THE FIRST ONE HUNDRED YEARS OF AMERICAN GEOLOGY

PUBLISHED ON THE FOUNDATION
ESTABLISHED IN MEMORY OF
PHILIP HAMILTON McMILLAN
OF THE CLASS OF 1894, YALE COLLEGE

THE
FIRST ONE HUNDRED YEARS
OF
AMERICAN GEOLOGY

BY GEORGE P. MERRILL

HEAD CURATOR OF GEOLOGY
UNITED STATES NATIONAL MUSEUM

(Facsimile of the Edition of 1924)

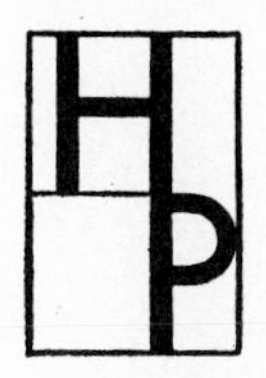

HAFNER PUBLISHING COMPANY
New York and London
1964

Reprinted by Arrangement

Printed and Published by
Hafner Publishing Company, Inc.
31 East 10th Street
New York, N.Y. 10003

Library of Congress Catalog Card Number: 64-20222

Lithographed in the U.S.A.
by Noble Offset Printers, Inc.
New York, N. Y. 10003

PREFACE.

EARLY American geology was preeminently a science of observation and deduction. Information on which to base theory and hypothesis was not available—indeed, did not exist. With the accumulation of recorded observations it became possible to carry conclusions beyond the point of mere observation, and the inductive method was evolved. Well toward the close of the period here to be passed under review, synthetic methods of research were introduced by which the attempt is being made to discover by actual experiment in the laboratory the correctness or falsity of deduction or of inductive reasoning.

The first century's progress was, however, due almost wholly to the accumulation of observed facts and the conclusions drawn directly therefrom. Scores of men, mostly untrained or self-trained, were observing and drawing inferences commensurate with their understanding. Comparatively few had sufficient mental grasp of the subject to form hypotheses and from analogies to draw general conclusions or formulate laws applicable to the larger questions of earth history.

In the following pages I have attempted to review almost step by step the work of these pioneers in a new field. Incidentally it has seemed desirable to touch upon the personal characteristics and methods of work of those thus engaged, and in order that the oft-seeming crudity of observation might be understood, to dwell upon the conditions under which the work was accomplished and the value of the conclusions as measured by the continually varying standard of the period. In brief, the method of treatment is here essentially the same as in my *Contributions to a History of American Geology*, published in 1904. Acknowledgments for assistance in one way or another are due, as before, to the late Professors W. P. Blake and W. H. Brewer and to Professor Charles Schuchert, Drs. R. S. Bassler, John M. Clarke, William H. Dall, A. H. Brooks, S. F. Emmons, A. C. Peale, and General A. W. Vogdes. For kindly criticism and suggestion, I am to no one more indebted than to Professor John J. Stevenson, and for assistance in procuring rare and unusual publications relating to the subject, to the indefatigable collector and enthusiast, the late Dr. W. S. Disbrow, of Newark, New Jersey.

THE

PHILIP HAMILTON McMILLAN MEMORIAL

PUBLICATION FUND

The present volume is the first work published by the Yale University Press on the Philip Hamilton McMillan Memorial Publication Fund. This Foundation was established December 12, 1922, by a gift to Yale University in pursuance of a pledge announced on Alumni University Day in February, 1922, of a fund of $100,000 bequeathed to James Thayer McMillan and Alexis Caswell Angell, as Trustees, by Mrs. Elizabeth Anderson McMillan, of Detroit, to be devoted by them to the establishment of a memorial in honor of her husband.

He was born in Detroit, Michigan, December 28, 1872, prepared for college at Phillips Academy, Andover, and was graduated from Yale in the Class of 1894. As an undergraduate he was a leader in many of the college activities of his day, and within a brief period of his graduation was called upon to assume heavy responsibilities in the management and direction of numerous business enterprises in Detroit; where he was also a Trustee of the Young Men's Christian Association and of Grace Hospital. His untimely death, from heart disease, on October 4, 1919, deprived his city of one of its leading citizens and his University of one of its most loyal sons.

CONTENTS

PAGE

Rogers on thermal springs. J. P. Couthouy on coral growth. W. Byrd Powell's work in the Fourche Cove region of Arkansas. State geological survey of New York. Work of Mather, Emmons, Vanuxem, and Hall. Sketch of James Hall. Hall's work in Ohio. His views concerning Niagara. Establishment of the geological survey of Canada. Issachar Cozzens on the geology of New York. Lawrence's geology of the Western States. Work of Ruffin and Tuomey in South Carolina. Dana's views on metamorphism. Beck and Dana on zeolite formations. Fremont's expeditions, 1842-1844. Association of American Geologists and Naturalists. Hitchcock on the trap tufa of the Connecticut Valley and on inclination of strata. Silliman on the intrusive trap of the New Red sandstone of Connecticut. H. D. Rogers on atmosphere of the coal period. C. B. Adams's work in Vermont. Sketch of Zadock Thompson. The Richmond bowlder train. Mather on the physical geology of the United States. Conrad's medial Tertiary of North America. David Christy's geological observations. Wislizenus's explorations in Mexico. Binney's views on the lóess. Rogers on the geological age of the White Mountains. Dana on continental problems. Dana's ideas on the volcanoes of the moon. Sketch of J. D. Dana. Tuomey's final report on the geology of South Carolina. Tuomey's survey of Alabama. Sketch of Tuomey. F. S. Holmes on the geology of Charleston and vicinity. Report of Messrs. Brown and Dickerson on sediments of the Mississippi River. Captain Stansbury's explorations in Utah. D. D. Owen's work in the Chippewa district and in Wisconsin, Iowa, and Minnesota. Early description of the bad lands. Agassiz's physical characters of Lake Superior. Sketch of Agassiz. Work of Jackson in Michigan. Foster and Whitney's work in Michigan. Hall's report on Lake Superior rocks. Desor on the drift. Dana and the Wilkes Exploring Expedition. Tyson's work in California. Higgins's and Tyson's work in Maryland.

CHAPTER V. THE ERA OF STATE SURVEYS, THIRD DECADE, 1850-1859. 293

Renewal of geological activity; new workers and new publications. H. D. Rogers on climate and salt deposits. Hubbard and Dana on subaerial erosion. Scope of a geological survey. Work on Millington and Wailes in Mississippi. Emmons's work in North Carolina. Norwood's work in Illinois. Hall's troubles with J. W. and J. T. Foster. Second attempt at a geological survey of Indiana. G. G. Shumard's work with Marcy's expedition. Trask's geological work in California. Swallow's geological survey of Missouri. Jules Marcou's geological map of the United States. Pacific Railroad surveys; work of Antisell, Blake, Marcou, Newberry, Schiels, Evans, Gibbs, and Swallow. Sketches of Antisell and Blake. Daniels's work in Wisconsin. Percival's survey of Wisconsin. Owen's survey of Kentucky. G. G. Shumard's work with Pope's expedition. Lieber's work in South Carolina. Dawson's work in Nova Scotia. Whitney's *Metallic Wealth of the United States*. Harper's work in Mississippi. Safford's work in Tennessee. Currey's *Sketch of the Geology of Tennessee*. Emmons's *American Geology*. Kitchell's geological survey of New Jersey. Emory's Mexican boundary survey. Dana on the geological history of the American continent; on the plan of the development of North America. The finding of *Paradoxides*. Lesley's *Manual of Coal*. Hitchcock's illustrations of surface geology. Joseph Le Conte on coral growth. Hall's geological survey of Iowa. Whitney's work in Iowa. Hitchcock's geology of Vermont. Geological surveys of Wisconsin, under Hall, Carr, and Daniels. Richard Owen's *Key to the Geology of the Globe*. Logan's proposed subdivisions of the Laurentian. The Ives expedition up

ILLUSTRATIONS

PLATES

FIGURES IN TEXT

Page

CHAPTER I.

The Maclurean Era, 1785-1819.

PROGRESS in early American geology rested largely in the hands of men of classical education who were engaged in the various pursuits of the so-called "learned" professions. It is to such men, for the most part untrained in observation, and who moreover were hampered by the mental attitude of the times toward investigation, that credit is to be given for inaugurating studies tending toward the solution of the story of the American continent and incidentally the earth at large. At the period set for the beginning of this history, books and libraries in America were few. In not a single university was geology taught as a science. There were no accurate maps, and topographic delineation was undreamed of. Neither were there railroad cuts nor deep well borings to give a clue to the earth's structure beneath the immediate surface. The country was largely a wilderness, and the information with which the geologist of today begins his career was uncreated. Naturally such as was available was almost wholly of European derivation; indeed, many of the workers had received what training they may have had in European universities.

We can here touch but briefly upon the condition of the science at the time this history opens. It was at best but an admixture of cosmogony and erroneous deduction, and is well set forth in the general and highly imaginative works of Liebnitz and Buffon, or the more specialized treatises of Guettard, Pallas, Demorest, and Werner, works the contents of which are admirably summarized by Lyell,[1] Geikie,[2] and Zittel.[3]

According to Dr. Geikie it is the Frenchman, Jean Etienne Guettard (1715-1786), who deserves special recognition "as the first to construct, however imperfectly, geological maps, the first to make known the existence of extinct volcanoes in central France, and one of the first to see the value of organic remains as geological monuments and to prepare detailed figures and descriptions of them. To

1 Principles of Geology, 9th ed., 1853.
2 The Founders of Geology, 1897.
3 History of Geology and Paleontology, 1901.

him also are due some of the earliest luminous suggestions on the denudation of the land by atmospheric agents." Though his work viewed in the light of today was full of error, and though he failed to recognize the connection between volcanoes and basalt, attributing to the latter an origin by crystallization from an aqueous solution, he is nevertheless regarded by Dr. Geikie as the founder of the science of volcanic geology.

The geological results obtained by the Russian Astronomical Expedition under Pallas in 1769, though important, could have had little or no immediate effect on American workers. They were first published in St. Petersburg, in German (1772-1776), and afterward translated into French. Pallas taught that a granite core was a constituent of all the great mountain ranges against which in a highly tilted condition rested the nonfossiliferous porphyries, serpentines, and schists. Upon these in turn rested the argillaceous schists and shales, and limestones containing marine fossils. The ranges lacking this granite core and composed mainly of schists and stratified rocks he believed to have been uplifted mainly through volcanic agencies, the eruptions of which were due to sulphur included in the schistose rocks. The remains of the mammoth and other gigantic mammals found in northern Siberia were thought to have been carried there by floods from the region of the equator. This last idea it will be noted later took firm hold on at least one American writer, and indeed was very generally adopted, the Irish mineralogist, Kirwan, being one of its most earnest advocates. During 1779-1796 the Swiss naturalist, De Saussaure, had published three quarto volumes giving the results of his observations in the Alps, and the Germans, Lehman and Fuchsel, during 1756-1770 had published important matter relative to the order of superposition of rocks. The two most prominent figures, and those whose views undoubtedly prevailed over all others in the latter part of the eighteenth century, were the German, Gustav Werner, and the Englishman, James Hutton. Indeed, our history opens in the very midst of a controversy between these two and their followers—between those who believed with Werner that the rocks of the earth's crust were the result of a gradual precipitation from the waters of a universal ocean, and who were not inaptly called Neptunists, and those who recognized the possibility of deep-seated igneous action, and were therefore called Plutonists.

Werner thought that all rocks had been formed through aqueous agencies as chemical or mechanical deposits, and in the same position as regards horizontality as now found, though where a dip of more than thirty degrees existed he thought it might be due to local disturbances. He divided all the known rocks into five classes, the oldest of which (the primitive) were direct chemical precipitates from the waters of a primeval ocean. They included not merely granites, but also gneisses, schists, and certain of the nonfossiliferous limestones. His second, or Transitional class, included schists, slate, graywacke, gypsum, and limestone, in part of chemical and in part of mechanical origin. In this group were found the first traces of organic remains. His third, or Flötz group, consisted of sandstones, limestones, chalk, coal, lignite, basalts, etc., which were accumulated as mechanical precipitates under various conditions of disturbance and repose. His fourth, or Transported class, consisted of water-accumulated and transported sands, clays and gravels, calcareous tufa, soapstone, etc. His fifth and last, lava and other volcanic ejectmenta which he believed to have been derived through the melting of other rocks by the combustion of subterranean beds of coal. He taught further that mineral veins were formed as open fissures in the earth's crust filled through infiltration of surface waters carrying the mineral matter in solution.

The views of Hutton, Werner's English contemporary, were as diametrically opposed as is possible. The oldest, or primitive rocks, which were considered chemical precipitates by Werner, he believed to be of igneous origin, as were other unstratified forms such as are known today under the names dolerite, diorite, greenstone, andesite, and basalt. The stratified rocks, on the other hand, were mechanical deposits, the materials of which were derived from the waste of the land transported by stream action and laid down on the sea bottoms of various geological periods. It followed as a necessary consequence that he believed them to have been laid down in essentially horizontal positions and, where now inclined, to have been uplifted through crustal movements. It was due to Hutton, more than to all others, that geologists were led to realize the wearing down of mountain ranges and continents through the action of weathering, and the truly enormous periods involved in geological time.

These brief and imperfect references must suffice for the present purpose. To attempt to review the teachings of others—of Humboldt, Cuvier, and the score of English and continental workers—would not merely exceed the available space of this volume, but would simply result in rewriting that which has been already ably done by the authors noted. Our interests now lie in the local applications of these seemingly crude and often impossible ideas.

It is, of course, impossible to state, in the majority of cases, whence any more recent writer may have drawn his inferences or inspiration. The literature of any given period may be known, but we have no means of knowing whether or not it was accessible, or, indeed, if he were aware of its existence. Discoveries in science, moreover, like inventions, are rarely or never the work of a single brain, but are matters of growth. They result from the gradual accumulation of facts or ideas, often seemingly wholly disconnected, but which some master mind takes at the proper time and molds to his uses.

Schöpf's Work on American Geology, 1787.

Fig. 1. *J. D. Schöpf.*

Johann David Schöpf's *Beitrage zur Mineralogischen Kenntniss des Oestlichen Theils von Nord Amerika und seiner Gebürge,* published in 1787, is commonly regarded as the first systematic work on American geology. Schöpf came to America as a surgeon to the Hessian troops during the war of the Revolution, and immediately after the treaty of peace in 1783 made a tour throughout the eastern states as far south as Florida.

As a foreigner, whose results were published in Germany, Schöpf's work, but for its date, would lie outside the limits of the present history. It may be said, however, that he noted the close similarity throughout all its parts of the flat lands (coastal plain) extending from the western end of Long Island to Florida, and that this was marked off from the hilly region to the northwest by a double series of waterfalls in all the rivers emptying into the Atlantic. Thus early was recognized the "fall line" as a physiographic feature of the American continent. A somewhat cautious attempt was made to show that these flat lands were simply more or less recent sea beaches, rising inland, and though he does not definitely speak of land elevation, he observed that all the streams traversing the plain "have excavated therein deep channels . . . which were formerly wider and hence shallower before shrinking into their present canals." On the other hand, however, he adds that if eastern America were so built up, *i.e.*, by the elevation of the coast, it would be in contradiction to the general theory "that the eastern shores of continents are being continually attacked, undermined, and reduced by the action of ocean currents flowing from east to west, or by tidal action." The northeast by southwest trend of the east coast he thought to be due to the action of the Gulf Stream. That both mountain regions and plains had been covered by oceanic waters was shown by fossils, but he found no means of deciding which remained the longer covered, though it was thought that the problem might be solved by measuring the thickness of the soil and its rate of formation.

The country lying between the Hudson River and Florida and the main mass of the mountains and the Atlantic he described as a level plain with a general incline toward the southeast, traversed in a northeast and southwest direction by great furrows, or depressions. These, in general, became narrower and the intervening hills higher toward the north. A puzzling, inexplicable feature was the absence of sediments over the front and lower ranges of mountains. Water formerly stood over the higher and inner mountain ranges, which are covered with their deposits. At whatever time this occurred, the front lower range must have been covered as well because of the condition of the general slope toward the sea. But—he asks—"if these higher ridges received a considerable addition of sand and clay, now occurring as whetstone . . . the question naturally arises, Why did not the same occur over the ridges situated in front of them?"[4]

There are several papers of a mineralogical nature, and by Ameri-

[4] Schöpf published also, the year following, his Reise durch einige der mittelern und sudlichen Vereinigten Nordamerikanischen Staaten, which contained many notes on the geology and physiography of the region traversed. This interesting work has been translated by A. J. Morrison, Ph.D., and published under the title Travels in the Confederation (1911).

Other important foreign treatises which were in existence at this time, or came into existence during the first half of the nineteenth century, and which, it is a fair and safe inference, had effect in molding American ideas, were: Werner's Plan of Succession of Rocks, 1787; Hutton's Theory of the Earth, 1795; Kirwan's Geological Essays, 1799; Playfair's Illustrations of the Huttonian Theory, 1802; Von Buch's Geognistische Beobachtungen, 1802; Jameson's Mineralogy, 1804; Bakewell's Introduction to Geology, 1813; Von Humboldt's Essay on the Trend of Rocks, 1822; Conybeare and Phillips's Geology of England and Wales, 1822; Buckland's Reliquae Diluviana (2d ed.), 1823; Humboldt's Essai Geognostique sur la Gisement des roches dans les deux hemispheres, 1823; Scrope's Considerations on Volcanoes, 1825; De la Beche's Geological Observer, 1830, and his Geological Manual, 1831; Lyell's Principles of Geology, 1830; De la Beche's Researches in Theoretical Geology, 1834 (an American edition in 1837); Buckland's Bridgewater Treatise, 1836; Murchison's Silurian System, 1838; D'Orbigny's Paleontology Français, 1840-1855; Humboldt's Kosmos, 1845; Bischof's Lerhbuch der Chemischen u. Physikalischen Geologie, 1846-1847 (English translation 1854).

can writers, of an earlier date than Schöpf's, which may be mentioned. In the *Memoirs of the American Academy of Arts and Sciences* for 1785 are to be found among others the following titles:

"An Account of the Oilstone found at Salisbury," by Samuel Webster; "Yellow and Red Pigment, An Account of, Found at Norton," by Samuel Deane; "An Account of Several Strata of Earth and Shells on the Banks of the York River, Virginia," by Benjamin Lincoln; and "Fossil Substance containing Vitriol and Sulphur, An Account of large quantities of, found at Lebanon, New Hampshire," by Jeremy Belknap.

Early Views on Volcanoes.

More instructive from our present viewpoint are two papers by David Jones and Caleb Alexander on the supposed volcanic nature of West River Mountain in the Connecticut Valley of Cheshire County, New Hampshire. Jones described in some detail the appearance of the mountain and the efforts of the natives, or *peasants*, as he called them, to discover thereon the gold which they imagined had been melted down to a solid body by the extreme heat of the eruption. The rock comprising the mountain was said to be in many places much burned, softened, and dissolved by heat, with cinders and melted drops adhering and hanging down like small icicles, somewhat resembling in color the cinders of a furnace or black glass. While convinced that there had been volcanic explosions in the mountain, he regarded such as having taken place at least fifty years earlier, while the volcano itself could not have been active perhaps "within the present nor the past century."

Alexander wrote much more confidently: "Once in winter there was an eruption. The years when the preceding eruptions happened I can not inform; the last was twenty-seven years since, which was the most violent eruption ever known in that place. It was toward the close of a dark evening when it was first perceived, being preceded by a louder noise than common; then directly was seen the fire, which was seen to burn for several hours."

After describing in some detail the appearance of the rock in the immediate vicinity of the spot from which the fire was supposed to have issued, he went on to say: "I am not able to determine whether there be anything of a sulphurous nature on this mountain, but this I dare affirm, that there have been several eruptions, but whether it may with propriety be called a volcano I know not." This determination he submitted as well he might, "to the judgment of gentlemen more acquainted with the nature of volcanoes than I can pretend to be."

In 1786 a Dr. William A. Baylies visited Gay Head, Massachusetts, and gave a somewhat detailed description of the same in Volume II of these Memoirs. That which most concerns us are the indications of supposed volcanic activity noticed here also. Referring to the well-known cliffs of variegated clays he wrote: "In fact it [*i.e.*, the volcano] had all the appearance of having blown out but a few days." That it was formerly a volcano was confirmed by further examination. "Large stones whose surfaces were vitrified, great numbers of small ones cemented together by melted sand and also cinders were to be seen in many places. . . . Besides there are very plain marks of four or five different craters." It is likely that the melted sand was but sand cemented by limonite, but what the crater-like forms could be it is at this date too late to say. Sufficient for the present that the entire locality is made up of stratified sands and clays of Cretaceous and Tertiary ages. The error was pointed out by Hitchcock in his final report on the geology of Massachusetts (1841).

Somewhat similar is an account given by a Dr. Greenway in the *Transactions of the American Philosophical Society* for 1793, of a volcano, or "Bursted Hill," on the Dan River near the North Carolina border of Virginia. The hill he described as formed of lava mixed with round white stones, there being large masses of the melted matter lying on the summit of the hill near the supposed place where the volcano burst forth. "It is the opinion of some that the hill has bursted twice, and that the second time it did not run with melted matter, as at the first eruption; but only threw out the large lumps of lava which appear on the top of the ground." Further descriptions of the character of the material leave no doubt but that the "lava" was here, too, simply an aggregate of siliceous pebbles cemented by a deposit of iron oxide in the form of limonite. Colonel George Gibbs, who visited the West River "volcano" in 1810 and made known the result of his observations through Bruce's *Mineralogical Journal*, came to a similar conclusion, finding no trace of any eruption or other signs of volcanic activity, the "lava" being hematite, or limonite, an oxide of iron.

Early Views on Earthquakes.

Earthquakes as well as volcanoes were naturally among the geological phenomena earliest to attract public attention, and naturally, too, the efforts to account for them were crude and labored since the actual source of either phenomena was not open to inspection: all was hypothetical.

The first account to which reference need be made is that of John Winthrop, professor of mathematics and philosophy in Harvard College, who, under date of November 26, 1755, delivered a "Lecture on Earthquakes," with particular reference to a series of shocks that had prevailed over New England the week before. The agents which could produce these movements, "which can heave up such enormous masses of matter and put into the most vehement commotion vast tracts of land," he conceived to be fire "and proper materials for it to act upon." The probable correctness of this view he felt to be strengthened by the fact that "those countries which have burning mountains are most subject to earthquakes; and that these mountains rage with uncommon fury about the time when the circumjacent countries are torn with convulsions." His account of the working details of these agencies is unquestionably a reflection of the then prevailing views of European geologists. He conceived a heterogeneous earth containing within it "many large holes, pits and caverns. . . . There are very probably long, crooked passages, which run winding through a great extent of earth and form a communication between very distant regions. Some of these cavities are dry and contain nothing but air, or the fumes of fermenting minerals; in others are currents of water." He conceived further that there was within the bowels of the earth an inexhaustible source of heat which, following Newton, might be "produced by the particles of different bodies rushing together, in virtue of their attractive powers." After enumerating a number of the bodies, as sulphur, iron, and water, which, when mixed, produce heat, he adds: "You have seen that there are in the bowels of the earth inflammable materials of various kinds and in large quantities; some in the form of solid or liquid bodies and others in that of exhalations and vapors; that there are also powerful principles constantly at work which are capable of inkindling these materials into an actual flame; and that the vapor generated from such flame will endeavor to expand itself on all sides with immense force. If now these inflammable vapors be pent up in close caverns, so as to find no vent till they take fire in any part, the same will spread itself, wherever it meets with materials to convey it, with as great rapidity, perhaps, as it does in a train of gunpowder, and the vapors produced from thence will rush along through the subterranean grottoes as they are able to find or force themselves a passage; and by heaving up the earth that lies over them, will make that kind of progressive swell, or *undulation*, in which we have supposed earthquakes commonly to consist; and will at length burst the caverns with a great shaking of the earth, as in springing a mine; and so discharge themselves into the open air."

Practically the same views were set forth by a Professor Samuel Williams also of Harvard in Volume I of the *Memoirs of the American Academy of Arts and Science* (1785), the title of the article being "Observations and Conjectures on the Earthquakes of New England." A list of recorded earthquakes is given with details of the accompanying phenomena, after which logically follows the chapter on "Conjectures and Causes." "From the phenomena and observations," he writes, "we may safely infer that the earthquakes of New England, whatever may have been the case elsewhere, have been produced by *something which moved along under the surface of the earth*. . . . I think we may lay it down as *a pretty certain fact* that the earthquakes of *New England* have been caused by something which has moved along under the surface of the country." This something, he goes on to state, "was probably a strong elastic vapor," an inference based upon sundry attendant phenomena among which "were some that preceded the earthquakes and looked like a *previous preparation*. . . . As though some grand fermentation was taking place in the bowels of the earth, the water in several wells and springs was uncommonly altered in its motion, colour, smell and quality." The noise or roar of the earthquake was further conceived to have been "such as might have been expected from a subterraneous vapor when fiercely driving along under the surface of the earth. When such vapors have a regular discharge through an aperture in the surface of the earth, they will vent themselves in copious effusions and exhalations and thus spend their force this way . . . but when the vapors are confined under the surface of the earth and have subterraneous passages, or proper strata for them to run in, by the volume of their expansion they will heave up the surface of the earth, and thus cause, not an instantaneous concussion, but a progressive swell or undulation of the earth, and this will be continued till the vapors thus confined find, or force for themselves a passage where they may burst from their caverns, and discharge themselves into the open air. And these are phenomena in all respects agreeing with those that have attended the earthquakes of this country." The origin of such a vapor is likewise traced and found to be due to the presence in the earth's crust of certain bodies like coal, fats, sulphur, etc., which generate or are easily turned to vapor, or as possibly due to fermentation of compound or mixed bodies. "But in no method is a more powerful vapor produced than by fire . . . and fire seems to be a fluid which is spread through almost all bodies whatsoever. It certainly exists in very large quantities in the bowels of the earth." The paper finally closes with an acknowledgment that a part of the account seems to be a matter of fact, and a part of conjecture, and he wisely adds. "Hypotheses may be of use to put us upon further enquiry, and a more critical examination, but are never to be received, any further than they are supported by proper evidence."

These views are not so widely at variance with those held by men like Humboldt and Newton as to allow their being pushed too lightly to one side. For their time they were as legitimate as those held today by the best of authorities upon the subject of radium. It is not strange, therefore, that as late as 1825 Dr. Isaac Lea, of Philadelphia, in an article in the *American Journal*, should accept them and at the same time attempt to show "the probable sources from whence such large supplies of combustibles are drawn, and to prove the identity of the volcano with the earthquake. . . . That these [the above described] channels have communication with the sea, there is no disputing; for volcanoes frequently throw up salt water and fish from the ocean," and moreover *all* volcanoes are situated near the sea. "Water, therefore, is an essential portion of the volcano and we may safely conclude that it is its most powerful agent." And this in defiance of the authority of a Dr. Stukely who, in 1749, had put forward the theory that they were caused by lightning!

The "astonishing elasticity" of water when "surcharged with caloric" is cited, and the source of the caloric assumed to be the internal heat of the earth, which is conceived, at a comparatively short distance below the surface, to be in a "state of ignition almost beyond our conception." The necessary combustible materials for keeping up the earth's internal fires he found in the roofs and walls of the branching underground channels aided by oxygen and water supplied by influx from the sea.

To the possible objection that rocks in themselves would not burn, he answered that, when such are decomposed, "their metallic bases, calcium, silicium, aluminum, magnesium, etc., are highly incandescent."

Maclure's Criticisms of Scrope and Lea.

Four years later (1829), Maclure, in discussing the theories of Poulett Scrope, took exception to the views of Lea, though he failed to offer anything satisfactory in return. He instanced the earthquake of 1811-1812 in the Mississippi Valley, in which only the alluvial formations had been disturbed, and questioned if such could not have been caused by the evolution of elastic gases arising from the fermentation of large masses of vegetable matter accumulated in the beds. If such were the case, however, he recognized the possibility of an increase in the frequency of earthquakes as fermentation went on, and evidently did not place a great amount of reliance even on his own theories, for he states, in referring to some of the ideas advanced by Scrope:[5] "All these speculations are out of reach of our senses and can be accounted only as amusement for the present."

But we anticipate.

Condition of Analytical Chemistry in 1789.

A paper by Robert McCauslin, communicated to the American Philosophical Society of Philadelphia in 1789 by B. S. Lyman and published in 1793, conveys a good idea of the crude condition of analytical chemistry at that date and enables one to appreciate the difficulties under which the mineralogists labored. In describing an "earthy substance" found near the falls of Niagara, he wrote:

> In order to determine the nature of this substance, I made the following experiments:
>
> Exp. 1st. I put an opaque piece, weighing fourteen grains, into the vitriolic acid diluted with three times its quantity of water, and let it remain there twenty-four hours, shaking it now and then. Not the least effervescence ensued, and on taking out the piece it weighed near one grain more than when it was put in, although care was taken to absorb the moisture which was upon its surface. This experiment was repeated with a shining piece, and with exactly the same result.
>
> Exp. 2nd. When put into vinegar it did not produce the least effervescence. The vinegar having stood upon it for some time, it was then poured off and spirit of vitriol dropped into it, yet not the least precipitation ensued.
>
> That I might not be led into error by the vinegar not being good of its kind, I repeated these experiments with chalk; and as both effervescence and precipitation took place it was evident there was no defect in the vinegar.
>
> Exp. 3rd. A small piece was exposed to the heat of a blacksmith's forge during fifteen hours. Upon taking it out and pouring water upon it no ebullition ensued; nevertheless it tasted like weak limewater; being then divided into two portions, a solution of mild fixed alkali was

[5] Scrope's Theory of Volcanoes appeared in 1825.

in it would recede from the center and rise till they arrived at that region of the air which was of the same specific gravity with themselves where they would rest; while other matter mixed with the lighter air would descend, and the two meeting would form the shell of the first earth, leaving the upper atmosphere quite clear. The original movement of the parts toward their common center would naturally form a whirl there, which would continue in the turning of the new-formed globe upon its axis, and the greatest diameter of the shell would be in its equator. If by any accident afterwards the axis should be changed, the dense internal fluid by altering its form must burst the shell and throw all its substance into the confusion in which we find it.

Again, in a letter to Mr. Bodoin, under date of 1790, he wrote:

Is not the finding of great quantities of shells and bones of animals (natural to hot climates) in the cold ones of our present world some proof that its present poles have been changed? Is not the supposition that the poles have been changed the easiest way of accounting for the deluge by getting rid of the old difficulty how to dispose of its waters after it was over? Since, if the poles were again to be changed and placed in the present equator, the sea would fall there about 15 miles in height and rise as much in the present polar regions, and the effect would be proportionable if the new poles were placed anywhere between the present and the equator.

Does not the apparent wrack of the surface of this globe, thrown up into long ridges of mountains, with strata in various proportions, make it probable that its internal mass is a fluid, but a fluid so dense as to float the heaviest of our substances? Do we know the limit of condensation air is capable of? Supposing it to grow denser within the surface in the same proportion nearly as we find it does without, at what depth may it be equal in density with gold?

Can we easily conceive how the strata of the earth could have been so deranged if it had not been a mere shell supported by a heavier fluid? Would not such a supposed internal fluid globe be immediately sensible of a change in the situation of the earth's axis, alter its form, and thereby burst the shell and throw up parts of it above the rest?

As if we would alter the proportion of the fluid contained in the shell of an egg and place its longest diameter where its shortest now is the shell must break if the whole internal substance were as solid and hard as the shell.

Might not a wave by any means raised in this supposed internal ocean of extremely dense fluid, raise in some degree as it passes the present shell of incumbent earth and break it in some places, as in earthquakes? And may not the progress of such wave and the disorders it occasions among the solids of the shell account for the rumbling sound being first heard at a distance, augmenting as it approaches, and gradually dying away as it proceeds? A circumstance observed by the inhabitants of South

America in their last great earthquake—that noise coming from a place some degrees north of Lima and being traced by enquiry quite down to Buenos Ayres, proceeding regularly from north to south at the rate of — [*sic*] leagues per minute, as I was informed by a very ingenious Peruvian whom I met in Paris.

William Bartram's Travels, 1794.

William Bartram in his *Travels in North America* makes sundry references to the rocks and minerals of the country passed over, which are of interest only in showing the condition of knowledge on geological subjects at the time. Bartram, it is true, was a botanist, but like most of his contemporaries, he dipped into all branches of science. Writing concerning middle Georgia, he says: "The rocks and fossils which constitute the hills of the middle region are of various species, as quartzum, ferrum, cos,[6] silex, galena, arena, ochra, stalactites, laxum, mica, etc. I saw no signs of marble, plaster or limestone; yet there are, near Augusta, in the forest, great piles of a porous friable white rock in large and nearly horizontal masses, which seems to be heterogeneous concrete consisting of pulverized sea shells with a small portion of sand; it is soft and easily wrought in any form, yet of sufficient consistency for constructing any building." In Cherokee County he noted that the cliffs seemed to be continually crumbling to earth in which he discovered "strata of most pure and clear white earth" (kaolin) and veins of "isinglas (*S. Vitrium Muscovitum*)" in flakes of such size as to be suitable for "lanthorns" or lights for windows.

Early Notices of Fossil Vertebrate Remains.

The finding of bones of gigantic extinct quadrupeds naturally excited much popular curiosity and scientific interest, and gave rise to crudities in reasoning such as today would excite ridicule from any but the most densely ignorant. The first mention upon record of bones of the American mammoth appear to have been in a paper by Reverend Cotton Mather, published in the *Philosophical Transactions of the Royal Society* for 1714. The bones were found in 1705 near Albany, New York, and included three teeth and a thigh (?) bone seventeen feet (?) in length. He considered these to be the bones of a human being and corroborative of the Scriptural account

[6] This word is assumed to be used in an algebraic sense (coss or cosa) meaning unknown. *Arena* is of course sand, and *laxum* probably any loose, unconsolidated material.

of a race of antediluvian giants.[7] Other elephantine remains were later found in Ohio and Kentucky and in 1793 a Mr. Robert Annan published in the *Memoirs of the American Academy* an account of bones found in 1780 some seventy miles north of the city of New York. The nature of the animal was to him problematical, "Certainly not a marine monster, for it lay above 100 miles from the sea; unless we can suppose that many centuries ago that part of the country was covered by the sea." From the appearance of the teeth it was judged to be carnivorous, though the possibility of their being those of an elephant was suggested.

Jefferson as a Paleontologist, 1797.

The ever busy and versatile Thomas Jefferson, when he came to Philadelphia to be inaugurated vice-president in 1797, brought with him a collection of fossil bones from the western part of Virginia, and the manuscript of a memoir upon them, which he read before the American Philosophical Society, of which he had been elected president the preceding year. The paper was published in the *Transactions* of the Society for 1799, as was also Dr. Wistar's more detailed description of the bones, which were those of *megalonyx*.[8] Again in 1801, when Congress was vainly trying to untangle the difficulties arising from the tie vote between Jefferson and Burr, when every politician at the capitol was busy with schemes and counter-schemes, Jefferson was corresponding with Dr. Wistar in regard to some bones of the mammoth which he had just procured from Shawangunk, in New York; and still again, in 1808, when the excitement over the embargo was highest, and when every day brought fresh denunciation of him and his policy, he was carrying on his paleontological studies in the White House itself. Under his direction upward of three hundred specimens of fossil bones had been brought from the famous Big Bone Lick of Kentucky, and spread in one of the unfinished rooms of the presidential mansion. The exploration of this lick, it should be noted, was made at the private expense of Jefferson, through the agency of General William Clarke, the western explorer.[9]

These several occurrences naturally excited the attention of the

[7] An Account of the Fossil Bones of the Great American Mammoth, by John Ware, Boston Journal of Philosophy, Vol. I, 1823-1824.

[8] A large sloth-like animal to which this name was given on account of the enormous claws.

[9] The Origin of the National Scientific and Educational Institutions of the United States, by G. Brown Goode.

curious, and numerous accounts of similar deposits began to find their way into print. In the *Transactions of the American Philosophical Society* for 1799 a Mr. George Turner gives an account of the finding in South Carolina of the teeth and portions of the skeleton of two "non-descript" animals, evidently mammoth and mastodon. One of these, he conceded, might have been herbivorous, but the second (the mastodon) must have been carnivorous. "With a body of unequaled magnitude and strength, it is possible the mammoth may have been at once the terror of the forest and of man." Under date of 1802, in the same publication, is given also an abstract of a communication received by Martin Duralde regarding the finding of bones supposed to be those of an elephant in the Opelousas country west of the Mississippi, *i.e.*, in Louisiana.

Peale's Account of the Mammoth, 1802.

Of greater interest, however, is a paper by the artist and naturalist, Rembrandt Peale, entitled "Account of the Skeleton of the Mammoth, a Nondescript Carnivorous Animal of Immense Size Found in America." This was printed in London under date of 1802, and dedicated to Sir Joseph Banks, Bart., president of the Royal Society. The particular skeleton described was exhumed in the vicinity of Newburgh, New York, in 1801. From the structure of the teeth and jaw Peale was led to consider the animal unquestionably carnivorous and he concluded his description with the devout quotation from Dr. Hunter: "If this animal was indeed carnivorous, which I believe cannot be doubted, though we may as philosophers regret it, as men we cannot but thank heaven that its whole generation is probably extinct."[10]

S. L. Mitchill on Geology of Eastern New York, 1798.

A medical journal is scarcely the place to which one would now refer for papers bearing upon geological matters. The early workers were, however, men of other of the "learned professions" whose writings found publicity through sources designed primarily for matter of quite a different character. In the early volumes of the *Medical Repository*, published in New York City (1797-1826) and edited by S. L. Mitchill, are to be found numerous articles of a geological nature, both original and by way of reviews, in which the

[10] See also The Newburg Mastodon, by Arthur Bibbins, Bull. Geological Society of America, Vol. XVIII, 1906. This contains a half-tone reproduction of a painting by C. W. Peale showing the work of exhuming in progress.

editor figured largely. Mitchill, who also became one of the most versatile of the scientific writers of his period, though not preeminent as a geologist, seems to have been early appointed a "commissioner" to investigate the geological conditions in eastern New York, and made his first published report in Volume I of the above-mentioned *Repository* (1798). He divided the area covered into the Granitic Tract, the Schistic Tract, the Limestone Tract, and the Sandstone Tract, noting in each the principal minerals, soils, and rocks. Incidentally he entered into an interesting discussion concerning the possible inorganic origin of limestones. "The calcareous earth is in some places filled with the remains of animals. . . . Although these materials are firmly compacted into stone, yet their shapes are often so entire and complete that the species may be exactly distinguished," although some were unlike any inhabiting the coasts at the present day. Discussing these he adds: "The opinion to which modern philosophy leans, that all the calcareous earth in the world is to be traced back to the collecting labor and accumulation of animals, appears doubtful, for this plain reason, that if lime is a primitive earth, or elementary substance, it must have existed in some form and some place before animals could have been furnished with it . . . and the Commissioner feels a persuasion that the real exuvial animals do not indisputably exist in large quantities . . . yet there are some appearances of shells, particularly, which are of a very doubtful or imposing nature." He then goes on to discuss the possibility of some of these "likenesses" having been produced in the crystalline processes incidental to limestone formation, without, however, committing himself absolutely. The possible fuel value of the peat deposits is taken under consideration, and the suggestion made that as wood grew scarce and the state produced no coal, sphagnum might be cultivated for fuel purposes in the wet grounds then lying waste. This was certainly a long look into the future.

Fig. 2. *S. L. Mitchill.*

In a later number of the *Repository* (1800) Mitchill described the geology of the region with particular reference to "Long or Nassau Island." Although the superficial—visible—portions of this island are now known to be all of loosely consolidated sands and gravels of glacial origin, Mitchill apparently thought otherwise.

From a survey of the fossils in these parts of the American coast, one becomes convinced that the principal share of them is granitical, composed of the same sorts of materials as the highest Alps, Pyrenees, Caucasus, and Andes, and, like them, destitute of metals and strata.

The occurrence of no horizontal strata, and the frequency of vertical layers, led him further to suppose that these strata were not secondary collections of minerals, but were certainly in a state of primeval arrangement.

The steatites, amianthus, schorl, feldspath, mica, garnet, jasper, schists, asbestos and quartz, must be considered as primitive fossils and by no means of an alluvial nature.

What inference remains now to be drawn from this statement of facts, but that the fashionable opinion of considering these maritime parts of our country as flats, hove up from the deeps by the sea or brought down from the heights by the rivers, stands unsupported by reason and contradicted by experience?

It is difficult at first thought to believe that this description was intended for the island now under this name. Nevertheless, further on he speaks of it as in former days separated from the mainland by a small river, and the strait which now divides them formed by successive inroads of the sea from the east toward the west, the small islands or shoals being residual remains of what was once upland of equal height with the main island but degraded by the leveling power of the waves.

The American Mineralogical Society, 1799.

Although the era now under discussion has been called the *Maclurean,* Dr. S. L. Mitchill was the most active and prominent participant during the earlier years. Indeed, through the medium of his journal, he for a time occupied in the geological and mineralogical field the position later filled by the elder Silliman. In Volume II for 1799 is noted the formation of the American Mineralogical Society, for "the investigation of the mineral and fossil bodies which compose the fabric of the globe; and more especially for the natural and chemical history of the minerals and fossils of the United States." The officers were: S. L. Mitchill, president; Solomon Simpson, vice-president; E. H. Smith, secretary; Edward Miller, treasurer. The prospectus issued indicated a still evident feeling of distrust of the peaceful intentions of neighboring tribes and more distant nations.

The American Mineralogical Society, instituted in the city of New York . . . earnestly solicit their fellow citizens . . . to communicate with them on all mineralogical subjects, but especially the following, viz.:

I. Concerning the stones suitable to be manufactured into *gun flints:* where these are found? and in what quantity? with example of the material for experiment.

2. Concerning native *brimstone* or *sulphur,* or the waters or minerals whence it may be extracted; a description of the tracts of country where sulphur or sulphur springs and ores abound; with probable estimate as to the practicability of collecting or extracting the sulphur economically; and with specimens of the ingredients for trial.

3. Concerning *saltpetre:* where (if at all) found in nature? of what extent? in what proportion? the mode of extracting or refining it? with specimens of the native earth or salt for examination.

4. Concerning mines and ores of *lead:* in what places? the situation? how wide the vein? in what kind of rock is it bedded? with pieces of the ore for assaying.

Specimens of ores, metals, coals, spars, gypsum crystals, petrifactions, stones, earths, clays, slates, chalks, limestones, marbles, and every fossil substance that may be discovered or fall in the way of a traveler which can throw light on the mineralogical history of America, examined and analyzed without cost; sufficient pieces, with the owner's leave, being reserved for placing in the Society's collection.

Specimens of all the above mentioned articles, or *information* concerning any of them may be forwarded (free of express) and will be thankfully received and duly attended to by the subscribers.

Notwithstanding the generous inducements offered, the society seems to have found little encouragement for existence, and no mention of it is made during the years that followed.

Mitchill appears to have been a man of extraordinary versatility. Living at a period when the sciences were in their infancy and it was possible for one mind to grasp a considerable portion of all that was known, he wrote and lectured intelligently on such a variety of subjects as would today only make one ridiculous. Even in his own time he did not wholly escape:

He was a polished orator, a versifier and a poet, a man of infinite humor and excellent fancy. His eccentricities furnished material for the wits of the day to fashion many a joke at his expense, over which no one laughed more heartily than himself. He was equally at home studying the Geology of Niagara or the anatomy of an egg; in offering suggestions as to the angle of a windmill or the shape of a gridiron; in deciphering a Babylonian brick or investigating bivalves and discoursing on conchology; and in advising how to apply steam to navigation or in disputing about the bible with his neighbor, the Jewish rabbi. He possessed a charm of manner and a magnetism of mind that was unusual and he did much to advance the public and private interests of America, and elevate our scholastic reputation in foreign countries.[11]

Wall of Supposed Human Workmanship, 1799.

Products of weathering or other natural causes often so closely simulate the work of human hands as to give trouble to the geologist and archeologist even at the present day, and cases like that mentioned below have come to the writer's attention through correspondence and the newspapers within a few years.

In the *Repository* for 1799 is found a letter from the Reverend James Hall to Professor James Woodhouse of the University of Pennsylvania, describing what is today easily recognizable as a small trap dike in granite, jointed and weathered into bowlder form masses closely imitative of human handiwork. Hall described it in detail and considered it "incontestably" proven artificial. Professor Woodhouse was not, however, so easily deceived, and in his reply detailed the now seemingly crude test performed by him to show the true nature of the material, and expressed himself "satisfied . . . that it consists of a mineral substance called basaltes and that it is a product of nature and not of art." The Reverend Zachariah Lewis brought the benefit of his theological training to bear upon the subject also and in the *Repository* for 1800 and 1802 gave the results of his investigations. With those who would attribute to it a natural origin he could not agree, and singularly enough was quite unable to appreciate the effect of differential weathering in removing the granitic rock on either side, leaving the wall-like dike standing in relief. "We cannot conceive it possible that a mass of melted substance, several hundred feet in length, twelve to fourteen in height and precisely twenty-two inches thick, should stand and cool in that position. The top would sink, and the sides spread; and this probably not equal in all its parts." For this and other equally good and (to him) valid reasons he rejected the generally adopted opinion of "philosophers" and concluded: "Thus it is evident, in the first place, that no substantial argument is offered to prove that this wall cannot be artificial; in the second place, that it has every possible appearance of being the production of art; and in the third place, that it cannot have been produced by nature in either of its ordinary modes of producing regular forms."

[11] Mag. of American History, Sept., 1886.

Seemingly this should have been considered as final, but doubters were in evidence and some twenty years later, in response to a request of Dr. Mitchill, a Mr. John Beckwith reinvestigated the subject, giving his results in the form of a letter printed in the *American Journal of Science* for 1822. While not wholly able to account for the conditions as found, Beckwith was, on the whole, inclined to believe the wall to be basaltic, though perhaps of aqueous rather than igneous origin. Multiplicity of observations have long since proven the correctness of the views put forward first by Professor Woodhouse.

Smith's Account of the Conewago Hills, 1799.

Turning once more to the *Transactions of the American Philosophical Society,* we find in 1799 Thomas P. Smith giving an account of "crystalline basalts" found in the Conewago Hills, near Elizabethtown in Pennsylvania. The "crystals" were described as generally tetrahedral and of very fine grain. The massive, non-columnar form was spoken of as amorphous, but it "has generally a very strong tendency to crystallize." Crystallized granite in "predominating tetrahedral forms" was also described. In the same transactions (1807) S. Godon made certain observations for a mineralogical map of Maryland, in which he noted the occurrence of gneiss and greenstone in the vicinity of Washington, and the finding of "fossil bodies," shells, and fossil woods in a ravine near Rock Creek Church. The city itself was rightly described as built on alluvial land, Rock Creek forming the boundary between the primary and alluvial soil.

In February of this same year B. H. Latrobe read before the society a paper describing the geographic distribution of the sandrock quarried at Aquia Creek on the Potomac, and used in some of the public buildings of Washington. The marked cross-bedding, so prominent a feature in the stone, he ascribed to wind action. A fact of more geological importance is, however, his recognition of the "fall line"—of the fact that a line drawn along the lower falls of our rivers is the ancient line of the seacoast from New York to the southwest, and indicates a higher ocean level of some 120 feet. Schöpf and other earlier writers, it will be remembered, recognized this line, though it does not appear that a like interpretation was put upon the phenomenon.

Appointment of Benjamin Silliman at Yale, 1802.

At this date, it is well to note again that none of the sciences were taught in the colleges and other institutions of learning in America or England.[12] Indeed, the general trend of public opinion was decidedly against the study of geology or the investigation of any question which might lead to the discovery of supposed inconsistencies in the Mosaic account of creation or to conclusions in any degree out of harmony therewith. The movement, therefore, by President Dwight in 1798 toward the establishment in Yale College of a department for the teaching of these subjects, was of the greatest importance and of far-reaching consequences. This culminated in 1802 with the appointment of Benjamin Silliman to the professorship of chemistry and natural science. Silliman was at that time about twenty-two years of age, only recently admitted to the bar, and was serving as a tutor in law, with not even the most rudimentary knowledge of the science he was to teach. He wrote of himself and of his qualifications:

> The appointment was of course the cause of wonder to all and of cavil to political enemies of the college. Although I persevered in my legal studies . . . I soon after the confidential communication of President Dwight [informing him of his probable appointment] obtained a few books on chemistry and kept them secluded in my secretary, occasionally reading in them privately. This reading did not profit me much. Some general principles were intelligible, but it became at once obvious to me that to see and perform experiments and to become familiar with many substances was indispensable to any progress in chemistry, and of course I must resort to Philadelphia, which presented more advantage to science than any other place in our country.[13]

To Philadelphia he accordingly went in the autumn of 1802, remaining for nearly five months attending the lectures on chemistry given by Dr. James Woodhouse in the Medical School of Philadelphia. His own first lecture at Yale was delivered April 4, 1804, while

12 Robert Jameson was appointed in 1807 to a professorship of natural history in the University of Edinburgh. Silliman, referring to conditions as they existed in London in 1805, states: "I found there no school public or private for geological instruction and no association for cultivation of the subject, which was not even named in the English universities." The Geological Society of London was not founded until November, 1807.

Woodward (Hist. Geol. Soc. of London, p. 81) states that the first mineralogical lectures at Oxford were delivered by Sir Christopher Pegge and later by Dr. John Kidd, but he mentions no date. Buckland was elected "reader in mineralogy" at Oxford in 1813. E. D. Clarke was appointed at Cambridge in 1808.

13 G. P. Fisher, Life of B. Silliman, New York, 1866.

in his twenty-fifth year, in a room in Mr. Tuttle's building on Chapel Street.

Few geological papers bear Silliman's name, and he is better known as a teacher and public lecturer. That which, however, has tended more than anything else to keep him in constant remembrance is his *American Journal of Science*,[14] founded in 1818, some eight years after the suspension of the *American Journal of Mineralogy*, to be noted later. The publication has continued down to the present day, and it is therefore the oldest and the most important geological periodical extant in America.

Silliman resigned his professorship in 1849 and died on December 24, 1864, having through his own efforts as a teacher, but more particularly through his personal influence as a writer and lecturer, done more to advance the science of geology than any other man of his day. Nevertheless, looking down the vista of a hundred years of advancement, the spectacle of this young lawyer secluding in his desk sundry books which he read privately in order to fit himself for the position of a professorship in Yale, is edifying, to say the least.[15]

Volney's Views of the Climate, etc., of the United States, 1804.

Although written by a foreigner, Count C. F. Volney's account of the geology and physiography of the United States cannot be passed over without a somewhat extended notice. Volney came to America in 1795 and passed three years in travel, visiting almost every part of the area comprised within what was then known as the United

[14] Known for many years only as Silliman's Journal.

[15] New Haven, June 29th, 1833.

FRIEND HALL:

. . . .

Prof. Silliman has a few days since finished his Mineralogical-Geological lectures which I have had the pleasure of hearing. You have doubtless read his Tour to Quebec, &c., and remarked the glowing stile in which much of it is written—it is not changed in his lectures any more than the subject requires & sometimes hardly that. Perfectly at home among the wreck and ruins of a former world, in either hand balancing a flood of waters & a lake of fire, before his respectable and attentive auditors of both sexes, he stands, like some kind but mighty spirit sent to instil into the minds of the rising generation the sublime and awful mysteries of the past creation, himself "filled to bursting nigh" with the majesty and grandeur of the subject; occasionally among other veterans making honorable mention of Prof. Eaton's name and services, yet not unfrequently dissenting to his opinions and more frequently adopting the opinions of his trans-Atlantic brethren, yet professing to be a genuine eclectic and unfettered by theories.

. . . .

(*Signed*) ABM. S.

PLATE II

BENJAMIN SILLIMAN

States. His work, *Tableau du Climat et du Sol des Etats Unis d'Amerique, etc.*, appeared in 1803 and was followed by an English translation printed in London in 1804, name of translator not given, and an American edition, from a translation by C. B. Brown, printed the same year in Philadelphia. However prejudiced Volney may have been, however biased in his statements in other lines, his expressed views on physiographical and geological subjects, taken in their entirety, were by far the best, with the possible exception of those of Schöpf, that had thus far appeared, and indeed in many respects were ahead of those expressed by some of his followers for many years. Singularly enough, however, he nowhere in his text refers to Schöpf's work. This notwithstanding the fact that he covered a portion of the same territory. Whether this omission is due to ignorance of the earlier publication or to a feeling of vanity and self-sufficiency on his own part, it is at this date impossible to state.

Fig. 3. *C. F. Volney.*

One chapter of Volney's work is devoted to the extent and geographical divisions of the United States. A few quotations from this are essential in order that one may appreciate the sparsely settled condition of the country as well as the difficulties under which the geologist worked. After referring briefly to the forested portion of the East, and to the western prairies, he says: "Such is the general aspect of the territory of these States. The picture is composed of an almost universal forest, varied and broken by five vast lakes or inland seas in the North; by immense natural meadows, or prairies, in the West; and in the center by a chain of mountains, whose ridges are parallel to the sea coast, at a distance of from 50 to 150 miles, and which turn, to the East and West rivers of a longer course, wider channel, and more ample stream, than we are accustomed to meet with in Europe. These rivers are broken into cataracts, from 20 to 140 feet in height, and enter the sea in mouths that expand into gulphs."

From a physiographical standpoint he recognized three grand divisions. First, the eastern country, placed between the sea and the mountains; second, the western country, situated between the mountains and the Mississippi; and third, the mountainous region itself, which is spread out between the two former ones. The mountains were described as beginning in lower Canada on the southern shore of the St. Lawrence and extending, with numerous bifurcations and ramifications, to the Mississippi and the mouth of the Ohio, where the great continental forest ends and the steppes or savannahs begin, which he compared with the steppes of Russia. The Atlantic country included what are now known as the Eastern and Southern Atlantic States, from Maine to Georgia. Throughout all this extent, the surface was described as "elevated but slightly above the sea, being more uniform and flat in the country south of Maryland than farther north." Long Island he considered to form the line of division between "the two kinds of surface, that to the northward being high and rocky," while that to the southward is "nothing but a flat of pure sand almost level with the ocean." "This sand has evidently been left by the sea, and is traced to a considerable distance inland. . . . As we approach the mountains we meet with a mixture of clay and gravel, which the waters have washed from the neighboring heights, and thus is formed a yellowish, poor, light soil, which prevails throughout the middle selvage of the Southern States. These lineaments are so clearly traced that we can not hesitate in considering Maryland, Pennsylvania, and New Jersey as gradually formed by the deposits of the Potowmack, Susquehannah, Delaware, and Hudson." The western region he appropriately termed the Valley of the Mississippi, because all the streams which traverse it are ultimately lost in that river. This valley he also divides into three grand districts, the surface features and soils of which he described with considerable detail, it even being noted that, throughout Kentucky and Tennessee, there were frequently observed "pits or funnels from 50 to 500 paces wide, and from 15 to 50 in depth, having at the bottom one or more holes or chasms, in which are swallowed up not only the rainy torrents from the neighboring heights, but even considerable brooks and rivers." These are obviously sinkholes, formed by the solvent action of the water, in a manner now, but evidently not at that time, well understood.

The mountain district he described as a belt or band computed at 1,200 miles in length, and varying in width from 90 to 150 miles, the greatest elevations of which would not exceed 4,000 feet, with the possible exception of Mount Washington, of which two estimates of 7,800 and 10,000 feet were given. Three principal ridges were distinguished in Virginia: first, the Blue Ridge; second, the North Mountain; and third, the westernmost, the Alleghany, a term believed to mean, in the Indian language, "endless," as previously noted by Schöpf and others.

During his journeys Volney collected specimens of the common minerals and rocks. Traveling mainly on foot, such were necessarily small, but when arranged and compared at Philadelphia with others presented to him, he was enabled "with the aid of some learned mineralogists," to draw up "a kind of physical geography of the United States," and to divide, "with sufficient certainty," the region between the Mississippi and the Atlantic Ocean, into five districts. First, the granite region; second, the grit or sandstone region; third, the calcareous or limestone region; fourth, the region of sea sand; and fifth, the region of alluvial or river-formed soil. Although noting the occurrence of calcareous rocks underlying Montreal, and similar materials along Lake Champlain and other localities mentioned, he nevertheless felt justified in making the somewhat remarkable statement that "all North America above Long Island is a rock of granite." The grit or sandstone of the Catskills formed the characteristic feature of the second region, which comprised the mountainous country of the Blue Ridge, the Alleghanies, the sources of the Kanawha, and generally all the southern chains as far as the angle of Georgia, and the Appalachi. The third or calcareous region comprised the western country beyond the mountains, extending, on the authority of MacKenzie, northwestward to the head of Sakachee River and the Stony or Chippewan Hills. The portion which he had himself seen lay between the Tennessee and the St. Lawrence, and was described as "an immense rock or bed of limestone disposed horizontally . . . and generally of a gray color." In this rock, particularly in Louisville and Cincinnati, he found abundant petrifactions, which, on his return to Paris, he submitted to Lamark, who reported upon them as referable to the genus of fossils, terrebratula. His conclusion regarding them was as follows: "After carefully considering these specimens, it is clear to me that the districts of North America which furnish them have formerly been covered by the sea, or at least they point out that portion of the surface which has once been the bottom and not the banks of the sea; for they all belong to the class of Pelagia."

From the fact that the western stratum lay nearly horizontal, while those in western Maryland and Pennsylvania were inclined, often at an angle as high as forty-five degrees, Volney was led to conclude that the Atlantic coast had been shaken and overturned by earthquakes. This view, he felt, was further substantiated by the fact that in the language of the aborigines of the East there were found words corresponding with earthquake and volcano, whereas no such terms could be found in the language of the western tribes. His "western tribes," it must be remembered, included only those of the region lying west of the Appalachians and east of the Mississippi.

The region of sand comprised all the maritime plains from Sandy Hook, opposite Long Island, to Florida. The interior boundary of this sand he described as a ridge or bank of "granitic talc," which extends as far as Roanoke, in North Carolina. In all its courses this border ridge is described as marked by cascades, which are the limits of the tidewaters. Here again was the fall line recognized. Between this bank and the sea, the surface, in a breadth of from 30 to 100 miles, he described as composed of sand, evidently deposited by the sea which once flowed at its foot. The region of alluvial or river-formed sand would appear to be the Mississippi Valley proper, including the delta. It is in this connection that is found the first reference to the extinct lakes of North America, as subsequently elaborated by Mitchill. "When we attentively examine the strata, we observe that the principal ridges of the Alleghany proceed in a line perpendicular to the course of the great rivers, which have been obliged to open themselves a way through the solid mass of these ridges. The more I consider the situation of the adjacent country, the more I am confirmed in the belief that the Blue Ridge was once entire, and by shutting the door against the river forced it to expand into considerable lakes. The numerous cross ridges which succeeded each other from Fort Cumberland can not fail to establish this lake westward of North Mountain. On the other side, the Valley of the Shenandoah and the Conigocheague would naturally form a lake between Chambersburg and Staunton. As the level of these heights, whence these two above-mentioned rivers take their course, is much below that of the Blue Ridge and North Mountain, the line of their summits may have formed the boundaries of this lake, which must have diffused itself southward as far as the great bow of the Alleghany. The two upper branches of the James River, barred up in like manner by the Blue Ridge, would augment this lake by all their waters, while northward the general level of the lake would find no obstacle to hinder its extension between the Blue Ridge and the Kittatinni, not only as far as the Susquehannah, but even to the

Schuylkill and Delaware." Aided by earthquakes, which he thought were once frequent about this maritime country, the waters, continually assailing and sapping the mountainous barrier, at length made themselves a passage through it. These, at first narrow and shallow, would speedily, by the action of mighty streams, be widened and deepened until at length the breach would extend from top to bottom, and the lake would be entirely drained. Through the rushing of the waters, consequent to this rupturing, the whole face of the lower country would be changed.

To the presence of these lakes Volney accounted the level or horizontal character of the strata throughout the western country. The theory would also, in his mind, account for the remains of trees and other vegetation, and even of animals, found embedded, and formed "a happy and natural explanation to the formation of those mines of coal which predominate in certain situations and districts."

He visited Niagara, and rightly attributed the formation of its channel to the wearing action of the water. "The chasm has been gradually worn away, from age to age, till it reached the place where the falls now appear. This operation has continued slowly but incessantly."

Although in the general matter of dynamics and physiography, Volney appears to have been somewhat in advance of the majority of men of his time, on the subject of earthquakes and volcanoes he was still sadly astray, as shown in his unquestioned adoption of the views of "Mr. Williams" (*ante*, p. 9).

Volney's observations, while traveling in Turkey and other eastern countries previous to his coming to America, had had a depressing effect upon his spirits and,—"saddened by the past and anxious for the future, I set out for a land of freedom to discover whether liberty, which was banished from Europe, had really found a refuge in any other part of the world." He was at first favorably impressed, but unfortunately "in the spring of 1789 there broke out so violent an animosity against France . . . that I was obliged to withdraw from the scene." Though developments in 1801 rendered atonement in part for the sufferings and suspicions to which he had been subjected, the incident seems to have left him permanently with the unfavorable impressions referred to by Mitchill in a long and critical review in his *Medical Repository* for 1805-1806.

This writer attributed Volney's somewhat peevish and hypercritical attitude to his unfortunate state of mind, "for he left France in a fit of disgust; he suffered in America two very severe attacks of malignant fever, five or six violent colds, and rheumatic affections which have proved incurable; he had been attacked in the newspapers as a suspected alien; and he seems to have departed from the United States in no better humor than when he entered them." It is added, however, "he has given a better history of our atmosphere and its currents, or winds, than any we have seen before. And, although we differ from him in some of his conclusions, candor obliges us to own that Mr. Volney possesses a happy talent for inquiry and observation."

Silliman's Mineralogy of New Haven, 1806.

On January 1, 1800, the Connecticut Academy of Sciences addressed a circular letter to every town in the state, containing subjects of inquiry arranged under thirty-two distinct heads, and requesting answers. To the fifth of these inquiries Silliman responded in 1806 with a sketch of the mineralogy of the town of New Haven. This was published in the memoirs of the Society in 1810, and is of interest, being the first attempt at a geological description of the region, as well as Silliman's first attempt at a geological survey.

With much laborious argument he showed that the town of New Haven is situated on an alluvial plain. East Rock he described as a whinstone trap, or basalt, identical with that from the Salisbury Craig near Edinburgh, Scotland. "The stone is reckoned among the argillaceous clays by some mineralogists and by others among the siliceous. The predominant ingredient is certainly silex or flinty earth, although when breathed upon it emits the smell of clay, which would induce one to refer it to the argillaceous family."

He would account for its presence on the supposition that it had "actually been melted in the bowels of the earth and ejected among the superior strata by the force of subterraneous fire, but never erupted like lava, cooling under the pressure of the superincumbent strata and therefore compact or nonvesicular, its present form being due to erosion."

Thus, at this early date and with a very limited amount of experience, Silliman was able to discriminate between effusive and deep-seated rocks. The rock resting upon the sandstone to the southeast of East Rock he found somewhat puzzling. "We must pronounce it granitic, although it is not a granite, and inclined to whin, although it is not a whinstone," he wrote, finally concluding that it formed a connecting link between granite and whinstone. Pine Rock and West

Rock were also identified as whin rock and basalt resting upon sandstone. Quartz and sandstone found at Westfield and at the Derby pike were referred to as "micaceous and magnesian schistus." He modestly concluded his paper by adding: "If there are errors [in the above] they are not the result of indolent and remiss inquiry, but of deficient information or erroneous judgment."

In these same *Transactions*, under date of 1808, Silliman, in conjunction with Professor Kingsley, gave an account of the meteorite that fell, in Weston, Connecticut, the year previous. This was the first really scientific description of the phenomena attending the fall of one of these bodies, and its mineral nature, to be given in America, and attracted widespread attention.

Mease's Geological Account of the United States, 1807.

A pudgy duodecimo volume of some five hundred pages, from the pen of a Dr. James Mease, bearing the date 1807, may be mentioned here, not so much on account of the scientific value of its contents as the pretentiousness of its title: *A Geological Account of the United States, Comprehending a short Description of their Animal, Vegetable, and Mineral Productions, Antiquities, and Curiosities.* The portion really geological comprises some thirty pages, though twenty-five more are given up to a catalogue of minerals. The geological part is acknowledgedly a compilation from the writings of Volney, Mitchill, and others. The country is divided into (1) granitic region, (2) region of sandstone, (3) calcareous region, (4) region of sea sand, and (5) region of river alluvions, the boundaries of each of which are given in some detail. The truly scientific (?) character of the work is shown by the following quotation taken from a description of the narrows of the Connecticut River.

> No living creature was ever known to pass through this narrow except an Indian woman, who was in a canoe, attempting to cross the river above it, but carelessly suffered herself to fall within the power of the current. Perceiving her danger she took a bottle of rum she had with her and drank the whole of it; then lay down in her canoe to meet her destiny. She marvelously went through safely, and was taken out of the canoe some miles below quite intoxicated.

Maclure on the Geology of the United States, 1809.

The year 1809 must ever be notable in the history of American geology, since it brought forth Maclure's *Observations on the*

Geology of the United States, with a colored geological map of the region east of the Mississippi. With the exception of Guettard's mineralogical *Map of Louisiana and Canada*, published in 1752, it was the earliest attempt at a geological map of America, and has caused its author to become known as the father of American geology and the William Smith of America.[16]

Maclure's personal history is not without interest in itself, and is worthy of note on account of his relation to American science and as illustrating the conditions under which a man at that date could rise to prominence in geological circles with little or no preliminary training. He was born at Ayr, Scotland, in 1763, and first came to America at the age of nineteen, with a view to mercantile employment, subsequently returning to London, where he commenced his career of commercial enterprise as a partner in the house of Miller, Hart & Company. He seems to have been successful, and accumulated a fortune. In 1796 he again visited America and, it is stated, took the necessary steps for becoming naturalized.

After retiring from business Maclure spent several years in England and on the continent of Europe, traversing the most interesting portions of the Old World from the Mediterranean Sea to the Baltic and from the British Isles to Bohemia. On returning to America he took up the important enterprise noted above. To accomplish this, in the language of his biographer:

> He went forth with his hammer in hand and his wallet on his shoulder, pursuing his researches in every direction, often amid pathless tracts

[16] See article by J. W. Judd, Geological Magazine, London, October, 1897, for account of Smith's work. Professor Judd believed the first geological map in existence to be one of Smith's, published in 1799. This bears the following title: A map of five miles around the City of Bath (England), on a scale of one inch and a half to the mile, from an actual survey, including all the new roads, with alterations and improvements to the present time (1799), printed for, and sold by, A. Taylor and W. Moyler, booksellers, Bath.

The map by Cuvier and Brongniart of the environs of Paris bears the date of 1809, which is the same as that of the first edition of Maclure's. Guettard's "carte Mineralogique" was published in a memoir on the minerals of Canada included in the Memoires de l'Academie Royale in 1756, though the map itself bears the date 1752. It is altogether in black and white and the locations of mines and useful minerals are indicated by symbols. The most striking feature is a broad *"Bande Marneuse"* along the Gulf and southeastern coast, which might at first sight be thought to indicate the coastal plain, but that after leaving Virginia it is greatly expanded and carried over the entire country between Lake Erie and western Maine, even crossing the St. Lawrence and extending up the valley of the Three Rivers. A portion of the "marneuse" (marly) material was evidently the blue-gray clay so characteristic of the river valleys of the glacial period.

PLATE III

WILLIAM MACLURE

and dreary solitudes, until he had crossed and recrossed the Alleghany mountains no less than fifty times. He encountered all the privations of hunger, thirst, fatigue, and exposure, month after month and year after year, until his indomitable spirit had conquered every difficulty and crowned his enterprise with success.

Encouraged by the liberal constitution promulgated by the Cortes, Maclure visited Spain in 1819 having in mind the establishment of a great agricultural school, "in which physical labor should be combined with moral and intellectual culture." Indeed, he seems to have here anticipated, in thought at least, the Morrill Bill and the agricultural colleges of the United States by nearly forty years. His efforts met with failure owing to the overthrow of the constitutional government. On returning to America in 1824 he was still intent on the establishment of such a school, and with this in mind purchased land and became connected with the communistic society at New Harmony, Indiana. Here too he met with disappointment and failure, and in 1827 withdrew, shortly after going to Mexico, where he died in 1840.

He was a liberal patron of science, and for twenty-two years, beginning with December, 1817, was president of the Philadelphia Academy of Natural Sciences, to which institution he subsequently donated his valuable private library and some $20,000 in money.

It was during his connection with the association at New Harmony that he is quoted as refusing to invest money in real estate in the city of Philadelphia, saying: "Land in the cities can no longer rise in value. The communistic society must prevail, and in the course of a few years Philadelphia must be deserted; those who live long enough may come back here and see the foxes looking out of the windows."

Opinions like this and those which follow are repeated here merely to show how impossible it was for any man to realize all that was in store for the United States. The possibilities of railroad transportation were scarcely dreamed of, and mountain barriers and desert plains were looked upon as natural boundaries between various peoples. Following Volney, he divided the country into three distinct and separate parts, differing materially from each other in relative situation and means of communication with the rest of the globe. The natural line of separation between the two main divisions was that formed by the Alleghany Mountains, which, from the poorness of the soil and the difficulty of access to market, he regarded as probably the last portion of the continent to become thickly inhabited.

To the west of this range was the Mississippi basin, which, in his view, was destined, on account of climate and soil, to become a country of immense agricultural possibilities. Traversed by only one great river, which is practically inaccessible to large ships of war, this section, inclosed by the Alleghanies on the east, the Rockies on the west, and the lakes at the north, he felt to be guarded against invasion by the comparative weakness of the mountain neighbors and the impossibility of attack by sea. It would therefore be given over to a nation of agriculturists, the rulers of which would be deprived by nature of even an excuse for keeping either a fleet or army such as had in the past always brought about the ruin of free and equal representative governments. "On this earth or in the page of history it is probable no place can be found of the same extent so well calculated to perpetuate a free and equal representative government as the basin of the Mississippi, both from its physical advantages and the political constitutions on which the state of society is bottomed."

The people of the Atlantic slope, he felt, would labor under entirely different conditions. With an extensive coast, accessible at all points, and with numerous rivers, they were liable to the depredations of a superior fleet and would naturally become a military people, involved in wars with European nations.

Maclure's observations, as already noted, were made in almost every state and territory in the Union, from the river St. Lawrence to the Gulf of Mexico. The memoir which embraced the accumulated facts was submitted to the American Philosophical Society and printed in their *Transactions* in 1809. The map is interesting not alone for its geological coloring, but as showing the paucity of knowledge regarding the physical features of the continent. (See Plate 1.) A continuous range of mountains was figured extending from northern Maine, along its western boundary, through eastern Vermont, western Massachusetts, across southeastern New York and New Jersey to Pennsylvania. More or less parallel ranges were figured extending down into northern Georgia, two of which turned toward the west, the more northerly terminating on the Ohio River just east of the Tennessee, and the other forming the divide between the Tennessee and the headwaters of the Tombigbee River. This last feature was, however, omitted in the second issue, published in 1817.

The classification of the formations adopted by Maclure in the later issue and as given below, was naturally largely Wernerian:

CLASS I. *Primitive rocks.*

(Sienna brown.)

(1) Granite, (2) Gneiss, (3) Mica slate, (4) Clay slate, (5) Primitive limestone, (6) Primitive trap, (7) Serpentine, (8) Porphyry, (9) Sienite, (10) Topaz rock, (11) Quartz rock, (12) Primitive flinty slate, (13) Primitive gypsum, (14) White stone.

CLASS II. *Transition rocks.*

(Carmine.)

(1) Transition limestone, (2) Transition trap, (3) Graywacke, (4) Transition flinty slate, (5) Transition gypsum.

CLASS III. *Flötz or Secondary rocks.*

(Light blue.)

(1) (*dark blue*) Old red sandstone, or First sandstone formation, (2) First or oldest Flötz limestone, (3) First or oldest Flötz gypsum, (4) Second or variegated sandstone, (5) Second Flötz gypsum, (6) Second Flötz limestone, (7) Third Flötz sandstone, (8) Rock salt formation, (9) Chalk formation, (10) Flötz trap formation, (11) Independent coal formation, (12) Newest Flötz trap formation.

CLASS IV. *Alluvial rocks.*

(Yellow.)

(1) Peat, (2) Sand and gravel, (3) Loam, (4) Bog iron ore, (5) Nagel-fluh, (6) Calc tuff, (7) Calc sinter.

His Alluvial class, it will be observed, occupied that portion of the Atlantic border beginning with Long Island and extending southward and westward to the western Louisiana line, comprising the beds now mapped by the United States Geological Survey as in part Cretaceous, but mainly Tertiary and Quaternary, and forming what is known, from a physiographic standpoint, as the Coastal Plain. The materials were described as mainly sands and clays, with considerable beds of shell deposits, and in New Jersey a greenish-blue marl (the Cretaceous glauconitic marls of recent workers), used as a fertilizer. There were also noted deposits of iron ore and ochre. His Primitive class included essentially the area mapped as Archean on the latest United States Geological Survey maps; the Transition, the narrow belt of sedimentaries along the Appalachian range, including the various horizons from Algonkian to Carboniferous; and the Secondary class, all that area to the west now known to be occupied mainly by Carboniferous and Silurian rocks with smaller areas of Algonkian and Cambrian. The red-brown sandstones (Triassic) of the eastern states were classed as Flötz or Secondary and called Old Red Sandstone. This sandstone he however separated by a deeper blue in the 1818 issue from the Secondary rocks on the western side of the range, because of its having a slight dip and agreeing, in the absence of organic remains and its relative position on the sides of many mountain ranges, with the Transition rocks. To the Transition beds he evidently referred all the crystalline limestones and dolomites (marbles) of western New England and the southern states, together with quartzites and graywackes. The roofing slates, now regarded as of Cambrian and Silurian ages, he classed as Secondary. The line between the Primitive and Transition may "perhaps be marked by the presence or absence of organic remains, or of aggregates of rounded particles the result of former decomposition, in part, by the more or less crystalline texture and its approach toward deposition."

To the northwest of the Transition belt lies an immense area of Secondary rocks, comprising, as above noted, the horizontal limestones and slates skirting Lake Champlain about Ticonderoga and Crown Point. There are also "immense beds of secondary limestones, of all shades from a light blue to black, intercepted in some places by extensive tracts of sandstone and other secondary aggregates," which "appear to constitute the foundation of this formation, on which reposes the great and valuable coal formation," which "extends from the headwaters of the Ohio in Pennsylvania, with some interruption, all the way to the waters of the Tombigbee." He noted that along the southeast boundaries of this formation, as on the fork of the Holston in Virginia and in Greene County and the Pigeon River region of Tennessee, gypsum, salt licks, and salt springs had been discovered. In his first map this was indicated by a line of green extending northeast and southwest entirely across the state. In the second he continued it northeast into New York, and he was led to the conclusion, since fully verified, that "we may hope one day to find an abundance of those two most useful substances (salt and gypsum), which are generally found mixed or near each other in all countries that have hitherto been carefully examined." He called attention to the presence of iron pyrites in the coal and limestone, of iron ores consisting principally of brown sparry and clay-ironstone, and of galena in the Mississippi Valley. On the Great Kanawha, near the mouth of Elk River, he noted the presence of "a large mass of black (I suppose vegetable) earth, so soft as to be penetrated by a pole ten or twelve feet deep. Out of the hole so made frequently issues a stream of hydrogen gas, which will burn some time;" and he queried "if a careful examination of this place would not throw some light on the formation of coal and other combustible substances found in such abundance in this formation." The occurrence of large detached masses of granite over an area from Harmony, in Indiana, to Erie, New York, and thence to Fort Ann, in some cases at least 200 miles from any known outcrops, was noted, but no suggestion made relative to their probable means of transportation.[17]

Godon's Mineralogical Observations, 1809.

In this same year S. Godon published in the *Memoirs of the American Academy of Arts and Sciences* a paper of twenty-seven pages on mineralogical observations made in the environs of Boston in 1807 and 1808.

As customary at that time, rocks and minerals were bunched together quite indiscriminately. He divided the minerals into two general groups, (1) *Simple minerals* and (2) *Aggregate minerals.* The simple minerals were then subdivided into *Acidiferous substances* (under which was placed carbonate of lime!), *Earthy substances, Combustible substances,* and *Metallic substances.* The aggregate minerals were classed under *Primordial soil* and *Alluvial deposits.* Under Primordial soil were placed all the primary and consolidated sedimentary rocks of the region. The nomenclature adopted, though somewhat cumbersome and awkward, was not more so than others since devised, the name of any rock being formed by adding the termination *oid* to that of the most characteristic mineral. Thus *amphiboloid, feldsparoid, argilloid,* were the names of

[17] A writer in the Edinburgh Review, commenting on this work, remarked that Mr. Maclure appeared to be thoroughly conversant with his subject, and while a disciple of Werner, his general views were nevertheless much more enlarged and philosophical than is usually met with in a geologist of that school, and further, that like most geologists who have had opportunities of extensive observation, he had found the theory of the Freyburg Professor (Werner) of very limited application. While, however, acknowledging the valuable information which the work contained, the writer was unable to bestow any praise on the manner in which the materials had been put together. "There is a great want of method and arrangement, for although the author has laid down a very good plan, he has not adhered to it, but has mixed up one part of a subject with another so as to cause considerable confusion; and were it not for the accompanying coloured map, it would often be very difficult to comprehend his conclusions."

rocks in which amphibole, feldspar, or clay formed the chief constituent. The rhyolites, quartz porphyries, and felsites of the region were classed as *simple petrosilex* and *compound petrosilex* and the Dorchester conglomerate as *wacke.*

The paper offers a striking illustration of the then prevailing lack of knowledge of the composition of rocks and of chemical methods. Thus an attempt was made at analyzing the "argilloid" by reducing 100 parts to a "subtile powder" and mixing it with an equal part of concentrated sulphuric acid. After standing fifteen days the undissolved portion was removed from the solution, washed, and weighed, whereby it was found to amount to 85 per cent. The solution was allowed to evaporate to dryness, and from the precipitated sulphate of lime thus obtained it was calculated that 5.5 parts out of the 15 soluble were of lime. Alumina and iron were determined by precipitation by ammonia—6.75 parts obtained. The liquid remaining from the alumina-iron filtrate was then dried and heated till the ammonia was driven off, the substance left then dissolved in water and allowed to crystallize, producing an admixture of "well-characterized sulphate of potash and sulphate of soda." "This analysis," while confessedly not of great accuracy, "is sufficient to establish the important fact of the existence of potash and soda as elements in some rocks in this part of the world." Truly an important discovery!

In the same memoirs and this same year Professor Parker Cleaveland, of Bowdoin College, perhaps all unintentionally, started the controversy relative to Glacial and post-Glacial uplift and depression by announcing the finding of fossil marine shells, belonging to genera still living, in deposits of sand and clay well above sea level at Brunswick, Maine.

Bruce's American Mineralogical Journal, 1810.

In January, 1810, there was established by Professor Archibald Bruce the *American Mineralogical Journal*, the first American publication designed primarily for geologists and mineralogists. The object in view, as announced was "to collect and record such information as may serve to elucidate the mineralogy of the United States, than which there is no part of the habitable globe which presents to the mineralogist a richer or more extensive field for investigation." The life of the journal was, however, short, the last issue bearing the date of 1814, and the whole number comprising but 270 pages.

The first paper was by S. L. Mitchill, then professor of natural history and botany in the University of the State of New York. This consisted of an annotated catalogue accompanying a suite of mineral specimens made during a tour to Niagara in 1809.

Among the other papers which followed, mention may be made of one by Colonel Gibbs on the iron ores of Franconia, New Hampshire; by Benjamin Silliman on the lead mines near Northampton, Massachusetts; by James Catbush on the blue earth of New Jersey; and by W. Meade on elastic marble, a "fossil of rare occurrence" found near Pittsfield, Massachusetts.

Fig. 4. *Archibald Bruce.*

The more important strictly geological papers were by Dr. Samuel Akerly and included a geological account of Dutchess County in New York, one on the improbability of finding coal on Long Island or in the vicinity of New York (Manhattan), and one on the geology and mineralogy of the island of New York (Manhattan). Akerly described the highlands of Dutchess County as consisting of granitic rocks, and the whole country north of the highlands as underlaid with primitive slate, most of the hills being composed of limestone. New York Island (Manhattan) he found underlaid throughout its northern part by Primitive rocks, granite, and limestone; the southern part, upon which the then existing city of New York was built, as composed of an alluvion of sand, stone, and rocks. This he considered a recent deposit "subsequent to the creation and even the deluge." He ascribed its deposition to a great flood of waters which he conceived to have at one time been confined to the region north of the highlands and which, by breaking through its southern barriers, had swept over the land carrying its load of "earth, sand, stones, and rocks" to be redeposited on Long and Staten islands and various other places, including the Jerseys. (See Mitchill's views, *ante*, p. 18.)

In the same journal Samuel Brown, of Lexington, Kentucky, gave a description of a cave on Crooked Creek, with remarks and observations on niter and gunpowder; Robert Gilmore, a descriptive catalogue of minerals occurring in the vicinity of Baltimore; while S. L. Mitchill proposed an amendment to Maclure's chart of the United States so far as it related to the character of the north side of Long Island, which he showed to be alluvial and not primitive, as stated, and Benjamin Silliman described the plain of New Haven as wholly alluvial and of very recent origin. This last paper is evidently a partial reprint of one offered before the Connecticut Academy of Sciences in 1810 and referred to elsewhere (p. 30).

Bruce was a practicing physician whose ruling passion is stated to have been a love for the natural sciences and for mineralogy in particular. He formed what was for the time a large and valuable cabinet of minerals and included among his personal friends and correspondents the celebrated French mineralogist, Hauy. He was ever ready to promote the interest of science, as may be inferred from his having established the *Journal.* Toward the close of his life his interest seemed to wane and at the time of his death in 1818, when but forty-one years of age, the publication was already so far in arrears that it was never resuscitated.

A paper by J. Corre de Sarra, the Portuguese minister then residing in Philadelphia, read before the American Philosophical Society in 1815, and published in their *Transactions* in 1818, is of interest as showing the condition of knowledge relative to so commonplace a phenomenon as that of rock weathering and formation of soils. His paper was entitled "Observations and Conjectures on the Formation and Nature of the Soil in Kentucky." He considered this soil the product "of the decomposition of an immense deposit of vegetables which the ocean had left uncovered by any other deposition."

The New Madrid Earthquake of 1811.

It would seem remarkable that no account is to be found in the early geological literature of a phenomenon so widespread and unusual as the earthquake of New Madrid, Mississippi, in 1811. Means of communication and dissemination of information were naturally slow, and there were in existence at that period few periodicals devoted to this and allied subjects. Nevertheless, the subject offers so inviting a field for speculation that it is difficult to account for its being ignored. The first reference I find bearing upon it (with the exception of that of Maclure, note on p. 11) is contained in a letter from Mr. L. Bringier to the Reverend E Cornelius, dated March 20, 1818, and by him transmitted to the editor of the *American Journal,* in which it was published in 1821.

The reference is brief and chiefly of interest from the fact that the writer was rightly opposed to attempting to account for the same through volcanic agencies. He, however, thought it might have been due to electrical causes.[18]

F. W. Gilmer's Ideas on the Natural Bridge, 1816.

It may be remembered that Jefferson, in his *Notes on the State of Virginia,* had described the now well-known Natural Bridge of Rockbridge County as spanning a gigantic fissure, the result of some great convulsion. In the Transactions of the American Philosophical Society for 1816 (1818) Francis William Gilmer had a paper on the same subject, illustrated by a full-page plate. The bridge was described in detail, but its formation was ascribed, not to a sudden convulsion, as argued by Jefferson, or to any extraordinarily sudden deviation from the ordinary laws of nature, but to the "very slow operation of causes which have always and must ever continue to act in the same manner." This cause he rightly considered to be the solvent action of meteoric waters on limestone. In this respect Gilmer, although scarcely known to geological science, was considerably in advance of the writers of his day.

Cleaveland's Treatise on Mineralogy and Geology, 1816.

With the exception of Maclure's *Observations on the Geology of North America,* undoubtedly the most important of the early publications was Parker Cleaveland's *Elementary Treatise on Mineralogy and Geology,* a work of upward of six hundred pages, with five plates of crystal drawings and a colored geological map.

For most of the geological observations and for the map Cleaveland was indebted to the *Observations* and to written communications from H. H. Hayden. The principal variation from Maclure's map lay in the adoption of the suggestion of S. L. Mitchill regarding the extension of the alluvial deposits on Long Island. It further differed from the 1817 edition in that a large portion of southern and western Maine, left blank by Maclure, was colored as occupied by Transition rocks.

The classification adopted was largely chemical, the minerals being divided into (1) classes, (2) orders, (3) genera, and (4) species, his definition of species being—"a collection of materials

[18] The bibliography relating to this earthquake is by no means scanty. A very full discussion of it is given by M. L. Fuller in Bull. No. 494, U. S. Geol. Survey, 1912.

which are composed of the same ingredients, combined in the same proportions."

In the tabular view given, all known minerals were grouped under four classes, as follows: Class 1, *Substances not metallic,* composed entirely or in part of an acid; class 2, *Earthy compounds* or stones; class 3, *Combustibles;* and class 4, *Ores.* As was the case with all writers of that day, basalt and several other compact rocks of indistinct mineralogical nature were classed as minerals. Thus is found under class 2, species 15, *porcellanite;* species 16, *silicon slate;* species 32, *emerald;* and species 34, *basalt.* This last was described as never crystallized but occurring "in large amorphous masses, but also under a columnar, tabular, or globular form" and passing insensibly into "greenstone, wacke, and perhaps clinkstone."

Not only was the mineralogical nature of basalt still little understood, but even its relationship to volcanic rocks as well. It was described as sometimes found in countries decidedly volcanic, but seldom near the craters of still active volcanoes; "on the contrary, it appears at the foot of volcanic mountains and sometimes almost surrounds them." Some of the most noted localities mentioned are the Giant's Causeway, island of Staffa, the Erzgebirge, Auvergne, etc. Its occurrence in the United States was thought to be doubtful, though he noted its reported occurrence on the Stony (Rocky) Mountains. It was regarded as of both igneous and aqueous origin.[19]

Anthracite, which forms species 6 under class 3, *Combustibles,* was described as "strongly resembling *coal,* from which, however, it materially differs." It was said to occur in Primitive or Transitional rocks, though sometimes connected with Secondary rocks. Obviously, therefore, it is stated, it "has not, at least in many cases, proceeded from the decomposition of vegetable substances which are not supposed to have existed when the primitive rocks were formed."

Pages 586 to 636 of Cleaveland's first edition included an "Introduction to the Study of Geology." Some of the statements here made

[19] The question of the origin of basalt was repeatedly solved (?) by the earlier geologists, each of the two schools—Neptunist or Plutonist—in its own way, and sometimes with an assurance that is today highly amusing. S. L. Mitchill in his Repository for 1799 describes some columnar iron ores which obviously owed their form to shrinkage through loss of water. "The specimens before me prove that argillaceous ore, which nobody has supposed to be a volcanic production, can take on a basaltic figure." This occurrence he thought settled the Neptunian-Wernerian controversy in favor of the former. "I am inclined to believe the weight of testimony is opposed to the formation of basaltic columns by fusiòn and the true manner in which many if not all of them have been produced is in the *Moist way.*"

PLATE IV

PARKER CLEAVELAND

are of interest: "Most of those extensive masses of strata with which geology is concerned are compound minerals, or aggregates, composed of two or more simple minerals mingled in various proportions and denominated rocks." While thus a division into species and genera may be possible with minerals, with rocks the case is quite different. "It is obvious that they can not admit of distinctions which are strictly specific." This observation holds good to the present day. His remarks on the position of the beds of rocks were no less interesting:

When primitive rocks are stratified the strata are seldom horizontal. On the contrary, they are often highly inclined and sometimes nearly or quite vertical. But whether these strata were originally inclined or whether, subsequent to their formation, they have been changed from a horizontal to an inclined position by the action of some powerful cause, is a question on which the most distinguished geologists are divided in opinion.

Again:

It is further evident that the higher the level at which any rock appears at the surface of the earth, the older is the rock; for it so declines as to pass under those rocks which appear at a lower level. The only exception to this general fact appears in those horizontal strata of secondary rocks which sometimes rest on the summits of high mountains.

The theory of the earth adopted by Cleaveland is apparently that of Cuvier. In his discussion of the origin of rocks and geological systems he was for the most part a follower of Werner, though he recognized some of the difficulties attendant upon the complete adoption of the Neptunian theory. "Though its general outlines may be correct, we are yet unable to give its details. It seems, however, to be rather encumbered with difficulties than absolutely confronted by existing facts." His views on volcanoes were largely a reflection of those advocated by his European contemporaries, and need but brief allusion.

Many parts of the external crust of the earth are subject to the action of subterraneous fire. In some cases these fires are comparatively mild and produce no important effects, excepting the destruction of the combustibles which feed them, and are nothing more than coal mines in a state of combustion. But in other cases these subterraneous fires rage with resistless impetuosity.

This book was hailed by a reviewer in the *Analectic Magazine* for 1817 as "one of those solid and judicious compilations that increase the mass of useful knowledge among us and promise to add to the best part of our literary reputation." Objection was, however, raised to attaching so much importance to the primitive form and structure of the crystal, under each species, since "the primitive forms of crystals are not of the greatest, but of the least, practical importance to a mineralogist." "Neither is attention paid to the primitive form in practice: we want to discover in a country we are exploring, gypsum or limestone. We are told, that the primitive form of gypsum is a four sided prism, whose bases are parallelograms with angles of 40 degrees 8 min., and 66 degrees 52 min., the sides of the base and the height of the prism as the numbers 12, 13, 32. Of limestone, we are told, that it is a rhomb with two acute angles, 78 deg. 28 min., and two obtuse angles, 101 deg. 32 min. The faces of this rhomb are inclined to each other, at right [*sic*] angles of 75 deg. 31 min. and 104 deg. 42 min. This very ingenuous, but too theoretical system of Hauy and his followers reminds us of the playful objection of *Simplicius* to the philosophical theories of *his* day. A man, says he, meets another in a lane; and wishes to inquire whether a friend of his has lately passed that way. Must he say, Friend, have you lately met a man going down the road? No! He should address him in the language of philosophy,—Animal, rational, visible, two-legged and without feathers, five feet eight inches high, have you lately met, during your peripatetic excursion, another animal, rational, visible, two-legged, and without feathers, six feet high? For practical purposes a small vial of acid and a blowpipe are of far more value than a goniometer."

The same reviewer remarks on that portion of the work dealing with basalt: "But it seems strange to say, that this mineral is never crystallized! Is not all columnar basalt crystallized? Is there more anomaly in the sides of a crystallized column of basalt than in a crystal of carbonate of lime? What definition of crystallization will exclude either the prismatic column or its articulations?"

Still another reviewer naïvely remarked that "in this age of book making it is no small negative praise if an author be acquitted of unnecessarily adding to the onerous mass of books." This writer incidentally commented upon Hauy's "curious discoveries regarding the six primitive figures which form the base of all crystals" and referred to it as "the most singular and acute discovery of our age." He adds, however, "There is a difference of opinion among mineralogists as to the practical use of crystallography in the discrimination of minerals."

With reference to the above it may be stated that Cleaveland sided with the French school in the opinion that the composition of minerals, so far as it is known, should be the basis of classification, but that when this is not known the external characters might be provisionally employed, and further, that while minerals might be arranged according to their internal composition they could be best described by their external characters.

He was so far disinclined to indulge in theory in teaching geology that he would describe a vein as *inclosed by* a rock rather than as *traversing* or *perforating* it, preferring a term which simply described the phenomenon without suggesting any theory as to the manner of formation. He is stated to have remarked once in reply to an inquiry that he did not believe that facts enough had been extended to warrant the sweeping generalizations of modern geology; that the more his knowledge of the facts of science increased the less confidence he had in any of the theories.

Cleaveland was a graduate of Harvard in the class of 1799, and later became professor of mathematics, natural philosophy, chemistry, and mineralogy in Bowdoin College, Brunswick, Maine. As with Silliman, he had at the time of his appointment (1805) received little or no training for the position he was to fill. For several years after graduation he seems to have been undecided between the profession of divinity and law, to neither of which was he strongly drawn. His interest in mineralogy was aroused more or less accidentally, his attention having been called to crystals of mica, quartz, and pyrite which were found by workmen engaged in blasting a ledge in the adjacent town of Topsham. On the not unnatural supposition that these might be of value they were referred to the college professor, who in this particular instance knew nothing specifically and had only a faint idea where to look for information. After perusing the mineral portion of Chaptal's *Chemistry* and making a provisional determination he packed a box of the specimens and sent it to Dr. Dexter, then professor of chemistry at Harvard, who verified the determinations and sent a small collection of named minerals in return. Thus began the career of one who was to take front rank in forwarding the science of mineralogy in America, and who afterward became an influential lecturer in medical chemistry.

In external appearance Cleaveland is said to have been stern and austere, and on any sudden provocation he was sometimes passionate and violent. But he was nevertheless large-hearted and possessed an exhaustless vein of kindly and generous feeling. For upward of fifty years he filled the position above outlined, at Bowdoin, even insisting on meeting his classes to the very last, dying, in fact, at the age of eighty years while preparing for his daily duties.

He is described by his biographer as possessing great natural ardor and activity, with a reputation for zeal, industry, and learning. To these statements certainly no one will take exception when the period, the magnitude of the work, and the manifold character of professional duties are taken into consideration. It must be remembered that his was the first attempt made in America at a systematic treatise on mineralogy, Shepard's book coming sixteen and Dana's twenty-one years later. The available foreign works were those of Jameson, Kirwan, Werner, and Brongniart; moreover, there were in America few, if any, important mineral cabinets.[20]

The work was received with great favor, two editions being issued, the second in two volumes bearing the date of 1822. A third was called for but never prepared, owing to the author's time being diverted by the chemical work of the newly organized medical department.

Second Edition of Maclure's Observations, 1817.

Not satisfied with his first edition, which, according to Marcou, was published during the author's absence in Europe, Maclure immediately set about obtaining the necessary information for a revision, and in 1817, after eight years of hard work, presented the amended memoir, which was republished in the *Transactions of the Philosophical Society* and issued also in the form of a separate volume of 127 pages, bearing the same date. The second issue differed considerably from the first, the most conspicuous features being the delineation of the mountain ranges and the correction of many minor details relating to the distribution of the various geological formations.

On Plate 2 of this issue were given five sections across the United

[20] The earliest mineralogical cabinet brought to America is stated to have been that of David Hosack. This was ultimately given to Princeton College, and it is stated that its identity has become completely lost. Other important cabinets were those of Archibald Bruce and George Gibbs. According to S. G. Gordon (American Mineralogist, Vol. IV, No. 2, 1919), the second cabinet brought to this country was that of Adam Seybert. This is still intact in its original condition among the collections of the Philadelphia Academy. Gordon also mentions the names of Thomas T. Smith, Sylvanus Godon, James Woodhouse, Girard Troost, Lardner Vanuxem, William Keating, and Thomas Nuttall, among the early mineralogists who doubtless possessed cabinets of more or less importance. Goode, in his Beginnings of American Science, mentions also the name of B. D. Perkins.

States from the Atlantic through the Appalachian regions and the Secondary rocks of the Mississippi, which were colored to correspond with the scheme shown in the map. (See Plate 1.) The first section extended from Camden on Penobscot Bay, Maine, to Oxboro, near Kingston, on Lake Ontario; the second from Plymouth, Massachusetts, to Cayuga Lake, New York; the third from Egg Harbor, New Jersey, to Pittsburgh, Pennsylvania; the fourth from Cape Henry to Arlington, Virginia; and the fifth from Cape Fear to Warm Springs, North Carolina.[21] The poverty of information relative to the region west of the Mississippi is shown in the statement that "the tops of the Stony Mountains are covered to a considerable extent with perpetual snows and pendent glaciers."

As was the case with a majority of the earlier surveys and geological textbooks, considerable attention was paid to the relation of the science to agriculture, and the two closing chapters were devoted to "Hints on the Decomposition of Rock," with an "Inquiry into the Probable Effects they may Produce on the Nature and Fertility of Soils." The kinds of soils resulting from the decomposition of various rocks were discussed with particular attention to their physical nature, but while occasional references were made to their content of lime, the alkalies, and other constituents, no chemical analyses were given, nor was their desirability apparently appreciated; nor does it seem to have been realized that throughout the entire glaciated area, as now known, there might be little connection between the soils and the rocks immediately underlying them.

J. F. and S. L. Dana's Geology and Mineralogy of Boston, 1818.

In this review of the early publications on the geology and mineralogy of New England mention must be made of J. F. and S. L. Dana's *Outlines of the Mineralogy and Geology of Boston and Vicinity,* which bears the date of 1818. This is noteworthy for containing a geological map of the area indicated, being antedated only a year by Edward Hitchcock's geological map of a part of Massachusetts on the Connecticut River and the various reprints of Maclure's, from which it was doubtless largely compiled. The classification adopted was that of Werner, and hence purely lithological. The various rocks enumerated and colored on the map were: I, *The*

21 These sections, as customary at that time, were printed in black and white and afterward colored by hand. As a natural result copies of the work are occasionally found in which the coloring has been omitted.

Primitive, including granite, argillite, primitive trap, porphyry, and syenite; II, *Transition,* including amygdaloid and graywacke; and III, *Alluvial,* including sand, pebbles, clay, and peat. No granite was, however, recorded as in place, that of Quincy and the areas west of north of Marblehead being called syenite. The argillite was considered the oldest rock occurring in the vicinity, and was represented as forming gently undulating eminences in Charlestown, Watertown, Chelsea, and Quincy. Greenstone, or primitive trap, was represented as occupying all of the Marblehead-Salem areas and large areas to the west, including Stoneham and Lexington. Porphyry, "a compound rock having a compact basis, in which are embedded crystals or grains of other minerals of contemporaneous formation," they found in Malden, Lynn, and Chelsea, while the large area beginning at the shore east and north of Lynn and extending southwestward as far as Malden, was colored as *petrosilex.* A wide strip, extending from just west of the Charles River to the coast, including Brookline, Roxbury, and Dorchester, was colored as *graywacke,* this being the conglomerate of later writers. On the northern and western edges of this are narrow belts colored as *amygdaloid,* the same being the rock later shown by Benton to be melaphyr, or ancient basaltic lava flow.

The minerals were divided into Class I, *Earthy fossils;* Class II, *Saline fossils;* Class III, *Inflammable substances;* and Class IV, *Metallick fossils.* Under Class I were included the phosphates and carbonates of lime, quartz, such silicates as mica, shorl, feldspar, garnet, epidote, the amphiboles, etc., and such compound substances as petrosilex, basalt, wacke, schaalstone, argillaceous slate, and clay. The second class, Saline fossils, included but a single species, sulphate of iron. Class III included hydrogen gas and peat, and Class IV sulphides and chlorides of copper, sulphide, oxide, and carbonate of iron, sulphide of lead, and oxide of manganese. The classes were subdivided into orders and the orders into genera, species, subspecies, and varieties. Thus novaculite was considered a subspecies of argillaceous slate, a species under Order II, Nonacidiferous substances of Class I, Earthy fossils. Altogether some 21 species of Earthy fossils were recognized, 1 of Saline fossils, 2 of Inflammable substances, and 8 of Metallick fossils. These were all described in detail, their physical characters and conduct before the blowpipe, and their place of occurrence. Perhaps no better means of showing the condition of the science at that time can be found than by reproducing here a page of the original descriptive matter under Class I, Order II.

SPECIES VII—WACKE.

Wacke, *Cleaveland,* p. 287. Wacce, *Jameson,* vol. 1, p. 376. Wacken, *Kirwan,* vol. 1, p. 223. Wakke, *Aikin,* p. 254.

External characters.

Its colours are grey and purple. Of grey, it occurs blackish grey and greenish grey; of purple, lavender purple. The colours vary much in their intensity.

It is dull.

It is amorphous and cellular.

It exhales a strong argillaceous odour when breathed upon.

It adheres to the tongue.

Its streak is greyish white, with a reddish purple tinge in some parts, and is dull.

It is moderately hard, passing to soft.

The fracture is from fine grained uneven to earthy; some specimens show a slightly slaty structure.

It is brittle.

It is easily frangible.

The fragments are indeterminately angular and not particularly sharp edged.

Its specifick gravity is about 2.88.

Chymical characters.

Before the blowpipe it melts into an opaque, semi-vitreous mass which appears porous when broken.

Geological situation and localities.

It occurs in beds in Petrosilex at Milton, and forms the basis of Amygdaloid at Brighton, Hingham, Newton, &c., and it is found also in rounded fragments at Needham, Newton, Brighton, &c.

Remarks.

This mineral sometimes much resembles ferruginous clay, and is intermediate between Clay and Basalt. It is very liable to decomposition, and when it forms the basis of Amygdaloid, by undergoing this change, it leaves the imbedded minerals projecting, or they fall out and leave the Wacke cellular.

This work was the subject of a scathing review in the *Analectic Magazine,* XIII, 1819, where the writers were accused of having borrowed at least three-fourths of their material "for the mere purpose of eking out the matter to the proper size of a *justum volumen* —of borrowing the most elementary ideas of the most common authors. . . . A student who has read either Cleaveland, Jameson, or Aikin, will find not a sentence that is new in at least nine-tenths of the book; which is in fact a disgraceful example of literary book-making, as it respects both the matter and the manner."

Mitchill on the Geology of North America, 1818.

In 1818 there was published by Kirk and Mercein in New York Robert Jameson's translation of Cuvier's celebrated *Essay on the Theory of the Earth,* and with it S. L. Mitchill's *Observations on the Geology of North America.*[22] It is the *Observations* alone that need now receive our attention, and this with particular reference only to what is said regarding the origin of the drift. Mitchill was one of the most prominent scientific men of his day, as has been repeatedly noted, had read and traveled extensively, and it may be safe to assume that the views held by him were supposed to rest upon a good foundation, though they were not wholly accepted, as is mentioned later. Many of his ideas, it will be noted, were strikingly similar to those of Volney (p. 24).

Mitchill taught that the Great Lakes were the shrunken representatives of great internal seas of salt water, which ultimately broke through their barriers, the saline lakes becoming gradually freshened by a constant influx of fresh water. The remains of the barriers which held back for a time this inland sea he thought to be still evident. One of them, he wrote, seemed to have circumscribed to a certain degree the waters of the original Lake Ontario and to be still traceable as a mountainous ridge beyond the St. Lawrence in upper Canada, northeast of Kingston, passing thence into New York, where it formed the divide between the present lake and the St. Lawrence, and, continuing to the north end of Lake George, apparently crossing the Hudson above Hadley Falls. Thence he believed it to run toward the eastern sources of the Susquehanna, along the Cookwago and Papachton branches of the Delaware, crossing the last-named a little north of Easton (Delaware Water Gap), the Lehigh north of Heidelburg (Lehigh Gap), and the Schuylkill northwest of Hamburg in Pennsylvania; continuing thence along to the north of Harrisburg, across the Susquehanna, in a southwesterly direction until it entered Maryland and passed the Potomac at

[22] There was manifested, particularly by the publishing houses of Philadelphia, an early tendency to reprint many of the English works on geology and mineralogy. This finds its most important illustration, from the present standpoint, in the reprinting of Frederick Accums's A Practical Essay on the Analysis of Minerals (Philadelphia, 1809); Cuvier's Essay, as noted above; Bakewell's An Introduction to Geology (New Haven, 1829); Cuvier's A Discourse on the Evolution of the Surface of the Globe (Philadelphia, 1831); and De la Beche's A Geological Manual (Philadelphia, 1832), etc.

Harpers Ferry into Virginia, where it became confounded with the Alleghany Mountains; thence gradually disappearing. Traces of it appeared to the westward, as at Cumberland Gap in Tennessee and the mountains west of Cape Girardeau beyond the Mississippi.

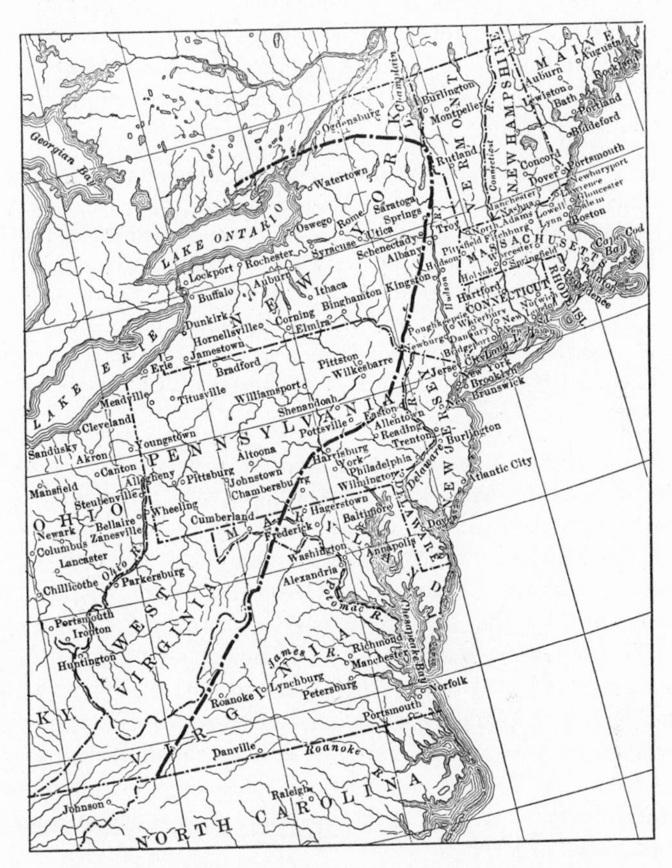

Fig. 5. *Map illustrating Mitchill's theory of barriers.*

It is evident that for a good part of its course, as traced, this barrier was but the Blue Ridge, while in eastern New York it was comprised mainly of the Catskills and Adirondacks.

To appreciate Mitchill's view, then, we have to imagine this now broken and gapped ridge as continuous throughout its whole extent, forming a vast dam holding back the waters of several salt inland seas to the northward, while the region to the southward was dry land. A time came, however, when the dams at various points proved too frail and gave way, the pent-up waters rushing through and carrying devastation with them like the waters from cloud-bursts or bursting reservoirs of today, but on a thousand-fold larger scale.

One breach was conceived to have been at the northeastern extremity of Lake Ontario. The Thousand Isles, to Mitchill's mind, "bear witness to the mighty rush of waters which thus prostrated the opposing mound and left them as scattered monuments of the ruin." "By this operation the water must have subsided about 160 feet," or to its present level.

> All the country on both the Canadian and Fredonian sides must have been drained and left bare . . . exposing to view the waterworn pebbles, the works of marine animals, their solid parts buried in the soil, their relicks bedded in the rocks, and the whole exhibition of organic remains formed in the bottom of such a sea as that was.
>
> Great masses of primitive rocks from the demolished mound or dam and vast quantities of sand, mud, and gravel were carried down the stream to form the curious mixture of primitive with alluvial materials in regions below.

A second breach was conceived to have been at the northern extremity of Lake George, whereby the lake was diminished to about its present size. A third was at Hadley Falls; a fourth at the upper falls of the Mohawk; a fifth was made by the pent-up waters of the Delaware above Easton (the Delaware Water Gap), Pennsylvania; a sixth by the Lehigh to the northwest of Bethlehem (the Lehigh Water Gap); a seventh by the Schuylkill; an eighth by the Susquehanna, and a ninth by the Potomac cutting its way through the Blue Ridge at Harpers Ferry. A second series of lakes and dams he conceived to have existed outside, *i.e.*, to the southward of those above mentioned. To the bursting of these he attributed many of the minor features of the present landscape. (See also Volney's views, p. 29.)

Little in the way of systematic geology is given, aside from the speculations above mentioned, though there are disconnected references to, and descriptions of, fossil remains and rocks found in various parts of the country. He, however, called attention to the possibility of the Great Lakes having formerly drained into the Mississippi, and to the gradual retreat of the falls of Niagara and the formation of the gorge through the undermining of the harder surface limestone, facts which seem to have been very early recognized. The work was favorably reviewed in the newly founded *American Journal of Science,* the writer incidentally remarking that he did not feel called upon to give an opinion regarding Professor

Mitchill's theories. A Mr. J. W. Wilson, whose name appears not elsewhere in this chronicle, did, however, venture to subject them to serious criticism. In this same *Journal* of 1821, he reviewed the theories of both Mitchill and Volney, and gave altitudes showing that the waters of the hypothetical lake could never have risen within 500 feet of the summit of the highlands at West Point without discharging over the tops of the mountains at Harrisburg, or through Lake Champlain into the Sorel and St. Lawrence before rising within 1,400 feet of Butlerville near West Point. These were good arguments, but explanations offered by Mr. Wilson were by no means satisfactory. "Is not that the best theory of the earth," he wrote, "that the Creator in the beginning at least formed it with all its grand characteristic features?" To this Silliman responded in a footnote: "The Creator undoubtedly brought the matter into being and established the laws which governed it; the operations of those laws then is always a fair subject for discussion and although it is the shortest it is not the most instructive course, to cut the knot when it may be untied."

The American Journal of Science, 1818.

The need of a periodical publication devoted mainly to geology and the physical sciences had early made itself felt. Bruce's *Mineralogical Journal*, already mentioned, which but partially filled the need, at best, had suspended publication in 1814, owing in part to the ill health of the editor and founder, and there seemed no prospect of resuscitation. It is stated by Silliman's biographer that the subject of a new journal was first broached by Colonel George Gibbs in 1817, and that the matter was fully decided after a consultation by Silliman with Dr. Bruce in the autumn of that year. The first number appeared in July, 1818, but a few months after Bruce's death. Thus came into existence the *American Journal of Science*, or *Silliman's Journal*, as it was long popularly known. The fact of its establishment is worthy of especial note as it was the first American journal to be devoted largely to geological subjects, although other sciences were by no means debarred. In fact, as noted by Silliman in his prospectus, the *Journal* was intended to embrace the circle of the physical sciences and their application to the arts and to every useful purpose.[23] From its earliest inception geological and mineralogical notes and papers occupied a prominent place in its pages, and a perusal of the numbers from the date of first issue down to the

[23] See Vol. L, 1846, of the Journal for a full history of its rise and progress.

present time will, alone, afford a fair idea of the progress of American geology. I shall note in the following pages many articles in this *Journal*, particularly the earlier numbers, as serving best to illustrate the condition of the science at the time of their publication.

John Kain on the Geology of Virginia and Tennessee, 1818.

Following out this plan the first paper to which attention need be called is one by Mr. John H. Kain, who wrote on the geology of the northwestern part of Virginia and the eastern part of Tennessee. This paper is apparently Kain's sole contribution to geology, but is of interest since, while dealing mainly with the mineralogical and lithological character of the region, it is nevertheless more comprehensive and more nearly systematic than anything that had preceded it. He noted that the region presented, with few exceptions, "none of the remarkable appearances which indicate the changes and convulsions which have been wrought by time, the great enemy of nature," the mountains seemingly having lain for ages in undisturbed repose, subject only to the erosive action of running streams. The rocks were described as Transition sandstones and limestones, the latter inclined at angles of twenty-five to forty-five degrees and bearing impressions of shells. The Natural Bridge of Virginia, he thought, "when viewed by a geologist," would probably be considered as a cave unroofed in all but one place, as he could not conceive of its having been made by a convulsion which had split and separated the rocks to a distance of fifty or sixty feet apart, leaving no indications of violence in the remaining span. This view is in accord with that expressed by T. W. Gilmer, noted elsewhere, though both opinions seem to have been overlooked or disregarded by the writer noted below. According to Glenn[24] this paper of Kain's contained the first distinct geological description of any part of the state of Tennessee, and gave the first definite information as to the use of the coal and mineral paint of the region.

The area included in Kain's observations as noted above, together with the adjacent portions of the Alabama and Mississippi territories, was afterwards visited by the Reverend Elias Cornelius and described in a subsequent number of the *Journal* for the same year. Cornelius would seem to have been a man of education and culture, but one whose professional training had quite unfitted him for the work of a geologist. In the course of his discussion he took occasion to refer to the Natural Bridge, which he described in some detail,

[24] Resources of Tennessee, Vol. II, No. 5, 1912.

but while he too dissented from those who had attempted to account for its formation through a convulsion of nature, in the sublimity of his faith he had "no difficulty even in supposing it to have proceeded from the hand of the Almighty as it is." He made no reference, however, to Kain's and Gilmer's papers. He referred briefly to the sandstones of Aquia Creek, used in the public buildings in Washington, and the calcareous breccia at Point of Rocks, Maryland, and noted that in crossing the country north of Virginia from the Atlantic to the interior he crossed successively the most important formations of the earth, from the most recent Alluvial to the oldest Primitive, and that the limestone of the mountain region was inclined at angles of twenty-five to forty-five degrees, while that to the west was nearly always flat or horizontal. Cumberland Mountain was described as of the table mountain type.

Edward Hitchcock's First Geological Paper, 1818.

This same year Edward Hitchcock, then a young theological student of twenty-five, but who was destined to be one of the most prominent figures of his time, made his appearance in the geological field through the presentation of a paper which, when all is taken into consideration, was really a remarkable one. Hitchcock came first into public notice in 1815 through some astronomical observations and corrections furnished Blunt's *Nautical Almanac.* His inclination, however, early took a geological turn, and throughout a prolonged period of activity, first as a clergyman and later as professor, president, and again professor in Amherst College, he kept himself ever prominently to the front.

The first state geological survey carried to completion, that of Massachusetts, 1830-1833, was primarily of his conception and was executed almost wholly through his efforts, as noted later (p. 142). He became, however, most widely known and is best remembered through textbooks, his work on the footprints found in the Triassic sandstones of the Connecticut Valley and his studies of the drift phenomena, in both of which last-named subjects he was a pioneer. Indeed, if one may be allowed to speak facetiously of so cultured and dignified a gentleman, he was America's first "superficial geologist."

The paper to which allusion is made above, *Some Remarks on the Geology and Mineralogy of a Section of Massachusetts on the Connecticut River, with a Part of New Hampshire and Vermont,* was published in the first volume of the *American Journal of Science.*

It is noteworthy on account of a geological map of the region, colored by hand,[25] and a transverse section of the rock strata from Hoosac Mountain to a point eleven miles east of the Connecticut River. The rocks were classed as Primitive, Secondary, and Alluvial, the older crystallines and the argillites being considered Primitive, and the traps and sandstones Secondary.

Eaton's Index, 1818.

The year was rendered further notable by the appearance in the geological arena of Amos Eaton, a man who, like Hitchcock, was destined to achieve a national reputation, but whose mental characteristics were as unlike as was possible among men in the same calling. His first geological paper occupied three pages of the *Journal* for the year under discussion. His first noteworthy publication was the *Index to the Geology of the Northern States*, which appeared in the form of a textbook for the geological classes at Williams College that same year. Eaton is described as a man of great force, untiring energy, and one of the most interesting men of his day. In 1816, at the age of forty, he abandoned the practice of law and went to New Haven to attend Silliman's lectures on mineralogy and geology. Subsequently he traveled many thousand miles on foot, throughout New England and New York, delivering in the principal towns short courses of lectures on natural history. In March, 1817, having received an invitation to aid in the introduction of the natural sciences in Williams College, his alma mater, he delivered a course of lectures in Williamstown. Such was the zeal at this institution, he wrote, that "an uncontrollable enthusiasm for natural history took possession of every mind, and other departments of learning were for a time crowded out of the college." In April, 1818, on invitation of Governor De Witt Clinton, he delivered in Albany, before the members of the State Legislature, a course of lectures on his favorite subject. Here was undoubtedly the beginning of the work which resulted in the establishment of the state survey in 1836.[26]

[25] The laborious and time-consuming work of coloring by hand each individual copy of the map in an edition even no larger than that of the Journal at this time can scarcely be appreciated by the modern worker.

[26] Eaton's activities were by no means limited to geology. In 1810 he had published an elementary treatise on botany, and in 1817, it is stated, the first Manual of Botany issued by an American author. The latter work passed through eight editions, the last, dated 1840, being a large octavo volume of 625 pages.

PLATE V

AMOS EATON

In the *Index* mentioned above, which has been pronounced "the first attempt at an arrangement of the geological strata of North America," the views expressed were largely tinged with Wernerism. They are reviewed in detail here, even when almost exact equivalents, on account of their local application.

He divided the rocks of the earth's crust into five classes: First, *Primitive;* second, *Transition;* third, *Secondary;* fourth, *Superincumbent;* and fifth, *Alluvial;* the body of the work, occupying pages fifteen to forty-one, inclusive, being given up to their description and geographical distribution.

Under the head of Primitive rocks he included *granite, granular limestone* and *quartz, gneiss, mica-slate, soapstone rocks, calcareous* and *granular quartz,* and *syenite.* These were barren of fossil remains and the oldest rocks to which human research had extended.

The Transition class included *metalliferous limestone, argillaceous* and *siliceous slate, graywacke slate,* and *rubblestone.*

The Secondary class included *red sandstone, breccia, compact limestone, gypsum,* and *rock salt;* the Superincumbent class, *basalt, greenstone, trap,* and *amygdaloid,* and the Alluvial class, *gravel, sand, clay,* and *loam.*

Illustrative of his ideas concerning the position of these rocks, the geological transverse section here reproduced was given, extending from Boston on the east to the Catskill Mountains on the west. In discussing the position of the gneiss of the Primitive series as it occurs between Dalton and Pittsfield, Massachusetts, he wrote: "It sinks laterally under the mica-slate to the west, and probably does not rise again until it reaches the

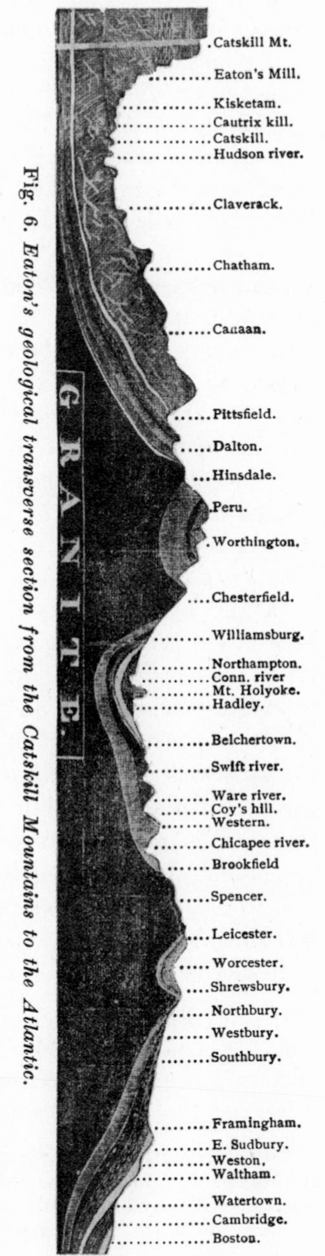

Fig. 6. *Eaton's geological transverse section from the Catskill Mountains to the Atlantic.*

continent of Asia." The soapstones and serpentines were thought to form one of the concentric coats of mechanically deposited materials, and their possible eruptive origin was not dreamed of.

In his Transition class he included the marbles of western Vermont, now considered to be of Lower Silurian and Cambrian age, and the roofing slates.

In the Secondary class were included the red sandstones of the Catskills (Devonian), as well as those of the Connecticut River (Triassic). Discussing the position of the Catskill stone, he wrote: "Bakewell removed this stratum from the Secondary class, where Werner placed it, to the Transition. He says this stratum terminates the series of transition rocks containing metallic veins and the more ancient organic relics. Had Bakewell ever visited Catskill Mountain he would undoubtedly have left the red sandstone where Werner placed it; for here the true old red sandstone of Werner contains the organized remains of at least one well-known phenogamous woody plant."[27] He also stated that this sandstone contained the petrified remains of the roots of the *Kalmia latifolia*, or common laurel! The gypsum and rock salt beds of New York and Pennsylvania, now considered of Salina age, he included in this Secondary class, and also the common compact limestone occurring in the western part of New York State.

In the Superincumbent class were included, as already stated, greenstone trap, amygdaloid, and basalt, which he considered varieties of one and the same rock. This assumption on his part is interesting in connection with the discussion of more recent years (since the introduction of the microscope into geology) regarding the now well-established relationship existing between basalt, melaphyr, and diabase. While these rocks were believed to be volcanic, their exact source was problematic. "On the Deerfield River the greenstone sinks down in a fissure in the red sandstone. . . . Bakewell would say here were volcanoes and here the melted greenstone was thrown up through the sandstone."

In discussing his Alluvial class, he wrote:

It is agreed by all geologists that all soils, excepting what proceeds from decomposed animal and vegetable matter, are composed of the broken fragments of disintegrated rocks. From this fact it is natural to infer that the soil of any district might be known by the rocks out of

[27] Concerning this Hall remarks (Geology of Fourth District, p. 6): "It is a remarkable fact that at this early period Mr. Eaton should have recognized the sandstone of the Catskill Mountains as the Old Red of Europe, which, now that we have identified its characteristic fossils, is proved to be true."

which it is formed; consequently, that rocks abounding in quartz would produce a sandy soil and those abounding in argillaceous slate a clay soil, etc. . . . This inference is certainly correct, but there is great difficulty in determining what rocks may have extended over any particular district and been entirely dissolved in former ages. Is there not good reason to believe that most of the strata now constituting Catskill Mountain . . . once extended over Massachusetts to the Atlantic Ocean?

That they once extended as far as Massachusetts and united to the same strata at Pittsfield and Stockbridge he thought beyond question, and he concluded that a mass of rock from 1,000 to 3,000 feet in thickness, from twenty to thirty miles in breadth, and perhaps eighty miles in length, had been dissolved and mostly washed down the Hudson River. The possibility of glacial drift was of course wholly unrealized, and in accounting for the masses of granite and syenite, weighing from one to fifty tons, which are scattered throughout the Connecticut River region he asks:

What force can have brought these masses from the western hills across a deep valley 700 feet lower than their present situation? Are we not compelled to say that this valley was once filled up so as to make a gradual descent from the Chesterfield range of granite, sienite, etc., to the top of Mount Tom? Then it would be easy to conceive of their being rolled down to the top of the greenstone, where we now find them.

In this year Eaton published also his *Conjectures Respecting the Formation of the Earth.* Still following Werner, he taught that the water now covering three-fourths of the earth's exterior was at one time thoroughly commingled with the solid materials constituting the globe in such a manner as to form a very thick paste. Such being the primary condition, the materials in this "globular mass of mortar" would soon begin slowly to settle together, the heaviest naturally at the center. Such settling naturally took place in inverse order of the specific gravities of particles, forming thus several concentric layers of metals, arranged like the coats of an onion. This settling went on until finally granite was deposited, several thousand years being considered necessary for the completion of all the strata. Neither animals nor vegetables were conceived to exist previous to the deposition of the granite, gneiss, mica-slate, and other primitive rocks, for no traces of their remains had been discovered in them.

As condensation went on, the water being disengaged from the inner deposits and forced gradually toward the exterior of the mass, the earth paste became more diluted, and a few zoöphytes, shell animals, and cryptogamous plants were created, as was shown by the finding of their remains in the Transition rocks overlying the primitive.

The continued dilution of the water by the deposition of the materials of the transition series was followed by the creation of several species of fish, and as the solution was thinner the deposits of the secondary rocks went on with considerable rapidity. During this time the red sandstone, compact limestone, and indurated marl were deposited, entombing incidentally individuals of the various plants and animals enumerated.

While this secondary formation was going on the internal heat of the earth immediately beneath the granite, by converting the water which remained in the subterranean interstices into steam, began to raise up the rocks of granite. The expansion of this steam found relief by forcing its way wherever the least resistance was presented, and as strata can be separated from each other easier than they can be broken through, the steam probably traveled laterally round the earth, separating the granite from the next stratum below. At length the force of the more highly rarefied steam became too great to be any longer confined within the coat of granite. It burst through at the weakest part and shot forth its craggy broken edges above the muddy waters which surrounded it.

Concerning the source of this heat, "whether it arose from the admixture and combustion of substances then abounding beneath the granite," or whether "it was excited by the concentric layers of metallic plates serving as a vast galvanic battery," Eaton was noncommittal.

The projecting edges of granite, together with the uplifted strata of transition and secondary rocks, formed the first islands and continents of dry land. Alluvial deposits had already commenced under water, and therefore parts of the raised islands and continents were prepared for the reception and support of plants and animals of the more perfectly organized structure.

These, however, were not the continents of the present day, which were conceived then to be at the bottom of the primeval ocean, but were rather the continents occupied by antediluvians, which are now in their turn probably at the bottom of the Indian Ocean. This land, he argued,

may have been supported by the meeting place of two vast segments of uplifted granite which contained beneath them an immense subterranean sea. Our present continent may now be supported in the same way and the meeting of the edges of segments form the granitic ridge which extends from Georgia to the Frigid Zone—that is to say, that which forms the Appalachian Mountain system.

In whatever manner the ancient world was supported, it is evident that when the wickedness of man drew down the vengeance of the Almighty, its foundations gave way and it sank to the bottom of the ocean never to be again uplifted.

Incidental to this catastrophe, he conceived there may have been formed a "vacuum wherein much water might subside"; or possibly several continents falling in contemporaneously, basins were formed

sufficient to hold all the waters which had hitherto covered the continents of our day; or, perhaps, the pressure at the outer margins of the falling continents might force up the granite, which raised our continents out of the ocean. . . . At any rate, the fountains of the great deep were broken up, and our continents, then at the bottom of the great deep, emerged into open day. While this tremendous crash of nature was going on, scales of various thicknesses from the various strata were shot up, detached and broken, which gave formation to our surrounding hills, the ragged cliffs of the Catskill and the bleak brow of the Andes. Some were formed at the bottom of the sea by volcanic fires; others have arisen from various causes since the great deep retired.

This order of creation he believed to be directly in accord with the account of Moses and the sinking of the ancient continent contemporaneous with the Noachian deluge. "When the fountains of the great deep were broken up and the bottoms of those fountains became dry land, the ancient world became itself the bottom of the great deep in its turn."

During this period all surviving animal life, both human and otherwise, was confined within the limits of Noah's ark, about one year elapsing from the time the sinking took place and the ark was floated before the new earth was sufficiently dry for the occupants to disembark. Grotesque as these views may seem today, they were not wholly unacceptable to other workers, if we may judge from current comment and criticism. Silliman, in his review in the *American Journal*, while questioning his conclusions on a few minor points, writes of it as "creditable to the author's industry and discernment, and as bearing every mark of verisimilitude."

American Geological Society, 1819.

The year 1819 was signalized by the organization, in the philosophical room of Yale College, of the *American Geological Society*, the first American society devoted to geological and allied subjects.

Though this continued in existence only until the end of 1828, it was productive of much good in stimulating workers throughout the country. Maclure was elected president, with Colonel Gibbs, Professor Silliman, Professor Cleaveland, Stephen Elliott, R. Gilmore, S. Brown, and Robert Hare, vice-presidents. Among the more prominent members were Akerly, Bruce, Cornelius, S. L. and J. F. Dana, Dewey, Eaton, Godon, Hitchcock, Mitchill, Rafinesque, Schoolcraft, and Steinhauer, while the names of Emmons, Harland, Lea, Morton, Troost, and Vanuxem appear among the younger and then less prominent workers.

The society published nothing, and aside from a few brief references in the *American Journal of Science*, has left little that is tangible to tell of its existence, though Eaton, in the second edition of his *Index to the Geology of the Northern States*, makes the following interesting comment concerning its personnel:

> The president of the American Geological Society, William Maclure, Esq., has already struck out the grand outline of North American geographical geology. The first vice-president, Col. G. Gibbs, has collected more facts and amassed more geological and mineralogical specimens than any other individual of the age. The second vice-president, Professor Silliman, his learned and indefatigable colleague, in these labors, gives the true scientific dress to all the naked mineralogical subjects, which are furnished to his hand. The third vice-president, Professor Cleaveland, is successfully employed in elucidating and familiarizing those interesting sciences; and thus smoothing the rugged paths of the student. Professor Mitchill has amassed a large store of materials, and annexed them to the labors of Cuvier and Jameson. But the drudgery of climbing cliffs and descending into fissures and caverns and of traversing in all directions our most rugged mountainous districts to ascertain the distinctive characters, number, and order of our strata has devolved on me. I make no pretensions to any peculiar qualifications other than that bodily health and constitutional fitness for labor and fatigue which such an employment requires.

Schoolcraft's Explorations, 1818-1821.

During the years 1818 and 1819 Henry R. Schoolcraft made a trip throughout what is now known as the lead region of the Mississippi Valley, and in November of the last-named year published, in the form of an octavo volume of some 300 pages, a book entitled *A View of the Lead Mines of Missouri; Including Some Observations on the Mineralogy, Geology, Geography, Antiquities, Soil, Climate, Population, and Productions of Missouri and Arkansas and Other Sections of the Western Country.* The work contained little of geological importance, the purport of the trip being mainly to study the lead deposits. He described the whole mineral country as "bottomed" on Primitive limestone, though he found quartz rock and later sandrock very common in the southern section of the Arkansas country. Secondary limestone was also met with, but was far less common than in Ohio, Indiana, Connecticut, and Illinois, the ore itself being found in the decomposition products from the Primitive limestone.[28] He mentioned the occurrence of granite in Washington and Madison counties, also greenstone, porphyry and iron ore, and correctly described the granite as being the only mass of its kind known to exist between the Primitive ranges of the Alleghany and Rocky mountains, being surrounded on all sides and to an almost immeasurable extent with Secondary limestone. He gave also a descriptive catalogue of the minerals found in the state. Among them mention was made of the flint from Girardeau County; several varieties of quartz, including the Arkansas novaculite; a red pipestone from the Falls of St. Anthony, which is evidently the catlinite of more recent writers, but which he called steatite; and other minerals, including baryte, fluorite, blende, antimony, native copper, etc. The micaceous iron ore of Iron Mountain, the coal found near Pittsburgh, Pennsylvania, and a few other minerals were mentioned.

Fig. 7. *Henry Schoolcraft.*

In 1820, acting under direction of Governor Lewis Cass, of Michigan, who was himself acting under authority of the then Secretary of War, Schoolcraft made a trip along the Great Lakes and to the sources of the Mississippi, the results of which, with general notes on the natural history of the region, were published the year following.

In 1821 he was a member of a second expedition authorized by the General Government to explore the central portions of the Mississippi Valley, the results of which were published in 1825. In 1822 he also reported to the General Government on the extent and value of the mineral lands on Lake Superior, and again in 1832 "resumed and completed" his explorations of the sources of the Mississippi, his results appearing in book form in 1855.

The expedition of 1820 (see map, Fig. 8) started at Detroit and made its way northward along the western shores of Lakes St. Clair

[28] In a letter to Dana, dated Jan. 31, 1841, he announced that further investigation had convinced him that this limestone belonged to the Transition series.

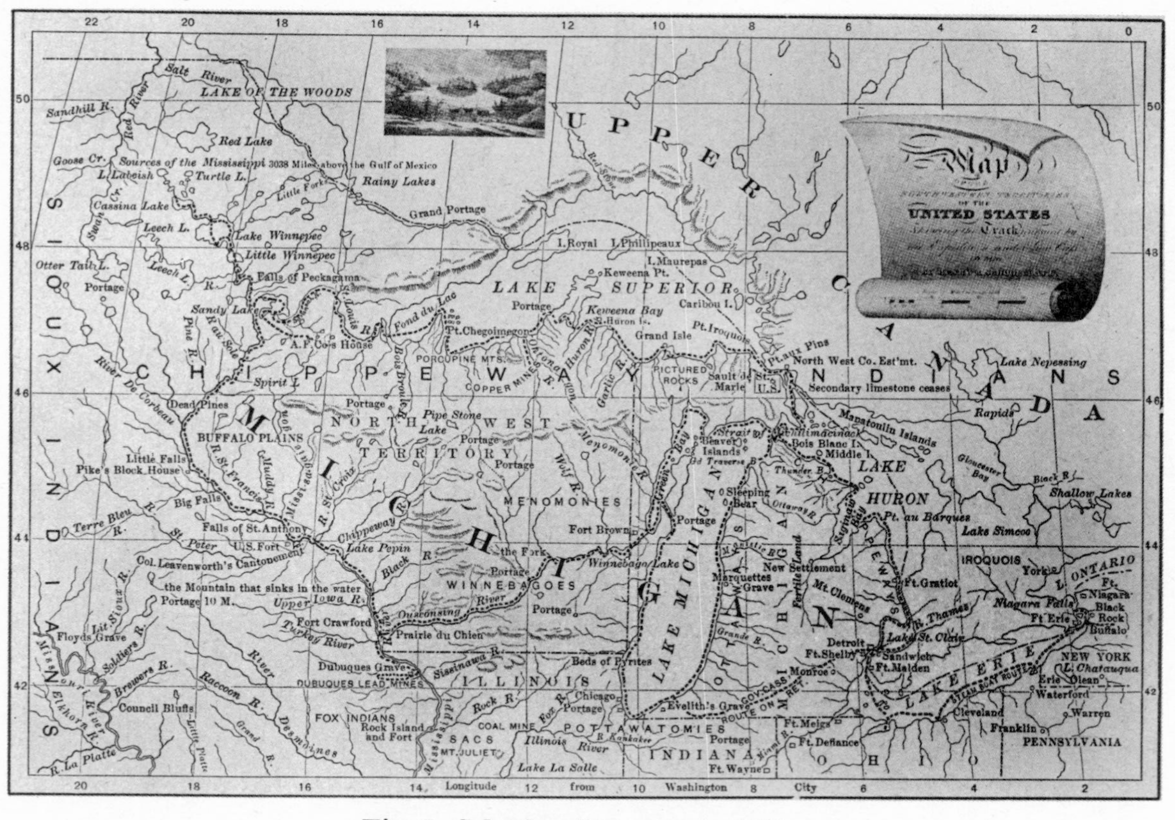

Fig. 8. *Schoolcraft's route in 1820.*

and Huron to the Straits of Michilimackinac, thence northward through St. Marys River and along the south shore of Lake Superior to Fond du Lac, up the St. Louis River, and down the Savannah to Sandy Lake and Lakes Winnipeg and Cassina farther north. The return trip was down the Mississippi to the lead region near Dubuque, Iowa, and thence northeasterly up the Wisconsin River to Green Bay, where the party divided, a portion going to Michilimackinac (now Mackinac), along the northern and eastern shores, and the rest keeping to the south, to Chicago, and eastward and north till the starting point was reached once more, one section of the party, with Governor Cass, leaving the lake near the southeastern end (at the mouth of the River du Schmein) and going overland to Detroit.

The narrative of the expedition of 1820 abounds with mineralogical and geological notes, which are in large part of a supposed economic nature. The occurrence of gypsum at St. Martins Island was noted, the island of Michilimackinac itself being of "transition and compact" limestone. A colored section was given showing the relative position of granite and overlying sandstone on Presque Isle (Fig. 10). This sandstone he described as overlapping the granite and fitting into its irregularities in a manner that "shows it to have assumed that position subsequently to the upheaving of the granite." The age of this sandrock he was unable to determine satisfactorily, though its position seemed to him to indicate a near alliance to the Old Red sandstone (Devonian). It may be added here that the correct position of this sandstone remained for a long time a matter of doubt and dispute but it is now considered to be of Potsdam Age.

He described the finding of the large block of native copper now in the National Museum in Washington, D. C., upon the bank of the Ontonagon River and offered some remarks on the probable origin of the same.

In his report to Secretary Calhoun under this same date, he dwelt in some detail upon the frequent occurrence of masses of copper in the drift, but stated that no body of metal sufficiently extensive to mine properly had thus far been discovered. He thought it probable, however, that a more intimate knowledge of the country's resources might result in the discovery of valuable ores in the working of which "occasional masses and veins of metal may materially enhance the advantage of mining." This prediction, it is scarcely necessary to say, has been abundantly verified.[29]

[29] An interesting account of the discovery and early—aboriginal—working of

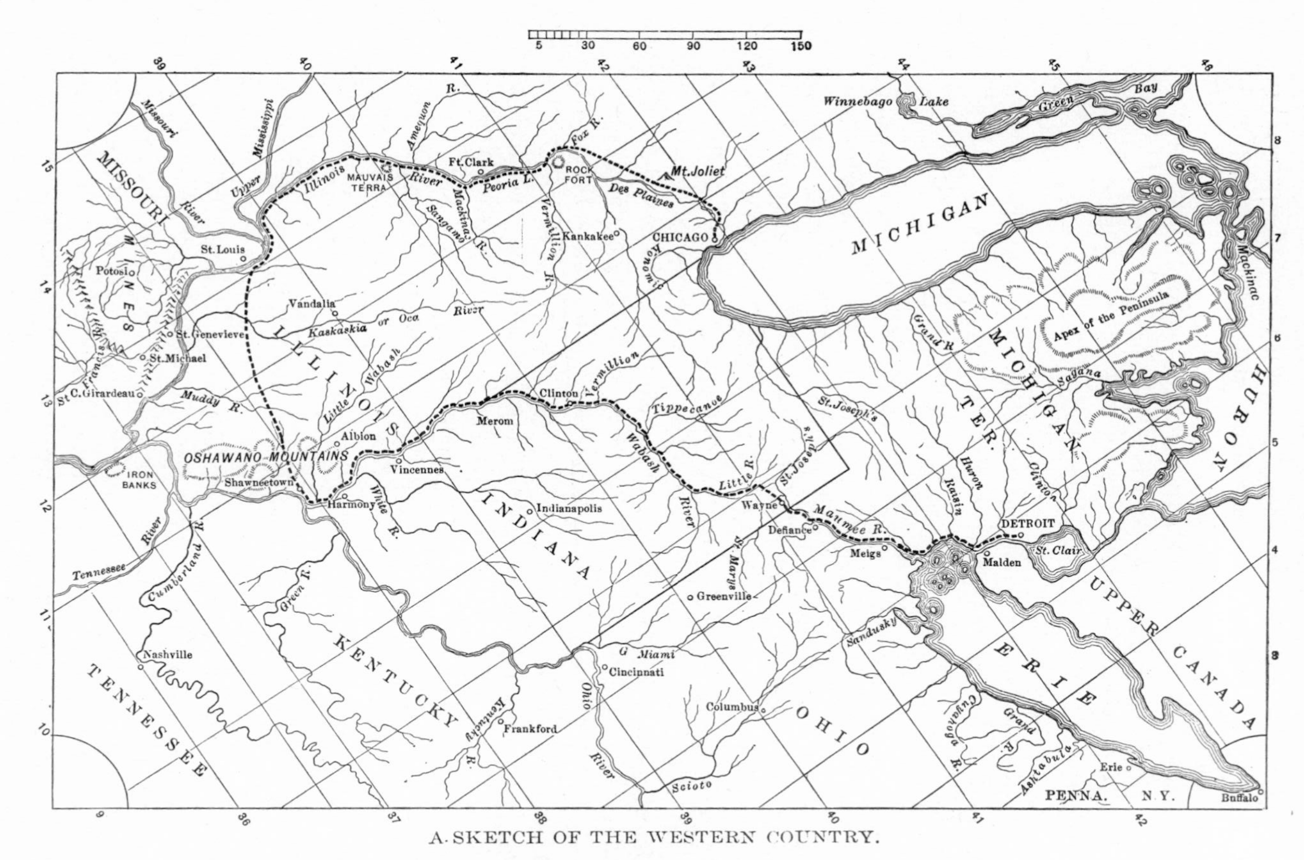

A SKETCH OF THE WESTERN COUNTRY.

Fig. 9. *Map of Schoolcraft's route in 1821 (1825).*

Writing on the prevailing theories as to the origin and distribution of metals and gems, he remarked, "There is no reason that can be drawn from philosophical investigations to prove that these substances may not be abundantly found in the climates of the north, even upon the banks of the frozen ocean," their distribution being apparently wholly independent of climatic conditions.

Fig. 10. *Schoolcraft's Section of Presque Isle.*

The Dubuque lead ore he described as occurring in detached masses in the ocherous alluvial soil resting upon a calcareous rock referable to the Transition class (in the revision of his work, published in 1832, he made this Carboniferous), and also in veins penetrating the rock. The relationship existing between the rock and the residual clay did not seem, however, to have been recognized.[30]

The presence of extensive beds of coal about forty miles southwest of Chicago, on the Fox River, was noted. The fact that bricks, made from clay occurring near Chicago, turned white was also remarked and explained on the ground that they were lacking in iron oxides.[31]

these deposits is given by H. V. Winchell in the Second Annual Report of the Proceedings of the Lake Superior Mining Institute.

30 Since this mode of occurrence was referred to by subsequent explorers, it may be well to state here that the ore was originally in the limestone, from which it was liberated by decomposition and left to accumulate in the residual clay, representing the insoluble constituents.

31 This has since been shown to be an error. These clays contain as much iron as others that may burn red. It is probable that, in the process of firing, this iron combines to form an iron-lime-magnesian silicate and is not oxidized to the extent of imparting the common brick-red color.

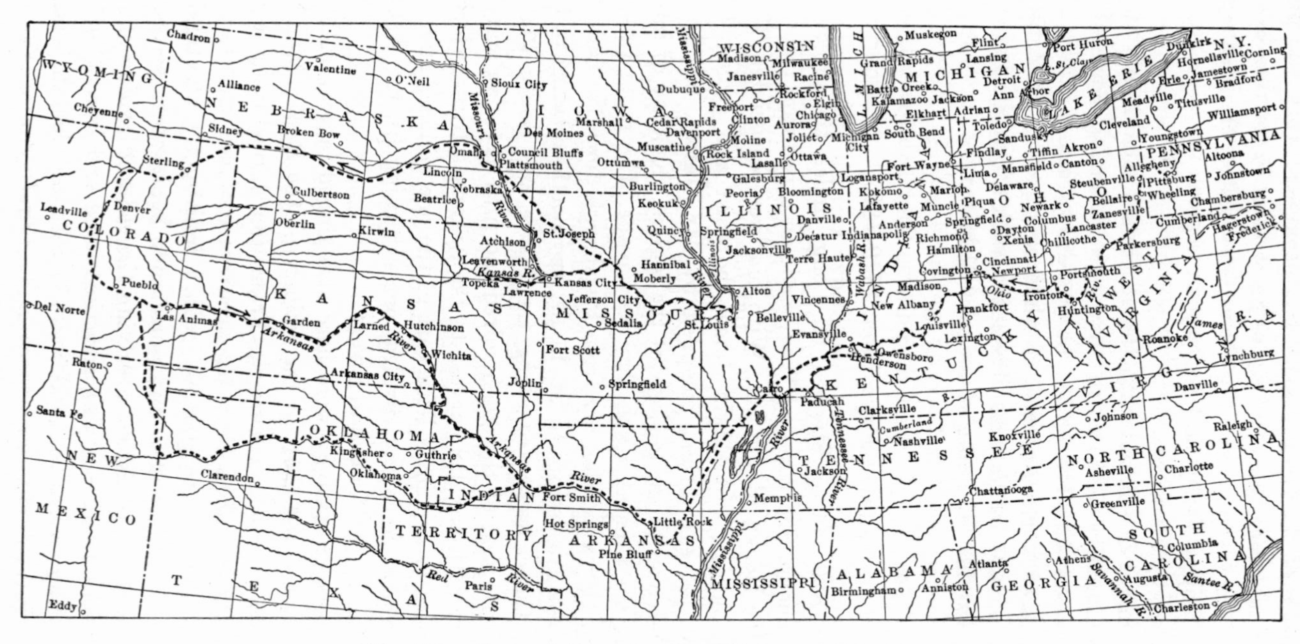

Fig. 11. *Map of Long's Expedition, 1819-1820.*

thought "to have been deposited at a very remote period when the waters of the primeval ocean covered the level of the Great Plain and the lower regions of the granitic mountains." The inclined position of these Secondary rocks in the immediate vicinity of the mountains was noted, and difficulty was found in accounting for the same. "Subsequent to the deposition of these horizontally stratified rocks," he wrote, "their position has been somewhat changed, either by the action of some force beneath the primitive rocks, forcing them up to a greater elevation than they formerly possessed, or by the sinking down of the Secondary, produced by the operation of some cause equally unknown." This matter he again referred to, thinking it possible, though scarcely probable, that the great and abrupt change in the inclination of the strata in the parts near the granite might be due to the gradual wearing away by the agency of rivers of some portion of the sandstone, and that those rocks now found in an inclined position were detached portions of what was formerly the upper part of the strata which, having been undermined on their eastern side and supported by the granite on the western side, had fallen into their present position. As noted elsewhere, other workers were finding difficulty in the solution of the same problem (see p. 27).

The presence of coal beds in the region of the Ozark Mountains was noted and, together with the associated limestones, identified as of Carboniferous age, as was also the limestone in the region of the lead mines.

An important suggestion was made relative to the possibility of obtaining water through bore holes sunk in the arid tract lying west of the Ozark Mountains. "It is not improbable," he wrote, "that the strata of many parts of this secondary formation toward its exterior circumference may vary from a horizontal to an inclined position, in consequence of which the water that falls in dew and rains in the hilly districts, becoming insinuated between curved stratifications, may descend toward the center of the formation under such circumstances as would insure its rising to the surface through well or bore holes sunk sufficiently to penetrate the veins."

The rocks of the Alleghany Mountains were classified as Granular Limestone, Metalliferous Limestone, Transition Argillite, and Sandstone. The report was accompanied by a volume of plates, which included geological sections on the thirty-fifth and forty-first parallels (Fig. 12). These were intended to form continuations of Maclure's third and fifth sections, already noted (p. 47).

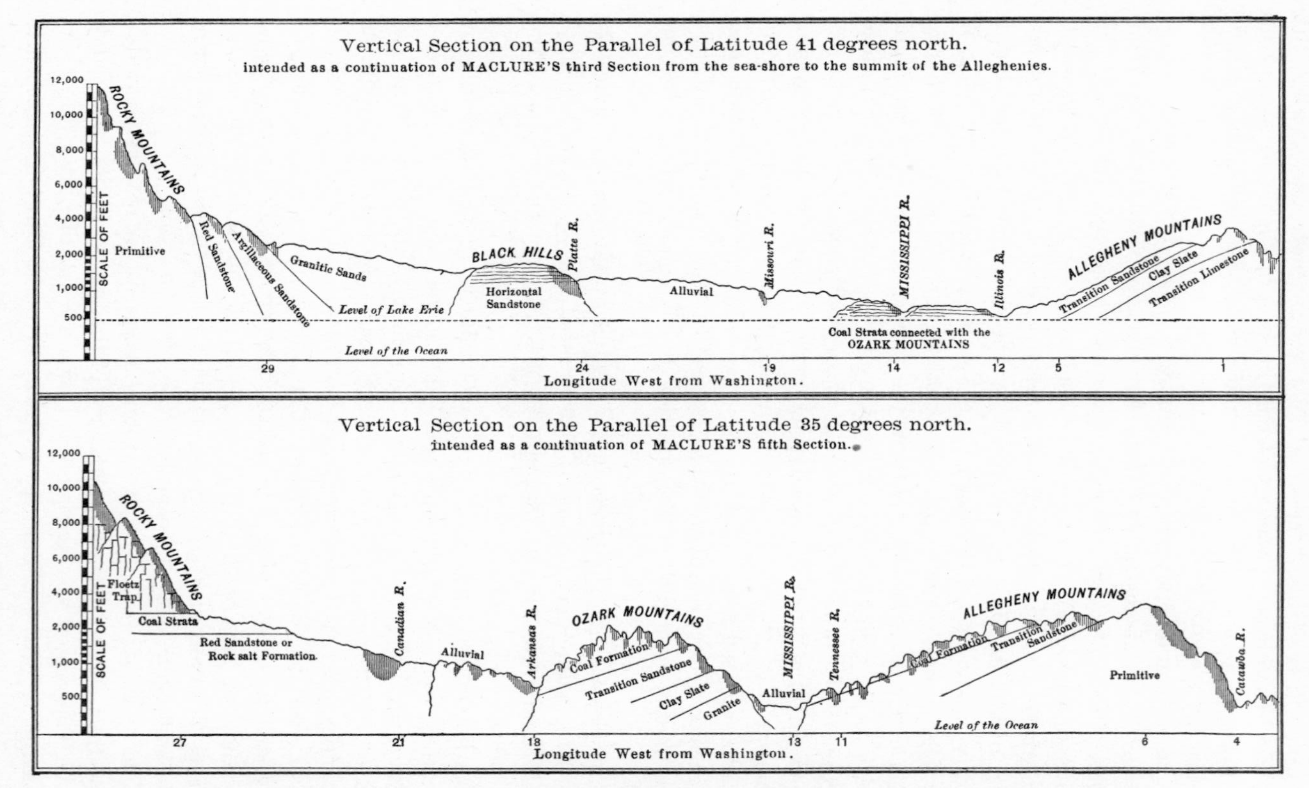

Fig. 12. *James's Geological Section from the Allegheny to the Rocky Mountains.*

Geological Notes by Silliman, Olmsted, and Atwater.

The second volume of *Silliman's Journal* (1820) had a long article by the editor—"Sketches of a Tour in the Counties of New Haven and Litchfield, Conn., with notices of the Geology, Mineralogy, and Scenery"—in which were described the various rocks passed over; among others the "primitive white marble" near New Milford. He described this as "a perfectly distinct bed in gneiss, which is found on both sides of it, and, of course, both above and below it," a sensible admixture of the two rocks being perceptible for some feet on both sides of the junction.

Just what Silliman's preconceived opinions on the subject may have been is not apparent, but he remarked in a satisfied way that the associations were exactly what he had anticipated. "The geological relations of the marble appear to have been perfectly distinct, . . . and give us new reason to admire that wonderful order and harmony, little expected by people in general, which are found equally in the mineral kingdom as in the animal and vegetable, and which afforded on analogical grounds the best reason to predict that the geological association of this marble would be found to be what it actually is.

"In the district are nearly all the important primitive rocks of Verneuil, while New Hampshire includes a considerable portion of his secondary formations."

In the same *Journal*, Denison Olmsted, then at Chapel Hill, in North Carolina, announced the discovery of a red sandstone formation in North Carolina, which he had traced through the counties of Orange and Chatham, with a breadth, in one instance at least, of about seven miles, and which he rightly conjectured might be a continuation of Maclure's "great sandstone formation."

James Pierce wrote rather prosaically on the geology and mineralogy of the Secondary region about New York, and Dr. John Bigsby, who later became prominent in Canadian geology, offered some remarks on the environs of the Carthage bridge over the Genesee River, giving a section showing the succession of the sedimentary strata to which only lithological names were applied, but with no suggestion as to their thickness and naturally none as to geological age.

Caleb Atwater also noted the occurrence of ancient human bones, together with those of a mastodon or mammoth in Ohio, which indicated to his mind that the whole country had at one time been covered with water, which had made it "one vast cemetery of the beings of former ages."

Akerly on the Geology of the Hudson River, 1819.

In August, 1819, Dr. Samuel Akerly, who is first mentioned on page 39, read before the New York Lyceum an essay on the geology of the Hudson River and vicinity. This was published the year following in the form of a small duodecimo volume of sixty-nine pages, accompanied by a colored section of the country from the neighborhood of Sandy Hook, New Jersey, northward through the Highlands in New York toward the Catskills. This section is mainly interesting for being the second attempt of its kind, being preceded by Eaton's transverse section from the Catskills to the Atlantic by only two years. The work consisted mainly of descriptive details relating to the lithological character of the various formations, and contained little else of value. Akerly was a disciple of Mitchill, whom he followed almost implicitly. He, however, questioned the "Old Red" age of the Connecticut River sandstone, on account of bones of land animals having been found beneath them at Nyack. Since, however, he gave credence to the reported finding of pine knots, earthen vessels, iron instruments, corn cobs, etc., under these same sandstones, and an iron instrument resembling a pipe in the anthracite coal of Rhode Island, little weight can be given even to his doubts.

CHAPTER II.

The Eatonian Era, 1820-1829.

ACCORDING to Amos Eaton, 1820 marked the close of the first era of American geology. Accepting this, it may well be called the Maclurean era. The second, including the decade 1820-1829, may with equal propriety be called the Eatonian era, since Eaton was the most prominent worker as well as most profuse writer of the decade. In so doing, however, we must not overlook the fact that Eaton was favored with unusual opportunities, owing to the munificence of the Honorable Stephen Van Rensselaer and that he himself would perhaps have called it the Rensselaerian era.

The era opened with promise, and though the results as apparent on paper were not great, yet much was actually accomplished. It was, so to speak, a transition period, one in which the possibilities of state and governmental surveys were seriously considered, and one, too, in which, so far as America was concerned, there was made the first systematic attempt at correlation by means of fossils. We meet, also, during this period the first really satisfactory suggestion as to the source of the glacial drift, the first recognition of overturned folds and the possibility of the repetition of strata through faulting and displacement. The cosmogonists had largely disappeared and in their place were men who had learned first to observe and then to draw conclusions according to their understanding of the observed phenomena.

Second Edition of Eaton's Index, 1820.

In 1820 Eaton published a second edition of his *Index* in the form of a small octavo volume of 286 pages. In this many of the earlier opinions were restated in his customary emphatic manner and with little, if any, modification. "I consider nothing in geology entitled to much confidence, which is purely theoretical. But I am willing to be held as pledged for every fact given as such in this Index; unless there are cases (which I hope are few) where I misjudged respecting the name or character of a mineral or stratum." His views relative to the origin of the continents were illustrated by the figure here

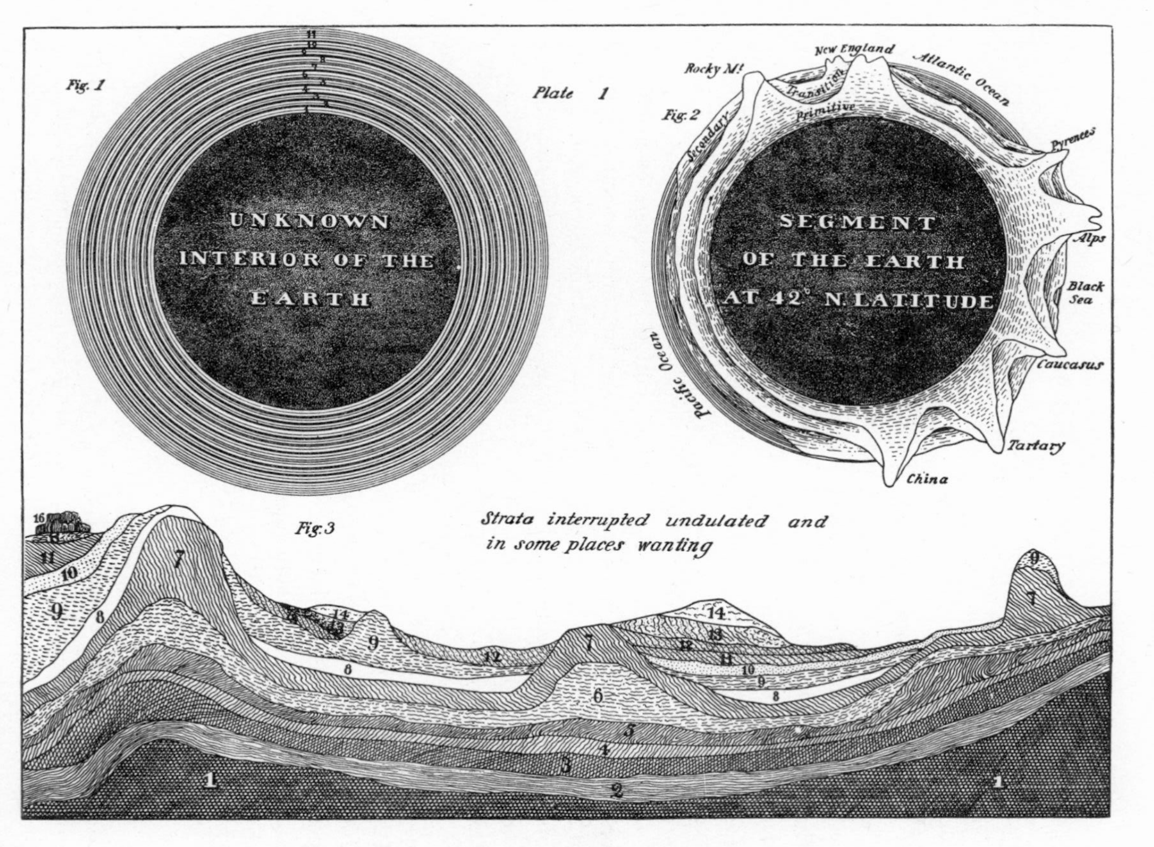

Fig. 13. *Eaton's Geological Segments, 1820.*

reproduced (Fig. 13) but the source of the energy which resulted in their elevation was still unexplained. He asserted "That the rents made by the grand explosion, which first upturned and disfigured the rocky crust of the earth, were in a north and south direction. That those crossing the forty-second degree of north latitude were principally made at the Pyrenees and Alps in Europe, Caucasus, Tartary, and China in Asia, Rocky Mountains and New England in America. . . . Whether this theory accords with the real origin of the present state of things or not is immaterial. It is introduced solely for the purpose of aiding the memory in studying the strata which we know do exist."

His views on the formation of caves in limestone were primitive and highly interesting:

When the waters of the ocean retired, the calcareous cement which now holds the shells together was in a state of soft paste. . . . After the waters retired the parts exposed to the sun's rays began to harden, contract, and crack into blocks. In some parts of the Heldeberg these blocks are of great extent, but I have seen acres of it where the stratum is very thin, checkered up into blocks from 2 to 10 feet square. Where the stratum is very thick and the fissures very long, large caverns were frequently formed, for the upper surface of the stratum was soon dried and indurated, while the whole remained soft a long time a few feet below the surface. If a stream of water happened to flow in the vicinity of the fissure, it would probably make its way into it and soon wash away the loose shells beneath the surface, which were merely enveloped in soft calcareous paste.

And further: "I have examined four of the largest caverns in the Heldeberg, and they all still exhibit conclusive evidence of their having been once in the state of mere fissures, and streams of water still traverse them all."

The majority of readers of today need scarcely be told that these limestone caves are formed wholly by a process of solution, by surface waters acting upon the strata only after they had reached, essentially, their present condition of induration and position above drainage level. The streams of water which are found traversing them are incidental and consequent rather than causative.

Concerning the formation of stalactitic iron ore, as exemplified in the Salisbury mine, he wrote:

These stalactites are always suspended from masses intermixed with the soil in such a manner that it is evident the iron was in a state of fusion when in contact with it. The foot, which still adheres to all stalactitic specimens, proves that the heat was continued after the ore was confined in its present state. If it was ever fused down from any rock, it must have been the same out of which the alluvion embracing it was formed. The eause producing such a high heat I shall not attempt to assign; but that the ore exhibits sufficient evidence of its having been recently fused, I believe no one can question who has ever inspected it in place.

Here also the reader of today scarcely need be told that all the peculiarities thus described were due not to heat, but to deposition from solution.

The inclined position of the crystalline limestone near the east line of Danbury, Connecticut, still confused him.

Here the layers dip to the west; but in West Stockbridge and Alford they generally dip to the east, though there seems to be no conformity in their direction. Had some force, applied at the eastern edge, raised these mountain masses from the horizontal toward the vertical position, leaving some inclined to the east and forcing others beyond a vertical position, they would have presented their present inclination.

This is the first suggestion of anything resembling an overturned fold which had thus far been made. He was correct in his observations, but the science had not sufficiently advanced to enable him to realize the possibilities. It will be noted, too, that he was working in a region the correct interpretation of which has been a source of dispute for three-quarters of a century. The glacial drift was, as in the first edition, still classed as alluvial, and the materials composing Long Island and the coastal plain were treated as having a common origin.

Eaton made use of Le Duc and Jameson's awkward name of *geest* to include the "most universal of all strata," which was found occupying "every inch of dry land which is neither naked rock nor covered with alluvion," and the character of which "is generally indicated by the rock upon which it lies and by those which have recently disappeared." Although thus defined, in his attempt to outline its distribution he failed to discriminate between the true residual material and that which was drift.

What he had to say on the subject of organic relics or fossils was merely an adaptation from Martin's *Systema reliquiorum.* He classified them under two heads, *petrifactions* and *conservatives.* The petrifactions or *substitutions*, as he sometimes called them, were "those relics which are entirely made up of mineral substances, which have gradually run into the places occupied by organized bodies, as those bodies decayed, and assumed their forms." The conservatives or preservations, on the other hand, were "those relics or parts thereof which still consist of the very same substances which originally composed the living, organized being."

These relics were named by annexing the termination *lithos*, a stone, to the scientific name of the living organism; as, for instance, a fossil fish would be *ichthyolithos*, though the English terminology he often modified from *lithos* to *lite*. Adopting this nomenclature, he grouped all organic relics or remains under nine heads, as follows: Genus I, *Mammodolite;* Genus II, *Ornitholite;* Genus III, *Amphibiolite;* Genus IV, *Ichthyolite;* Genus V, *Entomolite;* Genus VI, *Helmintholite;* Genus VII, *Concholite;* Genus VIII, *Erismatolite;* Genus IX, *Phytolite.*

Eaton's Survey of Rensselaer County.

In 1821 Eaton was employed by S. Van Rensselaer to make a geological and agricultural survey of Rensselaer County, New York. His report, printed in 1822, formed an octavo pamphlet of seventy pages, and was accompanied by a geological section, extending from the Onondaga Salt Springs across the county to Williams College in Massachusetts. The first thirty pages of the report were given up to discussions of the character and distribution of the various kinds of rocks and soils, and the remainder to methods of culture and an agricultural calendar.

The rocks of the county, as well as those of Washington County on the north and of Columbia on the south, were thought to belong chiefly to the Transition formations. Secondary limerock resting on the graywacke was found in Schaghticoke about four miles east of the Hudson and in the northern part of Greenbush. As to whether the argillite along the eastern margin of the county was Transition or Primary, he was in doubt. As with his contemporaries, he based his opinions largely upon lithological data, quite failing to realize that rocks of widely varying age may more or less resemble each other, according to local conditions and the amount of metamorphism they may have undergone. Passing westward from Williams College, the various rocks met with, as shown in his section and described, are (1) *Granular* and *Primitive Limestone*, (2) *Metalliferous* or *Transition Limestone*, (3) *Argillite*, (4) *Metalliferous Limestone* again, (5) a second band of *Argillite*, (6) *Graywacke* with sporadic patches of (7) compact *Secondary Limestone*, (8) *Secondary* or *Calcareous Sandstone*, and (9) *Argillaceous Graywacke*, semi-indurated *Argillite*, and *Clay-slate*. The first three were represented with steep westerly dips passing under the graywacke, which constitutes the principal formation.

J. B. Gibson on Trap Rock, 1820.

On November 17, 1820, the Honorable J. B. Gibson read a paper before the American Philosophical Society, which was published in the *Transactions* of that year, on the trap rocks of the Conewago Hills near Middletown, Dauphin County, and Stony Ridge near Carlisle, Dauphin County, Pennsylvania. He described the mode of weathering into bowlders, and rightly argued that such was due wholly to atmospheric agencies and did not indicate an original concretionary structure. He seemed to consider the columnar structure as due also to decomposition, mentioning as an example the columns of the Giant's Causeway, "which exhibit regular prismatic form only when it has long been exposed to the action of the atmosphere."

He did not consider the Conewago trap igneous, but that of Stony Hill presented to him an appearance decisively volcanic. This did not, however, necessarily mean that there may have once existed a crater here.

Like several other of the early workers he was not able to distinguish between forms due to crystallization and those due to a shrinkage of the mass through cooling or loss of moisture.

Nuttall's Observations, 1820.

The work of Thomas Nuttall merits attention here only on account of the time of its accomplishment. Nuttall was not a geologist, nor can he be considered an American. His principal work was of a botanical and ornithological nature, the paper, *Observations on the Geological Structure of the Valley of the Mississippi*, being his only contribution of note to the literature of geology. This was read before the Philadelphia Academy in December, 1820, and in its printed form occupies thirty-eight pages of the journal of the society.

Nuttall's travels took him along the southern shore of Lake Erie to Detroit; thence by canoe along the coast of Lake Huron to Michilimackinac; thence southwestward to Green Bay, Wisconsin, and by way of the Fox and Ouisconsin (Wisconsin) rivers to the Mississippi, near Prairie du Chien, and southward to St. Louis. This and other trips up the Missouri and Arkansas rivers gave him ample opportunity for such superficial observations as he was competent to make. These consisted largely of conjectures as to the geographic limits of the Secondary formations, which he found here to consist mainly of limestone strongly resembling the Mountain Limestone of Derbyshire in England, and he announced himself as "fully satisfied that almost every fossil shell figured and described in the *Petrificata*

Derbiensia of Martyn" was to be met with throughout the great calcareous platform of the Mississippi Valley. The limits and character of the "Ancient Marine Alluvium" and its fossil and mineral contents were discussed; the essay concluding with some observations on the Transition Mountains of Arkansas, with brief notes on the hone slate of Washita. C. R. Keyes seems very favorably impressed by Nuttall's work. He writes:

> Nuttall's paleontological correlations antedate by fifteen years Samuel Morton's similar efforts on the Tertiaries of our Atlantic coast, commonly regarded as the initial attempts in America along these lines. By two decades they were in advance of the first work of that pioneer paleontologist, Lardner Vanuxem. They anticipate by a full generation the famous investigations of Thomas Conrad and James Hall, of New York.[1]

Hayden's Essays, 1820.

In 1820 there was issued also H. H. Hayden's *Geological Essays.* These, dealing particularly with the Tertiary and more recent alluvial deposits of the Atlantic and Gulf coasts, forming what is now known as the coastal plain, are of especial interest. The work is verbose in the extreme however and more argumentative than logical. Indeed, were it not for its historical interest and for the light which is thrown upon the crudities of early observation and deduction, it would be scarcely worthy of consideration.

After referring to the geographical limits of this coastal plain, as defined and mapped by Maclure, and combating the opinions of previous observers, including Latrobe, Stoddard, and others, to the effect that it was formed by flood tides and the winds acting on materials cast up by the sea or through the transporting powers of the great rivers, Hayden proceeded to elaborate his own ingenious and wholly improbable theories. "Viewing the subject in all its bearings, there is no circumstance that affords so strong evidence of the cause of the formation [*i.e.*, the coastal plain] as that of its having been deposited by a general current, which at some unknown period flowed impetuously across the whole continent of America, and that from northeast to southwest." The course of this current he assumed depended on that of the general current of the Atlantic Ocean, the water of which rose to such a height that "it overran its ancient limits and spread desolation on its adjacent shores."

In seeking a cause for this general current the author referred,

[1] Century of Iowa Geology, by Charles Keyes. Proc. Iowa Acad. of Science, Vol. XXVII, 1920.

first, to the seventh chapter of the book of Genesis, "For yet seven days, and I will cause it *to rain* upon the earth forty days and forty nights, and *every living substance that I have made will I destroy from off the face of the earth.*" He then proceeded to show the inadequacy of this cause alone; the water being thus equally distributed over the ocean and the land, there "could be no tendency to cause a current in the former." Some other cause must therefore be sought, and fortunately his imagination proved equal to the task.

Accepting as probable the suggestion of "a writer of no common celebrity," to the effect that the cause of the general deluge was the melting of the ice at the two poles of the earth, and that this was occasioned by the sun deviating from the ecliptic, he proceeded to explain in his own way how this might be brought about, though acknowledging that no positive testimony could possibly be adduced to substantiate the fact.

Having admitted the possibility of the earth's changing its position so that the sun would pass "immediately over the two poles upon an unknown meridian," he showed that there would then result a rapid dissolution of the existing ice caps, such as would yield an ample supply of material, it being only necessary to give it direction. Considering as essential to the problem only the northern hemisphere, he remarked that from this polar cap there are but two outlets—the one into the Pacific Ocean through the narrow Bering Strait and the other through the wider channel between Greenland and Lapland into the Atlantic. Hence, when the melting ensued, by far the larger volume of water passed into the latter ocean.

> No sooner was this operation established, and this accession of strength and power thrown into the Atlantic Ocean in particular, than its tide began to rise above its common limits accompanied by a consequent current, both constantly increasing, the one in height and the other in rapidity, proportioned to the increase of power at the focus. . . . At the commencement of this frightful drama the current, it is highly probable, was divided by the craggy heights of Spitsbergen and a part thrown into the White Sea, while the other was thrown back upon the eastern and southern coast of Greenland, and from thence in a southwestern direction until it struck the southeastern coast of Labrador, along which it swept through the straits of Belle Isle, across Newfoundland, Nova Scotia, and along the Atlantic coast into the Gulf of Mexico. . . . In a short space of time, the southern and eastern coast of Labrador, over which this current was urged with increasing force, was desolated. The soil . . . was hurled adrift and . . . carried across the country into the Gulf of St. Lawrence, and across a part of New England into the sea, or general current of the ocean. Continuing to rise, the waters swept across

Davis Strait and rolled their tumultuous surges into Hudson Bay, embracing the whole coast of Labrador, while the unequal current of the St. Lawrence was *forced back and upward to its parent source*. . . .

At length the floods of the pole forming a junction with Baffins Bay and the Arctic Sea, defying all bounds, overran their ancient limits and hurled their united forces, in dread confusion, across the bleak regions of the north to consummate the awful scene. Thus lakes and seas uniting formed one common ocean which was propelled with inconceivable rapidity across the continent between the great chain of mountains into the Gulf of Mexico and probably over the unpeopled wilds of South America into the Southern Ocean. Fulfilled in this way were the awful denunciations of an offended God, by the sure extermination of every beast of the field and every creeping thing that creepeth upon the face of the earth.

To these causes he believed to be due not only the alluvial deposits of the coastal plain, including the delta of the Mississippi, but as well the nakedness of the rocks in Labrador and the northeastern portion of the continent, and the general phenomena of the glacial drift, the bowlders of the latter being conceived as transported by floating ice.

While all this was taking place on the American continent, the material supplied by the melting of the south polar ice cap was finding its way northward over Asia, carrying with it to northern Siberia the mammoth, rhinoceros, and other gigantic animal remains now there found. In this last, it will be observed, he followed the Russian explorer, Pallas.

The opinion held by many to the effect that the deltas of rivers were composed exclusively of alluvion brought down in the course of time and deposited at their mouths, he considered "a flagrant dereliction from truth and every principle of sound reasoning and established fact." Such he regarded as formed in part of natural alluvion of the rivers, in part of wind-blown sand and dust, and in part of waste from the fields due to cultivation and the cutting away of forests. Thus early he recognized the importance of man as a geological agent, and also that of the wind. He nowhere in the whole 150 pages of the discussion, however, recognized the now well-known fact that deltas are formed only at the mouths of rivers emptying into tideless seas.

Strangely enough, too, although one of the earliest to recognize the extent and importance of the alluvial formations, Hayden seems to have had very hazy notions concerning the origin of their materials. The belief held by many to the effect that every species of rock is liable to a slow but progressive form of disintegration and decomposition was to him likewise rank heresy, as "tantamount to a libel against the letter and spirit of Holy Writ." Not but that some rocks may indeed decompose, such as the "micaceous schistus," but "granite and other rocks of like nature, where the quartz, feldspar and mica are perfectly combined, are practically indestructible"; and the arguments he used to prove this are precisely those used today to prove exactly the opposite,—the rapidity of the destructive process,—that is, the evidence furnished by old stone monuments and buildings. Blinded by his religious prejudices and preconceived notions, he refused to accept proof of such decay, even when confronted by it in unmistakable forms, referring to such as but the "débris of the incompact or imperfectly formed mass that served as the covering, as it were, of the rocks, and which, being destitute of a cement, had fallen away to sand."

Even that "the soil which covers the face of the earth was produced by the disintegration of rocks" was to him an opinion unfounded both in natural as well as in moral philosophy, and betrayed a want of attention to the plans of the Omnipotent. "Who can or will contend that the mountains of our earth are becoming more and more depressed by the disintegration of the rocks of which they are composed? . . . Fortunately, however, it is not so. The Great Author of Nature intended it otherwise, and they are, and ever have been, the same in height, in all human probability, that they were from the commencement of time."

Hayden was the first to suggest the name *Ternary* for formations of Alluvial origin more recent than the Secondary. The last-named formations he considered "the results of a natural operation"; that is, natural deposits from water, probably in a state of perfect tranquillity. The Alluvial, on the other hand, were the results of "accidental operations."

These essays were very favorably reviewed by Silliman in the third volume (1821) of his journal, though the writer did venture to express the regret "that the respectable author had not pruned off from his style some redundances and inaccuracies of expression. These are, however, in a considerable degree veiled by a glow and energy of thought and language that evince a mind at once ardent and vigorous." Even the idea of the fusion of the polar ice cap was allowed to pass with no serious criticism though it was suggested that the flood of waters might have been produced through the expulsion of the same from cavities in the earth. J. E. De Kay, writing in 1828, ventured, however, to take exception to some of the expressed views regarding drift bowlders. He wisely suggested that since the speculative part of geology is but a series of hypotheses we should in every case admit that which explains the phenomena in the simplest possible manner. To his mind the simplest manner of explaining the presence of bowlders of Primitive rocks scattered over a Secondary or Alluvial region was to suppose that such had been, as igneous materials, extruded through all superincumbent strata, forming peaks which have since been destroyed by some convulsion of nature or by the resistless tooth of time, the bowlders being but fragments which had escaped destruction, though the place of their extrusion had become completely obscured.

Sketch of Hayden.

Fig. 14. *H. H. Hayden.*

Hayden's career was varied and interesting. Early thrown on his own resources, with an ardent desire for knowledge and craving for travel, he went to sea at the age of fourteen, making two voyages to the West Indies, after which he returned to his school and his books. When sixteen years of age he was bound out to an apprenticeship with an architect, with whom he served until he reached his majority, when he once more sailed for the West Indies and established himself at Point à Pitre in Guadeloupe Island, one of the Lesser Antilles. Ill health, however, finally drove him back to America, where he found employment in his profession in Connecticut and New York, with occasional intervals of teaching, in which, it was stated, he was very successful.

Through accident his attention was turned to dentistry, and with no preliminary training and only such knowledge as could be gained by reading, he established himself in the practice of that profession in Baltimore, somewhere about 1824, being then, it will be observed, between fifty and sixty years of age. In this work he is represented as being highly successful, rising rapidly in public confidence and holding the highest professional rank in the city. When the American Society of Dental Surgeons was established, he became its first president. Aside from his profession he devoted himself to studies in physiology and pathology, as well as geology.

Thomas Cooper's Ideas Concerning Volcanoes, 1822.

In 1822 Dr. Thomas Cooper, then president of South Carolina College, published in the *American Journal of Science* an article of nearly forty pages, giving his views on volcanoes and volcanic substances. He defined a volcano as a natural vent in the crust of the earth, made by subterranean fires to afford exit for gases, vapors, and solid substances that have been exposed to the action of intense heat in the bowels of the earth. The seat of the volcano he believed to be below or within the oldest granite. In action, the volcano is described as giving off smoke and flame derived from contact with coal strata, the eruption being usually accompanied by electric light, the source of which he acknowledged to be problematical. Compared with all this error, his recognition of the porphyries in the vicinity of Boston and the Triassic traps of the eastern United States as lavas stands out in marked contrast.

Sketch of Cooper.

Cooper, or "Old Coot," as he was called during the period of his activity at the South, seems to have been a queer character—"a learned, ingenious, scientific, and talented madcap," as Président Adams is said to have called him. He was not a geologist, excepting in books, but rather an educator and theorist. He was born in London in 1759 and educated at Oxford, where he paid chief attention to the classics, though his inclination was for the sciences. He came to America in 1795 and settled down for a time to the practice of law at Northampton, Pennsylvania.

A restless, aggressive spirit soon took him into the political field, where the violence of his newspaper attacks caused him at one time to be imprisoned and fined. After his release he became, first, land commissioner and then judge, being removed from the latter office for arbitrary conduct. He then turned his attention to chemistry and became in turn professor of chemistry in Dickinson College in Carlisle, professor of chemistry and mineralogy in the University of Pennsylvania in Philadelphia,[2] and professor of chemistry, mineralogy, natural philosophy and law (!) in the University of Virginia. The last-named position he held from 1817 to 1820, when he

[2] The following notice relative to Cooper's work appeared in the Analectic Magazine for October, 1817:

"About the middle of October Judge (Thos.) Cooper proposes to commence his

was forced to resign on account of the opposition on the part of the Presbyterians.[3]

From Charlottesville he went to Columbia, South Carolina, where he served as professor of chemistry (1820-1821) and then as president (1821-1834), being here again compelled to resign for his violent liberalism in matters relating to science and religion. Singularly enough, a committee appointed by the state legislature in 1831 to investigate his conduct with a view to his removal failed to make out a case, and the charges were dismissed.

Fig. 15. *Thomas Cooper.*

Cooper, as above noted, was not a geologist, as the term is now used, but comes in for recognition here on account of the prominent part he played in early educational movements relating to the introduction of the science in the universities. It was through his influence that the chair of geology at the South Carolina College was

Geological and Mineralogical Lectures in the University of Pennsylvania. They will consist of the following parts:

1. Introductory Lecture.
2. On the Globe of the Earth: on the general properties of Mineral Substances; specific gravity, hardness, fracture, chrystallization, colour, etc.
3. On the rocks called *Primitive,* and their component parts.
4. On the Substances found in Primitive rocks.
5. On the rocks termed *Transition,* and their component parts.
6. On the Substances found in Transition rocks.
7. On the rocks termed *Secondary.*
8. On the Substances found in Secondary rocks.
9. On Alluvial Formations.
10. On Basins. The great Basin of the Mississippi. The Basin at Richmond, Virginia. The Paris Basin. The London Basin. The Isle of Wight Basin.
11. On Volcanic Formations. On Floetz Trap.
12. On Organic Remains.

"This course of Mineralogy, which, as the reader will see, is very different in its outline from any hitherto attempted, will be illustrated by appropriate specimens, Judge Cooper's cabinet, being now the best adapted for the purpose, of any in the United States, Colonel Gibb's excepted. To which gentleman, and Mr. Maclure, Judge Cooper expresses his obligations for the kind assistance they have afforded him in this respect.

"This collection of between three and four thousand specimens, consists of his own collection; of the late Rev. Mr. Melsheimer's and of M. Godon's.

"It is expected the Course will occupy between two and three months, at three Lectures a week. Tickets 15 dollars."

[3] It is evident that Cooper could have done no teaching at Charlottesville, since the university was not opened to students until 1825. See Circular of Information No. 1, U. S. Bureau of Education, p. 106.

established, to which Vanuxem was called in 1821, and Le Conte in 1857. At the time he assumed the reins in Columbia, geology was taught at no other institution in America except Yale, and the only available textbooks were the reprints of Bakewell's geologies with Silliman's notes. To the latter, on account of their acceptance of Mosaic doctrines, Cooper took exception, and fiercely attacked them, first in lectures to his classes and subsequently in his pamphlet on *The Connection between Geology and the Pentateuch*, published in 1833. (See p. 157.)

Fig. 16. *Silhouette of Thomas Cooper.*

A man of powerful intellect, but a reckless busybody, bold and aggressive, he "walked roughshod over men's opinions and suffered the inevitable consequences." His personal appearance must have been as peculiar as his conduct. As described by one of his pupils, "He was less than 5 feet high and his head was the biggest part of the whole man. He was a perfect taper from the side of his head down to his feet. He looked like a wedge with a head on it."

Supposed Human Footprints in Limestone, 1822.

The credulity of even the scientific men of these early dates has been repeatedly referred to and is again illustrated by Schoolcraft's account of the finding of supposed human footprints in limestone belonging to the "Secondary" formation (Lower Carboniferous), on the west bank of the Mississippi River at St. Louis.

These (see Fig. 17) were described as the tracks of a man, standing erect, with his heels drawn in and his toes turned outward. The distance between the heels by accurate measurement was 6¼ inches, and between the toes 13½ inches. Attention was called to the fact that the impressions were not those of feet accustomed to a close shoe, the toes being too much spread and the foot flattened. The length of the foot was 10½ inches, width across the spread toes 4 inches, and at the swell of the heels 2½ inches. Public attention was first called to these prints by the Reverend Frederick Rapp, the head of the religious sect "Harmonites," who had them removed to his village of Harmony, and, it is said, taught that they were the impressions of the feet of the Saviour.[4]

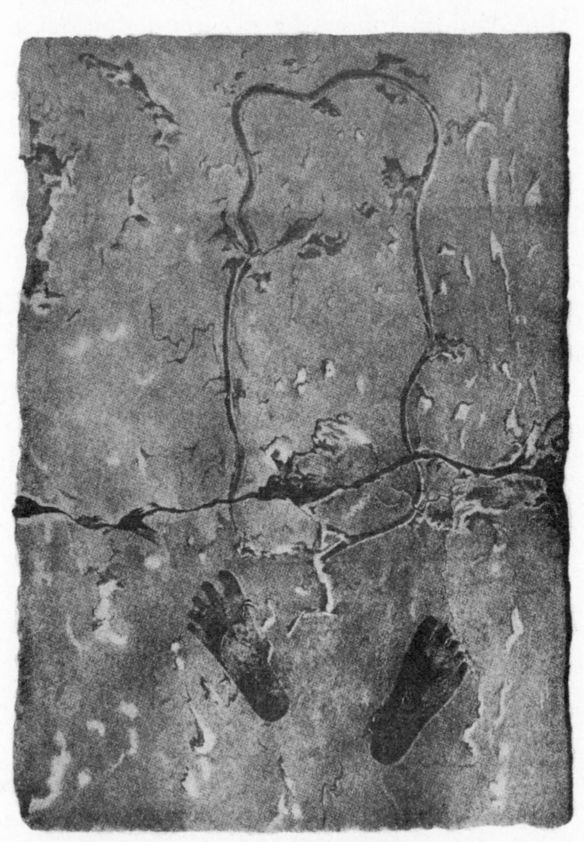

Fig. 17. *Supposed Human Footprints in Limestone.*

According to Schoolcraft, every appearance warranted the conclusion that the impressions were made at a time when the rock was soft enough to receive them by pressure, and the marks of the feet were natural and genuine. In this opinion, he stated, Governor Cass, of Michigan, coincided. He acknowledged, however, there were difficulties in the way of accepting this belief, and one of these was the absence of tracks leading to and from them. He could account for this only on the supposition that the toe prints might have pointed inland, "in which case we should be at liberty to conjecture that the person making them had landed from the Mississippi and proceeded no farther into he interior"!

Colonel Thomas H. Benton, of Missouri, to whom Schoolcraft wrote concerning the footprints, differed with him and thought them

[4] It would be interesting to know what has become of this block.

artificial, but his reasons therefor were not sufficient to convince Schoolcraft.

The matter was brought up again by David Dale Owen in 1842 in an article entitled "Regarding Human Footprints in Solid Limestone," which appeared in the *Journal of Science* of that year. Owen described the appearance of the tracks, and quoted the opinions of Maclure, Troost, Say, and Lesueur to the effect that they were of artificial origin. The English paleontologist, Mantell, was, however, inclined to the opinion of Schoolcraft. Owen himself considered them artificial for essentially the same reasons as those advanced by Benton, to the effect that, first, the footprints were not continuous, but isolated; second, this was but a solitary instance of human footprints in solid limestone; third, he could not conceive of the sudden consolidation of compact limestone after having received, while in a plastic state, such impressions; and, last, because of the age, nature, and position of the rock, and because no human remains had hitherto been discovered in any similar formation. He believed them to have been carved by aborigines with stone implements. In this he was doubtless correct.

Early Suggestions of Faulting, 1822.

Of interest at this time, as bearing upon the subject of faulting and displacement, is a paper in the same *Journal* by D. H. Barnes, containing a geological section of the Canaan Mountains in Connecticut, together with observations on the soils of the region. In the explanation of his section (see Fig. 18), beginning at the bottom, the beds are described as (1) clay loam with bowlders; (2) transition limestone; (3) white quartz, grading on one side into limestone and on the other into clay-slate; (4) slate; (5) graywacke slate. The strata then repeat themselves in the same order, the graywacke slate forming the summit of the knob. The entire formation is described as appearing "to have been broken off from the primitive tract on the east of it, and to have sunken down about one thousand feet perpendicularly." Referring to the beds and their associations on Hancock Mountain, he concluded that the strata on the top of the mountains might be considered as originally parts of the same bed now at the base of Canaan Mountain, whence they had "been disrupted by some mighty force," the eastern part remaining firm, while the western settled down to its present position. Referring to Maclure's map, he found that the two formations discussed "butt against each other in a line nearly straight for more than 300 miles." This he accounted for by supposing "that some mighty convulsion has rent asunder the continent from the St. Lawrence to the ocean," though what this force may have been he judiciously left for others to decide. He, however, thought it probable that it operated from beneath, "and that, after it had opened for itself a vent, and escaped through the rift caused by its action, the rock strata of the western part fell into the cavity which had previously contained the imprisoned agent."

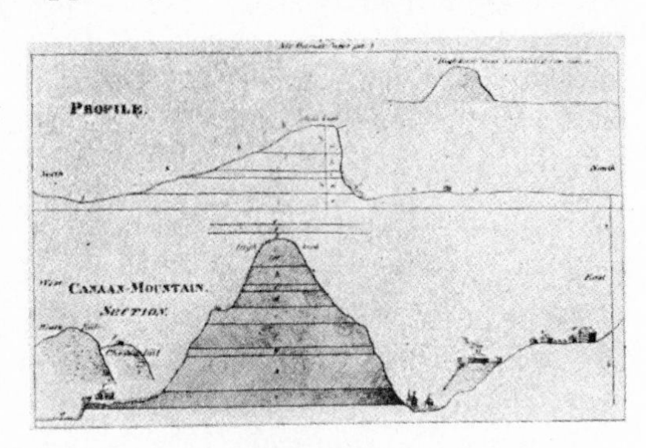

Fig. 18. *Barnes's Section of Canaan Mountain.*

Maclure's Conjectures, 1823.

In 1823, while in Madrid, Maclure had allowed himself to indulge in some interesting "speculative conjectures" concerning changes which he conceived might have taken place in the American continent. As he viewed it, the continent "east of the Stony Mountains consisted of a ridge of primitive mountains springing out of the great northern primitive formations, covered on the east and southeast by extensive beds of alluvion, apparently the deposition of the ocean, and on the west overlaid by Transition and Secondary, filling the immense basin through which the Mississippi now runs, with all its attendant streams." In imagination he could conceive of no period when the Alleghanies did not exist, but both analogy and reason pointed to a period when the mountains stood alone, surrounded by water and without their flankings of alluvial, Secondary and Transition materials. Thus far it is easy to follow him, but when we are asked to conceive of a time when the unbroken ridge of the Alleghanies on the east and south, with the Stony Mountains on the west, hemmed in an immense body of fresh water, which was drained only when the great rivers—the St. Lawrence, the Hudson, and Mississippi—had cut their way through, the pathway becomes more difficult.

Bigsby's Work in Canada, 1823.

This same year Dr. John J. Bigsby, a surgeon of the British army stationed in Canada, read before the Geological Society of London a paper, subsequently published in their *Transactions*, on the geology and geography of Lake Huron. He noted that the rocks of the north shore of the lake were mainly Primitive—granite, gneiss, basanite quartz rock, conglomerate, and greenstone. The other shores were described as occupied by Secondary rocks, frequently fossiliferous, which he thought formed part of an immense basin, "extending probably without interruption from the southern shore of Lake Winnipeg, spreading itself over the greater part of lakes Superior, Huron, and Simcoe, the whole of lakes Michigan, Erie, and Ontario, much of the western part of the State of New York, the whole of the States of Ohio, Illinois, Indiana, and Michigan, and the rest of the valley of the Mississippi." On Thessalon Island was found a new species of *Orthoceratite*, which, though described and figured, was not named. Drummond Island yielded also corals and a trilobite —*Asaphus platycephalus*. Naturally no attempt was made to determine the relative age of the rocks by means of these fossils.

In the year following, Doctor Bigsby had an article in the *Annals of the Lyceum of Natural History* (New York) on the geology of Montreal. He described the lowest rock of the region as a trap of a kind unique in the Canadas, and illustrating "in a beautiful manner the affinity existing between the formation of which it is a member and the primitive class in general." A horizontal shell limestone of a bluish-black color he noted as forming the floor of the plain surrounding the hill on which the city stands, and through this hill, as a center, the passage in all directions of a large number of trap dikes. The sandstone of St. Ann's he rightly described as underlying the limestone, and he noted the presence of fossil *Lingulæ*, *Terebratulæ*, *Trilobites*, and *Orthoceratites*.

In the consideration of the sands and gravels constituting the so-called alluvial, he followed the trend of opinion of his time, regarding them as products of the vast inland seas which succeeded the deluge. Officers of the British Army and Navy stationed in Canada often made, for the time, important observations which were published in the *Transactions of the Quebec Library and Historical Society*, and the *Quarterly Journal of the Geological Society of London*. The first mentioned society was founded in 1824, and the first volume of transactions bears the date 1829. It is here and in the succeeding volume (1830) that are to be found papers by Captain Bayfield of the Royal Navy, on the Geology of Lake Superior, and by Lieutenant Baddeley of the Royal Engineers, on the geology of the Labrador Coast, and an essay on the localities of metallic minerals in Canada. A catalogue of a mineral collection, comprising 1,310 numbers shows, for that period, a surprisingly long series.

Finch's Subdivision of the Tertiary, 1823.

The coastal plain of the eastern United States, which up to this time had been studied, or at least written about, only by Maclure and Hayden, was in 1823 made the subject of an essay by John Finch, of Birmingham, England, but residing temporarily in America. This appears to have been the first attempt at a subdivision of the Tertiary deposits and their correlation, and was, beyond question, the most important contribution to the stratigraphy of the region that had thus far appeared.[5]

From an examination of all available data as well as from a personal inspection of a part of the area, Finch was led to conclude that this formation, as existing in America, was identical and contemporaneous with the newer Secondary and Tertiary formations of Europe and other countries.

In this year Finch gave also a short sketch of the geology of the country near Easton, Pennsylvania, his paper being accompanied by a small colored map and a catalogue of minerals. According to J. P. Lesley, this was the first paper on the geology of Pennsylvania to appear in the pages of the *American Journal of Science.*

The paper on the coastal plain, above noted, is by all means the most important of Finch's geological publications, though in the *Journal* for 1826 he had a brief "Memoir" (3½ pages) on the red-brown sandstones of the Connecticut Valley and New Jersey, where he expressed the opinion that such should be considered as belonging to the New Red or variegated sandstones of Europe rather than to the Old Red, as contended by some. In this he was right.[6]

[5] According to Schuchert (Am. Jour. Sci., Vol. XLVI, 1918) Thomas Say, the biologist, was the first American to point out in 1819 the chronogenetic value of fossils in his Observations on some species of Zoöphytes, etc., stating that "the progress of geology must be in part founded on a knowledge of the different genera and species of reliquiæ which the various accessible strata of the earth present." (See further on p. 117.)

[6] Finch was author also of a volume entitled Travels in the United States of America and Canada (London, 1833), which contained a few notes on geology.

Olmsted's Work in North Carolina, 1823.

In the same year, Denison Olmsted, "a Connecticut school teacher," and student of Silliman's, was authorized by the president of the state board of agriculture to make a geological survey of the state of North Carolina. This act is sometimes referred to with sectional pride as being the first survey undertaken under state auspices. As a matter of fact, however, the work of Eaton in New York was the first sufficiently thorough and systematic survey to be dignified as a geological survey, though as this was done under the patronage of Honorable Stephen Van Rensselaer, it cannot be considered a public survey. Credit must be given to Massachusetts for the first geological survey made at the expense of the state, the same being begun under the direction of Reverend Edward Hitchcock in 1830, as will be noted later.

Fig. 19. *Denison Olmsted.*

Olmsted's first report, a pamphlet of forty-four pages, appeared under date of 1824 and was mainly of an economic character, dealing only with the distribution in the state of such substances as graphite, gold, coal, and building stones.

His second report appeared in 1827, but bore on the title-page the date 1825. He naturally dwelt upon the agricultural possibilities of the country and the suitability of its limestones and marls for fertilizing purposes. Some space was devoted to the great slate formation[7] and its included rocks and minerals. He noted that the "whole section of country from the great slate formation to the Blue Ridge" is granitic, containing numerous subordinate formations, such as mica-slate and greenstone, with beds of iron ore, etc., and described with some detail the Natural Wall of Rowan, which was supposed by early observers (see p. 21) to have been the work of human hands, but which he rightly ascribed to the natural jointing and decomposition of a basic igneous rock. The "peculiar assemblage of rocks that cross the Dan River at Buckingham" he classed as Transition.

[7] Olmsted's great slate formation lies west of and parallel with the freestone coal formation (Triassic) occupying more or less of the counties of Person, Orange, Chatham, Randolph, Montgomery, Cabarrus, Anson, and Mecklenburg and corresponds, therefore, to the Huronian of Kerr and recent workers.

In his "conclusion" he remarked that "the rocks are not, as in most other countries, particularly New England, exposed on the surface, but are very generally concealed by a thick covering of clay and sand." This he rightly regarded as having resulted from the decomposition of the rocks themselves, and not due to a deluge of waters, "as might at first be thought."

Sketch of Olmsted.

Olmsted was not primarily an investigator. He himself, as we are informed by his biographer, always regarded it as his more appropriate sphere of effort not so much to cultivate science as to teach and diffuse it. Graduated at Yale in 1813, and for a time employed as tutor there, he was appointed in 1817 to the chair of chemistry in the University of North Carolina. While here he conceived the idea of a state geological and mineralogical survey and laid the plan before the board of internal improvements in 1821, with the offer to himself perform the entire work gratuitously, asking only an appropriation of $100, to be afterwards renewed or not at the pleasure of the board, to defray his necessary expenses in traveling. This proposition was, however, declined,[8] only to be renewed a year or two later, with the result that in 1823 the assembly authorized the board of agriculture to have such a survey made and there was appropriated the sum of $250 a year, for a period of four years, to carry it out. The work thus inaugurated was interrupted in 1825 by his call to the professorship of mathematics and natural philosophy in Yale College, but was taken up again almost immediately by Dr. Elisha Mitchell and continued for two years longer (see p. 114).

The versatility of Olmsted is shown in the fact that while occupying this position at Yale he prepared in 1831 a two-volume work on natural philosophy, a textbook on astronomy in 1839, and became

[8] Referring to this failure to secure assistance, Olmsted, under date of January 9, 1822, wrote to a friend: "But the Legislature (the Senate, I mean) has, it seems, saved me any trouble on their account. I, however, feel most highly gratified and greatly encouraged at the handsome manner in which my proposition was treated by the Board of Improvement; and the readiness with which the resolution was adopted by the Commons inspires the hope that something may yet be accomplished at the public expense. But my feelings are too much interested in this project to yield to this failure; I hope to do something by my own exertions next summer, and trust the hospitality of the people of the State will make amends for my poverty. If I can live (as I think I can) on the charity of the people, I don't know what need I shall have of the public money; for Mr. M——— says he will lend me a horse."

well known throughout the scientific world through his papers on meteoric showers and the zodiacal light.

Of his personal character it is written:

> His uniform kindness and courtesy of demeanor and patience in imparting instruction, the excellent moral influence which he always exerted, as well by his consistent Christian example as by his personal counsels, the genuine friendliness of his disposition, and the unaffected interest which he always manifested in the welfare of his pupils—especially the readiness and fidelity with which he encouraged and assisted any who exhibited special fondness for the studies of his department—will not soon be forgotten by those who enjoyed the benefit of his instructions, and especially by those who were admitted to his closer friendship.

De Kay on Geology of Trenton Falls.

In the *Annals of the Lyceum of Natural History* (New York), Volume I, 1823, J. E. De Kay has some observations on trilobites, together with a short description of the geology of Trenton Falls, which are of interest. The rocks are described as of limestone, lying nearly horizontally, with numerous thin veins of argillaceous matter. After stating that organic remains furnish the most decisive evidence of the identity of different formations, he relegates them to the Transition class, "the Sub-medial of Coneybeare and Philips," the conclusion being based on the presence of the fossil *Calymene Blumenbachii.* Were it not for this, the slight inclination of the strata might "entitle it to be arranged as a part of the first floetz formation of Werner." This is of interest in connection with the observations of Finch, already mentioned, and those of Morton, later (p. 117).

Hitchcock's Geology of the Region Contiguous to the Connecticut River, 1823.

In this same year Edward Hitchcock published in the *American Journal of Science* a sketch of the geology, mineralogy, and scenery of the regions contiguous to the Connecticut River, the same being accompanied by a colored geological map embracing an area some 30 miles broad by 150 miles in length. The coloring and classification of the rocks were not strictly Wernerian, as might have been expected, but an attempt was made "to give every particular rock that position and extent on the map which it actually occupies on the earth's surface." The paper was devoted mainly to a discussion of the lithological nature and geographic distribution of the various rocks, which were classed as granite, gneiss, hornblende slate, mica-slate, talcose slate, chlorite, syenite, Primitive green slate, argillite, limestone, verd-antique, Old Red sandstone, Secondary greenstone, coal formation, and alluvion. Incidentally he discussed their possible origin and relationship.

The granite was described as occurring in beds, and thought to be Primitive, along with the gneiss, mica-slate and the greenstone to the west of New Haven, while that to the north and east was Secondary. The argillite was also considered Primitive, since it was highly inclined and destitute of organic remains. The red-brown sandstone now known to be of Triassic age he considered the equivalent of the Old Red sandstone of Werner and Cleaveland, though the possibility that a part of it, as at Chatham and Middletown, might belong to the coal formation was recognized.

He was inclined to agree with President Cooper as to the igneous origin of the greenstone, but committed himself only partially. He wisely added, "I cannot but say that the man who maintains . . . the original hypothesis of Werner in regard to the aqueous deposition of trap, will find it for his interest, if he wishes to keep clear of doubts, not to follow the example of Daubisson by going forth to examine the greenstone of this region, lest, like that geologist, he should be compelled not only to abandon his theory, but to write a book against it."

The finding at Deerfield of petrifactions belonging to the genus *phytolite* and to the species *lignite*, and agreeing with the petrifactions found in the Catskill red sandstones by Eaton and referred to "the tribe of Naked Vermes" was noted. Fossil bones of an animal some five feet in length were also mentioned as having been found in this sandstone at East Windsor. These were at first thought to be possibly human, but the unmistakable evidences of a tail caused this idea to be abandoned.

Dewey on the Geology of Western Massachusetts, 1824.

The following year Professor Chester Dewey came forward with a paper in the same journal, "A Sketch of the Geology and the Mineralogy of the Western Part of Massachusetts." This was likewise accompanied by a hand-colored geological map designed as a continuation of Hitchcock's and carrying the field of observations as far west as the Hudson River.

Working from the river eastward, he found a belt of Transition argillite succeeded by a broad belt of graywacke with included areas of the argillite; this followed by another continuous belt of the same, and then one of Transition limestone and a narrow one of Primitive argillaceous slate directly along the border line of New York and Massachusetts; beyond this, in Massachusetts, a broad belt of Primitive limestone with included areas of mica-slate, quartz rock, and Transition shell and compact limestone, which still farther to the east became the predominant rock, with a narrow area of gneiss, a small one of granite, and one of talcose slate.

Eaton's Survey of the Erie Canal, 1824.

The most pretentious piece of field work accomplished by Amos Eaton was that done under the direction of the Honorable Stephen Van Rensselaer in 1824, and comprised a geological and agricultural survey of the district adjoining the Erie Canal. The results of this survey were published in a volume of 163 pages, accompanied by a geological section extending from Albany to Lake Erie, and one by Edward Hitchcock from Boston to Plainfield, the two combined enabling Eaton to give a continuous section from Lake Erie to the Atlantic Ocean at Boston Harbor.

Eaton found along the line of the canal rocks belonging to the Primitive, Transition, and Secondary series. All these were described in considerable detail, but almost wholly with reference to their lithological features. When fossils occurred, such were noted, but were not utilized as aids to correlation.

The prevalent ideas, or at least *his* ideas, on chemical geology are well brought out in certain parts of this discussion. He recognized the fact that many of the valleys were excavated in the softer and more soluble rocks, like limestone, and that carbonic acid holds its base (the lime) with a tenure more feeble than that of the "common" acids; consequently he argued that when muriatic, or nitric acid comes in contact with limestone, it is immediately decomposed.

> We have vast quantities of muriate of lime in our wells, springs, etc., which is a very soluble salt. If nature has now, or formerly had, any method for presenting large quantities of muriatic acid to the lime rocks, they would of course be reduced to that soluble salt with great rapidity. Lime rocks would be rapidly dissolved, leaving valleys between those rocks which are subject to the ordinary disintegrating agents only. The valleys of Adams, Williamstown, Little Hoosick, etc., which are situated on limestone, could then be satisfactorily explained. If the common opinion that the ocean has stood over our continent be received as true, we have only to add one more conjecture to make out the requisite supply of muriatic acid; that is, we must suppose that the ocean at that time contained an excess of that acid.

And this in spite of admonitions by Van Rensselaer against indulging in theories! Eaton, however, denied any such propensity, claiming that he but traced "a few of nature's footsteps where the impressions still remain entire."

In discussing the saliferous rock occurring near Little Falls and its economic importance, he argued that the brine springs issuing from the same were "the daily productions of nature's laboratory." "We see the sulphur in iron pyrites taking oxygen from water, and thereby becoming sulphuric acid; we then see it uniting with magnesia, which is diffused in rocks, and thus forming Epsom salt. . . . We are all familiar, too, with the process of nature by which alum and copperas are made. Why may we not suppose that the two constituents of common salt (muriatic acid and soda) are in some state of combination in the rocks of the salt district, and that by some of those double decompositions with which nature is perfectly familiar salt is produced in the liquid state? May not this be the cause of the superior saltness of the brine springs of Salina over those foreign springs which are supposed to proceed from the solution of rock salt?"[9] It was in this work that Eaton introduced the term "Calciferous slate," thereby designating the slaty rock associated with gypsum and shell limestone which formed the principal portion of the ridge south from Oneida Creek to Pittsford.[10]

[9] A work written for children by one not a geologist can scarcely be expected to reflect the real condition of knowledge, even of the general public, at any given period. The following from The Child's Geology (1832) is nevertheless edifying:

"He who formed the earth and the sea, knew, that by saltness only, could the ocean be kept sweet. The saltness of the sea is absolutely necessary, to preserve it from putrefaction. So far from supposing that these masses of salt were abstracted from the ocean, it is more likely, that the saltness of the ocean is occasioned by stones of Rock-Salt at the bottom."

[10] The following reference to this survey appeared in the American Journal of Science, IX, 1825, and will serve to show how Eaton's work was regarded by his contemporaries. The note, it should be stated, was signed C. D. (presumably Chester Dewey):

"Mr. Jeffrey, the principal conductor of the Edinburgh Review, has obtained the opinion of Professor Buckland, the celebrated author of the *Reliquiæ Diluvianæ* (published in 1823), on the above work of Professor Eaton. In a letter to the Hon. S. Van Rensselaer Mr. Jeffrey has given the result. Mr. Buckland says that the 'author seems both to understand his subject and to have done his work carefully.' The work contains, indeed, abundant evidence of extensive and patient examination. This point will not be affected by the adoption or rejection of Mr. Eaton's peculiar views by our geologists. In some parts there is an evident improvement upon some of his previous publications on the geology of our country.

"Mr. Buckland makes some objection to the *style,* and complains of Mr. Eaton for 'affecting some needless novelties in technical language.' However true this

Views of Rafinesque, 1824.

The naturalist Rafinesque, as might have been expected, did not hesitate to express himself on subjects geological, his ideas, however, being a strange admixture of Wernerism and groundless imagery. It will be sufficient to note here only his "Annals of Kentucky," which appeared in Humphrey Marshall's history of that state, published in 1824, where, as Doctor Peter has succinctly remarked, "In only 26 duodecimo pages he gives the geological, ethnological, and historical annals of Kentucky from the first day of creation, according to Moses, down to the current year."

The geological history of the state he divided into six periods, the first being that of a general inundation, in which the "briny ocean covers the whole land of Kentucky and the United States, rising above 4,000 feet over the Cumberland or Sawioto Mountains and 5,000 feet over the limestone region near Lexington." The Organ and Mexican mountains alone in all North America rose above the water level.

Through a gradual decrease of the ocean and the decomposition and consolidation of its waters the various rock strata were deposited in the following order: (1) Limestone; (2) slate; (3) sandstone; (4) freestone; (5) grit; (6) pebblestone. These, he stated, were not always superincumbent or coexistent, though generally horizontal except the last four toward the Cumberland Mountains, which, having probably a granitic nucleus, have compelled the incumbent strata to become "obliqual" or inclined from ten to thirty degrees. In this, it will be noted, he showed an inability, common in his day, to comprehend the effect of subsequent uplift in tilting rock strata. By the operation of submarine volcanoes during this period the strata of coal, clay, and amygdaloid were formed and intermixed at various intermittent times with the other strata.

During the second period the Cumberland Mountains emerged from the sea, the waters of which sank to a level of 1,500 feet above those of today. The formation of the schistose rocks proceeded, vegetation began, and streams began to flow. During the third the level of the water was reduced to 1,100 feet and all the table-lands and highlands of Kentucky became uncovered. An inland sea still

charge may be, the censure is feeble when compared with the commendation contained in the previous quotation. In our country the work has been censured for this fault, and more particularly for the introduction of rocks or localities which do not belong to the district which is described. In this way unity is not preserved, and the continuity of the description is much interrupted. . . . It were to be wished that the common nomenclature of the rocks had been altered with a more sparing hand."

covered the Ohio limestone basin, extending from the actual mouth of Scioto River to that of Salt River. Land animals, insects, reptiles, birds, and quadrupeds were created during this period.

The fourth period was marked by a reduction of the level of the sea to 700 feet, and the limestone sea of Kentucky drained. Alluviums and bottoms began to form in the valleys and gulfs by the attrition of the strata and soil conveyed and deposited by the streams. Sinks and caves in the limestone region were formed. Lastly, Adam appeared in the Garden of Eden. This fourth period of Kentucky history thereby answered to the sixth day of the general creation.

The fifth period was that of Noah's flood, though Rafinesque acknowledged he failed to find any traces of such a violent convulsion in Kentucky. The ocean, which still bathed its western corner, subsided to about 300 feet above its present level and abandoned Kentucky forever. Strata began to consolidate, ponds and marshes decreased, animals multiplied, and vegetation overspread the soil.

The sixth period, that of Peleg's flood, was one of catastrophe. Great volcanic eruptions in Europe and America, with awful earthquakes, convulsed the Atlantic Ocean. During this period the Atlantic land disappeared, leaving only the volcanic islands—Azores, Madeira, Canary, and Cape Verde—to mark its position.[11]

Attempted Geological Survey of South Carolina, 1824.

In the acts of the General Assembly of South Carolina, passed in December, 1824, there occurs a clause appropriating $1,000 "for the salary of the Professor of Geology, and Mineralogy . . . and $500 for making a Geological and Mineralogical tour during the recess of the college, and furnishing specimens of the same." Presumably this movement was inspired by Vanuxem, then teaching at South Carolina College in Columbia, though definite proof is lack-

[11] Equally absurd speculations regarding early earth history were characteristic of this period. See Hayward's Natural and Aboriginal History of Tennessee, 1823. It may be admissible to refer also to a theory—scarcely geological—advanced about this time by a Captain J. C. Symmes, U. S. Army. Symmes argued that the earth is a hollow sphere, habitable within and widely open at the poles. So singular a burlesque was accepted by few except its author, who wrote, traveled, and lectured upon the subject until intercepted by death in 1829. Aside from his pamphlet, The Theory of Concentric Spheres, little now remains of his ideas except in the minds of the passing generation who may recall newspaper discussions of the "Symmes Hole Theory." A writer under the name of Seaborn (probably a pseudonym) subsequently (1820) wrote a small volume, Symzonia; A Voyage of Discovery, purporting to be the narrative of a voyage to the South Seas and into the earth's interior under the conditions described by Symmes.

ing. Reference in the bill is made to a "Catalogue of Mineral Specimens collected by a distinguished member of the faculty" as "furnishing the strongest assurance of his industry and science, and the richness of the country he has been directed to explore." The preamble is most amusingly verbose and grandiloquent, but seems to have failed to bring forth fruit. So far as can be learned Vanuxem did little more than make in 1826 a report to the newspapers, which was reprinted in Mill's *Statistics of South Carolina*, and later (1848) in Tuomey's final report. It consisted merely of a list comprising ten species of rocks and thirty of minerals, with the tests applied for their determination. Vanuxem resigned in 1826 and left the state.

The Long-Keating Expedition, 1823-1825.

April 25, 1823, Major S. H. Long, whose explorations of 1819-1820 have already been noted (p. 69), received orders from the War Department to make an expedition for a general survey of the country in the vicinity of the Great Lakes and the sources of the Mississippi; to prepare a topographic description of the same; to ascertain the latitude and longitude of all the remarkable points; to investigate its productions, animal, vegetable, and mineral, and to inquire into the characteristics and customs of the Indians. The route of the expedition (see map, Fig. 20) beginning at Philadelphia, was through Wheeling, West Virginia, to Chicago by way of Fort Wayne; thence to Fort Crawford and up the Mississippi to Fort St. Anthony, and the source of the St. Peters River; thence to the point of intersection between Red River and the forty-ninth degree of north latitude; along the northern boundary of the United States to Lake Superior, and thence homeward by the Great Lakes.

Although not intended primarily as a scientific survey, it was accompanied by Thomas Say, zoölogist, and William H. Keating, mineralogist. It was expected that Dr. Edwin James, who with Say had been a member of the expedition of 1819-1820, would be a member of this also, but through failure to connect with the party at Wheeling or Columbus it was deprived of his services.

An account of this survey, under the title of *Narrative of an Expedition to the Sources of St. Peters River*, etc., was prepared by Keating and published in two volumes in Philadelphia in 1824 and in London in 1825. Keating, it should be noted, was professor of mineralogy and chemistry as applied to the arts in the University of Pennsylvania, and though his published notes contain little or

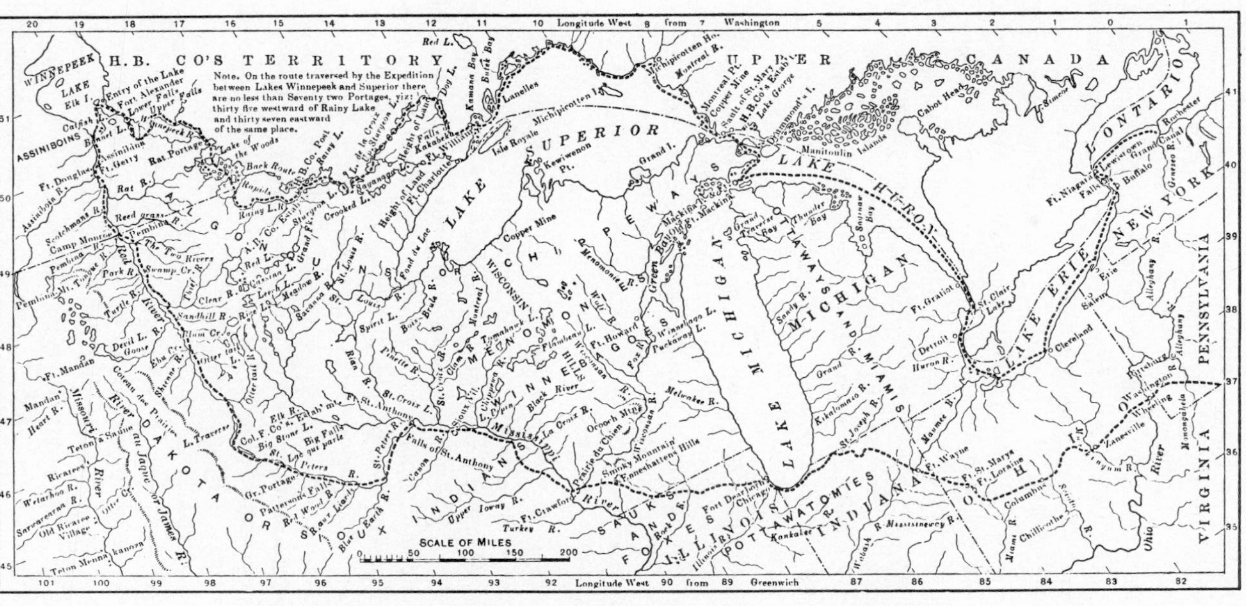

Fig. 20. *Map of route of Long's Expedition of 1823.*

nothing along the broad lines of geology, they are full of references which, at the time they were written, were of value.

He remarked on the disappearance of the Primitive rocks soon after leaving Philadelphia and the appearance of the Transition limestone, with occasional protrusions through this limestone of amphibolite. He also made reference to the red sandstone of Pennsylvania, New Jersey, and Maryland and the calcareous breccia found on the Potomac River. The occurrence of a white marble near Hagerstown and a secondary blue limestone in the vicinity of Cumberland was noted, and some attention was given to the coal formations in the vicinity of Cumberland and Wheeling. He noted further the occurrence of a limestone containing organic remains lying in a nearly horizontal position in the vicinity of Chicago, but, no superposition being visible, he was unable to determine its relative age.

Concerning the lead ores at Dubuque, and their apparent occurrence only in the alluvial soil, he wrote:

> From the specimens which we have seen, . . . we can not hesitate in asserting it as our opinion that no lead has yet been discovered on the Merrimeg or Mississippi in metalliferous limestones, but that wherever it has been found it has always been in an alluvium and never in regular veins or beds, nor even in masses which might be considered as coeval with the substances in which they are embedded.[12]

The Cambrian sandstone found between the Wassemon and the Wisconsin rivers he thought not older than the variegated or *Bunter sandstein* of Werner (Triassic), and the Lower Silurian limestone which lies above the sandstone, to correspond to the English Lias. The whole region comprising the headwaters of the Winnipeek River was looked upon as having been at a comparatively recent period an immense lake interspersed with innumerable barren, rocky islands, which had been drained by the bursting of the barriers which tided back the waters. The innumerable bowlders which he found covering the valley were regarded as due to the flood of waters caused by the bursting of these natural dams.

This was evidently a recognition of the extinct glacial Lake Agassiz, later described by Lieutenant G. K. Warren,[13] the history of which was worked out in detail by Warren Upham later.

Keating's ideas on the possible development of the mining industry were not of the most advanced type. Referring to the subject of the supposed source of the native copper reported by Schoolcraft (p. 65), he wrote:

[12] See explanation on p. 329.

[13] American Journal of Science, XVI, 1878, p. 417.

The question which appears to us of far greater importance is not where the copper lies, but what shall we do with it if it should be found. We are very doubtful whether any other advantage would result from it, at least for a century to come, than the mere addition in books of science of a new locality of this metal.

This in 1825! The development of the Michigan copper mines began about 1855, and up to 1889 the combined mines had produced over 1,000,000,000 pounds of the refined metal. The output for 1889 alone amounted to 87,455,675 pounds.

J. H. Steel on Overturned Folds, 1825.

In a "Notice of Snake Hill and Saratoga Lake and its Environs," by Dr. J. H. Steel, in the *American Journal* for this same year, is figured and described what is plainly an overturned fold. This is of interest, since while the nature of the fold was partially realized, the means by which it was brought about was considered as problematical. Similar phenomena had been noted by Eaton (p. 78) and by Maclure, but with nothing like as much detail. The writer says:

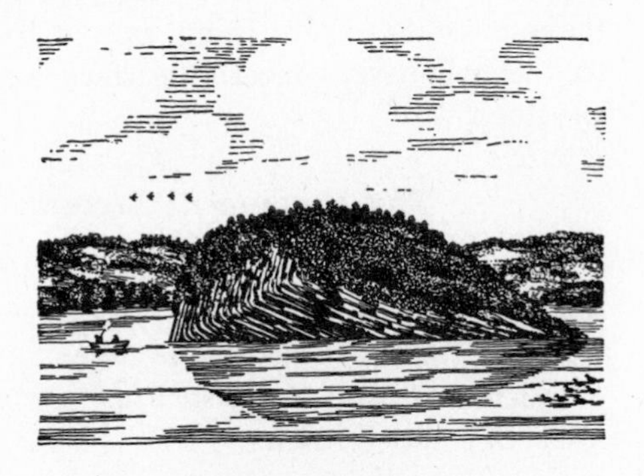

Fig. 21. *Steel's Sections of Snake Hill, Saratoga Lake.*

It is impossible to examine this locality without being strongly impressed with the belief that the position which the strata here assume could not have been effected in any other way than by a power operating from beneath upward and at the same time possessing a progressive force, something analogous to what takes place in the breaking up of the ice of large rivers. The continued swelling of the stream first overcomes the resistance of its frozen surface, and having elevated it to a certain extent, it is forced into a vertical position, or thrown over upon the unbroken stratum behind, by the progressive power of the current.

If it can be admitted that the operation of such a power did produce the effect here represented, it must have taken place before the materials of which the formation is composed had passed into an indurated state, as most of the strata remain unbroken, and, where the argillite has crumbled away, the curved part of the graywacke may be taken out entire, and some of them, which I now have in my possession, exhibit indentations and protuberances, particularly on their curved surfaces, evidently the result of friction while in a plastic state.

The description is of further interest when considered in connection with the dispute relating to priority of discovery which later arose between Messrs. Hitchcock and Rogers. (See p. 150.)

The fact that rock strata were not always found lying even approximately horizontal had been often noted, but no rational explanation suggested or, indeed, attempted. In this year, however, Maclure, in noting the fact that the Transition rocks dip, suggested the possibility of its being caused by their having been "disposed on the primitive, concerning which we can as yet scarce conjecture anything."

Van Rensselaer's Lectures on Geology, 1825.

The appearance in book form of Professor Jeremiah Van Rensselaer's lectures on geology, delivered before the New York Athenæum in 1825, affords an opportunity for summing up the condition of science at that date, such as was offered a few years earlier by Mitchill's *Observations*.

Van Rensselaer was not so much an original investigator as a student and teacher; hence the inference is fair that his work gives us a summary of existing knowledge rather than the details of his own observations or his individual views. He reviewed the opinions of the cosmogonists and theorists from Burnett, in 1680, down to Werner and Hutton, and referred to the work of his predecessors and contemporaries in America, concluding with a condensation from Cuvier's *Observations*, to the effect that, first, the sea had, at one period or other, not only covered all our plains, but remained there for a long time in a state of tranquillity; second, that there had been at least one change in the basin of the sea which preceded the present one; third, that the particular portions of the earth which the sea had abandoned by its last retreat had been laid dry once before and had at that time produced quadrupeds, birds, plants, and all kinds of terrestrial productions; and that it had been reinundated by the sea, which had since retired and left it to the possession of its own proper inhabitants. These facts, which had been proven through geological evidence, he thought to support the account of Moses, both agreeing, first, in the prevalence over everything else of water at the time of the creation; second, in the subsequent separation of the land from the water; and, third, in the eruption of the sea over the continent, the last corresponding to the Noachian deluge.

Van Rensselaer recognized the value of fossils in proving the identity of geological horizons, also the fact that organic remains have been deposited in successive generations and in such order that those of one bed bear a certain relation to each other and exhibit certain distinctive points differentiating them from those of earlier or later deposits, and that the greater the distance between the deposits the greater the difference between the contained fossils. This, I believe, is the second recognition by an American author of a now well-established principle.

The tendency to make sweeping generalizations founded upon purely local observations is noticeable in his writings, as in those of many of his contemporaries. In describing the gneiss he said:

It is the next rock to granite and occurs resting or lying upon it. When they are both seen in the same mountain its ledge is always the lower of the two. Mountains of gneiss are seldom so steep as those of granite, and the summits are not quite so peaked.

Or, again, in writing on the weathering of rocks:

The decomposition of granite is slow and when decomposed the unfriendly siliceous grains are easily washed away. There is neither vegetable nor animal matter in the compound; it does not absorb moisture, letting the moisture percolate, nor does it retain heat. The soil made from gneiss is not washed away so easily, and the mica yields more argillaceous matter. . . . Micaceous schist and argillite decompose more rapidly and form a better, though not a good, soil. . . . The rivers of primary districts have rocky beds and precipitous banks, etc.

Van Rensselaer was one of the first to recognize the necessity of exercising care in the selection of stones for building purposes:

We take our materials because they are near at hand, because they are cheap, and because others have taken the same, in preference to searching out others which . . . are more durable. Scarcely any one building in Europe or America of modern construction at the end of one thousand years will have one stone left upon another stone to denote the place where it stood, and the most splendid works of modern architecture are even now hastening to decay from want of attention to this subject.

This might with equal truth be written today.

His remarks on anthracite coal offer an interesting illustration of a disposition, still too frequently met with, to make the observed facts conform to preconceived opinions. Anthracite or native mineral carbon, or "blind coal," he wrote, is not, strictly speaking, a coal, though combustible.

> It occurs in the primary rocks, and hence it is obvious that it did not proceed from the decomposition of vegetable substances, since it is generally acknowledged that the primary rocks were formed previous to the existence of vegetation.[14]

An analytical key or a synopsis, given with a view to facilitating the study of geology, is of particular interest, since the methods of modern petrography have shown how absolutely impossible it is to classify rocks by a simple examination of the hand specimen and without the aid of a microscope. It is difficult to imagine a student in other than a thorough state of mental demoralization who should attempt to utilize this synopsis, an abstract of which is given below:

Synopsis of the Overlying or Superincumbent Rocks.

FIRST DIVISION.

Simple, or apparently so.

A. Wacke, of the German school. Resembles indurated clay, with an even and smooth earthy, or an uneven, somewhat granular fracture, and a shining streak.

a. Compact.

b. Cellular; but generally in that case partly amygdaloidal and appertaining to another division.

B. Indurated clay, more or less hard, with an earthy and dull fracture.

a. Compact.

This is different from the ferruginous clays found often with the trap rocks, which pass into jasper.

[14] Coal was at this date just beginning to receive consideration as of possible value for fuel. The American Journal of Science, Vol. XL, 1825, has the following editorial note relative to Pennsylvania anthracite:

"There can be no doubt that this coal will become an object of vast national importance. It is a very pure anthracite, but sufficiently combustible, *in a proper apparatus, and with proper treatment,* to maintain a constant and (if desired) intense fire, which burns without odour or smoke—is perfectly safe although left for the night and without watching, will continue till morning, and will then be found in a state of sufficient activity. It is admirable for halls, churches, and other similar places, where it is desired to throw into circulation a large volume of warm air; and we have it on the best authority that it is excellent for the smith's forge, and for other purposes of the arts."—Ed.

b. Cellular.

Like var. A, *b,* it is rarely cellular in large masses without also containing amygdaloidal nodules, when it passes to another division. The colors of this variety are usually ash or gray of different hues, or modifications of red, or brown, or purplish black.

Emmons's Manual of Mineralogy and Geology, 1826.

In 1826 there appeared a *Manual of Mineralogy and Geology,* designed for the use of schools and for persons attending lectures on these subjects, as also a convenient pocket companion for travelers in the United States of America, the same being from the pen of Dr. Ebenezer Emmons, destined later to act a very important part in American geological history, but at that time newly graduated, under Amos Eaton's guidance, from the Rensselaer Polytechnic Institute at Troy, New York.

Emmons at first studied, and later practiced, medicine, but in 1828 removed to Williamstown, where he had been appointed lecturer on chemistry. During the years 1830-1837 he also served as junior professor in the Rensselaer Institute, and in 1836 was appointed one of the four geologists of the New York state survey. In 1842 he became custodian of the state collections at Albany, and in 1843 engaged in investigations relating to the agricultural resources of the state, the results of which were published during the period 1846-1854 in the form of five quarto volumes. In 1851 he was appointed state geologist of North Carolina, a position which he retained until the time of his death.

As noted elsewhere, Emmons, during his work on the New York state survey, became involved in a discussion relative to the so-called "Taconic System," which lasted for nearly half a century and which undoubtedly seriously embittered the latter days of his life. The dispute, indeed, became at times so harsh that Emmons was practically ostracized by the scientific fraternity.

In 1851 he published a pretentious volume on North American geology, and, in connection with his work in North Carolina, two volumes on the agriculture and geology of the state. He died during the period of the Civil War, and his papers and notes are said to have become lost in the unsettled condition of the country which followed.

It is impossible here to go into a discussion of the true merits of the Taconic controversy, and the matter is reserved for a separate chapter. Hundreds of pages of printed matter have been published, covering a period of nearly fifty years, but the name has now disappeared from the maps and is of historical interest only.

Emmons's *Manual*, to which reference is above made, and which, as stated in the title-page, was adopted as a textbook at the Rensselaer school at Troy, was a small duodecimo volume of 230 pages. It had the distinction, so far as the mineralogical part was concerned, of being the second treatise of its kind written by an American and for American students, being preceded only by Cleaveland's *Mineralogy*, published in 1816. In geology it was preceded by the works of Cleaveland in 1816 and Eaton in 1818.

In Emmons's work the minerals were divided into four classes, those concerning which little was known, or which, one is tempted to add, did not yield themselves to this classification, being relegated to a common dumping ground—the appendix. The classes were subdivided, as Cleaveland and others had done before him, into orders, the orders again into genera, and the genera finally into species. His arrangement of classes was, however, somewhat different from that of any of his predecessors. Thus, the first class included those minerals, not metallic, which are "oxidable" or which are compounds whose bases are oxidable. It comprised native gases and liquids, sulphur and carbon, and the carbon compounds, but, singularly enough, no mention was made of graphite.

The second class included all minerals which are metallic or the bases of which are metallic; the third class all those which consist of an alkaline or earthy base in combination with acids, and the fourth those which consist of an earth or are compounds of earths with variable portions of alkaline and metallic oxides. In the seven sections into which this last class was divided he placed quartz in all its varieties, siliceous slate, pumice, obsidian, clinkstone, a large number of hydrous and anhydrous silicates, argillite, wacke, clay, phosphates, etc. But little attention was given to crystallography, and "silex, alumnie, and lime are considered as the oldest of the earths, as they enter into the composition of the primitive rocks." Altogether 297 species were recognized, 44 of which were mentioned in the appendix as little known. Many of these, such as pumice, obsidian, wacke, etc., cannot, of course, be properly considered as minerals. In spite of this unintentional exaggeration, it is interesting, for purposes of comparison, to note that in the 1892 edition of Dana's *Mineralogy* 824 species were recognized.

The part of Emmons's work devoted to geology was made up mainly of a "general description of North American rocks." The classification adopted was the same as that used by Eaton and need not be noted here further than to say that he included an argillite among his Primitive as well as Transition rocks, while Eaton limited argillite to the Transition series, though recognizing the possibility of a Primitive form.

Troost and His Work, 1826.

Among the early workers in stratigraphic geology, along lines laid down by Cuvier, Brongniart, and their successors, mention must be made of Gerard Troost, born in Holland in 1776 and dying in Nashville, Tennessee, in 1850. Like many naturalists of his time, Troost was a physician. He studied under Hauy in France and in 1809 was appointed by the King of Holland a member of the scientific corps to accompany a naval expedition to Java.[15] The English, then at war with Holland, prevented the sailing of the Dutch fleet, and Troost therefore embarked on an American vessel from a German port. The vessel was, however, captured by a French privateer, and Troost taken as a prisoner to Dunkirk. He was soon released and went to Paris, whence he sailed early in March, 1810, for Philadelphia. Java shortly after passing under British control, the projected expedition was given up, and he decided to remain in America.[16]

When the Philadelphia Academy of Sciences was established in 1812 Troost became its first president, holding the office for five years. With Owen, Maclure, and others, he joined the communistic society at New Harmony, Indiana, in 1825, but he removed to Nashville, Tennessee, in 1827, and the following year was elected professor of chemistry, geology, and mineralogy in the university of that city. In 1831 he became state geologist, holding the office till 1839, when it was abolished. Nine reports were made during his time, but seven of which were published. Prior to going to Tennessee his most important geological work was a survey in 1826, under the patronage of the Society for the Promotion of Agriculture, of the region about Philadelphia. His report comprised forty small octavo pages with a colored map of the region included within a half circle north of the Delaware River, having a radius extending a little beyond Chester, *i.e.*, some seventeen and a half miles.[17] As may be

[15] An absurd typographical error in my Contributions made this to read *Japan*.

[16] See article on Troost in Halsey's Life and Works of Philip Lindsey. I have decided to follow this rather than the account given in Appleton's Cyclopedia of American Biography.

[17] Marcou in his Mapoteca Geologica Americana, Bull. No. 7, U. S. Geol. Survey, 1884, expresses a doubt as to the existence of this work. The author, however, has a copy in his library.

readily supposed, fully three-fourths of the area was colored as gneiss, with narrower bands extending in a general way parallel with the river; in the northern part, of Primitive clay-slate and of limestone. Between the gneiss and clay-slate was a short, narrow belt of serpentine, and between the clay-slate and limestone one of eurite, and in the extreme northeastern portion of the sheet a band of Transition graywacke. Among the varieties of rocks mentioned, in addition to those enumerated and comprising subordinate formations, are diabase and pegmatite. The eurite was described as occurring north of the high ridge which separates the limestone from the granitic rocks and as being in every respect similar to that of Penig, on the Erzgebirge. "I was delighted," he wrote, "at meeting this rock for the first time on this side of the Atlantic. I imagined myself transported to the Erzgebirge, in Saxony, and remembered with renewed pleasure the father of geology, who made us acquainted with it." More than half of the paper, as might be expected when its date and the auspices under which the work was done are taken into consideration, was given up to a discussion of the physical and chemical properties of the soil.

Webster's Account of the Geology about Boston, 1826.

In the *Boston Journal of Philosophy* for the same year the ill-starred Professor J. W. Webster gave a somewhat detailed account of the geology of Boston and vicinity, describing the three hills, to which "Boston owes its ancient name and so much of its picturesque beauty," as being composed mainly of hard, compact clay with gravel and bowlders. Amister's Hill was described as composed largely of clay-slate, passing on the north into hornblende slate, the latter containing veins of "greenstone"; and Prospect Hill, in part of a greenish compact feldspar, which passes into clay-slate covered toward its northwest extremity by a mass of trap. "This hill," he wrote, "exhibits that gentle acclivity and rounded summit so common in the transition formations of the Wernerian school."

The Medford trap (diabase) he described as unfit for architectural purposes, owing to its rapid disintegration, a fact which has been many times commented upon in more recent years. The tendency manifested by the "greenstone" as a whole to exfoliate in bowlder forms with concentric structure, he correctly ascribed to weathering, as did Gibson writing several years earlier (p. 80).

He noted the occurrence of abundant joints in the conglomerate of Dorchester, but considered such as inexplicable with the geological

PLATE VI

GERARD TROOST

J. M. SAFFORD

information then available. An interesting light is thrown upon the lack of knowledge of the chemical composition of rocks at that time in the continuation of his paper the following year, in which he described this conglomerate—a highly siliceous rock—as passing into the overlying melaphyr, a basic igneous rock. In several places within the town of Brighton he thought to note the transition from one rock type to another. This deceptive appearance, it may be stated, is due to the fact that the melaphyr at the time of its extrusion was in a highly liquid condition and flowing out over the uneven surface of the conglomerate filled in all the inequalities, so closely welding itself as to form what was apparently one and the same mass. When subsequent erosion cut away a considerable portion of both rocks the appearance of isolated patches of melaphyr here and there on the eroded surface of the conglomerate was quite misleading. Some more recent observers have committed the same blunder of observation and faulty conclusion with far less excuse.[18]

Alanson Nash on Vein Formation, 1827.

In the *American Journal of Science* for 1827 Alanson Nash presented his first, last, and only geological paper that seems to have found its way into print—this relating to the lead mines and veins of Hampshire County, Massachusetts—and offered some interesting speculations as to their origin.

That they were not once open fissures filled from above, the fissures themselves being formed by the unequal subsidence of the earth's crust or through shrinkage caused by desiccation, according to the Neptunian theory, was to him evident for the following reasons: If the cavities were formed by desiccation and subsidence the veins would be widest at the surface and narrow as they descend, whereas, in fact, the very reverse is the case. If filled from above by mineral solutions which covered the globe, then he thought we ought to find beds of metallic matter in the valleys and plains also. Neither was he disposed to accept the views of the Plutonists, who regarded the veins as filled by "an injection from a fiery furnace below." Rather would he look upon them as contemporaneous both in formation and filling with the rocks in which they occur, being analogous to the granite veins of the same region. That the vein material did not adhere firmly to the wall rock, as is the case with the granite veins, was to him no argument against such a view. "In one case the vein is lapideous, in the other it is metallic; they are different."

[18] Proc. Bost. Soc. of Nat. Hist., Vol. XX, 1878-1880, p. 129.

Mitchell's Geological Work in North Carolina.

On the resignation of Olmsted, whose work in North Carolina has been mentioned on p. 94, Elisha Mitchell, also from Connecticut and a Yale graduate, was transferred from the chair of mathematics in the State University to that of chemistry and geology, and made an unsuccessful attempt at a continuation of the geological work. The following entry is found in his diary, under the date December 28, 1827: "The Geological Survey dies a natural death at the end of this year. There is no one who takes any interest in the business, nor, in the present state of the treasury did I find there was the least prospect in succeeding in any application to the legislature, and I therefore gave it up at once." Mitchell was a man of great culture and erudition, but had manifested no indications of a particular leaning toward geology until the transfer above noted, and today his fame probably rests more upon his knowledge of the physical geography and botany of North Carolina than of its geology. He was an enthusiastic explorer and collector, and finally lost his life on Black Mountain in 1857.[19]

Fig. 22. *Elisha Mitchell.*

His two most important papers on geology were published in 1828 and 1829, the first relating to the origin of the low country of North Carolina and the second to the geology of the gold regions. In 1842 he published a summary of his work in the form of a textbook for his classes, accompanied by a small geological map of the state, the only map of its kind thus far prepared.[20] A report on the

[19] Since known as Mt. Mitchell.

[20] There is, I am told, in the state library at Raleigh, N. C., a manuscript geological map of the state prepared by Olmsted during 1824 and 1825. It bears the following inscription: "This first attempt to sketch the geological features of North Carolina is respectfully inscribed to the Board of Agriculture. It claims to be merely an outline to be covered and filled up by succeeding observations. As such it is believed that the Board of Agriculture can make it of special service in prosecuting the Geological Survey, and that when published" (a few words here are illegible) "shows both Currituck Inlets, Roanoke Inlet, one in Oregon Inlet, Ocracoke Inlet, Drum Inlet, granite-gneiss, mica-slate, argillite, etc., sandstone, or coal formation, great slate formation, iron beds, limestone beds, plumbago or wacke, transition formations." It is not known if Mitchell was aware of this map, or utilized it in the preparation of his own. The state had of course been included in Maclure's map.

mineralogy of the state was also prepared in 1827 by his assistant, C. E. Rothe, of Saxony. A geological map of the eastern half of the state was said to have been made, but never published.

Mitchell, in his paper on the origin of the low country, took the ground that the various strata there found were formed in the bed of the sea and became dry land through the depression of the level of the ocean or the elevation of the land by a force operating from beneath. The shells found by him in these beds proved, in his judgment, that they were of comparatively recent origin, but just how recent he was not prepared to say. The presence of bones of elephants and mastodons, however, indicated to his mind an elevation prior to the Noachian deluge, to which catastrophe he evidently attributed their burial.

Mitchell's method of work was naturally radically different from that pursued by the geologist of today. Without accurate maps or means he traveled over the country apparently somewhat at random, sometimes on horseback and sometimes on foot, stopping where and with whom he could, and often under conditions of great bodily discomfort. The following entry from his diary[21] is characteristic:

> Saturday, passed on to Salisbury over a country which puzzled me and which I was prevented from examining fully by the rains: the soil is red; there is much black sand from hornblende where the water has washed. Toward the bridge the country becomes decidedly granitic. I believe this red soil can be produced by decomposition of hornblende rock. Not far from the river I saw a pile of rounded pebbles which I believed to be derived from the river, but which I now believe to be derived from the alluvial many miles below. The country between the ridge and Salisbury, and around the latter place, I do not fully understand. It may be genuine granite.

Incidentally it may be remarked that Salisbury lies in the midst of a wide granite belt through which have been intruded numerous dikes of dark, more basic rocks. And further, the red soil may be derived not only from hornblende rocks, but from rocks of almost any known type if decomposition has advanced sufficiently far.

Mitchell's Criticism of Olmsted and Rothe, 1829.

In 1829 Mitchell published in the *American Journal of Science* an article in which he questioned the correctness of the views of Olmsted and Rothe regarding the origin and occurrence of gold in the state.

21 Diary of a Geological Tour, by Dr. Mitchell, etc. James Sprunt Historical Monograph No. 6, Univ. of North Carolina, 1905.

Olmsted had described the metal as occurring in a "diluvial formation" consisting of mud and gravel, which was largely, though not entirely, limited in its distribution to an area of argillite from which he questioned if it might not have been derived, the gold itself having been separated from the diluvial material by streams which cut through it. Rothe, on the other hand, believed that the gold now found in the alluvial was derived from veins in granite, and spread over the country by a flood of waters breaking through the Blue Ridge and rushing in torrents over the entire gold-bearing region, though he had previously (1826) expressed the idea that the gold was derived from the bursting asunder of the gold-bearing veins by subterranean explosions and the gold thus scattered over the adjacent regions, some of it being carried down in the watercourses.

Mitchell's idea, which is undoubtedly the correct one, was that the gold occurred originally in veins and perhaps in part disseminated throughout the country rock, which was in itself in part Primitive and in part Secondary. From these rocks it was set free through atmospheric decomposition and subsequently distributed by gravity and running water. This paper is of further interest in that it contains a colored geological map of the gold region, the rocks being classified as Primitive, Transition or slate, Old Red sandstone, and Alluvium.

The Dark Ages of Geology.

The period that may with propriety be called the dark age of geology was that prior to the discovery and general recognition of the value of fossils in stratigraphy—the fact that the relative age or stratigraphic sequence of sedimentary rocks could be determined by means of the plant and animal remains they may contain.

According to Sir Archibald Geikie,[22] credit must be given to the Abbé Giraud-Soulavie for having first planted the seeds of stratigraphic geology. In a paper read before the Royal Academy of Science in Paris in August, 1779, Soulavie described the calcareous mountains of the Vivarais as made up of limestones belonging to five different epochs, the strata in each being marked by its own peculiar assemblage of fossils.

These views, though undoubtedly correct in the main, were not generally accepted even in France, a fact thought by Sir Archibald to be due mainly to the wretched style in which they were set forth, and the Abbé's fame has been eclipsed by his more brilliant suc-

[22] The Founders of Geology, p. 204.

cessors, among whom may be mentioned Desmarest, Rouelle, Lamanon, Cuvier, and Brongniart.

Cuvier, it will be recalled, was primarily a biologist, and may well be considered the first vertebrate paleontologist—the first to announce that the globe was once peopled by vertebrate animals of a type which have long since disappeared. In connection with Brongniart, he published in 1808 a memoir containing the results of their joint studies in the basin of the Seine. They showed that the formations there existing were arranged in a definite order and could be recognized by their lithological and paleontological characteristics. Although subsequent research has naturally tended to show that their observations and conclusions were not in all cases quite correct, it may, nevertheless, be said that they established on a basis of accurate observation the principles of paleontological stratigraphy, demonstrated the use of fossils for the determination of geological chronology, and paved the way for the enormous advances which have since been made.

First Attempt at Correlation, 1828.

When one recalls the avidity with which each new suggestion is seized upon by the student of today, it seems strange that so promising a field of investigation as was here thrown open should not have been immediately occupied. Nevertheless, it was not until 1828 that an American geologist took up the matter with an apparent full appreciation of its possibilities. In that year Dr. S. G. Morton read before the Academy of Sciences a paper based upon notes furnished him by Lardner Vanuxem, relating to a possible subdivision of the heretofore so-called Alluvial or Tertiary deposits of the Atlantic coast. The most important feature of this paper lay in its announced recognition of the value of fossils for purposes of correlation. With the exception of Professor John Finch, already mentioned, writers up to this time had very generally referred to these deposits as belonging to a single formation, either Alluvial or Tertiary, as the case might be. Vanuxem here asserted the existence of both Secondary and Tertiary formations, and showed "that the two may be at all times unequivocally identified by their

Fig. 23. *S. G. Morton.*

fossil remains." The relative geological position of the beds he gave as below:

Modern alluvial	Vegetable mold	7
	River alluvium	6
Ancient alluvial	White siliceous sand	5
	Red earth	4
Tertiary	Beds of Ostreae Mass of limestone, buhrstone, sand, and clay	3
Secondary	Lignite	2
	Marl of New Jersey	1

It was presumably because Vanuxem was on his way to Mexico, on private business, that he left so promising a lead to Morton, who, during the succeeding years, and particularly through the publication in 1834 of his *Organic Remains of the Cretaceous Rocks*, became recognized as the leading stratigraphic paleontologist and by some considered the founder of this particular branch of the science in America. (See further, on p. 121.)

Jackson and Alger's Work in Nova Scotia, 1828.

In 1828 Dr. C. T. Jackson, of Boston, in company with Francis Alger, published in the *American Journal of Science* a series of papers bearing on the mineralogy and geology of a part of Nova Scotia. These were accompanied by a colored geological map and, although mainly of a mineralogical nature, contained much which at the time was valuable. The map showed the distribution of the various formations (identified mainly on lithological grounds) of the northern half of the peninsula. A broad belt of Transition clay-slate was represented as extending from the Gut of Canso to St. Marys Bay. This was bordered on the northeast by a narrow bed of alluvium, the immediate border of the Bay of Fundy being occupied by trap rock or greenstone, as it was then called. A wide band of red and gray sandstone, alternating with shale and carrying beds of coal, occupied the region south of Minas Basin and all of the Cumberland County peninsula. The sandstone was described in considerable detail, and from its fossil contents it was judged to be a Secondary rock, although evidently older than the trap, again a recognition of the value of fossils for stratigraphic purposes.

The clay-slate of the South Mountains they thought to belong to the Transition class, since it dipped fifty degrees or sixty degrees to the northwest, while the sandstone dipped at an angle of only ten degrees or fifteen degrees to the north, to their mind clearly indicating the greater ntiquity of the first mentioned. The granite was undoubtedly of greater age than the clay-slate, since it contained "no relics of organized beings." They did not, however, consider it to belong to the oldest Primitive, since it showed at places a brecciated structure, contained black mica, and was "lacking in the metalliferous compounds and minerals which characterize the ancient formations." "It probably belongs," they wrote, "to the third or newest formation of Werner."

Concerning the source of the trap and its relationship to the sandstone it was stated:

The sharp fragments of the breccia and the breaking up of the strata also shew that the production of this rock or, rather, its nonconformable position on the sandstone strata, was effected suddenly. Whether it was ejected from the inaccessible depths of the Basin of Minas, or was thrown directly up through the strata of sandstone, we can not determine; but the occurrence of the trap only on the borders of the basin, which it almost surrounds, leads to the belief that this cavity was the crater, if it may be so called, from which, in former times, the trap rocks issued. The same remarks will apply to the whole North Mountain range, except that they probably originated from the unfathomable deeps of the Bay of Fundy, which is completely skirted on either side by trap rocks.

It will be noted here that Jackson for the first time cut loose from the Wernerian (Neptunian) doctrine. He realized this and stated that the evidence found convinced him of its insufficiency, and he was obliged to allow the superiority of the igneous theory as taught by Hutton, Playfair, and Daubeny.

Concerning the origin of the bed of hematite iron ore of the South Mountain region and its relation to the granites, they wrote:

Speculative geologists would doubtless regard the protrusion of the granite from the central regions of our globe as the cause of the disruption of the strata of clay-slate which was thus raised from the bottom of the sea, bearing with it the spoils of the ocean. The layers would thus be broken, their edges thrown up at an angle, and, by the contraction of the subordinate rocks, the superior strata being fixed, or the protrusion having carried the rocks so far as to poise the strata in a perpendicular position, a chasm would be formed into which the ore of iron was afterwards poured from above by a second submergence. From the similarity of fossils we should think the bed of iron ore must have been immediately formed after the disruption of the strata.

In 1833 a revised edition of the work was published in the Memoirs of the American Academy of Arts and Sciences. This was referred to by Featherstonhaugh in his journal as "the neatest and best executed work on geology which has been gotten up in the United States."

Sketch of C. T. Jackson.

Charles T. Jackson was born at Plymouth, Massachusetts, June 21, 1805; was graduated at the Harvard Medical School in 1829, and settled down to the practice of medicine in Boston in 1833, having spent a portion of the intervening years in Europe. He shortly, however, abandoned his medical practice in order that he might devote himself to chemical and geological investigations more to his taste. In 1836 he was appointed state geologist of Maine, and published during the three years he held that office three octavo reports comprising some 1,000 pages and an atlas of 24 plates. These volumes, while recording a large number of disconnected observations, contain nothing of striking interest or importance. They are devoted also largely to economic questions. It must be remembered, however, that the country at that time was largely a wilderness, without rail or carriage roads, and many of his journeys were made by canoes on streams and lakes, where exposures were few and far between.

In 1839, as state geologist, he made a survey of Rhode Island, and in the year following submitted the manuscript copy for his report, 1,000 copies of which were issued, constituting the first, last, and only official account of the resources of the state ever published. In 1840 he began work upon the geology of New Hampshire, having been appointed geologist for that state in September, 1839. This survey lasted until 1843, the final report appearing in 1844. It was while in the prosecution of this work that he made the discovery of tin ore at Jackson, and from it smelted the first bar of metallic tin produced in America.

In 1847 Jackson was appointed United States geologist to report on the public lands in the Lake Superior region, but spent only two seasons in the field, resigning for personal and political reasons in 1849.[23] It was through his instrumentality, however, in part, that the copper regions of Lake Superior were opened up. Other economic work of his which needs mention is that in connection with the dis-

[23] See further, on p. 278.

PLATE VII

C. T. JACKSON

covery and description of the emery mines at Chester, Massachusetts.

After withdrawing from the Lake Superior survey, Jackson devoted himself largely to laboratory work, having an office at 32 Somerset Street, Boston. As is well known, he was one of the claimants for priority in the discovery of the anæsthetic properties of ether. He was well and favorably known as a chemist and all-round naturalist, and is described as "an enthusiastic personage, a ready conversationalist, even eloquent in his speech and fond of story telling—a man of large stature, square shoulders, and massive head." As a geologist he was conservative almost to the point of obstinacy, as is shown by his steady adherence to the older forms of classification, though finding it necessary to depart somewhat from the ideas of Werner. He was but little given to theorizing, at least so far as is shown by his published works, and announced few, if any, new principles. His fame rests rather upon the extension of the geographic boundaries of our knowledge and the development of economic resources. According to Shaler[24] he was nevertheless a man of uncommon ability and wide learning in both chemistry and geology, but with an eager, most human and often pitiful hunger for applause; a man with a great deal of divining power but with a limited amount of field information, one not zealous of *seeing*, but "who could make safer inferences than any man I ever saw."

Alger was an iron and steel manufacturer, and was for a time interested in the zinc mines of Sussex, New Jersey. It is told of him that he would never part with his interests in any mine without reserving the right to all desirable mineralogical specimens that might thereafter be brought to light. His enthusiasm carried him to such extremes that finding a monster beryl—weight, five tons—at Acworth, New Hampshire, that could not be removed entire he bought the ledge where it lay, carefully had the stone cut away from around it, and considered it as forming a part of his cabinet. He died of pneumonia in Washington, D. C., at the age of fifty, while striving to perfect a shrapnel shell for the Federal Government.

Vanuxem's Views, 1829.

The matter of the relative age of rocks as indicated by their position with respect to horizontality, which had been discussed by

[24] Autobiography, p. 109.

Cleaveland in 1816, Maclure in 1825, and Jackson in 1828, was again taken up by Vanuxem in 1829. In a paper in the *American Journal of Science* in this year he called attention to several errors promulgated by American geologists, the first of which related to the existence of Alluvium and Tertiary rocks in the Southern Atlantic States, as he had previously announced in conjunction with Morton.

Of almost equal importance was the objection raised to the prevailing assumption that all of the so-called Secondary rocks were horizontal in position, or, on the other hand, that all horizontal rocks were, therefore, Secondary. He pointed out that rocks composed of mechanical particles when undisturbed would form horizontally lying masses, but that both uplifting and downfalling forces had existed and there was no certainty that such had acted in a uniform manner, giving rocks of the same age the same inclination. Therefore the position of beds as regards horizontality, he argued, could not be relied upon to indicate age. "The analogy or identity of rocks I determine by their fossils in the first instance and their position and mineralogical characters in the second or last instance." This is perhaps one of the most important generalizations that had thus far been made by any American geologist.

Sketch of Vanuxem.

Vanuxem was a Philadelphian by birth, but received his mineralogical and geological training at the School of Mines in Paris at the time when Brongniart and Hauy were both active. Graduating in 1819, he returned to America and assumed the chair of chemistry at Columbia College, South Carolina, resigning in 1826 to undertake some private mining work in Mexico. In 1830 he removed to a farm near Bristol, Pennsylvania, which continued to be his home during the rest of his life, though connected with the New York State survey during 1837-1843.

He is represented as a man of slight build, active and energetic, and with great powers of endurance; one who loved his work for the work's sake, and was always averse to receiving pay for his services excepting when circumstances rendered it absolutely necessary.

According to his biographer, he had the reputation of being visionary and full of untenable theories. Be this as it may, his published writings show no such failing, and were, for their time, remarkably free from error. There are few men of his day who saw more clearly or reasoned more correctly. As a geologist he ranks with those who have put forward new ideas rather than those who have extended geographic boundaries.[25]

First American Edition of Bakewell's Geology, 1829.

In 1829 there appeared the first American, from the third English, edition of Bakewell's *Introduction to Geology.* This was edited by Silliman, and was accompanied by the latter's *Outline of the Course of Geological Lectures Given in Yale College.* The reason for the reprint, as given by Professor Silliman in the preface, was that he might place in the hands of his classes a comprehensive treatise which they "would be willing to read and able to understand." The lecture notes, which merit our attention here, comprised 126 pages and may be reviewed in some detail as illustrating the character of instruction given at that time.

Silliman announced himself as being neither a Wernerian nor Huttonian, but simply a student of facts. The classification was, however, largely Wernerian, though he says, "It is one of convenience merely, and therefore there is no hesitation in deviating from it, or in substituting other views, when they appear preferable." The discoveries of geology he considered consistent with the biblical account, and "respecting the deluge, there can be but one opinion . . . geology fully confirms the scripture history of that event."

The earth he conceived as at an early stage covered superficially by a watery abyss, containing in solution acids and alkalies such as would augment its solvent powers. He regarded the solubility of all the existing elements forming the crust of the globe as clearly and actually demonstrated, but found a serious difficulty in attributing to the quantity of waters that now exist sufficient power to suspend all the materials of those rocks that bear marks of deposition from a state of chemical solution. Among other possibilities he, however, seriously considered that of a portion of the then existing waters having "been received into cavities of the earth, to await a future call to deluge the surface anew." From this solution the Primitive rocks—granite, gneiss, mica-slate, and some of the limestones—were thought of as having been deposited, the author thus far following Werner.

The question as to the origin of mountains and of the continents was, with him, a vexed one. "Some imagine that entire mountain

[25] According to Hall (Transactions of the American Association of Geologists, 1840-1842), Vanuxem was the first to point out the similarity of some of the western formations with those of New York, identifying the lower rocks of Ohio, Kentucky, and Tennessee with the Trenton limestone.

ranges and even entire continents have been raised by the force of subterranean fire," and he saw no inconsistency in admitting that both igneous and aqueous agencies might have been active in their production. But as to the source of the materials from which they were formed, he felt by no means clear. If supplied from regions immediately beneath, what fills the void? he asked. If it is arched over from side to side, what security is there that subterranean fires will not melt down the abutments and undermine the continent? But, whether the mountains were raised from below, left prominent by the subsidence of the contiguous regions, or were reared by accumulation, he regarded as immaterial for his present purpose. "It is agreed on all hands that they existed before the subsidence of the early ocean, whose retreat must of course have first exposed their summits."

Among the rocks belonging to the Transition period he included the crystalline limestones and dolomites of Bennington, Middlebury, and Swanton in Vermont; the graywackes of the Chaudiere Falls in Lower Canada, of Rhode Island, and of the Catskill Mountains; the breccia marble of Point of Rocks, Maryland; and the conglomerates of Dorchester and other localities in Massachusetts. These, as is well known, are now relegated to various widely separated horizons, extending from the Cambrian in the case of the Swanton marbles to the Triassic in the case of that of Point of Rocks.

The manner in which the fossil organisms were conceived to have become embedded in their various matrices is interesting:

> We can not doubt that the animals received their existence and lived and died in an ocean full of carbonate of lime, in solution, or in mechanical suspension, or both. When they died they of course subsided to the bottom and were surrounded, as they lay, by the concreting calcareous matter; . . . the interstices were filled by the calcareous deposit, and this being more or less chemically dissolved, produced a firm, subcrystalline mass, a section of which shows the animals sawn through.

Concerning the value of such remains, he was fully cognizant: "Fossil organic bodies contained in rocks are now considered as good indicia of the geological age and character of the strata in which they occur."

Silliman, in a paper of considerable length, in his *Journal* for January, 1829, had discussed the resemblance of trappean rocks to those of known volcanic origin, and writing with especial reference to phenomena exposed near Hartford, Connecticut, said: "Had the trap rocks been erupted into daylight like currents of lava, there would be no reason why they should not exhibit all the variety of appearance that belongs to lavas; but, if only forced through and among superior strata, or even if forced quite through them, but still remaining under the pressure of many miles of ocean, they would congeal under enormous pressure and of course would be long in cooling, and would in the main assume the stony or rocky, rather than the vitreous character." In consideration of these facts and the altered condition of the sandstone at contact with the trap he was strongly predisposed to regard the latter as of igneous origin, although holding himself in check, since "hosts of generalizations from few facts is a great evil in science."

Silliman's Views Regarding the Noachian Deluge.

As with others of his time and as above noted, Silliman was a believer in the Noachian deluge. Indeed, in point of detail, he outdid all others in his attempts at harmonizing apparently conflicting statements and ideas and wrote, as Huxley has aptly expressed it, "with one eye on fact and the other on Genesis." "There is decisive evidence that not further back than a few thousand years an universal deluge swept the surface of the globe." This deluge was considered quite distinct from the original, primeval ocean, and to have been brought about through direct intervention of the Creator for the purpose of punishing and partially exterminating the race. It was sudden in its occurrence, short in duration, and violent in its effects. In order to account for the biblical expression, "the fountains of the great deep were broken up," Silliman offered the suggestion that, contemporaneous with the forty days and nights of rain, a deluge of water burst forth from the bowels of the earth, whence it was forced by the sudden disengagement of gases. The presence of this subterranean water he had already conveniently accounted for by assuming that it was derived from the primeval ocean at the time it shrank away and left the dry land. Sufficient water to cover, when forced to the surface, the highest mountains, could, he calculated, be contained in a cavity the cubical content of which was only one two hundred and sixty-fifth part of the globe.[26]

Assuming that the antediluvian mountains were the same as today, but somewhat higher (say, 5½ miles), and accepting the fact that they were covered by the water, Silliman proceeded to show that, with a time limit of forty days, the water rose at the rate of a foot

[26] The idea that the waters of the flood were expelled from cavities in the earth's interior was quite prevalent at this time and continued plausible in the minds of many down as late as 1835. (See Kirby on the History, Habits and Instincts of Animals, etc., Bridgewater Treatise, Vol. I, 1835, p. 26.

in two minutes—*i.e.*, 30 feet an hour, or 181 feet in the time of a common flood or ebb tide, 362 feet in the time of the ordinary ebb and flow of the tide, or 726 feet in twenty-four hours.

The inequalities in the surface of the land would, however, increase even this rate, and a very graphic picture is drawn of "the inconceivably violent torrents and cataracts everywhere descending the hills and mountains and meeting a tide rising at the rate of more than 700 feet in twenty-four hours."

It is to such a catastrophe that he believed to be due the extinction of whole races of vertebrate animals, like the Siberian mammoth and others. That such would be amply sufficient no one will be likely to doubt.

CHAPTER III.

The Era of State Surveys, First Decade, 1830-1839.

THE decade beginning with 1830 stands out prominently as an era of public surveys. With the exception of the single immature attempt by Olmsted in North Carolina in 1824 no surveys at other than private expense had thus far been undertaken, though the subject had more than once been agitated. During the decade now under consideration, however, scarcely a year passed but witnessed the establishment of a state survey or the organization of an exploring expedition, to which a geologist was attached. There were established surveys in Massachusetts in 1830; in Tennessee in 1831; Maryland in 1831; New Jersey, Connecticut, and Virginia in 1835; Maine, New York, Ohio, and Pennsylvania in 1836; Delaware, Indiana, and Michigan in 1837, and in New Hampshire and Rhode Island in 1839.[1] In addition, the United States Government for the first time recognized the practical utility of the geologist by authorizing the surveys by G. W. Featherstonhaugh of the elevated country between the Missouri and Red rivers in 1834; of the Coteau des Prairies in 1835, and by D. D. Owen of the mineral lands of Iowa, Wisconsin, and Illinois in 1839.

The Wilkes exploring expedition, with J. D. Dana as geologist, was also organized and sent on its way in 1838. Beyond the limits of the United States Abraham Gesner was doing important work as provincial geologist in Nova Scotia and New Brunswick. Eaton and Maclure were still in the field, but new workers were rapidly forcing their way to the front, and the influence of the two pioneers was already on the wane. Other names which appear in this decade and grow to prominence are Timothy A. Conrad, James Hall, W. W. Mather, D. D. Owen, J. G. Percival, and H. D. Rogers.

Eaton's Geological Text-book, 1830-1832.

In 1830 Eaton published the first edition of his *Geological Text-book*, an octavo volume of sixty-four pages, accompanied by a

[1] For details of state surveys see Merrill's Contributions to the History of State Geological and Natural History Surveys, Bull. 109, U. S. Nat. Museum, 1920.

colored map giving a general view of the economic geology of New York and part of the adjoining states, the work for which had been under the patronage of the Honorable Stephen Van Rensselaer.

The book is worthy of a somewhat extended notice, since it affords an insight into the character of the instruction furnished students at the Rensselaer Institute, and also, since it was Eaton's second attempt at preparing a textbook. His first, it will be remembered, was his *Index to the Geology of the Northern States*, published in 1818, a second edition of which appeared in 1820. In these works he was anticipated only by Parker Cleaveland's *Text-book of Geology and Mineralogy*, which appeared in 1816, though the volume now under discussion was anticipated in 1826 by the work of Emmons.

Eaton's general views regarding the formation of the various geological deposits are summed up as follows: "The earth is composed of masses of rocks and detritus which are more or less extensive and uniform in their characteristic constituents." "These masses are mostly in regular deposits, and those of the same structure and composition regard the same order of superposition in relation to each other." A few of the outermost masses, having no reference to each other, he called "anomalous deposits." He divided the regular deposits into five classes, each of which "consists of three formations which are found to be corresponding equivalents in all the series. The lowest formation in each series is *slaty* or *argillaceous* and always contains beds of *carbon* in the state of coal, anthracite, or plumbago." The next is siliceous and destitute of beds of carbon, and the uppermost, also lacking in carbon, is composed chiefly of carbonate of lime.[2]

All the primitive formations, he taught, have been deposited in the form of concentric spheres like the coats of an onion. These contained the materials of which all other formations were afterwards made up. "Soon after these deposits were laid down they were broken up through several northerly and southerly rents by a very great force exerted immediately beneath the lowest of the primitive strata. In this semi-indurated and broken state materials were readily furnished for the outer strata." The source producing these rents, it will be remembered, was in his early works considered problematical (p.

[2] In an article entitled Geological Prodromus, published in the American Journal of Science during the latter part of the previous year, Eaton had given his readers warning of this proposed classification. "I intend to demonstrate, by reference to localities of easy access that;

"All geological strata are arranged in five analogous series; and that each series consists of three formations: viz., the carboniferous, the quartzose and the calcareous."

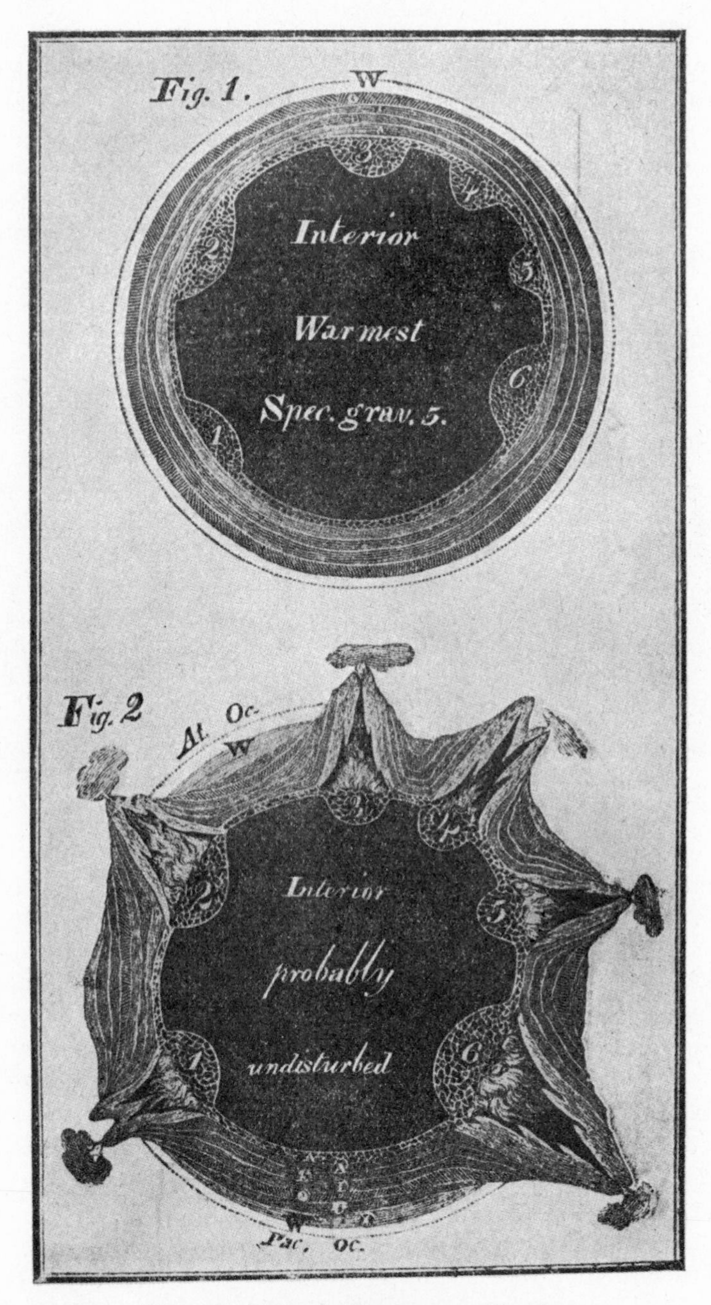

Fig. 24. *Eaton's Transverse Section of the Globe, 1830.*

77). In the work here reviewed he, however, solved the problem in a manner best understood by reference to Figure 24, which he claimed was an improvement on those given in the *Index*. As here represented the earth is cut into two parts at the forty-second degree of north latitude. Large bodies of combustibles of an undetermined nature, it will be observed, are conveniently stored under what are now the regions of maximum disturbance, as the Rocky Mountains, New England, Great Britain, the Alps, Pyrenees, Caucasus, and the

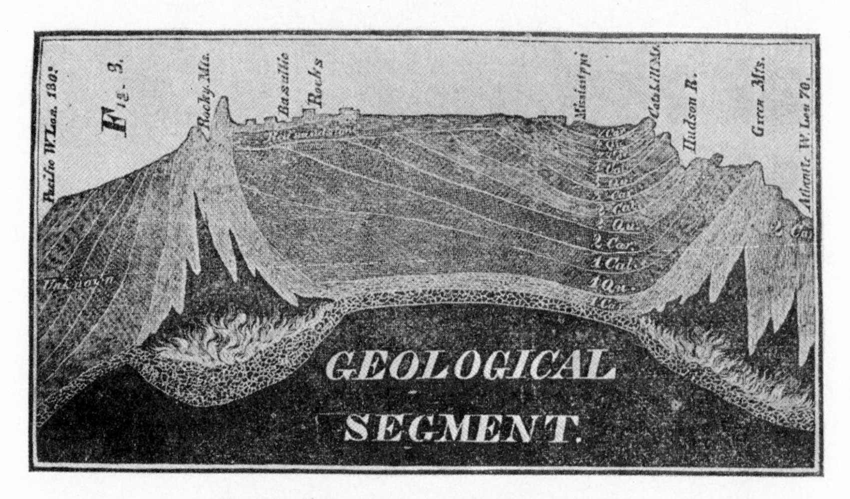

Fig. 25. *Eaton's Geological Segment, 1830.*

Himalayas. In the second figure combustion is supposed to have taken place, whereby an explosion was produced which burst through the Primitive and Transition series, and appalling indeed must have been the results.

A geological segment, also reproduced (Fig. 25), gives in greater detail a section across the American continent, showing the internal nucleus, the areas of combustible matter under New York State and the Rocky Mountains, and the alternation of the regular deposits. The combustible materials at the present time are supposed to be nearly exhausted, although still sufficient to cause ordinary earthquakes.

The superficial rocks of the crust, as shown in this section, were divided into four series, each of which was made up, in ascending order, of Carboniferous, Quartzose and Calcareous rocks, as noted above. The definitions of the various classes forming the Transition and Secondary remained much as in the early work, excepting that he divided his Class III into a lower and upper division and added a fifth series, the Tertiary, this last including those strata which contained remains of viviparous vertebrate animals.

The so-called anomalous deposits were those which have been produced at the earth's surface by fusion or disintegration of regular strata, and were divided into (1) *Volcanic*, (2) *Diluvion*, (3) *Post-diluvion*, and (4) *Analluvion*, the last including what are known as residual deposits—that is, those which result from decay *in situ*.

In the second edition of this work, published in 1832, the same general ideas were advanced concerning the formation and uplifting of the various rocks; a much more satisfactory chapter added on the character and objects of geology, and also a chapter on organized remains as auxiliaries in the determination of rock strata. In this respect this edition was decidedly in advance of the first.[3]

Many of the ideas put forward in the chapter on the character and objects of geology are, in the light of today, peculiarly interesting. The desire to harmonize all phenomena witnessed with the biblical account of the deluge was still manifested:

> Geological facts lead us to the history of created beings long anterior to written records. Such records may be erroneous, and we have no means of correcting them. But geological records are perpetual, unvarying, and can not be vitiated by interpolations or counterfeits. For example: The written history of the deluge might be varied more or less by erroneous copies and incorrect translations. But the geological records of divine wrath poured out upon the rebellious inhabitants of the earth at that awful period can never be effaced or changed.[4] These later records add to the Mosaic account, that even the antediluvial beasts of the forests and fens partook of the ferocious nature and giant strength of antediluvial man.

The bed of Lake Ontario Eaton regarded as formed by the rapid disintegration of saliferous and carboniferous rocks, while the beds of Lake Champlain and the Mohawk and Hudson rivers were thought to have originated, for the most part, through the disintegration of

[3] The edition of 1832 was issued and sold under direct supervision of the author. The following note in Eaton's handwriting relative thereto is in possession of the present writer:

"To General Howe (or whoever has received my geological text-book), deliver a copy of my text-book to Professor Silliman and another to Professor Tully on my account. Charge them to William S. Parker, and I will cause the same to be charged to me and also to be credited to you. July 24, 1832.

Amos Eaton.

N.B. I publish this edition myself, and have paid all cash, by way of trial."

[4] Eaton, however, failed to make allowance for the various interpretations that might be placed upon these records.

rocks at the line of contact between different formations. "The deep bed of the Hudson across the Highlands, however, may, without extravagance, be ascribed to the fusion by volcanic heat which produced the basaltic Palisades below the chasm. The same hypothesis may be well applied to the channel of the Connecticut River north of the northern line of Massachusetts, whence the volcanic lava flowed which now covers a series of basaltic prominences which form the northern part of Long Island Sound." And this as late as 1830-1832! Yet the Rensselaer school, where Eaton was teaching, was, and continued to be for many years, the chief training school for American geologists. Fortunately, his students were taught to *think* and not to blindly follow.

Eaton claimed at this date (1832) to have devoted more time and labor to American geology than any other individual, and credits General Stephen Van Rensselaer with having furnished the necessary facilities. "I made the first attempt," he wrote, "at a systematic arrangement of American rock strata." The various discoveries which he thought might be claimed as having been made under the auspices above noted are as follows:

(1) That each of the classes of rocks always begins with a carboniferous slate and terminates with calcareous rocks; having a middle formation of quartzose. This is apparently the first announcement of the now well-recognized principle of rhythmic or cyclic sedimentation, later elaborated by Professor J. S. Newberry (p. 470).

(2) The discovery of ferriferous stratum containing argillaceous iron ores, and which extends unbroken from near Utica to the extreme termination of Lake Ontario, in Upper Canada.

(3) That the bog ore properly belongs to the Tertiary formation.

(4) That talcose slate is the grand repository of hematitic iron ore, peroxide of manganese, and native gold.

(5) That the corniferous lime rock is the true carboniferous only.

(6) That crystalline granite is not entitled to a place among general strata, as it is never found other than as a bed or vein.

(7) That granular quartz rock and granular lime rock are entitled to a place among general strata.

(8) That all primitive rocks, excepting granular quartz and lime rocks, are contemporaneous.

(9) That there is evidence of a diluvian stratum having been deposited near the termination of the deluge, which formed a universal mantle about the earth.

In discussing the "regular" deposits, he wrote:

They exhibit grounds for conjecture, . . . if not absolute demonstration, that the surface of the earth has undergone five general modifications which no animals survived. Four of these modifications were followed by as many new creations of animals. Also that two new creations succeeded the final depositions of all regular strata. In the whole there appears to have been five creations of animals at least, and perhaps ten, since the primitive mass of earth was formed, and a long interval succeeded each creation.

In his chapter on soils he recognized and emphasized a principle now generally accepted, to the effect that the fertility does not depend upon ultimate chemical constitution, but rather upon physical properties; that a fertile soil should contain, first, sufficient stones and pebbles to keep it open and loose; second, sufficient clay to absorb and hold water in the right proportion; and third, sufficient fine sand to prevent the clay from baking in time of drought. That the fertility could not be told by a chemical analysis had been stated by him some years earlier, as follows:

Suppose, in one specimen, the soil, etc., should be quartz, in another feldspar, in another hornblende, in another sapphire, in another diamond, would there be any difference in the influence of the sand, etc., upon the productive quality of the soil on account of the different ultimate elements of which these different minerals are composed? Should they be so far decomposed, at some future period, as to become an impalpable powder, perhaps they may then differ in their influence upon vegetation. Perhaps we may foretell the future state of the country a century or two to come, where such extreme disintegration is effected. But the difference in the ultimate constituents can not possibly affect the question of fertility or barrenness *at the time the analysis is made*. For whatever effect can be ascribable to the one is equally a property of the other. They all hold water on their surfaces by the attraction of adhesion; they all keep the soil duly open and porous to give passage to the roots of vegetables; they all aid alike in bracing up plants and in keeping them in a fixed position, etc. Whatever is effected by one is effected by all—size, form, quantity, and all circumstances, other than their constituent elements, agreeing.

To this second edition was added a chapter of some eleven pages on the science of mining, one of eleven pages on geological localities, and five pages on localities of fossils and an equal number of plates. His gradually expanding views on the value of fossils are shown in his paper on "Geological Equivalents," of this year (1832). He here advanced the idea that the enumeration of the mineral constituents of rocks could never be satisfactorily applied for the determination of the relative position of strata, but that recourse must be had to the organic remains. "We find the same organized remains associated with equivalent strata in every part of the earth."[5]

If we are to judge from the preface of the edition of 1830, Eaton was by no means lacking in egotism, and had, at times, an unfortunate way of expressing his opinions, such as must have aroused antagonism in the minds of his collaborators. In referring to the work just mentioned, he wrote:

A *textbook* is too small a name for these days of puffing arrogance. But I propose to present all my *supposed* heresies to the geological fraternity in this form and under this title. And I beg the favor of the most rigorous criticism upon this book, small as it is. To stimulate men of science to the work of examination and of criticism, I will state that I intend to publish considerable in scientific journals, also a full system upon this plan. As I have had more than 7,000 pupils already (rather auditors), and shall probably have more still, it will be well for them "to be on the alert" if I am propagating errors. I am not in sport; I have, during the last fifteen years, traveled over 17,000 miles for the express purpose of collecting geological materials, the results of which are comprised in this little octavo pamphlet and exhibited in the accompanying map and wood cuts.

I may be accused of fickleness, on account of the changes which appear in every successive book I publish. I confess this is the ninth time I have published a geological nomenclature, and that I made changes in each of more or less importance. But I have always consulted my scientific friends, and every change was founded on new discoveries in "matters of fact."

And again:

[5] Several years later (*i.e.*, in 1839) in a letter to Silliman, he wrote:
"When I commenced my geological surveys, the application of organized remains for demonstrating strata was not studied in America. I had become acquainted with no method for determining the character of such strata, but that of tracing them separately through a vast extent of country, and then comparing their general characters. For this purpose I traveled some thousand miles at my own expense and with the liberal aid of students of Williams College, with Professor Dewey at their head, where I was employed more than a score of years since, by the authorities of the College, to introduce the natural sciences. Afterwards I traveled more than seventeen thousand miles on geologizing tours, at the expense of the Hon. Stephen Van Rensselaer, and I was always aided by several assistants and competent students. Had the application of paleontology been then as well understood as it now is, I could have settled the characters of most rocks as well in my closet by the aid of specimens. But it is a true remark in your last Journal, that strata must have been first settled according to the method to which I was compelled by ignorance to submit, before the service of organized remains could be successfully employed. In this country, no material progress had then been made in the study of organized relics; and even now, we have very few good paleontologists."

Students for whom this text-book is intended may feel no interest in anything *personal* relating to myself. But I will throw this paragraph in their way. I have been accused of arrogance for stating facts relating to American geology without formally bowing to European authorities. . . . I confess that this is a kind of "ipse dixit" text-book. It is so, because the plan does not admit of demonstration. In a future publication I intend to cite authorities from nature to illustrate my views. But I am prepared to abandon any of them, as I have frequently done heretofore, in cases of numerous errors, to which I am still subject.

Geology is a progressive science, and he who has any respect for his future reputation should be exceedingly cautious about committing himself on matters of fact or speculation. I confess that I have *most egregiously* violated this rule, but there are peculiar circumstances in my case, arising from my being "a hireling drudge" to the most munificent patron of this science (Stephen Van Rensselaer), which will palliate, at least, if not justify.

I despise arrogance; but I am within sixteen years of the "three score and ten," when the mind of man is averaged beyond the period of vigorous effort. About two score of these years have been devoted to natural science. I offer this as an apology for some dogmas forbidden to youth.

One can imagine him saying with Emerson, "What is well done I feel as if I did; what is ill done I reck not of."

Eaton was a close competitor of Schoolcraft in the matter of republishing, and a formidable rival of certain members of existing surveys. Silliman's patience evidently became exhausted, for he premises a half-page review with the remark:

The subject matter of this treatise has been published six times before the present edition, 1st in 1818 and 2nd in 1820, under the title Index to the Geology of the Northern States; 3rd in 1824 as a report of a Geological Survey of the Erie Canal, taken at the expense of the Hon. S. Van Rensselaer; 4th as a Geological Nomenclature; 5th as a Prodromus, presenting a new view of classification of rocks by series and formations; 6th as a Geological Text Book in 1830. Now (1832) it appears as a second edition of the last.

In 1837, when Eaton, though past his physical prime, should have been at his best mentally, we find Dana in a letter to Hall lamenting that "Eaton . . . has been so long before the world as an American Geologist that he deserves some commemoration from his associates in science. It is to be regretted that habits of inactivity and inaccuracy so strongly characterize his mental disposition, tending to efface a name that would otherwise stand high and shine bright in the historical tablet of geological science."

Monthly American Journal of Geology, 1831.

In July of 1831 there appeared the first number of G. W. Featherstonhaugh's *Monthly American Journal of Geology and Natural History*. This proved a short-lived but vigorously conducted journal in which essays on geology, as understood by the editor, occupied a leading part in each number. Troubles, financial and otherwise, beset the venture, and amongst the "otherwise" may be mentioned a lack of appreciation and support where he apparently had a right to expect both. "So that a work which has hitherto been stamped with general approbation, . . . has not been permitted to have a single subscriber in New Haven, the pretended seat of an American Geological Society," is his fierce plaint in the ninth issue.[6] A part of this apparent lack of appreciation may have, however, been due to a slight feeling of resentment on the part of the friends of Silliman and his *Journal of Science*, against a rival. Fisher in his *Life and*

[6] Featherstonhaugh's efforts seem to have been better appreciated abroad than at home, if we may judge from the following:

The editor lays before his readers, with a just pride and entire satisfaction, the following communication from the president of the Geological Society of London, and other distinguished naturalists:

London, June 18, 1831.

My Dear Sir: We, your undersigned friends in England, are happy to learn that you propose to establish a new periodical work in the United States, which, in embracing all subjects connected with the natural history of America, is to be specially devoted to the accumulation of geological facts and phenomena.

Knowing your zeal and ability, we have great hopes that a work so directed will meet with every encouragement in your country, and we are certain that it can not but be of service to the cause of science in general.

We shall at all times be desirous of aiding you with any communication in our power, and we subscribe ourselves,

Yours, very faithfully,

Roderick Impey Murchison,
President of the Geological Society of London.

Davies Gilbert,
Vice-President Royal Society.

W. D. Conybeare, F.R.S., F.G.S., etc.

A. Sedgwick, F.R.S., F.G.S., etc.,
Fellow of Trinity College, Cambridge.

Wm. Buckland, D.D., F.R.S., etc.,
Christ Church College, Oxford.

George Bellas Greenough, F.R.S., etc.

Charles Stokes, F.R.S., etc.

P.S.—I can not refrain, in particular, on my own part, from expressing the desire which I feel for the appearance of the proposed publication, as likely to conduce in the most important points to the effective progress of geology; to ascertain in detail the suite of formations, and the series of organic remains distinguishing them in a new continent, so widely separated from the old, and

Letters of Silliman, discussing the "dubious struggle for existence" which the journal underwent, speaks of a "discreditable effort" made by an individual to destroy and supplant it by a rival publication. This seemingly could only have reference to Featherstonhaugh.

In 1834 and again in 1835 Featherstonhaugh was appointed a government geologist, as noted later, but his work was necessarily superficial and is today of historical interest only. It is doubtful, however, if it merited the harsh criticism applied to it by Lesley (see p. 163). Personally he was an overfastidious Englishman, with an almost fanatical horror of tobacco and alcoholic liquors, or of obscene and profane language. The conditions of society and the domestic life of the common people as he found them during his work in the Southwest[7] were not so widely different from those existing two and even three generations later that serious exceptions could be taken to his criticisms by one going over the same ground. He had, however, in common with many of his countrymen, a lack of ability to accommodate himself to circumstances and this, together with an apparent hypercritical disposition, seems to have rendered him unpopular and perhaps did much to lessen his influence in scientific circles. He must, however, have been a man of serious purpose and indomitable perseverance to persist in his tours of exploration in the face of conditions which were—to him at least—simply intolerable.

Sketch of Featherstonhaugh.

Featherstonhaugh was a man of some means, and came to America when quite young, apparently through the mere love of travel. Being of good presence—standing more than six feet in height, well educated, and an accomplished musician—he easily procured admission into the best of society, married an American girl, and established

embracing such a range of various climate. So to compare the phenomena with those of Europe has ever appeared to me the most material desideratum in geology, for we may be sure that any analogies which are common to localities geographically so distant, and placed under physical conditions so distinct, are, in truth, analogies belonging generally to the whole globe; and thus we shall obtain data adequate for the foundation of a general geological theory.

Well acquainted with the attention you have paid to the formations on this side of the Atlantic, I am convinced that the execution of this task can not fall into more competent hands.

W. D. Conybeare.

G. W. Featherstonhaugh, Esq., Philadelphia.

[7] An itinerary of his survey of the elevated country between the Missouri and Red rivers, made in 1834, with many interesting and highly edifying remarks on personal experiences was published by Featherstonhaugh ten years later (1844) under the title of An Excursion through the Slave States, etc.

himself at Duanesburg, in New York, but was, apparently, never naturalized. He seems to have taken an active interest in agricultural, scientific, and political affairs, and, in company with Stephen Van Rensselaer, became one of the directors of the railroad from Albany to Schenectady, a charter for which was granted in 1826. The death of his wife and two daughters, however, caused him to leave Duanesburg and turn his thoughts toward exploration and science. He therefore removed to Philadelphia, where he established, in 1831, the journal above mentioned, and entered upon his career as a geologist.

Fig. 26.
G. W. Featherstonhaugh.

Somewhere about 1840 Featherstonhaugh returned to England, where, on account of his knowledge of American affairs he was appointed one of the commissioners to settle the international boundary dispute with Canada. Later he was made British consul for the Department of the Seine, France, and was the effective agent in bringing about the escape of Louis Philippe and his queen during the revolution of 1848.

Geological Survey of Tennessee, 1831-1835.

The year 1831 was, on the whole, singularly uneventful in geological history, but few papers of importance appearing, although undoubtedly a large amount of hard, preliminary work was being done.

Perhaps the most important event of the year was the establishment of the state geological survey of Tennessee, with Gerard Troost, then professor of geology and chemistry in the University of Nashville, but formerly of Philadelphia, at its head. This survey continued in existence until 1845, during which time nine annual reports were made, the first two of which were not published. The third report, made in October, 1835, comprised thirty-two octavo pages. The fourth report appeared in 1837, in form of a pamphlet of thirty-seven pages; the fifth in November, 1839, comprising all told seventy-five pages, including an appendix. It was accompanied by a colored geological map of the state, the first which had thus far appeared, and also a colored section across the entire state from Roane Mountain, in Carter County, to Randolph, on the Mississippi.

The region west of Tennessee was described as composed mostly of secondary strata, Cretaceous marl, greensand, and clay, horizontally stratified.

It should be noted that Troost, while seemingly recognizing the importance of the fossils carried by the rocks of various ages, did not at this date attempt to utilize them systematically as means of correlation. On page 18, however, he wrote:

> The lowest part of the stratum, where it is near the encrinital limestone, often contains members of Encrinites, and other fossils of that stratum, and is invariably of a calcareous nature. This shows that the formation of those siliceous strata is more or less contemporaneous with the encrinital strata.

Organization of the Geological Society of Pennsylvania, 1832.

In the spring of 1832 there was organized in Philadelphia, largely through the influence of Mr. Peter A. Brown, a Geological Society of Pennsylvania.[8] Of the seven active participants in the movement but one, Dr. Richard Harlan, achieved a more than local reputation in matters pertaining to geology. John B. Gibson became its first president. According to its constitution:

> The objects of the society are declared to be to ascertain as far as possible the nature and structure of the rock formations of the state; their connection or comparison with the other formations in the United States, and of the rest of the world; the fossils they contain, their nature, positions, and associations, and particularly the uses to which they can be applied in the arts and their subserviency to the comforts and conveniences of man.

This society continued in existence but four years, but served its apparent purpose in bringing about the establishment in 1836, of the state geological survey with H. D. Rogers at its head. A single volume, in two parts, of *Transactions*, numbering upward of 400 pages, tells the story of its brief existence. This contains papers by authors now for the most part little known. R. C. Taylor wrote on fucoids and the coal fields of Pennsylvania and Virginia, his papers being accompanied by numerous sections; James Dickson had an essay on the gold region of the United States, all comprised within the limits of sixteen pages; Jacob Green one on a sulphated ferruginous earth, and a description of a new trilobite from Nova Scotia. Others were by Richard Harlan, Gerard Troost, Thomas Clemsen, and H. Koehler. Few of the writers or active members achieved other

[8] The date 1834 given in my Contributions is an error. See Lesley's Historical Sketch of Geological Explorations in Pennsylvania, 1876.

reputation than that given by the single publication, though Taylor became an authority on coal, while Harlan and Troost later became widely known, the one as a vertebrate paleontologist and the other as a stratigraphic geologist. This, it is well to note, was the second geological society to be organized in America, the first being that at New Haven in 1819 (p. 61). The papers were mainly purely descriptive, but the illustrations, both colored and in black and white, were of excellent quality for their time.

Green's Monograph, 1832.

Fig. 27. *Jacob Green.*

The general condition of paleontological science at this time (1832), so far as it related to the trilobites at least, is shown by Jacob Green's *Monograph of the Trilobites of North America*, a duodecimo volume of ninety-three pages and one plate of ten figures. Thirty-three species were described. In a supplement bearing date of 1835 this number was increased to forty.[9] A series of casts, examples of which are still to be found in many of our educational institutions, accompanied the original publication.

The true nature of the *Trilobite* was at that time not well understood. Green himself seems at first inclined to agree with Latreille, who placed them intermediate between the Chitons and the Articulates, owing to their supposed footless condition.[10] Brongniart, De Kay, and others, it will be remembered, classed them with the *Entomostraca.*

A contemplated chapter on this branch of the subject remained unwritten, since the author felt that all matters of dispute were "put to rest by the late discovery of some living trilobites in the southern seas." Concerning the exact position in the zoölogical scale of these "living" forms there still exists some doubt. Dr. James Eights, who originally described them under the name of *Spæroma bumastiformis*, seems to have regarded them as isopods. Emmons, in his *Geology of the Second District of New York*, on the other hand, referred to them as trilobites, while Zittel, in the latest edition of his *Handbook on Paleontology*, refers them once more to the isopods.

9 Upwards of 1,500 are now (1923) known.

10 The trilobites were later shown by Beecher to have feet. Full details of these have been worked out by Walcott.

Green's ideas as to the value of fossils as indices of geological age were not as advanced as those of most of his colleagues or as his time would lead one to expect. After quoting some of the prevalent opinions, he added in a footnote:

> Nothing can be more opposed to true science than to pronounce on priority of formation or the comparative age of rocks from either their structure or the organic remains they present.

Concerning Brongniart's expressed opinion relative to their preponderating value over lithological criteria, he wrote:

> This seems to us to imply an admission that nothing definite can be inferred from the *nature of the rocks;* moreover, that between the nature of the rock and the organic remains there may be a palpable discrepancy. . . . The event has proved, from what we have already mentioned, that no evidence as to priority can be obtained from the nature of the fossil remains displayed in particular strata.

And this from a paleontologist!

Maclure's Geology of the West Indies, 1832.

Although the geological map referred to on page 32 was the most important of Maclure's works and the one upon which his fame as a geologist largely rests, there are at least two papers by him of later date that are worthy of consideration. The first of these bears the title *Observations on the Geology of the West India Islands from Barbados to Santa Cruz, Inclusive.* This was printed for the author in pamphlet form in 1832, while he was at New Harmony, Indiana. The mineralogical nature of the materials composing the islands was noted, as well as many of the structural features. He thought to divide the islands on geological grounds into two distinct classes, the first, or most eastern, comprising Barbados, Marie Galante, Grande Terre in Guadeloupe, Desirade, Antigua, St. Bartholomew, St. Martin, Anguilla, and Santa Cruz. These he found composed mainly of shell limestone or madrepore rock belonging to the transition class, though sometimes capped by secondary rocks. The second class, including the Grenadines, St. Vincent, St. Lucien, Martinique, Dominica, Basse Terre in Guadeloupe, Montserrat, Nevis, St. Christopher, St. Eustatia, and Saba, were more or less volcanic and probably thrown up from the bed of the ocean. Such an origin he felt to be proven by their containing intermingled masses of volcanic rock and coral or madrepore limestone. The idea that volcanic activity was dependent upon combustion, as the word is commonly used, still prevailed with him, and from the fact that the line of islands corresponded closely with the strike of the rocks composing them he thought that the seat of combustion probably existed in some substance running parallel to the general stratification. Further, since he found on the islands soufrières which deposited sulphur as well as formed alum, he concluded that sulphur was one of the combustible ingredients. In this he was not alone. The second paper, *An Essay on the Formation of Rocks, or an Inquiry into the Probable Origin of Their Present Form and Structure,* appeared in 1838. In this the rocks were divided into three classes: First, the Neptunian, including those plainly of aqueous origin; second, the volcanic; and third, those of doubtful origin, under which head he would place gneiss, mica- and clay-slate, primitive limestone, syenite, granite, greenstone, etc.

Conrad's Tertiary of North America, 1832.

In 1832 the melancholy Conrad began the publication of his work on the fossil shells of the Tertiary formations of North America. This marks, according to G. D. Harris, "the beginning of systematic research into this part of our continent's history."

The work was published in parts and in small editions, largely at Conrad's own expense, the numerous plates of illustrations being prepared, even to the drawing on stone, by himself. Nos. 1 and 2 appeared in 1832, and Nos. 3 and 4 in 1833. These editions early became exhausted, and in 1893 a new edition was prepared through the instrumentality of Professor Harris. It will be well to mention here that a second edition of his *Medial Tertiary Fossils* was brought out this same year by W. H. Dall, working in conjunction with the Wagner Free Institute of Philadelphia.

The work of 1832-1833 was devoted mainly to a description of the various fossil invertebrates belonging to the Tertiary formations and contained little of a strictly geological nature. He divided the Tertiary, in conformity with the English custom, into an Upper Marine, a Middle Tertiary, and a Lower Tertiary formation, and gave the geographic extension of each so far as it was known.

Hitchcock's Survey of Massachusetts, 1830, 1833, 1841.

In 1830 the legislature of Massachusetts passed a resolve authorizing the appointment of a surveyor to make a general survey of the commonwealth, and on the fifth of June following, on the recommendation of Governor Lincoln, further authorized the appointment of some suitable person to make, in connection with this survey, a geological examination of the region involved. On February 2, 1831, the limits of the geological survey were extended so as to allow the inclusion in the report of a list of the mineralogical, botanical, and zoölogical productions.

Professor Edward Hitchcock, then professor of chemistry and natural history in Amherst College, was selected as the "suitable person," and presented, early in 1832, a report of 72 pages, with a colored geological map, on the economic geology of the state. Of this 600 copies were ordered printed. In the early part of 1833 reports on the topographic geology (37 pages), scientific geology (430 pages), and the catalogues of animals and plants, with an appendix containing a list of the rocks and minerals described (248 pages), were likewise submitted.

The legislature promptly accepted these reports likewise and ordered the printing of 1,200 copies, including a reprint of the report first made. Altogether these formed an octavo volume of 692 pages, with numerous figures in the text, and a folio volume of 19 plates, including a colored map. This publication marked an epoch in American geological history, since it brought to a successful conclusion the first survey of an entire state at public expense.[11] It is true that a spasmodic beginning had been made in North and South Carolina, but it was so lacking in breadth of conception and failed so utterly in execution that it is only by courtesy that it can be considered as a *geological* survey. Under these conditions a detailed review of the contents of the reports will not be out of place.

In the consideration of the application of Beaumont's theory of mountains to those of Massachusetts, Hitchcock thought to find six systems of strata and contemporaneous uplift in the state. These systems as enumerated were: (1) The oldest meridional system, (2) the trap system, (3) the latest meridional system, (4) the northeast and southwest system, (5) the east and west system, and (6) the northwest and southeast system.

The Hoosac Mountains, considered under his oldest meridional system, he thought to be due to two different epochs of elevation, the last taking place after the deposition of the New Red sandstone, as was shown by the slight dip (fifteen degrees to twenty degrees) of the latter, the slates and gneisses of the range standing nearly vertical. The force which gave rise to the uplift he regarded as gravity

11 The total expense to the state of the three years' work, exclusive of cost of printing, was $2,030.

acting on the crust as the nucleus gradually shrunk away in process of cooling.

Not the least interesting chapters in the work are those relating to the unstratified rocks. Here, under the name of *greenstone*, he included the entire series of basic and intermediate eruptives now classed as melaphyrs, diabases, and diorites. Their mineral composition was stated, however, to be hornblende and feldspar, though the mineral identified as hornblende has been proven by modern microscopic methods to have been almost wholly augite. Columnar, compact, amygdaloidal, and porphyritic varieties were recognized.

Concerning the igneous origin of these rocks he professed little doubt, and he accounted for their appearance in the form of continuous sheets on the assumption that they were poured out, not from craters but from linear openings produced by the shrinkage of the earth's interior. Thus early was recognized the phenomenon of fissure eruptions, as later developed by Geikie and others.

Commenting on the occurrence of the greenstone in "veins" (small dikes), as at Nahant, he expressed the opinion that the slate in which these veins occurred could not have been solid at the time of the intrusion, this on account of the size of the fissures, which seemed to him to be too wide to have been formed by desiccation. Evidently the possibility of the formation of these fissures by dynamic agencies acting on consolidated materials was not then realized.

Under the name of *porphyry* he included the compact eruptive rocks long known as felsites and petrosilex, but now classed mainly as aporhyolites or quartz porphyries, such as form the cliffs at Marblehead, and are also found in other localities, as Hingham and Malden. Their porphyritic structure he thought could result only from igneous solution. The base of these porphyries he stated to be of compact feldspar, formed by the melting down of preexisting feldspathic rocks.

The slaty structure sometimes shown, which is now recognized as flow structure, he thought to represent the original structure of the slaty rock from which the porphyry was derived by fusion. This view is interesting when compared with those put forward nearly half a century later by one who, working under the erratic T. Sterry Hunt, argued in favor of the sedimentary origin of the entire series.

Hitchcock thought it probable that the mica-slate and gneisses of the Hoosac Range passed laterally into one another, owing to a decrease in the amount of feldspar in the gneissic rock. Such a transition seemed to him as possible, whether the rocks were to be regarded as direct crystallizations from aqueous solution or sedimentaries crystallized through the influence of heat. He was inclined himself to the belief that they were metamorphosed sediments, since he could conceive of no chemical process by which such a variety of minerals could have crystallized out simultaneously from solution. Such a crystallization, he argued, would be differential.

He utilized this same argument in writing on the origin of the granites. These he looked upon as resulting from the melting down of other rocks, gneisses being due to the more or less complete fusion and crystallization of feldspathic sandstone. The deeper lying portions, which were most highly heated, gave rise to granitic gneiss, and those further removed to the porphyritic, lamellar, and schistic varieties. It will be of interest to compare these views with those advanced by Clarence King nearly forty years later.

The siliceous conglomerate of Dorchester and Roxbury (now considered as of Carboniferous age) he classed under the name of graywacke as, indeed, he did every conglomerate, sandstone, and fragmentary formation that was older than the Red sandstone and Coal, thus avowedly following Humboldt.[12]

Under a misapprehension regarding the character of the amygdules, he included under the name of *varioloid wacke* the melaphyr of Brighton and elsewhere, as well as other altered Paleozoic lavas found at Hingham, Needham, and Saugus; also the siliceous or flinty slates of Nahant and jasper of Newport.

The red sandstones of the Connecticut Valley were rightly identified as belonging to the New Red, through the presence of "vertebral" remains found at East Windsor, Connecticut, and Sunderland and other localities in Massachusetts. He still, however, thought it not impossible that the lower portion of the beds might correspond to the Old Red sandstone of Europe. It will be remembered that in the *American Journal of Science* for 1823 he had argued that the upper beds belonged to the coal formation, on account of the included thin seams of bituminous coals.

The Tertiary formations were divided into two general groups, the most recent being made to include the beds of blue plastic clay alternating with layers of white siliceous sand, and the second division the plastic clays of Gay Head and Nantucket.

Naturally the drift so abundantly distributed over the surface was attributed to the action of the Noachian deluge. "That seen about Cape Ann can not fail to impress every reasoning mind with the conviction that a deluge of tremendous power must have swept

12 Humboldt's Essay on the Trend of Rocks in the Two Hemispheres appeared in 1822.

over this cape. Nothing but a substratum of syenite could have stood before its devastating energy."

One of the points which must impress the reader who is at all conversant with the work of modern physiographers is the inability of Hitchcock, as well as his contemporaries, to understand the relationship existing between a river and the channel in which it flowed. Discussing the course of the Connecticut River, he wrote: "The ordinary laws of physical geography seem here to be set at defiance," and "the geologist will be surprised to find it (the river) crossing the greenstone ridge" at its highest part, *i.e.*, through the gorge between Mounts Holyoke and Tom. The facts that he deduced from this and other like illustrations were that "surely the Connecticut River did not excavate its own bed, but the gorges through which it and the Deerfield and Westfield pass were excavated to a considerable depth before they began to flow."

The student of today need scarcely be reminded that the river did excavate its own channel and ran in essentially its present position before the warping which brought the trap ridge across its course took place.

Hitchcock was not a believer in the doctrine of uniformitarianism; indeed, while perhaps not an extreme catastrophist, he combated vigorously the views put forth by Lyell in his *Principles*,[13] and argued for the greater intensity of geological agencies in the earlier periods of the earth's history. Some of his reasons for this are worthy of attention. He believed the continent to have been elevated above the ocean, not little by little, but by a few paroxysmal efforts of volcanic force. Since this force, as acting during the past four thousand years, seemed too feeble to result in the elevation of a single mountain chain, so, he argued, it must have been more energetic in previous epochs.

The fact that the older rocks are more distorted and highly metamorphosed than the younger was thought to indicate also the greater intensity of the earlier agencies. Singularly enough, the near-shore origin of beds of conglomerate was not realized, the occasional occurrence of such among sedimentary rocks being thought due to the "occasional recurrence of powerful debacles of water," the like of which cannot be produced by any causes now in operation.

Concerning the origin of valleys there was manifested a great dearth of knowledge. "I am in doubt whether there is more than one valley in Massachusetts that is, strictly speaking, a valley of denudation," that being the one lying between Mount Toby and Sugar

[13] The first edition of this celebrated work appeared in 1830.

Loaf. The valleys in Berkshire and Worcester and possibly also that of the Merrimac River were considered primary, more or less modified by deluges and other abrading agencies.

Accompanying this report was a volume of nineteen plates, comprising a colored geological map of the state; nine general and special views; four of fossils; a map of the valley of the Connecticut; a map showing the direction of the strata; two plates of sections, three of which ran in an east and west direction and one in a north and south; and one "tabular view" of the rocks and their embedded minerals.

On the map the rocks were colored as belonging to six groups, as below. It will be noted that there is no attempt at other than a lithological classification until the New Red sandstone and Tertiary are reached.

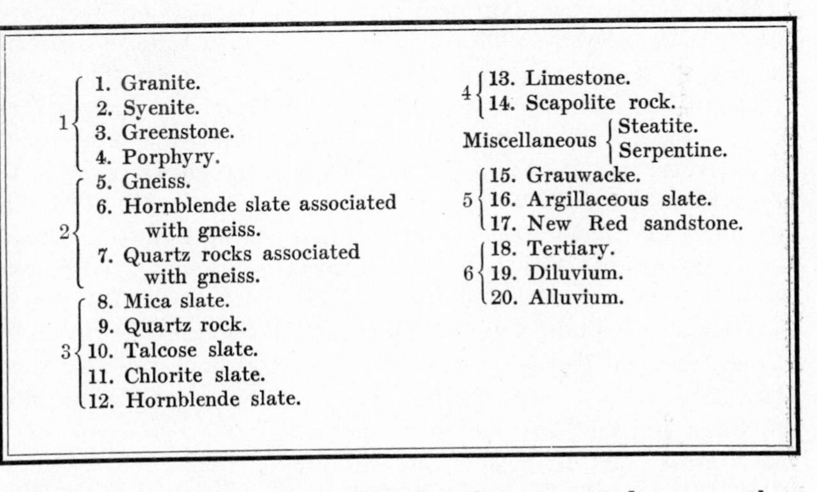

Group	Rocks
1	1. Granite. 2. Syenite. 3. Greenstone. 4. Porphyry.
2	5. Gneiss. 6. Hornblende slate associated with gneiss. 7. Quartz rocks associated with gneiss.
3	8. Mica slate. 9. Quartz rock. 10. Talcose slate. 11. Chlorite slate. 12. Hornblende slate.
4	13. Limestone. 14. Scapolite rock.
Miscellaneous	Steatite. Serpentine.
5	15. Grauwacke. 16. Argillaceous slate. 17. New Red sandstone.
6	18. Tertiary. 19. Diluvium. 20. Alluvium.

Deposits and mines of useful minerals were noted, among them being plumbago, coal, peat, lead, iron, copper, and manganese. The drawings of fossils and the general views were the work of Mrs. Hitchcock.

In 1837 Hitchcock was commissioned by Governor Everett to make a further geological and mineralogical survey in which economic problems should be of primary importance. His report appeared in the form of House Document No. 52, under date of 1838. In 1839 the legislature authorized the reprinting of all the reports, with such additions and corrections as the author chose to make. This reprint appeared in 1841 in the form of two quarto volumes of 831 pages, the first 126 of which are given up to a discussion of the soils, their kind, origin, and fertility.

Hitchcock here followed closely Sir Humphry Davy, Berzelius, and S. L. Dana, and dwelt particularly upon the *geine,* a problematic substance resulting from vegetable decomposition and supposed to contain all the essentials of plant food. This part of the work is of interest as containing C. T. Jackson's expressed and well-founded doubt as to the existence of such a body, and a long letter from S. L. Dana attempting to prove its chemical unity. Some twenty pages were devoted to fossil fuels, with especial reference to the anthracite of Worcester, Massachusetts, and of Rhode Island, and the bituminous coal of the New Red sandstone. A list of peat bogs, with reference to their possible availability for fuel, was also given. Building stones were discussed, and two pages devoted to the subject of rock decay and its probable cause. Metallic and nonmetallic ores were described.

Many of the views expressed in the early reports were repeated here with but slight modification. He still adhered to his six distinct systems of strata which were tilted at various epochs. The oldest, called the meridional system, embraced all the primary rocks lying between the valleys of the Connecticut and Worcester. The second, or northeast and southwest system, included the gneiss range in the southeastern part of Worcester County, the central part of Middlesex, and a part of Essex. The third, or east and west system, included the gneiss around New Bedford, the graywacke, and a large part of the syenite, porphyry, and greenstone of the state. The fourth, or Hoosac Mountain system, embraced all the rocks between the Connecticut and Hudson rivers, except the trap and New Red sandstone. The fifth included the New Red sandstone, and the sixth, or northwest and southeast system of rocks, mainly gneiss occurring in the southern part of the state bordering upon Rhode Island. In the geological map which accompanied this report he employed but six colors, which marked off the rocks of the state into six distinct groups, the members of each of which, with the exception of the fourth, were so nearly related that he thought they might be regarded as belonging to the same formation. In his first group he placed the granite, syenite, greenstone, and porphyry; in the second, gneiss and the associated hornblende slate and quartz rock; in the third, mica-slate, with the associated quartz rock and hornblende slate, talcose slate, and chlorite slate; in the fourth, limestone, steatite, and serpentine; in the fifth, metamorphic slates, graywacke, argillaceous slate, Coal Measures, and the New Red sandstone, and in the sixth group, Eocene Tertiary, diluvium, and alluvium.

The Connecticut River was described as flowing through a synclinal valley, which became nearly filled by the red sandstone deposit. This last, he thought, had since been eroded away, so that the valley as it now exists may "in some sense" be considered "a valley of denudation." Most of the valleys in Massachusetts he still thought due "to the elevation and dislocation of the strata." These are but reiteration of the views expressed in his first report.

He noted that the animal remains found in the older rocks differed most radically from existing species, and also that the organic remains found in northern portions of the globe corresponded more nearly to existing tropical species than did those now living in the same localities. Further than this, he announced that different species, genera, and families of animals began their existence at very different epochs in the earth's history, and that the same species rarely extended over from one formation into another. Notwithstanding this, though giving lists of fossils found in rocks of various horizons, he made no attempt to utilize them and thought correlation by such aids impossible. Following Phillips and Lyell in many of his statements, he yet announced as a principle that "Rocks agreeing in their fossil contents may not have been contemporaneous in their deposition," although they might not differ greatly in age. "From all that has been advanced, it appears that an identity of organic remains is not alone sufficient to prove a complete chronological identity of the rocks widely separated from each other; but it will show an approximate identity as to the period of their deposition, and in regard to rocks in a limited district it will show complete identity." Where both the mineral nature and the character of fossils were identical, identity or synchronism would be much more probable, but a want of such agreement, so far as it related to mineral character, was not considered fatal to the idea of synchronism.

He argued against the idea that the stratified primary rocks are merely detrital or fossiliferous rocks altered by heat, considering them rather products of both aqueous and igneous agencies when the temperature of the crust was very high and before organic beings could live upon it.

He noted that the dolomitic rocks seemed genetically related to the limestones, but his ideas as to the methods by which the changes had been brought about were naturally somewhat crude when considered in the light of today. From a study of the field relations of the dolomites of Berkshire County and other localities he came to the conclusion "that all the cases of dolomitization in Massachusetts occur either in the vicinity of a fault, or of unstratified rocks, or in the midst of gneiss, where the evidences of the powerful action of heat in the induration of the limestone and the obliteration of its stratification is as great as results from the presence of unstratified rocks." The actual dolomitization he thought to have been "probably produced by gaseous sublimations, chiefly of carbonate of magnesia, which penetrated the rock after it had been softened by heat so as to be permeable." This is probably a reflection of views earlier put forward by Von Buch with reference to the dolomites of the Tyrol. He considered it highly probable however that in some instances magnesian limestone had been deposited directly from thermal waters.

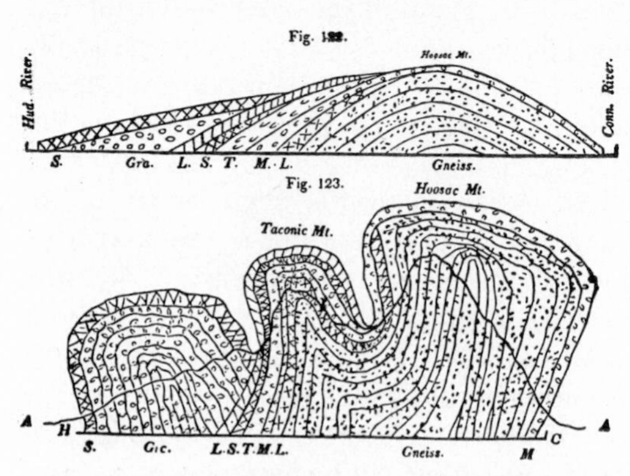

Fig. 28. *Hitchcock's illustration of folded axes.*

In seeking an explanation of the fact that in passing from the top of Hoosac Mountain toward the Hudson River one continually meets newer and newer rocks which appear to pass under the older—in other words, that there is inversion of the strata—he came to the conclusion, put forward with some hesitation, "that these rocks have actually been thrown over into an inverted position, or rather, have been so contorted by a force acting laterally that one or more folded axes have been produced."

The same feature had been mentioned in his report for 1835, but then appeared "too improbable to be admitted." His views on the subject were illustrated by a sketch, copied here in Fig. 28, the portion above the heavy continuous black line being conceived to have been removed by erosion. Later this observation was claimed to be the first recognition of the true character of overturned folds. This was, however, disputed by H. D. Rogers, who claimed to have made and published similar observations during the progress of the New Jersey survey in 1838-1839.

Below is the paragraph upon which the claim of Rogers was apparently based. The original statement is made in the course of a description of the various axes of elevation that have given rise to the more prominent physical features of northern New Jersey.

The movement which elevated Jenny Jump, seems to have been everywhere one of excessive suddenness and violence, as the strata along its anticlinal axis are not only frequently much dislocated and broken, but those lying immediately along its northwestern base are in several places thrown into an inverted posture, dipping not to the northwest, but in toward the base of the hill.[14]

The idea was, however, brought out much more clearly in a communication made by him before the Association of American Geologists and Naturalists in 1841, an abstract of which was given in their *Proceedings* and in the *American Journal of Science* for the same year. He then called attention to the "inverted dip" observable among the rocks of the Coal Measures (of Pennsylvania) and ascribed the same to a great force acting laterally, folding and crushing the axis so as to produce this inverted dip by tossing the strata many degrees beyond the perpendicular, thus producing the present apparent dip of the lower stratified or sedimentary rocks beneath the primary.

In view of the observations made by Eaton in 1820, Steel in 1825, and W. B. Rogers, in Virginia, in 1838, as noted elsewhere, the subject of priority is scarcely worth discussing.

Heavy demands have from a very early time been made upon electric currents to account for all manner of geological phenomena, from the formation of ore deposits to the production of slaty cleavage, or even the uplifting of mountain ranges. It is not strange, therefore, that Hitchcock should have felt justified in making a slight draft upon them to help him over some of his minor difficulties. The peculiar imitative and otherwise interesting forms assumed by the ferruginous concretions found in the Connecticut Valley demanded an explanation. "I know of no agent," he wrote, "that could have accomplished this [*i.e.*, the separation of the iron from the menstruum which held it in solution] except galvanic electricity," and again, "I strongly suspect galvanism to have been a chief agent in concretions of every sort."

In the report of 1833, and again in those of 1835 and 1841, Hitchcock called attention to the peculiarly flattened and otherwise distorted pebbles in the conglomerate at what is locally known as Purgatory, near Newport, Rhode Island. His description is important in view of his subsequent writings and the discussions which arose in connection therewith. The subject may, however, well go over until his later paper of 1861 (see p. 402).

This final report, coming at the time it did, naturally attracted much attention and favorable comment. From a brief preliminary review in the *American Journal of Science* the following paragraph is selected:

14 Geology of New Jersey, 1840, p. 55.

If we reflect that the vast mass of facts and information of various descriptions, and the reasonings and inferences contained in these columns, are all the result of untiring, nay, almost Herculean, efforts of an individual mind, continued among the harassments of constant professional duty during a period of ten years, we are encouraged to hope that we may yet see the day when the united efforts of our small army of working geologists now laboring in the common cause shall reduce the whole of our widespread territory to an intelligible perfect system.

Sketch of Hitchcock.

Hitchcock began his career as a teacher and clergyman though with decided leanings toward astronomy and the natural sciences. He has been eulogized somewhat enthusiastically by Lesley[15] as "a man of religion, a man of science; in both a docile student and an expert teacher; in both enthusiastic and self sacrificing; in both gentle, persuasive, affectionate, sympathetic; in both shackled by traditions which he both feared and hated to break, yet rigorously holding up his shackles and keeping abreast, and in some instances, ahead of the advancing age. . . . He was as a man, both timid and adventurous. Adventurous and progressive where he thought he could see his way; hesitating and submissive to authority when himself in the dark. And this composition of adverse habits, held in balance by circumstances, not by will nor genius, made him a representative man—a geologist in whose writings one can read the halting progress of American Geology—its ignorance of its own past history, its premature intuitions, its illbred waywardness and levity, its abortive investigations, its double-minded instability, its feeble conservatism, its energetic radicalism, the fertility of its fancy and the haziness of its judgment, its patience to wait and its power to work for what it is as ready to abandon in a moment for something new."

As a prolific writer Hitchcock was without an equal. He wrote five volumes and thirty-seven pamphlets and tracts on religious subjects; three volumes and as many tracts on temperance; fourteen volumes, five tracts, and some seventy-five papers on botanical, mineralogical, geological, and physical subjects and twenty-seven others, including a tragedy. His *Elementary Geology*, published in 1840, passed through thirty editions and then was rewritten.

[15] Biog. Mems. Nat. Acad. Sciences, Vol. I, 1877.

PLATE VIII

EDWARD HITCHCOCK

Nicollet's Work in America, 1833.

The explorations of Jean N. Nicollet in the Middle West need but brief mention.

Nicollet was born in France in 1790 and came to America in 1832 from financial considerations, having involved not merely his own fortune but those of friends in unfortunate speculations. His main scientific work up to this date had been of an astronomical nature, and it is perhaps natural that his attention in a new country should have turned toward exploration and map making.

From 1833 to 1838 he was engaged, with the cooperation of the War Department, in mapping the region of the upper Mississippi. From 1838 to 1841 he was employed by the Government in completing the surveys and maps thus begun. Other work of like nature was projected, but before plans were matured his health gave way and he died in 1843. His geological work, under the conditions mentioned above, was therefore of comparatively slight importance, excepting as calling attention to regions but little explored.

He claimed to have traced the Cliff limestone of Owen over a vast extent of country in the Mississippi Valley, and connecting his own work with that of Owen, Locke, H. King, and others, felt justified in assigning the Falls of St. Anthony as the northern limit of the formation. In this he was correct. He added to the knowledge of the geology of the region of the Sioux and Missouri rivers by bringing in important Cretaceous fossils, both vertebrate and invertebrate; he described the occurrence of the Indian pipestone (Catlinite) of the Coteau des Prairies and correctly ascribed the pseudo-volcanoes and their pumice-like product to burning lignite beds. Nicollet's work, and perhaps that of Featherstonhaugh, noted later, was presumably authorized in response to a resolution from the Committee on Military Affairs in the Congress of 1833. This resolution, it appears, inquired into "the expediency of authorizing the President to employ a suitable person in aid of the Topographical Bureau, to ascertain the mineralogy and geology of each of the several States of the Union, with a view to the construction of a mineralogical and geological map of the whole territory of the United States."[16] It may be added that, while the committee, in their report, seemingly favored the plan, they nevertheless refrained from urging its adoption to the point of making a specific appropriation to carry it out. "The committee is of opinion that all this can be effectually done without

[16] C. W. Keyes states that Nicollet's work of 1838 was in connection with the erection of the new Territory of Iowa, *op. cit.*, p. 413.

any increase of the annual appropriation under the Act of April 30, 1824, and by simply adding this important object to the duties of the Topographical Bureau." From which it is to be inferred that Congresses of that day did not differ greatly, in mental equipment, from those of the present.

Second American Edition of Bakewell's Geology, 1833.

The second American edition of Bakewell's *Geology*, from the fourth London edition, was printed, as was the first, in New Haven and under Silliman's supervision. The work needs notice, as, like the first, it was accompanied by an appendix giving a syllabus of his lectures at Yale, and showed his gradually expanding views and disposition to shake off the shackles of tradition, though still floundering in the deep waters of the Noachian deluge. Of particular interest are his remarks on the nature of geological evidence and its consistency with sacred history.

Silliman took the ground that there was no reason to believe that any part of the crust of the earth is now in the same condition as first created, every portion having been worked over in accordance with physical laws which are as much the Creator's work as are the materials upon which they operate. Unlike some of his predecessors and contemporaries, he did not, however, at this time feel it incumbent upon himself to close his eyes to, or even to distort, any evidence that might present itself in order that it might not conflict with the statement of sacred history. "Any attempt to disprove the truth or genuineness of the Pentateuch, and Genesis in particular, is wholly superfluous, and quite aside from any question that can in this age be at issue between geologists." "No geologist at the present day erects any system upon the basis of the scripture history." But admitting that the Mosaic account is genuine and true, he felt that one might with propriety compare geology with history and regard historical coincidence with observed phenomena as interesting, "because they are mutually adjuvant and confirmatory." As with other workers, including even some of the present time, he found most that is confirmatory in a study of the drift or "diluvial," which was ascribed to torrential action, and was perhaps concomitant with the universal deluge recorded in Genesis. After a somewhat prolonged discussion of the meaning of the term "days," as used in Genesis, he gave the following:

TABLE OF COINCIDENCES BETWEEN THE ORDER OF EVENTS AS DESCRIBED IN GENESIS AND THAT UNFOLDED BY GEOLOGICAL INVESTIGATIONS.

In Genesis.	No.	Discovered by geology.
Gen. I. 1, 2. In the beginning God created the heavens and the earth. And the earth was without form and void; and darkness was upon the face of the deep; and the Spirit of God moved upon the face of the waters.	1	It is impossible to deny that the waters of the sea have formerly, and for a long time, covered those masses of matter which now constitute our highest mountains;
3, 4, 5. Creation of light. 6, 7, 8. Creation of the expansion or atmosphere. 9, 10. Appearance of the dry land.	2	and, further, that these waters, during a long time, did not support any living bodies.—Cuvier's Theory of the Earth, sect. 7.
11, 12, 13. Creation of shooting plants, and of seed-bearing herbs and trees.	3	1. Cryptogamous plants in the coal strata.—Many observers. 2. Species of the most perfect developed class, the Dicotyledonous, already appear in the period of the secondary formations, and the first traces of them can be shown in the oldest strata of the secondary formation; while they uninterruptedly increase in the successive formations.—Prof. Jameson's remarks on the Ancient Flora of the Earth.
14 to 19. Sun, moon, and stars made to be for signs and for seasons, and for days, and for years.		
20. Creation of the inhabitants of the waters.	4	Shells in Alpine and Jura limestone.—Humboldt's tables. Fish in Jura limestone.—Humboldt's tables. Teeth and scales of fish in Tilgate sandstone.—Mr. Mantell.

TABLE OF COINCIDENCES BETWEEN THE ORDER OF EVENTS AS DESCRIBED IN GENESIS AND THAT UNFOLDED BY GEOLOGICAL INVESTIGATIONS.—*Continued.*

In Genesis.	No.	Discovered by geology.
Creation of flying things.	5	Bones of birds in Tilgate sandstone.—Mr. Mantell, Geological Transactions, 1826. Elytra* of winged insects in calcareous slate, at Stonesfield.—Mr. Mantell.
21. The creation of great reptiles.	6	It will be impossible not to acknowledge as a certain truth the number, the largeness, and the variety of the reptiles which inhabited the seas or the land at the epoch in which the strata of Jura were deposited.—Cuvier's Ossem. Foss. There was a period when the earth was peopled by oviparous quadrupeds of the most appalling magnitude. Reptiles were the lords of creation.—Mr. Mantell.
24, 25. Creation of the mammalia.	7	Bones of mammiferous land quadrupeds, found only when we come up to the formations above the coarse limestone, which is above the chalk.†—Cuvier's Theory, sect. 20.
26, 27. Creation of man.	8	No human remains among extraneous fossils.—Cuvier's Theory, sect. 32. But found covered with mud in caves of Bize.—Journal.
Genesis, VII. The flood of Noah, 4,200 years ago.	9	The crust of the globe has been subjected to a great and sudden revolution, which can not be dated much farther back than five or six thousand years ago.—Cuvier's Theory, 32, 33, 34, 35, and Buckland's Reliq. Diluv.

* Sheaths.

† One solitary exception is since discovered, in the calcareous slate of Stonesfield, in the bones of a didelphis, an opossum, a tribe whose position may be held intermediate between the oviparous and mammiferous races.

Buckland, the English geologist, in his *Bridgewater Treatise* (1836) refers to this as very interesting, but makes no further comment.

Professor Thomas Cooper, to whom reference is made on p. 86 and at this time connected with South Carolina College at Columbia, held radically different views and bitterly resented the insertion in a work intended as a textbook. "I am thus compelled to put into the hand of my classes . . . Dr. Silliman's geological doctrines—as different from what I have been accustomed to deliver as two opposite opinions could well be," he complained, and he proceeded, in a pamphlet of sixty-four pages, entitled *On the connection between geology and the Pentateuch* (1833), vigorously to set forth his reasons. "It is well for Professor Silliman," he wrote in his conclusion, "that his useful services to science have placed his reputation on a more stable foundation, than his absolute unconditional surrender of his common sense to clerical orthodoxy."

Perhaps that part of Professor Silliman's work which showed the keenest insight into geological problems is that relating to the subject of crystallization in rocks. The proximate causes of the phenomenon were recognized as heat and solution.

> There is no doubt that fire and water . . . have operated in all ages in producing mineral crystallization. Of these, however, fire appears to have been by far the most active, . . . and there is every reason to admit that even granite has been melted in the bowels of the earth, and therefore may crystallize from a state of igneous fusion. If this be true of the proper crystals of granite it may be also true of the imbedded crystals which it contains, and therefore of all other crystals. Those which contain much water of crystallization may present a serious difficulty, but perhaps pressure may have retained the water, and as the parts of the mineral concreted in cooling the molecules of water may have taken their place in the regular solid. Still we can see no reason for excluding water and other dissolving agents, acting with intense energy under vast pressure and at the heat of even high ignition, from playing a very important part in crystallization.

This, it will be observed, is essentially the aqueo-igneous theory of eruptive rocks, and could scarcely be improved on today. In this connection it is well to remember that Edward Hitchcock, in his *Geology of Massachusetts*, 1833 (see p. 145), inferred the igneous origin of granite from its crystalline structure, "since substances held in solution always crystallize in succession, while in granite we have a solid crystalline mass of three or four distinct substances which evidently crystallized contemporaneously." Silliman had previously (1830) committed himself to the igneous origin of the Connecticut traps or greenstones, basing his opinions upon the indurated and vesicular underlying sandstone.

Lea's Contributions to Geology, 1833.

The two most important original contributions of the year 1833 were Hitchcock's *Massachusetts*, already noted, and Isaac Lea's *Contributions to Geology*, the latter dealing with the Tertiary of Alabama, Maryland, and New Jersey. In Lea's work there were brought prominently forward for the first time, by an American writer, the striking changes that had taken place in the introduction of the Tertiary fauna and the close relationship existing between that and the fauna of the present day. Out of some 250 species of invertebrates found by him in the bluffs at Claiborne, Alabama, 219 were not referable to any known species, *i.e.*, were not found in any of the beds older than Tertiary and were new to science. It was in this work also that the undoubted presence in America of beds referable to the Eocene of Lyell was recognized, although the character of the fossils and the general position of the beds had been already noted by Conrad.[17]

Fig. 29. *Isaac Lea.*

Conrad on the Distribution of the Eocene, 1832.

The last-named, the year following, marked out the distribution of the Eocene in Maryland and noted the occurrence of Pliocene fossils overlying it at Vance's ferry on the Santee River. The Fort Washington beds, formerly thought by him to be Eocene, he now suggested might be more recent than Lea's Claiborne beds, and perhaps contemporaneous with the Miocene of Europe. This has since been shown to be an error.

[17] This publication of Lea's led to a misunderstanding with Conrad, or rather with Conrad's friends, Say and Morton, who felt that he was trespassing on Conrad's field. Dr. Lea, however, worked only on material he had received from Judge Tait, prior to Conrad's entry upon the field. (See Dall's Determination of Dates of Publication of Conrad's Fossils of the Tertiary Formations. Bull. Phil. Soc. of Washington, XII, 1893, pp. 215-240.)

Aiken on Mountain Uplift.

A paper by Professor William Aiken of Mount St. Marys College, Maryland, published about this time, is of interest for its bearing upon the prevailing theories on mountain uplift, and incidental phenomena. Aiken wrote on the geology of the regions between Baltimore and the Ohio River, and gave a cross section showing the kinds and inclination of the rocks.

Noting the reversal of the dip existing between Hancock and Cumberland, he wrote: "Beneath this space, then, we are authorized in concluding, the eruptive power that was instrumental in upheaving the Appalachian chain was most energetically exerted. This may be considered the true anticlinal region." He regarded the agent so efficient in throwing up mountain chains as igneous, "an opinion that gains confirmation, if any is needed, from the occurrence of thermal waters along the center line of the Allegheny region." It is apparent from this that he agreed with Daubeny and other European geologists as to the volcanic origin of these springs. Rogers's paper, ascribing them to the other causes, did not appear until eight years later (see p. 222).

The causes which would produce a marble from a limestone were acknowledgedly problematical, but, he added, "At a time when it is so fashionable to refer all troublesome facts in geology to that mighty cause, I may be excused for conceiving the possibilities of a blue limestone becoming a white one through the energetic agency of intruded igneous gneiss and sienite."

Ducatel and Alexander's Survey of Maryland, 1834.

In accordance with a resolution passed by the general assembly of Maryland, March 18, 1833, J. T. Ducatel, geologist, and J. H. Alexander, a civil engineer, were appointed to make a geological survey and new map of the state. This survey continued in existence until 1840, during which time three annual reports, one on the outlines and physical geography of Maryland, one on the new map, and one on the Frostburg coal formation, were issued, the last, however, not being an official publication.

The geological reports were given up largely to discussions of economic matters, such as the character of the soil and the occurrence of coal and ores of the metals. In the report for 1833 the occurrence of fossiliferous deposits in Prince George County; at Maryland Point, in Charles County; and on St. Marys River, in St. Marys County; also at Fort Washington on the Potomac, were noted, but none of the fossils were identified nor any suggestion made as to their probable geological age, although Conrad and Lea were both at work in the same field (*ante*, p. 158).

The rocks comprising the upper part of Cecil County, the greater portion of Baltimore and Harford counties, the upper districts of Anne Arundel County, and the whole of Montgomery County, were thought to be generally metalliferous and were grouped as Primary or Primitive. The serpentinous rocks were considered of importance, as likely to furnish the basis for the manufacture of epsom salts, a not uncommon opinion at that time. The frequent repetition in the early works of this idea sometimes leads one to wonder at the apparent need of so large a supply. On the map prepared under the direction of Mr. Alexander the lithological nature of the underlying rocks and the character of the soil were indicated by names, no attempt being made at coloring.

In 1835 Ducatel visited the eastern shore of the state, and in his report gave a very full account of the geography and agricultural condition of Dorchester, Somerset, and Worcester counties. He announced the presence of considerable deposits of greensand marl, which he believed would be of great value to the agriculturist. In 1836 he completed the geological survey of Calvert County and made extensive examinations in Anne Arundel, Prince George, Charles, and St. Marys counties, where he found further deposits of marl. In this same year he examined the coal and iron deposits of Allegany County and published a geological account and section of the beds.

In his report for 1838 he gave a general account of the mineral resources of Harford County and outlines of the geology of this and Baltimore counties, with a short treatise on the subject of lime burning.

This was considered by Ducatel to be the first attempt to connect a topographic and geological survey.

Sketch of Ducatel.

Ducatel was born in Baltimore, where his father, a Frenchman, conducted for many years a prominent pharmaceutical establishment. He received his early training in St. Mary's College in Baltimore, subsequently studying in Paris. In 1824 he returned to America, and shortly after entered upon a very successful career as a teacher of the sciences—chemistry, philosophy, mineralogy, and geology—serving in the Mechanic's Institute and the University of Maryland in Baltimore, and St. John's College in Annapolis.

Fig. 30. *J. T. Ducatel.*

In order that he might devote himself more fully to the work of the survey, he severed his connection with the institutions of learning about 1838. The move, however, proved an unfortunate one, as the survey was short-lived and he was thrown back upon his own resources.

In 1846 he visited, in the capacity of a geological expert, the Lake Superior mining regions. Through exposure while there he contracted a severe illness, which left him in an enfeebled condition from which he never fully recovered. He died from congestion of the lungs in 1849.

Few publications bear his name, though as a writer for the journals of the day he is said to have been quite prolific. His principal work was a *Manual of Toxicology.*

Featherstonhaugh's Survey, 1834.

In July, 1834, G. W. Featherstonhaugh, who was introduced to the reader on page 136, was authorized by Lieutenant-Colonel J. J. Abert, acting under instructions of the War Department, to make a geological and mineralogical survey of the "elevated country lying between the Missouri and Red rivers, known under the designation of the Ozark Mountains."[18] The survey was duly made and a report rendered bearing the date of February 17, 1835. This comprised altogether ninety-seven pages, the first forty-two of which were given up mainly to a discussion of general principles not germane to the report, but which, as indicating the condition of mind of the writer, are worthy of consideration. He regarded continents and islands as having originated through an expansive subterranean force, and believed the mineral veins to have been filled from below, rather than from above, as taught by Werner.

Granite was considered an igneous rock (designated "ignigenous").

[18] This appears to have been the first formal recognition of the science of geology by the General Government. See, however, footnote relative to Nicollet's work, p. 153.

Gneiss he recognized as often passing into granite, while some of the primary limestones he thought to have possibly "come from central parts of the earth, in a state of aqueous solution, and to have subsequently received their high crystalline character from being in contact with ignigenous rocks in an incandescent state." To the commonly received opinion, that all coal was of vegetable origin, he took exception, and thought that some beds "may have been the result of outpourings of bituminous matter." The wide geographical distribution of Carboniferous rocks west of the Appalachians and in the Mississippi Valley was noted, and a section given comprising 1,520 vertical feet of these rocks and overlying shale and millstone grit at Burkesville, Kentucky.

Passing to the specific part of the work, Featherstonhaugh discussed the occurrence of the lead ore (galena), its distribution in loose pieces in the soil, and its relation to the vein material in the solid rock, but the fact that this superficial layer was but the detritus from the decomposition of material of the same nature as the underlying rock does not seem to have been by him realized more than by his predecessors. He remarked, however, that the disseminated granules in the limestone at Isle La Motte furnished to the eye "sufficient proof that the stony and metallic matter were deposited at the same time," an idea not altogether different from that held today. Further on, however, he noted the occurrence of the ore in pockets or cavities filled with red clay as pointing "to a projection of this metallic or mineral matter from below."

He described the hematite deposit of Pilot Knob, Missouri, and correctly noted the improbability of finding precious metals in Arkansas in paying quantities. Magnet Cove he thought to have been perhaps "one of those extremely ancient craters that may have preceded those of which basalt and lava are the products." The crater form of the cone, it should be stated, while very suggestive of a volcano, is due to the weathering and erosion of intrusive igneous rocks of the nature of syenite.

Although Featherstonhaugh's standing with his fellow geologists does not seem to have been the best, some of his recorded observations certainly show a more philosophical mind and greater ability to grasp the broader problems of geology than do those of many of his contemporaries. Thus, discussing the muddy character of the Mississippi, he suggested that experiments might be conducted to show the amount of sediment annually brought down, whereby it might be possible to assign approximately a chronological period for the origin of the river, the commencement of the alluvial deposits, and the withdrawal of the ocean. This, so far as the present writer is aware, is the first suggestion of its kind to appear in the annals of American Geology.[19]

Commenting on the fact announced by him in 1828 and since, as he thought, confirmed, that, with the exception of the Tertiary and sub-Cretaceous beds of the coast, no rocks more recent than the coal-bearing series had been found in the United States, he suggested that the American continent might in reality be older than the European. This also is the first suggestion of its kind to be found in our literature.[20]

The report was accompanied by a colored geological section extending from the Atlantic at New Jersey to the Red River in Texas. On this all New Jersey and Delaware as far as Chesapeake Bay were colored as occupied by superficial sand and sub-Cretaceous beds; Maryland to the west of the Potomac mainly by Primary granites, gneiss, and slate, with a narrow band of Transition limestone; Virginia by Transition limestone overlaid in part by graywacke; Tennessee by graywacke overlaid at the Cumberland Mountains by Carboniferous limestone; Kentucky, Indiana, and Illinois by Carboniferous limestone; Missouri and Arkansas by Carboniferous limestone capped by calcareo-siliceous hills, and Arkansas, from a point about midway between the Fourche and Arkansas rivers, by graywacke capped by Old Red sandstone and occasional Tertiary deposits, with sub-Cretaceous beds beginning again at the Caddo River and extending nearly to the Red River in Texas, where they were covered by a ferruginous sandstone. It was a reprint of this section in the *Transactions of the Geological Society of Pennsylvania* that Lesley referred to in his *Historical Sketch of Geological Explorations* as "a rambling description and a worthless geological section across the continent from New York to Texas." Nevertheless the reviewer in the *American Journal* referred to it as worthy of consideration "on account of the wide range which it covers, the splendid features of the country and the scientific precision and perspicuity with which it is described."

19 Professor J. B. Woodworth, since the publication of my Contributions has called my attention to the following "cognate idea" expressed by Herodotus, Book II, chap. 11 (Rawlinson's translation): "Now if the Nile should choose to divert his waters from their present bed into this Arabian Gulf, what is there to hinder it from being filled up by the stream within, at most, twenty thousands years? For my part, I think it would be filled in half that time. How then should not a gulf, even of much greater size, have been filled up in the ages that passed before I was born, by a river that is at once so large and so given to working changes?"

20 See also p. 284.

In 1835 Featherstonhaugh, again under instructions from Lieutenant-Colonel Abert, made a geological reconnaissance of the region lying between the seat of government and the Coteau des Prairies, by way of Green Bay and the Wisconsin Territory. His report, issued in 1836, formed Senate Document No. 333, comprising 168 pages, with four plates of sections and diagrams. In this, as in his first, there is a large amount of preliminary matter of a very general nature. He noted that Washington and Georgetown were underlaid by gneiss, in which were perceived evidences of an "extensive anticlinal movement by which all the rocks along the entire length of the Potomac," as high up as the great bituminous field, had been affected, and that the true dip of the rocks was often "contradicted by the cleavage." This was a by no means unimportant observation. The erosive action of the river, as manifested by potholes in the schists at Great Falls, was dwelt upon and the Seneca sandstone and Potomac breccia were described, the former noted as often carrying anthracite and casts of calamites, but no suggestion as to the geological age of the beds was given. The Catoctin Mountains were described as "composed of primary slates, sandstones, and quartz having a northeast direction." Referring to the relationship existing between the Potomac breccia and the limestones, slates, and shales, he wrote:

> We thus have all the proofs that the Atlantic primary chain has come up from below through the limestone, triturating and breaking it up into fragments of every size, which were subsequently transported to the east of the chain by a current from the west, and deposited there, intermixed with the decomposed red shale.

He concluded, therefore, that this chain was elevated posterior to the deposition of the limestone, "which may be considered the equivalent of the lowest beds of Mr. Murchison's Silurian rocks." The unsymmetrical character of the folded sandstones and grits at Wills Creek, in Maryland, he described as affording evidence "that all the beds had been bent up by some action from below, and that from some inequality in the action or from some external cause the bed on which they lay, together with its associated strata, had collapsed toward the center in such a manner that they would appear to have been thrown up into a vertical position, if the uncurvated part had been concealed." The idea that mountain uplift might be due to a tangential rather than vertical movement had not as yet become prevalent.

In remarking on the constancy of the phenomena connected with the anticlinal arrangement of the whole series of Alleghany ridges, he ascribed their origin to an elevatory undulating movement, whereby "some parts of the strata were forced up into the anticlinal form, in a constant magnetic direction," the intervening distance between two axes or ridges being at the same time probably thrown down in a ruinous state. As the land arose and the waters retired the ruins would gradually be borne away, leaving the valleys as now existing.

The extensive bituminous coal beds of Maryland and Pennsylvania were described, the coal itself being considered due to plant growth *in situ*, and not to drifted material.

He thought to have found Carboniferous limestone near Navarino, Wisconsin, and also at the Falls of St. Anthony, in Minnesota. The lead-bearing beds of Dubuque, Iowa, were also judged, by their fossils, to be Carboniferous. In all this he was in error.

Rogers's Survey of New Jersey, 1835.

In 1835 there was organized under the authority of the legislature of New Jersey a state geological survey, of which Professor H. D. Rogers was made chief. The survey continued until 1839, the final report, a volume of some 300 pages and a colored geological map, bearing the date 1840, the first annual report being issued in 1836. In this final report Rogers argued that nearly the whole surface of the region occupied by the counties of Middlesex, Monmouth, Burlington, Gloucester, Salem, and Cape May was at some former period upon a level with the top of the surrounding hills, as shown by the finding of sandstone strata always at about the same elevation and in the horizontal position in which it was deposited. The hills he considered to be hills of denudation, that is to say, formed by the washing away or laying naked of the strata which formed the surface of the region about them; but, looking at the matter, as he did, through the eyes of a catastrophist, he was unable to say "from what quarter the mighty rush of waters proceeded which swept off so extensive a part of the upper rocks."

In view of the great difference of opinion which has existed and still exists in the minds of geologists with reference to the age of the white limestone near Franklin Furnace, New Jersey, it is not without interest to note that Rogers himself regarded it as having been originally the blue limestone of the district invaded at some period by mineral veins in a highly heated or molten state.

The formations of the northern portion of the state were divided into, first, a group of primary rocks confined to the islands and the vicinity of Trenton; second, a group of older secondary strata confined to the northwestern portions of Sussex and Warren counties, and third, a group of middle secondary strata lying in the broad belt of country between the southeastern foot of the Highlands and the boundary connecting Trenton and New Brunswick. With this third group he also connected the trap rocks of the region.

The great thickness throughout which this limestone had undergone crystallization, apparently from the heating agency of the dikes which traversed it, and the law which he traced in the development of some of the minerals, afforded, he felt, unquestionably strong support to the theory that gneiss and other primary strata had once been of sedimentary material converted by an intense and widespread igneous action into a universally crystalline condition. An important observation, though he was wrong in attributing the metamorphism to igneous action.

The presence of carbonate of copper diffused throughout the fissures of the shales indicated to his mind that a considerable portion of the metalliferous material, particularly the carbonate, had entered the strata in a gaseous or volatile condition and not in that of igneous fusion. The iron and zinc deposits of Sussex County, on the other hand, were "unequivocally genuine lodes or veins" filled with "matter injected while in a fused or molten state" and not beds formed contemporaneously with the surrounding rock. This is not in accordance with the present view, the Carbonate being universally recognized as secondary.

Views on the Gorge at Niagara, 1835.

It may be recalled that in 1820 Professor Bakewell of England visited the Falls of Niagara, and on his return published in the *London Magazine* a short memoir, in which he showed that the falls were once at Queenstown. In the autumn of 1846 he again visited the region and made the additional observation that the river at one time probably flowed through the ancient gravel-filled valley extending from the Whirlpool to St. Davids. In this, subsequent research has shown he was eminently correct. Nevertheless, his observations at the time were not wholly accepted, as is shown by the following.

In 1834 a Mr. Fairholme, writing in the London and Edinburgh *Philosophical Magazine* announced his agreement with Bakewell's views as expressed in 1820 and put forward at the same time certain of his own relating to the rate of back-cutting of the falls.

PLATE IX

H. D. ROGERS

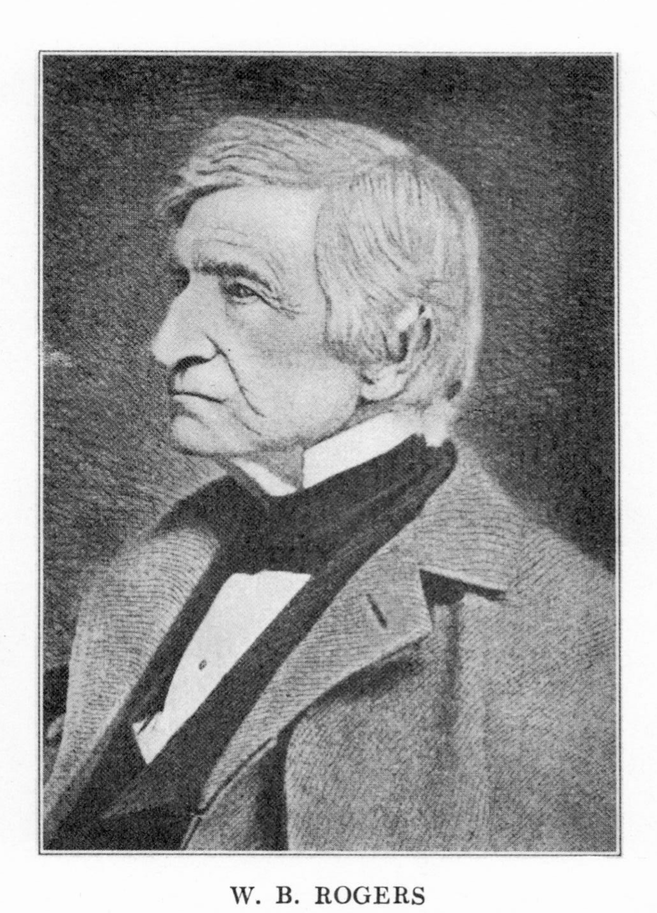

W. B. ROGERS

To both of these views H. D. Rogers, in the *American Journal of Science* for 1835, took serious exception. He believed that the channel below the falls had been formed in part as a diluvial valley and by some far-sweeping currents which denuded the entire surface of North America and strewed its plains and mountains with bowlders, gravel, and soil from the north. "The passage of such a body of water over the surface would deeply indent all the exposed portions of the land. Rushing in a descent from Lake Erie to Lake Ontario, from a higher to a lower plain, and across a slope like that at Queenstown, it would inevitably leave a long ravine."[21] Commenting on this, Silliman remarked in the same journal that an earthquake might possibly be instrumental in producing at once such a crack in the strata as would drain the lakes in a few days or hours, and to such an agency might be ascribed the channel in question.

Sketch of H. D. Rogers.

Rogers was born in Philadelphia in 1808, being one of four brothers,—James B., William B., Henry D., and Robert Empie,—all of whom rose to distinction as geologists or chemists. When not quite twenty-two years of age he became professor of chemistry and natural philosophy in Dickenson College, resigning in 1831 for the purpose of going to Europe to complete his studies. He returned to Philadelphia in 1833, and in the ensuing winter delivered courses of lectures on chemistry in the hall of the Franklin Institute.

In 1835 he was elected professor of geology and mineralogy in the University of Pennsylvania, and was also appointed by the legislature in the same year to make a geological survey of the state of New Jersey, as already mentioned. In 1836 he was made state geologist of Pennsylvania, and it is upon the work done in this connection that his reputation chiefly rests.

In 1857 he was made regius professor of natural history in the University of Glasgow, being the first American to thus receive a foreign appointment. As was the case with his brother, W. B. Rogers, he was a good lecturer, with quiet, gentlemanly bearing, never failing to make a favorable impression upon his audience, although by one of his Edinburgh pupils described as cold, impassive and self-contained, making few friends.

His work in Pennsylvania showed not merely administrative ability, but also the power of mastering the observations of his assist-

[21] Professor J. W. Gregory refers to this as Rogers's "blindest blunder." See also Hall's view, p. 236.

ants, and making therefrom important generalizations.[22] He was unquestionably the leading structural geologist of his time and was designated by the English geologist, J. W. Judd, foremost in the school of American orographic geology. That his generalizations and the theories he deducted will not in all cases hold today in no way reflects upon his ability. In judging his work, as, indeed, in judging that of any of his predecessors and contemporaries, one must take into consideration not merely the condition of knowledge at the time, but also the conditions under which they worked.[23]

J. G. Percival's Survey of Connecticut, 1835-1841.

One of the most unique figures in early American geology was that of James G. Percival. This man was born at Kensington, Connecticut, September 15, 1795, and graduated at Yale College in 1815, after which he studied medicine, receiving the degree of M.D. in 1820, and entering upon a troubled and, to his friends, troublesome career, which terminated only with his death in 1856.

He wrote poems, became editor of a newspaper, was a proof reader and assistant to Noah Webster in the preparation of his dictionary, and received a government appointment as surgeon at West Point and afterward with the recruiting service at Boston. But no form of practical work seemed suited to his taste, and he gave up position after position that he might devote himself to literature.

Peevish, often morbid and misanthropic to the point of insanity, and always complaining, truly such is queer material from which to make a geologist. "Slender of form, of narrow chest and with a peculiar stoop, a large, fine head, dark eyes, and inclined to sharpness of features; a wardrobe consisting of little more than a single plain suit—brown or gray—which he wore summer and winter until it became threadbare. He never wore gloves nor blacked his boots." Such is the picture held up to view by his biographer.

A geological and mineralogical survey of Connecticut being organized in 1835, Percival was given charge of the geology and C. U. Shepard the mineralogy. Shepard's report appeared in 1837. It comprised all told some 188 pages, but was not accompanied by a map, sections, or by figures of any kind.

[22] J. P. Lesley and others of his assistants found occasion to complain bitterly of his manner of "mastering" their work and giving no credit therefor (see p. 342).

[23] For summation of opinions regarding Rogers and his work, see address of Professor J. W. Gregory, before Glasgow Geol. Soc., Jan. 20, 1916.

Percival's report was long delayed, making its appearance finally in 1842. It would seem that the survey, when inaugurated, was expected to be but a superficial one, yet Percival was engaged upon it for five weary and laborious years, each year rendering his researches more minute, until he had collected over eight thousand specimens and made record of dips and bearings still more numerous. The legislators demanded a report, which was not forthcoming, and finally, in 1841, all appropriations were withheld and an abridged report published, much against Percival's wishes. This volume is beyond question one of the least readable of any issued by any state. A dry mass of lithological details, with little or no discrimination between important and unimportant matters—no theories nor generalities. No one for a moment will question that Percival had been, as he claimed, "laborious and diligent," yet the reader who searches from cover to cover for other facts than lithological characteristics, dip, and strike will search well-nigh in vain. The fact that the trappean outbursts were not in straight and continuous lines, but curvilinear in form and occupying a series of nearly parallel fissures, was practically the only break in the monotony of detail.[24]

He wrote to a friend:

I had twice surveyed the whole State on a regular plan of sections from east to west, reducing the intervals in the last survey to an average distance of 2 miles, thus passing along one side of each of the nearly 5,000 square miles of the State. . . . I had examined all objects of geological interest, particularly the rocks and their included minerals, with

[24] According to Dana, the work which he accomplished was, in the first place, an extended topographical survey of this portion of the state, and, secondly, a thorough examination of the structure and relations of the trap ridges, with also those of the associated sandstone. And it brought out, as its grand result, a system of general truths with regard to the fractures of the earth's crust, which, as geologists are beginning to see, are the very same that are fundamental in the constitution of mountain chains. For this combination of many approximately parallel lines of ranges in one system, the composite structure of the several ranges and the *en echelon,* or advancing and retreating arrangement of the successive ridges of a range, are common features of mountain chains. The earth's great mountains and the trap ranges of central New England are results of subterranean forces acting upon the earth's crust according to common laws. "The State of Connecticut, through the mind and labors of Percival, has contributed the best and fullest exemplification of the laws yet obtained, and thus prepared the way for a correct understanding of the great features of the globe."

It is told of Percival that one time when pursuing his vocation in the field he was held up by a countryman, who stated that he was a taxpayer and was, therefore, contributing towards his—Percival's—salary. He felt, therefore, that he had a right to know what was being accomplished by the survey. Percival regarded the man a moment, then putting his hand in his pocket withdrew it with a few cents which he handed over to the countryman with the remark, "Here is the amount of your contribution; I would rather remit than attempt to explain."

minute attention. I had scarcely passed a ledge or point of rock without particular examination.

Peevish and complaining, Percival evidently could not be made to understand why he should not be allowed to go on indefinitely, rendering a report when he himself should be satisfied of its correctness. With such a disposition the average country legislator naturally had no patience, and the abridged report, a volume of 495 pages, with an uncolored map, was finally printed in 1842, as already noted.

In 1853 Percival was employed by the American Mining Company in exploring the lead mines of Illinois and Wisconsin, and in 1854 was appointed state geologist of Wisconsin, as noted elsewhere. He died in 1856.

S. P. Hildreth's Work in Ohio, 1836.

In the *American Journal of Science* for 1836 S. P. Hildreth, later connected with the geological survey of Ohio, published what was, for its time, an important paper relative to the bituminous coal deposits of Ohio and the general geology of the Ohio Valley.

Although Hildreth was inclined to indulge in speculations founded upon scanty data, his paper is important for the numerous sections of the coal strata and as illustrating the condition of knowledge relative to both coal and petroleum.

He was one of the first to recognize the enormous amount of subaerial erosion that had taken place throughout the region, and that the Ohio River had carved out its own channel. He felt, however, that in times past the precipitation had been much greater than at present, and the abrasion of the surface by rain and torrents much more rapid. He thought to recognize in the Muskingum Valley Tertiary and Carboniferous rocks with New Red sandstone on the extreme southern border, and on the Clear Fork of the Little Muskingum the occurrence of a white limestone which was erroneously assigned to the Lias.

Considerable attention was given to the "muriatiferous" rocks, and he noted the outcropping of a five-foot bed of coal in the vicinity of the salt deposit as "evincing apparent design in Him who laid the foundations of the earth in the greater abundance of coal in those places where it would be the most useful," *i.e.*, in the evaporation of brine for making salt. The occurrence of petroleum springs on the Little Kanawha he thus describes:

By opening and loosening with a spade or sharpened stick, the gravel and sand . . . the oil rises to the surface of the water, with which the trench is partially filled. It is then skimmed off with a tin cup or some other suitable vessel, and put up in barrels for sale or domestic uses. In this way from 50 to 100 barrels are collected in a season, and much more could be gathered if demand required. In the adjacent hills is a thin bed of coal . . . but the source whence this petroleum flows must be deep in the earth, and the material which furnishes, vast in dimensions. The process is one of nature's hidden mysteries, carried on in her secret laboratory, far beyond the reach, and inaccessible to the curiosity of man.

The occurrence of gas springs was also noted, and one in the center of an open tract, "given to the public by the liberality of Washington," who "viewed it as an interesting natural phenomenon which no parsimonious individual ought ever to appropriate to his own benefit," is described in some detail and the same reverential spirit. "There appears to be no diminution in the amount of gas from its first discovery to the present time. The same Almighty and liberal hand, which furnished the perennial fountains with water, having also provided this gaseous spring with the means of an exhaustless supply."

His supposed discovery of the Liassic age of the white limestone, noted above, was later disputed by John Banister Gibson, chief justice of the supreme court of Pennsylvania, who claimed the credit for himself. Both were, however, in error, as shown by subsequent investigation.

Hildreth was a pioneer in science in this state. The contribution mentioned is one of his most important and probably the most important on the subject that had thus far appeared. He wrote also quite extensively on historical subjects and for some thirty-five years published a series of *Meteorological Registers.*

Mather's Geological Survey of Ohio, 1836.

In this same year there was organized the first geological survey of the state, and W. W. Mather appointed chief geologist. The survey lasted two years, seeming to have fallen through on account of local jealousies. Two annual reports were issued, of 134 and 286 pages, respectively, both bearing date of 1838.

Mather's assistants were the S. P. Hildreth just mentioned, who early resigned on account of ill health, J. P. Kirkland, John Locke, C. Briggs, Jr., J. W. Foster, and Charles Whittlesey, the last-named in charge of topographic work. Mather's own work was largely administrative and of an economic character, and he showed here as in his later work on the New York survey, a singular lack of discrimination as to the comparative value of the different subjects with which he had to deal.

In summing up the work of the assistant geologist, he traced the boundaries of the coal beds and estimated their thickness and probable yield. From the first season's work he had concluded that the state possessed sufficient coal "for every contemplated increase in population and manufactures for 2,500 years." At the end of the second season he had so far increased his already generous estimate as to state that "there is not only sufficient for domestic use for any reasonable time, but to supply the country around the lakes and throughout the valleys of the Ohio and Mississippi for as long a time as it is proper to calculate."

Hildreth's report must be read with a certain amount of allowance, since his ill health and consequent resignation precluded him from making certain possible modifications of his earlier statements. He described the occurrence of New Red sandstone, Lias, Oölite, etc., overlying the coal, and dwelt to a considerable extent upon the possible value of the buhrstone and also upon the salt springs.

Locke's report, comprising eighty-five pages in the second annual, was by far the most satisfactory, showing a much broader grasp of the general subject than that of any of his associates. He described the rocks below the coal formation as evidently having been deposited in the bed of a deep primitive ocean, though he failed to realize, or at least ignored, the value of the fossil remains in which the rocks abounded, his classification being wholly lithological. As pointed out by Orton, moreover, he failed to correlate the members of the series which he found with those of the same series elsewhere. He noted near the boundary of Ohio and Indiana a summit level and an anticlinal axis from which the strata dipped in opposite directions, eastwardly in Ohio and westwardly in Indiana. This is apparently the first recognition of what has since become known as the Cincinnati uplift.

He noted the immense amount of drift material and also the scratched and grooved surface of the underlying rock, describing it as "planished as if by the friction of some heavy body moving over it and marked by parallel grooves." He considered such to have been formed probably by icebergs floating over the terrace and dragging gravel and bowlders frozen into their lower surface. In this he followed the elder Hitchcock.

Briggs's reports, comprising some seventy pages of the first and second annuals, contained nothing of more than very local interest. A section was given showing the relative position and thickness of the strata in the counties of Wood, Crawford, Athens, Hocking, and Tuscarawas, but no attempt was made to refer them to any particular geological horizon, nor was the value of their fossils recognized. Indeed, there was but the briefest allusion to them excepting as curiosities. This seems the more remarkable, and in the case of Locke as well, when one considers that these regions are peculiarly prolific in invertebrate remains and the very important rôle they have been made to play since.

To Foster was assigned the geology of Muskingum County and parts of Licking and Franklin counties. He classified the rocks of the various formations of the region as Alluvium, Tertiary, Coal Measures, Fine-grained sandstone, Shale, and Mountain limestone, and noted the presence of remains of a mastodon and a casteroides in the alluvium.

Whittlesey's results consisted of, first, a brief pamphlet of some eight pages, relative to the general topography of the state, and second, a report of seventy-one pages, in which he gave geological sections of the rocks from Cleveland to the southeast corner of the Western Reserve, with miscellaneous notes on the order of the strata and on coal, limestone, iron ores, mineral springs, etc.

Sketch of Mather.

Mather was educated at West Point, and from 1829 to 1835 was assistant professor of chemistry, mineralogy, and geology in the Military Academy. While there he published (1833) a textbook on geology, a miniature sixteen-mo volume of some 139 pages, which seems to have been fairly well received at the time and is stated to have passed through several editions.[25]

He resigned his army commission in 1836 and gave up his time wholly to science. For a short time he was professor of chemistry and mineralogy in the University of Louisiana, but in 1836 became

[25] A copy of this book, in the possession of the present writer, has the following printed indorsement pasted on the inside of the front cover:

[Recommendation of Professor Silliman, of Yale College.]

DEAR SIR: You ask my opinion of the Elements of Geology, for the use of schools, by Wm. W. Mather. I think that it is a judicious, correct, and perspicuous work—containing, in a small compass, a solution of many of the most important facts and theoretical views in geology, and that it is well adapted to the object for which it was written.

Yours, respectfully,

B. SILLIMAN.

MR. WM. LESTER, JR.
NEW HAVEN, *June 18, 1834.*

connected with the Ohio state survey, and afterward with those of New York and Kentucky, as mentioned elsewhere.

His first geological paper seems to have been an "Illustration of a section through a part of Connecticut from Killingly to Haddam." This appeared in the *American Journal* in 1832. In 1834 he also published in the same journal a "Sketch of the Geology and Mineralogy of New London and Windham counties in Connecticut," the same being accompanied by a colored geological map.

He is described as a large and robust man, with great capacity for physical and mental labor—a man "equable in his disposition and gentle in his manners, considerate of others and just in his judgment of them, modest, but manly and self-reliant and thoroughly versed in the branches of science to which he devoted himself."

Cotting's Work in Georgia, 1836.

With a view of directing the attention of the legislature to the important subject of a geological and agricultural survey of the State of Georgia, the patriotic citizens of Burke and Richmond counties in 1835, at their own individual expense, directed Professor John R. Cotting

> to make a geological and agricultural survey of their respective counties. To examine all localities of limestone, marls, and all other minerals useful in agriculture and the arts. Also rocks that may be proper for the purpose of architecture, the construction of roads, railways, etc. To examine the water of springs and wells with regard to the salubrity or insalubrity of the same. To analyze the soils on different plantations in the two counties, with a view to their improvement. To illustrate the whole by drawings, diagrams, and a chart of the two counties, and to present a report of the same to his excellency the governor, in order that it may come properly before the two branches of the legislature should he deem the subject of sufficient importance.

In conformity with these instructions Cotting, in the latter part of March, 1836, began his survey, which was continued for eight months, the results appearing in the form of a duodecimo volume of 198 pages, unaccompanied by drawings or maps of any kind, since such could not be made within the state and the time assigned for publication would not permit of the originals being sent north.

The work was divided into three parts: (1) Topography, (2) Economic, and (3) Scientific geology. The ideas advanced were largely a reflection of those held by English and European authorities. The formations were divided into Tertiary, Secondary, Transition, and Primitive. Granite was recognized as an igneous rock, but it does not appear that its deep-seated origin and subsequent exposure by erosion were realized.

Many of these rocks [he wrote] occur not only together, forming a group or series, but are ejected in fragments through others and over incumbent strata to the highest series, assuming the appearance of having been once in a fused state. Thus fragments bearing all the features of having been fused or, at least, acted upon by fire or intense heat, occur scattered over the surface or imbedded in the strata of the Tertiary and Diluvial of this district, affording a demonstration that this region of country has been subjected to violent internal forces. The granite or syenite exhibits marks of some great force acting laterally and perpendicularly, which has rent the mass, heaved it up, and projected some of the fragments to a great distance.

Some interesting views on volcanic agencies were given in discussing the possible origin of the buhrstone, a chalcedonic rock carrying abundant casts of shells. The material as found occurred "only on the upper part of eminences and the edges of inverted cones." This he seemed to think offered a sufficient demonstration of its igneous origin, and he would account for its uplift on the theory of submarine explosions taking place simultaneously over a large extent of sea bottom. The vesicular character of the buhrstone he likewise considered indicative of its volcanic origin. "It is highly probable that the fused mass in that state was spread over the bottom of the then existing ocean and that these vesicles or holes were the effect of water converted into steam by the influence of the heat of the mass."

For the benefit of the general reader it may be well to state that the buhrstone is now known to be a product of purely chemical action, whereby the calcareous matter of a limestone is replaced by silica, and the vesicles, or holes, are largely cavities or casts of shell interiors.

Although no use was made of fossils in attempting to ascertain the relative age of the various rocks, he yet recognized the fact that they indicated a sedimentary origin of the beds in which they occurred. He wrote:

Geological investigation has led to the conclusion that there have been a number of deluges at different periods, or rather, that the oceanic waters have swept over the land, continued for an indefinite period, and then retreated several times, leaving their débris behind them. In no other place, perhaps, is the truth of the hypothesis better established than in this region, where fossil remains of different animals characterize different formations, as it is evident that these fossils must have been formed from animals who could not have existed under the earth, but on its surface and at the bottom of the then existing sea.

This well-meant attempt on the part of a few to arouse public interest proved only partially successful. The legislature of the following winter (December 7, 1836) adopted a resolution authorizing the governor to employ a suitable and well-qualified person to undertake the work and appropriated $10,000 to carry it on. Doctor Cotting[26] was thereupon appointed state geologist, but as the legislature of 1840 abolished the office nothing of value was accomplished.

Gesner's Work in Nova Scotia, 1836.

For but the third time in this history we are called upon to step beyond the limits of the United States, and for the second time upon the soil of Nova Scotia, a land later made famous through the labors of Logan and the elder Dawson. In 1836 there appeared a volume of 272 octavo pages, entitled *Remarks on the Geology and Mineralogy of Nova Scotia,* by Abraham Gesner. This had been preceded only by Jackson and Alger's work, and was for its time unquestionably a remarkable book. A geological map of the interior of the peninsula accompanied the volume. The southwestern border was colored as occupied by primary granites, gneiss, and mica-slate, this being succeeded by a wide belt (colored blue) of Transition slate, graywacke, and graywacke slate, and this, in its turn, by a broad band of sandstone (colored red) extending up as far as Westmoreland County, New Brunswick. The trappean area along the southwest shore of the Bay of Fundy was colored green. Important beds of iron ore were indicated in the Transition slate and of coal in the Red sandstone. These districts were described in considerable detail, and attention was called to the fact that the different formations corresponded in direction and general character with those of the United States.

Gesner's ideas regarding the uplifting of strata and the causes thereof were not at all in advance of his time. This is shown when, writing of the position of the slate, he said:

> The strata are variously inclined, and in some cases much twisted and broken; but generally they are so placed as to support the opinion that the primary rocks under their southern side have been uplifted by some violent and sudden movement which has thrown the neighboring slate in its present leaning, and often perpendicular, position.

The iron ores of the South Mountains he thought to be of aqueous origin, since they carried marine fossils. He was, however, unable to

[26] Erroneously stated as Dr. Little in my Contributions. G. P. M.

account for their presence. "From whence came these shells, and by what mighty convulsions and changes in this globe have their inmates been deprived of life and incarcerated in hard, compact, and unyielding rocks? By what momentous and violent catastrophe have they been forced from the bottom of the ocean, where they were evidently at some former period placed, to the height of several hundred feet above the level of the present sea, and even to the tops of the highest mountains?"

And further:

It is evident that the slate and ore containing the shells already mentioned were once at the bottom of an ancient sea. . . . By some mighty revolution, ground occupied by them has been uplifted and their native submarine possessions converted into slate, and even iron ore.

The magnetic character of some of this ore had been ascribed by Alger and Jackson to the influence of the intruded trappean rocks. To this Gesner did not agree, and called attention to the overlying position of the traps, indicating an extrusion at a later date than the formation of the New Red sandstone upon which they rest.

Notwithstanding the now seeming crudity of his notions as to the manner in which the fossil remains had become entombed, Gesner possessed very advanced ideas concerning their value for correlation purposes. This is shown in his discussion of the slate range extending from Yarmouth to the Gut of Canso. The fossils found therein, he felt, had an existence coeval with the original stratification.

They were inhabitants of the same age, enjoyed similar bounties, the same climate, and were companions at a period when the waters of the sea were as warm as those of the present tropical oceans; a fact easily proved by their organization and the beauty and delicacy of their shelly coverings. The corals, coraline sponges, and other vegetable productions of that period, although bearing a striking resemblance to those now flourishing in submarine situations, have nevertheless some peculiar characteristic features, distinguishing them from species of the same classes now inhabiting our shores, although their lineal descendants have long since passed away.

Gesner was an extreme catastrophist, and his ideas on the origin of the drift, as well as that of coal, were influenced more or less by the scriptural account of the flood. Discussing the fragments of slate and the masses of quartz rock and granite that were found scattered over the surface of the Red sandstone, and even entering into its composition at great depth, he argued that their shape demonstrated that they had been transported by the efforts of mighty currents. From this fact he conceived that similar causes had operated upon the surface of the earth at separate and distinct periods of time, one period having produced the ingredients of the newer rocks, which in their turn had been evidently denuded by the rapidity of overwhelming floods.

The giant bowlders, sometimes found on the very hilltops, he recognized as erratics, but could not believe them to be due to flood action. "They have doubtless been thrown upwards and left cresting the highest ridges, by volcanic explosions that have taken place since the general inundation of our planet." The general phenomena of the drift, however, he thought to have been almost certainly the effect "of an overwhelming deluge which at a former period produced those results now so manifest upon the earth. Not only hath the granite sent its heralds abroad, large blocks of trap are also scattered over the soil of Nova Scotia far from their original and former stations."

That coal is of organic origin he recognized, though as to the manner of its accumulation he was somewhat in doubt. He assumed that a part of it at least may have collected at the bottom of the sea, together with successive layers of sand and clay, and that the beds had since been uplifted by volcanic forces. The method of conversion of the organic matter into coal, he felt might also have been brought about through the intervention of volcanic forces.

In discussing other changes which have taken place on the surface of the earth, he again appealed to Genesis and queried if such might not

have been produced between that period when the globe was first created and the Noachian deluge, and might not many of those effects, the causes of which are now almost inexplicable, have been produced at that momentous period when the "windows of heaven were opened, and the fountains of the great deep broken up." . . . In no way can these phenomena be so satisfactorily accounted for and explained as by admitting the brief account of the creation of the world in the first chapter of Genesis; and that there is no necessity for making the world appear older than its date given by Moses.

Again:

The volcanic fires of the earth are gradually becoming extinct. They were evidently far more vehement in former ages than in the present day. Therefore, we have sufficient reasons to believe that from the creation of the world to the deluge great changes must have taken place upon the earth's surface. Who can clearly decide that the flaming sword which forever shut out our first parents from Eden's delightful garden was not a livid torrent of flame issuing from the ground polluted by sin?

Concerning the limestones of the coal formation along the Nepan River, he wrote:

On the surface of the limestone the detritus of the deluge forms a distinct covering; and, according to the opinion of some geologists, should not be considered in any way connected with the changes which have taken place in the strata beneath. But we would remark that although the beds of rounded pebbles and sand clearly demonstrate the effects of a flood, they can have no reference to the great geological catastrophe which ushered in that awful event. The depression of whole continents, the raising of the ocean's level bed, the distortion of strata previously horizontal, the elevation of mountains, and all those violent operations whereby the whole surface of this planet has been rent asunder, might have been the prelude to that overwhelming deluge, while the diluvial débris resulted from the action of torrents after the crust of the globe had been thus broken up.

Gesner's Work in New Brunswick, 1838-1843.

Two years later, *i.e.*, in 1838, Gesner was appointed provincial geologist of New Brunswick, a position he continued to hold until 1843, making during this time five brief reports, comprising altogether 440 pages. A partial geological map was prepared, but never published. A considerable portion of the reports was given up to a discussion of economic matters, particularly the coal beds, the value of which, as noted later, he overestimated. The first four reports dealt largely with that portion of the province south of the Miramichi, Nashwaak, and St. John rivers. The fifth and last (1843) described the country of the upper St. John and that between the headwaters of the Tobique and Ristigouche rivers.

A gradual advance is to be noted, particularly where Gesner in his second report subdivided his graywacke, which he thought comparable with Murchison's Silurian, into upper and lower members, of which the upper contained a fossil mollusk which he identified as a terebratulite, but which G. F. Matthew thinks to have been probably a somewhat poorly preserved *Orthis billingsi.* He also reported the finding of remains of conifers, calamites, and a *cactus* (?). The last Matthew thinks to have been some form imitative of a stigmaria. Though the full significance of these fossil remains was not appreciated by Gesner, yet he must be regarded as a pioneer in making known the flora of both the Cambrian and Carboniferous beds in New Brunswick.

In his fourth report he classed the graywacke with Sedgwick's Cambrian. About this time (1842) he seems to have become acquainted with Agassiz's glacial theories and referred to the expressed opinions of Lyell and Buckland with reference thereto, but thought that "the general occurrence of diluvial grooves and scratches upon the surfaces of the rocks . . . can not in general be explained unless by admitting that a great current of water has passed over the country from the north toward the south," though "it appears that ancient glaciers have had a powerful influence in the accumulation of those parallel mounds of erratic sand and gravel observed in different parts of these provinces, and likewise they may have been the means whereby bowlders of granite and other rocks have been scattered over the country."

Fig. 31. *Abraham Gesner.*

In his fifth report was given a general description of the coal fields of the province, which he estimated to embrace an area of 8,700 square miles, or nearly one-third the entire area of New Brunswick, making it one of the largest in the world. Through the influence of these reports much interest was excited in the public mind concerning the mineral wealth of the province, and a large number of mining adventures were undertaken with the expectation of immediate and favorable returns. Few of these were, however, successful, and the reaction in public feeling tended unjustly to throw discredit on Gesner's work and probably was influential in terminating his engagement, the survey being brought to an abrupt conclusion in 1843.

Sketch of Gesner.

Gesner was born of German-French ancestry in Cornwallis, Nova Scotia, and was by profession a physician and surgeon. For a long time after obtaining his degree he practiced his profession in the country bordering the shores of Minas Basin, making his visits on horseback, and often returning with saddlebags filled with specimens collected on the way. In 1838, two years after the publication just referred to, he was appointed provincial geologist of New Brunswick, and removed to St. John, where he established the Gesner Museum, afterward purchased by the Natural History Society of New Brunswick. This was an all-round natural history collection, many of the zoölogical specimens having been collected by Gesner himself, who was an ardent sportsman.

After the somewhat premature closing of the survey, Gesner left

St. John and returned to Cornwallis. During his residence here he wrote his works on *New Brunswick, with Notes for Emigrants*, and *The Industrial Resources of Nova Scotia*. Besides his medical practice and the scientific and literary work already mentioned, Gesner engaged in studies in applied science, particularly in electricity and chemistry, and in 1854 took out a patent in the United States for the manufacture of kerosene from coal and other bituminous substances. About this period he resided in Brooklyn, New York, and was connected with the company having works on Newton Creek near Perry Bridge. In 1861 he published a work on coal, petroleum, and other distilled oils. He returned to Halifax in 1863, where he died the following year.

He is represented as a man of medium height, with deep chest and square shoulders, black hair and eyes, and a face showing deep thought and strong intellectual power. He was unquestionably a man of great mental and physical energy and was noted for his deep piety, remaining all his life a firm churchman. It is told of him that, when connected with Guy's Hospital in London, if troubled by any physiological question which had become a subject of speculation, he always gave as his ultimate conclusion that "God made it so." This phrase became known as "Gesner's reason" and was so used among the students.

Shepard on Geology of Upper Illinois, 1838.

Fig. 32. *C. U. Shepard.*

Although C. U. Shepard's scientific work was mainly mineralogical, he occasionally contributed papers on areal geology. In the *American Journal of Science* for 1838, we find an article by him on the geology of upper Illinois. In this he described the extensive Kankakee swamp region south of Chicago, and the old beach lines above the present lake level. He also discussed the possible overflow of the lake in ancient periods to the southward, and dwelt to a considerable extent on the possibility of and the commercial advantages to be derived by uniting the waters of Lake Michigan and the Illinois River, forming thus a connection with the Mississippi.

The geology of the region was described as being exceedingly simple and uniform, the great rock formation being the magnesian limestone. The northern boundary of the coal formation he said he could not define with precision through lack of opportunity to explore it in detail. A geological section of the region between Fox River and Spring Creek accompanied the paper.

Opinions of the Reverend Samuel Parker, 1838.

Among the numerous laymen who strayed into the geological field mention should be made of the Reverend Samuel Parker, who made during the years 1835-1837 an exploring tour beyond the Rocky Mountains with the view, more particularly, of ascertaining "by personal observation the condition and character of the Indian nations and tribes and the facilities for introducing the gospel and civilization among them." Chapter XVI of the *Journal* as published is given wholly to geological considerations, and there are numerous brief references to the subject scattered elsewhere throughout the book. "The whole country, from the Rocky Mountains on the east and Pacific Ocean on the west, and from Queen Charlotte's Island on the north to California on the south, presents one vast scene of igneous or volcanic formation. Internal fires appear to have reduced almost all the regular rock formations to a state of fusion, and then through fissures and chasms of the earth to have forced the substances which constitute the present volcanic form." Whether intentional or otherwise the possibility of fissure eruptions as distinct from those of crater origin is here clearly recognized. As to the time when the volcanic fires were in operation his judgment was naturally somewhat warped, but he thought it to be nearly contemporaneous with the flood. "If these immense volcanic fires . . . had existed before the flood, and ceased their operations . . . before or at the time of the flood why do we not find here as well as in the higher regions of the north and in most other parts of the world, fossil organic remains?" He then goes on to discuss the occurrence of mammoth and other organic remains as reported from Siberia and their possible origin and comes back to the then prevalent belief that they were drifted there from the warm regions of the South. "No operations of existing seas or rivers account for such phenomena without the special interposition of divine agency. But when we refer them to one and the same cause [*i.e.*, Noah's flood] we find them rationally accounted for, without assuming that which is without proof. . . . Those who deny the inundation of the world by the deluge may be fairly called upon to account [otherwise] for the extermination of these animals and the condition in which they are found." The deep channel of the Columbia River afforded him, as did like phenomena in Massachusetts afford his contemporary, Hitchcock, food for speculation. "Undoubtedly the flowing waters have worn the rocks very considerably, and have produced changes, but upon no principles can it be supposed that they have produced so long and so deep a channel as the one through which the Columbia flows, and through such solid rock formation." In course of a discussion with several "literary gentlemen" of possible theories of origin it was suggested by one "that God had with his finger drawn a channel for the Columbia," and this was agreed upon as being the most rational explanation! Presumably no comment on this is necessary.

Mather's Reconnaissance of Kentucky, 1838.

In accordance with a resolution of the general assembly, approved February 16, 1838, W. W. Mather made a reconnaissance survey of Kentucky, devoting his attention mainly to the coal and iron ore resources of the state. His report of forty octavo pages appeared in 1839 and concluded with remarks on the beneficial results of a geological survey and outlining a plan for its continuance, a matter evidently contemplated in the original resolution. No action was taken, however, and the subject lay dormant until 1853 (see p. 321).

W. B. Rogers's Work in Virginia, 1835-1841.

On March 6, 1835, the general assembly of Virginia authorized the appointment of a "suitable person" to make a geological reconnaissance of the state with a view to its general geological features and to the chemical composition of its soils, rocks, ores, mineral waters, etc. Professor W. B. Rogers, then professor of natural history, philosophy, and chemistry in William and Mary College, Williamsburg, Virginia, and brother of H. D. Rogers, was selected for the position and entered upon his duties in 1835. He was assisted from time to time by George W. Boyd, Caleb Briggs, E. A. Aiken, C. B. Hayden, Samuel Lewis, J. B. Rogers, H. D. Rogers, R. E. Rogers, Thomas S. Ridgeway, and M. Wells. The survey continued in existence until the close of 1842, the act of authorization being then repealed by the legislature.

No provision was made for the preparation of a final report, although Rogers was ready to undertake the same, and the idea was not finally abandoned until as late as 1854. The necessary appropriation was not granted, however, and the seven brief annual reports submitted are all there is to show for years of careful and patient work under most adverse circumstances. No map and no sections were published at the time. In 1884, after Rogers's death, a reprint of all these reports together with sundry other of Rogers's papers was made under the editorship of Major Jed. Hotchkiss, a civil engineer, of Staunton, Virginia. They formed an octavo volume of upward of 500 pages. This was accompanied by a small geological map which Rogers had previously colored for Hotchkiss, and by numerous plates of sections.

Under the conditions enumerated above, it is not surprising that the reports contained little of more than local interest, and that the broader aspects of geology were barely touched upon. The region west of the Blue Ridge was described as occupied by fourteen groups of strata, which were designated by numbers, beginning with the lowermost. These all showed so general a conformity in superposition, and so remarkable a correspondence in their mineralogical and physical characters, as to clearly indicate, he felt, the propriety of regarding them parts of one great series accumulated over the widely expanded floor of the ancient ocean.

Little mention was made of the fossil contents of any of these beds, nor attempts at correlation or determination of their geological age. This seems the more remarkable when one considers the value attached to fossils by Rogers in his work on the Tertiary formations, where the inspection of a single shell, he claimed, would often enable one to pronounce upon the character of the stratum from which it was taken—that is, whether pertaining to the Eocene or Miocene.

The Massanutten Mountains were recognized as forming one great synclinal tract resting in a trough of slate. It is worthy of note, in view of the Rogers-Hitchcock controversy referred to elsewhere (p. 150), that in his report for 1838 Rogers described in considerable detail "the extraordinary phenomenon of inversion" presented throughout the long reach of North Mountains, whereby the entire series from formations II to VIII, inclusive, were made to lie conformably one upon the other, but in a reversed stratigraphical position; that is, the oldest above and the youngest at the bottom. But no attempt was made to account for the same.

In the report for 1839 the Tertiary marl south of the James River was described in detail and its Eocene and Miocene subdivisions recognized. In that for 1840 was given a list of the fossils found in the Miocene marls, and attention called to the discovery of a "remarkable stratum varying from 12 to 25 feet in thickness, composed almost entirely of microscopic fossils (Diatoms), and lying between the Eocene and Miocene, but referred to neither." This, the first discovery of its kind in the United States, was referred to as an "infusory stratum."[27] The coarse "Middle Secondary" conglomerates of Virginia, corresponding to the Potomac marble (Triassic) of Maryland, he thought to have been deposited by strong currents coming from the southeast, which laid down their load in a long, narrow trough extending from southwest to the northeast; such "being deposited in successive layers, commencing at its southeastern margin, would naturally assume the attitude of strata dipping toward the northwest."

When one considers the condition of the country at the time Rogers did his work—the lack of facilities for transportation, the entire lack of maps sufficiently accurate for purposes of plotting, the deep mantle of residuary material that nearly everywhere obscured the more solid rocks, and that there were no railroad cuts or other artificial exposures, such as exist today—one can but admire its accuracy. It is not yet too late to express regrets that the parsimony of the legislature should have stood in the way of the preparation of a final report.

Sketch of W. B. Rogers.

The first manifestations of a geological trend of mind on the part of Rogers appeared in 1833, when he set on foot inquiries relative to the greensand marl of Virginia. His first publication, as given in his biography, however, related to artesian wells. This appeared in the *Farmer's Register* of Richmond (1834-1835).

The poverty of his resources in 1835, while hopefully agitating the establishment of a survey for Virginia, cannot be better illustrated than by noting the request sent his brother Henry, in Philadelphia, for chemicals and apparatus to be used in the prosecution of his scientific work. The list comprised one platinum capsule, one pound of absolute alcohol (French), one-half ounce oxide of ammonia, one-half pound distilled muriatic acid (pure), one pound distilled nitric acid, one four-ounce vial of phosphate of ammonia, one foot small platinum wire for blowpipe.

He wrote:

My alcohol, with all the economy I have used, is almost exhausted. The

[27] In 1843 Tuomey announced the discovery of an "infusorial" stratum at Petersburg, in Virginia. This he referred to the lower portion of the Miocene.

gill which I had at the opening of the course has been used at least ten times in analysis, and though carefully distilled off in each operation, a portion is, of course, lost.

In August of 1835 Rogers was elected to the chair of natural philosophy in the University of Virginia, at Charlottesville, a position which he gladly accepted, being influenced, in part at least, by the more healthful climate of the place.

His first years here were, however, full of trial, owing to the dual nature of his duties—teacher and state geologist—and the lawless character of many of the students. Lacking, as he was, in physical stamina at the beginning, the trials as professor and the lack of appreciation of his work by the state legislature undoubtedly wore upon him severely and had to do with his comparatively early breaking down.

As early as 1846, in connection with his brother Henry, he formulated plans for the Polytechnic School in Boston, which place he felt persuaded was on all accounts the best suited for an institution of this kind. "I long for an atmosphere of more stimulating power," he wrote from Charlottesville, and with these thoughts in mind he resigned his professorship in 1848, but was induced to reconsider and remain for five years longer. On his final removal to Boston in 1853 he continued to take an active part in scientific and educational matters. The year 1859 found him again advocating his plan for an institute of technology, the matter being brought forward at this time in connection with the so-called Back Bay lands and their rapidly enhancing value, and in 1861 an act to incorporate the Massachusetts Institute of Technology passed the legislature and was approved by the governor. In 1862, an organization having been perfected, he was elected its first president, an office he continued to hold until forced to resign by ill health in 1870. In 1878, his health having improved, he was again induced to temporarily accept the presidency, holding the position three years, and being succeeded by General F. A. Walker.

Like other of the broader men of his time, Rogers was an all-round scientist, and wrote not merely on geology, but made observations on the aurora, experiments with reference to binocular visions, and other subjects.

His reputation as a geologist rests mainly upon his work in Virginia and that on the structure of the Appalachian Chain in connection with his brother Henry. He was, however, widely known as an educator and orator. To quote from one of his biographers:

The wide extent of his own studies and researches in mechanics, physics, chemistry and geology; his truly philosophical spirit, his unfailing courtesy and urbanity, his warm sympathies, his scientific enthusiasm, his commanding and stately presence, his rare gifts of expression, all combined to make him the ideal presiding officer. His introductions were most felicitous; his comments highly suggestive and inspiring; his summing up was always a masterpiece of discriminating and judicious reasoning, while, over all, his rich, tropical eloquence threw a spell as of poesy and romance, for to him the truth was always beautiful, and the most solid and substantial structure of scientific principle stood in his view against a sunset sky, radiant with a light which no painter's pencil ever had the art to fix to canvas.

Geological Survey of New York, 1836.

On April 15, 1836, there was authorized by the New York Assembly[28] a geological and mineralogical survey of that state. This led to an organization which has left a more lasting impression upon American geology than any that has followed or had preceded it. As fate ordained, the locality was one of the most favorable that could have been selected for working out the fundamental principles of stratigraphic geology; moreover, those appointed to do the work proved equal to the occasion. The New York survey gave to American geology a nomenclature largely its own; it demonstrated above everything else the value of fossils for purposes of correlation, and incidentally it brought into prominence one man, James Hall, who was destined to become America's greatest paleontologist.

To secure the greatest amount of individual freedom and to facilitate the work to best advantage, the state was divided into four districts (see p. 224), W. W. Mather being placed in charge of the first, Ebenezer Emmons the second, Timothy A. Conrad the third, and Lardner Vanuxem the fourth. This arrangement is stated[29] to have been made by Dr. Edward Hitchcock, who was first given charge of the survey of the first district but who early resigned for the purpose of undertaking a resurvey of Massachusetts. Nothing more remains to show what Hitchcock actually accomplished. C. B.

28 Agitation in favor of a state geological survey of New York began as early as 1827, but it was not until 1835 that the matter took definite shape. It was during these intervening years of agitation that a "well and favorably known citizen" proposed the following plan for carrying it out: "The geologist would travel his routes, having sent forward his appointment to be at certain places at set times, like the Methodist circuit rider. The people would come bringing their specimens. These he would label, pack and send to Albany for study." The work it was thought might thus be done in a short time and at slight expense!

29 Reminiscences of Amherst College, by Edward Hitchcock, 1863.

Adams was assistant geologist, but apparently resigned at the same time as did his superior officer. The survey continued in existence for five years, reports being issued annually in the form of assembly documents, the final reports appearing in 1842-1843 in the form of quarto volumes, comprising, all told, over 2,000 pages and 82 plates, sections, and maps.

The mineralogical and chemical work of the survey was placed under the charge of Dr. Lewis C. Beck, while Dr. John Torrey was made botanist and Dr. James E. De Kay given charge of the zoölogical department. At the end of the first season Conrad resigned, to become paleontologist of the survey, and James Hall, who had previously been an assistant to Emmons, was put in charge of the fourth district, while Vanuxem was transferred to the second. As with all the earlier organizations of its kind, agriculture and mining were considered subjects of primary importance. The results accomplished can, however, be best considered under the date of issue of the final reports (p. 224).

Geological Survey of Pennsylvania, 1836.

This same year witnessed also the establishment of a geological survey in the adjoining state of Pennsylvania, of which H. D. Rogers, at the time state geologist of New Jersey, was placed in charge. This, as planned, was expected to continue for a period of at least ten years, but was brought to an abrupt close in 1842, owing to the financial embarrassments of the commonwealth. Rogers, however, unwilling to relinquish the work in its incomplete form, continued on his own responsibility, and it is stated largely at his own expense, for three years longer, and in 1847 deposited in the office of the secretary of the commonwealth his final report, ready for publication. For reasons to be noted later, this was delayed until 1858.

During its period of existence there were issued six annual reports, the first bearing the date of 1836 and the last 1842. These were small octavos, destitute of illustration with the exception of a few outline sections, and of from 100 to 250 pages each. The results of the work were so completely elaborated in the final reports issued in 1858 that these preliminary publications have been almost completely lost sight of and are of merely historical interest. It is well to note that, owing to the large proportion of foreign population in the state at that date, two editions of the preliminary reports were issued—one in English and one in German. This dual publication was paralleled in the case of New Jersey, Ohio under Newberry, and Wisconsin under Percival.

PLATE X

TIMOTHY CONRAD

EBENEZER EMMONS

W. W. MATHER

LARDNER VANUXEM

As with the New York survey, the results of Rogers's work can be best summed up in a consideration of the final reports (see p. 375).

Wilkes Exploring Expedition, 1836.

After nearly ten years of agitation, there was passed by both houses of Congress, and approved by the President on May 14, 1836, a bill providing for an exploring expedition to the South Polar regions and the islands and coasts of the Pacific. This expedition, which finally sailed from Norfolk on August 17, 1838, and, from the name of its commander in chief, has come to be known as the Wilkes Exploring Expedition, had for its immediate object the increase of such knowledge as would be of interest and value to the whaling industry, which had suffered severely through loss of men and vessels in these remote regions. Research along other lines was purely secondary. Fortunately the scientific men of the day were fully alive to the possibilities offered, and, through their influence, an efficient corps of trained observers in various lines was permitted to accompany it. Of these only the geologist, J. D. Dana, who sailed on the *Peacock*, comes within the range of our present work, though incidental reference will be made to J. P. Couthouy, who sailed on the *Vincennes*. Inasmuch, however, as no tangible results from the expedition were made known until after the return of Dana, in 1841, the entire matter may also be dismissed here and taken up again under the latter date (see p. 286).

Jackson's Geology of Maine, 1836-1839.

In March, 1836, sixteen years after her establishment on a basis of independent statehood, the legislature of Maine authorized in stereotyped form the appointment of "some suitable person" to make a geological survey of the state "as soon as circumstances will permit," appropriating the sum of $5,000 to cover expenses. The legislature of Massachusetts authorized cooperation so far as it related to certain public lands which were the joint property of the two states. On June 25 a contract was entered into with Dr. C. T. Jackson for the carrying out of the same.

Jackson seems to have entered upon his duties promptly and energetically, making his first report, a volume of 116 octavo pages, in December, 1836. This was accompanied by an atlas of twenty-four plates. His results were apparently satisfactory to the legislature, the appointment being renewed the following year and $3,000 appropriated for expenses. Under this appropriation a second annual report of 168 octavo pages was forthcoming. This in its turn was seemingly satisfactory, for the geologist's salary was increased from $1,000 to $1,500 a year, while the sum appropriated for the carrying on of the work was made discretionary with the governor and council. Under such favoring conditions it is not strange perhaps that the state geologist became effusive and brought his third (and last) report up to 276 pages, with an appendix of sixty-four pages, containing a catalogue of the collections.

These reports, examined in the light of today, contain little which would be considered of geological importance. Jackson seems to have roamed somewhat at random over the state, with no apparent idea of the geological structure as a whole, and to have contented himself with making detailed notes on whatever was immediately at hand, regardless of its possible relationship to other formations at a distance, and with an eye particularly to economic questions.

It must be remembered, however, that the country was at that time largely a roadless, forested wilderness, and was, indeed is still, covered by lakes, swamps, and glacial detritus to an extent almost unknown and certainly unrealized outside of a few of the most northern states.

From the finding of marine shells at Lubec in layers of clay now some twenty-six feet above high-water mark, he rightly conjectured that the land had been elevated that amount within a comparatively recent period. He regarded the Devonian sandstone of the extreme eastern counties identical with the red Triassic sandstones of Nova Scotia, and seriously discussed the possibility of their containing bituminous coal. The slates of Piscataquis County were classed as Transition, and the possibility of their carrying anthracite coal was likewise discussed.

In his third annual report he noted the occurrence of fossil plants in the slates of Waterville and remarked it as a strange occurrence, since the rocks belong to the older Transition series.[30]

The probability of the occurrence of beds of coal at Small Point Harbor near Phippsburg was investigated. As a result it was announced that there was no possibility of such being included in rock of the nature there found, and that the coal sometimes thrown up by waves on the beach was presumably from English sources. So early a statement of this nature is interesting in view of borings for

[30] At the second session of the American Association of Geologists in Philadelphia, 1841, however, Professor O. P. Hubbard exhibited a specimen of this slate and was able to show by the aid of "Murchison's Silurian system" that they were not plant remains but annelid trails, a view which Jackson accepted.

coal which have taken place along this coast within a very few years.[31]

Jackson's views on the glacial deposits were naturally crude. The "horsebacks" (ridges of glacial gravel) were thought to be diluvial material transported by a mighty current of water. "It is supposed that this rushing of water over the land took place during the last grand deluge, accounts of which have been handed down by tradition and are preserved in the archives of all people. Although it is commonly supposed that the deluge was intended solely for the punishment of the corrupt antediluvians, it is not improbable that the descendants of Noah reap many advantages from its influence, since the various soils underwent modifications and admixtures which render them better adapted for the wants of man. May not the hand of Benevolence be seen working even amid the waters of the deluge?"

It is, perhaps, doubtful if the hard-fisted occupants of many of Maine's rocky farms would be disposed to take so cheerful a view of the matter![32]

However the work may impress the reader of today, it was considered by a reviewer in the *American Journal of Science* for 1837 as a "model of its kind. It has certainly not been surpassed by any similar effort in this country," and "The present sketch of Maine is a masterly production."

[31] "We ourselves happen to know of one instance, in which some persons in Maine, believing that they had discovered indications of a coal mine on the Kennebec River, actually sent a quantity of black tourmaline to Boston, which was exhibited by one of the principal coal dealers there as anthracite coal, and in a few days all the necessary implements for boring into a solid ledge of *granite* were prepared and sent to the spot. The exploration was not abandoned until the sum of two thousand dollars had been expended; and we believe the history of mining in this country does not afford an instance of more blind and determined disregard of the principles of science than this. Not many months afterwards, a person, probably on his own responsibility, visited Boston to obtain subscribers to stock in a new mining company, and brought with him specimens of gneiss and mica slate, in which he declared he found a bed of bituminous coal near the mouth of the Kennebec River. One or two persons known to the writer, became subscribers to the stock; and it was not until after one of them took occasion to visit the spot, that the gross fraud was detected. Pieces of coal were found there, but they came from Newcastle." (Am. Jour. Sci., Vol. XXXVI, 1839, p. 150.)

[32] New England would be desolate, if it were not for the diluvial débris scattered between the barren summits of the hills, which represents a soil susceptible of cultivation. The power of a deluge, capable of producing such a beneficial result, may be estimated by the force which has torn these masses of rock from their original position, and scattered them in profusion over the soil. They are monuments of Almighty Power. (Travels in the U. S. A. by I. Finch, 1833, p. 160.)

Jackson's Survey of Rhode Island, 1839.

In 1839 the general assembly of Rhode Island appropriated the sum of $2,000 to pay the expenses of a geological and agricultural survey and Jackson, fresh from his work in Maine, was placed in charge of this also. This, though lasting but a year, constituted the first, last, and only survey of Rhode Island carried on under state auspices.

Such a work was naturally productive of little of importance, and no new principles whatever were evolved. Aside from the gathering of a few facts of possible economic value, it resulted only in an extension of knowledge relative to the distribution of certain geologic groups. This, however, was a feature of nearly all the work carried on by Jackson. As further illustrating the condition of geological knowledge, the report is, however, worthy of consideration.

In his introduction he remarked on the attempt on the part of some geologists to abandon the name Transition and to group these rocks with the Secondary, according to the original schemes of the German geologist Lehman, and felt that a numerical division would doubtless be found preferable to any of the fanciful names proposed for some of its subdivisions. The names Cambrian and Silurian, as proposed in England, he thought, "will never be regarded in this country as appropriate terms for our rocks," yet they stand, today. He recognized the existence of contact phenomena, and also that the degree of crystallization and general structural features of igneous rocks are dependent upon conditions of cooling, though he considered pumice as formed when the fused rock, under little pressure, was brought in contact with water. The possibility of the cellular structure being due to expansion of vapor of water in the lava itself was not recognized. The hornblende rock, so extensively developed in Cumberland, Smithfield, and Johnston, was considered of igneous origin, and the suggestion made that its apparent stratiform structure might be due to an admixture of the argillaceous slate through which it was elevated.

The eruptive nature of the Cumberland titaniferous iron ore was also recognized, though naturally the fact that the rock was an iron-rich peridotite partially altered into serpentine, as later described by Wadsworth, was overlooked.

The origin of the drift was to him still obscure. He did not accept the theory of drifting icebergs, "nor can we allow that any glaciers could have produced them by their loads of sliding rocks, for in that case they should radiate from the mountains instead of following a uniform course along hillsides and through valleys." (See further, p. 622.)

The report was accompanied by six plates of Coal-Measure fossils and a geological map, and four cross sections.

Perhaps the most interesting and important part of the report from both economic and stratigraphic standpoints is that shown in his section extending from the state line on the east, northwesterly through Portsmouth to Cranston. Butts Hill, with its underlying coal beds, is shown occupying the center of a broad syncline, the coal appearing on both sides of the hill, near the water level. Concerning this Shaler writes: "It is, so far as I have learned, the first piece of work in which a great syncline was adequately determined from only limited outcrops, for the most of the area is covered by the waters of the great inlet known as Narragansett Bay. The constructive imagination . . . was clear in all his geological work which I have traced in the field."

Concerning the economic value of the Rhode Island coal for fuel purposes, Jackson seems to have had little doubt and he estimated that the beds could be made to yield 37,800,000 tons. He assured his readers, incidentally, that they must not place any value on opinions formerly held as to the difficulty of burning it, for the people had not at the time these views were expressed learned how to burn anthracite. It may be well to add that the Rhode Island coal, as well as that in the adjoining state, has never proven of value for fuel purposes, owing to the high degree of metamorphism to which it has been subjected and the consequent high initial temperature essential to its combustion.

Booth's Survey of Delaware, 1837-1838.

Geological work in Delaware at the expense of the state began and ended with the survey by J. C. Booth during 1837-1838.

A unique feature of the act establishing this survey was the requirement that an equal portion of the appropriation should be expended in each county, regardless of conditions. The clause was presumably inserted to allay local jealousies, but the absurdity of the same is, nevertheless, so great as to leave no occasion for comment. In the report, which appeared in the form of an octavo volume of 188 pages in 1841, the geological formations were divided into (1) *Primary*, (2) *Upper Secondary*, (3) *Tertiary*, and (4) *Recent*. The geographical distribution and lithological character of each was given in detail. The Primary included gneiss, feldspathean rocks (traps), limestone, serpentine, and granite; the Upper Secondary, the Red Clay Formation and the Green Sand Formation. The Tertiary was divided into the northern Tertiary, the southern Tertiary, the yellow clay formation of Appoquinimink Hundred, and the intermediate clays and sands. The Recent Formations were divided into the lower clays, the upper sands, and the river deposits. Attention 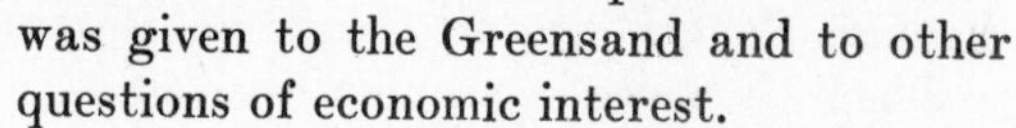was given to the Greensand and to other questions of economic interest.

Fig. 33. *J. C. Booth.*

The report as a whole represents the patient attempt of a man who, though thoroughly capable, was unversed in the broader problems of geology, who wrote down only what he saw and thought he understood. His remarks on the weathering of rocks are perhaps the most striking. The entire cost of this survey to the state was $3,000, of which sum Booth received $2,000 for his two years' work, the remainder going to defray cost of publication of the report and various incidentals.

Booth was primarily a chemist. His geological experience at the time of his taking charge of the Delaware survey amounted to but a year as assistant to Rogers in Pennsylvania. With the close of the work in Delaware he abandoned geology altogether and in 1849 became melter and refiner of the Philadelphia mint, holding the position until 1887.

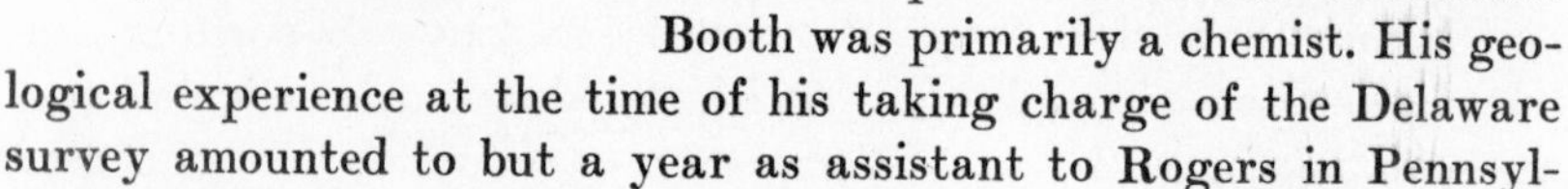

D. D. Owen's Survey of Indiana, 1837-1839.

In 1837 also there was organized the first state geological survey of Indiana, D. D. Owen, who had served as an assistant on the survey of Tennessee under Gerard Troost, being appointed state geologist. The life of the survey was limited to two years, and two reports were rendered, one of thirty-four pages, bearing date of 1838, and one of fifty-four pages, bearing date of 1839.[33]

[33] Reprints of these appeared under date of 1859. There are numerous references in the reports of 1838 and 1839 to sections and a geological map. In the reprint of 1859 entitled Continuation of Report of the Geological Reconnaissance of the State of Indiana, where reference is again made to this map, occurs the following footnote: "The original geological map here referred to was deposited in the state library but has not been published."

The essential similarity of the formations of Indiana with those of Ohio was recognized. The lowest lying, oldest rock, called blue limestone,[34] which he considered the equivalent of the Lower Silurian limestones of Europe, was described as forming, near the common boundary line between Indiana and Ohio, a kind of backbone which dipped gently in east and west directions, gradually disappearing in both states beneath a series of overlapping strata which, with one exception, have uniform characteristics. Here for the second time (see p. 172) is recognized the presence of the low swell known later as the Cincinnati anticline or uplift, which was subsequently identified by Newberry and Safford as a Middle Silurian emergence. This blue limestone was overlain by another limestone, thought to be the equivalent of the Cliff rocks (Niagaran) of Locke; this, in its turn, by a series of black slates, the equivalent of the Waverly sandstones of Ohio, and this in turn by a series of limestones in part oölitic and the equivalent of the Ohio conglomerate. This last, immediately upon which the coal formation rests at Oil Creek, in Perry County, he considered the uppermost member of a new series, to which he applied the name *sub-Carboniferous,* "as indicating its position immediately beneath the coal or Carboniferous group of Indiana," and "which merely indicates its position beneath the Carboniferous group without involving any theory."[35]

Immediately overlying this sub-Carboniferous (Mississippian) limestone was his *Bituminous Coal* formation, the latest and youngest of the series with the exception of the diluvium. He considered this formation a part of a great coal field which included nearly the whole of Iowa and Illinois and eight or ten counties in the northwestern part of Kentucky. Attention was called to the improbability of anthracite coal being found within the limits of the state.

Owen's recognition of the value of the fossil forms *Pentremites globosa* and *Archimedes* as horizon markers, as fixing the age of the oölitic limestone older than the Coal Measures, notwithstanding its lithological resemblance to the European Jurassic, and therefore marking the lower limit of the coal beds, was, for the period, an important generalization.

[34] This is the equivalent of the Cincinnatian division of the Lower Silurian of recent writers.

[35] The term was severely criticized as, taken literally, it would include every geological horizon beneath the Carboniferous. Later it was proposed to substitute the term *Lower Carboniferous,* but the present tendency is to do away with it altogether and to group the formations under the geographic term *Mississippian.*

D. D. Owen's Surveys of Iowa, Wisconsin, and Illinois, 1839-1840.

To properly appreciate much that is to follow, it must be remembered that, beginning with 1807, all government lands containing ores were reserved from sale and a system of leasing adopted. No leases were, however, issued until 1822, and little mining was done previous to 1826. For a few years, according to Whitney, rents for the mining lands were paid by the operators with comparative regularity, but after 1834, in consequence of the innumerable fraudulent entries of lands as agricultural which should, in reality, have been reserved as mineral, the smelters and miners refused to make any further payments, and the government officials were unable to enforce the claims. In consequence of these difficulties a resolution was adopted in the House of Representatives, on the sixth of February, 1839:

That the President of the United States be requested to cause to be prepared, and presented to the next Congress at an early day, a plan for the sale of the public mineral lands, having reference as well to the amount of revenue to be derived from them and their value as public property as to the equitable claims of individuals upon them; and that he at the same time communicate to Congress all the information in possession of the Treasury Department relative to their location, value, productiveness, and occupancy; and that he cause such further information to be collected and surveys to be made as may be necessary for these purposes.[36]

In accordance with this act Doctor Owen was appointed government geologist under direction of the General Land Office (James Whitcomb, commissioner) to make surveys in Iowa, Wisconsin, and northern Illinois. His first report, bearing date of April 2, 1840, was printed to form House Document No. 239, of the first session of the Twenty-sixth Congress.[37] It comprised, all told, 161 printed octavo pages, with 25 plates and maps, including a colored geological map and several colored sections. Fourteen of the plates were from Owen's own drawings. The district explored comprised an area of about 11,000 square miles lying in equal portions on both sides of the Mississippi River, between latitudes 41° and 43°, beginning at the north of

[36] For further information along these lines see J. F. Callbreath on Government Control of Minerals on Public Lands, Sen. Doc. No. 430, 1916; also A. H. Brooks on Applied Geology, Jour. Washington Acad. of Sciences, Vol. II, No. 2, 1912, pp. 19-48.

[37] A reprint or second edition of the work was brought out in form of Senate Document No. 407, 28th Congress, 1844.

PLATE XI

DAVID DALE OWEN

Rock River and extending thence north upward of 100 miles to the Wisconsin River.[38] (See Fig. 54.)

Owen's commission reached him at New Harmony, Indiana, on August 17, 1839. Its acceptance demanded that he explore "all the land in the Mineral Point and Galena districts which are situated south of the Wisconsin and north of the Rock rivers, and west of the line dividing ranges 8 and 9 east of the fourth principal meridian; together with all the surveyed lands in the Dubuque district," and complete his work before the approaching winter should set in. Concerning these somewhat remarkable conditions, Owen writes:

After duly weighing the nature of my instructions, estimating the extent of country to be examined, considering the wild, unsettled character of a portion of it, and the scanty accommodations it could afford to a numerous party (which rendered necessary a carefully calculated system of purveyance), and ascertaining that the winter, in that northern region, commonly sets in with severity from the 10th to the middle of November, my first impression was that the duty required of me was impracticable of completion within the given time, even with the liberal permission in regard to force accorded to me in my instructions. But on a more careful review of the means thus placed at my disposal, I finally arrived at the conclusion that by using diligent exertion, assuming much responsibility, and incurring an expense which I was aware the Department might possibly not have anticipated, I might, in strict accordance with my instructions, if favored by the weather and in other respects, succeed in completing the exploration in the required time.

I therefore immediately commenced engaging subagents and assistants and proceeded to St. Louis. There (at my own expense, to be repaid to me out of the per diem of the men employed) I laid in about $3,000 worth of provisions and camp furniture, including tents, which I caused to be made for the accommodation of the whole expedition; and in one month from the day on which I received my commission and instructions in Indiana (to wit, on the 17th of September) I had reached the mouth of Rock River; engaged one hundred and thirty-nine subagents and assistants; instructed my subagents in such elementary principles of geology as were necessary to the performance of the duties required of them; supplied them with simple mineralogical tests, with the application of which they were made acquainted; organized twenty-four working corps, furnished each with skeleton maps of the townships assigned to them for examination, and placed the whole at the points where their labors commenced, all along the southern line of the western half of the territory to be examined. Thence the expedition proceeded northward, each corps

[38] The original plates of this survey, bound up in one volume, are in the survey division of the U. S. General Land Office. The volume is marked D. D. Owen's Mineralogical Reports, Township Maps, Iowa and Wisconsin Territories.

being required, on the average, to overrun and examine thirty quarter sections daily, and to report to myself on fixed days at regularly appointed stations; to receive which reports, and to examine the country in person, I crossed the district under examination, in an oblique direction, eleven times in the course of the survey. Where appearances of particular interest presented themselves, I either diverged from my route, in order to bestow upon these a more minute and thorough examination; or, when time did not permit this, I instructed Dr. John Locke, of Cincinnati (formerly of the geological corps of Ohio, and at present professor of chemistry in the medical college of Ohio), whose valuable services I had been fortunate enough to engage on this expedition, to inspect these in my stead.

By the 24th of October the exploration of the Dubuque district was completed, and the special reports of all the townships therein were dispatched to your office and to the office of the register at Dubuque. On the 14th of November the survey of the Mineral Point district was in a similar manner brought to a close, and by the 24th of November our labors finally terminated at Stephenson, in Illinois, the examinations of all the lands comprehended in my instructions having been completed in two months and six days from the date of our actual commencement in the field. Also several thousand specimens—some of rare beauty and interest—were collected, arranged, and labeled.

The immediate result of this examination was to establish the fact "that the district surveyed is one of the richest mineral regions, compared to extent, yet known in the world." The area marked off as including the productive lead region, as shown by his map (see Fig. 54, p. 272) lay mainly in Wisconsin, but comprised also a strip of about eight townships on the Iowa side of the river and about ten townships in the northwest corner of Illinois, the ore-bearing strata being limited to a heavy-bedded magnesian limestone which, on account of its characteristic tendency to form perpendicular cliffs when subjected to weathering and erosion, he called the *Cliff limestone.* This he thought, from its fossil remains, to be the equivalent of the Upper Silurian, and perhaps a part of the Lower Silurian of Murchison. Though sparingly developed in the east, it "swells in the Wisconsin lead region into the most remarkable, most important, and most bulky member of the group. It becomes, as it were, the Aaron's rod, swallowing up all the rest," attaining a thickness of upward of 500 feet.[39]

[39] To James Hall's proposition to substitute the name *Niagara* for *Cliff,* Owen objected, claiming that the Niagara, as known in the East, represented only a part of the formation he was describing. In this last he was certainly right, more recent work having shown that his Cliff limestone included also beds belonging to what is now known as the Trenton period of the Lower Silurian (Ordovician).

PLATE XII

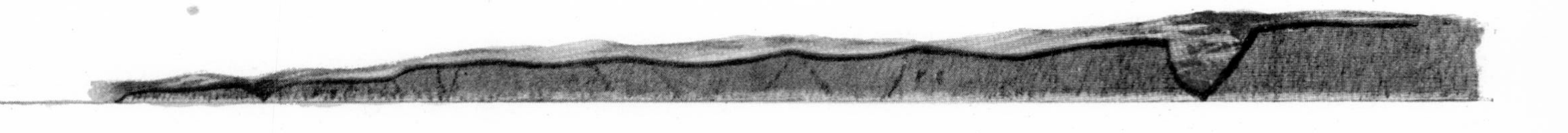

Section from the mouth of Lake St Croix to near the Great Bend, or Detroit of Lake St Croix.

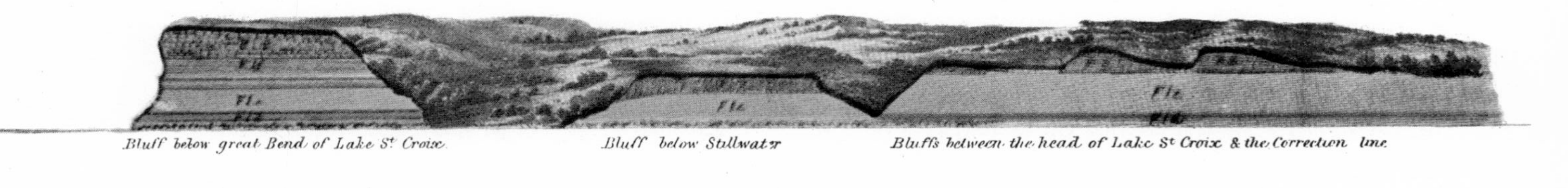

Section from the great Bend of Lake St Croix to the Correction Line

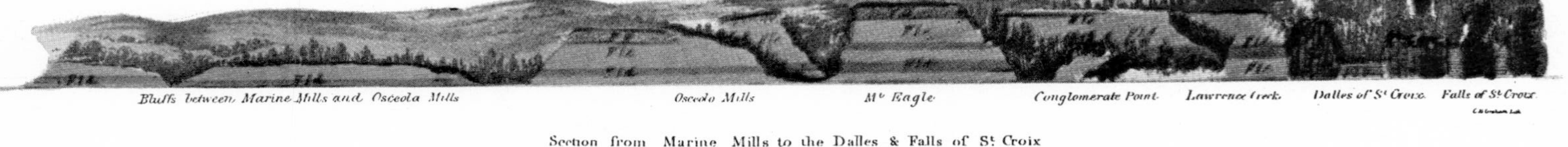

Section from Marine Mills to the Dalles & Falls of St Croix

Reduced copy of Plate IV in Owen's Report of a Geological Reconnaissance of the Chippewa Land District

The principal assistants on this survey were John Locke, who reported mainly on the barometrical observations, with particular reference to the measurement of altitudes and the making of geological sections, and a Mr. E. Phillips, who reported on the timber, soil, and productiveness of the region.

Sketch of D. D. Owen.

David Dale Owen was a son of Robert Owen, the well-known philanthropist and founder of the communistic societies at New Lanark, Scotland, and New Harmony, Indiana.[40] He first came to America in 1828, when seventeen years of age, but in 1831 returned to Europe, in company with H. D. Rogers, for the purpose of qualifying himself in chemistry and geology. Returning to America in 1832, he studied medicine at the Ohio Medical College in Cincinnati, whence he graduated in 1836. By one who knew him he is described as being very absent-minded, shy though amiable and of rather exquisite tendencies in his dress. That he was an artist of no inconsiderable ability is shown by the sketches of his numerous reports. The geological sections given in the report on the Chippewa Land District have never been equaled for picturesque effect. (See Plate 12.)

His earliest geological work was done in connection with Gerard Troost in Tennessee, and his first independent work that which has just been mentioned. Subsequently he made other surveys under the United States General Land Office, in which he showed exceptional administrative ability. Indeed, the organization and carrying out of the plan for a survey of the mineral lands of Iowa, Minnesota, and Wisconsin (in 1839-1840) within the short space of time and under the conditions imposed by Congress was a feat of generalship which has never been equaled in American geological history.

His life was one of unceasing activity and furnished one more illustration of the energy, persistency, and virility of the Scotch emigrants and their descendants in America. He died in the harness in 1860, while holding the position of state geologist of Indiana, and at the same time, and against the advice of his physician, attempting to finish the reports on the Arkansas survey of 1856-1857.[41]

40 Very interesting and entertaining references to the life of Owen, Maclure and other early American scientific men are woven into the story, Seth Way (Houghton, Mifflin & Co., 1917), by Mrs. Charles H. Snedeker, a granddaughter of Owen.

41 "Poor Owen is dead, suicide! His physician forced his way into his room and said 'Doctor, if you go on thus [*i.e.,* dictating from his sick chair to two

Conrad's Work on the Tertiary, 1839.

The problem of the Tertiary deposits of the southern Atlantic States, which up to date had been touched upon by Finch, Morton, Vanuxem, and Lea, was again, in 1839, taken up by T. A. Conrad. In the journal of the Philadelphia Academy of Sciences for this year Conrad described the Tertiary of Maryland as occupying all the tract south of an irregular line running from the vicinity of Baltimore to Washington, the Potomac being the western and the Chesapeake the eastern boundary. The deposit in the vicinity of Fort Washington on the Potomac, he now suggested, was probably contemporaneous with the London clay of England, *i.e.*, Eocene. This is the opinion held today.

The cause of the change in the character of animal life between strata belonging to the different animal horizons, was thought to be climatic. "Periodical refrigeration alone can explain the sudden extinction of whole races of animals and vegetables," he wrote.

It is well to note that in his *Geological Report for the State of New York*, Conrad had recognized the fact that in the earlier eras of the planet the temperature was uniform and the seas comparatively shallow. Hence, he thought, in the older rocks we should expect to find over the whole globe organic remains belonging to one group of species. Deep erosion and greater variation of temperature had caused more uncertainty in the Upper Tertiary formations.

In connection with Conrad's work, it may not be out of place to note that Professor J. L. Riddell this same year described the surface geology of Trinity County, Texas, and, basing his determinations mainly on the beds of lignite there found, identified the prevailing sandstone as Tertiary.

Incidentally, also, it may be stated that during 1840 James T. Hodge made a trip through the eastern portion of the southern Atlantic States and made extensive collections of fossils, which were turned over to Conrad for identification. Hodge's notes, as published in the Transactions of the Association of American Geologists, contain little of geological value, but the list of fossils comprised some 134 species, of which 32 were then new to science. All were of Tertiary age.

amanuenses] you will die in a week.' Owen quietly looked up and replied, 'I only want 13 days to finish it.' I have printed 90 pages of his Arkansas report for him here, but was not made aware of his dictation or I should have *im*proved it as I proved it." From letter of J. P. Lesley to James Hall, Dec. 4, 1860.

Sketch of Conrad.

Conrad was born in Philadelphia in 1803, and from early youth showed a decided taste for natural history studies, though for a time following the calling of his father—that of a publisher and printer. The work noted on page 200 was his second of geological importance, and was preceded also in 1831 by a paper on American marine conchology. Of his subsequent writings on conchology, paleontology, or general geology upward of twenty related to the Tertiary, and it is upon these that his fame chiefly rests, though in addition he described the fossils collected by the Wilkes Exploring Expedition, by Lieutenant Lynch's expedition to the Dead Sea, by the Mexican boundary survey under Lieutenant Emery, and was first geologist and then paleontologist to the New York state survey.

Personally Conrad was peculiar.

He wrote his letters and labels frequently on all sorts of scraps of paper, generally without date or location. He was naturally careless or unmethodical, and his citations of other authors' works can not be safely trusted without verification, and are usually incomplete. He had a very poor memory, and on several occasions has redescribed his own species. This defect increased with age, and, while no question of willful misstatement need arise, made it impossible to place implicit confidence in his own recollections of such matters as dates of publication. (Dall.)

He himself said in a characteristic letter to F. B. Meek, written July, 1863:

I go on Monday to help H. ferret out my skulking species of paleozoic shells. May the recording angel help me. God and I knew them once, and the Almighty may know still. A man's memory is no part of his soul.

Conrad was bitterly opposed to the doctrine of evolution, and predicted that Darwin's wild speculations would soon be forgotten. Every geological age came, according to him, to a complete close, and the life of the succeeding one was a wholly new creation. His feeling on this subject is well shown in a letter, also to Meek, dated February, 1869, where he says:

I sent you to-day a copy of my paper in Kerr's geological report. I perceive by Cope, Martin, and Hayden's researches that they have not found that phantom of the imagination, a formation between chalk and Eocene. The world has been so thoroughly harried by despairing development philosophers with so little result that they may as well say the "game is up," and not speculate on lost formations lying at the bottom of the ocean, a desperate expedient to save genetic succession. It is strange that geologists can not recognize an Azoic era after the chalk. I consider, in the light of paleontology, the advent of one only form of life in the beginning to be absurd. The history of Lingula ought to teach us the permanence of certain forms of life from the beginning, while thousands of others were created and died out.

Poor in health, given to melancholy and low spirits, he furnished to American geological history that which has a counterpart only in Percival, and like Percival, it may be added, he was sometimes given to rhyming, though unlike Percival, he was slovenly and careless in his work to a point beyond endurance.

His melancholy increased with age, and frequent ill health caused him at times to lose interest in every undertaking.

A period of moping would usually end in his writing some verses which nobody would praise, and this seemed sufficiently to nettle him, to rouse him thoroughly, and he would become again enthusiastic in the matter of shells and fossils. (Dall.)

It was presumably during one of these moping periods that he wrote again to Meek under date of October 24, 1864, as follows:

I am troubled with the cui bono malady, and frequently wish to be under the "clods of the valley." The idea of suicide haunts me continually, and I wish to get to work to banish the horrors, but whatever side I look at work it brings expenses I think I would regret. Singular inconsistencies of man. I don't expect to live much longer, yet shrink from expenses, but such is the hard mental twist that early poverty gives the mind. My blues have been considerably increased by the death of Doctor Moore,[42] who I loved more than most other men, and by the death of my nearest and dearest sister with whom I lived in Trenton. Now I must plunge into the marl beds to keep up a little time longer and hope to finish the Miocene figures before I go where no dreams of new discoveries will haunt me. Excuse this egotism; it is the last of it; but you now see why I do so little. Only one thing remains. I don't suppose you feel the want of a home, but I have felt it all my life, and the dreams of an Egeria have overtopped the dreams of science, so that in the midst of geological pursuits a horrid vacancy has yawned in my heart, and a grinning devil laughed over a drawer of fossil shells. A small Hamlet in science, I grow old and reform not.

Conrad's Ideas of the Drift, 1839.

Though primarily a paleontologist Conrad was sometimes drawn out of his chosen field by phenomena too obvious to be overlooked, and concerning the nature of which little was actually known, even by the best authorities. The occurrence of enormous bowlders in the

[42] Dr. W. D. Moore, of the University of Mississippi.

drift and resting often upon unconsolidated sand and gravel, naturally called for an explanation. That such could not have been brought into their present position through floods was to him obvious, neither could they have been floated by ice floes from the north during a period of terrestrial depression.

He assumed, rather, that the country, previous to what is now known as the glacial epoch, was covered with enormous lakes, and that a change in climate ensued, causing them to become frozen and converted into immense glaciers. At the same time, elevations and depressions of the earth's surface were in progress, giving various degrees of inclination to the frozen surfaces of the lakes, down which bowlders, sand, and gravel would be impelled to great distances. The impelling force might, in some cases, be gravity alone; but during the close of the epoch, when the temperature had risen, vast landslides—avalanches of mud filled with detritus—would be propelled for many miles over these frozen lakes, and when the ice disappeared the same would be deposited in the form of a promiscuous aggregate of sand, gravel, pebbles, and bowlders. The polished and scratched surface of the rocks in western New York he ascribed to the action of sand and pebbles, which were carried by moving bodies of ice, that is, apparently, to local glaciation.

George E. Hayes's Views, 1839.

The year 1839 seems to have been one of remarkable activity among American geologists. Moreover, the various publications began to show a greater variation in individual opinion and a disposition on the part of many to judge for themselves rather than follow too implicitly the opinions of others. Many of the ideas put forward were necessarily crude but furnish a very good insight into the gradual evolution of ideas upon a great variety of subjects.

George E. Hayes wrote upon the geology and topography of western New York and incidentally put forth several theoretical ideas worthy of mention, even though ill-founded. He took occasion to deplore the common custom of invoking the assistance of the Noachian deluge to account for such results as erosion and the distribution of the drift. He felt that "the condition of a continent gradually elevated from the ocean, whether by volcanic action or by the expansive force of crystallization, or by any other cause whatever, would be such as to account for all the geological phenomena hitherto attributed to the mechanical action of water." The formation of such terraces as Hitchcock had described in the valley of the Connecticut he thought might be due to the action of waves and tides.

It was idle to suppose that existing streams had carved out their own channels, and he ridiculed the idea that the falls of Niagara were once at Lewiston, seven miles below their present position, as had been contended, and rightly, by the English geologist, Bakewell.

He also decried all attempts at estimating the age of the falls or the time before they would so far cut back as to drain Lake Erie. It was his idea that the channel below the falls was cut while the rocks were being slowly raised above the level of the great inland sea and forming a limestone ridge or reef across which, at its lowest points, strong currents would alternately sweep during the ebb and flow of the tides. As the elevation progressed the currents would become more and more confined to the weakest and lowest points in the barrier, until in course of time the whole force of the conflicting currents would be concentrated at one point. "The power of the waves and the influx of the tide operating from below would be applied to the best possible advantage in tearing up the strata which most impeded their course, while the current, combined with the receding tide, would carry off the fragments. In this manner the valley of the Niagara was doubtless formed." (See Hall's views, p. 236.)

The lake beds of the region he imagined to have been formed by the unequal erosion along the edges of uplifted strata, Lake Erie lying at the junction of shale and limestone. He conceived a strong current to have set in through the Gulf of St. Lawrence before this limestone had become sufficiently elevated to shut out the sea from the basin now occupied by the lakes. This found its way through the valleys of the Mohawk and Hudson, dropping on its course the large quantity of bowlders foreign to the localities and now so plentifully distributed over the surface.

Jackson's Survey of New Hampshire, 1839-1840.

As early as 1823 Governor Woodbury, in his message to the state legislature of New Hampshire, had recommended an agricultural survey, having particularly in mind chemical investigations of the various kinds of soils. It was not, however, until 1839 that, under the earnest solicitation of Governor Page, a geological and mineralogical survey was organized, and Dr. C. T. Jackson, who has been already referred to as state geologist of Maine and Rhode Island, was placed at its head. Jackson entered upon his new duties on June 1, 1840, and devoted to the work the greater portion of the next three years, being assisted from time to time by J. A. Whitney, a recent recruit to the ranks of geological workers, M. B. Williams, W. F.

Channing, Eben Baker, and John Chandler. The survey continued for three years, receiving annual appropriations of $3,000 to cover expenses, Whitney himself as chief assistant receiving $3 a day. It is stated[43] that in order to circumvent the politicians who were insisting that all the minor salaried positions be filled by citizens of the state, it was arranged with the governor's assent to have no paid assistants whatever, Jackson to select trained men from his own laboratory who were to volunteer their services, trusting that in the end the legislature would appropriate sufficient funds to indemnify. A partial compensation was later provided, from which Whitney received the sum mentioned.

As an outcome of the survey there were issued two annual reports dated 1841 and 1842, and a final report in 1844, the last in form of a quarto volume of 384 pages and 11 plates, including two of colored sections and an uncolored geological map of the state on which the various rock types were indicated by figures. An ideal section from Nova Scotia westward into Pennsylvania represented the White Mountains with a core of granite with strata of gneiss dipping regularly east and west on either hand. The schistose rocks overlying the gneiss he colored as of Cambrian age. These in their turn were, with the exception of the interpolation of the Green Mountain quartz rock on the west, figured as flanked on either hand, east and west, first by Silurian and then Coal Measures strata. Elsewhere in the report he referred to the regular east and west dipping of the beds in Maine and Vermont as indicating an anticlinal axis in New Hampshire which he thought to have been brought about by the primary rocks having been driven up through them like a wedge. This "ideal section" was later shown by C. Hitchcock to have been too largely ideal, since not only did the mountains have no granite core, or axis, but there was no such regularity of east and west dipping strata, the beds being often overturned, affording repetitions of the older strata.

He noted further that the apparent dipping of the Vermont marble beds under the Green Mountains might be due to "a folding or doubling back of the strata, which may curve around and pass down toward the shores of Lake Champlain, where the same strata may be composed of marine shells," and also that the secondary rocks must at some time have been horizontal, forming a continuous deposit extending entirely across the region now occupied by the White Mountains. If, however, the secondary strata were to be restored to their original horizontal position, he found an insufficiency

[43] Life and Letters of J. D. Whitney, p. 40.

of materials to cover the gap left by the removal of the primary rocks composing these mountains. This he would ascribe to the breaking up of the secondary rocks at this point through the sudden elevation of the primaries and a more or less altering of the same by heat.

It was Jackson's idea, further, that highly inclined stratified rocks could have been made to assume that position only by some distorting cause other than aqueous action. This cause he conceived to be "a deep-seated power residing in the interior of the crust of the globe," the power itself being furnished by the "great caldron of molten rocks and pent up gases and steam, there being no more difficulty in our conceiving of the adequacy of this force than in the prodigious power of steam in moving the enormous engines" daily seen in operation. He conceived that a better idea of this power might be gained if one "look to the dimensions of the earth's great boiler, and consider the comparative thinness of its sides." Volcanoes he devoutly referred to as safety valves which prevented a general upbursting of volcanic fires, and "hold all in the most perfect order, and preserve the earth in safety."

His views on the glacial drift would seem to have undergone no appreciable change since 1839, although perhaps somewhat differently expressed in the later reports. He wrote of this drift as due to the "ocean waters and seas of ice from the polar regions having been hurled with violence over the surface of the northern hemisphere during a period of shallow subsidence." The drifting of bowlder-laden icebergs would account for the glacial striæ, but the glacial theory of Agassiz he looked upon as quite insufficient and absurd. It is to be noted, however, that his only conception of glaciers was that of mountain glaciers of the Swiss type, the one-time presence of which was disproved by the nonradiating character of the drift and striæ. This glacial flood to which he now appealed, however, was not considered contemporaneous with the Noachian deluge, but as having occurred before the advent of man.

The coal beds were considered as products of plant growth in low, moist, and warm bogs, which were at no great height above sea level, and which were frequently subjected to slow and gradual submergence and reelevation; in this he followed the authorities of the day.

Houghton's Work in Michigan, 1837-1841.

In 1837 there was organized a state geological survey of Michigan, the sum of $3,000 being appropriated to carry out the work of the first year, the amount being increased to $6,000, $8,000, and $12,000, respectively, for each of the three succeeding years, which limited the life of the organization.

Douglas Houghton, then not thirty years of age, was made state geologist. During his term of office there were issued five annual reports, the last bearing the date of 1842. The chief assistants were Bela Hubbard and C. C. Douglas.

Fig. 34. *Douglas Houghton.*

The survey continued its existence until 1841, when it expired through lack of funds, no further appropriations being granted.[44]

Houghton was not, however, content to drop the work, and in 1844, in connection with W. A. Burt, devised a plan for connecting the linear surveys of the public lands of the United States with the geological and mineralogical survey of the country. This was fully set forth in a paper prepared and read by him before the Association of American Geologists at Washington in that year. Under the recommendation of the General Land Office, Congress appropriated funds for the purpose, and Houghton was appointed to undertake the work. According to the arrangement thus agreed upon, Burt was to take charge of running the township lines of the Upper Peninsula, the subdivisions to be made by deputy surveyors, and Houghton was to have the directorship. The rocks crossed by lines were to be examined, and observations made as to the general geological and topographical features of the country. The system had been fairly organized and the field work of one season nearly completed when on the thirteenth of October, 1845, he met his death by drowning during a snowstorm, while making his way in an open sailboat along the west shore of Lake Superior.

Sketch of Houghton.

Houghton graduated in 1828 at what was then America's chief training school for geologists, the Rensselaer Polytechnic Institute, and was almost immediately appointed assistant to Professor Eaton. In 1830, having but just attained his majority, he delivered a course

[44] For details see the author's Contributions to a History of State Geological and Natural History Surveys, Bull. 109, U. S. Nat. Museum, 1920, pp. 158-203.

of lectures at Detroit on chemistry, botany, and geology, an example no man of several times his age would care to follow today unless quite regardless of reputation. But such was the condition of knowledge at that time. In 1831 he entered upon the practice of medicine, and soon after received the appointment of surgeon and botanist to the expedition under Schoolcraft for the discovery of the sources of the Mississippi. His public career as a scientific man began, therefore, with that date.

Fig. 35. *Bela Hubbard.*

He is described by his biographer[45] as a man of rather small stature, with hands and feet small and delicately formed; head large and well developed; with blue eyes, well sheltered underneath massive brows, and showing upon his ears, nostrils, and mouth scars from an accidental explosion of powder, which took place during some of his boyish experiments. In early life he had suffered severely from a hip disease, which had left one leg a little short, though not sufficiently so as to cause him to limp. His temperament was warm and nervous; his movements quick and earnest. Young, ardent, and generous to a fault, he seems to have made friends wherever he went, and soon became the most prominent and popular man in the state.

Everywhere his ability and energy were acknowledged. No name throughout the distant and rural districts was so often uttered. His daring, his generous acts, his good humor, his racy stories were repeated everywhere. . . . Every man seemed to feel a pride in the growing celebrity of Houghton, and the familiar epithets "The little doctor," "Our Doctor Houghton," "The boy geologist of Michigan," became common throughout the State.[46]

[45] Alvah Bradish, Memoir of Douglas Houghton, Detroit, 1889.

[46] T. B. Brooks, after spending some years in the same field, wrote: "Anyone . . . who will peruse the mss. notes left by Dr. Houghton will be convinced that his views regarding the geology of the older rocks were far in advance of his time and such only as geologists years afterwards arrived at, and those which are but now, thirty years after he recorded them, universally accepted."

CHAPTER IV.

The Era of State Surveys, Second Decade, 1840-1849.

THE fever for state surveys, so prevalent during the last decade, would seem to have very quickly subsided, since during the period now under consideration such were established only in Alabama, South Carolina, and Vermont. Governmental surveys were also few, being limited to those by D. D. Owen in the Chippewa land district, and Jackson, Foster, and Whitney in the Lake Superior region.

The cause of this sudden cessation is not quite apparent. Nine of the twenty-six states forming the Union at the beginning of the decade had, so to speak, escaped the contagion, and only the three above mentioned succumbed during the period, leaving Arkansas, Georgia, Illinois, Kentucky, Louisiana, Mississippi, and Missouri still open to attack. It is possible, and perhaps probable, that the period of great financial depression beginning in 1836 may have had influence, but as nine of the sixteen surveys inaugurated during the decade 1830-1839 were established either during 1836 or the three years immediately following, this is perhaps open to question. An important factor may have been the lack of geologists to agitate the subject and carry on the work, nearly every man of prominence and experience being engaged in organizations already under way. The period was, nevertheless, one of importance, one of manifest results rather than of organization and preparation.

The single event of greatest consequence during the decade was the appearance of the four quarto volumes constituting the final reports of Mather, Emmons, Hall, and Vanuxem of the New York survey, an event which would, however, have been paralleled by the reports of Rogers in Pennsylvania but for the dilatoriness of the state authorities. The volume of literature was naturally greater than at any previous period, since it included many of the reports of organizations of the previous decade, as those of Percival in Connecticut (1842), Booth in Delaware (1841), Jackson in Rhode Island and New Hampshire (1840), (1844), and Rogers in New Jersey (1840). The establishment of a geological survey of Canada in 1841, the coming of Lyell to America in the same year, the publication of his *Travels* in 1844, and the coming of the Swiss glacialist and zoölogist, Agassiz, and the French paleontologist, Verneuil, were events the importance of which it is impossible accurately to estimate. The most noted of the American participants in the events of the decade, as will be seen, were H. D. Rogers, then in full vigor; Edward Hitchcock, mature and conservative; D. D. Owen, by all means the leader in reconnaissance work; and W. W. Mather. Joseph Leidy began the extensive series of papers and monographs on fossil vertebrates which soon gave him world-wide recognition, James Hall was rapidly forging to the front among invertebrate paleontologists, while J. D. Dana, in the full flush of early manhood and fresh from the experiences of the Wilkes expedition, began the important series of papers dealing with the grander problems of earth history which soon placed him foremost in the ranks of American geologists and, indeed, among the geologists of the world.

Among other names which were to appear prominently will be found those of E. T. Cox, Ebenezer Emmons, J. P. Lesley, F. B. Meek, B. F. Shumard, Michael Tuomey, and J. D. Whitney.

Geology, at the opening of the decade, had found a place in the curricula of the leading colleges of the land, as at Bowdoin, in Maine; Amherst and Williams, in Massachusetts; Yale, in Connecticut; the Rensselaer Institute, in New York; the University of Virginia, at Charlottesville; the University of North Carolina, at Chapel Hill; and the College of South Carolina, at Columbia.

The Smithsonian Institution came into existence also during this period. The National Institute for the Promotion of Science and the Society of American Naturalists and Geologists were formed at the very beginning of the decade, the last-named in 1847, to be merged into the American Association for the Advancement of Science. The new organization held its first meeting in Philadelphia on September 20, 1848. It is of interest to note that of the fourteen officials, including the standing committee, ten were geologists or paleontologists, and the first paper to be read as reported in the Proceedings was by Peter A. Brown on "The Fossil Cephalopodes Belemnosepia." Among other papers were one by Lieutenant C. H. Davis on "The Geological Action of the Tides," one by the Rogers Brothers on "The Decomposition of Rocks by Meteoric Waters," and one by Agassiz on "The Terraces and Ancient River Bars, etc., of Lake Superior." This is however anticipating, and we must return to our chronology.

Agassiz's Glacial Theory, 1840.

In 1840 an immense stride in the study of the drift deposits was made through the publication in Europe of Louis Agassiz's *Études sur les Glaciers*, a work comprising the results of his own study and observations combined with those of Von Charpentier, E. T. Venetz, and F. G. Hugi. The work was published in both French and German and brought to a focus, as it were, the scattered rays with which the obscure path of the glacial geologist had been heretofore but dimly illuminated. But libraries in America were few and far between, the workers were poor, and many remained long in ignorance of the existence of the treatise or gained but a partial and imperfect knowledge of its contents through hearsay or brief reviews in periodicals. It was Agassiz's idea, based upon observations in the Alps and the Juras and what was known regarding existing conditions in northern Siberia, that at a period geologically very recent the entire hemisphere north of the thirty-fifth and thirty-sixth parallels had been covered by a sheet of ice possessing all the characteristics of existing glaciers in the Swiss Alps. Through this agency he would account for the loose beds of sand and gravel, the bowlder clays, erratics, and all the numerous phenomena within the region described, which had heretofore been variously ascribed to the Noachian deluge, the bursting of dams, the sudden melting of a polar ice cap, or even to cometary collisions with the earth.

Agassiz's views were favorably received by the majority of workers in Europe and Great Britain, though there was naturally a highly commendable feeling of caution against their too hasty acceptance. As a reviewer in the *American Journal of Science* put it:

> These very original and ingenious speculations of Professor Agassiz must be held for the present to be under trial. They have been deduced from the limited number of facts observed by himself and others and skillfully generalized; but they cannot be considered as fully established until they have been brought to the test of observation in different parts of the world and under a great variety of circumstances.

The effect of this publication became soon apparent. The subject is one, however, that can be best treated in a special chapter and may be passed over here (see p. 615).

R. C. Taylor and the First Geological Model, 1841.

At the April, 1841, meeting of the Society of American Geologists and Naturalists Mr. R. C. Taylor exhibited and described in detail a model of the western part of the southern coal field of Pennsylvania. This he stated to be the first model constructed in the United States. Such being the case, it is worthy of remark that he considered the then customary (and to some extent still prevailing) habit of exaggerating the vertical relief as "a hideous burlesque upon the actual aspect of the district represented or, rather, misrepresented," and that he had himself "for some time ceased to make any such difference between the horizontal and vertical scale." Taylor was particularly active during the period 1831 to 1839; his papers, relating almost wholly to economic problems, appearing mainly in the *Transactions of the Geological Society of Pennsylvania.* His most pretentious publication was that entitled *Statistics of Coal,* a volume of 754 pages relating to the coal fields of the world, published in 1848. "I have two undertakings on my hands," he wrote Hall in 1843, "of some magnitude. One of them is the ascertainment and a concise notice of every deposit of coal and lignite formation upon our globe. . . . I pursue the inquiry as to coal everywhere; and the map in illustration will elucidate some useful facts and principles which as yet are not manifest. . . . A friend of mine in this city has lately published a pamphlet suggesting that coal formed a certain zone round the globe. As he theorized before getting together the facts and details it may be very probable his scheme will not fit. This, of course, can only be acquired, as I am now doing by an accumulation of adequate data." Again: "My other undertaking is somewhat colossal, for it comprehends the entire illustration of the present territorial globe. To do this requires a vast deal of labor and hard reading."[1]

Certainly an ambitious undertaking.

Visit of Charles Lyell, 1841.

In 1841-1842 Charles Lyell, the English geologist, visited America. He traveled extensively over the eastern part and in 1845 published, both in England and America, in the form of two attractive and readable volumes, an account of his "travels," together with geological observations. As he had received a generous welcome from American geologists, and as they undoubtedly had discussed freely with him their problems, it is perhaps not strange that a rumor, shortly after his return, to the effect that he was to take this action should have aroused some feeling and sharp criticism. "I have condemned unqualifiedly such a course in Mr. Lyell—a course which I should not have anticipated, and which from my intercourse I

[1] See Appendix, p. 667.

thought him incapable of," wrote one of them.[2] In the autumn of 1845 Lyell came again to America and subsequently (in 1849) published his *A Second Visit to the United States.* Both books contained much relating to geology, as well as other matters such as would naturally be of interest to a foreigner, but the fears on the part of sundry Americans, lest their own work should be anticipated, were apparently groundless.

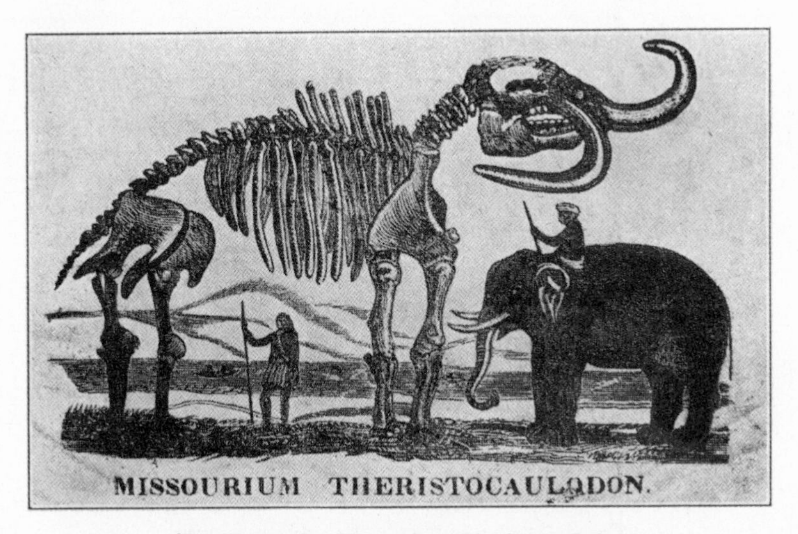

Fig. 36. *Koch's Missourium Theristocaulodon.*

Koch's Discoveries in Vertebrate Paleontology, 1840-1845.

Mention has already been made in these pages of the finding of bones of giant mammals and the crude inferences drawn therefrom. It remained, however, for a foreigner, a Dr. Albert Koch of Germany, to outdo his predecessors in flights of the imagination or Munchausenian exaggeration, whichever view one may choose to take of the matter. Koch was evidently a man of some education and during a period of several years, beginning about 1835, was engaged, apparently of his own volition, in "making researches of animal organic remains in the far west of the United States of N. A., and particularly in the state of Missouri." As might have been expected he found numerous remains of elephantine animals and finally, in 1840, discovered in the Osage country the skeletons upon which,

[2] Appendix, p. 668.

together with those of the Zeuglodon to be described later, is based his claim to the immortality conferred by these pages. His *Description of the Missourium Theristocaulodon or Missouri Leviathan* (*Leviathan Missouriensis*) was published in pamphlet form in both English and German and is a strange admixture of credulity and ignorance coupled with textbook knowledge of anatomical details. He here described and figured what was unmistakably a member of

Vol. I.
New York, Saturday, September 20, 1845.
No. 38.
THE SKELETON OF THE GREAT SEA-SERPENT;
Or, Hydrargos Sillimanii.
Discovered in Alabama in January, 1845, by the German Naturalist, Dr. Albert Koch, and now exhibited in New York.

Fig. 37.

the mastodon family, but with a vertebral column extended out of all proportion by the insertion of bones belonging to several individuals, with the height exaggerated by a false position of the limb bones, and with the tusks curving out horizontally (Fig. 36), the last-named feature being considered sufficiently characteristic to warrant the specific name *Theristocaulodon* or sickle tusk. The dimensions of his restoration as given were, length thirty feet, height fifteen feet. Anatomical details convinced him that the habits of the living animal were similar to those of the hippopotamus of the present day, and from comparisons with the accounts given in the forty-first chapter of Job, he announced, after examining the subject in all its bearings: "I have come to the conclusion that the leviathan here alluded to is none more than the Missourian before described."[3]

With so mobile an imagination it was not difficult to account for the elimination of so terrific a phase of nature's handiwork. The principal instrument to bring this about "was a certain comet that came in contact with our globe," causing a vividly described period of devastation and converting "the garden of delicious fruit trees and blooming flowers . . . into a gloomy forest of thorn and thistles."

Koch's second Munchausenian essay was made in 1845 when he announced finding in Washington County, Alabama, the skeleton of an enormous reptile-like animal "without exception the longest of all fossil skeletons found either in the old or new world." Here again he showed remarkable powers in the process of reduction from many to one and produced a form 114 feet in length which he estimated to have reached a probable limit of 140 feet in life. This he named *Hydrarchos Harlani.*[4] It was shortly shown, however,[5] that here too he had included in the vertebral column members of more than one individual as well as from several and even distant localities, and in addition quite mistaken the zoölogical affinities. The fairly complete skeleton now on exhibition in the National Museum at Washington (under the name *Basilosaurus cetoides*) is but fifty-five feet in length and unmistakably mammalian, though whether most nearly related to the seals or whales is an open question.

Troost's Work in Tennessee, 1841.

Troost's sixth annual report as state geologist of Tennessee (see *ante*, p. 138), a pamphlet of forty-eight pages, was made in October, 1841; his seventh, of forty-five pages, in 1843; the eighth, of but twenty pages, in 1845; and the ninth and last in 1848. During the interval between his fifth and sixth reports he had become acquainted with the work of Sedgwick and Murchison in England, and the result is a slight modification of his views regarding the formations of the state. He now announced the conclusion that those rocks west of the line which separated Tennessee from the state of North Carolina, and previously classified by him as Transition under the names of graywacke, etc., belonged to Sedgwick's Cambrian system and, he

[3] I am informed by C. Davies Sherbon that this skeleton is in the British Museum of Natural History.

[4] Also called *Hydrargos sillimani.* Renamed *Zeuglodon* later by the English anatomist, R. Owen.

[5] Jour. Acad. Nat. Sciences of Phila., Vol. I, 1847-1850, pp. 5-17.

added, "Geologists will rejoice that henceforth the name of graywacke will be doomed to oblivion." Unfortunately Troost's *ipse dixit* was not final, and the word continued in use, illy defined and obnoxious, down to a very recent period. The range known by the name of Bays Mountains, which extends into Tennessee for about 100 miles in a northeast and southwest direction, he believed to terminate this system toward the west. A few pages farther on he, however, thrust a doubt on this determination by writing:

> I consider . . . that Bays Mountain forms the upper part of the Cambrian system or that it perhaps belongs to the Old Red Sandstone; that hence toward the north of that mountain chain another formation commences, which is the Silurian system of Murchison.

The oölitic limestones and siliceous strata toward the west of the Cumberland Mountains he thought to rest upon a continuation of the Silurian system, more generally known under the name of "Mountain limestone." This he made to extend over the whole of middle Tennessee to a few miles east of the Tennessee River.

In his seventh annual report he announced that the region about Nashville was composed of strata belonging to the Silurian system, and mentioned the fossils which he considered characteristic of the lowest strata. In this report he gave a detail colored geological map of Davidson, Williamson, and Maury counties. The eighth and ninth annuals were, as noted, very brief and contained little of public value.

Troost's collection of fossil Echinodermata, containing all his types and species described in what is known as Troost's list of 1850, together with the unfinished manuscript relating thereto, passed after his death in 1850 into the custody of the Smithsonian Institution, by which it was handed over first to Louis Agassiz, and several years later to James Hall, to be put into shape for publication. For reasons unknown to the public this work was delayed year after year, until, after the death of Hall in his turn (in 1894), it once more came into the possession of the Institution.[6] In the meantime all that was new had been elsewhere described. Thus, through sheer neglect, Troost lost much of the credit to which he was justly entitled.

Conrad's Work in Florida, 1842.

In 1842 an expedition under a Captain Powell was sent out by the Secretary of the Navy to survey Tampa Bay, Florida. Through the

[6] It has since been studied and described by Miss Elvira Wood in Bull. No. 64, U. S. Nat. Museum, entitled "A Critical Summary of Troost's Unpublished Manuscript on the Crinoids of Tennessee."

influence of the National Institute in Washington, T. A. Conrad was allowed to accompany the party for the purpose of examining into the geology of the region. Conrad remained on the coast some three months, his results being published in the *American Journal of Science* for 1846.

From paleontological evidence he was led to consider the limestone of the Savannah River in Georgia, between Savannah and Shell Bluff, as Upper Eocene, and thought that very probably the prevailing limestone of Florida would be included in this division. He also found evidence to lead him to believe that a considerable post-Pliocene elevation of the whole Florida peninsula had taken place, and that the Florida keys were a product of this movement. In this he was only partially correct.

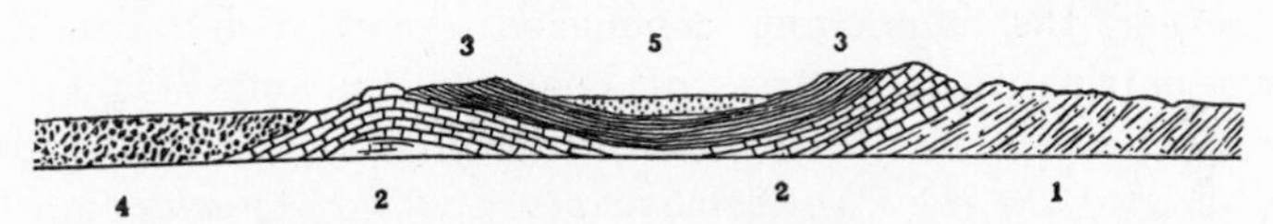

Fig. 38. *Owen's General Section across the Western States.*

D. D. Owen on the Geology of the Western States, 1842.

At the fourth annual meeting of the Association of American Geologists and Naturalists, D. D. Owen read a paper on the geology of the Western states, including under this designation the region of the Ohio Valley. Abstracts of this paper appeared in the *American Journal of Science* for 1843. At the November (1842) meeting of the Geological Society of London, Mr. Lyell presented for Owen the same paper, which was afterward published in the *Journal* of that society under date of 1846. It seems proper, therefore, that it should receive consideration here, particularly since Owen attempted a correlation of the American rocks with those from Europe. As published, the paper was accompanied by a colored geological map of the Ohio Valley and the generalized section shown here in Fig. 38. The attempts at correlation, it should be noted, were founded on paleontological data. The limestones immediately underlying the coal (3 of the section) he considered the equivalent of the European Carboniferous. These rested upon a series of rocks, the upper 100 feet of which Hall had assigned to the Devonian. The middle and lower beds Owen thought to be the equivalent of the English Upper Ludlow, as defined by Murchison. Below this he found shales which he believed equivalent to the English Lower Ludlow and the New York Marcellus shale, while the shale-bearing rock he thought the equivalent of the Wenlock formation of Murchison.

The fundamental rocks of the Ohio Valley he made the equivalents of Murchison's Lower Silurian, the lower 75 to 100 feet corresponding to the Llandeilo flags and the rest to the Caradoc sandstones. The corresponding formations of New York appeared to be the Trenton limestone and shale, representing the older series, and the Salmon River and Pulaski sandstone, the rest.

H. D. and W. B. Rogers on Appalachian Structure, 1842.

Turning for the moment from these minor details, the enumeration of which is seemingly essential to our method of treatment, we will next review the important conclusions reached by the Rogers brothers during their surveys of Pennsylvania and Virginia. In a paper read before the Association of Geologists and Naturalists in 1842, entitled *On the Physical Structure of the Appalachian Chain as Exemplifying the Laws which Have Regulated the Elevation of Great Mountain Chains Generally*, they called attention to the fact that the chain consists of a broad zone of very numerous ridges of nearly equal height, characterized by their great length, narrowness, steepness of slope, evenness of summit, and remarkable parallelism. They divided the chain on topographical and structural grounds into nine distinct divisions, consisting of alternating straight and curved portions, and noted the remarkable preponderance of southeastern dips throughout its entire length, the general trend of which was northeast and southwest. This was particularly characteristic of the portion along the southeastern or most disturbed portion of the belt, but toward the northwest dips in the opposite direction became less steep and more numerous.

They accounted for the phenomena of these dips by assuming them to be due to a series of unsymmetrical flexures, presenting in most instances steeper or more rapid arching on the northwest than southeast side of every convex bend; and, as a direct consequence, a steeper incurvation on the southeast than northwest side of every concave turn; so that, when viewed together, a series of these flexures has the form of an obliquely undulating line, in which the apex of each upper curve lies in advance of the center of the arch. On the southeastern side of the chain, where the curvature is most sudden, and the flexures are most closely crowded, they present a succession of alternately convex and concave folds, in each of which the lines of greatest dip on the opposite sides of the axes approach to parallelism and have a nearly uniform inclination of from forty-five degrees to sixty degrees toward the southeast. This they described as a doubling-under or inversion of the northwestern half of each anticlinal flexure. Crossing the mountain chain from any point toward the northwest, they noted that the form of the flexures changes; the close inclined plication of the rocks producing their uniformly southeastern dip gradually lessens, the folds open out, and

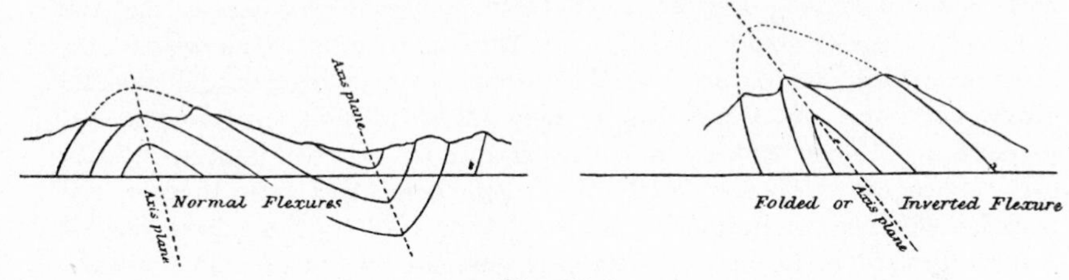

Fig. 39.

the northwestern side of each convex flexure, instead of being abruptly doubled under and inverted, becomes either vertical or dips steeply to the northwest. Advancing still farther in the same direction into the region occupied by the higher formations of the Appalachian series, the arches and troughs grow successively more gentle and the dips on the opposite side of each anticlinal axis gradually diminish and approach more and more to equality, until, in the great coal field west of the Alleghany Mountains, they finally flatten down to an almost absolute horizontality of the strata at a distance of about 150 miles from the chain of the Blue Ridge or South Mountain. (See Fig. 39.)

This work, so far as it related to the structure of the chain, has been improved upon in recent times only by the discovery of enormous overthrust faults in the southwestern portions. Their ideas regarding the origin of the chain have not, however, so successfully withstood the work of the modern physicists. They assumed that this difference in the dips was due to a combined undulatory and tangential movement of the earth's crust, which was propagated from the southeast toward the northwest; that is, they thought the various ranges composing the chain to be actually stiffened waves or billows of crustal matter comparable to the waves of less viscous material, like water or lava. These views may be best understood by direct quotation:

We assume that in every region where a system of flexures prevails the crust originally rested on a widely extended surface of fluid lava. Let it be supposed that subterranean causes competent to produce the result, such, for example, as the accumulation of a vast body of elastic vapors and gases, subjected the disturbed portion of the belt to an excessive upward tension, causing it to give way at successive times in a series of long parallel rents. By the sudden and explosive escape of the gaseous matter, the prodigious pressure, previously exerted on the surface of the fluid within, being instantly withdrawn, this would rise along the whole line of fissure in the manner of an enormous billow and suddenly lift with it the overlying flexible crust. Gravity, now operating on the disturbed lava mass, would engender a violent undulation of its whole contiguous surface, so that wave would succeed wave in regular and parallel order, flattening and expanding as they advanced, and imparting a corresponding billowy motion to the overlying strata. Simultaneously with each epoch of oscillation, while the whole crust was thus thrown into parallel flexures, we suppose the undulating tract to have been shoved bodily forward and secured in its new position by the permanent intrusion into the rent and dislocated region behind of the liquid matter injected by the same forces that gave origin to the waves. This forward thrust, operating upon the flexures formed by the waves, would steepen the advanced side of each wave precisely as the wind, acting on the billows of the ocean, forces forward their crests and imparts a steeper slope to their leeward sides. A repetition of these forces by augmenting the inclination on the front of every wave would result finally in the folded structure, with inversion in all the parts of the belt adjacent to the region of principal disturbance. Here an increased amount of plication would be caused, not only by the superior violence of the forward horizontal force, but by the production in this district of many lesser groups of waves interposed between the larger ones and not endowed with sufficient momentum to reach the remoter sides of the belt. To this interpolation we attribute, in part, the crowded condition of the axes on the side of the undulated district which borders the region where the rents and dikes occur, and to it we trace the far greater variety which there occurs in the size of the flexures.

The date of the Appalachian uplift was considered subsequent to the formation of the Coal Measures, as the final paroxysmal movement which terminated "in that stupendous train of actions which lifted the whole Appalachian chain from the bed of the ancient sea."[7]

[7] A writer (not named, evidently an editorial) in the Am. Jour. Sci., Vol. XXXI, 1837, p. 290, argued that the main chain was due to electrical causes. "If the earth contain, as is probable, a series of immense galvanic batteries, producing the results of electric and magnetic phenomena, the identity of the element being inferred then it may be assumed that this principle acting by such vast machinery and on so grand a scale is adequate to the explanation required." The general north and south extension of this power, proving the

H. D. Rogers on the Formation of Coal, 1842.

At this same meeting of the Association H. D. Rogers gave a paper on the origin of the coal beds. He here took the ground that the singular constancy in thickness of the Pittsburgh bed and its prodigious range were strongly adverse to any theory of accumulation by drift as held by many workers, both European and American. It was his idea that the area now covered by the coal was once an extensive flat, bordering a continent, beyond which lay a wide expanse of shallow but open sea; that these low flats were occupied by peat bogs derived from and supporting a luxuriant growth of stigmaria, while along the land margin and the drier areas were conifers, lycopods, and tree ferns. The stigmaria would thus form a uniform mass of pulpy peat admixed with leaves and other easily transported débris of the trees along the shores, but free from trunks and coarse branches.

To account for the shaly and sandy partings of the beds and their numerous impressions of plants, he imagined a sudden sinking of the land, producing thus a tidal wave from the open sea, carrying destruction to the forest growths, and on its return dragging back and spreading out over the sea bottom (and hence over the accumulated peat) sand, gravel, and silt, together with floating trunks of trees, all of which go to make up the roof slate and other partings in the beds. A period of tranquillity followed; another bed of peat accumulated, and so on. His theory differed from those of Buckland, Beaumont, Lyell, and others in excluding from the coal-making materials the trunks and branches of the more woody trees and the catastrophic interruptions in the growing processes due to earth sinkings and earthquakes.

He further accounted for the difference in volatile constituents between the coal of the eastern and western beds on the supposition that the former were debituminized by the action of steam and gaseous matter emitted through the crust of the earth and by means of the cracks and crevices formed during the undulation and permanent bending of the strata, which resulted in the formation of mountain ranges.[8]

accumulation of forces in that direction, might, he thought, in one age, through the sudden expansion of the elements, throw up the precipitous barriers of the mountains of Chile, and in another, under different combinations, the Rockies.

[8] A most excellent review of the opinions of Rogers and other early and later writers is given by J. J. Stevenson under the title Origin of the Pennsylvania Anthracite, Bull. Geol. Soc. of Am., Vol. V, 1894, pp. 39-70.

W. B. Rogers on Thermal Springs, 1842.

W. B. Rogers also read at this meeting a paper on the connection of the thermal springs in Virginia with the anticlinal axes and faults. He recognized the work of the European geologists in connecting similar phenomena with areas of terrestrial disturbance, and showed that, with but few exceptions, the thermal springs of the Appalachian region issued from the steep dipping or inverted strata on the northwest side of the anticlinals and this he thought might be laid down as the general law of their position. These views he regarded as in harmony with those of Arago and Bischoff. From those of the English geologist Daubeny they differed, seemingly, in that the latter regarded such springs as indicative of volcanic eruptions going on in a covert and languid manner under certain parts of the range.[9]

J. P. Couthouy on Coral Growth, 1842.

In 1842 there appeared in the *Proceedings of the Boston Society of Natural History* a paper by J. P. Couthouy, one of the naturalists of the Wilkes Exploring Expedition, in which it was shown that the distribution of coral growth in the ocean was limited by temperature. This paper brought forth very promptly a reply from Dana, who, while agreeing as to the facts and theories, claimed that Couthouy had simply borrowed the idea from him; that the explanation "was originally derived from my manuscript, which was laid open most confidentially for his perusal while at the Sandwich Islands in 1840." The assertion was naturally denied by Couthouy with equal promptness and the charge was subsequently withdrawn, though it must be confessed that the evidence, so far as it appears in *print*, favors Dana's claim.

Dana, it should be remarked, was led, from his observations while on this expedition, to agree with Darwin as to the formation of atolls (annular coral reefs), but could not agree with him regarding regions of subsidence and elevation. He found nothing which to him supported the idea that islands with barrier reefs were subsiding, while those with fringing reefs were rising.

[9] We may anticipate here and state that C. T. Jackson, in 1859, in a paper before the Boston Society of Natural History, noted that thermal springs were generally found along lines of disrupted strata and near the junction of sedimentary and eruptive rocks. Vanuxem, in the discussion which followed, agreed in thinking that the springs were genetically connected with faulting, but suggested, further, that there was a connection between certain rocks and the springs which issued from them; that the rocks generally insulated their own waters.

Byrd Powell's Work in the Fourche Cove Region of Arkansas, 1842.

In the summer of 1842 Dr. W. Byrd Powell, of Little Rock, passed, of his own volition, some six weeks in a study of the geology of the Fourche Cove region, his results being published in a pamphlet of twenty-three pages and map, by the Antiquarian and Natural History Society of Arkansas. The Doctor at the time of his writing occupied the position of lecturer on phrenology (!) and geology in the Medical College of Louisiana. Whatever may have been his qualification in the first-named subject, as a geologist he was no whit above the majority of workers of his day. The rocks of the cove and their association were described in considerable detail, the prevailing syenite being called granite, although he noted that it contained no quartz. He differed with Featherstonhaugh, however, in assuming, and wrongly, that the rock was primitive and not intrusive. For a supposed transitional form of greenstone into basalt he adopted De la Beche's name *Cornean*. The other rocks noted were basaltic clinkstone, amygdaloid, gneiss, etc. Some of his observations are open to question. Less than a gill of mica from the gneiss, he announced, yielded him half an ounce of metal, which he considered mostly iron. He further remarked that the hornblende rocks of the region contained a large percentage of iron, which was chiefly, if not wholly, native; also that they contained native lead and a little silver!

State Geological Survey of New York, 1836-1843.

The organization of the geological survey of the state of New York in 1836 has been already noted. Annual reports were issued regularly during the period of its existence, but with one or two exceptions the results may be summed up as below in a review of the final reports issued in quarto form in 1842-1843.

MATHER'S WORK IN NEW YORK.

The first district, to which Mather was assigned, embraced twenty-one counties in the southeast corner of the state as shown on the accompanying maps (Fig. 40). His report, the most voluminous of the series, comprised 639 pages with 45 plates, in part colored, and a geological map. Mather thought to have recognized in the district assigned to him seven periods of elevation and disturbance of the strata. These he described as follows:

First, one preceding the laying down of the earliest fossiliferous rocks, which gave rise to the primary Highlands of the Hudson and the primary mountains of Saratoga and Washington counties. On this he found the Potsdam and Calciferous sandstones and black limestone resting unconformably.

Second, one which preceded the deposition of the Helderberg

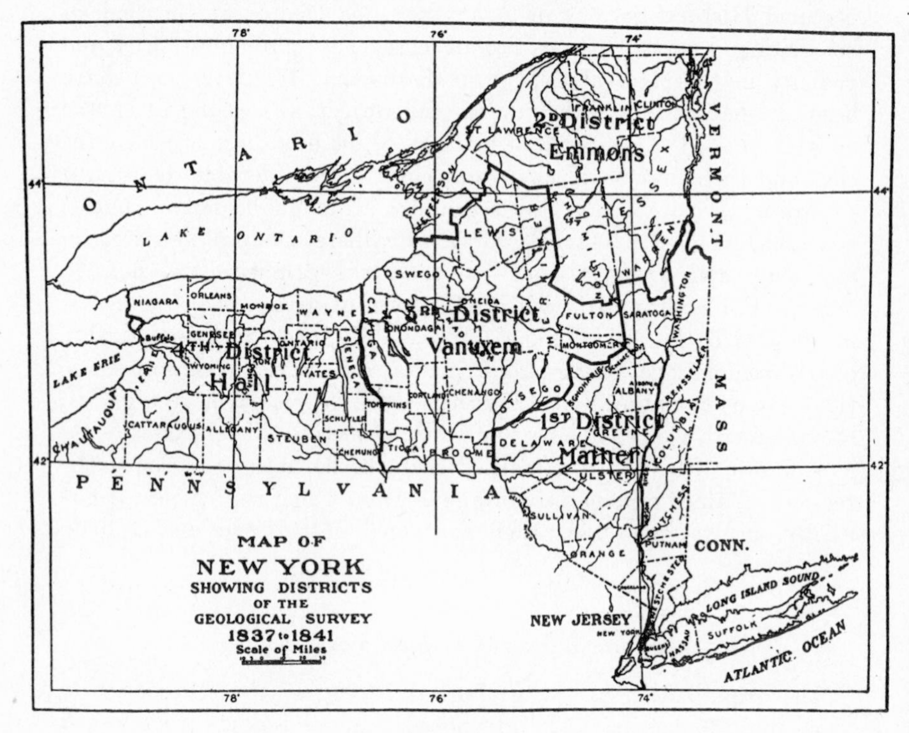

Fig. 40.

limestone, which he found lying unconformably on the upturned earlier rocks. This manifested itself mainly in the Hudson Valley and included the Highlands and Westchester County.

Third, one subsequent to the deposition of the rocks of the Ontario division, the effects of which he thought to recognize from Green Pond Mountain through Bellevale Mountain, Goose Pond, Sugar Loaf, the Skunnemunk Mountains, and along the Shawangunk Mountains.

Fourth, one subsequent to the deposition of the Catskill division and even the coal formation. This disturbed all the preexisting rocks in New York east of the Delaware. To it he ascribed the bending and wrinkling of the coal formations in Pennsylvania, Maryland, and Virginia. The era of disturbance was regarded as between that of the coal and the New Red sandstone.

Fifth, a period of disturbance at the end of the deposition of the Long Island division and before the deposition of the Quaternary.

Sixth, a period between the drift and Quaternary by which the land, without deformation, was lifted to about the level of the Quaternary formations.

Seventh, a period after the deposition of the Quaternary by which a large part of the land was lifted 100 feet above the level of the oceanic waters.

He thought, also, to recognize four periods of metamorphic agency, though, singularly enough, his metamorphism was that of contact by igneous masses only, and hence his periods those of igneous intrusion. To the then ordinarily accepted causes of folding, elevation, and depression,—as steam and elastic vapors in a high state of tension, the contraction of the earth by secular refrigeration, or the undulatory action of the fluid interior, combined with tangential force—Mather added a fourth. He conceived that, as the earth is a cooling and shrinking body, revolving with increasing rapidity owing to its gradually diminishing diameter, the ocean, being fluid, would not immediately partake of this increase but, as it were, lag behind and, therefore, as the earth revolved from west to east, a current of greater or less strength would set to the westward over the entire ocean, but most strongly at the equator. The inertia of the solid mass of the interior of the globe would also cause it to press to the westward with a power dependent upon the rapidity of rotation. Given, then, lines of weakness, where yielding would take place, motion and distortion or elevation might follow. If the interior were fluid and the solid exterior floating upon it, a change in the velocity would produce still stronger effects and changes of latitude of masses of the earth's surface would result. Evidences of this were seen, he thought, in the wrinkled and folded strata.

The large amount of drift scattered over the central and northern Mississippi region he ascribed incidentally to ice-laden currents from Hudson Bay and the polar seas which, flowing over the northern part of the United States, would be met by the warm waters of the Gulf Stream, causing them to deposit their loads. The warm current flowing northward would be superimposed on the cold current, the latter continuing southward beneath it, transporting the finer materials such as now occupy the lower Mississippi Valley. It may be added that Mather showed a somewhat amusing disposition to overload and overwork the Gulf Stream. Not merely did he ascribe the glacial drift and the New Red sandstone deposits to this agency, but he would have the entire sedimentary series of the New York system dependent on the same causes. "And we can perceive no region from which such a vast amount of mineral matter can have been abraded and washed away . . . unless it has been brought by the equatorial and polar currents in their ceaseless flow through all time since the ocean has occupied the surface of the earth. . . . And these depositions seem to have been formed in a kind of eddy produced by the meeting of the flow of warm waters of the Gulf Stream through the Mohawk Valley with the polar current through the Champlain and Hudson valleys." To the presence of this cold polar current on the east he attributed also the comparative paucity of fossil remains in these rocks when compared with those same horizons to the west, and to the same joint action of polar and equatorial currents, the formation of the coal beds. The equatorial current, he argued, performed a circuit around from New Mexico along the Rocky Mountains, a part flowing into the polar sea and Hudson Bay and the remainder through the northern part of the United States and southern part of Canada, where it again divided, one part flowing over and through the St. Lawrence Valley and the remainder over the Mohawk Valley and along the Blue Ridge around once more to the Mississippi, where it would rejoin the same stream, a polar current meantime flowing through the St. Lawrence and Hudson valleys to the valleys of the Red River of the North and the Mississippi. The meeting of these currents at particular points would produce eddies and consequent stagnation. The transported organic matter, becoming water-logged, would sink to the bottom. This he thought might account for the presence of tropical vegetation in the polar regions.

The white, red, mottled brown, and blue clays and variegated sands of Long and Staten islands he regarded as the geological equivalents of certain beds of New Jersey underlying the drift and Quarternary and of Cretaceous age. The materials he thought to have been derived from the breaking down of the gneissic, granitic, and other crystalline siliceous rocks which extend parallel to the Atlantic chain from Georgia to Maryland and reappear again in Connecticut and Rhode Island, extending probably into Massachusetts. To him it bore evidence of transport from the southwest, and he believed the Gulf Stream, as before, to have been the transporting agent, the velocity of which was checked by the southward-traveling polar current. These latter currents sinking to the bottom would not be conducive to growth of animal life, and hence the beds are comparatively nonfossiliferous.

Mather believed that during the Quaternary division of time a vast inland sea occupied the basin of the St. Lawrence and Hudson valleys, in which the clays and marls were laid down, the water having only a very moderate flow; that subsequently the country was raised *en masse* to a height of from 300 to 1,000 feet. This elevation he conceived to have been sudden, causing strong currents to flow through the channels communicating with the ocean, depositing sand, gravel, and bowlders in their eddies. (See further on p. 252.)

He also believed that the limestones, that are frequently crystalline white and variegated marbles in the western part of Vermont, Massachusetts, and Connecticut, were metamorphosed forms of the Mohawk limestone and Calciferous sandstone; that they were, in short, rocks of the Champlain division, but much more highly altered and modified by metamorphic agencies than Taconic rocks. It should be said that, while using the term "Taconic," Mather stated distinctly that "in describing the Taconic rocks separately I have yielded to the opinion of my colleagues who have considered them as interposed between the Champlain division and the primary. I can discover no evidence of any such interposition, but consider them as rocks modified by metamorphic agency and intermediate in their characteristics between the unchanged rocks of the Champlain division and those still more altered and crystalline along the eastern line of New York and in the western part of Vermont, Massachusetts, and Connecticut." (See also under The Taconic Question, p. 594.)

The trap rocks of the Hudson Palisades he looked upon as ancient lavas that had flowed through the rocky fissures in dikes while this part of the continent was still beneath the waters of the ocean. These and the associated red sandstones he rightly believed to be of the same age as those of Connecticut, Virginia, and Nova Scotia, though he was in error in thinking the Lake Superior red sandstone to be contemporaneous.

EMMONS'S WORK IN NEW YORK.

Emmons, who had charge of the second district, presented his results in a quarto volume of 437 pages, with 17 plates and colored sections. It was during this survey that he proposed the name *New York Transition* system for the stratigraphic series of rocks extending from the primary up to and including the Old Red sandstone. Subsequently this name was shortened by the omission of the word *transition*, and in this form was adopted by the other members of the survey. He classified the rocks of the system in ascending order as, first, *Champlain division;* second, *Ontario division;* third, *Helderberg division;* and, fourth, *Erie division.* The first, or Champlain, division included the Potsdam sandstone, the Calciferous sandrock, the Chazy and Birdseye limestone, the Isle la Motte marble, Trenton limestone, and Loraine shales; the second, or Ontario division, the Medina sandstone, green shales, and oölitic iron ore, now known as the Clinton ore, the Onondaga salt and plaster rocks, and the Manlius water lime; the third, the Pentamerus lime, the Oriskany sand, Schoharie grits, and Helderberg limestone; the fourth, the Marcellus and Hamilton shales, the Tulley lime, Genesee slate, and the Ithaca and Chemung slates and grits.

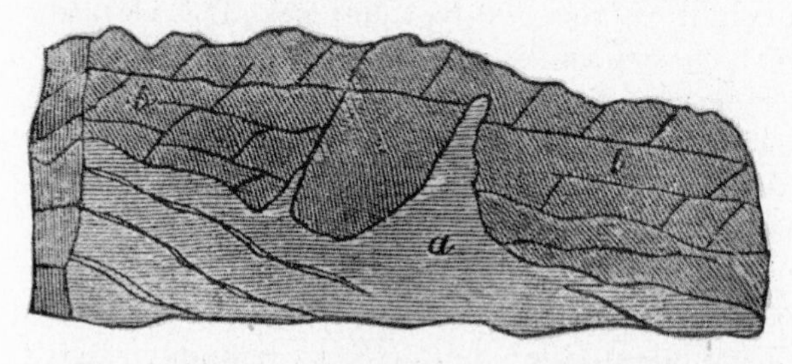

Fig. 41. *Emmons's Illustration of Igneous Nature of Limestone.*

He regarded the Transition rocks of Essex and St. Lawrence counties equivalent to the graywacke series of European authorities. The so-called Primitive limestone—the peculiar, coarsely crystalline, serpentinous limestone occurring in Essex County—he conceived might be an eruptive rock in which the carbonic acid had been retained by the pressure of superincumbent masses, the experiment of Sir James Hall bearing him out in this. The occurrence of plumbago he thought also favored the igneous origin of the limestone, since plumbago is so often produced in smelting furnaces.

Considerable space was given up to discussions of economic problems, particularly those of the iron ores. A point deserving particular attention, though not made prominent in the report, is that relating to the occurrence and origin of the magnetic iron ores. In describing these and their association with the igneous (hypersthene) rocks, he wrote: "The rock which incloses the ore is clearly unstratified; from which fact we are also to infer the igneous origin of the inclosed mass of ore. We are clearly driven off from every other mode of formation." But he adds as a cautionary clause, "The subject, however, must remain open to future investigation."

Emmons was, at this time, a catastrophist, even to the extent of asserting that the deep canyons in the Potsdam sandstone, like the Au Sable Chasm, were "opened by some convulsion of nature."

It was while connected with the New York state survey that he conceived the idea that a series of obscure rocks, forming a belt some fifteen miles wide along the western border of Berkshire County, and lying along both sides of the Taconic Mountains, were distinct from any of the so-called Primary rocks, and lay below those of the New York system. He therefore proposed, on stratigraphic grounds, to raise them to the dignity of a system by themselves, which should be called the Taconic system. Inasmuch as the controversy which arose over this new system raged vigorously for nearly half a century, the subject is considered worthy of a special chapter. (See p. 594.)

VANUXEM'S WORK IN NEW YORK.

Vanuxem's report on the third district, a volume of 306 pages, consisted of little more than an uninteresting account of the geographic distribution of the various rocks and their fossil contents. In fact, it fell short of what one would be led to expect from a perusal of his earlier papers. His district, it should be noted, comprised the counties of Montgomery, Fulton, Otsego, Herkimer, Oneida, Lewis, Oswego, Madison, Onondaga, Cayuga, Cortland, Chenango, Broome, Tioga, and the eastern half of Tompkins.

The views given on the origin of ore deposits are worthy of a moment's notice. In discussing the uplift on East Canada Creek, where the upturned edges of the Utica slate and Trenton limestone abut against the Calciferous with only a small trappean dike and vein of lead ore intervening, he found it easy to predict that, were the ore derived from the visible upturned rocks, the quantity would be small. But, if independent of them, proceeding from something like the foundations of the deep, the result could not be so easily foretold. It, *i.e.*, the vein, "is at the junction of dissimilar masses, hence electrical results could take place; where different exudations by the layers could unite, chemical action would result, and union with the former would produce galvanism; these conditions are requisite for the formation of metallic veins. . . . The bottom of the uplifts must extend to the point of fluidity, therefore, ejections from that part may exist in the fissure, though it is more probable that it was filled by lateral infiltration from the masses between which it exists."

As a matter of petrographic interest, it may be well to note that the serpentine described by Vanuxem, first in his third annual report (1839) and last in his final volume (1842), occurring near Syracuse, and concerning the origin of which he was then somewhat in doubt, was in 1844 described as an igneous rock, and as the only mass of the kind in the vicinity of the Salt Spring. This fact seems to have been quite overlooked by Williams in 1887, when, working with all the appliances of modern petrography, he showed the rock to be an altered peridotite.

In view of its enthusiastic acceptance by Conrad (see p. 255), it may be well to refer back to a theory of elevation given by Vanuxem in his annual report of 1837, but omitted from the final volume. "Knowing that the effect of crystallization is expansion; that to this property water, in freezing, bursts our conduit pipes, splits our trees and rends our roofs; as a consequence, too, crystallized bodies float upon their fluid material; an expansion of a crust of a few hundred miles in thickness from the known expansion of many mineral substances in crystallizing gives us an elevation greater than the heights of any known mountain, and to this force, as a primary cause, we owe our uplifted rocks, our mountain chains and parallel fissures." It is difficult to believe that a man of Vanuxem's caliber should not have realized that all minerals do not—like water—expand on crystallizing.

HALL'S WORK IN NEW YORK.

To James Hall was assigned at the close of the first season the fourth district in the state survey as already noted (p. 188). This comprised fifteen counties and included all that part of the State lying west of the parallel of Cayuga Lake and between Lake Ontario on the north and Pennsylvania on the southeast (see map). The area includes the least disturbed portions of the state, those portions where the rocks lie in a beautiful consecutive series, extending from the Medina up to and including the Chemung and affording means unrivaled for the unraveling of the stratigraphy of the continent. It is doubtful if, at the time this district was assigned to Hall, he himself realized its possibilities—certainly his collaborators did not. Indeed, at the time of Vanuxem's transfer and Hall's promotion the region was regarded as one of little promise and was willingly relinquished to the youngest and least experienced of the force.

Hall's report, appearing in 1843, comprised 683 octavo pages, with 19 plates, including a colored geological map of the state and the United States as far west as the Mississippi River.

He adopted the term *New York system*, as suggested by Emmons, to include all the oldest fossiliferous rocks of the state, excluding the Chemung and Catskill. As defined by him it would be the equivalent of what was embraced in the *Transition* of Werner. It likewise included Sedgwick's Cambrian, Murchison's Silurian, and the Devonian of Phillips, omitting the Old Red sandstone.

Concerning this he wrote:

> Nowhere is there known to exist so complete a series of the older fossiliferous rocks as those embraced within the limits of the State, and for the reason that in New York, where the means of investigation are best afforded and where the whole series is undisturbed, there is manifested the most complete and continuous succession, showing but one geological era for the deposition of the whole. In that era the earth first witnessed the dawn of animal life and ages of its greatest fecundity in marine organisms and the approach of the period when it became fitted to support a vegetation so luxuriant and universal, of which no modern era has afforded a parallel.

Precedence, it should be noted, was given in this report to paleontological characters over all others in distinguishing the sedimentary strata, but the fact that lithological characters must not be wholly disregarded was recognized, a matter to which he had previously called attention (p. 235).

> Changes in the lithological features of a rock . . . which may render observations unsatisfactory or doubtful are usually accompanied by greater or less change in the nature of the fossils. In no case, therefore, are to be overlooked either of the three important facts and characters, viz.: Lithological character, order of superposition, and nature of the contained fossils.

Hall was also at this time evidently a catastrophist and thought the drift soils, terraces, and the deep valleys and water courses to be due to the violent action of water, which may have been caused in part by a sudden submergence and the rapid passage of a wave over its surface. His views were, indeed, in many respects, little, if any, in advance of those held by Volney and Mitchill twenty-five years earlier. With them he conceived of an inland sea bounded by and held back by the Canadian Highlands on the north, the New England range on the east, the Highlands of New York and the Alleghanies on the south, and the Rocky Mountains on the west. These presented a barrier of from 1,000 to 1,200 feet above the level of the ocean until broken through by the St. Lawrence, the Susquehanna, the Hudson, partially by the Mohawk at Little Falls, and perhaps also by the Connecticut.

> But, to whatever cause we attribute the phenomena of the superficial detritus of the fourth district, the whole surface has been permanently covered by water, for it seems impossible that partial inundations could

have produced the uniform character and disposition of the materials which we find spread over the surface.[10]

He apparently failed at first to realize the efficacy of subaerial erosion, and thought that the immense amount of denudation which had taken place in his portion of the state could only have been accomplished beneath the sea, when it entirely covered the surface of the country and was subject to tides and currents like those of the present ocean.

Thus we may conceive this whole extent of country to have been submerged beneath the ocean for a long period; and that in its subsequent elevation it has been washed by the advancing and retiring waves, which have worn the deep indentations in the limestone cliffs and broken up the edges of the strata.

It may be well to state here that although the survey as an organization came to an end with the publication of the reports above noted, the work was by no means completed, and, by subsequent enactment of the assembly, Hall and Emmons were authorized to continue, under salary, until 1844. The time was subsequently extended until 1847 and again until 1849. The work grew in proportion to the opportunities offered, however, and finally the survey was the subject of a legislative investigation with the usual disastrous results. There was no question but that good work was being done, but, as it seemed to the layman, overdone, particularly when the tendency to superillustrate certain species of fossils was considered. It was shown in one instance that 8 species of fossils had been illustrated by 117 figures, one *Spirifera radiolarius*, having been figured 27 times, the drawings and engravings for which cost $526.50. Again, it was shown that the fossil *Pentamerus galeatus* was pictured by each of the four geologists and by a total of 74 figures, the drawings and engravings for which cost $174, in addition to the $150 for printing, or $324 for the one species.

It would be interesting to note how many times and in how many works these same species have since been figured.

Sketch of James Hall.

Hall was born at Hingham, Massachusetts, in 1811, and studied under Eaton at the Rensselaer Polytechnic Institute, graduating in 1832, after which he served for a short time as librarian, and was then appointed assistant to Eaton, at a salary of $600 a year.

[10] Appendix, p. 733.

PLATE XIII

JAMES HALL

His first systematic work in geology was done under the patronage of Stephen Van Rensselaer in St. Lawrence County. With the organization of the state geological survey in 1836 he was appointed assistant to Emmons, but, after the withdrawal of Conrad, was placed in charge of the fourth district—

> the level, uninteresting western portion of the State, which he was told was good enough for a young man of twenty-five. The region was not the western New York of to-day; roads were less numerous and less carefully made; exposures were rare and poor. It was necessary to wade along streams for miles to gain fragments which were to be pieced into tentative sections; the people were suspicious, fearing some new scheme for increasing the taxes; but none of these things moved him; as in later years, difficulties only increased his determination. So his is the only one of the four final reports which deals broadly with the problems of the young science, and, though upon the contemned fourth district, it is the only one which has endured with authority and become a classic in geological literature. (Stevenson.)

The final withdrawal of Conrad from the paleontological work also left Hall almost sole master of the field, which he for a time filled so completely that he came to regard it as his own, and resented fiercely any intrusion upon his domain. This resentment and his desire to gain priority in all matters relative to Paleozoic paleontology caused him at times to resort to methods so questionable as to rouse the antagonism of nearly every paleontologist in America,[11] and subjected him to criticism of the severest kind. Undoubtedly a portion of the hatred and opposition to him which early became manifest was the outgrowth of that peculiar jealousy which the poet Riley has so aptly set forth in his humorous rhyme:

> And I've known some to lie and wait,
> And git up soon and set up late,
> To find some fellow they could hate
> For going at a faster gait.

Much of it was, however, apparently justified. His disposition to down an opponent through sheer weight of personal authority rather than by proof or argument sometimes resulted in placing him in laughable and awkward positions before his audience, from which he always emerged unembarrassed. It is told of him that at the Buffalo meeting of the American Association for the Advancement of Science a member rose to question a statement of his to the effect that three species of Spirifer characterized three zones of the Che-

[11] See Meek's comments relative to Hall's work in Iowa.

mung formation, the gentleman affirming that, so far from this being the case, the three species could often be found upon one slab, indicating that they belong to the same zone. This Hall emphatically denied, and declared that if such a slab could be shown him, *i.e.*, with the three species associated, he would eat it. The next day such a slab was actually produced, though it is not upon record that he was called upon to redeem his promise.

Fig. 42. *James Hall.*

He was a man of tremendous physical energy, whom no amount of opposition could overthrow, and on the failure of the legislature in 1850 to make the necessary appropriations for a paleontological survey of the state he carried it on for a time at his own expense, even when this involved the sacrifice of his private means.

Stevenson tells us that the fundamental feature of Hall's character was "child-like simplicity united to self-confidence and indomitable energy." With reference to the last-mentioned qualities no one will take exception.

> Knowing what he wanted he took a direct line, with little regard for anybody or anything which might be in the way to oppose. . . . He deceived his opponents by always telling the truth, something strange to politicians, but in time they came to understand him well, and strong men sought combat simply to measure strength, as in gladiatorial contests of olden time. Almost invariably he was victorious, but victory was often worse than defeat, for it converted into life-long enemies men who before had been merely indifferent. . . . He held his place for almost two-thirds of a century through no favor of man, but solely because he refused to be displaced.

For the benefit of those who, after the science of paleontology was well upon its feet, were disposed to claim it as a branch of biology, it may be well to remark that with Hall the problems of geology were always uppermost.

> His quartos on the New York paleontology are his monument, and the casual observer is liable to see in him a biologist rather than a geologist, but until his later years he was a geologist. His studies were from the standpoint of one seeking to determine relations between the physical and biological conditions in order to solve problems of correlation. The great problems of geology, not those of biology, were uppermost in his

mind until less than twenty years ago. His presidential address to the American Association for the Advancement of Science in 1857 (see p. 384), was so far in advance of the time as to be thought not merely absurd, but mystical, yet to-day it is recognized as one of the most important contributions to one of the most difficult problems in physical geology. Even in his later years, when biological problems had assumed their proper importance for him, he would have resented an intimation that he was any less of a geologist than before.

Fifteen quarto volumes, comprising 4,539 pages and 1,081 full page plates of fossils, stand as an enduring monument to his industry, a record which never has been, and presumably never will be, surpassed in the annals of American geology.[12]

Hall's Work in Ohio, 1841.

In 1841 Hall, fresh from his work on the geological survey of New York, made a tour through the then western state of Ohio to the Mississippi River, with a view of furthering the work of Vanuxem in 1829 relative to the identity of certain of the western formations with those of New York. As a result, he claimed that the rocks seen near Cleveland were "perfectly identical with those of the middle portion of the Portage group," and that the rocks of the Chemung group, now known to be Waverly, appeared near Newbury and Akron. The Cliff limestone of the Ohio geologists he considered an equivalent of the Helderberg series of New York, or at least of the Onondaga and Corniferous members.[13] Perhaps the most important of his generalizations was the following:

> From the facts here stated, the conclusion seems unavoidable that the character of fossils is, or may be, as variable as lithological characters; in fact, that the species depend in some degree upon the nature of the material among which they lived. Fossil characters, therefore, become of parallel importance to the lithological; and, in order to arrive at just conclusions, both must be studied in connection, and localities of proximity examined. In the case of the Hudson River group of shales and sandstones, in passing from New York to Ohio the lithological character is

[12] Professor Charles Schuchert informs me that if to the above are added his miscellaneous publications in octavo the number of pages of text published during his life may safely be set down as not less than 10,000. Large as are these figures, they were, however, exceeded by the celebrated Bohemian paleontologist, Barrande, who issued during his life 18 quarto volumes, comprising 5,568 pages of text and 798 plates of fossils, besides leaving manuscript and material for at least 10 volumes more, of which four have been issued since his death.

[13] At that time, the Corniferous, Schoharie, Cauda-galli, Oriskany, and the Onondaga, or Salina and Waterlime, were considered subdivisions of the Upper and Lower Helderberg.

almost entirely changed, and at the same time also the most prominent and abundant fossils are unlike those of the group in New York. More careful examination, however, reveals the fossils which characterize this group at the East, and also at the same time some obscurely similar lithological characters. Similar lithological changes, accompanied by like changes in fossils, occur in more limited districts within the State of New York. Without desiring to diminish the value of fossil characters as means of identifying strata, it must still be acknowledged that similar conditions in the bed of the ocean, and, apparently, similar depth of water, are required to give existence or continuation to a uniform fauna; and when we pass beyond the points where these conditions existed in the ancient ocean, we lose, in the same degree, the evidences of identity through successive depositions, often of very different nature; yet, at the same time, these may not have had a very wide geographical range.

He wrote further:

One of the most interesting changes in the products on going westward is the great increase of carbonate of lime, and the diminution of shaly and sandy matter, indicating a deeper ocean or greater distance from land. The source of the calcareous deposits is thus shown to have been in that direction, or in the southwest, while the sands and clays had their origin in the east, southeast, and northeast, producing a turbid condition in the waters of these parts during long intervals, and the formation of chemical deposits. In New York we are evidently upon the margin of this primeval ocean, as indicated in the character of the deposits as well as organic remains; the southwest unfolds to us that portion where greater depth and more quiet condition prevailed.

This is apparently the first clear enunciation by an American of what are now generally recognized as well-established principles.

Hall's Views Concerning Niagara, 1842.

It was but natural that Hall, while geologist of the fourth district of New York, should write concerning Niagara and its gorges. The views of Bakewell, Rogers, and Hayes have been already noted in these pages. Hall in his state report (1841), but more in detail in the *Boston Journal of Natural History* for 1842, covered the ground in a manner much more thorough and scientific than any of his predecessors.

He announced at the outset his disagreement with Professor Daubeny,[14] who considered that the terrace or escarpment at Lewis-

[14] Charles Daubeny of Oxford, England, who in 1838 had delivered a memoir on the geology of North America before the Ashmolean Society of Philadelphia. This was printed at Oxford, the following year, in form of a pamphlet of seventy-three pages.

ton was produced by a fault,[3] as he found no evidence whatever of even the smallest disturbance.

He noted the abrupt change in the direction of the stream at the Whirlpool, and also the gravel-filled channel extending from this point to St. Davids, but failed to realize that this may have been the one-time channel of the Niagara. He felt, rather, that this ravine was excavated by the power of the waves of the sea, aided probably by a stream which may have been of very insignificant proportions. The fact was, however, recognized that the river was at that time carving out its own channel, and the existing gorge from the falls to Lewiston was due to this cause.

Establishment of a Geological Survey of Canada, 1842.

In 1841, after several years of agitation, there was established a geological survey of Canada, of which W. E. Logan was put in charge in the spring of 1842, the sum of £1,500 being appropriated for the purpose of carrying out the work. During the first season Logan spent several weeks in examining portions of the coal fields of Nova Scotia and New Brunswick, and made his famous section of the Coal Measures at the South Joggins, which gave the details of nearly the whole thickness of the coal formation, some 14,500 feet, including 76 beds of coal and 90 distinct Stigmaria under-clays. Logan remained at the head of this survey until 1869, when he resigned, to be succeeded by Selwyn (see p. 445). During this period he submitted sixteen reports, dealing mainly with stratigraphic and economic subjects, and in 1863 a summation of his results under the caption of *The Geology of Canada* (p. 411).

Issachar Cozzens's Geology of New York, 1843.

Among the early members of the New York Lyceum was one Issachar Cozzens, a chemist and geologist, who has left as his main claim to recognition here an octavo volume of 114 pages and 9 plates, entitled *A Geological History of Manhattan or New York Island*, together with a map of the Island and a suite of *Sections, Tables, and Columns for the Study of Geology, particularly adapted to the American Student*. This was published in 1843. The map and sections were all hand-colored, the latter somewhat gorgeously, and included, aside from those relating to New York Island proper (Fig. 43), sections of Staten Island; one across the Palisades on the west side of the Hudson River; one from Stony Point, on the Hudson

River, through Dunderberg Mountain in New York; one from Brenton Reef to Portsmouth, Rhode Island, and one of Niagara Falls, the latter showing the origin of the falls through the gradual undermining of the softer shales and the breaking down of the harder limestones above.

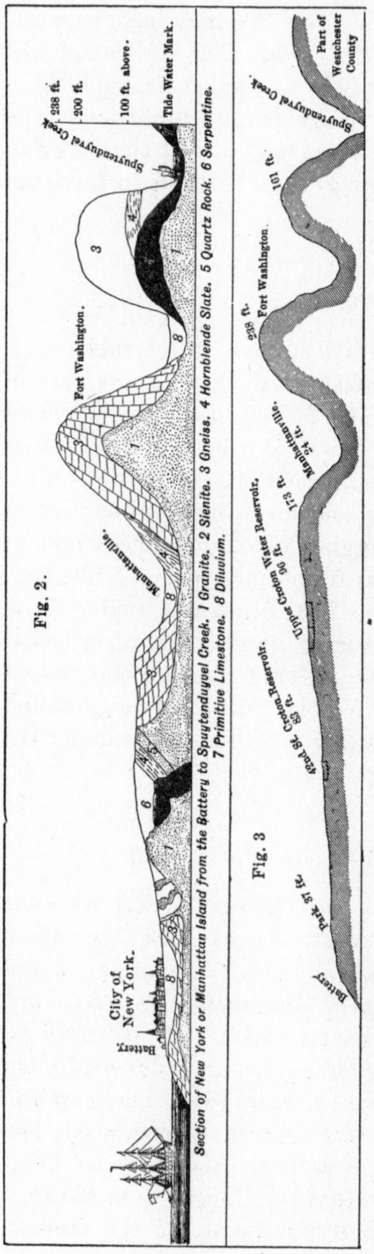

Section of New York or Manhattan Island from the Battery to Spuytenduyvel Creek. 1 Granite. 2 Sienite. 3 Gneiss. 4 Hornblende Slate. 5 Quartz Rock. 6 Serpentine. 7 Primitive Limestone. 8 Diluvium.

Fig. 43. *Issachar Cozzens's Profile of the Island of New York.*

The rocks of Manhattan Island proper were classed as (1) *granite*, (2) *syenite*, (3) *gneiss*, (4) *hornblende slate*, (5) *quartz rock*, (6) *serpentine*, (7) *primitive limestone*, and (8) *diluvion*.

In the various sections given, the vertical distances are grossly exaggerated, and the rocks piled on top of one another without any evident consideration of their original position or thickness, nor is there, in the text, any indication that the author considered at all the problems relating to original horizontality, uplift, and erosion. The red sandstone associated with the New Jersey traps (Triassic) he considered Old Red, but the impressions of fossil fish that had been reported as found therein he thought came from "one of the upper members of the Coal Measures which lie above it." The Triassic conglomerate of Maryland and Virginia was also incidentally referred to the Coal Measures. This seems a little strange when we consider the previous work of his contemporaries, Redfield and Hitchcock.

It is stated in his preface

that the work was first undertaken for his own amusement and study, and it is possible that neither his influence nor attainments can be gauged by this one publication.[15]

Lawrence's Geology of the Western States.

In 1843, also, there was published by Byrem Lawrence a geological map and descriptive pamphlet of the western states, which at that date included nothing west of the Mississippi Valley. This was designed for popular use, and was claimed to be the first attempt of its kind. "The West," he wrote, "has no mountains nor even hills, except such as have been produced by the action of running water. . . . The rocks here have never been broken by violence nor tilted up as in the East, but lie apparently horizontal in the sides of the Hills." The drift he described in some detail, its source being recognized and its cause attributed to water and ice. "It is probable there may be some analogy between the whole matter and the icebergs of the present day."[16]

Ruffin's and Tuomey's Work in South Carolina, 1843-1844.

In 1843, too, the noted Edmund Ruffin, agricultural surveyor of South Carolina, published a report on the commencement and progress of his survey, in the form of a pamphlet of 175 pages. This naturally related largely to agricultural matters. It contained, however, a list of the invertebrate fossils of the state and numerous analyses of the marls. He noted that nearly the entire country above the lower fall line of the rivers was occupied by granitic rocks and their residual soils. The first attempt at a systematic *geological* survey of South Carolina was inaugurated in 1844 with the appointment of Michael Tuomey as state geologist. The first report, an octavo pamphlet of forty-five pages with an appendix, was submitted in November of the same year. This was devoted mainly to economic

[15] In the Annals of the Lyceum of Natural History for 1867, Mr. R. P. Stevens had a paper, read in 1865, on the past and present history of the geology of New York Island. In this, after reviewing the work of H. H. Hayden, Maclure, Akerly, and Cozzens, he gave the results of later observations by himself in the 75 to 100 miles of artificial sections exposed since Cozzens's time. He showed the island to be a portion of the mainland of Westchester County cut off by a profound fault. The rocks enumerated were chiefly gneiss, which he considered metamorphosed Taconic, cut by granite, and interbedded with limestone.

[16] Lawrence is credited in Darton's bibliography of American geology with but two very brief geological papers—"On Coal in Arkansas" and "The Geology of Arkansas"—published in New Orleans in 1851 and 1853, respectively.

considerations, and, aside from a few notes on rock weathering, contained little of general interest. His final report, which appeared in 1849, will be noted later.

Dana's Views on Metamorphism, 1843.

Prior to the introduction of the microscope and the study of rock structures by means of thin sections (about 1870) views regarding metamorphism, particularly such as are due to shearing stresses in the older crystallines, were naturally somewhat vague. With rocks too fine of grain to allow a determination of their structure and mineral composition by the unaided eye, the mode of occurrence only could be relied upon to give a clew to their origin, and it was not until the thin sections made it possible to sometimes trace a distinct gradation from schistose and foliated rocks of decidedly metamorphic aspect into massive eruptives of the same ultimate composition that a feeling of doubt began to arise in the minds of observers regarding the assumed sedimentary origin of the gneisses and crystalline schists.

It was during this prematinal period, as Rogers might have termed it, that J. D. Dana came forward with a suggestive paper in the *American Journal of Science* on the analogies between the modern igneous rocks and the so-called Primary formations and the metamorphic changes produced by heat in the associated sedimentary deposits. The conclusions arrived at were based upon observations made during the Wilkes Exploring Expedition, and the exciting cause of a paper at this time would appear to have been the somewhat varying views recently put forth by Lyell.

Dana argued: (1) That the schistose structure of gneiss and mica-slate was not necessarily an evidence of sedimentary origin; (2) that some granites having no trace of a schistose structure may have had a sedimentary origin; and (3) that the heat producing metamorphism was not applied from beneath by conduction, but was rather due to heated waters of the ocean which permeated the rocks. As confirmatory of the first he called attention to the parallel arrangement of the minerals and the consequent platy structure, with a tendency to split along certain lines, which were sometimes found in volcanic and other igneous rocks. The possibility of a schistose structure due to dynamic (shearing) causes was not, however, suggested.

His argument for the possible metamorphic origin of certain granites was based upon the fact that some of the basaltic tuffs observed by him in the Andes and in Oregon had become so indurated that their fragmental origin was almost wholly obscured. He felt that if rocks of this type could be so remodeled or rehardened as to be scarcely distinguishable from the parent rock, sedimentary deposits of granitic material might undergo a like change.

His argument against metamorphism by dry heat was enforced by calling attention to the low conducting power of stone, which is such that heat, even to the point of fusion, may be transmitted only a few inches. "Lavas may be heated to a red heat within a yard of the surface and still be so cool above that the bare foot may walk upon them." He believed that subterranean waters or water on sea bottoms might become so heated by "volcanic fires" as, on permeating the rocks, to bring about the metamorphism. This, of course, would mean that the gneissic rocks were not necessarily deep seated (hypogene), as held by Lyell, but analogous to other rock formations deposited and solidified at or near the surface.

Beck and Dana on Zeolite Formations.

The geological chemists, as might be expected, had their troubles in accounting for what they observed. Professor L. C. Beck, the mineralogist of the New York survey, in 1843 described various phenomena in connection with the igneous rocks, dwelling particularly upon the presence in them of sundry hydrous minerals as zeolites and serpentine. He noted that the latter occurred at times under such conditions that its igneous origin was evidently assured. "But," he added, "if this is the correct view, how happens it that while all the minerals of the granite, gneiss and limestone are destitute of water, the serpentines are almost always loaded with that substance?" The problem he thought might perhaps be solved by an investigation into the composition of the minerals of trappean and particularly those of the zeolite group. From the fact that volcanic products were often of a hydrated nature he evidently felt that the presence of these minerals was not paradoxical, though he seemingly failed to recognize the fact that they were in themselves secondary, *i.e.*, not products of the primary cooling of the molten magma. In his final summing up he becomes almost prophetic, when we consider the modern ideas of vein formation and magmatic differentiation. "Now I think it conceivable that the character of the igneous eruptions may have been connected with circumstances attending the different depths to which the refrigeration, and consequently the solidification of the crust may have extended. When the granitic deposits had been partially solidified, fissures and cracks in the crust would be followed by injections of the same mineral ingredients, in some instances perhaps sparingly mixed with those below. Hence the formation of true granite veins with their accompanying minerals during such a state of things might be easily accounted for. But, as the solidification extended toward the interior, the erupted matter would exhibit a different aspect, owing perhaps in part to the new agencies which were brought into action, but chiefly to a real difference in the mineral matter or composition, which we have reason to believe exists in different parts of the central nucleus, or at different distances from the surface." Such an idea would require but little modification to make it applicable to the origin of the "veins" of granitic pegmatite.

This subject was brought up again by J. D. Dana, in a paper read at the meeting of the Association of American Geologists in 1845. In this paper Dana brought out the distinction, overlooked by Beck, between the essential minerals and those occupying seams and cavities in rocks and which he rightly considered as secondary. To the source of the materials forming these secondary minerals he devoted especial attention. They could not have proceeded from vapors attending the eruption and, further, zeolitic minerals, he pointed out, had been found under such conditions as to prove that they "actually result from decompositions and recompositions" of preexisting minerals, and their formation in the cavities, either crevices or vesicles, he argued was due to infiltration, and he felt that available evidence pointed to heated solutions, although recognizing the fact that in some special cases, as the formation of zeolitic stalactites in caves, ordinary temperatures might have prevailed. These views he stated further were not wholly original with him, but were in accordance with those previously announced by European geologists. It would appear, too, that they were essentially the same as those advocated by him two years earlier at the Albany meeting of the Association.

Fremont's Expeditions, 1842-1844.

The expeditions of Fremont to the Rocky Mountains, Oregon, and northern California were not accompanied by a geologist and the few observations in the field were of little value. Attempts were made at collecting plants, fossils, and other natural history objects, the transportation of which was often accomplished with great difficulty and sometimes accompanied by disaster, as shown by the following transcript from his report:

Unhappily, much of what we had collected was lost by accidents of serious import to ourselves, as well as to our animals and collections. In the gorges and ridges of the Sierra Nevada, of the Alta California, we lost fourteen horses and mules, falling from rocks or precipices into chasms or rivers, bottomless to us and to them, and one of them loaded with bales of plants collected on a line of two thousand miles of travel; and, when almost home, our camp on the banks of the Kansas was deluged by the great flood which, lower down, spread terror and desolation on the borders of the Missouri and Mississippi, and by which great damage was done to our remaining perishable specimens, all wet and saturated with water, and which we had no time to dry.

Notwithstanding all the difficulties, a small series of fossil plants were brought east and placed with Professor James Hall for identification. These were referred by him to Upper Jurassic formations. For more than forty years these specimens remained lost and forgotten until, in 1887, they were rediscovered among the collections of the National Museum. Still more recently they have been restudied and described by Dr. F. H. Knowlton and by him referred to the frontier subdivision of the Upper Cretaceous.

Association of American Geologists and Naturalists, 1844.

The Washington meeting of the Association of American Geologists and Naturalists, May, 1844, was of exceptional interest and importance. The retiring president, H. D. Rogers, gave a long and exhaustive address on the "History of the Recent Labors of American Geologists," and a summation of the present condition of geological research in the United States. Among the numerous shorter papers offered only those of Hitchcock, Silliman, and Rogers can here be noted.

HITCHCOCK ON THE TRAP TUFA OF THE CONNECTICUT VALLEY AND ON INCLINATION OF STRATA.

Professor Hitchcock described a trap tufa differing from "common trap" in being conglomerate and apparently carrying organic remains. He believed that this was produced before the main ridges of trap along the Connecticut River, by precursory outbursts of pumice, scoria, ashes, and melted matter falling over the bottom of the ocean, where it became admixed with sand and gravel. After this layers of sandstone accumulated over it, and finally the main ridges of trap were protruded through the strata, tilting them up.

The dip of the sandstone in the valley he thought to be due in part to elevation by the protrusion of the trap and in part to the lifting and lateral movements of the adjoining primary ridges. He felt, however, that the sandstone might have been originally deposited on a slightly inclined plane. This matter of the inclined position of strata, how much of it was due to original deposition and how much to upheaval, it may be remarked, was one which seems often to have troubled the earlier geologists, and also one which, so far as the literature shows, was never solved by itself, but incidentally, and in connection with other problems, where it was gradually lost to sight. Cleaveland, it will be remembered, in his work on geology and mineralogy, 1816, expressed a doubt as to whether the inclined position of strata was original or due to some powerful cause which had elevated them from horizontality. Maclure in 1825 suggested that the dip, so far as the Transition rocks were concerned, might result from their having been laid down on a primitive floor, concerning the position of which nothing was known. Hitchcock, with his usual caution, felt that both causes might have been instrumental in producing the dip observed.

SILLIMAN ON THE INTRUSIVE TRAP OF THE NEW RED SANDSTONE OF CONNECTICUT.

At the same session Benjamin Silliman, Jr., presented a report *On the Intrusive Trap of the New Red Sandstone of Connecticut.* His conclusions were to the effect that the sandstone was let down from suspension in water in the inclined position which it now occupies and that it had suffered no change in dip, excepting in immediate connection with the injection of the trap rocks. He considered it probable that the strata were deposited by a primeval ocean current setting from the southwest and west. Subsequent to this accumulation and consolidation the lower primary rocks were disrupted, the igneous rocks injected through the fissures and distributed along the lines of least resistance in the sedimentary strata—that is, along and up the plane of the dip, thereby lifting the strata parallel to the dip from the beds on which they had before reposed. At the same time there were produced in the upper strata fissures and transverse cracks which were filled with the molten trap. This injection he thought was probably continued during a long period, but all referable to the same geological epoch and anterior to the elevation of the strata in which it occurred.

After the deposition and injection ceased and the elevation of the present continent had commenced, denudation set in, induced by a northerly current, itself due to the flowing off of the oceanic waters incidental to this elevation. By this denudation the soft shales and other materials were removed and the trap ridges developed. Thus simply, deposition, intrusion, uplift, and erosion were all accounted for.

H. D. ROGERS ON THE ATMOSPHERE OF THE COAL PERIOD.

Early speculations regarding the origin of the earth, though largely fanciful, were founded only too frequently upon the brief outline of a series of events as chronicled in the first chapter of Genesis. Theories regarding earth development and its probable destiny other than its catastrophic annihilation through divine wrath, as a punishment for sinful man, were slow to appear. When such did appear, however, they were founded upon a much more scientific basis.

That a mutual reaction was going on constantly between the superficial portions of the earth and its surrounding atmosphere was doubtless realized by all of those who wrote on the weathering of rocks and the formation of soils; but, so far as known to the present writer, H. D. Rogers was one of the first to show by direct calculation that the earth had been in the past, as at present, robbing the atmosphere of some of its constituent parts and gradually storing them up in its solid crust, although Vanuxem in 1827, while connected with South Carolina College, had pointed out the probable change in the atmosphere during geologic time through the absorption by the earth of nitrogen and oxygen, and also the probability of a warm, moist climate during the period of coal formation. At the Washington meeting of the Association, Rogers submitted a communication on the probable constitution of the atmosphere at the period of the formation of the coal. He estimated that the amount of carbon existing in the atmosphere today in the form of carbonic acid would be only sufficient to furnish through vegetable action about 850,000,000 tons of coal, while the probable quantity of this substance in existence, all of which must have been elaborated from the ancient atmosphere, was nearly 5,000,000,000 tons;[17] that is to

17 Now estimated at more than one thousand times that amount for America alone.

say, about six times what the present atmosphere would produce. So great a reduction in the carbonic acid, implying a corresponding augmentation of oxygen, he felt to be a matter of great interest in geology, as showing that every modification in the constitution of the air had adapted it to the development of animals progressively higher in the scale of organization. Singularly enough the paper seems to have excited little interest at the time.

C. B. Adams's Work in Vermont, 1844.

Of all the New England States Vermont was the last to become infected with a desire for a geological survey, and it was not until 1844 that the final steps were taken which resulted in the appointment of Professor C. B. Adams as state geologist. Adams was educated at Phillips Academy and Amherst College, became a tutor and for a brief period assistant to Professor Hitchcock in his work of organizing the geological survey of New York. When appointed state geologist of Vermont he selected Zadock Thompson his chief assistant, Denison Olmsted, Jr., and later T. Sterry Hunt aiding in mineralogy and chemistry. Reverend S. R. Hall, of Craftsbury, was employed to look after the agricultural features. Up to March, 1848, the survey continued in a fairly prosperous condition, but at the session of the legislature for this year appropriations were for some reason withheld and the work stopped. Four reports in all were issued, the last a mere pamphlet of eight pages.

The purpose of the survey, as stated in the first report, was to collect and analyze the soil, the simple minerals, both of economic and scientific importance, and to make investigations into the character and limits of the geological formations. The reports, on the whole, were extremely fragmentary. Owing to the death of Adams before the manuscript of the final report was prepared, a large part of his work was lost. This was due mainly to the fact that he took his notes in a "peculiar shorthand, which could be read by no one but himself." As suggested by Professor Thompson, it would have required more labor to decipher his notes than to go over the ground anew.

The second annual report was prepared—as was not infrequently the case at that date—in the form of a general treatise on geology, and began with an elementary chapter which dealt with all subjects relating to geological phenomena, whether applicable to the state or not. The views advanced were those given by the textbooks of the period and need not be mentioned in great detail. An ideal section

(Fig. 45) from the first report is, however, reproduced here as illustrating equally well the rough method of picture making with which the early workers had to content themselves and the prevailing ideas relative to the earth's crust and interior. He would, indeed, be regarded as a novice in geology today who would not at first glance inquire what supported the mass of solid material represented as resting upon the molten lava, yet this question does not seem at this time to have even presented itself for consideration. A little space was devoted to the subject of the connection between geology and the Bible, to meet the wants of the "many well-meaning persons" who were disturbed by a supposed incongruity between the principles of geology and the Mosaic narrative. A very abstruse explanation of concretionary structures was offered, Adams failing, as did many of his contemporaries, to discriminate between concretionary structure as it is now understood, and certain forms due to contraction in cooling assumed by igneous rocks. He looked upon columnar structure in basaltic rocks as a peculiar form of concretion caused by lateral pressure such as might exist between spheres compressed at the sides.

Fig. 44. *C. B. Adams.*

Agassiz's glacial theories had been apparently without effect, even if they were known, and icebergs were still considered sufficient to account for all the phenomena of the drift.

In 1847, the year before the closing of the survey, Adams had

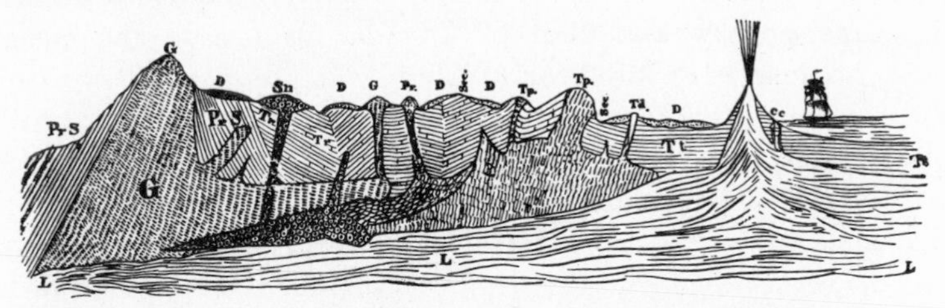

Fig. 45. *Adams's Ideal Section of the Earth's Crust.*

accepted a professorship at Amherst, and in 1853, in conjunction with Alonzo Gray, published an *Elements of Geology*, a duodecimo volume of 350 pages. Here his geological work seems to have ceased.

Indeed, from the date of the abandonment of the state survey he devoted himself mainly to zoölogy, and particularly to conchology, making large collections especially rich in West Indian and Central American forms. While at St. Thomas in 1853, he contracted yellow fever, from which he died. He is described as a man of sturdy build, medium height, with large, black eyes and black hair; a man of tremendous physical endurance, knowing neither fear nor what it was to be tired; but, withal, of a quiet and self-contained demeanor.

Sketch of Zadock Thompson.

Fig. 46. *Zadock Thompson.*

Thompson, Adams's chief assistant, is represented to us as a man poor in this world's goods as well as in general health, and modest almost to a fault—one who from childhood had shown a passion for writing books. His first publications were almanacs, which he himself sold while traveling on foot about the country. It is told of him that when at one time interrupted by his clerk with an inquiry as to the prediction that should be made for the weather during a midsummer month in the forthcoming issue, he replied somewhat testily, "Say 'Snow about this time.'" So it appeared in the printed almanacs, and to the astonishment of all, including both printer and author, snow did fall in Vermont that year in the month of July.

The history of the man is interesting. He graduated from the University of Vermont at the age of twenty-seven. He published an arithmetic, a geography and map of Canada for use in the common schools; became in 1832 editor of the *Green Mountain Repository;* wrote a history of Vermont, which appeared the same year; studied theology, and took deacon's orders in the Protestant Episcopal Church in 1836. He preached for a time and then returned once more to his book writing, first publishing in three volumes a *Natural, Civil, and Statistical History of Vermont;* then a textbook on the geology and geography of Vermont; finally becoming an assistant to Professor Adams on the state survey in 1845, when his geological work really began.

"Tall, angular, of a very quiet and sedate, yet very pleasant, manner, a man of most amiable and sweet temper, loved by all who knew him, and respected for his sound sense and accurate judgment." Such is the picture given us of him, who certainly was one of the remarkable men of his times although his name does not stand the highest in the annals of geology.

The Richmond Bowlder Train, 1845.

Under the caption of *Description of a singular case of dispersion of blocks of stone connected with the drift in Berkshire County, Massachusetts,* Dr. Edward Hitchcock described in the *American Journal of Science* for 1845 a remarkable train of glacial bowlders extending from Fry's Hill in the Canaan Mountains of New York, for a distance of some fifteen or twenty miles, southeasterly into Massachusetts. (Fig. 47.)[18]

Though forming three somewhat meandering trains extending from the hill above mentioned, through the adjoining valley and upward over an elevation of 800 feet at the state line, across the Richmond Valley and over Lenox Mountain, 600 feet in height, to and over Beartown Mountain, 1,000 feet in height, the lithological nature of the bowlders was such that they could all be traced to a common source.

Naturally, so striking a phenomenon excited investigation, and, naturally, too, Hitchcock, in the then existing condition of knowledge regarding glacial transportation, found difficulty in accounting for it. He recognized the similarity of the trains to the lateral glacial moraines described by Agassiz in his *Études sur les Glaciers,* which had appeared five years previously, but could not conceive of a glacier traveling directly across the intervening ridges, even were the mountains in the vicinity of sufficient altitude to give rise to the same. Neither did the consideration of river drift or floating ice afford him a satisfactory conclusion. "In short, I find so many difficulties on any supposition which I may make that I prefer to leave the case unexplained until more analogous facts have been observed."

Unsatisfactory and apparently unimportant as the paper may, at first thought, seem, it is mentioned here on account of the extraordinary explanation of the phenomenon offered by the Rogers brothers three years later. According to their description the trains started each from a particular depression in the summit of a high ridge, in Canaan, New York. Taking a direction S. 35° E., they

18 This remarkable train of erratics was first brought to public attention by a Dr. S. Reid, who described it in the columns of the Berkshire Farmer of 1842.

cross the higher ridges and their intervening valleys for distances of from ten to twenty miles. Neither train is more than 300 or 400 feet in breadth and the distance between them not more than half a mile. The transported blocks are of all sizes up to twenty feet in diameter, sharply angular, free from scratches and all lithologically identical with the ridge from whence they started. After exhibiting to their own satisfaction the inadequacy of either the iceberg or the glacial hypothesis to account for their production, the authors, in a paper before the Boston Society of Natural History in 1848, attempted to show how all the phenomena might be explained on the theory of a sudden discharge of a portion of the Arctic Ocean southward across the land. They discussed the important functions of the "wave of translation," showed its surpassing velocity and great propulsive power, and traced the influence of vehement earthquakes near the pole in dislodging the northern waters and ice and maintaining in the rushing flood these vast and potent waves. They then suggested that, at a certain stage of the inundation, the ice, previously floating free, might impinge with irresistible violence against the tops of submerged hills, and that the Canaan Mountain stood precisely in the position to take the brunt of the ice-driving flood as it swept down the long, high slope of the distant Adirondack and across the low, broad valley of the Hudson.

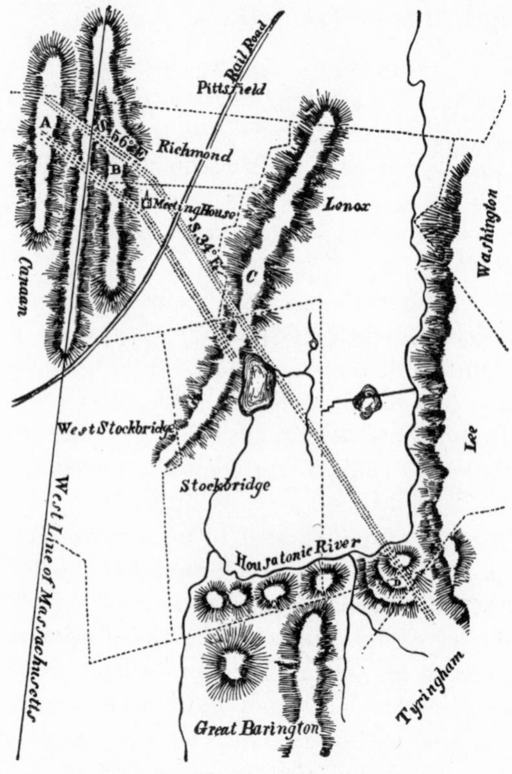

Fig. 47. *The Richmond Bowlder Train.*

They then proceeded to show how, at the instant when some enormous ice island struck the crest of the mountain and scooped the trench which is there visible, a great vortex was produced by the obstruction thus suddenly thrown in the path of the current, which, endowed with an excessive gyratory velocity, sustained and carried forward the greater part of the fragments. As in the instance of the waterspout and the whirlwind, the whirlpool would gather into the rotating column the projected blocks and strew them in a narrow path along the line of its pendent apex.

The paper terminated with an application of this idea in detail, to the explanation of each important feature of the trains; to their deflection from a straight line; the intermission in the bowlders at certain places in the train, and the fact that some of the blocks had been violently broken at the moment previous to their final deposition.[19]

Fig. 48. *J. B. Perry.*

Truly there were catastrophists in those days.

We may anticipate here by stating that at the meeting of the American Association for the Advancement of Science in 1870, Professor J. B. Perry read a paper on glacial phenomena, in which he referred to these trains incidentally, and argued that they were part of the true glacial drift—morainal material deposited by the melting of the ice sheet of the glacial epoch. Again, in 1878, Mr. E. R. Benton studied the trains in considerable detail, and in the *Bulletin of the Museum of Comparative Zoology* published a detailed description and map of the region, from which the one given herewith (Fig. 47) is reproduced. Benton took the ground that the bowlders were deposited by the ice sheet of the glacial period at the time of its final melting, the direction of the striæ on the underlying rocks being essentially the same as that of the bowlder trains. In connection with his investigations, Benton wrote to Professor W. B. Rogers concerning the views above advocated by him, and received in reply the letter from which the following abstract is taken:

19 In 1852 the English geologist, Lyell, visited this region in company with James Hall, the former's views on the subject being published in 1855 in the Proceedings of the Royal Institution. He regarded the hypothesis of glacial transportation as out of the question, and apparently also that put forward by the Rogers brothers, for he made no mention of it. His own idea was that the large erratics had been transported to their present resting place by floating coast ice.

At that time [*i.e.,* when this paper was written] paroxysmal dynamics had still many advocates, and the attempted explanation may be interesting as a specimen of the bold type of speculation in which some of the early geologists sometimes ventured to indulge. But, for myself, I may say that long years of observation and study make me more distrustful of our knowledge of causes and more willing, in geology as in other things, to labor and wait.

Mather on the Physical Geology of the United States, 1845.

In the *American Journal of Science* for 1845, W. W. Mather, whose work in New York and Ohio has been mentioned, published a paper on the physical geography of the United States in which he still further elaborated some of the interesting and rather unique ideas regarding the origin of the secondary rocks and the elevation of islands, continents, and mountain chains which he had previously put forth (p. 225). These may be referred to in considerable detail. He again argued on the apparently safe basis that the earth is a cooling body, contracting, and hence undergoing an increased velocity of rotation upon its axis. The oblateness of its spheroidal form, due to the increase of centrifugal force, would therefore induce a flow of water from the polar to the tropical regions, and as the earth revolved from west to east these currents from the poles would bend more and more to the westward as they advanced to lower latitudes. On the other hand, the water in the tropics, being gradually expanded through heat, would tend to flow off toward the polar regions.

These currents he thought to have operated throughout all the time since the ocean occupied its present bed, and to them he would ascribe the mechanical distribution of many of the sediments making up the Secondary rocks. Considering only those that constituted the Equatorial, the Gulf Stream, and the Labrador Current in the Atlantic, he thought to show that within the great eddy due to their interference were deposited all the materials constituting the immense mass of sedimentary rocks between the primary ranges north of the Great Lakes and the Gulf of Mexico, and between the Blue Ridge and the Rocky Mountains.

The red sandstone formations extending from Carolina to Stony Point on the Hudson he believed to have been formed through the transporting power of the Gulf Stream, its northward extension having been cut off by a polar current flowing through the Champlain and Hudson River valleys. The origin of the loosely consolidated material now classed as glacial drift was also referred to the same agency. The coal beds of the eastern and central portions of the United States he looked upon as formed at the bottom of the ocean in which great eddies occurred, and where plants brought from the tropics and other sources would float and circle about until they sank. He noted that these beds were based upon a sandstone which at the outcrops on the edges of the coal basins was sometimes a conglomerate or coarse pudding stone, while through the center of the basins it was much finer. This fact denoted to his mind a stronger current on the exterior of the coal basins than within its area.

Concerning the regions over which the polar currents were supposed to have flowed, and from whence, as a consequence, were derived the materials for the sedimentary rocks, he could say little from actual observation. Drawing for his materials mainly on the writings of travelers, though acknowledging that such might be faulty, he nevertheless pictured a scene of barrenness of the entire region north of the St. Lawrence and Great Lakes and from Newfoundland to the Stony Mountains such as is equaled only by that given by H. H. Hayden in his attempt to account for the deposits of the coastal plain (p. 81).

Although the quotations from travelers lack that accurate examination that is necessary to a determination whether the surfaces thus described have been exposed to the action of violent and long-continued currents, yet they have their weight. When considered in connection with the effects of known physical causes, it is rendered more than probable that the currents under consideration have flowed from the polar regions toward the equator and from the tropics toward the poles when this continent was beneath the ocean, and that the matter which composed the deposits of the sedimentary rocks of the United States was washed away by these great equilibrating currents from the bed of the ocean, from reefs, islands, and coasts, and finally deposited from suspension over the great area where we now find it exposed to observation.

The probable cause of the elevation of the land he thought to be secular refrigeration accompanied by contraction, whereby a gradually accumulating tension was finally overcome by a paroxysmal yielding of the solid strata. At the same time a sudden increase of the velocity of the earth's rotation would result, and this in turn would increase the flow of the ocean currents as above mentioned.

In attempting to account for the overturned character of the folds in the mountain chains of the eastern United States, he seems to have actually outdone the Rogers brothers in fecundity of ideas. He wrote:

Paroxysmal elevation and the action of inertia offer a satisfactory explanation of the folded axes and eastwardly dip. . . . Suppose the sudden elevation of a mountain mass one mile in height, or more distant from the axis of rotation than it was before its elevation. It would still retain the linear velocity it had when a mile nearer the axis of rotation, while the proper linear velocity at this increased distance would be $\frac{3.1415}{24}$ miles, or 690 feet greater per hour than that which it had before its elevation. Inertia would therefore cause the mass at the top to press to the westward with a force proportional to its mass and the above-mentioned velocity, and at intermediate heights with a proportionally less momentum. If the strata be capable of yielding, they must, when elevated in highly curved wrinkles, tend to fall over westward as a consequence of the influence of inertia and the revolution of the earth from west to east on its axis.

Such views from a man of Mather's training and experience and at this late date seem extraordinary, to say the least.

Commenting further on the effects of centrifugal force, particularly upon bodies of different densities, he argued that as such is greatest under the equator, any subterranean force tending to elevate portions of the earth's surface by their elastic tension would here be most effective. In this way he accounted for the supposed fact that most of the highest mountains are found within the tropics. Reasoning along the same lines, he conceived it possible that fractures might be expected to develop in the direction of circles parallel to the equator at a distance intermediate between it and the poles, where the curvature resulting from the revolution of the earth would be greatest.

Conrad's Medial Tertiary of North America, 1845.

In continuation of his work begun in 1832, T. A. Conrad in 1845 issued his volume on the *Medial Tertiary Formations*, the general plan of the work conforming closely to those which have already received consideration.

The nature of the country, its strata, and fossils were described in detail. The most interesting part of the paper, from the present standpoint, lies in his remarks on the theory of elevation and the age of the deposits. He thought the formation of each Tertiary division and the final annihilation of its fauna to have been a very gradual process, taking place in quiet waters and having no connection with volcanic agencies or any violent movements of the earth's crust, but merely depending on changes of temperature.

In discussing these temperature changes he showed some peculiar ideas on the conductivity of rock masses and the effects of cooling. The cold, penetrating deep into the igneous rocks beneath the surface, he thought, would result in a maximum amount of crystallization and expansion,[20] producing thereby a slight elevation of the crust. Since the cold penetrated gradually, he argued that the elevation would also be gradual, more appreciable, of course, during epochs of unusual cold. In brief, crustal elevation, he would have us believe, was induced not by volcanic agency, but by "the all-powerful and pervading influence of crystallization in the primary rocks." In this he followed what he called the "sublime" theory advanced by Vanuxem (see p. 230). The idea that the abrupt change in the character of the animal life at the close of each geological epoch by a sudden fall of temperature was, however, avowedly the theory of Agassiz.

David Christy's Geological Observations, 1845-1848.

In connection with work pertaining to the Liberian Colonization Society, in 1845-1848, David Christy traveled extensively throughout the eastern and middle United States. Being a man of active mind and with a love for the sciences, he made many geological observations which were first embodied in a series of letters to Dr. John Locke and published in the *Cincinnati Gazette*. These were afterward issued (in 1848) in pamphlet form, some seventy pages, with five plates of fossils and three colored geological sections: the first from Mine La Motte and Pilotknob, Missouri, to Hollidaysburg, Pennsylvania; the second from Lake Erie to Pensacola Bay, and the third from Richmond, Indiana, through Oxford, Ohio, to Beans Station, Tennessee.

It is impossible to say what portion of the information given in these letters can be claimed to be the result of original investigation or observation. As stated by Locke in his introductory note, Christy "has studied the magnificent pages of Nature . . . with so much success that he has availed himself of a vast amount of knowledge not found in any ordinary printed volume. He has studied the fossil remains, so abundant in our rocks, until he is enabled to identify strata at remote geographical distances, and has referred the geological formations to the Blue Limestone of Cincinnati as a kind of zero, informing us whether the rocks at any place are above or below

[20] As a matter of fact, cooling and crystallization in a rock magma would result in contraction.

that zero." Locke further states that he knew of "no other individual who has actually drawn approximate sections of the strata from the Atlantic to Iowa and from Lake Erie to the Gulf of Mexico; most of this work being the result of his own observations."

Fig. 49. *David Christy.*

In a letter addressed to M. de Verneuil in 1847 (and afterward printed and bound up with the above-mentioned pamphlet), relative to the erratic rocks of North America, Christy showed himself to have been a catastrophist. He believed that there had been two periods of elevation of the land in the history of the continent, with an intervening submergence. The second and last emergence he considered to have been rapid and to have given rise to the swift currents which were instrumental in producing the drift.

Wislizenus's Explorations in Mexico, 1846.

In the spring of 1846, F. A. Wislizenus,[21] a "German by birth and an American by choice," left St. Louis, Missouri, with the intention of making a tour through northern New Mexico and upper California for the purpose of examining into the geography and general natural history of the country. "The principal object of my expedition was scientific. I desired to examine the geography, natural history, and statistics of that country, by taking directions on the road with the compass, and by determining the principal points by astronomical observations. I made a rich collection of quite new and undescribed plants. I examined the character of the rocks, to gain insight into the geological formations of the whole country. I visited as many

[21] "There is also a Thuringian who looks like an American, but is none, as the poet sings:

> " 'Thüringens, Bergezum Exampel, geben
> Gewächs, sieht aus wie Wein
> Ist's aber nicht'

yclept Wislizenus, who hath wandered in New Mexico and written a book and is very full of prickly pears, burs, and cacti überhaupt. Him also will I move to communicate with you. I wish you would sometime hither and see these men whom you would find not good naturalists merely, but accomplished and agreeable persons." From letter of G. P. Marsh to S. F. Baird, 1848.

mines as possible, and analyzed some of the ores. I made barometrical observations, to ascertain the elevations above the sea. I kept meteorological tables, to draw general results from them for the climate, its salubrity and fitness for agriculture, and took memoranda in relation to the people—their number, industry, manners, previous history, &c."

Fig. 50. *F. A. Wislizenus.*

Unfortunately for his intentions, war between the United States and Mexico broke out while he was within the jurisdiction of the Mexican Government, and he was compelled to remain in the State of Chihuahua for a period of six months or until the arrival of the American troops, when he accepted a position in the medical department of the army and returned with it by way of Monterey to the States.

His opportunities for observation were, naturally, much less than was at first anticipated, and his geological report, amounting to but five pages, was published with 137 pages of itinerary and meteorological tables as Senate Miscellaneous Document No. 26 of the Thirtieth Congress, first session. In this he noted that the bluffs on the Arkansas some 341 miles from Independence were of a grayish conglomerate limestone, containing fossils which "seemed to belong to the Cretaceous formation." Near Las Vegas he found a dark blue limestone with casts of *Inoceramus*, which were also relegated to the Cretaceous series. Near El Paso he noted the presence of a limestone containing the fossil coral *Calamopora* and a bivalve shell of the genus *Pterinea*, which, as a consequence, he considered to belong to the Silurian formation. The presence of numerous eruptive rocks was noted and an outline map published, in which the lithological nature of the rocks was indicated. The map comprised the area lying between the Arkansas and the Rio Grande rivers, but extended south and west of the latter as far as Monterey and Chihuahua.

Binney's Views on the Loess, 1846.

The fine siltlike character of the superficial deposits of the Mississippi Valley had frequently been noted by the various geologists who passed over the region, but it remained for Amos Binney, the conchologist, to give the first reasonably satisfactory account of their probable origin. At a meeting of the Boston Society of Natural History in April, 1846, he exhibited a collection of fossil shells from the so-called Bluff formation at Natchez, on the Mississippi, and announced the belief that the formation was analogous to the Loess of the Rhine Valley and a result of fluviatile action rather than attributable to the glacial drift.

Rogers on the Geological Age of the White Mountains, 1846.

In the *American Journal of Science* for the same year W. B. Rogers had a paper on the "Geological Age of the White Mountains of New Hampshire," in which he announced the finding of a fossiliferous strata involved in the uplift, and affording, as he believed, convincing proof that the region now occupied by the mountains was overspread by the ocean at an era as late as the Matinal (L. Silurian) or more probably Levant (U. Silurian) period. He thought, further, to have found proofs of two different epochs of elevation, the latest of which probably was a part of the general movement which lifted the whole Appalachian chain. Whether the earlier accorded with the close of the Matinal or early Levant only further research would tell.

> If we conceive the whole tract from the Upper Hudson and Lake Ontario, eastward to and including Maine, to have emerged from the waters into permanent land during the close of the Matinal period and the first stages of the Levant while the still wider spaces to the S.W. and N.E. remained undisturbed for the reception of later strata, we shall be able to interpret many important facts in the geological structure. . . . The suggestion we have made, that the Primal and Matinal rocks of the White Mountains emerged from the waters in the Levant period, and were elevated into anticlinal and synclinal flexures with a different strike from those of the more extensive crust undulations of the late Carboniferous date, offers a natural cause, we think, for the superior elevation of their outcrops in this mountain chain, compared with their height in the Green Mountains and other districts where only one series of axes upon a large scale is discernible.

J. D. Dana on Continental Problems, 1846-1849.

During 1846 and 1847 J. D. Dana, who from the beginning of his career had shown a capacity for the broader, more profound questions of geology, came forward with a series of papers dealing with earth history which were of fundamental importance. These, it should be stated, were based largely upon experience and observations while with the Wilkes Exploring Expedition (1838-1842), the substance of which naturally found its way into print through the medium of the *American Journal of Science* prior to the issue of the final report in 1849.

In the first paper, dealing with the origin of continents, Dana considered the earth as a cooling globe and showed that the areas now constituting the continents were those first "free from eruptive fires," and hence the first to cool and solidify. Progressive cooling and consequent contraction produced the depressions forming the ocean basins. The continents, then, were due not to uplift, as had been commonly taught, but rather to a retreating of the ocean waters into a gradually deepening trough. It is this paper which is commonly referred to as containing the first announcement of the theory of the permanence of continents. In fact, however, such a construction is largely a matter of inference and it was not until the publication of the volume on the Wilkes Exploring Expedition in 1849 that this idea was clearly set forth. In the *Journal* for 1847 Dana brings up these problems under the caption of *The Geological Results of the Earth's Contraction* and *The Grand Outline Features of the Earth.*

In these several papers he accepts the prevailing theory that the earth was once in a condition of fluidity, and that the now oceanic areas were at one time the most intensely heated portions of the crust and, as above noted, had as a result undergone the greatest amount of contraction on cooling. Further, that the mountain ranges and main fissures along the ocean borders were due to this same contraction. This theory he felt did away with the almost preposterous though prevalent idea that continents and mountains had been lifted by a force acting from beneath, a force which could not be satisfactorily located and accounted for; a theory which did not account for the mountains retaining their position even would it offer satisfactory explanation of their production.

He showed, further, that the folding and faulting of the strata as described by the Rogers brothers in Virginia and Pennsylvania could be readily accounted for on the theory of a force acting laterally, and that such folds would have their steepest incline on the side farthest from the source of movement and would also be most abrupt nearest this source. The fact that such results were not in all cases uniform he conceived to be due to variation in the thickness of the beds, to a want of uniformity in the materials, and to inequality in the action of the force upon the different parts of the line along which it operated. The idea that mountains might be produced by tides and paroxysmal movements beneath the crust (as advocated by H. D.

Rogers) was set aside on the ground that such should have occurred at earlier periods, and, further, it would not account for the principal ranges in the east and west of the continent.[22] The geological epochs were announced in this same paper as perhaps due to the alternation of prolonged periods of quiet with those of more or less abrupt change, due to contraction.

In the second paper referred to, on *The Grand Outline Features of the Earth*, Dana argued "that the great chain of mountains as well as of islands, are interrupted ranges consisting of overlapping lines, either straight or curved, and that curves constitute an essential feature of the system." He showed that throughout, northwest and northeast lines are prevalent, and that these lines are usually curved, instead of conforming to the direction of a great circle. He apparently accepted the theory that the course of mountain ranges, islands, and coast lines is attributable to the courses of former fissures in the earth's surface, and, in discussing the electrical, contractional, and other theoretical causes, suggested that if, as claimed, curves of magnetic intensity on the earth's surface are approximately isothermal lines, they must also be lines of equal cooling and hence of equal contraction and tension. But whatever may have been the origin of the fissures, he thought there could be no doubt but that a kind of cleavage structure, or at least capability of fracturing easily in two directions, was given the crust during its formation.

The conclusions which appear to flow from the facts that have been presented are as follows:

That the general direction and uniformity of the grand outline features of the globe may be in a great degree the simple effects of the earth's cooling, this operation resulting in (1) solidification, and under the circumstances, whatever they were, an attendant jointed structure or courses of easiest fracture, in two directions nearly at right angles with one another, both varying according to the rates of cooling in different parts; and (2) occasioning tension in the crust through the contraction going on beneath, with some relation to circular areas, but especially to large compound areas, which tension caused ruptures conforming or not to the lines of jointed structure according as the force of tension acted in accordance with this structure or obliquely to it. (3) The age of mountains can not, therefore, be determined necessarily by their courses; a different direction in a particular region in different ages is not improbable, since

[22] The general theory of changes of level by expansion and contraction, and the resultant rise of continents, was not wholly original with Dana, somewhat similar views having been advocated by the English geologists, Babbage and De la Beche, and the French geologist, Prevost. It was, however, here first clearly set forth and in a manner to attract attention. The idea of the relative *permanence* of the continents and oceanic basins was apparently Dana's.

the same contracting area might exert its horizontal force in somewhat different directions at different epochs, or other such areas might co-operate and exert a modifying influence; and at the same time an identity of direction for different ages was to have been expected.[23]

Dana's Ideas on Volcanoes of the Moon, 1846.

In a letter to James Hall, dated August 19, 1846, Dana wrote: "I shall offer something on the moon at the next meeting of the Association; we may soon expect that it may be annexed and it is important that its resources should be exploited."[24]

In this paper, which was published in the *American Journal of Science* for that year, Dana compared the lunar craters with those of the earth's surface and in particular with that of Kilauea, with which he found a striking resemblance. As Kilauea was simply a boiling volcano and built up no cinder cone, he argued that a similar condition of affairs may have existed in the moon. He went on to state, after enumerating the resemblances: "We may therefore say unhesitatingly, without fearing an impeachment of our sobriety, that the moon's volcanoes are in fact volcanoes, although the craters would receive comfortably more than a score of Etnas." As to whether or not these volcanoes were still active he found no proof, but as there was no water in the moon, which is assumed to play so important a part in the igneous phenomena of the globe, he was in-

[23] "*Permanency of continents and oceans.*—It is now more than fifty years since James D. Dana began to teach that the rising continents and the sinking oceanic basins have been, in the main, permanent features of the earth's surface. He did not mean, however, that the continents have always had essentially the same shape, elevation, and areal extent that they have today. Still, Dana did not fully appreciate the amount of continental fragmenting that has taken place in the course of geologic time, though he clearly pointed out the foundering of Australasia, speaking of it in his famous *Manual of Geology* (page 797) as 'a fragment of the Triassic world.' The teachings of Dana as to the permanency of continents and oceanic basins have been accepted in some form by all geologists and lie at the basis of all zoogeography and evolution as well. In Dana's time and to some extent even today geologists are swayed by the Wernerian or Neptunian theory of earth history, which postulates a gradual emergence of the land out of the decreasing hydrosphere through loss of water by crustal absorption. Now, however, geologists are holding more and more to the hypothesis that the earth periodically shrinks, and each time it does so some parts or all of the continents rise more or less; but that in the main there is subsidence of the ocean bottoms equal in amount to the rising land-masses, that the water of the hydrosphere is constantly increasing in amount, and that even though the continents are in the main permanent, yet they are partially breaking down into the oceanic basins." Schuchert, Amer. Jour. Sci., Vol. XLII, August, 1916, p. 93.

[24] The annexation of Texas, it may be recalled, had been accomplished the year previous and the accession of the adjacent Mexican territory was imminent.

clined to believe that sulphur might have been instrumental in promoting some of the observed changes.

In this connection he brought up for consideration the idea, discussed elsewhere, of volcanic regions presenting a center of solid feldspathic rocks, unstratified and compact, while the exterior consisted mainly of basaltic lava, and arrived at the conclusion that the actual mineral nature of the igneous rocks depended not only on heat and pressure, but also on rate of cooling.

Sketch of J. D. Dana.

James D. Dana was born in Utica, New York, in 1813, and was, therefore, practically contemporaneous with James Hall. He became a student of Professor Silliman's in 1830 at Yale College, leaving in 1833, somewhat in advance of graduation, to avail himself of an offer to cruise in the Mediterranean, as instructor in mathematics to the midshipmen in the Navy.

The first paper in his bibliography was published in 1835 and gave an account of Vesuvius as seen by him during this trip in 1834. In 1836 he returned to New Haven, remaining for two years, the latter part of the time again an assistant to Professor Silliman. It was during this time, scarcely four years after his graduation and when but twenty-four years of age, that he brought out his first *System of Mineralogy*, a volume of 580 pages, and a most remarkable work for the time and the conditions.[25]

From 1839 to 1842 Dana served as geologist and mineralogist on the Wilkes Exploring Expedition, and for the first thirteen years after his return his chief energies were devoted to the study of the material collected on the expedition and to the preparation of his reports, of which two—the volume on geology, 756 pages, 5 maps, and 21 plates (1849), and the one on crustaceans, 1,620 pages, with an atlas of 96 plates (1854)—are monumental.[26] His labors, however, were not limited to the reports, for during the same period he prepared and issued three editions of the *System of Mineralogy*

[25] The System has now gone through six editions, though after 1868 the work was done mainly and finally wholly by his son, E. S. Dana, professor of physics and curator of the Mineralogical Collection in the same university. The last edition, that of 1892, comprises 1,104 royal octavo pages, with over 1,400 figures in the text.

[26] Yet in 1837 he wrote to Hall: "I feel but poorly prepared for the laborious and responsible duties that will be required of me in the Southern seas, and should never had the presumption to propose myself for the situation. I, however, yielded to repeated solicitations (though at first refused), trusting to unremitting exertions and industry to supply the many deficiencies. I fear, however, that I am venturing beyond my depth."

PLATE XIV

JAMES DWIGHT DANA

(1844, 1850, 1854) and two editions of the *Manual of Mineralogy* (1848, 1857), besides writing numerous papers for scientific periodicals.

In 1846 Dana became an editor of the *American Journal of Science,* associated with Professor Silliman, who had founded it twenty-eight years before. His labors in connection with the *Journal* continued until the close of his life. In 1850 he was appointed professor of natural history in Yale College, and in 1864 the title was changed to that of professor of geology and mineralogy. His duties as instructor, however, he did not take up until 1855, but after this date, with some interruptions due to ill health, his active connection with the college continued until 1890. It is perhaps well to add that just before his appointment to Yale in 1850 he had been invited to a similar position at Cambridge, Massachusetts, in connection with Harvard College, but by the prompt action of a generous friend in the Yale faculty in providing the necessary funds he was induced to remain in New Haven and accept the Silliman Professorship.

Fig. 51. *J. D. Dana.*

In 1859 long-continued overwork brought a breakdown of serious character and from which he never fully recovered, and although later some degree of health came back, he was always subject to the severest limitations until the end of his life. Only those immediately associated with him could appreciate the inexorable character of these limitations and the self-denial that was involved, not only in restricting work and mental effort, but also in avoiding intercourse with other men of science and friends in general, in which he always found the greatest pleasure. Little by little the power for work was restored and by husbanding his strength so much was accomplished that, besides other writing, he was able to bring out in 1863 the first edition of his *Manual of Geology,* and in 1864 the *Text-book of Geology,* and four years later his last and most important contribution to mineralogy, the fifth edition of the *System.* This last great labor, extending over four years, was followed by a turn of ill health of an alarming character and from which restoration was again very slow.

The years that immediately followed were filled with the same quiet labor, geological investigations in the field, the writing of original papers and books, the editorial work of the *Journal,* and his duties as a college instructor.

They were remarkably productive years, notwithstanding the difficulties contended against, notably his renewed illness in 1874 and 1880. A large number of important papers were published, chiefly in the *Journal.* New editions of the *Manual of Geology* were issued in 1874, 1880, and 1895; of the *Text-book* in 1874 and 1883; while a new geological volume, called *The Geological Story Briefly Told*, was issued in 1875, and one on the *Characteristics of Volcanoes* in 1890, after his second visit to the islands. A second edition of his *Corals and Coral Islands*, the first edition of which appeared in 1872, was also brought out in 1890.

According to Le Conte it was Dana who first studied geology from the standpoint of the evolution of the earth through all time. For him geology was the development of the earth as a unit. He first made geology a philosophic history.

But it was not as an investigator and writer only that Dana achieved success. As a teacher he seems to have won the respect and regard of all with whom he came in contact, and to have left on the minds of students—even those who had no taste for geology—a lasting and favorable impression. Many of his sayings in class lectures were epigraphic: "I think it better to doubt until you know. Too many people assert and then let others doubt." Again, "I have found it best to be always afloat in regard to opinions on geology."

Nor can we regard him as merely a geologist. His work on crustaceans, comprising 1,620 pages, with an atlas of 96 plates (1854), shows that equal success could have been attained in the biological field had he chosen to follow it. The mental vigor and staying powers of the man were simply extraordinary, and it is not too much to say that he stands out head and shoulders above all his contemporaries, if not above all who preceded him.[27] His interest never flagged; no problem was too large for him to grasp, no detail too small for his consideration.

Tuomey's Final Report on the Geology of South Carolina, 1848.

It will be remembered that in 1844 Michael Tuomey was appointed state geologist of South Carolina, a position which he continued to

[27] "Dana has gone to Europe for the benefit of his health. Prof. Silliman writes me they have heard from him several times—he is improving rapidly, but he had to run away from Paris—the savants ran after and feasted him, and kept him up so late at nights that he could get no repose. He deserves all the honors and attentions he received, for he is not only a great man, but a *true* and a *good* man. No scientific man from America ever before created such a sensation in Paris." (F. B. Meek in letter to Hayden, 1859.)

hold until his appointment to the chair of geology, mineralogy, and agricultural chemistry in the State University of Alabama.

A preliminary report has already been noted (p. 239). The final report appeared in the form of a volume of 293 pages in 1848. This, with the exception of the volume on the fossils of South Carolina, of which he and Doctor Holmes were joint authors, was the most systematic and pretentious of his publications.

This survey, like many of its predecessors elsewhere, was undertaken with a view to developing the agricultural resources of the state. The condition of the public mind toward pure science at the time is well reflected in the almost pathetic postscript of Mr. Tuomey to his preface, in which he simply states that, while the report was passing through the press, he was informed that the plates containing figures of fossils had not been considered essential by the committee on publication and were therefore omitted! The work comprised: (1) An introduction of 59 pages, given up, as was so often the case with these early reports, to general geological considerations and as applicable to any other region as to that at hand, and (2) the report proper, comprising seven chapters (234 pages) and an appendix of 56 pages, containing a catalogue of the fauna of the state. It also contained a geological map of the state, the first to be issued.

The introductory chapters are interesting only as reflecting the condition of the science at the time. Tuomey estimated that our actual knowledge of the earth's crust extends to a depth of fifteen miles, measured from the tops of the highest mountains. He adopted the Kant-Laplacean theory of the origin of the earth, and recognized slaty cleavage as distinct from joints and stratification, regarding it as resulting "from the tendency of the simple component substances of the rock to arrange themselves in crystalline forms at a time when the semifluidity of the mass permitted a certain degree of motion among its particles." The crystalline structure of the gneisses and other metamorphic rocks he conceived to result from a degree of subterranean heat which, although intense, did not destroy the lines of stratification or bedding which they received at the time of their deposition. In his table showing the order of superposition of fossiliferous rocks, he divided them into: (1) The *Recent* or *post-Pliocene period;* (2) the *Tertiary;* (3) the *Newer Secondary* or *Cretaceous;* (4) the *Middle Secondary,* including the Wealden and Lower Lias with the intermediate beds; (5) the *Older Secondary,* including the Triassic; (6) the *Newer Paleozoic,* extending from the Magnesian Limestone or Permian down to and including the Lower

Carboniferous shales; (7) the *Middle Paleozoic*, including the Devonian; and (8) the *New York System* of Upper and Lower Silurian, including the Chemung and Champlain Division and intermediate formations.

In the part of the report relating to the geology of the state proper attention was called to the extensive decomposition of the surface rocks and the production of bowlders through decomposition along joint planes. The occurrence of the flexible sandstone, or itacolumite, was noted, and it was rightly remarked that "the flexible portions of the rock seem to be in the incipient stages of disintegration." The gold and iron mines received a considerable share of attention, and a map was given showing in color the deposits of magnetic, specular, and limonite ores, and the limestone of the York and Spartanburg districts.

In the line of purely physical local geology, perhaps the most interesting part of the report lies in his discussion of the possible subsidence of the coast going on at the present time. He showed, to his own satisfaction at least, that the presence of stumps of trees below, or partly below the present sea level was due to the gradual undermining of the sand and mud flats on which they grew and the consequent settling of the stumps, retained in an upright position by their wide-spreading roots, and not to a subsidence of the coast, as taught by Lyell, Bartram, and later by Lieber and Cook.[28] The origin and nature of the soils, the mining and preparation of fertilizers, including lime burning, the washing and milling of gold ores, and kindred subjects, were touched upon. Many soil and fertilizer analyses were given, the latter and their discussion being largely by Professor C. U. Shepard, then professor of chemistry in the Medical College of the state, at Charleston. The fossils were identified by Conrad and Gibbs; as was natural a considerable portion of the work was given up to a discussion and description of the Tertiary deposits, which cover more than half the area of the state.

Tuomey's Survey of Alabama, 1848-1850.

Prior to 1848 the professor of geology in the University of Alabama had been required to spend a portion of his time, not exceeding four months of each year, in making geological explorations. In January, 1848, the general assembly passed a resolution appointing Tuomey, then recently elected to this professorship, state geologist,

[28] Tuomey's view was upheld by Professor N. S. Shaler as recently as 1870, Proceedings of the Boston Society of Natural History, XIII, p. 228, and also by Suess in his Face of the Earth (Eng. Translation), Vol. II, p. 471.

and requiring him to make a report to the general assembly at each of its biennial sessions. Thus simply was inaugurated the first systematic geological survey of the state.

Under this law, Tuomey's *First Biennial Report*, in form of an octavo volume of 176 pages and a colored geological map, appeared in 1850. In this the rocks were divided into those of (1) the *Primary*, (2) *Metamorphic*, (3) *Silurian*, (4) *Carboniferous*, (5) *Cretaceous*, and (6) *Tertiary systems*, the last named being represented, so far as then known, only by the Eocene. Much of the report was taken up with economical considerations, particular attention being paid to coal and iron.

The second report of the survey appeared under the editorship of J. W. Mallet, the publication having been delayed through the procrastination of the public printer and Tuomey's death, which occurred in March, 1857.

Tuomey recognized on Marble Creek in Limestone County a blue limestone which was a continuation of the Silurian rocks in Tennessee. The Devonian rocks were represented by black slate found on the principal streams flowing from the north into the Tennessee between Flint River in Madison and Shoal Creek in Lauderdale County. He regarded the divisions of the Carboniferous made by Troost as sufficiently characteristic in north Alabama to be retained.

The clastic material—that is, the loosely consolidated sands and gravels—which he mapped as extending across the middle of the state and north of the verge of the Cretaceous beds, and colored as post-Tertiary, was referred to as belonging to the drift, although having little resemblance to that of the North. "If the southern drift be at all connected with that of the north," he wrote, "it may be explained by supposing that the northern glaciers suddenly melted, and that the water thus liberated in immense volume took a southern direction, carrying with it the débris torn from the surface over which it passed until it met the Tertiary sea, upon the shores of which its burden was deposited." This theory, he felt, would sufficiently account for that enormously long ridge of drift extending parallel with the Atlantic coast, for the moment the current entered the Tertiary sea its velocity would be checked and the greater part of the transported detritus deposited. The reader will here recognize an old idea, but slightly modified.

Commenting on the fact that these beds contained no fossils, he wrote:

> The only way by which I can account for this . . . is by supposing that, before the drift period, the bottom of this sea had been elevated and

converted into dry land, and that at the commencement of the drift period a depression of the land took place; that the time between the influx of the sea and the deposition of the drift was too short for marine animals even to have commenced a colonization, and that the land was again elevated into its present position and subjected to long-continued denudation, which produced its present configuration; that after this elevation the rivers excavated their present channels.

Tuomey was assisted in this work by Oscar M. Lieber, geologist, and J. W. Mallet, chemist.

Sketch of Tuomey.

Tuomey was born in Ireland in 1805, and went to England when about seventeen years of age, shortly afterward coming to America. He first settled near Philadelphia, going thence to Maryland, where he served for a time as a tutor in a private family. He entered the Rensselaer school in New York, and after graduation went south, serving first as an engineer on a railway in North Carolina and afterward as a teacher in Virginia. His love for natural history led him to make collections of fossils, which brought him in contact with Sir Charles Lyell, James Hall, J. D. Dana, and others, and he shared with Rogers the honor of discovering the "infusorial" (diatom) beds near Richmond and Petersburg.

Fig. 52. *Michael Tuomey.*

In 1844 he became state geologist of South Carolina, as already mentioned, and in 1847 was appointed professor of geology, mineralogy, and agricultural chemistry in the University of Alabama; in 1848 receiving in addition the appointment of state geologist. As a teacher Tuomey is represented as possessing in a remarkable degree the faculty of interesting the students, those, even, who took no particular interest in the subject-matter he taught. His native Irish wit did much to render his lectures entertaining, especially to those who were not the victims of it. He was particularly unmerciful in his rebukes and exposures of shams and affectations.

His most important publications were his official reports on the geology of South Carolina, and his joint report with Dr. F. S.

Holmes, of Charleston, on the Pliocene fossils of that state, which appeared in 1855.

F. S. Holmes on the Geology of Charleston and Vicinity, 1849.

The withdrawal of Tuomey from the field in South Carolina incidental to his removal to Alabama left the position of state geologist vacant until the appointment of Oscar Lieber, in 1855. In the meantime F. S. Holmes, working privately, gave in the *American Journal of Science* for 1849 a brief paper on the geology of the vicinity of Charleston, which may be noted here, with the preliminary remark that this was Holmes's first venture in the geological field. In this paper he called attention to the fact that the city of Charleston was built upon geological formations presumably identical in age and in other respects similar to those upon which London and Paris are located, *i.e.*, upon the Eocene. The adjacent sea islands he thought to have been formed through the action of the ground swell of the ocean and the streams flowing down from the interior during the time when the land was gradually emerging from the sea.

He agreed with Tuomey in taking exception to the then generally received opinion that the sea was advancing upon these shores, having been led by his own observations to the conclusion "that if the ocean does wash off portions of the shore at one exposed point it deposits the same at no great distance upon another." The supposed indications of subsidence, such as the stumps and roots of trees now below the level of high tide, he accounted for on the supposition that outer sand barriers, which had prevented the ingress of salt waters, were gradually removed, allowing the waves to wash away the fine silt and mud between the roots of the trees, thus permitting them to sink into it and become embedded.

Holmes classified the formations met with as post-Pliocene, Miocene, and Eocene, the first mentioned, where observed, resting directly upon the Eocene. The fact that the Miocene was covered with diluvial or alluvial material he accounted for on the supposition that during the deposition of the post-Pliocene the Miocene areas were above water, or had been denuded of their post-Pliocene covering previous to the deposition of the alluvial or diluvial sands and clays.

Report of Messrs. Brown and Dickerson on Sediments of the Mississippi River.

It is in this connection that attention may well be called to a report made to the American Association at the Philadelphia meeting

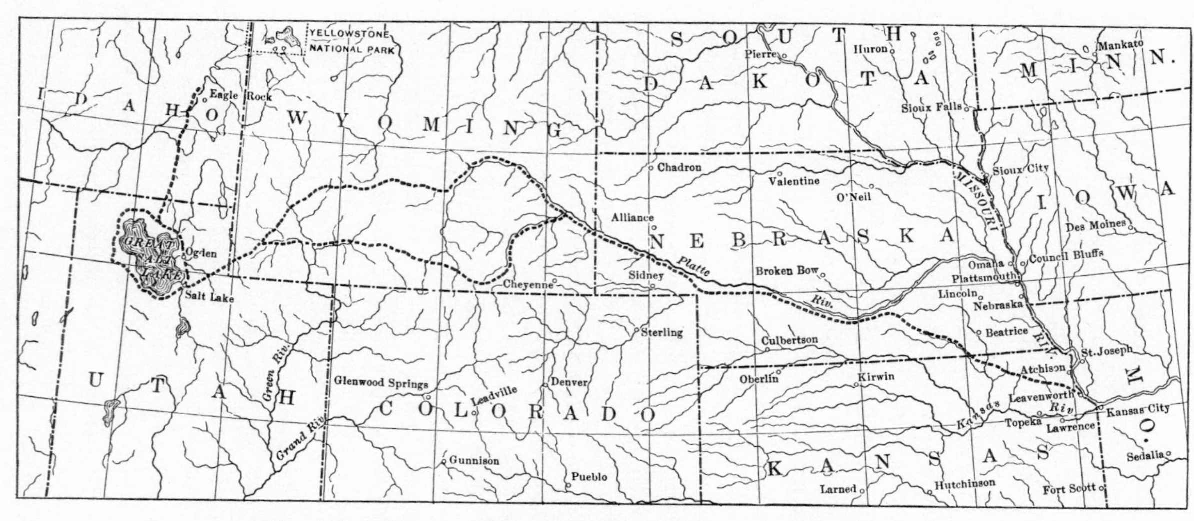

Fig. 53. *Map Showing Stansbury's routes in 1849-1850.*

in 1848 by Messrs. Andrew Brown and M. W. Dickerson on the sediments of the Mississippi River. In this report, which covered a period of observation of eighteen to twenty years and dealt mainly with the annual rate of discharge of both water and sediments into the gulf, it was stated that in the work of making excavations in New Orleans, erect tree trunks with roots as they grew were found buried at a depth of nine feet below sea level. On the not improbable assumption that this subsidence was still going on and would continue at the same rate for an indefinite period, the time, it was estimated, may yet come when the city itself will sink to below the level of the ocean.

Captain Stansbury's Explorations in Utah, 1849.

In 1849 Captain Howard Stansbury, of the corps of topographical engineers, acting under authority of the War Department, explored the valley of Great Salt Lake. The expedition left Fort Leavenworth May 31, 1849, taking a northwest direction and striking the Platte River at Fort Kearney, proceeding thence along this and the North Fork to the Sweetwater, and thence across South Pass to Fort Bridger and Salt Lake City. The return trip was made by a more southerly route, through Bridger Pass, striking the old route again at Fort Laramie, and thence back to Fort Leavenworth, the latter point being reached the sixth of November, 1850 (see Fig. 53).

No geologist accompanied this expedition, though a small series of rocks and fossils which were collected were reported upon by James Hall. Forms belonging to the Cretaceous, Carboniferous, and Silurian or Devonian ages were recognized.

D. D. Owen's Work in the Chippewa District and in Wisconsin, Iowa, and Minnesota, 1847-1851.

In 1847 D. D. Owen was again employed by the Treasury Department, under immediate supervision of General Land Commissioner R. M. Young, to make surveys in the Chippewa district of Wisconsin and northern part of Iowa. The region lies between 43° and 47° north latitude and 89° and 94° longitude west of Greenwich, embracing about 46,000 square miles and, as shown on the map (Fig. 54), comprising that portion of the country "lying chiefly east of the upper Mississippi above Lake Pepin and extending north to Lake Superior." Incidentally, there was included a portion of Iowa "stretching north from the northern boundary of the geological survey of 1839 as far as the St. Peters River, and also a tract of country north of the Wisconsin River."

As in Owen's previous survey, the question to be decided was a practical one and the time limited, only the summer and autumn of 1847 being devoted to field work. The report, printed in form of Senate Document No. 57 of the first session Thirtieth Congress, bore the date April 23, 1848. It comprised 134 pages, with one geological map, 23 lithographic plates from drawings by Owen, and 13 colored plates of sections. Some of these last were beautiful combinations of sections and perspective landscapes, and gave at a glance a general idea of the surface features as well as the character and dip of the underlying rock masses, such as have not been excelled. (See Plate 12, from his section 4.) Even when one considers that, as Owen states, the working time of the members of his corps was from twelve to fifteen hours a day, still it is remarkable that so much was accomplished and presented in such good form. Though a detailed geological survey was made of only about thirty townships west of the fourth principal meridian on Black River and sixty townships on the St. Croix, sufficient data were obtained to enable him to lay down with approximate accuracy the general bearing and area of the principal formations of over two-thirds the area above noted. He showed, also, that both the upper and lower magnesian limestone, were lead-bearing, and that there existed in Wisconsin two if not three trap ridges similar to those of Michigan, which "hold out a prospect of productiveness."

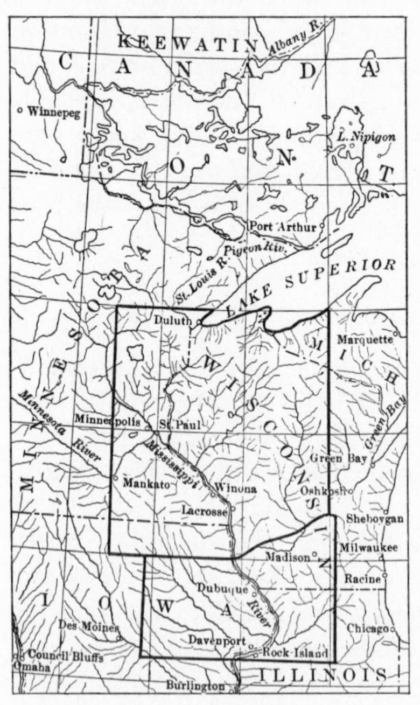

Fig. 54. *Map of Area Surveyed by D. D. Owen in 1839 and 1847.*

In this report, too, he first announced that the upper Mississippi country north of the Wisconsin River was based upon magnesian limestones which were older than the lowest formations of the valley of the Ohio, a portion of them being contemporaneous with the Calciferous group and Potsdam sandstone of New York. He noted also that the Falls of St. Anthony are receding and were probably at one time at a point near Fort Snelling.

Concerning the red sandstones of the south and west shores of Lake Superior, he wrote, "There is strong presumptive evidence that they were deposited subsequently to the Carboniferous era." His evidence, though, was admittedly weak and later was acknowledged to be wholly misleading and insufficient. That Jackson fell into a similar error is noted later.

Fig. 55. *Trapdikes on Lake Superior.* (*After Owen.*)

This preliminary reconnaissance ended, Owen was instructed to make a survey of the Northwest Territory, embracing chiefly Wisconsin, Iowa, and portions of Minnesota. For reasons of his own, the area to receive consideration was so far extended westward as to permit the sending of Dr. John Evans into the *Mauvaises Terres*, or Bad Lands of the upper Missouri, in Nebraska. The manuscript of his final report of this work was submitted in 1851 and published under date of 1852 in the form of a quarto volume of 628 pages of text, with 15 plates of fossils, 19 folding sections, and a geological map. The illustrations of the fossil remains are of particular interest, being medal-ruled on steel, the first of their kind produced in America.

On the title-page of the volume appeared the cut, here reproduced (Fig. 55), which so long did duty in the textbooks of Le Conte and others.

Owen regarded the gypsum deposits of Dubuque as due to original deposition at the bottom of an ocean, the sulphate of lime having probably been derived from submarine sources during the formation of the bed. This view is somewhat remarkable when one considers that of the total lime salts in solution in sea water, 90 per cent occur in the form of sulphates and would be deposited as gypsum during the ordinary processes of evaporation.

His views on the origin of the drift were in accordance with those of the leading authorities of his day. The large bowlders he thought to have been deposited by floating ice and drifted by currents from the north while the country was depressed. The opinion which he had previously expressed (in 1848) concerning the age of the Lake

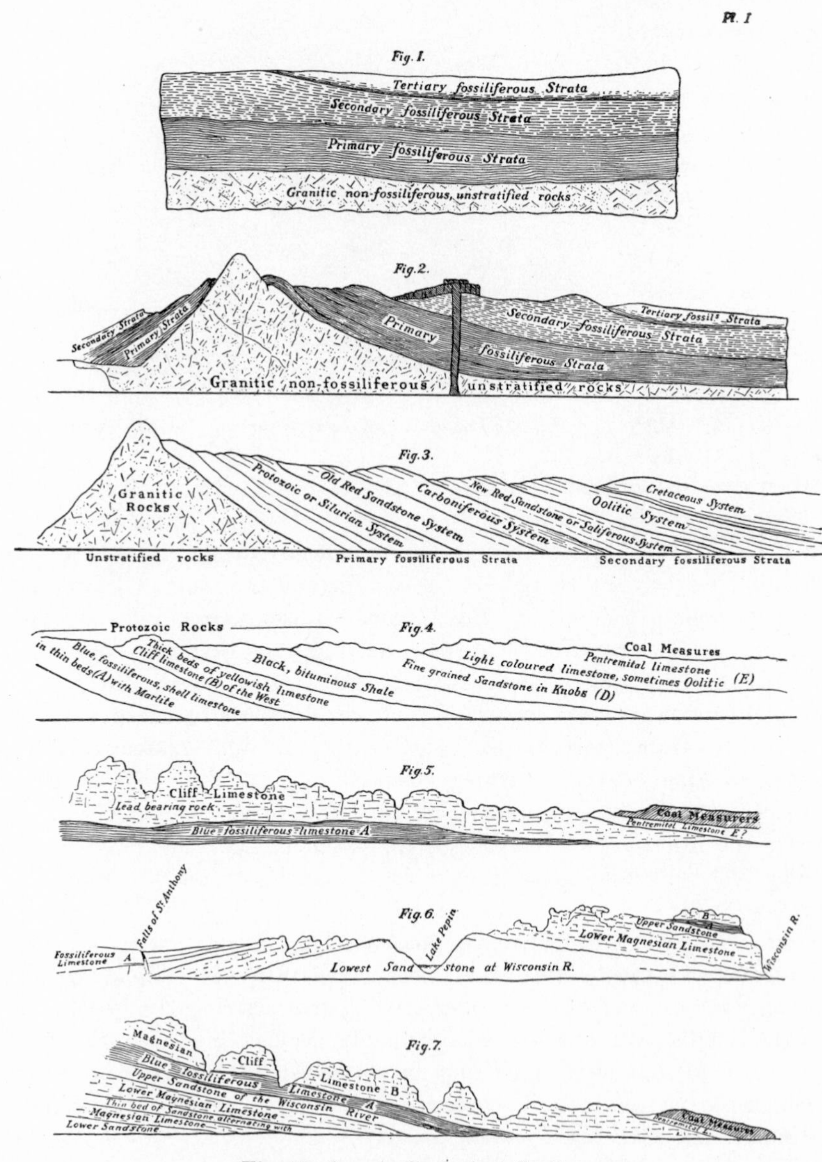

Fig. 56. *Owen's Geological Sections.*

Superior sandstones was in this final report retracted, and he relegated them, on stratigraphic evidences only, to the Potsdam formations, which is in accordance with the prevailing opinion at the present time. Of greater importance was the fact, first announced in his report for 1847-1848, but here brought out in detail, that underlying his lower magnesian limestone (Chazy) there were at least six different trilobite-bearing beds, separated by from 10 to 150 feet of intervening strata. Previous to this no remains of this nature had been reported from any American strata older than the Canadian period of the Lower Silurian. These trilobite-bearing strata, it should be noted, were found resting immediately upon the primal rocks and hence formed the true base of the zoölogical series in the Mississippi Valley.

The Bad Lands.

The *Mauvaises Terres*, or as now known, the Bad Lands, were at that time known only to Indian traders and a few explorers. It is not strange therefore that Evans, in his portion of the report, should have designated them as depicting

one of the most extraordinary and picturesque sights that can be found in the whole Missouri country. . . . From the uniform, monotonous, open prairie, the traveller suddenly descends, one or two hundred feet, into a valley that looks as if it had sunk away from the surrounding world; leaving standing all over it, thousands of abrupt, irregular, prismatic, and columnar masses, frequently capped with irregular pyramids, and stretching up to a height of from one to two hundred feet, or more.

So thickly are these natural towers studded over the surface of this extraordinary region, that the traveller threads his way through deep, confined, labyrinthine passages, not unlike the narrow, irregular streets and lanes of some quaint old town of the European Continent. Viewed in the distance, indeed, these rocky piles, in their endless succession, assume the appearance of massive, artificial structures, decked out with all the accessories of buttress and turret, arched doorway and clustered shaft, pinnacle, and finial, and tapering spire.

One might almost imagine oneself approaching some magnificent city of the dead, where the labour and the genius of forgotten nations had left behind them a multitude of monuments of art and skill.

On descending from the heights, however, and proceeding to thread this vast labyrinth, and inspect, in detail, its deep, intricate recesses, the realities of the scene soon dissipate the delusions of the distance. The castellated forms which fancy had conjured up have vanished; and around one, on every side, is bleak and barren desolation.

Then, too, if the exploration be made in midsummer, the scorching rays of the sun, pouring down in the hundred defiles that conduct the wayfarer through this pathless waste, are reflected back from the white or ash-coloured walls that rise around, unmitigated by a breath of air, or the shelter of a solitary shrub.

The drooping spirits of the scorched geologist are not permitted, however, to flag. The fossil treasures of the way, well repay its sultriness and fatigue. At every step, objects of the highest interest present themselves. Embedded in the débris, lie strewn in the greatest profusion, organic relics of extinct animals. All speak of a vast fresh-water deposit of the early Tertiary Period, and disclose the former existence of most remarkable races, that roamed about in bygone ages high up in the Valley of the Missouri, towards the sources of its western tributaries; where now pastures the big-horned *Ovis montana,* the shaggy buffalo or American bison, and the elegant and slenderly-constructed antelope.

The work of studying the vertebrate fossils brought back by Evans from the region thus described fell to Dr. Joseph Leidy of Philadelphia. Among them was the now well-known Oreodon, an animal with grinding teeth like the elk and canines like those of omnivorous, thick-skinned animals, belonging, as was then thought, to "a race which lived both on flesh and vegetables and yet chewed the cud like our four-footed grazers." This was the first systematic account published of the Bad Lands fossils and it might not unjustly be considered as marking the beginning in America of studies in vertebrate paleontology, though Leidy had, in 1847, published an important memoir on the fossil horse, and fragments of the jaw of an enormous pachyderm from the region had been described by Dr. H. A. Prout of St. Louis in the *American Journal* for the same year.

Agassiz's Physical Characters of Lake Superior, 1848.

In 1848, while occupying the chair of zoölogy and geology at Harvard University, Louis Agassiz, in company with Jules Marcou and a party of students, undertook an exploration of the Lake Superior region, the results of which were published in 1850, under the caption of *Lake Superior; its Physical Character, Vegetation and Animals, compared with those of other and similar regions.* Marcou would have us believe that this volume marked an epoch in natural history publications in America, this mainly on account of the superior style of its illustrations. Certainly there was much to justify the claim.

The country was nearly everywhere roadless, and transportation by water possible only through the aid of birch-bark canoes. As might be expected, the purely geological observations were of little value, excepting so far as they related to glaciation. Agassiz argued that the form of the lake was due to "a series of injections of trap dikes of different characters, traversing the older rocks in various directions." He found six systems of these dikes, to the trend of which the various lake shores in a general way conformed. The relationship of the various copper deposits he attempted to explain on the somewhat remarkable, as well as ingenious, assumption that the material had been poured out in a melted condition, and, cooling quickly, remained in the native state, offering to the agencies of change a relatively small surface exposed in proportion to its mass. At a distance from the main mass, where the ejections were small with relatively large surfaces exposed, they became more or less completely changed into oxides, sulphides, carbonates, etc. The reader need scarcely be reminded that authorities today hold quite a different view, and regard the copper as having been precipitated by reduction to a native state from salts held in permeating solutions.

Naturally, Agassiz's views on the glacial phenomena of the region are of paramount interest. These are, however, set forth elsewhere (p. 615).

Sketch of Agassiz.

Louis Agassiz was born in Motier, Switzerland, in 1807, and came into world-wide notoriety through his works on fossil fishes and his enthusiastic exploitation of the glacial discoveries of Hugi, Venetz, and Charpentier, while in Neuchâtel. He came to America to better his finances in 1846, and after delivering courses of lectures before the Lowell Institute in Boston and in other of the eastern cities, accepted, in 1847, the professorship of geology and zoölogy in Harvard University. Agassiz was more biologist than geologist, and his services to the latter branch of science after coming to America were more as a teacher and through arousing public interest than by research. In these lines he has never been excelled. Though prone to jump at conclusions, as shown in his hasty assumption that the bowlders of decomposition found by him in Brazil were drift bowlders and indicative of a former glacial period in that latitude, his wonderful faculty of teaching others to think, rather than blindly to follow, justly gave him a reputation rarely attained, and perhaps never exceeded, by men of his calling.

Work of Jackson in Michigan, 1847.

In accordance with an act of Congress approved March 1, 1847, Dr. C. T. Jackson was appointed by the then Secretary of the Treasury R. J. Walker, to make a geological survey of that portion of Michigan lying south of Lake Superior and north and northwest of Lake Michigan. As in previous operations of like nature by Owen, the object of the survey was to ascertain what of the lands could be classed as mining lands and what agricultural.

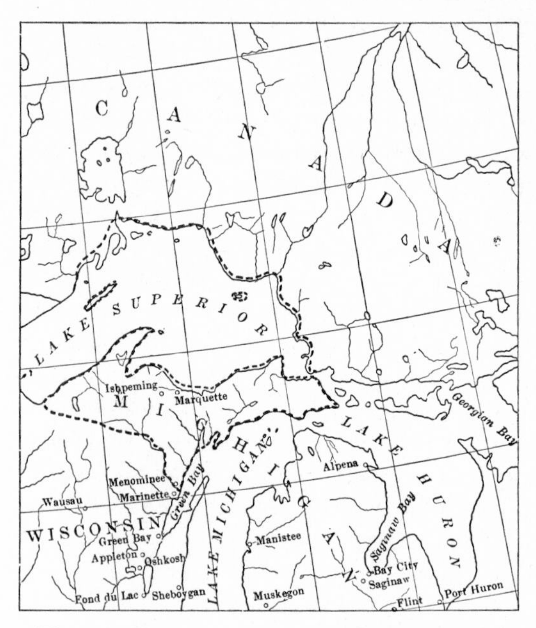

Fig. 57. *Map of Area Surveyed by Jackson, Foster and Whitney, 1847-1849.*

Concerning the personnel of this survey it is stated[29] the party varied somewhat from year to year, there being always at least two first assistant geologists, each at the head of a party and responsible for his own district. Jackson himself did less field work than his assistants, being occupied with looking after appropriation bills in Washington.

The assistants were J. D. Whitney, W. D. Whitney for a single season, Walcott Gibbs, Charles A. Joy, John Locke, William Francis Channing, and John Wells Foster. In general, besides Jackson and his two assistants, there were five or six other geologists, mineralogists, naturalists or surveyors on salary, with the necessary accompaniment of packmen, rodmen, and cooks, the latter being mostly Indians and Canadians, with a varying number of beginners who served without pay for the sake of experience.

The difficulty of surveying in an unbroken wilderness of thickets, marshes, and lakes, was complicated by the usual congressional difficulties at Washington. Congress was slow in passing the appropriations. Moreover, the Michigan congressmen felt that no outsider

[29] Life and letters of J. D. Whitney.

could do justice to the mineral resources of their state, and only the utmost efforts on the part of the head of the survey brought about the defeat of an amendment to the appropriation bill which would compel him to reside in the state, have all the chemical work done at Detroit, and employ as assistants only "practical" men acquainted with woodcraft, and citizens of Michigan. There was, moreover, much hostile criticism of the survey and personal opposition to Jackson on the part of mine owners and others on the ground. Jackson himself was not at his best, and it seems probable that the malady which years later sent him to end his days in a hospital for the insane had already gained a hold upon him. Be this as it may, the criticisms of the survey became so outspoken that in the spring of 1849 both Foster and Whitney resigned. An investigation by the Department at Washington followed with the result that Jackson also was allowed to retire, and the completion of the survey left to his two former assistants who were reinstated.[30]

Jackson's report was published as House Document No. 5, Thirty-first Congress, first session, 1849. It comprised upward of 560 pages, with 19 plates, geological maps of Keweenaw Point and Isle Royal, and three sections of mines. The eruptive rocks of Keweenaw Point were described as having been intruded through linear caverns or fractures in the superincumbent rock which they had frequently overflowed, so as to rest unconformably on their strata. Here is again recognized the possibility of fissure eruptions as distinctive from the crater eruptions of modern volcanoes.

The red sandstones were erroneously relegated to the New Red series, the decision being based upon a bed of limestone carrying the fossil *Pentamerus oblongus*, which was found in the midst of the sandstone near Anse.[31] The native copper, he thought, with Agassiz, to have come as such from the molten interior of the earth.

Jackson, it is well to note, was opposed to the principle of the reservation of mineral lands by the general Government. He wrote:

It may be useful to the public to cause geological and mineralogical surveys to be made for their information, but I am satisfied that the reservation of mineral lands is a great evil to the country, and that the Government never can derive revenue from such sources, while the re-

[30] Appendix, pp. 669-671.

[31] The early attempts at correlation of the various beds of red sandstone in the eastern United States were based largely on lithological grounds, and created a great amount of confusion. Jackson changed his mind at least twice as to the age of that of Perry, Maine, and as noted above at one time considered that of Lake Superior Triassic, though according to Whitney he never saw it in place. Rogers would not admit of the Triassic age of either, but thought them to be

striction most seriously embarrasses the settlement of newly acquired territory. The above remarks are applicable to the whole copper region, and I would not advise the reservation of any part of it as mineral land.

Foster and Whitney's Work in Michigan, 1849-1851.

The work begun by Jackson was, as above stated, continued by Messrs. J. W. Foster and J. D. Whitney. Their reports were issued in octavo form, Volume I, on the copper lands, constituting House Executive Document No. 69, first session, Thirty-first Congress, 1850, and comprising 224 pages, with 12 plates and a facsimile of a map of Lake Superior made by the Jesuit missionaries in 1671. Volume II, or Part 2, on the iron region, appeared as Senate Executive Document No. 4, special session, March, 1851. This comprised, all told, 406 pages, with 33 plates. A colored geological map of the area surveyed accompanied the report. Some of the more important items noted are as below:

Fig. 58. *John W. Foster.*

They stated that Lake Superior occupies an immense depression which has been for the most part excavated out of the soft sandstone of the Potsdam age. The configuration of that portion of the lake lying west of longitude 88° was deemed due to two axes of elevation extending in parallel lines from the northeast to the southwest, which upraised the sandstone, causing it to form a synclinal valley.

This, it will be noted, is radically different from the idea put forward by Agassiz (p. 277).

The conglomerate composed of rounded fragments of jaspery rock, so abundant throughout the south shore copper region, they regarded as formed by friction and testifying to the "intensity of

Silurian or Devonian. B. F. Shumard in a letter to Meek, dated May, 1850, expressed the belief that the Lake Superior stone was of Potsdam age. "Almost every one," he wrote, "who has examined this stone has his own particular views in regard to its age. Agassiz has recently affirmed that it is the 'New Red' but his opinion is without doubt erroneous. I have not changed my opinion in relation to it, viz.: that it is a formation either equivalent to the Potsdam sands or beneath it." Whitney, in summing up matters in 1860, stated that Owen finally came to regard this stone as of Potsdam age, as did both himself and Foster. This is the view held in the later reports of the state survey of Michigan, though the Perry beds are considered Devonian.

the force with which the eruptive rocks have been propelled from the interior through the earth's crust," the detritus having been redistributed by water, following in this the teachings of Von Buch. The graphic account of the conditions under which it was supposed to have been formed is worthy of being reproduced entire:

We may suppose that at one time all of this district [*i.e,* the copper district] formed a part of the bed of the primeval ocean. Adopting the theory of a cooling globe, we may further suppose that the waters were in a heated condition and differed essentially in chemical composition from those of the present oceans. The earth's crust was intersected by numerous powerful fissures, and the communication between the exterior and interior was unobstructed. Volcanic phenomena were much more frequent and exerted on a grander scale. Each volcanic paroxysm would give rise to powerful currents and agitations of the water, and their abrading action in detaching portions of the preexisting rocks, and depositing them in beds and layers on the floor of the ocean, would operate with greater intensity than at the present time. We can trace the remains of one volcanic fissure extending from the head of Keweenaw Point, in a southwesterly direction, to the western limits of the district, and of another, in a parallel direction, from the head of Neepigon Bay to the western limits of Isle Royale. Along the lines of these fissures existed numerous volcanic vents, like those observed at this day in Peru, Guatemala, and Java, which were characterized by periods of activity and repose. From these vents were poured forth numerous sheets of trap, which flowed over the sands and clays then in the progress of accumulation. During the throes and convulsions of the mass portions of rock would become detached and rounded simply by the effects of attrition, and jets of melted matter be projected as volcanic bombs through the air or water, which, on cooling, would assume spheroidal forms; while other portions of the rock, in a state of minute mechanical division, would be ejected in the form of ashes and sand, which, mingling with the water, would be deposited as the oscillations subsided among the sands and pebbles at the bottom of the sea. During the whole of this period of volcanic activity the sands which now form the base of the Silurian system were in the progress of accumulation and became mingled with these igneous products. The level of the sea, as evidenced by the ripple marks, was subject to repeated alterations; sometimes it rose so shoal that the marks of the rippling waves were impressed on the sands, at others it sank to unfathomable depths.

In the process of consolidation the rocks became traversed by numerous fissures, and the water, charged with lime, was forced in like jets of steam, filling them with materials different from the inclosing mass. In this way the pores in the conglomerate and the vesicles of the amygdaloid were filled.

The formation of the copper and silver ores was considered as caused probably by electro-chemical agencies. To again quote their exact words:

> The existence of two metals side by side, like copper and silver, each chemically pure and capable of being alloyed in any proportions; the accumulation of the latter near the cross courses or at the junction of two mineral planes; the changes in the metallic contents of lodes in their passage through different rocks, and the parallel arrangement of the earthy gangues, all seem to indicate the existence of electrical currents during the period of their formation.

They were disposed to look upon the iron ores as of purely igneous origin "in some instances poured out but in others sublimed from the interior of the earth. . . . During this [*i.e.*, the Azoic] period, the interior of the earth was the source of constant emanations of iron, which appeared at the surface in the form of a plastic mass in combination with oxygen or rose in metallic vapors, or as a sublimate, perhaps as a chloride; in the one case it covered over the surface like a lava sheet; in the other it was absorbed into the adjacent rocks, or diffused through the strata in the process of formation." In addition to these original sources they recognized the igneous rocks rich in ferruginous silicates, from which under the forces of denudation the ferruginous particles were derived and swept into the inequalities of the surface to form lenticular masses, subsequently covered by other materials. When the siliceous matter had become impregnated with the metallic matter, through some unexplained means indefinitely stated as "segregating forces," the whole mass assumed a banded structure. The beds were later elevated or folded, and thus originated the jaspery iron ores that have since afforded abundant opportunities for study and material for entire monographs without a complete solution of the question. By those best qualified to judge, however, these ores are now considered almost wholly of secondary origin and due to a redeposition of iron leached by surface waters from overlying rocks and concentrated at lower levels.[32]

Concerning the origin of the drift and the phenomena of the groovings and striations of the rocks in the regions, they were still somewhat in the dark. The position of bowlders resting on stratified deposits of sand and clay was thought to be antagonistic to the theory of a general ice cap similar to that of the circumpolar region.

[32] There were no available works on chemical geology at this date. The first volume of Bischof's Lehrbuch der chemischen und physikalischen Geologie was issued in Bonn in 1847, but could have had little effect on American workers. The English translation of the completed work appeared in 1854.

It was felt, rather, that such might have been transported by floating ice (not icebergs) in the same manner that bowlders are even now each spring transported from the borders of the northern lakes and rivers and dispersed over the adjacent swamps and lowlands.

The slates of the region were looked upon as probably originally beds of volcanic ashes subsequently consolidated—a by no means improbable theory.

The igneous rocks were classed as dolerite, anamesite, and basalt. The sandstone, which occupies almost exclusively the bed of Lake Superior and occurs in isolated patches along the shore and on the islands, they rightly classed as Potsdam, differing in this respect with Jackson, who considered it as New Red, and from Locke and others, who thought it to be the equivalent of the Old Red or Devonian sandstone of Europe.

It was in this volume that they introduced into American literature the term *Azoic*, intending thereby to include a body of strata in part at least of sedimentary origin which, while lacking in fossils, did not necessarily preclude the existence of life at the time of their formation but which everywhere underlaid stratified rocks containing the lowest known forms of animal life. It included the *Metamorphic Series* of the Canadian and English geologists, which were subsequently designated by Logan under the term Laurentian. The term Azoic, it may be well to state, was provisionally adopted, though in a slightly different sense, by Dana and used in the first edition of his *Manual* (1863), but afterward replaced by the term Archæan (*Am. J. Sci.*, 8, 1874). The failure on the part of other geologists to adopt the term in its original sense seems to have aroused the ire of Whitney and was perhaps the chief motive in causing him, in collaboration with M. E. Wadsworth, to bring out in 1884 his *Azoic System and its Subdivisions*, in which, with characteristic vigor and terseness, he expressed the opinion that the "chances of our having at some future time a clear understanding of the geological structure of northeastern North America would be decidedly improved if all that had been written about it was struck out of existence." As to how we were to arrive at our present knowledge without this great mass of preliminary, and undoubtedly often erroneous, writings, nothing is stated, however.

In the volume devoted to general geology is an important chapter on elevation and parallelism of mountain ranges, and an evident tendency manifested to accept the theories of Elie de Beaumont. "We are disposed to regard these axes of elevation not simply as irregular lines of limited extent, but as possessing much uniformity in direction, and traversing entire continents; but we admit that, in the present state of geographical and geological knowledge absolutely certain conclusions cannot be attained." It is a little amusing to note in the conclusions of this chapter a claim to originality in the idea that the so-called new world was probably the older of the two continents, since Featherstonhaugh, whose work Whitney had condemned by wholesale, had advanced the same opinion in 1834 (see p. 163).[33]

Hall's Report on Lake Superior Rocks.

The work on the Paleozoic rocks, given in Part II of these reports, was done by James Hall, of New York. The limestones first seen upon St. Marys River Hall regarded as identical with the Chazy, Bird's-eye, Black River, and Trenton limestones of New York. The Cliff limestone of Owen he designated the *Galena* limestone, which he erroneously considered a distinct member of the Lower Silurian system, not recognized in the East.

In his chapter on the parallelism of the Paleozoic deposits of the United States and Europe, Hall called attention to the fact that the simplest principles of elementary geology teach us that sedimentary beds, having the same thickness and same lithographical characters, cannot have spread over an area so wide as that now included between the European and American continents. All sedimentary deposits must vary in character at remote points as the physical conditions of the ocean cannot be presumed to have remained the same over a wide extent of surface. Under such circumstances, absolute parallelism is not to be sought for or expected. Calcareous deposits, as would naturally be supposed, have been found to be more persistent and more uniform in the character of their fossil contents; but these, over some portion of their extent, have often been invaded by argillaceous and arenaceous sediments, and the fauna is found to be in a greater or less degree influenced by such circumstances.[34]

[33] Apparently an unsuccessful attempt was made by Whitney for a new and better edition of the report than that issued by Congress. "You remember that after that abominably printed document came out we got a small appropriation [$1,500] to enable us to make some further examinations with the understanding that we were to have the privileges of having a final report printed like Owen's, by Lippincott & Company. When I came back from the Lake, Foster and I went on to Washington and endeavored to carry that arrangement through, but found the Commissioner suddenly converted over to the idea that the public printer could do the work better than anybody else and so we had to give up the idea of doing anything else than hanging round the public printer and having our illustrations executed in the usual congressional documental style."

[34] Appendix, p. 671.

Desor on the Drift.

The drift phenomena of the region surveyed by Foster and Whitney were described by E. Desor, who divided the deposits into four classes, as follows, beginning with the lowermost:

1. A layer of coarse, pebbly loam called "coarse drift."
2. Clay resting either on the coarse drift or directly on the rock.
3. A deposit of sand, gravel, and pebbles, irregularly stratified, resting upon the clay or the bed rock.
4. Isolated bowlders scattered over the whole region.

While disclaiming any intention of giving a general theory for the causes and origin of the drift, he argued that the phenomena he described indicated neither paroxysmal agencies nor the operation of any single cause, however long continued. "They disclose a long series of events which have resulted from causes highly diversified, and as yet imperfectly known. Three periods are recognized in the history of the drift of the Lake Superior region: (1) 'The period of the grooving and polishing of the rocks,' which 'must be considered as the dawning of the drift epoch'; (2) a period of comparative quiescence, extending over a long period of time and during which the stratum of red clay was deposited. This is the second era of the drift. (3) The overlying stratum of sand and gravel presumably formed by water, but at higher levels than the clay and indicating a still further depression of the land. This period characterized by intervals of agitation and repose." The transportation of the bowlders, he thought, took place at the close of the drift epoch.

In another article in the *American Journal of Science* for the same year (1852), concerning the post-Pliocene of the Southern States and its relation to the Laurentian of the North and the deposits of the Mississippi Valley, Desor attempted to account for the large bowlders in the drift of Long Island by means of ice rafts. Those in the post-Pliocene of the Southern States were doubtfully referred to water action only.

Desor, according to Marcou, was a German, who had come to America as private secretary to Agassiz, and whatever views he had on glaciation or other scientific subjects may be regarded as mainly absorbed rather than learned from observation. Through becoming overpresumptuous he had a falling out with his chief, which resulted in his discharge in 1848 and return to Germany in 1852, passing thus beyond the limits of our field of research.

Dana and the Wilkes Exploring Expedition.

J. D. Dana, as already noted, served as geologist in the United States Exploring Expedition under Captain Wilkes during 1838-1842. The results of his observations during this time are embodied in the tenth volume of the reports of the expedition, a royal octavo of 756 pages, with a folio atlas of 21 plates issued in 1849. Many of the conclusions given in this volume were first published in the columns of the *American Journal of Science*, and have already received attention.

One of the earliest results of Dana's work, as here chronicled, was the establishment of the principle that temperature influenced the growth and distribution of corals. A claim to priority in this discovery was made by James P. Couthouy, and brought about a somewhat bitter personal controversy between the two authors. This has been alluded to elsewhere (p. 222). The existence of harbors about the coral-bound reefs was attributed largely to the action of tidal and local marine currents, though the presence of fresh-water streams it was thought might have contributed toward the same end.

The then popular theory of the formation of coral reefs and atolls through the gradual subsidence of volcanoes, the crater corresponding to the lagoon and the rim to the belt of land, was rejected, though he believed that, beyond question, a subsidence had taken place throughout a large part of the Pacific, and hence that subsidence must form a part of any true theory of their origin.

He believed that the atoll once formed a fringing reef about a high island. This, as the island subsided, became a barrier reef, which continued its growth while the land slowly sank. The area of waters within finally contained the last sinking peak, which itself ultimately disappeared, leaving only the barrier at the surface and an islet or two of coral in the inclosed lagoon.

These were essentially the views put forward independently by Darwin. More recent work by the younger Agassiz and others has shown them to be not wholly correct.

From the actual extent of the present coral reefs and islands Dana inferred that the whole amount of high land lost to the Pacific by subsidence was at least 50,000 square miles, probably more, though he would not go so far as to conclude that a continent once occupied the place of the present ocean, or indeed of a portion of it.

In the discussion of the Hawaiian volcanoes it was noted that no apparent connection existed, so far as indicated by the phenomena of eruption, between Mount Loa and Kilauea, sixteen miles distant and more than 10,000 feet lower. It was therefore concluded that the two conduits, which he assumed were once connected by a fissure, had become isolated through the solidifying of the lava between them, each conduit being possibly a separate branch of some deep-seated channel.

The wide difference in the height of the columns of lava in these two volcanoes so near together, caused him also to question the statement so commonly made by writers of that day to the effect that volcanoes were the earth's safety valves.

Assuredly, if, while Kilauea is open on the flanks of Mount Loa—a vast gulf 3¼ miles in diameter—lavas still rise and are poured out, Kilauea is no safety valve even to the area covered by this single mountain alone.

The conclusions based upon the study of these volcanoes were to the effect that:

1. The majority of the Pacific volcanic summits were formed from successive eruptions of molten rock, alternating sometimes with cinder or fragmentary ejections.

2. The eruptions are, in general, the result of a rising or ascent of the lavas, owing to the inflation by heat of such vaporizable substances as sulphur and water, the overflow or lateral outbreak taking place in consequence of the increased pressure from gravity and from the elasticity of the confined vapors, the contraction of the earth's surface being no more necessary for an eruption than the contraction of the sides of a pot of water to make it boil.

He concluded further that volcanic action usually proceeded from fresh water gaining access to a branch belonging to some particular outlet or vent, and not to a common channel at greater depth. The lack of sympathetic action between two neighboring vents was thus explained on the ground that the union of their channels "took place far below the level to which the waters that ordinarily feed the fires gain access."

To the elevation theories of volcanic craters advocated by Von Buch he took exception, as he did also to the theory of Bischoff, who appealed to the internal igneous fluids for the source of volcanic action.

The highly feldspathic, coarsely crystalline, and solid centers of certain volcanic mountains, contrasted with the more vesicular and less dense outer portions, he rightly ascribed to slowness in cooling, the central mass being protected on all sides from the external air. Incidentally he discussed a problem which has become known to modern petrographers as that of magmatic differentiation. He argued that, given a large crater like that of Kilauea, the rise of the lavas through the center would be accompanied by a descending current along the side, though of less distinctness. The essential constituents of a rock, for example, being augite and feldspar, wherever the temperature of the liquid mass became sufficiently lowered, there the feldspar would commence to solidify or would slowly stiffen in the midst of the fluid material made up of the other ingredients. Under these conditions the ascending vapors would urge the feldspar upward much less freely than the more liquid part of the lava, for the latter would yield more readily to the inflating vapors and thus become lighter and rise to the surface. This process, going on throughout the whole progress of the cone, would keep the center feldspathic below a short distance from the summit. The residue from the feldspathic crystallization, consisting of ferruginous silicates, would be brought upward in the form of a frothy scoria which must on either side, in part, return to supply the place of the ascending current. On cooling, then, the more basaltic portions would constitute this exterior descending part. Thus, a feldspathic center and basaltic flanks would be the result of one and the same process. This feldspathic center, further, by being inclosed within a thick covering of rocks, would cool slowly, forming, perhaps, disseminated crystals in the earthy base, or, if cooling sufficiently slowly, a crystalline granular mass like granite or syenite.

He recognized the fact "that particular rocks have no necessary relation to time, excepting so far as time is connected with a difference in the earth's temperature or climate and also in oceanic or atmospheric pressure, for, if the elements are at hand, it requires only different circumstances as regards pressure, heat, and slowness of cooling to form any igneous rock the world contains."

The date of the beginning of volcanic activity on the Hawaiian Islands he placed as far back as the early Carboniferous or Silurian epoch, and believed: (1) That there were as many separate rents in the region as there are now islands; (2) that each rent was widest in the southeast portion; (3) that the southernmost rent was the largest; and (4) that the order of extinction of the volcanoes was as follows: 1, Kauai; 2, Western Oahu; 3, Western Maui (Mount Eeka); 4, Eastern Oahu; 5, Northwestern Hawaii (Mount Kea); 6, Southeast Maui (Mount Kale-a-kala); 7, Southeast Hawaii (Mount Loa).

From the general arrangement of the islands in the Pacific as a whole and the phenomena connected therewith, the conclusion was reached that the Hawaiian group originated in a series of rents or ruptures seldom continued at the surface for a long distance, but frequently advancing successively, one after the other, causing the resultant islands to appear in the form of a curve rather than a straight line, and, further, it was announced that: (1) While straight ranges are of occasional occurrence, curved ranges are still more common; (2) curvature may arise either from a gradual change of trend in the subordinate parts or from the position of these parts in a series; (3) the same great chain may change its direction sixty degrees or more, and consequently (4) the course of a mountain chain can be no evidence of its age. In this it will be observed he differed radically from Elie de Beaumont.

Dana noted further that the Pacific islands were arranged mainly in two systems of linear groups nearly at right angles with each other, the linear groups being based on a series of ruptures instead of a single uninterrupted fissure. The prevailing uniformity of trend of these fissures or ruptures he believed to be due both to the nature of the crust fractured and the direction of the fracturing forces. He accepted the doctrine of an earth cooled from a state of fusion, and regarded the influence of electric currents on the position of continents in process of formation as an established fact. An outer crust having once formed, the deep-seated crystallization would go on at a rate inconceivably slower, and circumstances would be favorable for a coarse crystallization of the material below and for the operation of electrical currents. The rupturing force he believed to be contraction caused by cooling. He argued that a cooling globe incrusted over by refrigeration while contraction was still going on beneath would, like a Prince Rupert's drop, be in a state of tension. Such a tension is bound to produce fractures and displacements, the direction of which would depend on rate of cooling in different parts and on the change in the earth's oblateness accompanying a diminution of its diameter. He concluded, from the absence of volcanoes in the interior of continents, that these portions of the globe cooled first and became solid; the intermediate portions cooling later and at a less rapid rate contracted most, since the crust was here thinnest. The oceanic areas would therefore be gradually subsiding and the tension increasing; moreover the tension, from its nature, would be exerted nearly horizontally. He inferred, therefore, that the subsiding oceans have produced the mountains of the continents, and that the oceanic and continental areas have never changed places, and he saw no reason for appealing to an incomprehensible subterranean force for the uplifting of the mountain chains or the continents. Such may have been "only a result on the whole of the deepening of the ocean's bed. It is obvious . . . that the earth has reached its present condition by gradual progress from a state of prolonged igneous action through epochs of increasing quiet, interrupted by distant periods of violence, to the present time, when even the gentlest oscillations of the crust have almost ceased." These ideas here somewhat elaborated he had previously put forward in the *American Journal of Science* as noted elsewhere (see p. 259).

In his discussion of the origin of the coal beds of New South Wales, Dana concluded "that the layers of the coal series were probably deposited by fresh-waters during the different stages of annual floods and wider deluges occurring at more distant periods; that the subsidence, which may have been gradual during the coal deposits, finally submerged the whole."

Another important observation bearing on the same subject was made when writing on the geology of Luzon.

> One of the interesting points about this Lake [Laguna de Bay] is the fact that vast quantities of plants live on its surface and pass down the river into the bay, carrying along great numbers of fresh-water snails of different species. Here we have, therefore, fresh-water shells and vegetation which is not marine accumulating under salt water, for they sink after a while and must become buried in the mud of the bottom, along with the remains of marine life. This floating vegetation illustrates a theory with regard to the vegetation of the coal beds.

Tyson's Work in California, 1849-1851.

The gold excitement of California in 1848-1849 drew attention to a region the geology of which was practically unknown. Fremont's expedition to California and Oregon in 1843-1844 was not accompanied by a geologist, and the few fossils collected were described by James Hall. *Some Notes on the California Gold Region*, six pages only, were given by C. S. Lyman in the *American Journal of Science* for 1849, while J. D. Dana had touched upon the subject during his return overland after the disaster to the Wilkes Exploring Expedition.[35] The main gold-bearing area was, therefore, practically an unknown land. The appearance in 1850 of Philip T. Tyson's *Geology and Industrial Resources of California* was consequently important. Tyson seems to have gone to California in 1849 as a private citizen, but so great was the demand for information concerning the region

[35] It will be recalled that the ship *Peacock* on which Dana sailed was wrecked at the mouth of the Columbia River.

that on his return he made a report to Colonel J. J. Abert, which was printed as a Senate document the year following.

In this report Tyson gave eight sections across the gold country, two of which extended from the coast to the Sierras. These were published as mere outlines, showing the direction of the dip of the rocks, but with no pretense to scale. He described the western flank of the Sierras as consisting of a vast mass of metamorphic and hypogene rocks, stretching from the Sacramento Valley to the axis of the mountains. The metamorphic rocks, mainly slates, contained the veins of auriferous quartz, through the breaking down of which had been derived the gold found in the gravels of the ravines.

Fig. 59. *P. T. Tyson.*

Making all due allowance for Tyson's laudable desire to check the wide and rapidly spreading excitement, bordering almost upon insanity, caused by grossly exaggerated accounts of the richness of the mines, still it would seem as if he were overzealous. Certainly he underestimated their value, to his own detriment and that of others. But it must be remembered that at that time and in that remote region, deposits, either placer or in veins, which could today be worked profitably, were valueless. He warned prospective investors that the large bodies of gold-bearing quartz found on the surface would if followed downward be found to be "nothing more than descending veins securely held between solid rocks, and that the cost of mining such was enormous, whilst the chances were almost wholly against their containing gold in proportion that would pay expenses." Indeed, he considered the prospect of a profitable mining of the veins as "altogether too remote and uncertain to be relied on."

Higgins's and Tyson's Work in Maryland, 1848-1858.

During the session of 1847-1848 the legislature of Maryland passed an act providing for the appointment and commission of a "person of ability, integrity, and suitable practical and scientific attainments," who should act as agricultural chemist for the state. These requirements seem to have been met in the person of Dr. James Higgins, who received the appointment and held the office until 1858, during which time he issued five reports. The office was not really a geological one, and the matter is mentioned here as bearing upon the subject only indirectly. During the session of 1858 bills were brought before the legislature to have the title of the office changed to *geologist* and, again, to *chemist and geologist*. Both, however, failed.

Higgins was succeeded in 1859 by the Philip Tyson above noted, whose first report of 145 octavo pages and appendix of 20 pages was issued February 14, 1860. Like the reports of his predecessor, this was given up very largely to a discussion of agricultural questions, but contained chapters on the "Minerals Comprising the Rocks" of the state; "The Mineral Character of Rocks"; "The Consideration of the Rocks as Grouped into Geological Formations," and also their "Geographic Distribution in Maryland"; and on "Chemical and Physical Geology," in which the question of the origin of soils through rock weathering was discussed.

The main interest in the work, from the present standpoint, lies in the colored geological map and sections which accompanied it, and which had the merit of being the first special map of the state, the area having, of course, been included in the general maps of Maclure and others. The various formations were classified according to the scheme of Rogers on the map of Pennsylvania, and a table given showing which of these were found within the state limits. The second report, which appeared in 1862, comprised ninety-two pages, and was given over almost wholly to a discussion of economic questions, including the soils and ores, coal, marbles, clays, etc. From the presence of fossil cycads, found associated with the iron ores, Tyson was disposed to consider these and the clays in which they occur as belonging to the oölitic period.[36]

[36] L. F. Ward, in his paper in the Nineteenth Annual Report of the U. S. Geological Survey, 1897-1898, considered them Cretaceous (Potomac).

CHAPTER V.

The Era of State Surveys, Third Decade, 1850-1859.

THE period of financial depression which proved so disastrous to the state surveys during the last decade at last ran its course. Several new states had in the meantime been added to the Union, some of which showed commendable promptness in authorizing geological surveys. New organizations were thus formed in fourteen states, eight of which had made no previous attempt. These eight, in alphabetical order were California, Illinois, Iowa, Kentucky, Mississippi, Missouri, Texas, and Wisconsin. Six states for the second time undertook the work—Michigan, New Jersey, North and South Carolina, Tennessee, and Vermont. The National Government was also active, the most important undertaking being the surveys in connection with proposed Pacific railways. In addition to these, Captain R. B. Marcy made a survey of the Red River region of Louisiana, Major W. H. Emory one of the Mexican boundary, and Colonel Pope one into the arid region of New Mexico along the thirty-second parallel. To each and all of these expeditions geologists, or at least naturalists, were attached. In the British provinces Logan's survey was doing good service, while Dawson, alone and unofficially, was working in Nova Scotia.

This was an era of publication, not merely of state survey reports, but of books and general treatises. Emmons's *American Geology*, Dawson's *Acadian Geology*, Hitchcock's *Geology of the Globe*, Lesley's *Manual of Coal* and his *Iron Manufacturer's Guide*, and Whitney's *Metallic Wealth of the United States* were among the more important productions. The publication of by far the greatest importance of this decade was, however, the long-delayed report of the Pennsylvania survey, to which allusion has been made elsewhere—two ponderous quartos which were truly epoch-making, although much of their contents that had at the time been considered new or important had found its way into print elsewhere. The publication in Europe of Murchison's *Siluria*,[1] the ninth edition of Lyell's *Prin-*

[1] This work met with a very favorable reception in America (as indeed it should), if one may judge from the extensive notices of the same by B. S. and J. D. W. in the American Journal of Science for 1854 and 1855.

ciples, and F. Roemer's *Die Kreidebildungen von Texas* were also matters worthy of note. Hitchcock's *Surface Features* belongs to this era, and marked the beginning of systematic study along lines of physiography. Among the names of new workers will be found those of W. P. Blake, J. W. Dawson, Leo Lesquereux, Oscar Lieber, William Logan, J. G. Percival, J. S. Newberry, J. M. Safford, G. C. Swallow, Alexander Winchell, and A. H. Worthen.

H. D. Rogers on Climate and Salt Deposits, 1850.

At the August, 1850, meeting of the A. A. A. S., H. D. Rogers pointed out the "intimate connection between the present basins of salt water and the existing distribution of the earth's climate." Assuming that the original source of the salt of the ocean was in the chlorides of the volcanic rocks of the earth's crust whence it was removed by the decomposing and leaching action of meteoric waters, he compared the gradual accumulation of saliferous matter in the lakes of arid regions with the salines of earlier geological ages, and believed them due to similar causes and conditions. "Following up the same general fact of the incessant solution of the rocks, we behold, in the great sea itself, a basin like the other salt ones which has no outlet for its surplus supplies but back again into the atmosphere by evaporation. Looking then at the primeval condition of an atmosphere of aqueous vapor, and at the state of the ocean just after the period when the earth's general temperature had ceased to be incompatible with the liquid state of water, it was a fresh ocean and not a salt one." This is an interesting forerunner of the form of speculation which led Professor Joly of Dublin, in 1899, to calculate the age of the earth from the rate of leaching out of the sodium during the weathering of rocks and its accumulation in the ocean.

Hubbard and Dana on Subaerial Erosion, 1850.

Two other papers of this year bearing on rock decay and erosion are worthy of consideration. Professor Oliver P. Hubbard, of Dartmouth College, during a study of the trap dikes of New Hampshire, noted that the same could be traced continuously across the country, at varying levels, from mountain top to valley bottom. From this fact he rightly argued that the valleys had been carved out through decomposition and erosion since the dikes were formed. The difference in altitude at the various outcrops gave, then, a measure of the

PLATE XV

J. G. PERCIVAL

EUGENE W. HILGARD

amount of erosion. In a previous paper (1837) on "The Mineralogy and Geology of the White Mountains of New Hampshire," Hubbard had recognized the intrusive nature of both granite and "trap," and also the efficacy of frost and moisture in promoting rock disintegration and decomposition.

Fig. 60. *O. P. Hubbard.*

Dana's observations, though of a somewhat different nature, were none the less interesting. In writing on denudation in the Pacific, he took the ground that the ocean is powerless to excavate valleys along the coast, and that the deep valleys like those of Tahiti are due to subaërial decomposition and erosion, an observation no geologist of today would venture to doubt.

Scope of a Geological Survey.

The limitations of the geological surveys have as a rule been loosely drawn, and even today they are made to include researches along all manner of scientific and industrial lines. Rarely, however, has a plan been suggested in which more was required with less prospect of adequate return than that of a "gentleman" from Mississippi in a letter to James Hall, under date of June 19, 1847. This is worthy of reproduction almost entire.

Being a citizen of Mississippi by adoption, and feeling a lively interest in her success, I have with others laboured to do her good. As one of the means of benefitting her, there has been some exertion made to get up a geological survey, in union with a perfect agricultural report, but our Legislators are unwilling. This failure has induced individual action, and with the view of making it effective I beg your views. To understand what manner of information we want, I beg to inform you that we propose raising a contribution of $2500. or $5000., to be placed in the hands of one or more trustees for the purpose of insuring the services of the best geologist that can be procured, the amount being given him as a bonus to survey the state, and we only require a publication in good style, he receiving the profits and the bonus, unless the State should see cause to purchase, when the Surveyor shall be paid liberally and (I prefer) he retain the $2500. as a present from the subscribers, or the amount be returned to subscribers as they may determine on prior to making any survey. It may be better to commence the survey with one of the best counties, and publish the report, through the sales to pay for prosecuting the work, or some other mode as may be preferred by a majority of the parties so subscribing.

I prefer the selection of a competent *gentleman,* put in his hands our cash, whether $2500. or more, and require of him to do his best until he completed the state, and let him make the best bargain he could with the state, we paying him our mites as inducement to do the work. In connection with this we would require of him to analyze soils in any locality for the work, but when done for individuals he should receive a fair compensation, to be as the other his own property. I think this latter would give $1000. per year.

Now Sir, could anyone be induced to take hold of such a work, and who would be likely to do so? Who would you recommend—this may be a delicate question, but please pardon me and attribute to my anxiety to get the best man available. Writing to you must show you that I prize your labors. What would be the best plan to carry out this project? What would you, or others, be willing to undertake at, per year, and how long probably would it take? Would $2500. with all perquisites and all resources, pay you or any other thoroughly competent gentleman?

I have started this matter, and when our subscribers meet for organization I wish to have all information before them, yet I will faithfully keep to myself anything that I am required to do, thus placing all fear of saying too much aside.

I wish when the Geological Survey of each county is made to have at the same time a thorough agricultural survey made, such as number of acres cultivated—in cotton—corn—oats—potatoes—or wheat; how many acres owned. How many slaves, number of hands over 12, ages from 1 to 5; 5 to 10; 10 to 15, to 20 or to 60. How many negro children born, proportion of sexes and ages on each plantation; number of work horses proper; brood mares; costs ———; stock in numbers and kinds; wool and bacon made; return of crops per acre, per hand; best prices; etc., etc.—in short every species of information.

Such a work would meet the views of any planter in this state, and would be purchased by others, as it would be a complete history of the agriculture of Mississippi.

Be pleased to make any suggestions that your knowledge may make useful to one who you will see has none of these matters at heart.

I write in haste, having much of it to do.

. . . .

Please answer and give all the information that is desirable, and believe me to be with great respect,

M. W. P.

What may have been Hall's reply to this extraordinary letter, the writer has no means of ascertaining, but nearly half a century later in answer to a query from B. W. Kumler, regarding the duties of a state geologist, he wrote:

I have received your letter of the 7th inst. asking me to give you some information as to the duties of a State Geologist.

I may say briefly that the duties of a State Geologist are to examine and determine the kinds and character of rocks occurring within the state and the limits of each one of the formations; that he should lay down upon a map the limits and extent of each one of the several formations within the state; to describe the same in all their characters and also to describe their contained mines and minerals and fossils; and to communicate such description to the people of the state. The natural channel for such publications being the reports made through the Legislature. His first duty is what I have already stated, to determine the nature and limits of each one of the geological formations. It will happen in all states or large areas of country that some one or more of these several geological formations will prove productive of minerals or ores, while other portions may be barren of such products, and it is the duty of the State Geologist to ascertain these facts that he may himself give special attention to the formations which promise mineral resources, and also to be able to advise other persons where they may expend their energies in search of minerals with some prospect of success, and on the other hand to advise them against expending money in geological formations barren of metalliferous ores or other economic products.

These in a general way are the duties of a State Geologist but in some parts of the country there may be special reasons why he shall give his time and attention largely in one direction; as for example a special investigation may be made as has already been done in some of the states regarding the sources of water and the area of a country likely to give artesian wells. In the newer states many of these points will be brought to the attention of the geologists by the expressed wants of the people, but he should not forget that the first element is the determination of the nature and extent of the geological formations which constitute the underlying rocks of the country and to determine their relations one to the other, and the nature of their mineral and fossil contents.

Millington and Wailes's Work in Mississippi, 1850-1852.

Whether it was or was not due to the efforts of the "citizen by adoption" and author of the first letter above quoted, is not apparent, but as a matter of history, an agricultural and geological survey of Mississippi was inaugurated in connection with the State University through an act of the legislature approved March 5, 1850. Under this Dr. John Millington was appointed professor of geology and agriculture, in connection with the professorship of chemistry, which he already held. Professor Millington, however, being by age disqualified for field work (he was over seventy), relinquished the situation the latter part of 1850 without having made a report, and in 1852 was succeeded by Professor B. L. C. Wailes. Wailes made but one report, this an octavo volume of 356 pages and an appendix. Of this, only pages 207 to 288 deal with matters strictly geological. The work is by no means of a high order and made no permanent impression upon the science either in the state or country at large. The brilliantly colored sections are crude and the language pedantic. In fact, there is scarcely a single original observation which can be unhesitatingly accepted, owing to the general air of unfamiliarity with the subject that is apparent throughout.

He made no attempt to classify the rocks he described otherwise than Cretaceous, Tertiary, and Quaternary, and inferentially classed among the latter the sandstones of the Grand Gulf group, which he mentioned as overlying the Diluvial gravel. He, however, traced correctly, according to Hilgard, the northern limits of the Grand Gulf formation from the Mississippi across the Pearl River to Brandon, and described its occurrence in the southwestern Mississippi.

E. Emmons's Work in North Carolina, 1851.

In 1851 Ebenezer Emmons, heretofore connected with the survey of New York, was appointed state geologist of North Carolina, a position he continued to hold until his death in 1863, though the work of the survey came to a close in 1860, owing to the Civil War. Emmons was assisted by his son, Ebenezer Emmons, Jr., and during his period of office issued five reports, the first, bearing the date of 1852, forming an octavo pamphlet of 181 pages. It related principally to the geology of the eastern counties and the coal. Emmons recognized the fact that the coal-bearing rocks of North Carolina were not of the same age as those of Pennsylvania, but were presumably the same age as those of the Richmond fields, which at that time were thought by W. B. Rogers to be Liassic, though Emmons rightly questioned if the saurian remains found in the sandstones did not point to their possible Permian age; he added they might be Permian or Triassic.

> The meager list of plants and animals . . . furnish only grounds for conjecture to what age the formations belong. My opinion, derived from all the facts and circumstances known to me, inclines me to adopt the belief that it is the upper New Red sandstone. Still, if the Richmond coal basin is of the same age as the coal rocks of North Carolina, geologists will be disposed to place the series along with the Oölites or Lias, as Profs. W. B. Rogers and Sir Chas. Lyell have done.[2]

[2] At a meeting of the Boston Society of Natural History on January 4, 1854, Professor W. B. Rogers summed up the evidence regarding the red sandstone of

Emmons in this work rejected the old Wernerian classification of (1) Primitive, (2) Transition, and (3) Tertiary, and introduced that given in tabular form below. It seems to have met with little favor, and was not, so far as the present writer is aware, adopted in a single instance elsewhere.

EMMONS'S ROCK CLASSIFICATION, PROPOSED IN 1852.

I. Pyrocrystalline—crystallized by the agency of fire. Primary of authors.
II. Pyroplastic—molded by fire. Ancient and modern volcanic rock of authors.
III. Hydroplastic—molded by water. Sediments of authors.

The first class is divided into two sections:

1. Unstratified pyrocrystalline, as granite, hypersthene rock, pyrocrystalline limestone, sienite, magnetic iron ores.
2. Stratified pyrocrystalline gneiss, mica slate, talcose slate, and hornblende steatite.

The second class is divided into two sections, also:

1. Modern pyroplastic rocks, lavas, tuffs, pumice, and all the products of volcanoes, which are cooled in the air.
2. Ancient pyroplastic rocks, the ancient lavas, cooled under water, basalt, porphyry, and green stone.

The third class is divided into systems, most of which are admitted by geologists of the day.

The systems belonging to the class of hydroplastic rocks, the consolidated and loose sediments, are exhibited in the following table:

I. Tertiary system: 1, Postpliocene; 2, Pliocene; 3, Miocene; 4, Eocene.
II. Cretaceous system: 1, Upper Cretaceous, including the true chalk, with flints; 2, Lower Cretaceous, including the green sand, iron sands, etc.
III. Wealdon, unknown in the United States.
IV. Oölite and Lias.
V. New Red Sandstone or Trias: 1, Upper; 2, Middle; 3, Lower.
VI. Permian system.
VII. Carboniferous system.
VIII. Devonian system.
IX. Silurian system: 1, Upper; 2, Lower.
X. Taconic system.

Virginia and North Carolina, and found it "to confirm the conclusion of their Jurassic date." The fossils thus far formed in the more western red sandstone

His second report, comprising 351 pages, was issued in 1856. It related chiefly to the geology and mines of the so-called midland counties. In 1858 his *Report on the Agriculture of the Eastern Counties* was issued. About one-third of this, notwithstanding its title, was given up to paleontological matters. In 1860 two more reports, both short, appeared—the one devoted mainly to agriculture and the other a special report on the swamp lands.[3]

In the report for 1858 Emmons announced a principle which has since been enunciated in somewhat different words by our most eminent authority on soils. This, in his own words, is as follows:

> In the examination of soils the physical properties require as much attention as the chemical, for in order that a good chemical mixture of elements may be fertile they should possess a certain degree of adhesiveness or closeness which will retain water.

Much attention was given to the marls of the state with reference to their availability for fertilizing purposes, though he recognized the fact that, unfortunately for the best interests of agriculture, they were of too low grade to bear transportation to distant points.

Norwood's Work in Illinois, 1851-1857.

A geological survey of Illinois was organized under an act of the general assembly approved February 17, 1851. This act authorized the governor, auditor, and treasurer to employ a competent geologist, who should be required to make "a complete geological and mineralogical survey of the entire territory of the State," and provided that

belt and its extension through Pennsylvania and New Jersey showed this also to be Jurassic, a little lower probably than that of Virginia and North Carolina. He felt that there was little doubt but that the same conclusions would apply also to the sandstone of the Connecticut Valley. Later, still other and more interesting fossils must have been found, for Whitney, writing regarding the August, 1856, meeting of the American Association, says: "Emmons had a magnificent collection of fossils—saurians and the like—from the so-called New Red sandstone of North Carolina. It was the most curious and striking set of fossils ever got together in this country, as was declared by the paleontologists present."

[3] In connection with this survey, there was published in 1861 by Edmund Ruffin a work of nearly 300 pages, entitled Sketches of Lower North Carolina. This was devoted mainly to soils and the swamp lands of the Atlantic coast, the geology being largely a reflection of the prevailing opinions of the time. This work has been largely overlooked by bibliographers and is not mentioned even by Holmes in his history of the state surveys.

the said geologist should proceed to ascertain the order, successive arrangement, relative position, dip, and comparative magnitude of the several strata or geological formations in the State, and to search for and examine all the beds and deposits of ores, coals, clays, marls, rocks, and such other mineral substances as may present themselves; to obtain chemical analyses of the same, and to determine by barometrical observations the relative elevation of the different portions of the State.

He was also required "to procure and preserve an entire suite of the different specimens found in the state, to be preserved and properly arranged in a cabinet, and placed in the office of the secretary of state in the state capitol"; and eventually "to publish with the final reports a geological map of the State."

Under the provisions of this act Dr. J. G. Norwood was appointed state geologist, and entered upon his duties in October, 1851, by an examination of the formations exposed in the bluffs of the Ohio River between Shawneetown and Cairo.

The act appropriated the sum of $3,000 a year for carrying on the work. In 1853 the amount was increased to $5,000, with $500 additional for topographical purposes. Up to January, 1857, some $27,000 had been expended, though without visible results.

The usual trials of a state geologist seem to have come across the path of Mr. Norwood and perhaps some that were unusual. Be that as it may, a committee was appointed to investigate the conditions of the office. They reported that in their belief the money had been well expended, a large amount of work had been done, and well done, and they recommended an appropriation of the sum of $6,500 to enable him to complete a report on the economic resources on which he was engaged.

The recommendation seems, however, to have failed of its purpose, since the only outcome was a small octavo pamphlet of less than 100 pages devoted wholly to a discussion of the coals of the state. This bore the date of 1857, and was his last as well as first report. It was accompanied by a small, one-page, colored geological map.[4]

Presumably, all members of the legislature were not as favorably disposed toward Norwood as were the members of the investigating committee, for the following year (1858) he was supplanted by A. H. Worthen, who had been connected with James Hall on the state survey of Iowa and who was endorsed by Hall for the position in Illinois.

[4] In 1855 Norwood, in connection with Mr. Henry Patton, published in the Journal of the Academy of Natural Science a paper comprising some seventy-seven pages of text relating to the work thus far accomplished.

Hall's Troubles with J. W. and J. T. Foster, 1851.

Inasmuch as some confusion still exists in the minds of many regarding an episode in American geology, due to a similarity of names, space may well be given here to an explanation.

It appears that J. W. Foster, one of the Lake Superior geologists working with Whitney and Jackson, somewhere about 1851 compiled a geological map of the United States, a copy of which was sold by him to the United States General Land Office. This was compiled in part from a map made by Hall, with additions from surveys of Alabama, South Carolina, Tennessee, and Texas in the South and by Owen in the West. Hall was very indignant on learning of this, claiming that he had been at work upon the map since 1848 and had no intention of allowing Foster to copy it except for his own use and with no expectation of his claiming it and other compiled additions as his own. It is apparent that considerable ill feeling existed for a time between the two most concerned. The second episode also centered about a map. It appears that a certain *J. T.* Foster, who figures not otherwise in geological history, compiled what Whitney designated as "a ridiculous attempt at a geological chart," at about this same period. An attempt was made to introduce this into the schools of New York. To prevent this Hall wrote Agassiz requesting his opinion of it, which opinion, expressed plainly, Hall published. Whereupon Foster entered suit. *J. W.* Foster, Whitney, Dana, and Agassiz all testified to the truth of the criticism and the case was dismissed. It created considerable interest in scientific and educational circles at the time, and the two cases as outlined above are often confounded.[5]

Second Attempt at a Geological Survey of Indiana, 1852.

In 1852 the question of the reestablishment of the geological survey of Indiana was agitated, but nothing came of it until the following year, when, acting under a recommendation of Governor J. A. Wright, the legislature made a small appropriation to become available in January, 1854. With this for a beginning, the governor appointed Dr. Ryland T. Brown state geologist. The venture was, however, short-lived. Brown made but one report, which the legislature refused to publish, and at the same time refused further ap-

[5] Appendix, p. 673.

propriation for continuing the survey, the main reason given for such action being an overvaluation of the Indiana coals on the part of Brown. The report, it should be mentioned, was subsequently published in that of the board of agriculture.

G. G. Shumard's Work in 1852.

In 1852, in compliance with a resolution of the United States Senate, Captain R. B. Marcy was directed to proceed "without un-

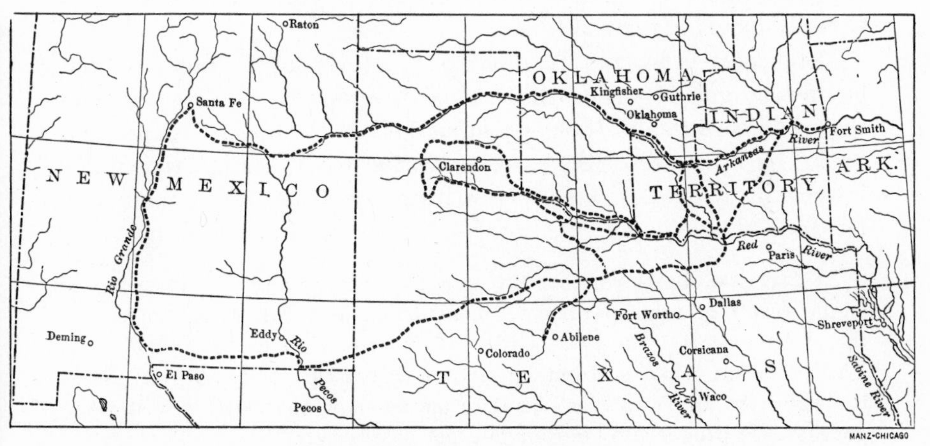

Fig. 61. *Map of Routes of Marcy and Shumard, 1849-1852.*

necessary delay" to make an examination of the upper Red River and the country immediately bordering upon it from the mouth of Cache Creek to its sources. The officers of the expedition, which left Fort Belknap, Texas, on the second of May, consisted of Captain Marcy, in charge; Brevet Captain George B. McClellan, U. S. Engineer; and Dr. G. G. Shumard; the last named, in addition to his duties as surgeon, acting as naturalist. The report submitted formed an octavo volume of 320 pages with 65 plates, of which 25 may be considered as geological, 6 of them being of invertebrate fossils. The geological part proper comprised but 32 pages, which were by President Edward Hitchcock, of Amherst College. The invertebrate fossils were described by B. F. Shumard.

G. G. Shumard's report consisted of a brief summary of the geology of a portion of northwestern Arkansas, which he considered essential to a proper understanding of that which was to follow, followed by memoranda of the various types of rocks, minerals, and formations passed over each day. It was accompanied by a colored geological section from Fort Belknap, Texas, to Washington County, Arkansas, showing a granitic axis flanked on either side by Carboniferous limestone, in its turn overlaid by Coal Measures, and with Cretaceous deposits between Fort Wichita and Cross Timbers in Texas.

In the paleontological report by B. F. Shumard there were described Carboniferous fossils from Washington and Crawford counties, Arkansas, and Cretaceous forms from Fort Wichita and the Cross Timbers regions of Texas.

Hitchcock, in his report, dwelt particularly on the possible Carboniferous age of various beds of coal, reported by Shumard, and the economic value of the gypsum, as well as the ores of copper and gold. His reference to the canyons of the Red River is particularly interesting in view of his early writings regarding the Connecticut (p. 146):

> You seem in doubt whether this gorge was worn away by the river or is the result of some paroxysmal convulsion. You will allow me to say that I have scarcely any doubt that the stream itself has done the work. The fact that when a tributary stream enters the main river it passes through a tributary canyon seems to me to show conclusively that these gorges were produced by erosion and not by fracture.

Work of J. B. Trask in California.

In March, 1853, by resolution of the senate of the state of California, Dr. John B. Trask was called upon to furnish what information was in his possession on the geology of the state. Trask was a medical practitioner, and an active member of the California Academy of Sciences. His geological publications up to date had dealt only with earthquakes and paleontological subjects. In answer to the call of the senate he promptly submitted the matter issued in the form of an assembly document of thirty pages entitled *On the Geology of the Sierra Nevada, or California Range.* This seemed to have met with approval, and on May 6 following, a joint resolution was passed authorizing further work in the Sierra Nevada and Coast Range mountains. As a result of this and subsequent acts there were issued reports in 1854, 1855, and 1856.

The classification adopted by Trask was that of the older geologists of his time, the rocks being divided merely into *Primary, Secondary,* and *Tertiary.* Although fossils were found, he did not attempt to describe them in his first report, and seemed in most cases to fail to realize their value as indicative of any particular geological horizon. His ideas regarding metamorphism were somewhat crude, as is shown by the following: "It is a well-known fact that intrusive dikes of trap in passing through limestone will change the calcareous formation to true talcose rocks."

Fig. 62. *J. B. Trask.*

He divided the state into three divisions: First, a primary or central district, included between latitudes 38° 30′ and 40° north; second, the northern district, included between 40° and 42° north; and third, the southern district, included between 38° 30′ and 36° north. The rocks of the primary district, he stated, are for the most part primitive, being composed of granite, porphyry, trap, and other allied rocks, of which serpentines form the one important part. The sedimentary rocks of the district were divided into first, argelite (*sic*) slates; second, conglomerates; and third, sandstones, the last named being regarded as probably of Miocene age.

The rocks in the northern district were described as of essentially the same character, a few minor differences only being noted. The southern district contained many rocks in common with the other two, although there were in addition many basaltic areas. He found a primary limestone in the area between the American and Merced rivers, and evidence of at least three successive periods of upheaval in this portion of the continent.

In his second report he described the geology of the most elevated portions of the counties of Butte, Sierra, Yuba, Nevada, and Placer; also the more southern counties of San Francisco, Santa Clara, Santa Cruz, Monterey; the north part of San Luis Obispo, Tulare, Mariposa, Tuolumne, Alameda, Contra Costa, and San Joaquin. During the months of October and November he visited Nevada and Calaveras counties. He pointed out the presence of post-Pliocene fossils in argillaceous sandstone of the Coast Mountains between Point Pinos and Nacimiento River. He divided the Tertiary rocks as below:

Period.	*Group.*	*Where found.*
Eocene	Middle	Calaveras County, at Murphys and other localities. Bones of extinct animals, etc.
Miocene	....	North and south of San Francisco in the Coast and Monte Diablo mountains. Consisting of marine shells, with most of the species extinct.
Pliocene	Lower	Coast mountains and Gabilan Spur. Also in cavern deposits in Calaveras County.
Post-Pliocene	....	Southwest of Monterey. Marine shells, all of existing species.

He had no hesitation in saying that no coal would be found in any part of the Coast Range south of the thirty-fifth parallel of north latitude, though the presence of Carboniferous limestone in Shasta County led him to express a hope that the desired material might yet be found within the limits of the state.[6] In his report for 1855 the predominating fossiliferous rocks of San Luis Obispo and Santa Barbara counties were considered of Miocene age. The San Bernardino Mountains were described as made up for the most part of primitive rocks, granite forming by far the larger part of their highest ridges and peaks.

In his final report (1856) on the geology of northern and southern California he considered a portion of the counties lying in the Coast Mountains, north of the Bay of San Francisco. He found it to be made up, for the most part, of mountain ridges having precipitous flanks, with deep, rugged, and in many cases, almost impassable canyons, though between the mountain ridges there were valleys, some of considerable magnitude. The rocks, back from the coast were mainly post-Tertiary lavas with smaller areas of primitive and fossiliferous rocks. The character of the soils and climate, as related to agricultural possibilities, was touched upon and about half of the pamphlet of sixty-six pages was given up to a description of mines.

In this connection it is well to note that, at the request of A. D. Bache, of the Coast Survey, W. P. Blake prepared in 1855 a brief paper on the physical geology and geography of the coast of Cali-

[6] These hopes were partially realized, lignite coal, but of Cretaceous age, having been found and to some extent worked in Alameda, Amador, Contra Costa, Fresno, Kern, Monterey, Orange, Riverside, and San Bernardino counties.

fornia from Bodega Bay to San Diego. Blake thought the sedimentary rocks of Punta de los Reyes to be probably of Miocene age, the sandstone at the entrance of San Francisco Bay to be Tertiary, and the metamorphic rock of the peninsula to be an altered sandstone. The serpentine of Lime Point was shown to be eruptive, as was also the granite of Cypress Point and the Bay of San Carlos, and younger than the conglomerate.

Swallow's Geological Survey of Missouri, 1853.

This same year (1853) there was organized a geological survey of Missouri with G. C. Swallow at its head,[7] and A. Litton, F. B. Meek, F. Hawn, and B. F. Shumard, assistants. Five reports were published. The first, bearing the date November 10, 1853, consisted of but four pages. The second, dated 1854, comprised over 400 pages, including the reports of the assistants above mentioned. None of the sections or maps given were colored, nor were any new principles or striking features brought out. The work as a whole is an extremely uninteresting array of details, essential only to the geographical extension of geological knowledge. Some space was devoted to economic geology, a consideration of soils, etc. Dr. Litton's report consisted almost wholly of details of lead, copper, and iron mines. Perhaps the most important economic item in Meek's report was his calling attention to the limited area of the coal beds, which lie in narrow basins in the encrinal limestone.

Fig. 63. *G. C. Broadhead.*

The third annual report of progress (6 pages) appeared in 1857, the fourth (14 pages) in 1859, and the fifth (19 pages) in 1861. G. C. Broadhead, who later was himself state geologist, became an assistant during 1857-1861, J. G. Norwood in 1858-1861, Henry Engelmann in 1857, and John Locke in 1860.

During the period which intervened between the publication of the second report and the stopping of the survey in 1861, a large portion of the state had been visited by members of the corps, and full reports were written on Cape Girardeau, Perry, St. Genevieve,

[7] Dr. Henry King was candidate for this position but failed to secure the appointment.

Jefferson, Crawford, Phelps, Pulaski, Laclede, Wright, Ozark (including Douglas), Clark, Morgan, Miller, Saline, Chariton, Macon, Randolph, Shelby, Osage, and Maries counties.

After the survey had been discontinued the legislature authorized L. D. Morse and G. C. Swallow to publish all the results of the work of the previous seven years, but the project was abandoned on account of the expense, and not taken up again until 1870 (see p. 459).

Jules Marcou's Geological Map of the United States, 1853.

In 1853 Jules Marcou, a French geologist who had come to America in 1848, published a geological map of the United States and the British Provinces of North America, with ninety-two octavo pages of explanatory text, two geological sections, and eight plates of the fossils which characterize the formations. The map, which was colored by hand, extended as far west as the one hundred and sixth meridian, and the geological sections, not colored, extended, the first from the Atlantic Ocean at Yorktown, Virginia, to Fort Laramie, Wyoming, and the second from Lake St. Johns in the Hudson Bay Territory to Mobile, Alabama.

The author adopted the formation names established by M. de Verneuil and Murchison, in order to render possible a satisfactory comparison with the existing geological maps of England, Germany, Russia, Scandinavia, and Bohemia. The divisions Cambrian and Taconic, however, were not recognized, since, in the opinion of the compiler, they "ought to disappear from geological classification, for they give two names to the metamorphic rocks, of which they are integrant parts, in all regions where these beds have been observed." This statement regarding the Taconic is a little surprising when one considers the part Marcou played in the subsequent controversy (see p. 604).

Perhaps the most striking feature of the map, at least that in which it differed the most from those of later date, lay in the enormous development of the Devonian, which was made to occupy a large portion of southern New York, as now, and to extend thence in a continuous, broad, but gradually narrowing band southwesterly to near Tuscaloosa, in Alabama. A continuous belt was also indicated as extending from the east side of the south end of Lake Michigan westerly nearly to the Missouri River. The areas of brown Triassic sandstone of the eastern states were colored yellow (Keuper), and the belt west of Richmond and one near Raton Mountains, in New

Mexico, as Lias or Jurassic (blue). A band of Keuper was also represented extending from the eastern end of the south shore of Lake Superior quite across the Coteau des Prairies to the ninety-eighth meridian. A continuous band, colored as of granite porphyry, syenite, greenstone, gneiss, mica-schist, etc., was represented as extending from below the twenty-fourth parallel in Mexico northward to the great bend of the North Platte in Wyoming, where it was overlain by Silurian, Devonian, Coal Measures, and Keuper strata, the last flanked on the north by a broad band of "copper trap." Here again the series extended westerly to the edge of the map and northeasterly to a point about midway between Mandan and the Little Missouri River in Dakota.

Fig. 64. *Jules Marcou.*

This attempt on the part of Marcou was certainly commendable, requiring courage as well as judgment. Unfortunately, he does not seem to have used discretion in all cases in the selection of his authorities, and made altogether too sweeping generalizations, often in direct contradiction of available facts made known by other workers. Nevertheless Agassiz in a letter to Hall wrote: "He [*i.e.*, Marcou] is not the man to be misled by a charlatan. I know him to be a good observer and his papers show an admirable power of generalization, for which he has not received the acknowledgment he truly deserves. Nothing has been noticed in his publications except mistakes of detail."

In 1855, 1858, and again in 1872, Marcou published in Germany and France other editions of his map, in which he comprised the whole country from the Atlantic to the Pacific Ocean. Here again a striking feature is the enormous area of country west of the Missouri at Iowa and extending almost to Great Salt Lake and the Colorado River in Arizona, colored as Trias with broad intercallations of Jurassic. This is the same in the issues of both 1855 and 1858, although in the map of 1853 he had colored the same area (at least as far as this map extended) Cretaceous, Tertiary, and Quaternary. In the edition of 1855 the entire west coast of California south of Humboldt and as far down as Monterey was colored as occupied by eruptive and metamorphic rocks, while broad belts in the interior, along the courses of the main rivers, were colored Tertiary. In the edition of 1858 this was in part corrected, the Tertiary being extended to the coast.

The various editions of these maps, on account of the errors mentioned and numerous others perhaps even less excusable, though less conspicuous, were severely criticised by American geologists, particularly harsh reviews of the maps being given by James Hall and Dana in the *American Journal of Science* for 1854 and 1855 and by W. P. Blake in the same journal for 1856.

Marcou complained bitterly of the manner in which his work was received and, could one accept the accuracy of his observations, would seem justified. Unquestionably however he set too high a value on his own work and was inclined to disregard much that was good in others unless corroborative.[8]

The Pacific Railroad Surveys of 1853-1856.

Section 10 of an Act of Congress approved March 3, 1853, reads as follows: "And be it further enacted That the Secretary of War be, and is hereby authorized under the direction of the President of the United States to employ such portion of the Corps of Topographical Engineers, and such other persons as may be deemed necessary, to make such explorations and surveys as he may deem advisable, to ascertain the most practical and economical route for a railroad from the Mississippi River to the Pacific Ocean, and the sum of one hundred and fifty thousand dollars or so much thereof as may be necessary, be and the same is hereby appropriated out of any money in the treasury not otherwise appropriated, to pay the expense of such explorations and surveys." Under this act several parties under direction and command of the Army Topographic Corps were sent out, to nearly every one of which were attached a mineralogist and geologist, physician, naturalist, and sundry civil assistants in the capacity of engineers and draftsmen. The geologists thus employed were, in alphabetical order, Thomas Antisell, W. P. Blake, John Evans, Jules Marcou, J. S. Newberry and James Schiel.[9]

The published results of these explorations, geological and otherwise, form the thirteen quarto volumes of Pacific Railroad Reports,

[8] Appendix, pp. 675-681.

[9] It may be well to state here that only those military expeditions are mentioned in the following pages that were accompanied by a geologist or "naturalist" and made geological observations or at least collections which were later reported on. A full list of western exploring expeditions from 1857 to 1880, under government auspices is given in Part III, Vol. I, Reports of Wheeler Surveys West of the 100th Meridian.

so well known to all naturalists. Collectively the geological portions amounted to about 1,000 printed pages.

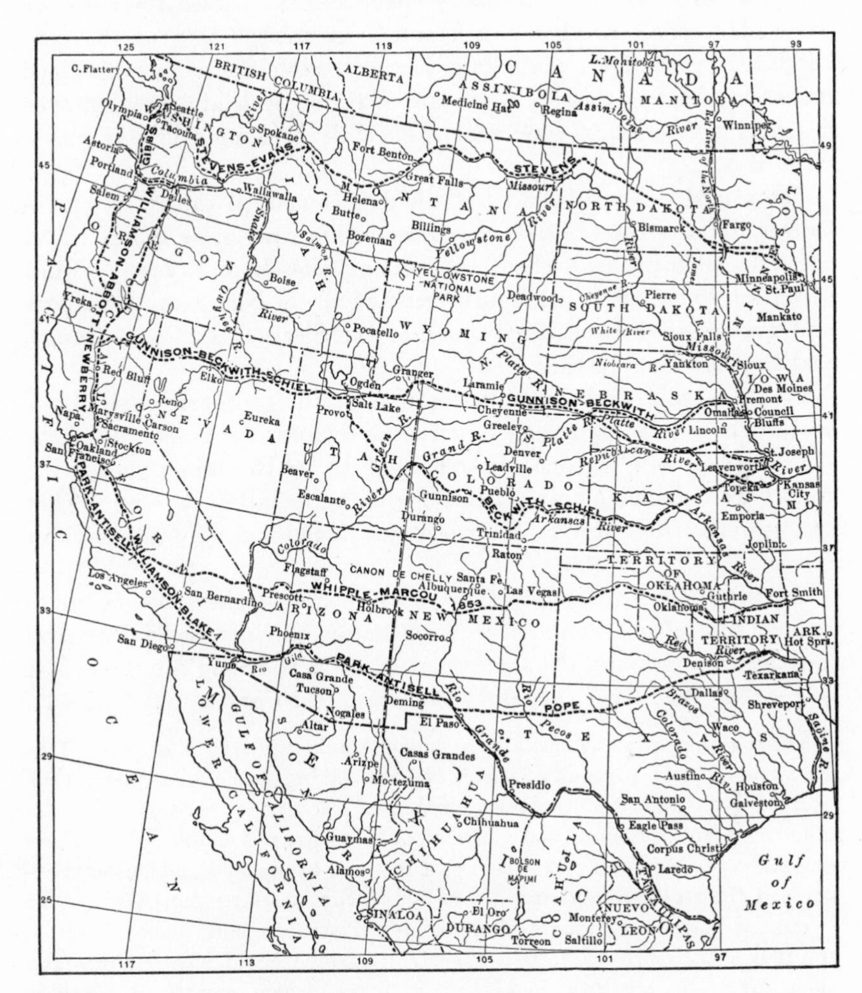

Fig. 65. *Map showing principal routes of exploring parties in connection with the Pacific railroad surveys.*

The Work of Antisell.

Thomas Antisell accompanied the party under command of Lieutenant J. G. Parke, surveying the route from San Francisco southward along and near the thirty-third parallel. (See Fig. 65.)

Antisell recognized the post-Miocene age of the final uplift of the

Coast Range and thought that the elevating force must have taken place from two points, one in the north and one in the south, and that the forces became gradually spent as they passed toward each other. He conceived that this resulted in an uplifting of the consolidated crust of the state at either end, while the center remained quiescent, causing thereby a rupture of the superficial strata, or even a depression below sea level near the middle, forming thus San Francisco Bay.

Influenced to some extent by Elie de Beaumont's theory of mountain uplift and the relative age of mountain ranges, as indicated by their parallelism, he, without committing himself in any way, called attention to the north and south trend of all the New Mexican ranges and northwest and southeast trend of the Sierras, the Coast Range, and the ranges of Nevada.

No Paleozoic rocks were recognized in the southern part of the state. The sandstone underlying the Carboniferous limestone of the Gila region, New Mexico, was thought to be Devonian, though no fossils were found by which this might be definitely proven. What fossils were collected were described by Conrad and were all Tertiary forms.

As with Hitchcock and other of the early workers, he failed to realize the full capabilities of river action during gradual crustal movement, and found the same difficulty in accounting for the course of the Gila that Hitchcock did with that of the Connecticut (p. 146).

> During the series of elevations which finally uplifted this range (the Pinaleno) to its present altitude, the upheaving force must have been exerted even upon the southern portion of the range, raising the tableland of northern Sonora and Chiricahua to so great a height. This strain may have produced a fissure from east to west, or cracked and perhaps depressed the strata along parallel 33°, and thus enabled the Gila to take that as a permanent course. Some such catastrophe must have occurred, for it is scarcely probable that the river unaided could have cut through such lofty hills and hard rocks as it appears to have done in its passage through these mountains, running as it does at right angles to the strike of the ranges.

The report was accompanied by numerous sections in black and white through the coast ranges, and a colored geological map of the region extending from San Francisco to Los Angeles; also by a colored map and section of the region from the Rio Grande (then known as the Rio Bravo del Norte) to beyond Maricopa Wells (112° 30′).

Sketch of Antisell.

Antisell was born in Ireland, of French Huguenot parentage, in 1817, and came to America for political reasons in 1848, where he practiced medicine until 1854, though in the meantime holding the position of lecturer on chemistry in a number of colleges, including those of Woodstock, Vermont; Pittsfield, Massachusetts; and at the Berkshire Medical Institute. Prior to his American experiences, though by profession a surgeon, he had manifested a lively interest in matters pertaining to other sciences, and had published in 1846 a duodecimo volume of eighty-four pages on Irish geology and a manual of agricultural chemistry, prepared with especial reference to the soils of Ireland. His work with the Pacific Railroad Survey, above noted, comprised his only geological investigations after coming to America.

Fig. 66. *Thomas Antisell.*

In 1856 he was appointed an examiner in the Patent Office, but resigned to enter the Union Army as surgeon in 1861. In 1866 he became chief chemist of the Department of Agriculture, and in 1871 was appointed by the Japanese Government an expert in chemical technology in Tokyo. He returned to the United States in 1877 and resumed his former position in the Patent Office, where he remained until 1891.

Work of W. P. Blake.

To W. P. Blake fell the lot of accompanying Lieutenant Williamson, whose route also led from San Francisco southward along the San Joaquin River, through San Francisco Pass to the Mohave River, and eastward along the thirty-second and thirty-third parallels to the mouth of the Gila. The region, as pointed out by S. F. Emmons,[10] is not one from which definite geological data could be obtained, the rocks, with the exception of recent and Tertiary formations, being barren and classed as metamorphic and eruptive. He however noted the widespread occurrence of Tertiary beds about San Francisco and in southern California and the occurrence of eruptive serpentinous rocks. He also made many interesting minor

10 Presidential address, Geological Society of Washington, 1896.

observations on the polishing and grooving of hard rocks by wind-blown sand and other desert phenomena. Mountains of the isolated or "lost mountain" type were described, and the fact that the Colorado Desert was below the level of the sea noted. The age of the coast mountains was determined as at least post-Pliocene, and the impregnation of the rocks of the region with gold thought to have been contemporaneous with the uplift.[11]

To Blake, too, fell the study of the collections made by Pope's expedition along the route near the thirty-second parallel. It was the original intention of Pope that this work should be done by Marcou, but owing to the decision of Davis, referred to elsewhere, the collections were returned in part "in a confused condition and with many of the labels displaced" and given to Blake.

Captain Pope's route extended from Preston on the Red River in a southwesterly direction to the Pecos River, and thence nearly due west to the valley of the Rio Grande at El Paso and Dona Ana in New Mexico, crossing the Organ, Huerco, and Guadalupe mountains. The result of this expedition was to establish the Carboniferous character of the limestone of the Organ Mountains, the prevailing granitic type of the Huerco Range, and the presumable Carboniferous age of the limestone and sandstone of the Guadalupes (see also Shumard's observations, p. 371). The underlying formations of the Llano Estacado were judged to be probably Cretaceous and not Jurassic, as mapped by Marcou. For the geology of the region between Llano Estacado and Preston, Blake drew largely on Shumard's publications.

The gypsum beds were thought to have originated through the action of percolating waters containing sulphuric acid derived from the decomposition of pyrites on the underlying limestone.

Sketch of Blake.

Blake came first into public notice through his connection with this survey. His work here was necessarily in the nature of reconnaissance. On his return east he was for two years editor of the *Mining Magazine* of New York, a now extinct publication, and in 1862-1864 served as a government geologist in Japan. In 1865 he became professor of geology and mineralogy in what is now the University of California, at Berkeley. In 1873 he was sent to Vienna

[11] Blake had a number of copies of his report bound up separately under the title Report of a Geological Reconnaissance in California, etc., 1858. This has led to some confusion among bibliographers.

in connection with the proposed United States Centennial Exposition at Philadelphia and on his return was commissioned by the Smithsonian Institution to prepare for the exposition an exhibit illustrating the mineral resources of the United States. That this work was well done no one acquainted with conditions and results can question. The collection, after the close of the exposition, formed, together with materials donated by foreign governments, the basis of the geological collections in the New National Museum at Washington. In 1894 Blake was appointed director of the Arizona School of Mines at Tucson, a position he continued to occupy until his death in 1910. Personally, he is stated to have been an ideal teacher, and to have had no mercy on indolence or false pretense, but his genial and unaffected manner and interest in his own work as well as that of others made him friends wherever he went.

Fig. 67. *W. P. Blake.*

He was fully six feet in height and during the latter part of his life, when crowned with a mane of snow-white hair, of very striking appearance.

Marcou's Work in Connection with the Pacific Railroad Surveys, 1853.

Marcou received his appointment in May, 1853, with orders to join Lieutenant A. W. Whipple in the exploration of the route near the thirty-fifth parallel. This lay by way of Fort Smith, Arkansas, westward through New Mexico, south of Albuquerque and Santa Fé to Los Angeles, California, and occupied the period between June 2, 1853, and March 26, 1854. On returning to Washington, June 1, 1854, Marcou, on the plea of ill health, asked permission to work up his results in Europe; and, acting apparently under a misunderstanding, packed his own collections, together with others made by the party under Captain Pope along the thirty-second parallel, and made preparations to leave the country on September 27. This procedure seems, however, to have displeased the then Secretary of War (Jefferson Davis), who insisted, under date of September 25, that the report be made in this country. It being then too late, as claimed, for Marcou to change his plans, he was allowed to depart, but instructed under threat of prosecution to turn over all collections and notebooks, the property of the Government, to the accredited representative of the United States at Paris. Naturally this treatment aroused considerable indignation on the part of Marcou, though it is doubtful if he was justified in his subsequent proceedings. Such of the collections and notes as were returned were placed in the hands of W. P. Blake to be worked up, but, while there is no reason for criticising the latter, the report was not all that could have been desired.[12] It is a painful fact, moreover, that none of Marcou's types were returned to America, but were, by him, distributed to English and continental museums, one of the collections being now in the possession of the Geological Society of London.[13]

Before leaving for Europe, however, and before becoming aware of the intentions of the Secretary of War, Marcou made two brief reports or summaries—one of the route traversed by himself, and the other on the materials collected by Captain Pope's party. These were published in House Document No. 129 and were republished, together with other papers by the same author, in Zurich in 1858, under the title of *Geology of North America, with Two Reports on the Prairies of Arkansas and Texas, the Rocky Mountains of New Mexico and the Sierra Nevada of California.* The latter work was accompanied by a geological map of the United States and a section across the country from Fort Smith to the Pacific Ocean, the same being a reprint of a map published by him in 1855 in the *Bulletin of the Geological Society of France* and in the *Annales des Mines.* (Referred to on p. 309.) It also contained a colored map of an area from thirty-five to seventy-five miles in width across New Mexico along the thirty-fifth parallel, bearing date of 1857, and a reproduction of Maclure's map, the original of which bore date of 1809.

In this summary Marcou identified certain horizontal beds overlying the Carboniferous limestones as Triassic, a formation which on the basis of his geological map of the United States, published in Boston in 1853, above noted, he claimed to have been the first to recognize in the West. At Pyramid Mount, in the Llano Estacado, he claimed to have found Jurassic rocks overlying conformably the Keuper (the first discovery of rocks of this horizon in America), and at Galisteo, in New Mexico, the White Chalk division of the Cretaceous. From the fact that the Cretaceous beds lay unconformably upon the upheaved Triassic and Jurassic beds, he concluded that the

[12] Marcou's original notes and Blake's translation of the same were published in parallel columns in the third volume of the Pacific Railroad Reports.

[13] Appendix, p. 675.

same were deposited after the principal dislocation of the Rocky Mountains, which took place at the end of the Jurassic period.

The Sierra Nevada he rightly inferred to be of more recent uplift than the Rocky Mountains, probably dating from the end of the Miocene or Pliocene, and the Coast Range from the end of the Eocene.

Newberry's Pacific Railroad Work.

J. S. Newberry was geologist to the party under Lieutenants Williamson and Abbot, surveying routes in northern California and Oregon. Newberry regarded the sandstone in the vicinity of San Francisco as probably Miocene, and the serpentines as of igneous origin—in this agreeing with Blake. The rocks at Arbuckle's diggings, near Fort Reading, were considered Cretaceous. He did not agree with Trask in thinking that the Sierra Nevada terminated at Lassen's Butte, or that the Coast Mountains, when continued northward, formed the Cascades of Oregon and Washington. He considered Mount Shasta a part of the Sierra Nevadas (as did also King, later), which were themselves probably of greater antiquity than the Coast Range Mountains. In this he was right. He thought to have discovered evidences of a continuous ice sheet in the region of the Three Sisters and Mount Jefferson in the Cascades. The canyons, as of the Columbia, he rightly regarded due to stream erosion, not to rifts by volcanic action. The few fossils collected were described by Conrad and all as Miocene species.

Schiel's Work.

The survey along the thirty-eighth and forty-first parallels was first placed under the direction of Captain Gunnison, who, however, lost his life in a fight with the Indians in Sevier Lake Valley in 1853. Lieutenant E. H. Beckwith was then placed in charge, with Dr. James Schiel as geologist. The route lay through Kansas, up the Arkansas River, and across the front range to the valleys of the Grand and Green rivers, south of Salt Lake; thence across the Humboldt Desert and Sierra Nevadas to the Pitt and Sacramento rivers in California. The report contains scarcely anything of value from a geological standpoint, being mainly mineralogical and lithological, though a few invertebrate fossils are described and one fossil fish, the first named being mainly specimens of *Productus*, *Terebratula*, *Inoceramus*, and *Gryphea*. Nothing was said concerning their probable geological age.

The writer was told by the late Professor W. H. Brewer that he—Brewer—received an appointment to accompany the Gunnison party as geologist, but through a delay in the delivery of letters missed connections, and did not go. Gunnison himself did not favor the appointment of a geologist, but preferred having "some doctor who knew something about geology." The entire party was massacred by Indians, with the exception of four Mormons. This raised a suspicion that Mormons were involved in the affair, and an investigation was ordered. The Mormon authorities, however, promised to see that the Indians were properly punished and went through the form of arresting the guilty ones and confining them, pending a trial. All were allowed (?) to escape before the trial came off, and nothing was done.

Work of Dr. John Evans in Washington and Oregon Territories.

Dr. John Evans, whose early work in the Bad Lands is noted elsewhere, reported to Governor I. I. Stevens on the geology of the northern route. The manuscript of this report, it is stated,[14] was lost in transit. This, however, was an error. The report was made which, as appears from correspondence, reached the Government Printing Office in Washington some time in 1860-1861, only to be withdrawn and lost for a period of over[15] fifty years. From the incomplete manuscript now available, it is shown that Evans began his work at Vancouver on the Columbia River in 1854 and proceeded northward as far as the Lewis River. An unfortunate hiatus covering the period from June 29, 1854, to August 4, 1855, here occurs. On this date he left Fort Colville on the Dalles of the Columbia River, passing northeasterly along what is now very nearly the line of the Northern Pacific Railroad as far as Pend d'Orielle, returning a little to the east of that line as far as Willow Creek, a small stream west of the junction of the Umatilla River with the Columbia. A second hiatus occurs covering the period until June 29, 1856, when he is found on Elk Creek, a tributary of the Umpqua River some 150 miles to the south. This creek was followed downward to the Umpqua River and thence he proceeded by boat to its mouth and down the coast to Empire City, which was reached on July 2. July 18 following, he proceeded thence northeasterly and easterly across the Elk and Sixes

[14] P. 230, Vol. I, Reports of Explorations and Surveys for a Railroad Route to the Pacific Ocean, etc.

[15] This manuscript was recovered in part, in New Orleans in 1917 by D. I. Bushnell, Jr., by whom it was deposited in the archives of the Smithsonian Institution.

rivers as far as the Coquille and Willamette, passing down the latter to Vancouver. The geological notes with all sections missing are of little moment, consisting mainly of references to the lithological character of the rocks observed with little concerning dip or strike, and nothing relative to their fossil contents.

George Gibbs's Work.

George Gibbs accompanied a party under the command of Captain George B. McClellan and made reports on the country lying on Shoalwater Bay and Puget Sound, and also the central part of what was then the Territory of Washington. His published notes contained little of geological interest. Gibbs, it should be stated, while active in scientific matters, did little work along the broad lines of geology. His bibliography consists, in addition to the above, of three brief notes, altogether comprising less than a printed page. He thought to have discovered that the paving blocks—pebbles—on Waverly Place in New York City had through the weight of traffic yielded as if plastic, so that the concavity of one fitted into a corresponding convexity in its neighbor. He also called attention to the "glades" in Oakland, Allegany County, Maryland, which he believed to have been the seats of ancient glaciers.

Fig. 68. *George Gibbs.*

Swallow's Survey of the Southwest Branch of the Pacific Railroad in Missouri, 1859.

In this connection mention should be made of the survey by Swallow of that portion of the railroad line built under state auspices and extending from St. Louis westward to the state line and known as the Southwest Branch. The report comprised less than 100 pages, and is of interest on account of the colored geological map of the region traversed. It dealt mainly with economic questions.

Daniels's Work in Wisconsin, 1854.

Under an act of the Wisconsin legislature, approved March 25, 1853, Edward Daniels was appointed state geologist, but at the end of the year was superseded by J. G. Percival, whose work in Connecticut has already been referred to (p. 168). His first and only report, bearing date of 1854, consisted of a small octavo pamphlet of eighty-two pages, devoted largely to a consideration of the lead fields. One geological section in black and white was given, extending across the lead mines from the Mississippi River opposite Dubuque to the Blue Mounds.[16] The rocks were classified, beginning with the uppermost, as (1) *Coralline beds,* (2) *Gray limestone,* (3) *Blue limestone,* (4) *Buff limestone,* (5) *Sandstone,* (6) *Lower Magnesian limestone,* and (7) the *Sandstone of the Wisconsin River.* A vertical section was also given showing the succession and relative thickness of the rocks underlying the lead regions. In this last the gray limestone was given as the prevailing "surface rock of the mines, containing veins of lead, and, in its lower beds, zinc and copper." He found evidence, as he thought, to strengthen the conjecture of Owen to the effect that the Lower Magnesian limestone would be found to contain lead ore in workable quantities.

Aside from the above-noted publication, Daniels was the author of an article on the iron ores of Wisconsin, in the report of the geological survey of Wisconsin for 1857 (sixty-two pages), and a brief note regarding the lead district of Wisconsin, which appeared in the Proceedings of the Boston Society of Natural History in 1854.

According to Hall, Daniels was the first to point out in 1855 the unconformity of the western Coal Measures with the older rocks, though J. W. Foster in 1856 published a section showing a similar unconformity, the discovery of which he credited to Norwood.

Percival's Geological Survey of Wisconsin, 1854.

J. G. Percival, who succeeded Daniels, received his appointment in August, 1854, and served until his death, which occurred on May 2, 1856. Two reports, in the form of octavo pamphlets of about 100 pages each, were issued as the result of his work. The first of these was published under his own supervision; the second after his demise. There were two editions, one in English, and one in German. As with his previous work in Connecticut, these reports are extremely prosy and made up largely of very minute descriptions of the lithological nature of the various rock formations of the state, their geographic distribution and relative position. Fossils were mentioned only occasionally, and, otherwise than his reference to them as primary and

[16] In some of the volumes of the issue this section was colored, formations Nos. 1 and 2, sienna brown; No. 5, blue; No. 6, red brown; No. 7 and the lowest-lying sandstone of the Wisconsin River, yellow.

secondary rocks, there is no suggestion as to their probable geological age, with the exception of the reference in the second report to the fact that the so-called "Mound limestone" had been considered from its fossils the equivalent of the Niagara limestone. The character of the ore deposits, and the position, number, and character of the veins were noted with great detail, and in a single instance he indulged in a little speculation regarding the origin of the ore:

> The appearances seem no less to indicate the origin of the mineral and the accompanying ores from beneath, probably from the primary rocks underlying the lowest secondary; and that they rose in such a condition that they were diffused through a certain definite extent of the materials of the rocks, and then segregated in their present form, and this along certain lines which have determined their arrangement.

Owen's Geological Survey of Kentucky, 1854-1860.

April 8, 1854, D. D. Owen was appointed state geologist of Kentucky, a position which he continued to fill until the time of his death in 1860, though in the meantime having been appointed state geologist of Indiana also. During this period there were issued four large octavo reports devoted mainly to economic matters, to descriptions of the coal, iron ores, building stones, and other useful minerals, and comprising, all told, upward of 2,000 pages, with sections, maps, and plates of fossils. He was assisted at first by Dr. Robert Peter, chemist, and Sidney S. Lyon, topographer; later Leo Lesquereux, paleobotanist, E. T. Cox, geologist, and Joseph Lesley,[17] topographer, were added to the force.

Owen divided the formations of the state as follows, beginning with the uppermost: (1) *Superficial deposits,* (2) the *Coal Measures,* (3) *sub-Carboniferous limestone,* (4) *Black lingula shales,* (5) *Gray coralline falls limestone,* (6) the *Chain coral* and *Upper Cliff limestone,* and (7) the *Blue, Shell,* and *Bird's-eye limestone,* the last named being the most ancient of any yet found within the state limits.

A local epidemic of milk sickness in cattle was examined into in considerable detail, and the conclusion reached that the disease was "intimately connected with the geological formation," and probably due to the presence of soluble salts of aluminum, iron, and magnesia, produced in the shales by the decomposition of iron pyrites. An amusing tendency to magnify the importance of minor matters was shown in the suggestion that the animals at pasture may have be-

17 An uncle of J. Peter Lesley of the Pennsylvania survey.

come weakened and peculiarly susceptible to disease through breathing, while feeding with the nostrils close to the ground, an atmosphere deficient in oxygen, the abstraction of this element being due to the oxidation of the pyrite in the surface rock.[18]

Concerning the oölitic structure of the sub-Carboniferous limestone, he wrote that the structure "seems to have been formed in eddies where the water circled round in spiral or funnel-shaped currents which kept particles of fine sand revolving in such a manner that they acquired concentric calcareous coatings, until, having attained the size of fish roe, their gravity was sufficient to overcome the power of suspension of the rotary currents, when they sunk to the bottom." Inasmuch as this oölitic structure is now known to prevail over many hundred square miles of area, this explanation seems scarcely sufficient.[19]

Peter's work on the chemical composition of the soil was very thorough, and, taken all in all, the work done by this survey on problems relating to the soil was of greater importance than any previously accomplished. The method followed by Peter has, however,

[18] This reminds one of Kirwan's attempt to account for the longevity of the ancients, as chronicled in the scriptures. He conceived the atmosphere of the early geological period to have been greatly enriched in oxygen thrown off by the rank growths of vegetable matter "without any proportional counteracting diminution from the respiration or putrefaction of animals," which were multiplying much more slowly. But these conditions were reversed by the flood, which destroyed all vegetation and left the surface of the land "covered with dead and putrefying animals and filth which copiously absorbed the oxygenous part of the atmosphere and supplied only mephitic and fixed air." Hence the constitution of men must have been weakened and the lives of their enfeebled posterity gradually reduced to their present standard. S. L. Mitchill in his Medical Repository of 1799 expressed equally interesting views. His conclusions, based upon observations relative to the prevalence of epidemics in sundry regions, were to the effect that countries abounding with calcareous earth were mostly free from such ravages by reason of the power which that material possesses of neutralizing the acids of putrefaction, the conclusion being founded upon the supposition that the "septic gas" or "effluvium" to which the epidemics were probably due was of an acid nature. The suggestion found acceptance in at least one mind,—that of Hippolyto I. Da Costa, of Lisbon, Portugal, who in a long letter to Mitchill, printed in the Repository for 1800, described in detail the unsanitary conditions there existing although the city was as a whole unusually healthy. He attributed this to the fact that the city was paved and built largely of marble or limestone and the interiors quite generally whitewashed. "I confess Sir I cannot find any explanation of this phenomenon if I reject the theory of the influence of calcareous earth upon septic fluids, and I well comprehend the reason of it if I admit it." Thus are the modern sanitary engineers and bacteriologists outdone in their own field.

[19] On pp. 433-437 of Vol. IV of these reports is given an "Explanation of Plates" intended for four plates of fossils which so far as can be learned were never published.

been proven to be of comparatively little value, the fertility of a soil, as announced years earlier by Emmons, being dependent more upon its physical than its chemical properties.

G. G. Shumard's Work in Connection with Pope's Expedition, 1855.

In 1855 G. G. Shumard accompanied, in the capacity of geologist, the expedition under command of Captain Pope for the purpose of boring artesian wells upon the desolate western plains along the line of the thirty-second parallel. Papers based upon his observations and containing descriptions of fossils appeared in the *Transactions of the St. Louis Academy of Sciences*, I, 1856-1860, though, so far as can be ascertained, no final report ever appeared.

Shumard announced the finding of Permian fossils about thirty miles above the mouth of Delaware Creek, in the country lying between the Rio Pecos and the Rio Grande. At Delaware Creek the oldest rocks were Cretaceous, overlaid by Quaternary. The Guadalupe Mountains he described as consisting of white, gray, and bluish-black limestone, containing fossils, some of which appeared to belong to the Coal Measures and others to the Permian, the beds being finally set down as of Permian age. The Sierra Alto was shown to have a granitic nucleus, from which the stratified rocks dip away on either side.

Silurian rocks were found in the El Paso Mountains, and the Jornada del Muerto was shown to be a small trough composed mostly of limestone, sandstone, and shales, covered to a depth of five or six feet with loose detritus. The upheaved edges of these underlying strata formed the mountains on either side. The Organ Mountains were shown to be of limestone, belonging to the Coal Measures.

Lieber's Work in South Carolina, 1855-1860.

By an act of the legislature of the state of South Carolina in 1855 Oscar M. Lieber was appointed director of the geological, mineralogical, and agricultural survey of the state, a position which he held up to 1860. During this period he made four annual reports, the first bearing date of 1857 and the last 1860. This, it will be remembered, was the third survey of the state undertaken at public expense, the first being purely agricultural, under the direction of Edmund Ruffin, and the second geological and agricultural, under direction of M. Tuomey.

In the reports of Lieber, matters of both theoretic and economic nature received attention, though an undue amount of space was perhaps devoted to a discussion of the itacolumite, which he considered to be a true sandstone. He differed in this respect with Mr. Tuomey who classed it under *quartz rock*, but which from his description was evidently a highly quartzose mica-schist, passing on either hand into a compact quartzite or arenaceous sandstone. Its flexibility he thought to be due to its fineness of grain and admixture of mica or talc, delicately laminated structure, and a certain degree of compactness in the constituents of each lamina rather than decomposition. He discussed the origin of the rock, and compared it with the itacolumite of Brazil in its relation to the diamond.

A peculiar tendency on the part of Lieber to use outlandish terms, particularly Brazilian Portuguese, is manifested in this report. Thus he mentions finding the "*tapanhoancanga*, or *Canga*, which sometimes passed into a reddish *quader*-sandstone." He also wrote of the "*oryctognostic*" composition of the rock, and designated a prospector as a "*costeaner*."

He classified the igneous rocks of the state, commencing with the newest, as:

Trachytic rocks:
- Eurite and quartz porphyry (?).
- Coarse trachyte of eastern Lancaster.
- Domite, phonolith.

Trappean rocks: Diorite and Diorite slate, soapstone (?), Talcose trap (?).

Granites:
- Melaphyre, Egeran (?).
- Aphanitic porphyry.
- Coarse-grained granite, etc., of Taxahaw (syenite).
- Other granite and gneiss.

(The marks of interrogation with some of these denote that their exact relative position is not established.)

Lieber's views regarding mineral veins and ore deposits were largely influenced by those of European authorities, but, nevertheless, he entertained certain independent ideas which at this date are instructive. He believed with Werner that the veins were filled crevices, but thought them to have been filled from below rather than above; that is, the minerals constituting the veins were derived from the interior of the planet and brought to the surface by mineral waters or steam, where they were deposited chiefly in the crevices themselves, the surplus only distributing itself among the surface waters, whence it was afterward precipitated.

PLATE XVI

OSCAR M. LIEBER

EUGENE ALLEN SMITH

He argued against the idea that they had been filled by the leaching action of water permeating the rocks on either side, saying:

Entirely the reverse is, however, the case. Thus, minerals which belong to the veins but which are found in the country were in reality derived from the former; the vein crevices were the first reservoirs, and the few scattering particles of the minerals of the veins which we find in the adjacent country rocks found their way into the latter by elimination or secretion and by sublimation from the surcharged vein crevices.

Concerning the variation in vein structure and the changes of the contents of veins at various depths, he wrote:

Except in those few cases where the enrichment of a vein is due to extraneous or superficial causes, we may safely say that veins gain in the quantity of their metalliferous contents as we leave the surface; that is to say, the lodes increase in diameter and in the compactness of their ores.

Thus do we find one who was the best authority in the state promulgating a theory now known to be wholly erroneous and misleading.

The question of the origin of gold received consideration, and the opinion was expressed that "the gold in some igneous rocks with us and elsewhere may have been brought up with the fluid masses from beneath, but it may also have been imparted by the solution of auriferous sedimentary rocks traversed at the period of injection." While noting that dikes exerted a conspicuous influence on the discernible presence of gold, he believed that these did not impart the metal, or, if they did, only in exceptional cases, and that as a rule certain sedimentary rocks must be regarded as the true source of the gold into which it was infused at the time of their deposition, though by what means or from what source he found it impossible even to guess.

Each of the reports was accompanied by two or more colored geognostic maps, in which the various formations were grouped according to their lithological nature rather than according to their geological age.

J. W. Dawson's Work in Nova Scotia.

Aside from the work of Alger, Jackson, and Gesner, already noted, no geological work of importance had been carried on in Nova Scotia previous to the decade now under consideration. In 1842 J. W. Dawson, a native of Pictou, and destined to become a prominent figure in matters relating to provincial geology and education entered the list of workers with a description of a geological excursion to Prince Edward Island, which was published in *Hazard's Gazette.* From this beginning he developed into one of the most prolific writers and for over half a century not a year elapsed without the appearance of one or more papers from his pen.[20] A large proportion of these related to the Coal Measures of Nova Scotia and their included plant and animal life. As early as 1853 he announced the finding in the sandstone of a cast of an upright Coal Measure tree trunk, the shell of a land mollusk and fragmental skeleton of an air-breathing animal, a reptile or batrachian, to which last he gave the name *Dendrepeton Acadium.* Subsequent exploration brought to light other remains which were described in detail first in the *Canadian Naturalist* of 1863 and later in book form, under the title *Air Breathers of the Coal Measures.*

As noted above, a majority of Dawson's papers dealt with problems involved in the Coal Measures and the formation of coal. He felt that the evidence as he found it in Nova Scotia indicated the one-time occurrence of a long succession of oscillations between terrestrial and aquatic conditions which were not accompanied by any material permanent change in the nature of the surface of the land and its organized inhabitants. He conceived that the elevation and depression of large areas were not absolutely contemporaneous, but such that by a mere change of place one could have passed from a coal swamp to a modiola lagoon or a tidal sand flat.

One of the most important, indeed the most important, of Dawson's early publications, was issued in 1855, a small octavo volume of 388 pages, entitled *Acadian Geology.* This passed through several editions, the fourth and last of which, a volume of some 800 pages, appeared in 1891. The first edition contained a colored geological map of the province, and gave in a condensed form what was known of the geology of the peninsula. A discussion of the views held, and as subsequently modified, may well be reserved until we come to the edition of 1878 (see p. 489).

In the *Quarterly Journal of the Geological Society* for 1859 Dawson had an important paper in which he announced that the vegetable matter of the coal beds consisted principally of Sigillariæ and Calamiteæ. With these, however, he found intermixed remains of many other plants.

The structure of the beds as he found them conformed to the view that the materials were accumulated by growth in place without

[20] Dawson's bibliography contains upward of 360 titles, of which at least nine were books.

any driftage of materials. The rate of accumulation he thought to have been very slow, the climate of the period being such that the true conifers grew not more rapidly than their modern congeners. Making all due allowances, he felt safe in asserting that every foot of thickness of pure bituminous coal implied the growth and fall of at least fifty generations of Sigillariæ and, therefore, an undisturbed condition of forest growth enduring through many centuries.

In this same journal for 1865, under the title of "On the Conditions of the Deposition of Coal," he again referred to the subject. Writing with reference only to the beds of Nova Scotia and New Brunswick and their associated sediments, he showed that the coarser matter of the Carboniferous rocks was derived from the neighboring metamorphic ridges, but much of the finer material was probably drifted from more distant sources. He thought there was no good reason to doubt that in the Carboniferous period the greater part of the Laurentian and Silurian districts of North America existed as land. Considering the relative position and lithological nature of the beds of the Carboniferous period, and comparing them with those of other periods of the Paleozoic age in eastern America, he thought to find indications of the existence of periodic cycles such that similar beds were deposited at corresponding periods in each, the parallelism being tabulated as below, the several formations being arranged in descending order.

Character of group.	Lower Silurian.	Upper Silurian.	Devonian.	Carboniferous.
Shallow, subsiding marine area, filling up with sediment.	Hudson River group.	Lower Helderberg group.	Chemung group.	Upper coal formation.
Elevation, followed by slow subsidence, land, surfaces, etc.	Utica shale.	Salina group.	Hamilton group.	Coal Measures.
Marine conditions; formation of limestones, etc.	Trenton, Black River, and Chazy limestones.	Niagara and Clinton limestones.	Corniferous limestone.	Lower Carboniferous limestone.
Subsidence; disturbances; deposition of coarse sediment.	Potsdam and Calciferous sandstones.	Oneida and Medina sandstones.	Oriskany sandstone.	Lower Coal Measures and conglomerate.

J. D. Whitney's Metallic Wealth of the United States, 1854.

While nearly all the state and land-office surveys organized up to 1850 had economic ends in view, and while, further, numerous papers of a more or less economic character had appeared from the hands of various writers from time to time, yet literature relating to American ore deposits remained scant and of an extremely unsatisfactory character.

The appearance in 1854 of Whitney's *Metallic Wealth of the United States* marked, therefore, an important epoch. The work comprised upward of 500 octavo pages, and though written with especial reference to the ore deposits of the United States, contained references to those of all the principal foreign localities as well and remained the standard work of reference up to the time of the appearance of F. Prime's translation of Von Cotta's *Letine von den Erzlagerstätten* (1869), and Kemp's *Ore Deposits of the United States* in 1893.

Many of the principles set down by Whitney have, naturally, been shown to be erroneous, as he himself lived to realize. Thus, regarding the occurrence of gold, he wrote: "In general it may be said that the older the geological formation the greater the probability of its containing valuable ores and metals." And further: "There is room for doubt whether the great gold deposits of the world did not originate exclusively in the Paleozoic strata, since we are not aware that the rocks which have been proved to be of Azoic age have been found to be auriferous." In this, it will be noted, he followed the teaching of Murchison.

To the reader of today, it will seem scarcely possible that, at the date of Whitney's writing, there were no mines worked for silver alone or silver and lead in the United States, the supply of the metal coming almost wholly from the native gold of California. Argentiferous galena from the Washington mine in Davidson County, North Carolina, had been smelted to some extent, but work was suspended at the time of his writing. How little the silver resources of the West were realized is shown by his comment on the rapid increase of the gold output from year to year and the comparative decrease in that of silver.

Silver is, in a geological point of view, the metal best adapted for a standard of value, since, possessing all the valuable qualities which make gold suitable for that purpose, it is not liable to those fluctuations in its production to which this latter is exposed. There is no discovery of a new continent to be looked forward to whose mines shall deluge the world with silver, and any increase in the amount of this metal produced must come chiefly from the working of mining regions already known.

Yet he lived to see the annual output of American silver become so great as to practically remove it from the list of precious metals and cause its rejection for all but subsidiary coinage. He recognized the fact that none of the deposits of lead in the Mississippi Valley could be considered as coming under the head of true veins (*i.e.*, fissure veins), and that the productive deposits did not generally exceed a hundred feet in thickness in the Galena (Lower Silurian) limestone immediately overlying the Trenton. No ore of consequence was known to occur in the so-called "Blue Limestone" (Trenton, Bird's-eye, Chazy, and Black River), and it was not considered probable that the fissures would ever be found to extend through the intervening sandstone into the Lower Magnesian beds. Sinking through the sand into these latter beds was, therefore, considered as mere random exploration and a foolish enterprise. On several of these points he further committed himself while connected with the geological surveys of Iowa and Wisconsin (see pp. 349, 355).

He classified the ore deposits as (1) simple alluvial deposits which were recognized as residual from the decomposing limestone; (2) deposits in vertical fissures which had a very limited longitudinal extent; and (3) deposits in flat sheets. All of these were regarded as deposits from aqueous solution in depressions of the surface or in vertical fissures of the nature of gash veins produced by the shrinkage of the calcareous strata.[21]

These views differ from the more modern in that it is now deemed probable that the ores were first precipitated in an ocean in which they were held in solution as sulphates and reduced to sulphides through the decomposing organic matter; that subsequently, through the action of percolating surface waters, the same were once more oxidized, segregated in the fissures, and reduced a second time to the condition of sulphides. The productive capacity of the mines in the Mississippi Valley he thought to have even then reached its maximum.

Whitney recognized the eruptive nature of the Iron Mountain masses of Missouri, and regarded the hematite ores of the Lake Superior region as also of an eruptive nature (see p. 282), though

[21] Dr. Henry King, in a sketch of the geology of the Mississippi Valley, read before the Association of American Geologists and Naturalists in 1844, and printed in the American Journal of Science for that year, argued that the ores of zinc, copper, and lead occurring in the Cliff limestone were deposited contemporaneously with the inclosing rock.

the work of later geologists has shown them to have had a chemical origin. Concerning the specular and magnetic iron ores of New York, he thought to have found evidence of direct eruptive origin as perhaps less conspicuous, many of them exhibiting appearances of secondary action, such as might have been brought about by volcanic agencies and powerful currents which swept away and abraded portions of the original eruptive masses, "rearranging their particles and depositing them again in the depressions of the strata." The lenticular beds of ore occurring parallel with the stratification were particularly referred to as originating in this manner.

A year later Whitney had in the *American Journal of Science* an article on the changes which take place in the structure and composition of mineral veins, which is of interest in connection with the recent revival of the subject of secondary enrichment of ore deposits. He here described the now well-known copper deposits of Ducktown, Tennessee, and the superficial alteration of the vein matter, and called attention to the fact that the black ore, then being mined, was formed by a process of natural concentration by surface water, which was constantly decomposing the material above the permanent water level and carrying it downward to the point where it was stopped by the solid portion of the vein. By this means a large portion of the copper once disseminated throughout perhaps 100 feet of vein stone had become concentrated into a thickness of perhaps two or three feet. In this it is to be noted he was one of the first to recognize the now well-known principle.

Harper's Work in Mississippi, 1854-1857.

In June, 1854, Lewis Harper, or properly, Ludwig Hafner, was elected by the trustees to the chair of geology in the University of Mississippi, which act, under the law, constituted him also state geologist.

In 1855 he was relieved of a portion of his duties as instructor in the university in order that he might personally take the field, and an allotment of $1,000 was granted for the employment of E. W. Hilgard, then fresh from his studies in Heidelberg, as an assistant. The two worked together during the season of 1855, but the dual nature of Harper's duties proved unsatisfactory to the university trustees and in 1856 the law was repealed, Harper retiring as state geologist but Hilgard retaining the position of assistant. During the legislative session of 1856-1857 a bill was passed, however, providing $3,500 for the publication of the second annual report. This contained a colored geological map of the state, which, though less detailed, corresponded in a general way with that published later (1860) by Hilgard. Indeed, there is every reason for supposing that the map itself was compiled largely from Hilgard's notes. The work, like its predecessor, had little influence upon geological thought, and was, if not ignored, unfavorably reviewed by the journals of the day.[22]

Concerning it Hilgard later remarks:

It need only be said that it is a literary, linguistic, and scientific curiosity, and probably unique in official publications of its kind. It was the labored effort of a sciolist to show erudition, and to compass the impossible feat of interpreting and discussing intelligently a considerable mass of observations, mostly recorded by another, working on a totally different plan from himself.

The same bill provided for a complete separation of the survey from the university, provided for the establishment of an independent organization, and Hilgard was made state geologist at a salary of $2,000 a year.

Safford's Work in Tennessee, 1854.

In 1854 the general assembly of Tennessee passed an act creating for a second time the office of geologist and mineralogist of the state. J. M. Safford, then holding the chair of chemistry, geology, and natural history in the Cumberland University at Lebanon, was appointed to the position, continuing to serve until the outbreak of the Civil War, when the work was abandoned. During this time he published two biennial reports, the first, in 1856, comprising 164 pages, and the second, in 1857, comprising but 11 pages. A final volume on the geology of Tennessee, comprising 550 pages with 7 plates of fossils, appeared in 1869.

The first report, known under the title of *Geological Reconnaissance of Tennessee,* was presented to the general assembly in

22 "This Geological Report on Mississippi is very unequal in its different parts and requires a careful revision before it can become good authority. Many of the sections have a fantastic boxing off of layers which is quite unintelligible to us. Certain rocks are pronounced to be Carboniferous, because the genus *productus* occurring in them is not known to exist, the author says, in older strata. Some peculiar concentric aggregations in clay are attributed to whirlpools in the waters. The work argues against the subdivisions of the Eocene proposed by Conrad on very insufficient data and an evident want of appreciation of the subject. There are errors, also, in the identification of the Cretaceous, Tertiary, and post-Tertiary beds which betray insufficient observations and an imperfect acquaintance with the science." (Am. Jour. Sci.)

1855, and published in 1856. It was given up largely to a somewhat popular treatise on the economic geology of the state, with much interesting stratigraphical matter.

He discussed the occurrence of the ores of iron, copper, lead, zinc, and even gold and silver, and included also a brief chapter on

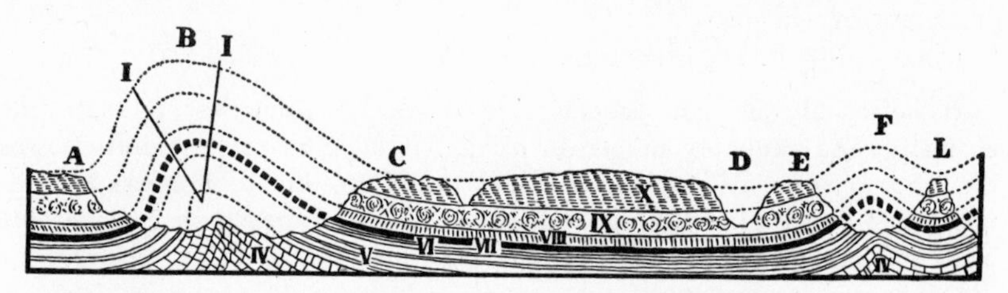

Fig. 69. *Safford Section from the Cumberland Table-land to the Eastern Base of Lookout Mountain.*

"Aluminum or the Metal from Clay." "Stone coal" and the marbles received a share of attention, as well as various minor minerals. From a strictly geological standpoint the matter given in chapters 5 and 6—the closing chapters of the report—was of greatest importance. Here he dwelt on the general geological structure of the state and gave the now well-known Cumberland section, reproduced with the original legend in Fig. 69, and also the ideal cross section illustrating the geological structure of East Tennessee shown in Fig. 70. He pointed out the striking parallelism of the valleys and ridges and the occasional vertical position of the beds, as well as the frequent repetition of the same series of beds, due to folding and faulting.

Fourteen geological formations were recognized, extending from the mica-slate group of metamorphics of Azoic age to the Alluvial, post-Tertiary series. The much discussed and still enigmatical Ocoee conglomerates were included in his semi-metamorphic series at the top of the Azoic. Troost, it will be remembered, put them at the base of his Cambrian.[23]

The importance of fossils for purposes of correlation he at this time distinctly recognized. "Every formation has, in great part, its own species of fossils. Most of those found in one do not occur in any other. Upon this fact depends the great utility of fossils. They furnish, when known well enough to be recognized, unmis-

[23] The Ocoee is regarded today as mainly of pre-Cambrian age, although a portion of what has in the past been called Ocoee may belong to a Carboniferous horizon.

takable evidence of the geological position, and hence the general character of the formation in which they occur."

This report showed on the part of Safford a thorough insight into the intricacies of the structure of the state and an ability to grasp the salient features and master the broader problems in a manner not realized by most of his contemporaries and many of his successors.

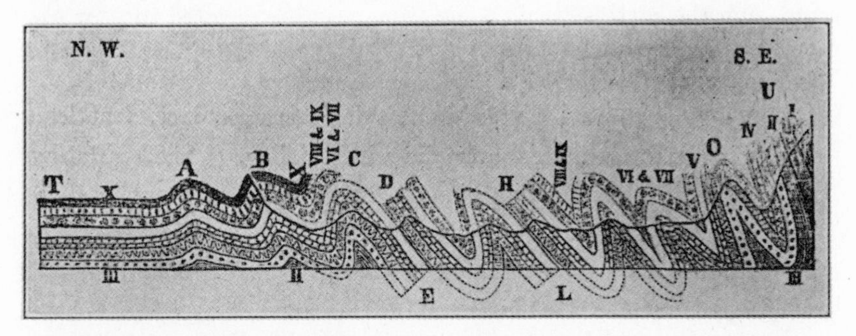

Fig. 70. *Ideal Section, illustrating the geological structure of east Tennessee.*

T, A, B, The Cumberland Mountain or Table-land. At T, the formations are horizontal; at A, they rise in a moderate fold, as at Crab Orchard; at B is a more abrupt fold, which, by partial denudation, leaves a crest like that of Walden's Ridge, in the southern part of Roane, etc.; U, Unaka Mountain.

The second biennial report, which appeared in 1857, gave simply a brief account of the objects and utility of the survey, with a general summary of what had been accomplished.

R. O. Currey's Sketch of the Geology of Tennessee, 1857.

In this connection Richard O. Currey's *Sketch of the Geology of Tennessee* may be mentioned briefly. Currey had been a pupil of Troost and later a professor of chemistry and geology in East Tennessee University. His geological writings are few and limited mainly to such publications as the *Southern Journal of Medical and Physical Sciences*, the *Nashville Banner*, and *The Virginias*. The sketches above referred to were first published in 1853 in the *Nashville Banner*, and subsequently brought out, under the title quoted, in form of a booklet of 128 pages, accompanied by a reprint of Safford's map in black and white. It contained little more than a description of the economic products of the state and their distribution.

Emmons's American Geology, 1854.

In 1854 Dr. Ebenezer Emmons began the publication of a work on American geology, the intention being, as announced in the "proposals," to bring it out in four or five parts, each of which should contain about 200 pages of matter. The first part, containing 194 pages, was issued at Albany in 1854. It was devoted mainly to a discussion of the general principles of geology, and of the rocks composing the earth's crust. Some 70 pages were devoted to "An Application of Geological Facts and Principles to the Business of Mining." Part II comprised 250 pages and 18 plates, and was given up to an exposition of the author's views on the Taconic and Lower Silurian systems as developed in America and England. Part VI, the last ever issued, was devoted mainly to a discussion of the supposed Permian and Triassic beds and their included fossils.[24]

This work, as outlined, was by far the most pretentious of its kind that had thus far appeared in America, antedating Dana's *Manual* by some ten years, and having for its predecessors only the American reprints of Bakewell's *Introduction,* Buckland's *Bridgewater Treatise,* and De la Beche's *Manual,* besides the smaller textbooks of Eaton and Emmons, published in the twenties. There was, therefore, ample room for a work of this nature, but it is doubtful if the occasion was propitious or Emmons the man for the task. Too much of the work (122 pages) was given up to a defense of the author's own peculiar views, which were not in all cases the best, or most generally accepted. His style was poor, often lacking in perspicuity, and in many ways he laid himself open to savage onslaughts of criticism.[25] Unjust as much of this criticism may have been, Emmons, it must be confessed, showed singular ignorance of, or indifference toward, much of the work that had been done in the West and Northwest, and his ideas on the early history of the globe were not the most advanced. Heat was considered at first the predominant and active element. As the activity of fire diminished, that of the antagonistic

[24] The statement made in the author's Contributions to the effect that but two parts of this work were issued, was an error. As some confusion exists as to the number of volumes and parts issued, it may be well to state that Vol. I, as commonly found in libraries, bears the date of 1855 and comprises the Parts I and II mentioned above. In 1857 he published another "part," which it was his intention should be Part VI of his third volume. He, however, died without completing the work. In 1875 all were reprinted, each as a separate volume, Part VI bearing on the title-page the legend Part III, but on page 2 the proper designation, Part VI. And thus the poor fellow's troubles were handed down to the bibliographers!

[25] See Hall's 10-page review (unsigned) in the American Journal of Science, XIX, May, 1855.

element, water, increased. The first was paroxysmal in its action, the latter constant. In America he found the evidences of aqueous agencies on a grand scale, but "volcanic fire" seemed so far a thing of the past that it was "impossible to obtain specimens [*sic*] even for laboratory illustration."

Fire or heat, he wrote, acts in four ways: (1) In the elevation of areas by expansive forces beneath the crust; (2) by the transference of fused matter from the interior to the surface; (3) by producing areas of subsidence, caused by this loss of matter from beneath; and (4) by the elevation or depression of areas simply through expansion and contraction of strata by heat and cold. The internal heat he thought effective in sustaining "that degree of temperature which is best fitted to the organic and structural conditions of living beings" on the earth.

The classification adopted was essentially the same as that put out in his North Carolina report (p. 299) and need not be repeated here.

The pyrocrystalline rocks, however, he thought to be a result of the primary consolidation of the earth's crust. The fact that some of these occurred in dikes was due to the fissuring of the earlier cooled portions of the crust by shrinkage and the forcing up through these fissures of the lower, still unconsolidated portion. He imagined that the age of rocks might be deduced from their crystalline state, the older having been subjected to greater heat and hence becoming more perfectly fused and most highly crystalline on cooling. The granites were, therefore, the first products of cooling and the oldest of rocks.

Concretions were looked upon as symmetrical bodies due to "a force by virtue of which the molecules are really transferred to central points about which they accumulated." The parallel planes of the rhombic forms in limestones and slaty rocks, and jointed structure in general, "admits of the same explanation," as did also the concentric weathering shown by massive rocks. Further than this, he thought that the phenomena of rift and grain, in a massive rock was identical with cleavage in a simple mineral. Such rocks "are not only composed of crystallized minerals, but they are crystalline in the mass."

The serpentine of Syracuse, New York, he considered a magnesian rock altered by proximity to eruptives, though that described by Jackson on Deer Isle, Maine, he seemed to think might have been erupted in the condition of serpentine. The occurrence of a sinter-like material associated with a serpentine near Macon, North Carolina, on the other hand, suggested to him the possibility that the material might in this case be a hot spring deposit. It may be well to note here that Hall in a discussion relative to the report on the mineralogy of New York, held at a meeting of the Boston Society of Natural History in 1844, had mentioned the occurrence of a knob of serpentine discovered by himself in the neighborhood of Syracuse, and spoke of it as being "the only mass of igneous rock known in the vicinity of the salt springs." Emmons still adhered to the idea of the igneous origin of the crystalline limestones of St. Lawrence County, as announced in his New York report. For the first time in an American textbook there was recognized the relationship existing between conglomerates and sandstones as shore deposits and the pelagic, calcareous, and argillaceous materials laid down at greater depths.

Kitchell's Geological Survey of New Jersey, 1854-1856.

In 1854 Dr. William Kitchell was appointed superintendent of the geological survey of New Jersey. He continued in office until the close of 1856, during which time he made three annual reports, being assisted in the geological work by George H. Cook and in the topographic by Egbert L. Viele and Thomas B. Brooks.[26] It was expected

[26] The field orders given Brooks are highly entertaining in this day and generation. The following is an exact transcript furnished by his son, the well-known A. H. Brooks, of the Alaskan division of the U. S. Geol. Survey. It may be well to remark further that in return for all this Brooks was paid $60 per month.

THOMAS BENTON BROOKS, ESQ.

Sir:—In addition to your duties as assistant engineer, you are hereby instructed to make the following observations and investigations in that section of the State over which your labors extend.

1. To describe the *Physical Features* of each township—embracing all ranges of mountains, hills, elevated lands and valleys; all isolated mountains, hills and elevated lands, their extent, height and course; their declivity—whether steep and rugged, abrupt and precipitous—whether cultivated or uncultivated, covered with rocks in place or with drift—as sand, gravel, boulders, etc.

Note particularly the most elevated mountains and hills, and describe the scenery from their summits—its extent, background, foreground, etc.

Describe all rivers, streams, lakes, ponds and springs; their origin and direction, extent, depth, velocity of current, availability for water power, by what animals inhabited, their abundance, etc.

2. Examine the geological formations—viz. the different varieties of rocks; their strike, dip, axes, and as far as possible define their boundaries on the map.

Search for, and examine carefully all *fossil remains, metalliferous deposits* and *mineral localities.*

Note particularly the indications of magnetic iron ore by the attraction of the magnetic needle.

3. *Specimens.* Collect three specimens of all characteristic rocks, ores, minerals, etc. Collect all fossils—large and small—take notes on them and pack them according to the printed directions.

Collect specimens of fish of all kinds, reptiles, birds, insects, etc. Carefully

that Conrad would take charge of the paleontological work, but failing health forcing him to resign, it was placed in the hands of James Hall. Up to the time of the discontinuance of the survey, however, Hall appears to have done little. The vertebrate fossils collected were placed in the hands of Joseph Leidy. Mr. Henry Wurtz served as chemist and mineralogist.

Such part of the work as was done directly by Kitchell contained little that was original, though as noted by Clarke he was the first to point out that the algal growth, *Chara*, played an active part in the precipitation of calcium carbonate and the formation of fresh water marls. His sudden death, in the midst of his active duties, is perhaps responsible for so small a showing.[27] The stratigraphic work was largely that of Cook. Considerable attention was paid to the greenstone marls, their distribution and chemical composition, although other economic questions were taken up, particularly those relating

preserve, label and take such notes on them as are directed in the accompanying pamphlets.

4. Carefully locate on the map all mines, mineral deposits and localities of fossils, together with the names of their owners, also obtain *correctly* the names of all mines, furnaces, forges, and quarries—the length of time they have worked —the use or disposal which has been made and is now made of their productions— collect historical data of them from their commencement to the present time, and all other matter of interest and importance.

5. Keep a register of the weather including storms, thunder showers, and prevalent winds—their direction and severity. Make thermometric observations in appropriate places three times daily at 6 A.M., 12 M. and at 6 P.M.

Note localities where lightning has struck, together with its history. Make inquiries in regard to the number of times it has struck in the vicinity, together with the *particulars* of each.

6. Ascertain the principal crops raised and the average of each per acre. What manures are used, etc.

7. *Journal, notes and reports.*

Keep a journal of each day's operations and investigations. Page and number or letter each note book, and each specimen to correspond therewith, so that reference may be made to specimens from the note book and to the note book from the specimens.

All specimens collected while engaged in the service of the state are to be submitted to the Superintendent of the survey.

All reports are to be rendered to the superintendent, with the exception of those required by E. L. Viele, State Topographic Engineer.

Report a synopsis of investigations every two weeks by letter to the superintendent of the survey at Newark.

(*Signed*) WM. KITCHELL,
Superintendent of the Geol. Surv., N. Jersey.

Newark, N. J., May 19, 1855.

27 According to Dr. Cook (Geol. of N. J., 1868, p. xiv) Kitchell's plans for the survey had been laid out on a comprehensive scale, but his notes and papers were found to be "unarranged for publication," though they were largely used in the preparation of the report.

to the iron ores. The magnetic ores were regarded as deposited contemporaneously with the sedimentary rocks in which they were inclosed, and the white crystalline limestone of the highlands was classed as Azoic.

In the third annual report, printed in 1856, Cook called attention to the swamp lands and buried timber, and also the apparent gradual encroachments of the sea in Cape May County, and stated it as his opinion that the sinking of the coast was going on at the rate of about two feet in one hundred years.

The results of these observations were afterward reprinted with some additional matter in separate form, entitled *Geology of the County of Cape May, State of New Jersey*, bearing the date of 1857. This contained a colored geological map of the county and catalogues of the zoölogical and botanical collections and of a few fossil invertebrates.

The survey was brought to a close on the first of May, 1856, owing to the fact that the state did not feel able to provide an appropriation for its continuation.

Emory's Mexican Boundary Survey, 1855-1856.

The surveying party organized under Major W. H. Emory, for the purpose of fixing the Mexican boundary in accordance with the treaty of 1854, was accompanied by Dr. C. C. Parry in the capacity of geologist and botanist and Arthur Schott as an assistant. The observations made by these two men were largely physiographical and mineralogical, with numerous notes of a lithological nature. Collections of fossils were fortunately made, which were referred to Messrs. Hall and Conrad for determination. The published report, the work of all the above mentioned, comprised but 174 pages of text, with 21 plates of fossils. It was, however, noteworthy in containing a colored geological map of the Mississippi Valley and country to the west as far as the Pacific coast, prepared with the assistance of J. P. Lesley. This has the distinction of being the earliest colored map of the region published by the Government. It is of further interest for showing how little was definitely known of the region. The mountain ranges were colored as metamorphic, often flanked by more or less sinuous, narrow bands of Cretaceous and Carboniferous rocks. Large areas of igneous rocks were represented occurring in the extreme northwest (California, Oregon, and Washington), but the interior, the Great Basin region, was left almost entirely blank—geologically a *terra incognita*.

The fossils, collected largely from Texas, were almost wholly Tertiary, Cretaceous, and Upper Carboniferous forms. A few Lower Silurian forms were figured and named by Hall, but no descriptions were given.

Dana on the Geological History of the American Continent, 1855.

It is a pleasure to turn from these seemingly petty but necessary details and disconnected observations to papers showing a broader outlook and of a more philosophical bearing.

In his presidential address before the American Association for the Advancement of Science in 1855 Professor Dana dwelt on the geological history of the American continent as understood up to that time. "Geology," he announced, "is not simply the earth's history and may or may not have been the same in distant places. It has its more exalted end—even the study of the progress of life from its earliest dawn to the appearance of man; and instead of saying that fossils are of use to determine rocks we should rather say that the rocks are of use for the display of the succession of fossils."

He called attention to the fact that each continent has had its own periods and epochs, but that the subordinate divisions of time were not universal. "We should study most carefully the records before admitting that any physical event in America was contemporaneous with a similar one in Europe. The unity in geological history is in the progress of life and in the great physical causes of it, not in the succession of rocks." He remarked on the comparative simplicity of the continent, both in form and in structure, its outline being triangular, the simplest of mathematical figures, and its surface only a vast plain lying between two mountain ranges, and further, that the bordering heights were proportional to the size of the adjacent ocean, a matter which was elaborated in the paper on the plan of development in the geological history of North America. This principle, he announced, was of wide application, unlocking many mysteries in physical geography, and he felt it pointed to the conclusion that the subsidence of the ocean basins had determined the continental features and that both results were involved in the earth's gradual refrigeration and consequent contraction. America had thus the simplicity of a single evolved result in contrast with the complexity of that of Europe.

He then passed briefly in review the succession of epochs in American geological history, accepting apparently the results of the

Rogers brothers on the structure of the Appalachians and the general subject of mountain making. He noted that through the Devonian and Silurian periods the seas over the American continent were in at least twelve distinct epochs swept of all or nearly all existing life, and as many times repeopled; also that the continent had never been the bed of a deep ocean but was rather a region of comparatively shallow seas and, at times, emerging land, its general outlines being marked out in the earliest Silurian. This last view, he remarked, was that urged by De Verneuil and appears to have been the prevalent opinion of American geologists at that date. During the first half of the Lower Silurian era the whole east and west were entirely covered with the sea, the depth of which may at times have been measured by thousands of feet but not by miles. During the Potsdam period the continent was just beneath or at the surface. In the Trenton period the depth was greater, giving purer waters for abundant marine life. Limestones were being laid down over what is now the great Mississippi Valley, while sandstones and shales were as generally forming in the East. He recognized a period of violent disturbances toward the close of the Lower Silurian era, succeeded by slow oscillations through the Silurian, Devonian, and Carboniferous ages, gradually increasing their frequency throughout the last, and ultimately raising the land into the long series of folds characterizing the Appalachian system. This epoch of maximum disturbances he thought should be interpreted as one of slow measured movement of inconceivable power pressing forward from the ocean toward the northwest; for the rocks were folded up without the chaotic destruction that sudden violence would have been likely to produce.

The question of the existence of a distinct Cambrian system he thought had been decided adversely by the American records, the fossil remains thus far found belonging to the Lower and Upper Silurian. The term Cambrian, if used at all, he thought should be made subordinate to Silurian.[28] The Taconic system of Emmons, as noted elsewhere, belonged with the upper part of the Lower Silurian.

Concerning the origin of the American coal fields, he agreed with H. D. Rogers in thinking that the condition of a delta or estuary for the growth of coal plants was out of the question. He took occasion to add that the idea that the rocks of our continent had been supplied with sand and gravel from a continent now sunk beneath the ocean was wholly improbable. There were no facts to prove that such

[28] He later changed his views on this point and adopted in company with other geologists the terms Cambrian and Lower Silurian (now Ordovician) for the lowermost divisions of the Paleozoic series.

a continent ever existed and the whole system of progress was opposed to it.

Dana on the Plan of Development of North America, 1856.

In an article on the "Plan of Development in the Geological History of North America," published this same year in the *American Journal,* Dana further elaborated these views, again calling attention to the fact that the greater mountain ranges border on the greater oceans in both Americas, and also to the apparent fact that volcanic activity, or at least the evidence of heat, is greatest along the coasts bordering on the greater oceans. Assuming the typical form of a continent to be a basin with borders of mountains facing the oceans, the heights of the mountains and volcanic activity proportional to the size of the oceans, and that volcanoes characterized oceanic islands and not continental interiors, he showed that the extent and position of oceanic depressions have, in a great degree, determined the features of the land; that oceanic depression and continental elevation have both been in progress with mutual reaction from the beginning of the earth's refrigeration, and that the original V-shaped character of the North American continent was due to forces acting from the two oceans. "Contraction was the power, under divine direction, which led to the oscillations of the crust and the varied succession in the strata."[29]

These views, it will be observed, are largely in accordance with those advanced in his *Wilkes Exploring Expedition Report* in 1848 and again emphasized his ideas on the permanence of continents and ocean basins.

The Finding of Paradoxides.

Prior to 1856 the age of the slates of Braintree, Massachusetts, had been problematical, though by most of the geologists of the time they had been considered to be Primary or possibly Transitional. In 1851 or thereabouts a Mr. Eliphas Haywood, while blasting out some of the material for underpinning, found therein fragments of trilobites. These were shown to a Mr. Peter Wainwright, who recognized their value and brought them to the attention of Professor Rogers, who in his turn brought them before a meeting of the Boston

29 It was presumably this phrase that caused one of Dana's contemporaries to make the following sarcastic reference to the publication, in a letter to Professor James Hall: "I have seen Dana's pamphlet. Natural history is one thing and religion is another, and I can't see why geology should be mixed with religion more than with obstetrics."

Society of Natural History in August, 1856. These were later identified as belonging to Barrande's *Paradoxides* and as perhaps identical with Green's *Paradoxides harlani* which had been described from a specimen, source unknown, in the cabinet of Mr. Francis Alger. Rogers therefore felt justified in referring the slates to the horizon of the lowest Paleozoic group, or about that of the Primal rocks, Potsdam sandstone, and the Protozoic sandstone of Owen. "Thus for the first time we are furnished with data for fixing conclusively the Paleozoic age of any portion of this tract of ancient and highly altered sediments, and what is more, for defining, in regard to this region, the very base of the Paleozoic column, and that too by the same fossil inscriptions which mark it in various parts of the old world."

Fig. 71. *Paradoxides Harlani.*

The beds, it may be noted, are, in the latest edition of Dana's *Manual*, referred to the Middle Cambrian and therefore lie below the Potsdam.

Lesley's Manual of Coal, 1856.

The most important work of the year on economic geology, and perhaps the most important thus far to appear, with the exception of Whitney's *Metallic Wealth of the United States*, already noted, was Lesley's *Manual of Coal and its Topography*, issued in 1856—an octavo volume of 224 pages. The scope of the volume was, however, not quite what one would be led to expect from its title, and its possible usefulness marred by the irrelevant personal claims of the author. Not confining himself to the subject of coal and its topography, he entered freely into the subject of mountain structure, with especial reference to the Appalachians, the formation of valleys, theories of the drift, topographical drawing, and kindred subjects. Lesley was, it should be remembered, an assistant on the survey under Rogers in 1839-1851. Reference to his work while on the survey is therefore to be expected, though it must be confessed one's expectations are somewhat exceeded when he claims, on his own behalf and that of Messrs. Whelpley and Henderson—also assistants on the Rogers survey—to having been the first to unravel the intricacies of Appalachian structure. "Years of patient toil," he writes, "it cost us to unfold the mysteries of the Pennsylvanian and Virginian range."

This same claim, in an even more aggressive form, he put forward again three years later, in his *Iron Manufacturer's Guide*, which appeared under the authority of the American Iron Association, of which he was secretary, in 1859. These attacks brought forth an emphatic rejoinder from R. E. Rogers (a brother of H. D. Rogers), which was printed privately in form of a pamphlet of twenty-two pages in 1859.

The merits of the case cannot here be decided, but it would seem most probable that this offers but another illustration of the experiences of every executive who has planned, directed, and rendered possible the work of subordinates, only to find in the end that the value of his instrumentality is quite underestimated, and all credit claimed by him to whom opportunity was given.[30]

Dr. Edward Hitchcock's Illustrations of Surface Geology, 1856.

In 1856 Dr. Edward Hitchcock came once more to the front through the medium of the *Smithsonian Contributions to Knowledge* with a paper of some 150 royal quarto pages and 12 plates. In this he considered all changes, as of erosion and deposition, which had taken place since the close of the Tertiary period to belong to the "alluvial formation," and due to causes still in operation. The products of these changes he classed as (1) drift unmodified and (2) drift modified, under the latter including such deposits as beaches, ancient and modern; submarine ridges; sea bottoms; osars; dunes; terraces; deltas; and moraines. The drift proper he regarded as a product of "several agencies—icebergs, glaciers, landslips, and waves of translation"—which, though more active in the past than now, are still at work. The sandy and gravelly plains (the "overwash" of modern glacial geologists) and the low ridges of New England were thought to

30 "The point in the volume that will excite most remark, is the claim advanced in behalf of Mr. Whelpley and Mr. Henderson, of having first unraveled the Appalachian Mountain system. Professor H. D. Rogers published his views on the subject at considerable length in the Transactions of the Geological Association of 1842, having presented them to the Association at the meeting in that year. Mr. Lesley makes no allusion to the paper of Professor Rogers, and does not mention his name in connection with the subject. 'Years of patient toil,' he says, 'it cost *us* to unfold the mysteries of the Pennsylvanian and Virginian range,' including himself with the two persons just mentioned. These gentlemen had been assistants of Professor Rogers previous to the publication of that paper, and the facts which they observed, were then collected. To substantiate such a claim it is necessary to prove that Professor Rogers was dependent on the suggestions of these gentlemen for the theory he had advanced, and as preliminary, to settle the legitimate relations between an assistant and the superintendent in a Geological survey." (Am. Jour. Sci., Vol. XXII, 1856, p. 302.)

represent old sea bottoms, to be "explained only by the former presence of the ocean above them, with its tides and currents."

The terraces of the Connecticut River were described in great detail, with numerous references to those of other regions, both at home and abroad. The chief agency in the formation of these appeared to him to have been water. Moraine terraces, however, demanded the action of stranded icebergs in addition (see further on p. 547).

The illustrations accompanying this paper were beautifully executed, perhaps the best, both those in color and in black and white, that had thus far been produced.[31]

A second work of Hitchcock, published this same year and needing a brief reference, was his *Geology of the Globe*, a small octavo volume of 136 pages, 6 plates of fossils, and 2 colored geological maps, one of the United States and Canada and one of the world.

The source of inspiration was evidently Boue's treatise published under the auspices of the Geological Society of France. On the maps the various formations, indicated by color, were: (1) Hypozoic and metamorphic strata, with granite, syenite, and some porphyries; (2) Primary fossiliferous strata to the top of the Carboniferous; (3) Secondary strata; (4) Tertiary strata; (5) Alluvial deposits; and (6) Volcanoes and igneous rocks of the Alluvial and Tertiary periods. Comparatively few areas were left uncolored—areas scarcely larger than those left as unknown or uncertain on the map published in 1884 by the official survey of the United States. When one thinks of this he can but smile at the remark of the author when referring to the coloring: "Nor can I doubt that it is done approximately correct." The work was necessarily largely a compilation and contained little original matter.

Joseph Le Conte on Coral Growth, 1857.

With the exception of papers by Dana and Couthouy scarcely anything of consequence on the subject of corals, geologically considered, had thus far appeared from the pen of American writers. At the meeting of the American Association for the Advancement of Science in 1857 Joseph Le Conte, then professor of chemistry and geology in the University of South Carolina, made his first appearance in the geological arena, with a paper, subsequently published in the *American Journal of Science*, in which he argued that the penin-

[31] Hitchcock had presented a paper covering much the same ground before the August, 1849, meeting of the American Association for the Advancement of Science.

sula of Florida owed its southern continuation to a double process of coral-reef growth and sedimentation from the Gulf Stream. It was his idea that the sediments brought down by the Mississippi sank to a depth beyond observation before reaching the Keys, but were gradually deposited along the inner side of the curve formed by the current as it passed around the point. When the bottom should have become sufficiently elevated by this process, a coral reef would begin to form. This process he thought to have gone on until the channel had become narrowed as now, and the rush of water, together with the depth, practically precluded further deposition. Waves throwing broken material upon the beaches would serve to fill the old channels and convert them into dry land. In brief, "the peninsula and Keys of Florida have been the result of the combined action of at least three agencies. First, the Gulf Stream laid the foundation; upon that corals built up to the water level; and, finally, the work was completed by the waves."

Le Conte further conceived that the rate of flow of the Gulf Stream has been gradually accelerated by the narrowing of its channel, and hence had made itself manifest at a greater distance into the Atlantic, thus bringing about a gradual amelioration of the climate of England.

Hall's Geological Survey of Iowa, 1855-1858.

In 1855 the first state geological survey of Iowa was organized, and James Hall of New York made state geologist. The organization continued through 1858. The reports of the survey appeared in the form of two octavo volumes, of which the second was given up wholly to paleontology. This formed what the reviewer in the *American Journal of Science* described as "doubtless the best contribution yet made to our knowledge of the crinoids and other echinoderms of the Carboniferous system." Hall was assisted by J. D. Whitney, who served as chemist and mineralogist, and by A. H. Worthen, F. B. Meek, R. P. Whitfield, and Benjamin S. Lyman.

Hall noted the rapid westward thinning out of the various beds of sedimentary rocks, including those of the Hudson River, Medina, Clinton, Niagara, and Onondaga salt group, and Catskill Mountain group, and made the following comment:

> This remarkable fact of the thinning out westerly of all the sedimentary formations points to a cause in the conditions of the ancient ocean, and the currents which transported the great mass of material along certain lines, which became the lines of greatest accumulation of sediments, and

consequently present the greatest thickness of strata at the present time. It is this great thickness of strata, whether disturbed and inclined as in the Green and White mountains and the Appalachians generally, or lying horizontally as in the Catskill Mountains, that gives the strong features to the hilly and mountainous country of the East and which gradually dies out as we go westward just in proportion as the strata become attenuated.

Again, with reference to the subdued topographic features of the West, he wrote: "The thickness of the entire series of sedimentary rocks, no matter how much disturbed and denuded, is not here great enough to produce mountain features."

Fig. 72. *Hall's Section of Cincinnati Uplift. a, a, Axes of Silurian Limestone. b, Shales and Sandstones of the Coal Measures.*

In this, it will be observed, is given the gist of his theory of mountain formation, as announced two years later.

Numerous changes in nomenclature and important correlations were made, or at least attempted. The Lower Magnesian limestone of Owen he considered identical with the Calciferous of New York, and gave the name Galena to the limestone immediately succeeding the Trenton, which Owen had called Upper Magnesian. The "Coralline and Upper Pentamerus beds of the Upper Magnesian limestone" of Owen he identified as Niagara, and proposed the name Leclaire to designate a gray to whitish limestone lying next above the Niagara and outcropping near the town of that name. The Coal Measure rocks he showed to have been deposited after the disturbances producing the low folds in the Silurian beds, illustrating their relative position by the section here reproduced. (Fig. 72.)

The encrinital group of Burlington and that of Hannibal, Missouri, thought by D. D. Owen to be distinct, were considered by Hall to be identical, and the entire series of Carboniferous limestones were regarded as "successively deposited in an ocean, the limits of which were gradually contracted upon the north while at the south the conditions were becoming more and more favorable to the development of this kind of deposition and to the support of the fauna which abounded throughout the period, until both culminated in the great limestone formation of Kaskaskia," the Archimedes limestone of Owen. He also noted that in the section exposed at Burlington the passage from the Chemung (Devonian) to the Burlington (Carboniferous) limestone was so gradual in both physical aspect and the generic and specific characters of the fossils, that no greater change could be observed than between any of the subordinate groups. This is worthy of note in connection with the discussions arising later over the reports of C. A. White dealing with the same subject (see p. 433).

In discussing the relationship of the limestone and coal, Hall quoted from his report on the geology and paleontology of the Mexican boundary, elsewhere alluded to. With reference to this he wrote:

> The conditions favorable to the production of an extensive deposit of marine limestone are not such as usually accompany the production of coal. In the present instance the ocean, depositing the great limestone formations previous to the coal period, occupied to a great extent the present area of the Coal Measures which succeeded in the valley of the Mississippi. . . . We begin thus to comprehend the truth, that during the period of the great coal formations of the West, in which these calcareous deposits were in course of formation—that during the oscillations which we know to have occurred throughout the coal period there was a time when the whole area became depressed so as to allow the waters of the western ocean to flow over all the Coal Measure region; or, at least, as far northward and eastward as the northeastern part of Ohio, and from hence it derived the limestone under consideration.

The series of beds immediately underlying the Carboniferous rocks at Burlington, and now considered as Kinderhook (sub-Carboniferous), Hall relegated to the Chemung; this proved to be an error, and one that it required many years to eradicate.

It was noted that Potsdam sandstone seemed likely to be found strongly developed in the Rocky Mountain region.

> Subsequent to this period, however, every sedimentary formation indicates the proximity of land on the east. The great thickness of strata, coarse materials, and numerous fucoids of the Hudson River group in its eastern extension indicate proximity to land, or the course of strong currents, while in the west the formation dies out in some inconspicuous fine shaly and calcareous beds, which, both in nature and condition of the material and in the fossil contents, indicate great distance from land and a quiet ocean.[32]

[32] The volumes of this work now in the library of the National Museum were the personal property of F. B. Meek, and many of the statements made by Hall are savagely criticised by the former in marginal notes. Thus, with reference to Hall's statements as to the Potsdam, Meek notes: "It was quite safe for him to make this prediction, when he knew at the time he was writing it, I had identified Potsdam fossils in Hayden's collections from the Black Hills. These fossils were then in Professor Hall's house—he had seen them and heard me say I regarded them as Potsdam species."

Still jealous of his discoveries of supposed Permian fossils in the West, Meek also bitterly criticised Hall's reference to the occurrence of Permian and Jurassic

Hall argued that the treeless character of the prairies was due mainly to the character of the soil, and inferred, on what seemed to him to be reasonable grounds, that the soil itself originated in form of an almost impalpable sediment, which gradually accumulated on the bottom of an immense lake once occupying the region. This will be recognized as the view now commonly accepted for the origin of the loess, although the glacial origin of the material was not suspected.

Whitney's Work in Iowa, 1858.

As already noted, Hall was assisted in his work on this survey by J. D. Whitney, who served in the capacity of chemist and mineralogist, though, as Hall could be on the ground but a small portion of the time, often the *acting,* but never the *real* head of the organization. A good deal of attention was given by Whitney to the chemical nature of the coals and the distribution and mode of occurrence of the lead and zinc ores. There was, however, in his report very little

rocks in the Rocky Mountain region, and added in the margin: "This was all intended to bear the date of 1857, although every word of it was written after the publication of our [*i.e.,* Meek and Hayden] Permian and Jurassic discoveries in 1858."

Again on page 142, Hall stated: "In the early part of 1857 Mr. Worthen placed in the hands of the writer some peculiar fossils collected several years since in Illinois and supposed to be from the Coal Measures. These, however, were at once recognized as of peculiar forms, differing from Coal-Measure fossils, and a farther examination proves them to be of Permian types, and closely allied to British species." To this statement Meek has added the following marginal note: "At the time he obtained these fossils from Mr. W. he told Mr. W. he regarded them as lower Cretaceous forms." Also "If he had stated when this 'later' examination was made, it would have been all right; but he leaves it to be inferred that it was sometime in 1857, when, in fact, it was not until after the publication of our paper on the Kansas fossils on the 2nd of March, 1858; and yet he intended at that time that this report should bear the date of 1857."

On page 144 Hall suggested the probability of finding in the West beds of the age of the Jura or Oölite of Europe. To this Meek has added the following: "How wonderfully sagacious he was to make this prediction, when he knew we had Jurassic fossils then in his house (which he had seen) from the Black Hills, Nebraska! Yet what is here written is intended to date in 1857."

Again, Hall said: "Thus far the collections made in the explorations across this western country have brought us no true Jurassic fossils, and it is only in the far north and upon the Pacific coast, as well also as in the southern extremity of the continent, that we have the evidences of the existence of these rocks below the Cretaceous formation." To which Meek appended the following: "At the very time when Professor Hall wrote those words some of Doctor Hayden's Jurassic fossils from the Black Hills were lying in a tray within fifteen feet of him, in an adjoining room. I had called his attention to them and told him I was satisfied they must be Jurassic forms."

It is the old story of jealousy and heartburning which seems an almost invariable outcome of any attempt at mutual collaboration.

of a theoretical nature, and speculations regarding origin were almost entirely lacking. He thought the iron ore of Jackson to have originated, in part at least, through the decomposition of nodules of iron pyrites which were distributed irregularly through the rock. In other cases he thought it to be a deposit from springs, the original material, however, having been derived from the decomposition of pyrite, as before.

He noted that where the Galena limestone had a maximum thickness, the lead deposits were limited to the central and lower portions, and never penetrating the underlying Blue (Trenton) limestone, and he had satisfied himself that, in a great majority of cases at least, the deposits diminished in productiveness rapidly after passing below a limit perhaps fifty feet above the base of the Galena. When, however, the Galena limestone was diminished in thickness, he noted that the lead deposits were found in lower and lower positions until, at last, the bottom of the Blue limestone was reached, where the Upper sandstone entirely cut off the ore, there having never been an instance, so far as he has ascertained, in which a crevice had been worked in that rock.

He found reason, therefore, to differ with Owen in his statement to the effect that when a mine was sunk through the "Cliff" limestone to the "Blue" beneath, the lodes shrank to insignificance.

As to the possibility of the lead-bearing crevices extending downward indefinitely and the probability of deep mining ever being likely to prove profitable, Whitney wrote, "The question can be answered unhesitatingly in the negative," and he went on to state that the most profitable method of mining would be in the form of horizontal excavations or drifts and not by means of vertical shafts.

These deposits, it should be noted, he described as occurring in gash veins. The mineral deposits of the Northwest in general were classified under the heads of surface deposits, vertical crevices, and flat sheets, as in his work on the metallic wealth of the United States.

Perhaps the most peculiar idea advanced in any part of the report is that regarding the origin of the siliceous matter of the Potsdam sandstone.

It has been generally assumed that such sandstones were originally formed by mechanical agencies, the material being supposed to have gradually accumulated from the grinding down of previously existing quartzose rocks. The facts collected, however, seem rather to point to chemical than mechanical causes. . . . We can hardly understand how such an amount of quartzose sand could have been accumulated without its containing at the same time a considerable quantity of detritus, which could be recognized as having come from the destruction of the schistose, feldspathic and trappean rocks, which make up the larger portion of the Azoic series wherever it has been examined. The uniform size of the grains of which the sandstone is composed and the tendency to the development of crystalline facets in them are additional facts which suggest the idea of chemical precipitation rather than of mechanical accumulation.

The explanation of this error lies in the fact that the grains of which this sandstone is made up have been cemented by a secondary deposition of interstitial silica, so deposited as to convert each granule into a more or less perfect crystal, of which the original sand grain forms the nucleus. It is a case of secondary growth and enlargement, a phenomenon now well known to every petrographer, but undreamed of at the time of Whitney's writing.

Although it was generally understood that the life of the survey was to be limited and but one report called for, yet it would appear from correspondence that Hall in 1860 was asking funds for continuing the work and even in 1863 asking to be reimbursed for the sum of $2,300 expended without authority, but trusting to the future favorable action of the legislature. Nothing more was done, however, until 1866.[33]

Hitchcock's Geology of Vermont, 1856-1861.

In 1856 Professer Edward Hitchcock was appointed state geologist of Vermont, a position which he continued to hold until 1860, at which time he submitted a final report, which appeared in 1861 in the form of two quarto volumes of 982 pages, accompanied by 3 maps, one colored to show the geology and two showing the distribution of the terraces and beaches, and 35 plates of scenery, fossils, sections, etc. Hitchcock was assisted by his sons, Edward Hitchcock, Jr., and C. H. Hitchcock, and by Albert D. Hager, in whose charge was placed the final publication of the report.

The main objects of the survey, as outlined in the introduction, were first, to gain such a knowledge of the solid rocks of the state as to be able to delineate them upon maps and sections; second, to study the loose deposits lying upon the solid rocks and trace out the changes which the surface of the state had undergone; third, to collect, arrange, and name specimens of rocks, minerals, and fossils from every part of the state for the state cabinet; fourth, to obtain a full collection, for the same cabinet, of specimens valuable from an eco-

[33] See p. 433.

nomic standpoint; and fifth, to identify the metamorphosed rocks of the state with those which had not been thus changed.

The systematic geology was in immediate charge of the elder Hitchcock; the economic in that of A. D. Hager, and the chemistry, of C. H. Hitchcock. The Reverend S. R. Hall aided in the preparation of a report on the northern part of the state and its agricultural possibilities. The paleontological work was done by James Hall, and the paleobotanical by Lesquereux. Hitchcock (the elder) recognized at this date that each rock formation was characterized by its peculiar group of fossils not found in any other; so that a paleontologist, on seeing a specimen, could usually tell from which of the series it came, thus indicating a great advance over any of his previous reports.

His views on metamorphism and the production of schistose and slaty cleavage were still, however, in an immature state. It is to be noted, however, that all through his work he is far from being dogmatic, usually contenting himself with a discussion of, and full quotations from, the expressed opinions of others, stating his reasons for adopting or rejecting any one, as the case may have been.

Referring to the agencies of metamorphism, and particularly the chemical changes involved in the "transfer of ingredients from one part of a mass to another," he wrote, "We know of no other agency by which this could be produced, except by galvanism." To the same agency he would ascribe the production of cleavage, foliation, and joints. He nevertheless recognized the possibility of the metamorphism of sediments through pressure, and dwelt in considerable detail upon the elongation and flattening of pebbles in the conglomerates near Newport, Rhode Island. The jointing of these conglomerates, as already stated, he regarded as due "to some polar force such as heat or galvanism. Mere shrinkage could not have separated the pebbles so smoothly, much less could a strain from beneath have thus fractured them. . . . A mere inspection of the rock in place will satisfy anyone that no mechanical agency is sufficient to explain these phenomena." Yet today "mere inspection" will satisfy anyone that mechanical agency is amply sufficient. (See also p. 403.)

He regarded the feldspar in all of the crystalline rocks as the result of a process of metamorphism:

Silicates probably furnished the ingredients, which, being abstracted by hot water, left the excess of silica in the form of quartz, and formed the feldspar and mica to fill up the interstices. The feldspar, which has converted the cement into gneiss, could have no other origin, and this fact, in connection with all the rest which have been adduced, forces a presumption that feldspar in nearly all the crystalline rocks, stratified and unstratified, is a product of metamorphism.

Further on, he wrote:

Metamorphism furnishes the most plausible theory of the origin of the Azoic stratified rocks, which are mica, talc, and hornblende schists, gneiss, serpentine, white limestone, etc., such as cover a large part of Vermont.

The influence of water as an essential constituent of trappean magmas was evident, since the intrusion of the material, often in thin sheets scarcely thicker than writing paper, could not be explained on the theory of fusion from dry heat alone. "By means of water the materials could be introduced wherever that substance could penetrate."

His ideas concerning the origin of the drift had been considerably modified since his earlier reports. Their discussion however is reserved for a special chapter (p. 615).

On lithological grounds the alluvial deposits were classified as (1) *Drift and* (2) *Modified drift*, and on chronological grounds into four periods, in each of which the continent was differently situated in respect to the level of the ocean. These were, first, the *drift period*, when the continent was under water at its greatest depth; second, the *beach period*, when it began to emerge; third, the *terrace period*, when it rose to nearly its present situation; and, fourth, the *historic* or *present period*. The phenomena of the modified drift as terraces and beaches were worked out in detail in this report by C. H. Hitchcock.

In the chapter on hypozoic and paleozoic rocks the belief was expressed that the rocks of Vermont had been thrown into a succession of folds while in a semiplastic condition by a force from the direction of the Atlantic, and that their crests have been subsequently denuded, some 10,000 vertical feet having disappeared from the surface of the Shelburne anticline. The strata of those folds with westerly dips, on the western side of the anticlines, he regarded as occupying a normal position; while those with easterly dips, on the western side, had been inverted, so that "though we cross an uninterrupted succession of easterly dips in going eastward we cannot infer that we are constantly meeting with older and older rocks, and therefore that mere superposition would not justify one in deciding upon their relative ages." It is here that the idea of mountain formation by tangential pressure rather than from direct uplift from beneath is first made manifest.

Concerning the position of the much-disputed red-sand rock, it was remarked that without an exception it rested upon the Hudson River group, and the stratigraphical evidence showed it to be of the age of the Medina sandstone or Oneida conglomerate. In this he was in error, the red-sand rock being now conceded to belong to the Lower Cambrian. He also thought the black slate beneath the red-sand rock to belong to the Hudson River group, although Emmons had described it as Taconic and thought to recognize a Potsdam sandstone in the vicinity of Whitehall, in this respect antagonizing the views of Adams, as announced in his second report.

The name Æolian limestone was given to the marble beds of Dorset Mountain, which had been renamed Mount Æolius. Its exact geological position was not determined, although it was thought that it might be as new as the Carboniferous. In this he was mistaken.

The clay-slate directly overlying the calciferous mica-schist was considered probably of Devonian age, and the calciferous mica-schist itself thought to have been originally a limestone formation charged with a good deal of silex and with perhaps silicates and organic matter, a great deal of the lime having been abstracted by carbonated water and the rock converted into a schist by metamorphism.

The numerous beds of serpentine which were found in the state limits were regarded as having been derived mainly from beds of hornblende schist and diorite, and much of the granite considered as recent, at least, as the Devonian period.

It seems almost extraordinary that, even at this date, there were believed by Hitchcock, as well as by others, to be many well-authenticated instances on record in which toads, snakes, and lizards had been found alive in the solid parts of living trees and solid rocks, as well as in gravel deep beneath the surface. In these cases it was assumed that the animals "undoubtedly crept into such places while young, and, after having grown, could not escape. Being very tenacious of life and probably obtaining some nourishment occasionally by seizing upon insects they might sometimes continue alive even after many years."

Geological Surveys of Wisconsin under Hall, Carr, and Daniels, 1857.

Under date of March, 1857, an act of the Wisconsin legislature provided for a geological and agricultural survey of the state, which was to be under the joint control of James Hall, Ezra S. Carr, and Edward Daniels. This joint leadership seemed, however, to have proven unsatisfactory, and by a second act, approved April 2, 1860,

James Hall was appointed principal to the commission. This last organization proved also to be short-lived, the survey being discontinued in 1862, the legislature even going so far as to refuse to refund to Hall the money actually expended by him or to pay his own salary.[34]

The actual work of the survey was not begun until 1858, when it is stated Hall and Carr employed at their own expense Charles Whittlesey to explore the country between the Menominee and Oconto rivers, and in the spring of 1859 an engagement was entered into with J. D. Whitney to make a survey of the lead region. The work of both Whittlesey and Whitney was completed and an act passed and approved April 15, 1861, authorizing the publication of one thousand copies of what was expected to form the first volume of the report. The next year, under the excitement of the war, the legislature repealed the act authorizing the survey. Carr and Daniels abandoned the field, but Hall, contending that he had a contract under seal with the Governor according to provisions of the law, claimed that the legislature could not annul the same. He continued his labor and completed that portion of the work which had been assigned to him in his original division among the three commissioners. Manuscript for a second volume of the report, it is stated, had in 1875 been ready for publication more than a dozen years, and Hall had made repeated applications to the legislature for compensation, but in vain.[35]

The single royal octavo volume of 455 pages, of which 72 were by Hall, and 9 plates, a colored geological map of the lead region, and a diagram showing the position of the ore crevices, appeared in 1862 as the only result of the organization.

In his portion Hall repeated his attempt at correlation of the western with the eastern formations. In addition he called attention to the driftless area of the state, and to the fact that, over the whole area south of the Wisconsin River, the Galena limestone and a por-

[34] Appendix, p. 681.

[35] Hall to Dana, May 24, 1862:

". . . One great motive for me to engage in the Wisconsin Survey was to enable Whitney to bring out his Report on the Lead Region, the map of which had been long before begun, and at one time we expected to bring it out in the Iowa Survey. The Wisconsin Survey has cost the expenditure of much time uselessly, and the unfortunate organization, which brought the Survey into disrepute, has been the cause of most of this annoyance and waste of time. Geological Surveys in the west where one has to go a thousand miles to begin his work are not sinecures. . . ."

See also Appendix, p. 683.

tion of the underlying Blue limestone had been removed by decomposition and solution to a depth of some 350 to 400 vertical feet.[36]

Whitney, in his portion of the report, showed, as in his previous work (p. 349), that the lead-bearing fissures of the Northwest were limited to one set of strata, and that there are but rarely any evidences of dislocations or faults. Hence he was forced to the conclusion that the ore-bearing crevices originated through the action of some local cause or forces which were limited in their field of action to a comparatively narrow vertical range. In short, he conceived the lead-bearing fissures to have been originally closely allied to what are called joints. How far shrinkage and crystalline action may have been concerned in the production of these, he was unable to say, but "it will be sufficient at this time to recognize the fact that either or both these causes united would have produced a tension in the mass of rock which would lead to the production of fissures in two different directions." He felt, further, that the direction of the fissures might have been influenced, if not absolutely determined, by an elevatory movement of the region in which they are developed.

He looked upon the veins as having been filled from above, the various metals having been in solution as sulphates in oceanic water at the time of the deposition of the lead-bearing strata, the precipitation as sulphides being brought about through the agency of sulphureted hydrogen gas, which was liberated by the decomposing organic matter. This is substantially the view held by most authorities today. He thought, further, to recognize a general law connecting the lack of mineral deposits in the nonfossiliferous strata with their abundance in those containing organic remains. As in his previous work (*Metallic Wealth of the United States*), he differed with both Daniels and Owen as to the metalliferous character of the Lower Magnesian rocks, thinking it quite improbable that they would be found sufficiently rich in ore to be worked economically.

No report of Whittlesey's work seems to have been published at the time, though a paper presented before the Boston Society of Natural History in June, 1863, was probably based upon the ob-

36 It is stated that Whitney was greatly displeased at Hall's action in inserting these 72 pages in his report and having it issued under his name. He felt that it should have been put out as a separate volume. To satisfy himself on this score he purchased from Hall "a small number of copies" and had them bound up with the omission of Hall's preliminary chapter, "for distribution among my personal friends and those especially interested in the subject." In the preface of this special issue he states that when the MS. was handed in, in October, 1860, he had supposed it would be so published. The printing, it should be stated, was not begun until after Whitney had left to take charge of the California survey, and the proof revisions were left to his brother, W. B. Whitney.

servations made. This paper was devoted to a history of the mining operations on the Penokee mineral range and its geology. As a provisional arrangement of the rocks of the district, he proposed the following, in descending order; Formation No. 1, *Potsdam sandstone;* Formation No. 2, *Trappose,* in two members; Formation No. 3, *Hornblendic;* Formation No. 4, *Siliceous,* two members; Formation No. 5, *Granites* and *Syenite* of Central Wisconsin. A considerable number of chemical analyses were included, though what purpose they were intended to serve is not apparent. An incident, unintentionally amusing to those conversant with the careless manner in which Whittlesey prepared his manuscripts, is given. In explanation of certain typographical errors he wrote: "In the Chippeway language the name of iron is *Pewabik;* and I thought it proper to designate the mountains where this metal exists in quantities . . . as the Pewabik Range. The compositor, however, transformed the word into *Penokie,* a word which belongs to no language."[37]

It was extremely unfortunate, in view of the difficulties that beset the geological survey, that Whittlesey should himself have been intractable. In every instance there seems to have been a falling out between him and his superiors or collaborators, which can apparently be traced only to a "cantankerous" disposition.

Nevertheless, Winchell states[38] that "Colonel Whittlesey was a good geologist. His studies were chiefly structural and stratigraphical, and his field work was almost always upon unfossiliferous rocks; but, when in Ohio, he gave due attention to the collection of organic remains. . . . He appreciated their significance in geological investigation." He was a man of greatly varied career. Educated at West Point and graduating in 1831, he resigned at the close of the Black Hawk War and opened a law office in Cleveland, Ohio, becoming also editor and part owner of the *Whig and Herald.* In 1837, however, he accepted an appointment under Mather as topographer, geographer, and structural geologist. In 1844 he made an agricultural survey of Hamilton County, Ohio, and later served as geolo-

[37] The compositor could scarcely be to blame if we are to judge from the following. Whitney to Hall, June 24, 1860:

"Day before yesterday came a roll by express which when first opened I supposed to be a manuscript in some unknown tongue, sent by mistake to me instead of my brother, who is in the deciphering line of business. On closer inspection, however, I gradually came to the recognition of the fact that it was in the English language, and written in the best and plainest Whittleseyan hand. I will study on it at intervals and give you my results. It seems to be, so far as I can yet make out, an attempt to show that Mr. Foster's observations on the Menominee River were incorrect as far as naming the rocks were concerned."

[38] Am. Geologist, Vol. IV, 1889, p. 263.

gist for a private company exploring for copper on the upper peninsula of Michigan. From 1847 to 1859 he was with the government surveying parties in the Lake Superior region and Wisconsin. With the outbreak of the Civil War he reenlisted, but resigned again on account of his health in 1862. He made explorations in the upper Mississippi region, became archeologist and historian and was one of the chief supporters of the Western Reserve and Northern Ohio Historical Society.

Richard Owen's Key to the Geology of the Globe, 1857.

Richard Owen's *Key to the Geology of the Globe,* issued in 1857, while he was professor of geology and chemistry in the University of Tennessee, is worthy of note for the reason of containing the first presentation of the theory of an original tetrahedral form of the globe put forward by an American writer. Owen conceived that "the earth in some of its former geological epochs occupied a smaller volume than before the whole of the present superficies emerged from the ocean, and than it did before some of the later successive layers were deposited on the earlier formations." Further, that the original internal nucleus was of the form of a cube or spherical tetrahedron, and that the rocks of the different successive geological periods will be found less and less dense in structure as they leave the north pole and approach the equator. When these beds were upheaved, "the edges of the formations appear to have been brought to the surface along concentric lines, which are parts of great circles intersecting each other in such a manner as to form equilateral spherical triangles on the earth's surface, each angle or intersection being equidistant from our present north pole." The distribution of coal, the metals, and other mineral products he conceived to correspond with the lines marking these spherical triangles.

The forces instrumental in bringing about these upheavals were supposed to have originated through the internal fluid materials being thrown into periodical waves by the attraction of the sun and moon, creating thereby electrical disturbances, etc. Indeed, the entire earth he seemed to regard as a gigantic magnet, made so by the heat of the sun, an idea which has been as yet by no means abandoned.

Before this separation took place, the layers composing the South American continent were supposed to have rested on the layers of submerged Africa, Australia to have been superimposed upon Arabia, and North America over a portion of Europe.

Owen further called attention to the fact that, if the north pole were elevated 23½° above the horizon and the globe then revolved, the western coast lines of the chief continental masses would be brought successively to the horizon, proving their parallelism; if depressed to a similar extent the parallelism of the eastern coasts would be similarly proven. In other words, he showed that the coast lines of the continents, as well as of many of the islands, tended to conform to the axis of the ecliptic, and he regarded this angular distance of 23½°, which marks the northward extension of the sun in summer, as a natural unit of measure in the structure of the earth.

Many of the ideas put forward were fanciful in the extreme.

Our planet, perhaps, typifies an ovule from the solar matrix. In its earlier igneous chaotic state it bore analogy to the yet undeveloped amorphous structure of vegetable ovules and the animal ovum. Like them, it had at an early period a nucleus, on which after a time air and moisture deposited additional materials, derived from the matrix. At a later period yet a part of these same materials were carried in mechanical mixture, partly in chemical solution, to promote the development of later formations, forming new continents, etc., just as a portion of the seed (the albumen) and the food-yolk of the egg go to nourish the expanding germ.

The separation of continents typifies the propagation of offshoots, or artificially by cuttings in plants, and seems to resemble the fissiparous mode of reproduction observed among the lowest animals. In some of the earlier cataclysms we have the type of the ruptured Graafian vesicles, while at a final convulsive deluge, the period when the Western Continent and Australia were detached, and when possibly the moon as a terrestrial ovule was thrown into space, we readily recognize the type of ruptured pericarpal dissemination of seed in the vegetable world, of completed incubation and parturition in the animal kingdom. . . .

It is by no means contended that this earth is a monstrous organism, with all the parts and properties of a plant or animal; but simply that in it we have everything developed according to the same laws and plan pervading the rest of creation; that in it we see foreshadowed the type of those future forms and changes which organic bodies undergo by the assimilation of these very inorganic materials.

In addition to the above there was a large amount of irrelevant matter, including chapters on pathologic, therapeutic, and ethical geology and advice to young men. By a peculiar coincidence, at the Montreal meeting of the American Association for the Advancement of Science in August of this same year, Professor Benjamin Pierce, of the United States Coast Survey, brought forward ideas almost identical with Owen's, so far as relates to the parallelism of the coasts, and felt disposed to regard such conditions as evidence that the sun, by influencing in some way the cooling of the crust, had determined its grander outlines.[39]

Logan's Proposed Subdivision of the Laurentian, 1857.

At the Montreal meeting of the American Association for the Advancement of Science for this same year, Logan, referring to previous statements by himself relative to a possible subdivision of Azoic rocks, as given in his annual reports, described in some detail a series or group of clastic rocks, consisting of siliceous slate and slate conglomerates, holding pebbles of crystalline rocks and sometimes showing ripple marks, which he had traced along the north shore of Lake Huron from Sault Ste. Marie for 120 miles, and which had also been noted in other parts of the Dominion by other observers.

As seen by him in the Lake Huron area the formation was some 10,000 feet in thickness and plainly of later age than the gneisses, as well as of distinct lithological character. For this he then formally proposed the name Huronian, although the term had been previously used in the same connection by Murray in 1848 and Hunt in 1855. To the portion of the Azoic immediately underlying the Huronian he gave the name Laurentian, after the Laurentide Mountains.

At this same meeting he further suggested a possible subdivision of the Laurentian series, in which the calcareous rocks should be considered as distinct from the siliceous. This announcement and the subsequent discussion led to a mass of literature which is perhaps second only to that involved in the Taconic controversy, although, fortunately, without any of its bitterness. Indeed, the entire matter is strongly suggestive of Emmons's Taconic, even if the rocks described could not be accurately relegated thereto.

The principal participants in this discussion, which is even now scarcely at an end, were, aside from Logan, R. D. Irving and Alexander Winchell, though many others took part. Later it was apparently shown that under the name Huronian two distinct formations were comprised, to the older of which A. C. Lawson in 1886 gave the name Keewatin.[40]

39 These ideas were not wholly new. Those desiring a historical review of the subject, with an account of recent advances and theories will find the same in a paper by J. W. Gregory, The Plan of the Earth and Its Causes, first printed in the Geographical Journal for March, 1899, and reprinted the same year in the Appendix for the Report of the Smithsonian Institution. Owen's book was briefly noted in the American Journal of Sciences. Two or three pages of the matter were quoted, and dismissed with "Remarks on these passages are unnecessary."

40 For a history of the discussion, see A Last Word with Laurentian, by Alexander Winchell, Bull. Geol. Soc. of America, II, 1891, pp. 85-124.

The Ives Expedition up the Colorado River, 1857-1858.

During 1857-1858 Professor J. S. Newberry accompanied as geologist the corps of topographical engineers, under the immediate direction of Lieutenant J. C. Ives, on an expedition up the Colorado River from its mouth to a point called Fortification Rock, north of the thirty-sixth parallel. The main object of the expedition was to ascertain the navigability of the Colorado, with especial reference to the transportation of supplies to the various military posts in New Mexico and Utah. Newberry entered the country from San Francisco by way of San Diego and Fort Yuma. The party ascended the river in an iron steamer constructed for the purpose in Philadelphia as far as the Great Bend above the mouth of Black Canyon, and returned thence to the Mojave Valley, south of Pyramid Canyon. Thence the homeward route lay eastward overland to Fort Leavenworth by way of Sitgreaves and Railroad passes in the Black, Cerbat, Bill Williams, and San Francisco mountains, Salt Springs of the Little Colorado, northward to the Moqui Pueblos, and eastward to Fort Defiance, Santa Fé, and Las Vegas (see map, Fig. 73). In the latter part of the trip the route lay along that traversed by the surveys for a railroad route to the Pacific under Lieutenant Whipple, with which Jules Marcou was connected as geologist.[41] Newberry's report, Part III of the report of Lieutenant Ives, comprised 154 octavo pages, with 3 plates of fossil invertebrates and plants. Made necessarily as a hasty reconnaissance, it nevertheless contains interesting generalities. He rightly regarded the canyon of the Colorado as one of erosion, but conceived that in earlier times the river filled to the brim a series of isolated basins formed by the various mountain chains and interlocking spurs of the synclinal trough through which the river runs, and that during the lapse of ages "the accumulated waters, pouring over the lowest points in the opposing barriers, have cut them down from summit to base," thus forming the canyons. The massive walls of Pyramid Canyon, with their capping of stratified gravel, he considered conclusive proof of the former existence here of an unbroken barrier "stretching across the course of the Colorado and raising its waters to an elevation of at least 250 feet above their present level." The idea of a possible uplift across the river's course, an uplift so slow as to be counteracted by erosion, as afterward taught by Dutton, was undreamed of, and the mental restorations of part surface contours given are at times startling for their magnitude.

[41] The return route, it will be noticed, does not differ greatly from that now followed by the A. T. & S. F. R. R.

PLATE XVII

JOHN STRONG NEWBERRY

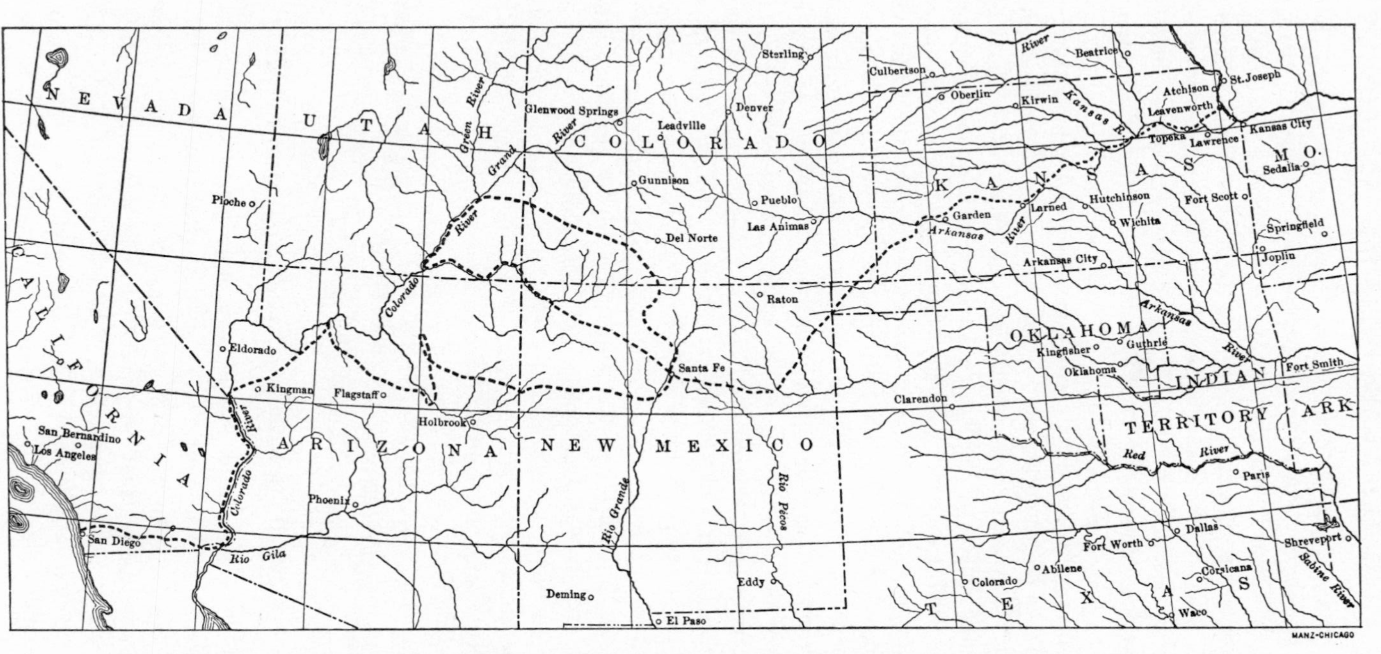

Fig. 73. *Map showing routes traversed by Newberry with Lieutenant Ives and Captain Macomb.*

He gave a section of the Grand Canyon, showing the bottom granite overlain by Potsdam sandstone, and this by Silurian (?), Devonian (?), and Lower Carboniferous (?) rocks, and finally, Upper Carboniferous limestone. This was the first section of these beds published. The report also contained several graphic views of the canyon from sketches by F. W. Egloffstein, but which have been largely overlooked since the more recent work of Powell, Dutton, and Holmes.

The source of the materials which make up the sedimentary rocks of the West, he thought, could not have been derived from the emerged surfaces east of the Mississippi, but were rather "formed by the incessant action of the Pacific waves on shores that perhaps for hundreds of miles succumbed to their power, and by broad and rapid rivers which flowed from the mountains and through the fertile valley of a primeval Atlantis." The outlines of the western part of the North American continent, to his mind, were approximately marked out by groups of islands, broad continental surfaces of dry land, and areas of shallow water from the earliest Paleozoic times.

As already stated, a portion of Newberry's return route lay along that previously (in 1853-1854) traversed by Marcou, and it is interesting to compare the views of the two. He questions Marcou's determination of the age of the formations near Partridge and Cedar creeks, near Bill William Mountain, Marcou regarding them as Devonian and Lower Carboniferous (Mountain limestone), while Newberry thought them to be not older than Carboniferous. In like manner he questioned the Permian age of the Canyon Diablo beds, rightly identifying the same as Carboniferous, and the red sandstone overlying this limestone as likely to prove Permian or Triassic. The yellow sandstone along the Rio Grande was thought to be Cretaceous rather than Jurassic.

Newberry's Work with the Macomb Expedition, 1859.

The Arizona trip, as above outlined, was accompanied by its full share of discomforts, as may be readily imagined by one at all conversant with the nature of the country, and Newberry's letters to collaborators in the East were often highly entertaining.[42] Undeterred by recollections of the past, Newberry in 1859 again took the field, this time in connection with a topographic party under Captain J. N. Macomb. The party left Santa Fé about the middle of July

[42] Appendix, p. 684.

and, crossing the Rio Grande del Norte, followed up the valley of the River Chama, finally leaving it at the dividing ridge between the waters of the Gulf of Mexico and those of the Gulf of California. From here they struck across the headwaters of the San Juan River, passing along the junction of the Grand and Green rivers. Thence they proceeded southward to the San Juan River, which they followed up as far as Canyon Largo, passing thence down the valley of the Puerco to the old pueblo of Jemez, and thence easterly back to Santa Fé. The route, it will be perceived, thus covered part of the territory previously gone over in connection with the Ives expedition.

The report, though written in 1860, was not published until 1876, owing to disturbances incidental to the Civil War. It comprised an octavo volume of 148 pages, with 8 full-page plates of fossil plants and invertebrates, and a map of the route traversed. Some of the conclusions arrived at during the work of the Ives expedition were confirmed by the more recent work. Perhaps the most interesting of the results as given related to the orographic movements attendant upon the elevation of the Rocky Mountains. His conclusions were to the effect that:

First, the Rocky Mountains existed, in embryo at least, previous to the deposition of the older Paleozoic rocks, the presence of the upheaved Potsdam sandstone in the Black Hills region showing conclusively that this part of the country was buried beneath the waters of the primeval ocean.

Second, volcanic activity, which began as early as the Middle Tertiary, continued even into the present epoch.

Third, previous to the deposition of the Lower Cretaceous strata the central portion of the continent was above the ocean level, the main portion of the Cretaceous sediments being deposited during the period in which a subsidence of several thousand feet took place.

Fourth, at the close of the Cretaceous age a period of elevation began, which continued to the drift epoch. This was succeeded by a period of depression and again by one of elevation.

Fifth, the great elevatory movement of the Rocky Mountains took place between the close of the Cretaceous period and that of the Miocene Tertiary.

Sketch of Newberry.

Newberry was born in Connecticut, but his parents moving to Ohio when he was but two years of age, he passed his boyhood in what was then the western frontier. His father, in 1828, then living at Cuyahoga Falls, opened up the coal mines at Tallmadge, making the first systematic attempt at introducing coal for fuel along the lake shore. The abundant and beautiful fossil plants found in the roof shales of these mines undoubtedly did much toward turning the young man's thoughts toward science, and a visit from James Hall in 1841, while he was but nineteen years of age, still further stimulated his interest. He graduated at the Western Reserve College in 1846, and afterward studied medicine, receiving the degree of M.D. in 1848. Two years were then passed in study abroad, during which time he made his first scientific publication—a description of a fossil-fish locality at Monte Bolca, Italy. In 1851 he returned to America and began the pursuit of medicine at Cleveland, but although acquiring a large practice he was soon induced to give it up and enter upon a wider field of usefulness. In 1855 he became connected with Lieutenant R. S. Williston's survey in California, and later with that of the Ives expedition to explore the Colorado, as well as with that of Captain J. N. Macomb to explore the San Juan region of Colorado and adjacent portions of Arizona, Utah, and New Mexico, as noted. In 1857 he became attached to the Smithsonian Institution in Washington and also accepted a professorship in Columbian University in the same city.

During the Civil War he was attached to the Sanitary Commission, and in 1866 became professor of geology and paleontology in the School of Mines of Columbia College, New York, with which institution he remained connected until the time of his death in 1892. In 1869 he assumed charge of the geological survey of Ohio, as noted elsewhere. That the survey, in common with most others of its kind, came to an untimely end was no serious reflection upon Newberry, though, as his biographer remarks, a mistake was undoubtedly made in postponing the economic portions of the work until the last, and thereby arousing the antagonism of the rural members, one of whom is quoted as having remarked that too much money was devoted to clams and salamanders, referring of course to his Volume IV on the zoölogy and botany of the state. Newberry also did a large amount of paleontological work in connection with the surveys of Illinois, New Jersey, and the United States surveys under Hayden and Powell. His most important works were in connection with those of Illinois and Ohio, and two monographs published by the United States Survey, one on the *Paleozoic Fishes of North America*, and the other on the *Flora of the Amboy Clays of New Jersey*, the latter being edited by Dr. Arthur Hollick after Newberry's death.

He is described as a man whose personality inspired confidence in others, though possessing little of what is known as executive ability—a man of kindly, cheerful disposition, and whose desire for fame never went so far as to cause him to assume credit for others' work.[43] A pleasing writer and conversationalist and accomplished musician, he made many friends and retained to the end the respect and love of all who knew him.

His shortcomings were known to his friends as well as were his excellencies; he was impetuous and sometimes he was severe, possibly unjust, in judging men or in dealing with them. But of bitterness he knew little; of forgiveness he knew much. His defects were those of a strong man; in many they would have been sources of weakness, but somehow they seemed to make his friends stand more firmly by him.

Newberry's work received prompt and ready recognition both at home and abroad. He was a charter member of the National Academy of Sciences; foreign correspondent of the London Geological Society, receiving the Murchison medal in 1888; president of the American Association for the Advancement of Science in 1867; president of the New York Academy of Science for twenty-five years; vice-president of the Geological Society of America, and president of the International Congress of Geologists in 1891, although then too ill to attend its meetings.

Work of David Dale Owen in Arkansas, 1857-1860.

On April 20, 1857, D. D. Owen, while his Kentucky work was still unfinished, was appointed state geologist of Arkansas, the appointment to take effect "from and after the first day of October, 1857," at which time it was supposed the Kentucky work would be brought to a close. Owen selected for his assistants, E. T. Cox and Leo Lesquereux, with Joseph Lesley for topographer, and Robert Peter and W. Elderhorst, chemists. Two large octavo volumes, comprising together some 689 pages, were issued in 1858 and 1860, the first having been printed apparently at Little Rock and the second in Phila-

[43] "Expedition and exploration are of inestimable value, as giving one capital to work with—both of material (matter) and thoughts—but for a man who has a good wife and pleasant home, three darling boys, as I have, absence from them for any purpose however pure and noble is only exile, and fame gained only at the price of continued absence is not worth its cost. If I must choose between the two, I say give me domestic happiness and let fame go to the winds." (Letter N. to H., December 21, 1860.)

delphia.[44] They were illustrated with colored lithographs and engravings after originals in Owen's well-known style.

On paleontological evidence the zinc and lead-bearing rock of northwestern Arkansas, *i.e.*, the 300 feet of Magnesian limestone and silico-calcareous rock that underlie the marble strata, forming about 250 to 300 feet of the lower and main body of the ridges of Marion County, were put down as of Lower Silurian age, and as in all probability belonging to that subdivision known as the Calciferous sandrock of the New York system. The horizon of the "marble limestones" of northwestern Arkansas he was unable to fix definitely, but seemed to think it probable they would prove to represent the Onondaga of New York.

Concerning the origin of the lead ores, he wrote:

> My impression is that the lead ore once occupied these north and south crevices, and was subsequently removed, in part or in whole, into its present bed by transportation, analogous to that known to mineralogists under the name of the pseudomorphous process, by which one mineral is removed while another takes its place, assuming often the form of the first mineral instead of the usual form belonging to itself. . . . That it should have been deposited like a limestone or sandstone is altogether improbable, and contrary to the usual nature of such ponderous and difficultly soluble minerals.

The cause of the hot springs was considered to be the internal heat of the earth. Not that the waters came actually in contact with fire, but rather that they were completely permeated with "highly heated vapors and gases which emanate from sources deeper seated than the water itself."

The white novaculite from the vicinity of the hot springs he believed to be of the age of the millstone grit, and "once a simple ordinary sandstone. From the state of ordinary sandrock it has been altered or metamorphosed into this exquisitely fine material, not as I conceive, by contact with fire or igneous rock, but by the permeation of heated alkaline siliceous waters." Through this permeation he conceived that "the particles of solid rock have been gradually changed from grains of quartzose sand to impalpable silica, and the greater part of the oxide of iron, manganese, and other impurities carried out in solution from the pores of the rock, leaving nearly chemically pure silica behind." Such changes are practically inconceivable, and the best authorities today regard the novaculites as siliceous replacements of limestone.

[44] The editing of the second volume was done by J. P. Lesley, after Owen's death. See also p. 199.

A large number of soil analyses were made by Doctor Peter "with a view to settle that very important question" whether such can be of utility to the agriculturist in showing the relative fertility of soils and the loss sustained by cultivation. In Owen's opinion such utility was established "in the most incontrovertible manner." Later writers and investigators have, however, found occasion to take a different view of the matter and most of them could agree with the ideas expressed by T. Sterry Hunt in a letter to Hall written in 1862. "You ask me about soil analyses. I have no time for six months for any kind of investigations in the laboratory, and if I had, I hope I am too wise and too honest to waste any money in such things. I am wholly of Prof. Johnson's opinion as to their worthlessness."

According to Branner the appropriations for Owen's survey were vigorously combated by some of the legislators, one of whom, in his attempts to defeat the bill, introduced the following amendment:

Sec. 12. The same amount which is appropriated to the State geologist shall likewise be appropriated to a phrenologist, . . . and a like amount to an ornithologist and their several assistants, who shall likewise be appointed by the governor, and shall continue in office fifty-four years . . .; and the secretary of state shall forward one copy of each report to the governor of each State in the Union except such as may be known to be black republican governors; also, one copy to the Queen of England and to the Emperors of France and Russia; also, a copy to the Queen of Spain: provided that government will sell Cuba to the United States on reasonable terms.

Sec. 14. It shall be the duty of the phrenologist to examine and report upon the heads of all the free white male and female citizens in the State, and their children, except such as may refuse to have their heads examined.[45]

Hilgard's Geological Survey of Mississippi, 1857-1860.

Lewis Harper, whose work in Mississippi is referred to on p. 330, was succeeded in the office of state geologist by E. W. Hilgard, under whose efficient direction there was brought about the first and only survey of the entire state. His *Report on the Geology and Agriculture of the State of Mississippi* appeared in 1860. It contained a systematic and fairly detailed account of the various formations, the classification and terminology adopted being that of Safford and Tuomey.

The Orange sand of Safford was shown to characterize the greater

[45] Journal of Geology, II, 1894, p. 826.

part of the surface of the state, and he thought it proven beyond question that its deposition had taken place in flowing water, the general direction of the current of which was from north to south. By far the greater portion of the state was shown to be occupied by deposits of Tertiary age, leaving out of consideration the stratum of Orange sand which covered a large part of the actual surface.

Hilgard's work had an agricultural bearing and, as may be readily inferred from the work of more recent years both in the Mississippi Valley and California, had to do largely with soils, their original physical and chemical constitution, fertility, and methods of rejuvenation.[46]

The Kansas Permian Controversy, 1858.

At a meeting of the St. Louis Academy of Sciences, held on February 18, 1858, G. C. Swallow presented a paper announcing the finding of Permian fossils among some collections submitted to him for identification by Major F. Hawn. This, as being the first apparently authentic announcement of the occurrence of Permian rocks on the American continent, excited a great deal of interest and started a somewhat bitter personal controversy between Hawn and Swallow and Meek, which may be referred to in some detail.[47]

Hawn, it appears, was a civil engineer engaged in the linear survey of Kansas, whose interest in matters geological led him to make extensive collections of fossils. These, for purposes of identification, he divided, sending the supposed Cretaceous forms to F. B. Meek, then at Albany, New York, and the Carboniferous to G. C. Swallow, at Columbia, Missouri, the particular collection referred to in this controversy being made at the Smoky Hill Fork of the Kansas River.

The collections received by Meek were of such a character as to lead him to think that the beds from which they were taken might be Triassic or Permian, though perhaps belonging to the Upper Coal Measures. He immediately wrote Hawn, asking that he obtain for him other materials and, if possible, get from Professor Swallow all

[46] For details of this work see my Contributions to the History of State Surveys, Bull. 109, U. S. Nat. Museum, 1920.

[47] It should be noted that Marcou claimed priority in the discovery of Permian (Dyas) beds in America, while connected with the Pacific Railroad survey in 1853. (See his American Geological Classification and Nomenclature, Cambridge, 1888.) The correctness of his paleontological identification was questioned by Meek, White, Hall, and Newberry.

of those forwarded to him which were not Carboniferous. To this Hawn readily agreed and wrote as follows.

Professor Swallow certainly will not attempt to interfere with you in this matter. He knows perfectly well the relation existing between us and expressed himself gratified that I was furnishing you with important information in furtherance of your investigation. Furthermore, he has not the data to establish a relation between the several points under review by you. I merely sent him the Carboniferous fossils for classification and comparison with those of the Missouri collection, that a parallel may be established in my further operations.

Again, under date of September 5, 1857, Hawn wrote to Meek:

I wrote to Professor Swallow . . . and requested him particularly to send you all the fossils that were not Carboniferous, as they were intended for your use. I hope he will comply with my request, as I shall not have time to go to see him.

And still again, under date of September 9, of the same year:

Should this formation turn out to be as you anticipate, new and important, will you discuss the details in an article for my contemplated work on the mineral and agricultural characteristics of K. T., and describe the fossils that are new?

The following year, under date of January 4, he wrote:

I have a letter before me from Professor Swallow in which he thinks the specimens sent him from Smoky Hill Fork are Carboniferous. Therefore suppose he has not sent you any fossils. Of course, I should not set up my limited knowledge in opposition, but I certainly have forms unknown to me and not occurring in Carboniferous rocks of Missouri which I examined.

Under these conditions it is difficult to understand why Hawn should have later entered into a partnership with Swallow for the working up of the collections and the publication of the results, and it is perhaps not to be wondered at that Meek should have denounced the proceeding somewhat harshly in a paper prepared under the joint auspices of himself and Hayden and read before the Albany Institute on March 2 of that year.[48] Hawn, in a subsequent letter, excused himself on the ground that Meek's letters had led him to believe that he had abandoned the investigation for want of time or for want of confidence in the final result, and that consequently he brought the

[48] See also Appendix, p. 685.

matter to the attention of Swallow with the urgent request that the collections be worked up immediately, and at the same time notified him of Meek's suggestion regarding their Permian nature. Swallow himself, after the mischief was done, wrote a very conciliatory note to Meek, which seems, however, to have been far from healing the breach between the parties most concerned.

The facts of the matter, so far as they can be made out, both from the publications and the letters now in the archives of the Smithsonian Institution, seem to show that, beyond question, Meek was the first to recognize the possible Permian nature of the fossils in question; that he so notified Hawn, and supposed the matter was so arranged that he could work the materials up at his leisure. In the meantime Hawn, as suggested in his letter, thinking that Meek had laid the matter aside, brought it to Swallow's attention, who published the matter with almost unseemly haste in the *Transactions*,[49] as well as in the local newspapers of the day. In addition, he gave the substance of the discovery in a letter to Dana, dated February 16, which was published as an appendix of the *American Journal of Science* for March of that same year.[50] In a later paper in the same journal he stated very frankly that "so far as I know, Mr. Meek first discovered the Permian character of the Kansas fossils, and communicated to Major Hawn his impressions on the 3d of September, 1857. He also mentioned his discovery, as I am informed, to some friends at the Smithsonian Institution on the 17th of January, 1858, and communicated the same to Prof. Leidy on the 16th of March. Major Hawn frankly declares that his first impressions that the rocks might be Permian are due to Mr. Meek." So ended the matter so far as priority of discovery was concerned. It did not, however, end the discussion as to the real equivalency of the beds in question.

[49] Swallow's paper begins with the following statement, which illustrates his anxiety to claim priority in a discovery, the importance of which was greatly overestimated: "In presenting the following paper to the scientific world, we feel it incumbent upon ourselves to state that it has been prepared in great haste, in the midst of other pressing duties," etc.

[50] Apparently inserted at the last moment before the number was bound up, and too late to be noted in the table of contents. The letter reads as follows: "I have just finished the examination of a collection of fossils from Kansas Territory made by Major Hawn who was formerly connected with our survey. The larger part of the collection is from the Upper Coal Measures; but by far the most interesting part cannot be referred to the Carboniferous, or to any other formation known to exist in the west. From the beds in doubt there was but one known Carboniferous species *Terebratula subtilita* of Hall. It is quite certain they are not Cretaceous. After a somewhat careful comparison with the Permian fossils of Rublit, we are satisfied that they are Permian."

Shumard on Permian in New Mexico.

Fig. 74. *B. F. Shumard.*

On the eighth of March of this same year B. F. Shumard announced, at the Academy of Sciences at St. Louis, that certain fossils which his brother, G. G. Shumard, had brought from the white limestone of the Guadalupe Mountains of New Mexico were, at least in part, identical with the Permian fossils of Kansas. This is referred to on page 314. Subsequently the supposed Permian was recognized in Illinois by Hall from fossils sent to him by Worthen, and by J. G. Norwood, the latter announcing his discovery in a letter to B. F. Shumard dated March 31, 1858, and later published in the Transactions of the St. Louis Academy of Sciences.

Meek and Hayden on Triassic in Kansas and Nebraska.

In the *American Journal of Science* for May of the year following (1859), Meek and Hayden had a paper on the so-called Triassic rocks of Kansas and Nebraska, in which they expressed the opinion "that the entire series from near the top of the lower Permian sections of Swallow's and Hawn's down even lower than the horizon where they drew their line between the Coal Measures and the Permian, should be regarded most nearly related to the Carboniferous and might well be called Permo-Carboniferous." This view for some time prevailed, but recent work has shown that beds, the equivalent of the true European Permian, are present in Kansas, Nebraska, the Guadalupe Mountains, and perhaps in other parts of the West as well.

B. F. Shumard's Geological Survey of Texas, 1858.

Information regarding the geology of Texas up to 1858 had been gradually accumulating through the work of geologists connected with the various exploring expeditions, and more particularly through the publications of Ferdinard Roemer, a German who had come to America in 1845 for the purpose of geological explorations, and who passed a year and a half in active work in that state. The most important of his publications, the outcome of this labor, was his *Die Kreidebildungen von Texas, und ihre organischen Einschlüsse,* Bonn, 1852—a large quarto of 100 pages of text and 11 plates of invertebrate fossils. In 1858 the first attempt at a systematic survey under state auspices was attempted, and B. F. Shumard, whose paleontological work in connection with explorations under Captain R. B. Marcy and Colonel Pope have been already noted, was appointed state geologist. Shumard served, through much trial and tribulation, only to 1860, when he was suspended for political reasons[51] and Francis M. Moore, who had been one of his assistants, appointed in his place.

But one report was issued, a pamphlet of seventeen pages, bearing the date 1859. It very briefly outlined the work accomplished, laying considerable emphasis upon the presence of coal and other economic products, including petroleum. G. G. Shumard was assistant geologist, W. P. Riddell, chemist, and A. Roessler, draftsman.[52]

Safford on the Unconformability between the Lower and Upper Silurian Formations in Tennessee, 1858-1859.

In a letter to J. D. Dana, and by him printed in the *American Journal of Science* for April, 1858, Professor Safford, then at Cumberland University, Lebanon, Tennessee, called attention to an important and previously unrecognized unconformability existing between the Lower and Upper Silurian formations in Tennessee, indicating, as he felt, the one-time presence of an island some eighty to ninety miles in diameter in the Upper Silurian and Devonian seas, and occupying a central position with reference to the present limits of the state. This he thought to have been elevated at the close of the Lower Silurian period and not again depressed until the beginning of the period marked by the deposition of the Upper Devonian black shale. Later, in 1859, he pointed out the fact that the crinoidal and variegated marbles and ferruginous limestones of eastern Tennessee were originally deposited in the form of long, narrow belts, stretching in a northeast and southwest direction entirely across the state. This striking feature was accounted for on the supposition that the mate-

[51] See Appendix, p. 686.

[52] G. G. Shumard's notes were later (1886) brought together by H. P. Bee, commissioner of insurance, etc., and published under the title A Partial Report on the Geology of Western Texas, forming a volume of 144 pages.

rials were deposited in troughs, formed by earlier oscillations or corrugations of the earth's crust, the conditions thus confirming the views held by Dana as to the early Silurian age of the beginnings of the Appalachian oscillations.[53]

H. D. Rogers's Final Reports on the Geology of Pennsylvania, 1858.

The geological survey of Pennsylvania, established under H. D. Rogers in 1836, came, it will be remembered, to an untimely end in 1842. Rogers, however, unwilling to relinquish the work in its unfinished condition, continued it at his own expense until he was able in 1847 to make his final report to the office of the secretary of the commonwealth. Here, for some reason unknown to the present writer, the manuscript was allowed to lie until the spring of 1851, when appropriations were made for revising it and bringing it up to date. The appropriations were continued until 1855.

Mismanagement of the funds, for which, it is said, Rogers was in no way responsible, still prevented the publication of the report, and it was not until 1858 that it was finally issued, and then only under a special contract between the author and the state, whereby the latter was to furnish the sum of $16,000 and Rogers was to receive 1,000 copies of the book and the original manuscript. The work was issued by W. Blackwood & Sons, London and Edinburgh, and J. P. Lippincott & Company, of Philadelphia, and was in the form of two quarto volumes containing 1,631 pages, with 23 full-page plates, 18 folded sheets of sections, and 778 figures and diagrams in the text.

This was an epoch-making work, and beyond question the most important document on the geology of America that had appeared up to date, with the exception of the final reports of the New York surveys, issued in 1842-1843. From these it so differed, however, as to be considered quite by itself. In the New York survey stratigraphy

[53] These views are again advanced by Ulrich and Schuchert (Report of the New York State Paleontologist for 1901, pp. 633-663), who state that several folds were developed or older ones accentuated at the close of Beekmantown (Calciferous) time. Between these folds of the southern Appalachians during later Lower Siluric times was the "Lenoir Basin," "containing several disconnected longitudinal folds high enough to affect the direction of currents and consequently the character of the sediments, and, in a smaller degree, faunal distribution." The variegated marbles are in the western or "Knoxville trough" of this basin.

based upon organic remains was ever uppermost. With Rogers's reports quite the reverse was true, the physical side preponderating over everything else. While the value of fossils in determining the relative age of strata was recognized, yet out of the entire 1,631 pages but 20 are devoted to invertebrate paleontology and 47—by Lesquereux—to a description and discussion of the fossil flora of the Coal Measures. Naturally, many of the results given in these reports and the opinions expressed had found their way into print several years earlier, through the publications of the American Association of Geologists and Naturalists and other media. Especially was this the case with the results relative to Appalachian structure.

Rogers was aided by a considerable corps of assistants, many of whom afterwards became noted in the same lines of work. Among these were James C. Booth, S. S. Haldeman, and J. Peter Lesley.

The first volume of these reports treated of the metamorphic rocks and the Paleozoic strata. The second began with a discussion of the coal basins of the state, to which over 600 pages were devoted. Some thirty pages were devoted to the rocks of the Mesozoic red sandstone series, which was followed by a discussion, first, upon the igneous rocks and minerals with special bearing upon their economic value; second, the conditions of the physical geography during the laying down of the Paleozoic strata of the United States; third, the organic remains of the state; fourth, the laws of structure of the more disturbed zones of the earth's crust; fifth, the classification of the several types of petrographic structure illustrated in the Appalachians; sixth, the coal fields of the United States and British provinces, with their chemical and physical characters; seventh, the method of searching for, opening, and mining coal, as pursued in Pennsylvania; eighth, foreign coal fields and coal trade and statistics of the iron trade.

The nonfossiliferous rocks underlying the old Paleozoic were classified as *Hypozoic* or true metamorphic and *Azoic* or semi-metamorphic, the Hypozoic including the true gneisses and crystalline schists, and the Azoic or semi-metamorphic strata "various coarse talcoid and chloritic schists, semi-porphyritic, arenaceous grits and conglomerates, and jaspery and plumbaginous slates," carrying veins and dikes of a metalliferous character.

The difficulty in at all times separating the Azoic from the overlying Paleozoic was recognized, and also the fact that it might at times grade into it without a break. Also was recognized the fact that a portion of his Lower Paleozoic was metamorphic or semi-metamorphic. These views are interesting when one considers how emphatically Rogers combated Emmons's Taconic system.

The belt of metamorphic and semi-metamorphic strata extending from Newfoundland to Alabama was designated the *Appalachian* or *Atlantic Belt* and the one extending westward from the north coast of the St. Lawrence Gulf to the Great Lakes the *Laurentide*, owing to the fact that the latter was well developed in the Laurentide Mountains.

The semi-metamorphic Azoic rocks were thought to be richer in minerals than the true gneisses, and to these were referred the schists of the Atlantic coasts, bearing lead, copper, zinc, and iron, and the auriferous quartz veins of California.

The Paleozoic formation proper, which was estimated to cover probably one-half of the area of the United States and to have a total thickness of from 30,000 to 35,000 feet, was divided into fifteen distinct series or sets of formations, "extending from the deposits which witnessed the very dawn of life upon the globe to those which saw the close of the long American Paleozoic day." The names assigned by him to these formations were regarded "significant of the different natural periods into which the day divides itself, from the earliest dawn to twilight." These were, beginning with the oldest, *Primal, Auroral, Matinal, Levant, Surgent, Scalent, pre-Meridian, Meridian, post-Meridian, Cadent, Vergent, Ponent, Vespertine, Umbral,* and *Seral*—signifying the Dawn, Daybreak, Morning, Sunrise, Ascending day, High morning, Forenoon, Noon, Afternoon, Waning day, Descending day, Sunset, Evening, Dusk, and Nightfall. These terms, based on time, he thought preferable to the inexpressive ones, mainly of a geographic character, then in vogue. It should be stated, however, that they were never generally adopted.

The Primal, Auroral, and Matinal were considered the approximate representatives of Sedgwick's Cambrian; the Levant, Surgent, Scalent, and pre-Meridian, near representatives of the English Silurian, beginning with the Upper Caradoc sandstones. He recognized both a physical and paleontological break in the succession of strata at the contact of these two great divisions. At the base of the Carboniferous he thought to recognize another break correspondingly sharp.

The entire Paleozoic system was thus divided into three natural divisions, as follows:

Division	Periods	Equivalent	Formations
Upper Paleozoic	Seral Umbral Vespertine	Carboniferous	Coal Measures. Middle Carboniferous. Lower Carboniferous.
Middle Paleozoic	Ponent Vergent Cadent	Later Devonian and Carboniferous.	Catskill. Chemung and Portage. Genesee, Hamilton, and Marcellus.
	Post-Meridian Meridian	Older Devonian	Upper Helderberg. Oriskany.
	Pre-Meridian Scalent Surgent Levant	Silurian	Lower Helderberg. Onondaga Salt Group and Niagara Limestone. Clinton. Medina.
Lower Paleozoic	Matinal Auroral Primal	Cambrian	Trenton Limestone, Utica and Hudson River slates. Black River, Chazy limestone, and Calciferous sandstone. Potsdam.

The sediments making up these Paleozoic strata were, according to his views, derived from a land area now occupied by the Tertiary seaboard plain and the waters of the Atlantic Ocean. This deposition was preceded by a wide movement of depression, which began in the south or southwest and permitted the waters of the Appalachian sea to occupy what is now the upper half of the southern Atlantic slope. There was, however, left above water to the southeast of the present Atlantic plain, a large tract of continent or great chain of islands which, with numerous fluctuations in their limits, remained as such down to the close of the Carboniferous period.

It was considered as "susceptible of demonstration that the various coal basins, bituminous and anthracitic, of Pennsylvania, Maryland, Ohio, Virginia, Kentucky, and Tennessee were originally united," "the whole as one great formation." This is essentially the view held today. The structure of the coal rendered "it apparent that no irregular dispersion of the vegetable matter by any conceivable mode of drifting . . . could produce the phenomena which they exhibit," and he could not conceive of "any state of the surface adapted to account for these appearances, but that in which the margin of the sea was occupied by vast marine savannahs of some peat-creating plant, growing half immersed on a perfectly horizontal plain, and this fringed and interspersed with forests of trees, shedding their leaves on the marsh." In this he agreed with Beaumont, though the evidence was of different character. He further believed that the coal material became finally engulfed through earthquake action, the sea receding and then returning once more laden with detritus, carrying everything before it and reaching far inland. Thus the entire marginal forest growth might be uprooted and as the sea again retreated the material would be spread broadcast and mixed with coarse rock detritus. When finally the earthquake paroxysms ceased and the sea became quiet, the fine silt in suspension was deposited, forming thus the soil for another growth.

The regular decrease in the amount of volatile bituminous matter in the coal as one passes from west to east was accredited to the action of the "prodigious quantity of intensely heated steam and gaseous matter emitted through the crust of the earth, by the almost infinite number of cracks and crevices which must have been produced during the undulation and permanent bending of the strata. The coal in the east basin would thus be effectually steamed, and raised in temperature in every part of the mass would discharge its bituminous matter in proportion to the energy of the disturbance."

Rogers, it should be noted, found in the American Carboniferous no recognizable base defined by organic remains alone, and in his classification relied mainly on the suddenness of the change from marine to terrestrial forms and the rapid appearance of that amazing vegetation characteristic of the coal period.

He conceived the Connecticut red sandstone to have been deposited on sloping shores within a narrow estuary, its greatest depth near its eastern margin, the material itself having been derived from the west; and he agreed with Emmons in regarding it of Jurassic and Oölitic age.

One of the most important chapters in the work related to types of orographic structure and the physical structure of the Appalachian chain, which was worked out in conjunction with his brother, Professor William B. Rogers. This formed the substance of a paper read before the Association of American Geologists at its third session in April, 1842, and already noted under that date (see p. 218). It is well to recall, however, that, under the title of *Dislocation of an Anticlinal Axis Plane—Uninverted Side of Wave Shoved Over the Inverted,* he described and illustrated the overthrust faults of the modern geologist, and remarked on their misleading character, owing to the resultant dipping of younger strata beneath those which are older.

The origin of valleys occupying the crests of anticlinal ridges, which he designated as "valleys of elevation," he rightly described as due to the carving force of waters. The mountains themselves were thought to have been elevated through and above the ocean by successive but sudden earthquake movements; the water, as a consequence, was propelled in gigantic billows and rushing sheets transversely across the anticlinal ridges, scooping them out where weakest into the form of terraced oval valleys.

Slaty cleavage was looked upon as due in part to movement, and in part to the action of heat waves traversing the rocks. In proof of this were cited his own observations to the effect that, in gneisses lying in an undisturbed horizontal position, the foliation is almost invariably coincident with the stratification, the heat producing it having flowed upward, invading stratum after stratum. Where granite occurs invading gneissic strata, the foliation is parallel to the plane of outflowing heat. He thought to discover two general laws: First, that the planes of foliation "are parallel to the waves of heat which have been transmitted through the strata," and second, "the foliation is parallel or approximately so to the cleavage, when both occur in the same rock mass."

A singular and unaccountable discrepancy in the work of Rogers lies in his almost completely ignoring the latest glacial views as advocated by Agassiz and others. The fact that the greater part of the work was completed and the manuscript prepared prior to 1848 can scarcely be considered a sufficient reason, since the years 1851-1855 were devoted to its revision and bringing it up to date. Moreover, his mention of the fact that the marine Pleistocene formation of Canada had been designated Laurentian clays by Desor (in 1850), shows that he was at least conversant with the literature of the period; the fact that he had considered, if not fully comprehended, Agassiz's views, is shown by a brief paragraph, in which he described and figured drift striæ seen on an exposed surface of Umbral sandstone on the south side of the Wyoming Valley.

His ideas regarding vein formation and the origin of quartzite were quite extraordinary. Thus, speculating on the wide distribution of the Potsdam sandstone, its remarkable uniformity and purely quartzose nature, he wrote:

> May we not conjecture that this . . . was supplied from the great dikes and veins of auriferous quartz, which . . . issued in a melted condition through the rents and fissures of the crust over all the region of the Atlantic slope? . . . Outgushing bodies of this quartz mingled with volcanic steam, and suddenly chilled and pelted upon by cold and heavy rains, may have been granulated into sand, as would occur with heated unannealed glass, and then washed in copious streams into broad, shallow, and tide-moved sea, and there gradually dispersed and precipitated.

That the sand grains were crystalline, and not glass seems to have been quite overlooked.

From the above, as from other of his writings, it will appear that Rogers was a catastrophist. Further than that, viewed in the light of today, many of the conclusions which he drew from observed phenomena and the theories advanced are strikingly absurd for a man of his learning. This is particularly true with reference to his ideas on the formation of anthracite (p. 376), the origin of valleys (p. 220) and mountain chains (p. 377), as well as that of the Potsdam sandstone just mentioned. Indeed, his entire work well illustrates the peculiar, uneven make-up (if one may be allowed the expression) of some of our best workers. "Great men, of great gifts you shall easily find, but symmetrical men, never."

The work was favorably reviewed in the *American Journal of Science* for November, 1859, almost the only unfavorable criticism relating to the deficiency of paleontological work and the geological nomenclature:

> The author has left this great department [of paleontology] of the survey to future workers. This being so, the author has hardly a broad enough basis for the institution of a new system of nomenclature and of subdivisions for the Paleozoic formations, and especially for diverging in these respects from the New York survey, in which the subdivisions had been founded upon a thorough study of the organic remains. The names of these subdivisions, Auroral, Matinal, Levant, Surgent, and so on, can not be proved to be better than those before adopted. They are founded on the idea of a Paleozoic day, which has had no existence except in the fancy of the writer. This unfortunate framework, about which Professor Rogers has clustered his facts, is no serious impediment to the geological reader who has a key at hand for comparison.
>
> The work is a great one, worthy of the State which authorized the survey. It contains a vast amount of information in all its departments, and will ever rank among the most important of the reports of the geology of the United States.

Lesley, the director of the second survey, notwithstanding sundry severe criticisms of Rogers's methods noted elsewhere, felt called upon to speak favorably of the work in its entirety. In his report to Governor Hartranft in 1874 he wrote:

> The accuracy of most of the statements made by the geologists, whose reports are condensed and consolidated in the final report of Prof. H. D. Rogers, published in 1858, is a matter of constant remark and admiration with the associates now in the field . . . their views were broad, their isolated observations numerous and exact, but their districts never were *accurately surveyed* by them, nor could be.

Amos H. Worthen's Work in Illinois, 1858-1875.

As previously noted (p. 301), J. G. Norwood, state geologist of Illinois, was succeeded in 1858 by Amos H. Worthen.[54] Under the latter's direction the survey lasted until 1875, when active field work was discontinued, owing apparently to an indisposition on the part of the legislature to provide the necessary funds. Six volumes of reports had been issued up to this time, and by the aid of subsequent special appropriations, two more were completed, the last bearing the date 1890, and on the title-page the name of Joshua Lindahl, state geologist, and Worthen as director. The total published results of this survey amounted to upward of 4,000 pages of text and 197 full-page plates of fossils.

Worthen was aided at various times by J. D. Whitney, Leo Lesquereux, Henry Engelmann, J. S. Newberry, F. B. Meek, H. C. Freeman, H. M. Bannister, H. A. Green, James Shaw, G. C. Broadhead, Orestes St. John, and E. T. Cox. The work of Whitney naturally related to the mining problems of the state, Lesquereux to the paleobotany, Meek to the invertebrate paleontology, Newberry to the vertebrate paleontology, and the others mentioned to general stratigraphy.

In the first volume Worthen divided the sub-Carboniferous into five groups: The Chester, St. Louis, Keokuk, Burlington, and Kinderhook, the term Chester being used in place of the Kaskaskia of Hall, and the St. Louis including the Warsaw of Hall. The name Kinderhook, it may be stated, was first proposed in 1861 by Meek and Worthen, and was intended to include all the beds from the base of the Burlington to the top of the black shale. The blue, green, and chocolate-colored shales immediately underlying the Kinderhook group in western and southern Illinois he considered Devonian.

Whitney, in this report, made the Galena limestone the "sole depository of lead in western Illinois," a view not quite in agreement with that expressed with reference to the Iowa and Wisconsin fields. Two maps were given, one geological of the northwest corner of the state, and the other showing a diagram of the lead-bearing crevices near Galena. The origin of these lead-bearing crevices "seems to be the same cause by which what are called joints by geologists have been formed in almost every variety of rock occurring in large homogeneous masses, and especially where a decided crystalline texture

[54] Worthen was "backed" for this position by Dana, Agassiz, and Hall. Edward Hitchcock, whose influence was asked, had, it seems, already endorsed a Mr. McChesney. The last-named figured quite prominently in the matter, but was evidently purely a politician.

PLATE XVIII

H. ENGELMAN

J. G. NORWOOD

A. H. WORTHEN

F. B. MEEK

LEO LESQUEREUX

exists in them." The course of the main set of fissures he thought might have been determined by the axis of upheaval, by which the whole region had been slowly elevated along the north boundary of the district, the metals themselves having been held in solution in oceanic waters and precipitated through the agency of decomposing organic matter. This is essentially the view held today, with this difference: Most geologists believe that the ore was originally disseminated throughout the limestone, and that it has become segregated in veins and pockets through the leaching action of meteoric waters.

Lesquereux, in connection with his paleobotanical work, took occasion to announce his adherence to the view that the coals were the result of the accumulation of sphagnous mosses, in place, *i.e.*, were not drift. He also introduced a chapter on the origin of prairies, in which he argued "that all the prairies of the Mississippi Valley were formed through the slow recession of sheets of water of varying extent, whereby the existing lakes were gradually transformed into swamps and bogs and ultimately into dry land. The black surface soil of the prairies he thought to be due to the growth and decomposition of bog vegetation, confervæ, etc." With this view Worthen, in the fifth volume of the reports, did not wholly agree. No one theory, he thought, was sufficient to explain all the phenomena noted, though the chief cause of the treelessness of the prairies he felt to be due to the character of the soil itself. The loess was regarded as of fluviatile origin. Concerning the origin of the drift as a whole, he wrote:

> Thus it will be seen that the first and greatest of the drift forces was the glacier; then the floating iceberg and ice field produced their results, carrying the large bowlders from place to place and dropping them over the ice-cold seas; and lastly, the wave and current force of water, after the ice had in part or altogether melted, leaving the loose clays, sands, and subsoils, substantially as we find them.

In his second report, published in 1866, Worthen adhered to the determination first published in the *Transactions of the Philadelphia Academy of Sciences*, 1865, to call by the name *Cincinnati group* the rocks included by Hall under the name of *Hudson River*. This volume was devoted wholly to paleontology, and contained articles by Newberry, Meek, Worthen, and Lesquereux. Newberry, in his work on the fossil fishes, accounted for the abundance of some of these forms in certain strata by a sudden introduction of "heated waters or noxious gases" in the Carboniferous seas where the forms lived.

The third volume, which appeared in 1868, was given up to a discussion of the geology of the various counties, with paleontology by Meek and Worthen, and the fourth volume, which appeared in 1870, was devoted quite largely to paleobotany by Lesquereux, vertebrate paleontology by Newberry and Worthen, and descriptive geology of the various counties by Worthen, Bannister, Bradley, and Green.

Sketch of Worthen.

Worthen was born in Vermont in 1813, and educated in the common schools and the local academy. In 1834 he emigrated to Kentucky, and in 1836 removed thence to Warsaw, Illinois, where he made his permanent home. Until 1855 he was engaged in mercantile pursuits, but devoted all his spare time to a study of the local geology, to which he was attracted by the abundant fossil remains for which the region was noted.

At the time he began his work satisfactory textbooks were few and the difficulties which he and others of his time encountered may be best understood when it is remembered that the work was undertaken more than sixty years ago, when railroads were practically unknown, when postage on a single letter cost twenty-five cents, and when, moreover, money was scarce and labor cheap. There being no overland freight or express lines, all his exchanges of specimens with friends in the East were made by means of Mississippi River steamboats between Warsaw and New Orleans, and sailing vessels between New Orleans and Boston. Often months would elapse between the time of his sendings and the return of exchange material.

In 1851 Worthen first began attendance upon the meetings of the American Association, and in 1853, as noted, took part in the survey of the state of Illinois, under the direction of J. G. Norwood. From 1855-1857 he was assistant to James Hall on the survey of Iowa, and in 1858 was made state geologist of Illinois in place of Norwood, as already mentioned.

Worthen's own labors related principally to the Carboniferous rocks, and to him belongs the credit of being the first to work out the true relations of the various subdivisions of the Lower Carboniferous system in this section.[55]

Personally, as we are told by his biographer, he was of manly

[55] "Worthen makes no pretensions to a thorough knowledge of chemistry (which department of the survey he has placed in the hands of a competent man at Chicago), but I do know there is no living man better acquainted with all the details of western geology than he is. That is, I mean, with the formations that occur in the western states. Hall got his classification of the subcarboniferous rocks, bodily from Worthen." Meek to Hayden, March 15, 1861.

presence, kindly, candid, and of unpretentious manner, impulsive and generous to his friends, charitable even to those with whom he had little sympathy, but uncompromising in his love of justice and scientific truth. His thorough interest in his work is shown by his persistent continuation of the same under the most adverse conditions. Again and again his work was in danger of suspension by the threatened failure of the necessary appropriations by the legislature, and more than once they were so far reduced that only the most careful management averted disaster. Once during 1875-1877 the appropriations were allowed to entirely fail, but he continued his work without compensation and with such sincerity of purpose that they were resumed by the next legislature.

Sketch of Lesquereux.

Lesquereux, whose paleobotanical work has been on several occasions described in these pages, and to whom we shall again have occasion to refer, was born at Fleurier, Switzerland, in 1806, and came to America with Arnold Guyot about 1848. His early scientific studies were on living plants; but later, and particularly after coming to America, he turned his attention almost exclusively to fossil forms. His work on the coal plants of Ohio, Pennsylvania, Illinois, and Arkansas began in 1850, his papers appearing in the survey reports of all these states. His work on the coal flora of Pennsylvania was particularly valuable, forming what was at the time the most important work on Carboniferous plants published in America. He became early connected with the Hayden surveys, and to the time of his death, in 1889, was actively engaged in the study of the materials collected by members of this organization. Deaf from early manhood, a foreigner, with but poor command of English, he labored under enormous disadvantages. To an interviewer he once remarked:

> The science student's life is absorbed with grave and serious truths; they are naturally serious men. My associations have been almost entirely of a scientific nature. My deafness cut me off from everything that lay outside of science. I have lived with nature, the rocks, the trees, the flowers. They know me. I know them. All outside are dead to me.

Hind's Work in the Winnipeg Country, 1858.

In April, 1858, Henry Youle Hind, of Toronto, was authorized by the provincial government of Canada to explore the region "lying to the west of Lake Winnipeg and Red River, and embraced (or nearly so) between the rivers Saskatchewan and Assiniboine, as far west as

South Branch House on the former river." He was directed further to procure all the information in his power respecting the geology, natural history, topography, and meteorology of the region. The work was accomplished between June 14 and October 31, the results being published in 1859 in form of a quarto volume of 201 pages, with a colored geological map, 2 plates of fossils, and other maps, figures, and sections. The region was described as occupied by Laurentian gneisses to the east of Lake Winnipeg, succeeded to the west by Silurian, Devonian, and Cretaceous formations. The Cretaceous fossils were described by F. B. Meek and the Devonian and Silurian forms by E. Billings.

Hall's Views on Sedimentation and Mountain Building, 1859.

Hall's principal contribution to strictly physical geology related to the accumulation of sediments and the formation of mountain chains. The first brief announcement of this was made in his reports on the geology of Iowa (p. 346). In 1857 he brought the matter before the public once more in his address before the American Association for the Advancement of Science, at Montreal. This address was, however, not printed at the time, and it was not until two years later that he formulated his views and gave them in extension in Part VI of the third volume of the *Natural History of New York*. In 1869, in an address on the Geological History of the North American Continent, delivered before the Albany Institute of New York, he reiterated many of the opinions previously announced, and, finally, in 1882 brought out the original address of 1857, and this, it is said, without revision. Whatever changes or additions it was found desirable to make were added in the form of supplementary notes. The subject may, therefore, be conveniently reviewed at this date (1859).

Hall had shown that one simple and intelligible sequence of strata from the Potsdam sandstone to the end of the Coal Measures covered, with slight exceptions, the entire country from the Atlantic slopes to the base of the Rocky Mountains, and that the sedimentary deposits of the entire period were much thicker in the East and for the most part poorer in calcareous matter than in the West; that an approximate measurement of all the strata along the Appalachian chain gave an aggregate thickness of 40,000 feet, while the same formations in the Mississippi Valley measured scarcely 4,000 feet. In short, there was certainly in the Appalachians more than ten times the thickness of the same series in the West. Still there were no mountains of this altitude; that is to say, there were no mountains the altitude of which equaled the vertical thickness of the strata composing them, and he affirmed that the line of greatest accumulation of sediments was along the direction of the Appalachian chain; that the chain was itself due to the original deposition of material and not to any subsequent action or influence breaking up and dislocating the strata of which it is composed.

Discussing the cause of the folding and plication, he referred to the fact previously recognized by Herschel to the effect that sea bottoms, when loaded by accumulated sediments, undergo a process of subsidence which may cause an elevation of the adjacent continental areas, a principle which was then becoming generally recognized and which has since become known under the name of *isostacy*. When, then, these sediments were spread along a belt of sea bottom, as they were originally in the line of the present Appalachian chain, the first effect would be to produce a yielding of the earth's crust beneath and a gradual subsidence. Proof of this subsidence was furnished by the great amount of material and the evidences of shallow water accumulation; it was impossible, he argued, to suppose that the sea had been originally as deep as the thickness of these beds, *i.e.*, 40,000 feet.[56]

The line of greatest depression would, therefore, be along the lines of greatest accumulation. By such a process of subsidence the lower side of the beds would become gradually curved and stretched, and there would follow, as a sequence, rents and fractures. On the surface above, which would be contracted horizontally by such subsidence, there would be produced wrinkles and foldings of the strata. Into the rifts formed below it was conceivable there might rush fluid or semifluid material, producing what are now evident as trap dikes.

The sinking of the mass would produce a great synclinal axis, and

56 The idea that the earth's crust, or indeed the entire sphere, possessed a certain amount of elasticity, that certain areas might become depressed through loading, or again elevated on relief from load, was by no means a new one. In notes made by Lewis Evans in 1755 accompanying a map of the Middle British Colonies of North America (ed. of 1776) there occurs the following: "It is easy to show the earth and sea *may* assume one another's places, but positively to assert *how that actually happened* in times past is hazardous; we know that an immense body of water is contained in the great lakes at the top of this country, and that this is dammed and held up by ridges of rocks. Let us suppose these ridges broken down by any natural accident, or that in a long course of years a passage may be worn through them, the space occupied by the water would be drained. This part of America, disburdened of such a load of waters, would of course rise, as the immediate effect of the shifting of the centre of gravity in the globe at once or by degrees, much or little, accordingly as the operation of such event had effect on that centre. The directly opposite part of the earth would, as part of the same effect sink and become depressed and liable to be

within this axis, whether on a large or small scale, would be produced numerous smaller synclinal and anticlinal axes. The greater amount of compression above or stretching below along the line of maximum thickness of the sediments would account for the gradual decline toward the margin of the major syncline or the evidences of fracture and distortion. This, he thought, afforded a partial explanation of the fact that mountain elevations in disturbed regions bear, in their altitude, a much smaller proportion to the actual thickness of the formation than do the hills in undisturbed regions; and further, that since in the formation of an anticline the beds are weakest at the ridge and become more liable to denudation, such are often worn down to form low ground or even deep valleys, while the synclinal arches, being protected in the downward curvings of the beds, may remain to form the prominent mountain crests, as is observable in the southern Appalachians. It nowhere seemed to him that folding or plication had contributed to the altitude of the mountains, but rather that the more extreme the plication the greater had been the general degradation of the mass wherever subjected to denuding agencies.

The chief elevation of the Appalachian chain, he argued, was continental and not of local origin, and the present mountain barriers to him were but the visible evidences of the deposits upon an ancient ocean bed, while the determining cause of their elevation existed long anterior to the production of the mountains themselves.

At no point nor along any line between the Appalachian and Rocky Mountains could the same forces have produced a mountain chain, because the materials of accumulation were insufficient, and though we may trace what appears to be the gradual subsiding influences of these forces, it is simply in these instances due to the paucity of the material upon which to exhibit its effects.

deluged without any apparent reason discoverable in those parts for such a change."

Incidental to this he pictures a result of the building of isthmian canals which has either been shown groundless or overlooked by modern engineers!

"We know from observation how much higher the Atlantic ocean is than the Pacific, and how it is piled up against the American coast on the western shore of the Gulf of Mexico, driven thither by the trade winds, and the attraction of the moon and sun. Let us suppose it possible that a passage might be forced through the Isthmus of Darien, or some other part of America between the tropics; these waters then would pour down from this height, and be discharged through this passage, instead of running back through the Gulf of Florida; the height of the Atlantic would be lower between the tropics, and the level of the Pacific ocean would rise; the centre of gravity of the earth would shift, and there would be few places on the earth but what would perceive the effect, although none would be able to conceive the cause that did not know the particular event of this passage being opened."

Referring to the amount of metamorphism which the rocks had undergone, Hall thought that we must look to some other agency than heat for the production of the phenomena, and that the "prime cause must have existed within the material itself; that the entire change was due to motion or fermentation and pressure aided by a moderate increase of temperature, producing chemical change." Just what is meant by this it is difficult to say, but, inasmuch as Hall seems to have been in consultation with T. Sterry Hunt, it is safe to assume that it was intended to include all possible causes which future investigation might show to have been operative.

These views seem not to have been favorably received at first by the American geologists, and were facetiously referred to by Dana as proposing a system of mountain making with mountains left out. To this Hall very justly replied that he had not intended to offer any new theory of elevation, nor to propound any principle as involved beyond what had been suggested by Babbage and Herschel. What he did intend to imply was that mountain formation was due to sedimentary accumulation and subsequent continental elevation. T. Sterry Hunt, coming to Hall's defense, argued that his views were largely in harmony with those previously maintained by masters in the science of geology, De Montlosier, as early as 1832, having declared that the great mountain chains of Europe were but the remains of continental elevations which had been cut away by denudation, the foldings and inversions in the structure of mountains being looked upon as local and accidental.

According to Meek, J. S. Newberry, then fresh from his work with the Ives and Macomb expeditions, took occasion to refer to the subject in a paper read before the American Association at Newport (1860). Unfortunately the paper seems never to have been printed, though Meek states that he completely demolished Hall, and that all the leading men in attendance sided with him.

To the views as outlined Hall apparently adhered to the last. To a letter of inquiry from G. L. Vose questioning why, if his theory were correct, the mountain folds should be steeper on one side than on the other, Hall replied under date of November 5, 1864:

The depression arising from accumulation will be the greatest in the center of the mass, or along the line of greatest thickness, thus probably together with a general contraction of the crust causing the plication. It appeared to me that the longest slopes would be toward the center of the great synclinal, while the outer slopes would be shorter and steeper from the general movement toward the center, this movement being in fact equivalent to a pressure from the northeast against the base of the synclinal.[57]

Notwithstanding this arrangement it may be found that I have invented an explanation for a condition which is far from universal, for I do not think that the difference in the slopes occurs with that regularity which we have been taught to believe nor do I think it possible, theoretically, that such regularity could exist when we consider the inequalities of the thickness and nature of the sediments, and sometimes the sudden changes noticed in the material of these sediments. . . . If I can sustain the great principle which I advocate, viz.—that mountains are not produced by upheaval, but by accumulation and continental elevation, I shall feel that I have done something to advance the science of geology in true principles. I feel quite sure of its ultimate adoption, because I *feel quite sure* that it is the only true explanation, the only mode of making mountain ranges, for they cannot be made without materials and no imaginary upheavals will ever explain their existence.

Seventeen years later, in response to a letter from Clarence King, he wrote:

In relation to my views of mountain-building, I do not know that my views have undergone any material change. I cannot believe that mountain ranges exist without, in the first place, *accumulation* of sediment—which must be along coast lines—or perhaps I should not take it for granted that you believe all our formations of sedimentary origin. But when our oldest Laurentian granites show that they were originally conglomerates, and that the pebbles of these conglomerates are *stratified* or *laminated* rock, it is going a long way back with sedimentation.

Then we know that these accumulations did not always take place in a deep sea, for in the semi-metamorphic rocks we have ripple marks, fucoids and mud cracks at various depths through 20,000 feet of thickness. I believe, therefore, that these great accumulations produced a depression of the crust. I do not intend to ignore contraction from cooling, but really this seemed to me to be recognized as an elementary principle, and it did not occur to me at the time to use the arguments. I believe you can have no great amount of depression without producing numerous minor foldings as you may see within every synclinal fold in a gneiss rock.

Mountains are not elevated as ranges of mountains, but as part of the continental movement. Erosion has taken place along the weakest lines, the anticlinals. The synclinals have been protected and remain as the final results.

I was led to this conclusion by a study of the New England ranges—the Adirondack (Laurentian) and the Appalachian as far south as Virginia—and a comparison with the thinning expansions of the same forma-

[57] Appendix, p. 688.

tions in the west. I have seen too little of the Rocky Mountains to be entitled to an opinion, but at a point known as Bear Mountain, I believe, forty miles west of Laramie, and where the snow remains till August, I found the mountain mass regularly stratified and dipping to the westward away from the eastern escarpment.

I can only say therefore that I have seen nothing to change in any material degree my general views of mountain building. Were I to review and re-write my views I might present some points more clearly, but the hoped for time has not yet come to me, and I must trust to you and to others to interpret with what limits you will the expression of views perhaps upon too limited an observation and acquired amid the more serious labor of tracing out and identifying by their fossil contents the widespread palaeozoic formations of the United States on the east of the Mississippi River.

In a letter to J. Dorman Steele, dated May 24, 1877, Hall stated his position as follows:

In my preface to Volume 3 is made for the first time, so far as I know, the proposition that the newer mountains were higher than the older; that a part of this was due to the rising mass being loaded with a greater accumulation of all the later formations. It is probably due also to the much longer period of erosion which had operated on the older ranges. Agassiz's reasoning on this point is fallacious because *mountain upheaval* is *continental elevation,* and mountain elevation depends, as I have clearly shown (Volume 3), on the thickness of the strata, for certainly mountains cannot stand on nothing. You may uplift, upheave, convulse and cataclysm to all eternity but you cannot make mountains without the thickness of strata. (See further p. 465.)

Engelmann's Work in Utah, 1859.

In 1859 Henry Engelmann of St. Louis accompanied, as geologist, an expedition commanded by Captain J. H. Simpson, organized for the purpose of opening new wagon routes for military purposes across the Great Basin of Utah. Engelmann's report, accompanied by one of Meek's on the invertebrate fossils, formed Appendices I and K of the general report of the expedition which, owing to the outbreak of the Civil War, was not published until 1876. Originally, as it would appear from the text, the manuscript was accompanied by a geological map and profile, though such do not seem to have been reproduced. No explanation is offered, but it is probable that developments during the long interval of delay were such as to make them of doubtful value, and they were suppressed. Engelmann's observations began with the country in the immediate vicinity of Leavenworth. He noted the presence of rocks of Carboniferous and Permo-Carboniferous ages along the Republican River, and of Cretaceous and Tertiary deposits farther west. In the district leading from the eastern foothills of the Rocky Mountains to the divide between the waters of the Atlantic and Pacific he thought to recognize rocks of Silurian and probably Devonian, Carboniferous, Permian, Triassic, Jurassic, Cretaceous, and Tertiary ages.

CHAPTER VI.

The Era of State Surveys, Fourth Decade, 1860-1869.

THE period of the Civil War naturally might be expected to be a period of uncertainty and inaction in matters relating to the sciences.

In all the seceding states work then in progress was brought abruptly to a close, and in several of them—as Missouri, North Carolina, and Texas—the records became so far lost or ruined through neglect as to make them well-nigh valueless for future reference. Throughout the North the results were less disastrous, though even there work was in some instances temporarily discontinued, owing to the failure of legislatures to make the necessary appropriations. In three instances, however—in California, Maine, and New Jersey—surveys were established in the very midst of the threatened disaster.

With the passing of these years of turmoil, active work was begun once more in states where it had been but temporarily suspended, and in others new organizations were authorized, as in Kansas in 1864, Iowa and North Carolina in 1866, and Indiana, Louisiana, Michigan, and Ohio in 1869. A futile and ill-considered attempt at a state survey of Nevada was also made in 1866, but, fortunately for the reputation both of the science and the individual, no one was found to undertake the work under the conditions proposed.[1] W. E. Logan continued his work as provincial geologist of Canada, and Murray and Richardson were at work in Newfoundland.

The literature of the decade was scarcely as voluminous as in either the preceding or the one following. By far the most important, when all is taken into consideration, was the first edition of Dana's *Manual*—a work of 798 pages, which at once took its place as the leading authority on matters pertaining to American geology, a place which it has continued to hold through all its four editions down to the last (1895). In addition to this may be mentioned

1 The sum of $6,000 is stated to have been appropriated, with the stipulation that the survey be completed in eight months. For details, see Bull. 109, U. S. Nat. Museum, p. 291.

Logan's summation of the geology of Canada (1863) and Cooke's of New Jersey (1868).

This was largely an era of new workers, or, at least, new leaders. Eaton was gone; D. D. Owen died at its very beginning; the elder Hitchcock had largely ceased his labors; while the Rogers brothers had both become absorbed in the work of teaching and administration, and no longer took an active part. Dana was undoubtedly the leading figure, while James Hall, by sheer physical energy, dominated in matters paleontological. Many new names and faces appear. Among them may be mentioned those of T. B. Brooks, Elkanah Billings, Robert Bell, E. T. Cox, E. D. Cope, F. V. Hayden, C. H. Hitchcock, T. S. Hunt, W. C. Kerr, W. E. Logan, Alexander Murray, Edward Orton, Raphael Pumpelly, Carl Rominger, N. S. Shaler, C. A. White, and Alexander Winchell.

As a whole, the decade was one of extension of knowledge of geographic boundaries rather than one of discovery or announcement of new principles. Of all subjects, that of glaciation received, perhaps, the most attention. The line of separation between this and the decade to follow must be quite arbitrary, since several important surveys were organized during 1869 and continued well on into the seventies.

R. Thomassy's Practical Geology of Louisiana, 1860.

At the beginning of this era there appeared R. Thomassy's *Geologie Pratique de la Louisiane*, a small quarto volume in French of 263 pages, with 6 plates. Why the word "pratique" should have been introduced into the title is not apparent, a large proportion of the work being given up to geographical and meteorological or physiographical matters, and the really geological portion limited to a description and discussion of the lower Mississippi, its delta, and attendant phenomena. Its appearance seems to have excited little interest, not even being noted in the *American Journal of Science*, and the few original ideas advanced were referred to by Hilgard and other subsequent writers only to show their erroneous nature.

Thomassy dwelt in considerable detail upon the absorption of the waters of the Mississippi by the porous terranes above New Orleans and their consequent diminution in volume seaward. The "mud lumps," so common in the lower reaches of the river, he thought due to mud springs, having their source at a somewhat higher level on the land and opening upward in the bed of the stream.

The name *mud lump,* it may be well to state, is applied to the numerous small islands lying near the main outlets of the passes of the river. They obviously were built up from the bottom, but their exact mode of formation has been open to question. E. W. Hilgard in 1871 showed with seeming conclusiveness that they were composed of clay laid down on the bottom in a condition of loose consolidation and later forced or squeezed out from under the crest of the gradually accumulating sands forming the bar.

North Island, or Petite Anse, as it is more commonly known, was considered by Thomassy to be of volcanic origin, and he thought to have discovered traces of the violent corrosive action of thermal waters and acids on the rock fragments which he conceived to have been ejected from the depths of the Gulf through explosive action. Richard Owen, it may be mentioned, while serving in the Federal Army and stationed at New Iberia, Louisiana, studied these same deposits cursorily, and came to the conclusion that the island was not volcanic. (See also Hilgard's views, p. 468.)

Little can be learned definitely regarding Thomassy. It is stated by Veatch[2] that in the latter part of 1862 or early in 1863 a Frenchman by this name undertook the manufacture of salt at Bistineau in the northern part of the state. He was a man of somewhat fastidious taste and regarded with disfavor by the local people. After he had leveled off a tract of land and was about to begin operations, a party of men arrived on this ground from Arkansas, and disregarding his rights and protestations, proceeded to dig wells for themselves, threatening that if he did not "dry up and leave the country" they would hang him. He seems to have heeded the warning, and nothing more was heard of him.

Winchell's Work in Michigan, 1859-1861.

In 1859, largely through the efforts of Alexander Winchell, then professor of physics and civil engineering in the state university, there was established for the second time a geological survey of Michigan. Of this Winchell was appointed director, making his first *Report of Progress*—an octavo volume of 339 pages—in 1861. In this he called attention to the futility of efforts then being made to produce salt in the vicinity of Grand Rapids, and fully anticipated the development of the same industry in the Saginaw Valley, his explorations enabling him to locate the salt beds at a depth of 650 feet. Attention was given, also, to the occurrence of gypsum, coal,

[2] Report on the Geology of Louisiana, Part VI, 1902.

iron, and other economic products and to the geographic distribution of the various formations throughout the state.

The so-called Marshall sandstone (Lower Carboniferous) he considered on paleontological grounds to be above the beds of the Hamilton group. He found evidence which led him to conclude that the Ohio and Michigan coal basins were never continuous, as had been heretofore asserted, and, indeed, that the geological column in the latter state had been built up quite distinct and independently from that in adjacent regions. He could find no parallelism between the Carboniferous limestones and those lying farther to the west, and all the evidence indicated to him that these deposits were laid down in an isolated basin cut off from that of Ohio to the south throughout the entire period from the Helderberg to that of the drift. In consequence of the outbreak of the Civil War no appropriations were made for the continuation of the survey after 1861.

Sketch of Winchell.

Winchell was born in the town of Northeast, Dutchess County, New York, in 1824, and graduated at Wesleyan University, Middletown, Connecticut, in 1847. His scientific tendencies are said to have manifested themselves at a very early age, although he showed no marked preference for any branch of study, unless it was mathematics, in which pursuit he seems to have been little short of precocious. Immediately after graduating he entered upon a career of teaching and lecturing, which kept him prominently before the public for over forty years.

His first public geological lectures were given at Pennington Seminary, Alabama, in 1849. In 1850 he assumed charge of an academy at Newbern, Alabama, but finding the conditions were not what he had been led to expect, he resigned, and in the spring of 1851 opened the Mesopotamia Female Seminary at Eutaw, in the same state. Finding, however, that he was illy adapted to the successful management of a southern female institution, he gave up this position to accept the presidency of the Masonic University at Selma. While engaged in the work of presenting the claims of this university before the people in the southern part of the state, he made extensive geological tours throughout the region and brought together large collections in natural history. In November, 1853, he was elected to the chair of physics and civil engineering in the University of Michigan, at Ann Arbor, and entered upon his duties in January, 1854.

In 1859 he was commissioned state geologist by Governor Moses

Wisner, as above noted, holding the position for two years, when the survey was abolished through failure on the part of the legislature to make proper provisions for its continuance. In 1869 the survey was reorganized and Winchell again appointed director. He assumed for himself the personal investigation of the Lower Peninsula.[3]

Owing to hostility to the management of the survey, aroused, it is stated, by personal enemies of Winchell, it appeared likely that the appropriations for carrying on the work would again fail. Hence he resigned the position in 1871.

In 1873 he resigned also his professorship at the university, accepting the chancelorship of Syracuse University, New York, which he held, however, only until June, 1875. In this latter year he was offered the professorship of geology, zoölogy, and botany in Vanderbilt University, at Nashville, Tennessee, but did not see fit to accept, although he did subsequently fill a three months' engagement there. In May, 1878, he took leave of this university also and the chair was abolished, owing to some foolish differences of opinion that had arisen between himself and Bishop McTyeire, who took exception to his stand in reference to pre-Adamites and evolution.

In June, 1879, Winchell was again called to Ann Arbor, being offered the chair of geology and paleontology in the State University. This position he accepted and continued to hold until the time of his death. In March, 1887, he was offered the position of state geologist of Arkansas, but refused.

Winchell's work as a geologist, as may be readily imagined from this sketch, was of a more or less intermittent nature, and it is as a teacher, public lecturer, and popular writer on scientific subjects that he is best known. The advanced stand which he took regarding evolution brought him in conflict with the religious element, particularly at Vanderbilt University, as already noted.

He was for a time connected with the Minnesota geological survey, and was one of the original promoters of the *American Geologist* and the Geological Society of America. As stated by his biographer:

> His largest educational field, however, was the public platform. Here he was under no constraint by reason of youthful auditors. No limits were set to his rhapsodic scientific eloquence; no courteous regard for the amenities of possible professorial etiquette hampered the free flow of his criticism, or the exultant prophecy of the betterments of the future. . . . Himself a working geologist in the field, he was well acquainted with

[3] Winchell's report on the geology of the Grand Traverse Region in 1871, seems to have been made independently of any organized survey.

geological methods. A teaching geologist in the university, he was skillful in imparting his own knowledge and in training others to habits of observation and investigation.

As a writer of books Winchell will be remembered for his *Sketches of Creation,* 1870; the *Doctrine of Evolution,* 1874; *Pre-Adamites,* 1880; *Sparks from a Geologist's Hammer,* 1881; *World Life,* 1883, and his textbook, *Elements of Geology,* 1886. Of these his work on World Life is undoubtedly the one showing the greatest amount of research and thought, and was at that time probably the only one in the English language covering in a systematic manner the entire field of world history.

Work of D. D. and Richard Owen in Indiana, 1859-1862.

In 1859 D. D. Owen, for the second time, accepted an appointment as state geologist of Indiana, but with the stipulation that, until his surveys in Arkansas and Kentucky, upon which he was then engaged, were completed, most of the field work should be performed by his brother Richard. Owing, however, to his death in 1860, the work fell wholly upon the brother, who succeeded him in the office. The results of the survey are comprised in an octavo volume of 364 pages, issued in 1862. Like other of this author's writings, it is prolix and uninteresting, differing in this respect in a marked degree from those of David Dale.

The importance of fossils was recognized, and he accepted it as an "unquestioned truth that a certain vertical range or ascertained thickness of fossiliferous rock is characterized by the organic remains of plants and animals, differing more or less from the plants and animals in the rocks above as well as those in the rocks below the given layers or strata." The rocks of the state were all classified under: (1) Lower Silurian; (2) Upper Silurian; (3) Devonian; (4) sub-Carboniferous; (5) Coal Measures, and (6) the Drift or Quaternary. Some twenty-five pages were given up to soil analyses by Robert Peter, and seventy to a report on the distribution of the geological strata in the Coal Measures of the state, by Leo Lesquereux. No maps or sections accompanied the report, which, as a whole, is singularly lacking in interesting or instructive matter. Joseph Lesley served as topographer.

With the outbreak of the Civil War Owen resigned to take command of a company of volunteers, and this, the third attempt at a geological survey of Indiana, came to an end.

Ideas of Rogers and Agassiz on Subsidence and Deposition, 1860.

Prevailing ideas regarding conditions of sedimentation have been frequently referred to in these pages, but with more particular reference to the position of the sea floor—whether inclined or horizontal—upon which the sedimentary beds were deposited. A new phase of the question was brought to light during a discussion between W. B. Rogers and Louis Agassiz at the meeting of the Boston Society of Natural History, March 21, 1860.

Rogers argued that the strata between Lake Ontario and the Pennsylvania coal region were deposited on gradually subsiding sea bottoms. On this supposition only could he account for the relative position of the beds and their very great thickness. Had they been formed as argued by some on a gradually rising bottom, the older deposits would crop out at the higher levels and the successively later ones at lower levels. Agassiz, on the other hand, maintained that there was no subsidence during the deposition of the New York strata, and that the facts indicated just the reverse, *i.e.*, an upheaval. During the upheaval, he argued, the level of the sea might be actually less, from the contraction of the earth while cooling, but in consequence of this contraction the ocean would always remain at a certain depth, sufficient for the deposition of the thousands of feet of strata. The study of the fossils he argued was also opposed to the theory of subsidence and denudation, since those of the Primary were never found carried into the Secondary beds.

At the April 4 meeting the discussion was resumed, when Rogers took the ground that an assumption of an upward movement of the sea bottom carried with it the admission of an original depth so enormous as to be incompatible with the accumulation of the material of the earlier strata, unless, indeed, the strata were supposed to be formed exclusively within a moderate distance of the shore.

> We might imagine a series of strata to be successively laid down in a gentle slope approximately parallel to that of the ancient sea bottom, each terminating against this surface without being continued into the profounder depths beyond, and we might suppose the floor to be rising in the region of this accumulation at such a rate as to bring successive tracts, farther and farther from the ancient shore line, within limits of depth admitting of mechanical and organic deposition; but in such circumstances of formation these earlier strata, instead of extending, as they are believed to do, almost continuously over the whole ocean floor, would be seen to terminate at no great distance from the original shore line, but

abutting against the bottom at the places where the depth had set a limit to their accumulation.

No hypothesis of a secular rising of the sea bottom, he therefore argued, could explain the formation of the Appalachian Paleozoic deposit. They indicated, rather, a long period of subsidence of the ocean floor, varied by many and long pauses of upward oscillations.

Agassiz, in reply, admitted the probable shallowness of the ocean in which the strata were deposited, and also that during a local upheaval of the shore the whole sea bottom was probably subsiding, the subsidence being due to shrinkage, caused by the cooling of the earth's crust. This view was accepted by Rogers as amounting to a virtual disclaiming of the theory as first advanced.

C. A. White's Views on Geological Equivalency, 1860.

It will be remembered that in 1858 Hall, in his Iowa report had commented on the gradual transition of the Devonian into the Carboniferous beds. At the June, 1860, meeting of the Boston Society of Natural History Dr. C. A. White, later state geologist, made his first contribution in a paper entitled "Observations upon the Geology and Paleontology of Burlington, Iowa, and Its Vicinity." In this he referred to the "imaginary line" separating the Devonian from the Carboniferous rocks, "a line where Devonian species ceased only to predominate, and upward from which the Carboniferous species flourished in full force." While not proposing to discuss the question of the actual equivalency of these rocks with those of New York, he wrote:

> Admitting that some of the species found in the lower beds have been identified with those of the Chemung group of New York, it settles beyond question their geological equivalency, but does not necessarily prove that they were contemporaneous. Indeed it seems probable that they were not so, by an interval of time that it would take the species to migrate that distance. May it not therefore be inferred that the species originated at the east, and were migrating westward during the time that the bottom of the Chemung seas was sinking and receiving upon it the deposits of the Old Red sandstone—thus making these Devonian rocks equivalent to the Chemung of New York and contemporaneous, at least in part, with the Old Red of the Catskill mountains?

In the discussion which followed, W. B. Rogers, after calling attention to a similar difficulty in separating the Devonian from the Carboniferous experienced by foreign geologists, remarked to the effect that such a gradual change or commingling of races was but a natural result of accumulation of strata during a long period of comparative repose, and that the abrupt changes often found were the memorials of the disturbing and destroying agencies to which its living races have been successively exposed. He argued, further, that all attempts at synchronizing distant deposits must be limited by vague results. Even when the correspondence of fossils would seem to mark a simultaneous origin, the possibility of migration and the long lapse of years which may have been required for the extension of a living race into distant parts should not be lost sight of. "We cannot accept," he said, "as an absolute law in paleontology the principle that like organic types are excluded from reappearing in a given region after having once, and for a long time disappeared." The doctrine of colonies, he thought, was a striking instance of the departure from such a law.[4]

On the other hand, it was argued by Agassiz, two years later, that there may be synchronism of deposits without identity of fossils. This was shown by the present distribution of animal life. "If at the present epoch the fauna of America and Australia should become fossilized, there would not be the slightest resemblance between the representative species of the two continents. The paleontologist must be ready to admit that very different fossil fauna may be contemporaneous, and that their difference does not necessarily imply a distinct zoölogical age."

Early Views on Petroleum and Natural Gas.

Although the occurrence of petroleum, or coal oil, and natural gas had been known for many years and oil had been obtained from springs and by distillation from various natural bituminous products, it was not until the completion of the drilling of the Drake well at Titusville in Pennsylvania, in 1859, that the possibility of obtaining a commercial supply of oil direct from wells began to be appreciated, and geologists to turn their attention to questions relating to its origin and occurrence. Among the earlier writers on these subjects only T. Sterry Hunt, E. B. Andrews, and Alexander Winchell need be here referred to.[5]

[4] The "doctrine of colonies" was put forward by the Bohemian paleontologist, Barrande, to account for the occasional presence of a later fauna in the midst of an older one, as of Silurian forms in Ordovician rocks. Those desiring will find this theory well elaborated in the introduction of Nicholson and Lydekker's Manual of Paleontology, pp. 60-62.

[5] For a detailed history see Early and Late History of Petroleum by J. T. Henry, Philadelphia, 1873.

Hunt, writing in the *Canadian Naturalist* for August, 1861, and with reference to the Canadian oil, said:

> These wells occur along the line of a low, broad anticlinal axis which runs nearly east and west through the eastern peninsula of Canada, and bring to the surface in Enniskillen the shales and limestones of the Hamilton group which are there covered with a few feet of clay. The oil doubtless rises from the Corniferous limestone which . . . contains petroleum; this being lighter than water, which permeates at the same time the porous strata, rises to the higher portion of the formation which is the crest of the anticlinal axis where the petroleum of considerable area accumulates and slowly finds its way to the surface through vertical fissures in the overlying Hamilton shales, giving rise to the oil springs of this region.

This, so far as the writer has information, is the first enunciation of the anticlinal reservoir theory, and taken in connection with Hunt's other writings, as those relating to the origin of the bituminous matter in the dolomite underlying the city of Chicago, would, if accompanied by sufficient field observations, have placed him in the front rank of petroleum authorities.

In the December number of the *American Journal of Science* for the same year E. B. Andrews, of Marietta College, gave the results of his observations on the geological relations and distribution of oil in the Ohio fields, and likewise showed that the oil and gas had accumulated in quantity only in regions where the strata had been disturbed and fractured, as along the crests of anticlinals. The chief point of difference in the views of these two writers would appear to have been that Andrews considered the crevices in the shattered rock to form the reservoirs, while Hunt *apparently* regarded the oil as diffused throughout a porous sandstone.

Neither of these papers seems to have attracted the attention deserved, and little more of importance appeared in the literature until Alexander Winchell in 1865, writing with particular reference to the oil field of Michigan, called attention to the equally essential need of an impervious covering for the retention of the oil.

> Wherever the oil producing shales are exposed to the air or are covered by a porous medium, the product of the slow spontaneous distillation going on escapes into the atmosphere and is lost. Where the shales are covered by an impervious layer . . . the oil and gas elaborated are retained in the rocks, filling cavities by driving the water out by elastic pressure and saturating porous strata embraced in the formation.

Later, in 1873, this branch of the subject was discussed by Newberry and by Lesley and his assistants in connection with the surveys of Ohio and Pennsylvania. It remained, however, for Edward Orton and I. C. White to develop a practical working hypothesis relative to the origin and natural storage conditions of both petroleum and gas, but this at a date so recent as to be quite beyond the limits of this review.

Work of Moore and Buckley in Texas, 1860-1867.

The removal of B. F. Shumard from the position of state geologist of Texas, in 1860, as noted on p. 372, was followed by the immediate appointment of his successor, Dr. Francis Moore, "an honorable and cultured gentleman of much executive ability," a one-time newspaper editor, but, so far as can be learned, never a geologist.[6] In April, 1861, the survey was, however, suspended by an act of the legislature, and both Shumard and Moore requested to report on what had been accomplished. With this request neither party complied, Shumard going immediately to St. Louis, where he died in 1869, and Moore leaving for the North on the outbreak of the Civil War, where he, too, died in 1864.

The notes of these short-lived organizations seem to have passed into the hands of S. B. Buckley, a botanical collector under the Shumard régime, who returned to Austin after the close of the war and succeeded in having himself appointed state geologist. In this capacity he issued one report, bearing date of 1866 and comprising some eighty octavo pages, dealing with generalities and matters pertaining to the agricultural resources of the state, but containing little or nothing of geological value and but few references to what had been done by his predecessors. Buckley's period of rule was short, coming to an end in 1867.

No attempt at resuscitation was made until 1870, when the governor, under authority from the legislature, appointed John W. Glenn state geologist. Glenn, however, found the position uncongenial and resigned the following year, to be succeeded in 1874 by Buckley once more. The latter held the office but two years, when the

[6] Moore, as I am informed by Professor Wm. B. Phillips, was editor of the *Houston Telegraph.* In 1840 he had published a small volume entitled Map and Description of Texas, Containing Sketches of its History, Geology, Geography and Statistics. This was printed by H. Tanner, Jr., of Philadelphia, and Tanner & Sturnell of New York. Shumard, in a letter to F. B. Meek, complained bitterly of his treatment, claiming that he had been superseded while absent in the field, that no intimation of a change had been given him, and that he was not even allowed the privilege of making out a report covering his two years of service.

governor, becoming convinced (truthfully, it is to be feared) that the survey promised to be of no practical benefit to the state, vetoed the bill appropriating moneys for its support.

During this second term of office Buckley issued two reports, under dates of 1874 and 1876, respectively, comprising altogether some 220 pages. The first contained a brief history of the past surveys of the state and a somewhat partisan account of an unhappy disagreement between Shumard and himself. Both the statements and the manner in which they are presented are such as to prejudice the unbiased reader against Buckley, and this in particular when one reflects that there was no apparent occasion for the publication in an official report of matters of this nature. Buckley was replied to vigorously by A. R. Roessler, who had also been one of Shumard's assistants.

The main portion of both reports was given up to a discussion of economic problems, and little that was new geologically appeared. He noted the eruptive nature of Pilot Knob, but was unable to determine whether the period of eruption antedated the Cretaceous or not. Much space was devoted to a discussion of soils and crops, together with notes on the fauna and flora.

Hitchcock's Views on the Metamorphism of Conglomerates, 1861.

In his report on the geology of Massachusetts (1833), and again in 1835 and 1841, Edward Hitchcock had called attention to the flattening and distortion of pebbles in a conglomerate near Newport, Rhode Island, as already noted. As time went on the importance of the subject, particularly in the light of subsequent discoveries, seems to have grown upon him, and in a paper in the *American Journal of Science* for 1861, and in his reports on the geology of Vermont for that same year, the ground was gone over in great detail and his gradually expanding views were fully elaborated.

In his report of 1833 the conglomerates were described as composed of elongated rounded nodules of quartz rock passing into mica-slate, with a cement of talcose slate, the quartz pebbles with their longest diameters uniformly parallel, the entire mass, pebbles and all, being divided by fissures as perfect as if "cut through by the sword of some Titan." Later, while engaged in the work of the Vermont survey, he found analogous conglomerates along nearly the whole western side of the Green Mountains, and in 1859, in company with his son, C. H. Hitchcock, he again visited the Newport locality in search of facts to aid him in solving the Vermont problem. His results were given before the American Association for the Advancement of Science that same year, and the subject again brought before the Association by C. H. Hitchcock in 1860.

The facts brought out and presented showed the pebbles to be (1) much elongated in the direction of the strike of the beds; (2) that, while flattened, this feature was not so striking as their elongation; (3) that they were often indented by one being pressed into another; (4) that they were often bent, sometimes in two directions; and (5) that they were cut across by parallel joints, or fissures, at intervals of from a few inches to many feet. These facts led him to conclude that (1) the rock had once been a conglomerate of the usual character, and had undergone metamorphism whereby the cementing material had become crystalline and schistose and the pebbles elongated, and (2) the pebbles themselves had been in a more or less plastic state; otherwise any attempt at change in form would result only in fracture and comminution. The clean manner in which the pebbles were cut by joints was thought also to indicate a condition of plasticity. The flattening and distorting force was thought to have operated laterally, and the jointing to be due to some polarizing force acting upon soft materials, a simple inspection of the rock in place being sufficient to satisfy anyone that no mechanical agency would alone be sufficient to produce the phenomena. Applying the same method of reasoning to the conglomerates of Vermont, he came to the conclusion that these, too, had undergone metamorphism, giving rise to schists and gneisses and that, indeed, granites and syenites might result from the metamorphism of stratified rocks. The chemical details as worked out would naturally not hold in their entirety today, nor would the idea of a polarizing force be longer seriously considered; but the fact remains that this paper, as a whole, marked a long stride in advance along the line of metamorphism, and for its time was comparable in its importance with the later work of Lehmann on the crystalline schists of Saxony.[7]

[7] B. S. Lyman brought the subject up for discussion before the Association at the Buffalo meeting in 1866 and contended that the apparent flattening was due to the schistosity of the rocks from which the pebbles were derived and not to plasticity.

The subject was in 1868 again taken up by Professor G. L. Vose, whose conclusions, based on a study of conglomerates in the vicinity of the Rangeley Lakes, were largely confirmatory of those of the Hitchcocks. He was not, however, disposed to regard the pebbles as having been in a plastic state, but rightly contended that under the forces that had there prevailed rigid pebbles might be bent and flattened in the manner described.

Natural History Survey of Maine, 1861-1862.

By an act of the legislature approved March 16, 1861, Ezekiel Holmes, of Winthrop, Maine, was commissioned, under the direction of the board of agriculture, to make a scientific survey of the state. C. H. Hitchcock, then of Amherst, Massachusetts, was commissioned geologist, while George L. Goodale served as botanist and chemist, C. Houghton as mineralogist, A. S. Packard, Jr., as entomologist, and C. B. Fuller as marine zoölogist. Others, including G. L. Vose and Dr. N. T. True, rendered assistance at various times.

Fig. 75. *C. H. Hitchcock.*

Two brief seasons were devoted to field work, the results of which appeared in the reports of the board of agriculture for 1861-1862, the strictly geological portions being limited to some six hundred and odd pages. The work done under these conditions was necessarily somewhat disconnected. Northern Maine was still practically a wilderness; there were no maps and absolutely no railroads. A geological map in black and white was, however, prepared of the region north of Calais and the forty-fifth parallel, and also one of the eastern portion of the state included between St. George and Belfast.

The fossiliferous rocks were rightly considered to be mostly Paleozoic and probably all lying below the Carboniferous series. The red sandstone of Perry, the problematic horizon of which had been discussed by others, as previously noted in these pages, was "indisputably Devonian," the statement being based upon expressed opinions of Jackson, Rogers, Newberry, and Dawson. It was noted that the fossiliferous marine clays, which were supposed to be of the same age as the similar deposits along the St. Lawrence and the Champlain valleys and referable to the Terrace period, sometimes underlay a coarse deposit referable to the unmodified drift. Without committing himself definitely on this point, Hitchcock suggested the possibility, therefore, of a recurrence of the drift agencies—that is, a period of second drift. This, so far as the present writer has information, is the second suggestion by an American of such a possibility. It seems, however, like the first, to have been quite overlooked (see p. 488).

The quartz rock in the vicinity of Rockland, Thomaston, and Camden was relegated to the Taconic period of Emmons, and the associated limestones were looked upon as contemporaneous with the Stockbridge limestones of Emmons, or, what is the same thing, the Æolian limestones of the elder Hitchcock.

J. P. Lesley on Mountain Structure, 1862.

In May, 1862, J. P. Lesley read a paper before the American Philosophical Society describing the structure of the Alleghany Mountains. In this he assumed that the rocks of the Blue Ridge Range, on the eastern side of the valley, were a prolongation of the Green Mountains of Vermont, and consisted, therefore, of the Quebec group, or Taconic System. In this he followed the Rogers brothers, as he acknowledged. He accounted for the change in the drainage of the New River, which breaks into the Appalachians, by assuming a structural change in the geology, most of the mountain valleys north of this being unbroken anticlines and synclines, while those south of it were monoclines bounded by immense faults or downthrows. The Appalachians of southern Virginia and eastern Tennessee as viewed by him are grouped in pairs by faults, the fracturing being in parallel strips from five to six miles wide, each strip being tilted easterly so that the upper edge of one strip, with its Carboniferous rocks, abuts against the bottom or Lower Silurian edge of the strip next to it. The Paleozoic zone, therefore, included between the Great Valley and the backbone escarpment, is occupied by as many pairs of parallel mountains as there are parallel faults, and as these lie in straight lines at nearly equal distances from one another the mountain ranges run with great uniformity side by side for 100 or 200 miles until cut off by cross faults, or by a change in the courses of the principal faults.

J. P. Lesley on Uplift and Erosion, 1863.

In a paper before the American Philosophical Society the year following on "A Remarkable Coal Mine or Asphalt Vein in the Coal Measures of Wood County, (West) Virginia," Lesley was led into a discussion of uplift and erosion, and expressed sundry other ideas worthy of note.

It will be remembered that Rogers held the opinion that the great amount of erosion which had manifestly taken place in the Pennsylvania Appalachians was cataclysmic—the result of a great rush of waters over the face of the continent at the time of the upheaval.

This view Lesley was at first inclined to accept, and even as late as 1864 confessed himself as not entirely convinced that such a cataclysm was not necessary in order "to explain the earlier and perhaps the larger part of the whole phenomena." Nevertheless, in assuming that the valley found cutting across the above-mentioned vein had been carved out since the vein filling took place, he found reason for rejecting both the idea of cataclysmic erosion dating from the time of the uplift, and that of the secularists who regard the present face of the country as but the latest phase of an infinite series of oceanic degradations still in progress, neither would he accept Rogers's view or theory of a "pulsating planetary lava nucleus" to account for the phenomena. He adopted rather the idea of a lateral thrust caused by shrinkage of a cooling globe and that of a succession of denuding actions of unknown force and indefinite number. "A homogeneous element with sufficient force, acting either by one or by repeated blows," would bring about the present condition of affairs.

No one will deny that water, if obtained in sufficient quantity at a sufficient velocity, would be such an agent. In the acknowledged instability of the crust of the earth, in its acknowledged less stability in ancient times than now, we find the possibility, nay, we feel the certainty, that the oceans have at times been launched across the continents, and we need nothing more to satisfy all the conditions for an explanation of Appalachian topography.

One great obstacle, he thought, in the way of advance in topographical science among geologists "has been an innocent ignorance of the titanic postulates upon the ground, and therefore an inability to reconstruct in imagination the awful vaults of rock which have been removed from over at least 50,000 square miles of the surface of the United States merely along the one belt of the Appalachian Mountains, between the coal area and the Blue Ridge range."

However this may have been, Lesley himself could not be accused of any such "innocent ignorance" and consequent impotency of imagination. This was shown also when speculating on the character of the folds of the Appalachians, as controlled by the roughness of the old surface of the more or less disturbed and eroded sedimentaries and the thinness of the newer formations, whereby there was a tendency for more or less hitch and catch below and crack and shove above. "As a whole, the plicating energy must have acted with a steady evenness of thrust, which carried up the anticlinal waves of the crust unbroken, and in some cases to a height of between 5 and 10 miles above the present surface level."

Truly a strange admixture of views. As a modern catastrophist he was equaled only by Clarence King.

Whitney's Geological Survey of California, 1860-1869.

In 1860 J. D. Whitney, to whose work frequent reference has been made, was appointed state geologist of California, in which capacity he served until 1874, when the survey was discontinued. According to the original act, he was required to make an accurate and complete geological survey of the state and to furnish a report containing a full scientific description of its rocks, fossils, soils, and minerals, and of its botanical and zoölogical productions. W. H. Brewer was principal assistant in botany and agricultural geology, William Ashburner in mining, and W. M. Gabb in paleontology. Clarence King rendered assistance in a volunteer capacity.

Notwithstanding the apparently favorable conditions under which the survey was organized, its life was short and the results far from what both workers and the public had a right to hope. It early became evident that there was a lack of sympathy between the legislature and the director. This is in part accounted for below, and is made evident in the annual report and public lectures delivered before the assembly. Failure on the part of this body to pass the necessary appropriation bill in 1869 was overcome through the effort of J. D. Whitney, Sr., who advanced the funds for the season's expenses. He was subsequently reimbursed, but the act of March 25, 1870, appropriating money for but two years' work in completing the survey and publishing the results brought all to an end, though some of the data obtained was worked up and published by Whitney, later (see p. 492).

The reports of this survey are embodied in three volumes on geology and paleontology published by the state and two volumes on the auriferous gravels, published by the Museum of Comparative Zoölogy at Harvard after Whitney's retirement. It was announced in the statement of progress for 1872-1873 that a geological map of the whole state had been colored, but it seems never to have been issued.

The first volume or report of progress and synopsis of field work from 1860 to 1864 appeared in 1865. This was a quarto of 498 pages. It contained a great amount of descriptive matter relating to the areal geology of various parts of the state, particularly of the Coast Ranges and Sierra Nevada, with a chapter on the mining regions.

Whitney concluded, from the discovery of a single shell in the rocks of Alcatraz Island, that the so-called San Francisco sandstone was of undoubted Cretaceous age. The serpentines of Mount Diablo and the San Francisco peninsula he considered metamorphic sediments—sandstone—a mistake which was later repeated by Becker in his volume on the quicksilver deposits of the Pacific slope.

He was decidedly pessimistic regarding the probability of the occurrence of petroleum on the Pacific coast, and unhesitatingly discouraged the promoting of enterprises of this nature. With reference to the region south of the Bay of Monterey, he wrote:

> As the bituminous shales are everywhere turned up on edge and have no cover of impervious rock, the inference is unavoidable that flowing wells or at least those delivering any considerable quantity of liquid petroleum, can not be expected to be got by boring to any depth. The probabilities, at least, are against it.

When one reflects that the output of the California fields in 1911 amounted to upward of 81,000,000 barrels, he is led to question the infallibility of human judgment in these matters. Such an error was, however, pardonable; first, in that he based his opinion largely upon the reports of assistants and, second, in that the presence of asphaltum was regarded, at the time, as precluding the existence of petroleum. Antagonism from promoters whose efforts at selling stock were checked by these reports undoubtedly influenced the legislature in refusing appropriations. "Petroleum is what killed us. By the word 'petroleum' understand the desire to sell worthless property for large sums, and the impolicy of having anybody around to interfere with the little game."[8]

All those chains or ranges of mountains in California were considered to belong to the Coast Ranges which had been uplifted since the deposition of the Cretaceous beds; those which were elevated before the Cretaceous to belong to the Sierra Nevada. The slates of the western slope in Mariposa County were identified as of Jurassic age, and the Calcareous slates of Plumas County as belonging to the Triassic. The limestones in the Gray Mountains had been referred by Trask to the Carboniferous. With this Whitney agreed.

The peculiar dome-shaped concentric structure of the granite in the Sierras was dwelt upon with considerable detail, and the curved structure of the sheets was thought to have been produced by the contraction of the material while cooling or solidifying. The Yosemite Valley itself he thought due to a differential movement, the half dome seeming beyond a doubt to have been split asunder in the

[8] See further, Bull. 109, U. S. Nat. Museum.

Chester Averill, William More Gabb, William Ashburner, James Dwight Whitney.
Charles F. Hoffman, Clarence King, William Henry Brewer.

MEMBERS OF THE GEOLOGICAL SURVEY OF CALIFORNIA, 1860

middle and one-half gone down in what he called "the wreck of matter and the crash of worlds." In other words, he considered the valley due to the downward drop of an enormous fault block. Later investigators have been inclined to regard the valley as due merely to river erosion, facilitated by the vertical jointing of the rocks.[9]

The first volume of the paleontological reports appeared in 1864. This comprised 243 pages, with 32 full-page plates of fossils, the Carboniferous and Jurassic being described by F. B. Meek, and the Cretaceous and Triassic by W. M. Gabb. Concerning the work thus far done, Whitney wrote:[10]

> Perhaps the most striking result of the survey is the proof we have obtained of the immense development of rocks, equivalent in age to the upper Trias of the Alps, and paleontologically closely allied to the limestones of Hallstadt and Aussee, and the St. Cassian beds, that extremely important and highly fossilliferous division of the Alpine Trias.

And further on:

> Enough [fossils], however, have been found to justify the assertion that the sedimentary portion of the great metalliferous belt of the Pacific coast of North America is chiefly made up of rocks of Jurassic and Triassic age. . . . While we are fully justified in saying that *a large portion of the auriferous rocks of California consist of metamorphic Triassic and Jurassic strata,* we have not a particle of evidence to uphold the theory . . . that all or even a portion are older than the Carboniferous. . . . We are able to state . . . that this metal [gold] occurs in no inconsiderable quantity in metamorphic rocks belonging as high up in series as the Cretaceous (p. 261).

Subsequent to this, apparently,[11] W. P. Blake, geologist of the California State Board of Agriculture, made a claim to having, in 1863, found on the American River a fossil ammonite, which he thought established the Secondary age of the gold-bearing strata. It was claimed by Brewer, however, that Blake did not know for a certainty if the specimen were found in place or, indeed, if it were an ammonite or ceratite; also that the fossils were not found so early as the date claimed by Blake (1863). Brewer is very emphatic in his statements to the effect that Whitney was the first to announce the Jura-Trias age of these rocks, and the published reports seem to bear him out.[12]

[9] See H. W. Turner, Proc. Calif. Acad. of Sci., Geology, I, No. 9, 1900.
[10] Amer. Jour. of Sci., XXXVIII, November, 1864, pp. 256-264.
[11] See Amer. Jour. of Sci., XLII, 1886, p. 114.
[12] Appendix, p. 689.

The second volume of the paleontological reports, published in 1869, comprised 299 pages, with 36 full-page plates, and was given up wholly to descriptions of Tertiary and Cretaceous fossils by Gabb. In his introductory note Whitney reiterated his statement above quoted concerning the age of the gold-bearing rocks and the absence of rocks older than Carboniferous, not merely in the state, but west of the one hundred and sixteenth meridian.

The topographic work of Hoffman, in connection with this survey, has been claimed verbally, by Brewer, to have been considerably in advance of any heretofore undertaken. He introduced a system of rough triangulation well adapted to the needs of the survey but his contours were still indicated by hachures. This point seems worthy of note in connection with the discussions which arose later over the consolidation of the various western governmental surveys.

As to the exact cause of the failure of the survey under Whitney, opinions may differ. It was a by no means unusual fate in the history of state surveys and reasons of one kind and another are easy to find. Taken all in all that given by Rossiter Raymond[13] seems in the light of present knowledge to best fit the case.

It happened that, when the question of a further appropriation was pending, the only report which had been issued by the Survey (Whitney's) was a volume on paleontology; and an opponent of the appropriation carried the House with him by simply reading random extracts from that dry and technical treatise, as samples of the character of the work which had been done at the public expense up to that time. The appropriation was refused; and the valuable work of Bowman and others, on the old-river channels of California and their gold-bearing gravels and cements, was thereby barred from publication for several years. For this result, Prof. J. D. Whitney, the distinguished head of the Survey, has often been blamed, on the ground that he expended money and time in a preliminary topographical and geological survey, without attacking problems of immediate industrial interest. Personally, I think there is some foundation for this criticism. Prof. Whitney, with a lofty and serene regard for the logical sequence of science, and an equal disregard for the clamor of industrial interests, had begun his work with the topographical reconnaissance necessary as a basis for accurate geological deductions and correlations; and, in the course of this preliminary labor, his field-parties had made incidentally many interesting palæontological observations, undoubtedly significant in their bearing upon the geology of the State. Prof. Whitney had also started investigations of more immediate and evident practical importance; but unfortunately, in his plan of a permanent and monumental scientific achievement, these were not of

[13] In a footnote to a Biographical Sketch of J. D. Hague, Bull. No. 26, Am. Inst. Min. Engs., 1909, p. 113.

prime importance, and were advancing slowly. Probably the thought never occurred to him that it would make any difference what he published first, as the fruit of his work for the State; and thus he made the profound mistake in policy of issuing, merely because it was ready, a learned book on palæontology for the benefit of a limited outside public of specialists, and to the profound dissatisfaction of the people who had paid him, and were, reasonably or unreasonably, expecting something else for their money.

Whitney was a strong, aggressive man of striking personality, and unquestionably the best-trained man of his time in his particular calling. His work in the Lake Superior region, and the upper Mississippi Valley, together with his *Metallic Wealth of the United States*, were powerful stimulants to truly scientific methods of the study of ore deposits and did much to put the calling of the mining geologist on a higher and legitimate plane. He was outspoken and healthfully harsh in his criticisms of poor work and as a writer of vigorous English equaled only by Lesley.

Logan's Geology of Canada, 1863.

As previously noted (p. 237), Logan, during his period of service as provincial geologist, submitted sixteen reports dealing mainly with stratigraphic and economic subjects. In 1863 he brought forward a long promised volume—*The Geology of Canada*—a large octavo of 983 pages, accompanied by an atlas containing a colored geological map and sections.

Logan was assisted during the early part of his work by Alexander Murray and James Richardson, geologists; Elkanah Billings, paleontologist, and T. Sterry Hunt, chemist. Later Robert Bell, Edward Hartley, Thomas Macfarlane, Charles Robb, H. G. Vennor, and others of less prominence, were added to the force. A portion of the paleontology was assigned to James Hall, and the nomenclature adopted was essentially that of the New York survey.

The volume, as would naturally be expected, comprised detailed descriptions of the various formations included within the dominion; their geographical distribution and fossils. Special chapters, largely by Sterry Hunt, included descriptions of the minerals, waters, rocks, and economic products. A "supplementary" chapter included a discussion of the drift. No consolidated stratified rocks were recognized of later age than Carboniferous (the Bonaventure Series). Laurentian rocks consisting mainly of highly feldspathic gneisses (anorthosites) with bands of crystalline limestone were recognized as occupying by far the larger portion of the area, at least 200,000 square miles. These are overlaid in part by the slate conglomerates and intrusive "greenstones" of the Huronian, the geographic distribution of which had not been made out in detail, and these again by the various members of the Paleozoic series. In an earlier paper in the *American Journal of Science* (May, 1862), Logan had assumed that during the Potsdam period the older rocks forming the coast of the Lower Silurian sea extended under comparatively shallow waters easterly and southeasterly from the region of the St. Lawrence and Ottawa to the fault between Gaspé and the Mohawk and southwesterly as far as Alabama. All around this shallow sea they descended abruptly into deep water at a possible angle of forty-five degrees. The forces producing corrugations in the earth's crust he thought might have been applied in the form of lateral pressure, and the corrugated area, for physical reasons, would be limited to the slope between shallow and deep water, the buttresses of gneiss marking the limit of disturbance.

Overlying the graptolitic shales of the Utica and Hudson River formations in the vicinity of Quebec, there was found a conformable series of sandstones, shales, and conglomerate limestones, which were considered, notwithstanding their position, older than the Hudson River group and to which he applied the name *Quebec group.* This was again subdivided into an underlying green sandstone series called the *Sillery formation,* and an overlying *Levis formation.* These were supposed by both Logan and Billings to be mainly contemporaneous with the Calciferous and Chazy groups of the New York geologists, but more recent investigations by Selwyn and Ellis have shown that while the Levis beds are Calciferous in the lower parts, the Sillery is probably all Cambrian (Dana), the Quebec, as a whole, being a northern continuation of the Taconic series which, however, Dawson later denied. Be this as it may, the advisability of retaining the name has been much questioned and given rise to numerous controversies not at all unlike those relating to Emmons's Taconic. The matter is well summed up by J. W. Dawson in the *Canadian Record of Science* for 1890, which, though extending beyond our time limits, may be referred to. In this article retention is defended on the ground that the formation so called differs radically lithologically and in structure from its equivalent farther west, being indeed in part a great Paleozoic bowlder formation. The application of local names for local developments of particular series, he argued, is always to be commended. It is scarcely necessary to add that, holding to these views, it was necessary for him to uphold Emmons in his unfortunate warfare. By a singular error, to which Marcou was prompt to call attention, this Quebec group on the map referred to above was made to include the Chazy and Calciferous formations below the Trenton.

It is worthy of note that, while his successors have been almost invariably opposed to the idea of the origin of the drift and rock striations through a glacial ice sheet, Logan evidently committed himself to the theory. Concerning the origin of the lake basins of western Canada he wrote:

> These great lake basins are depressions, not of geological structure, but of denudation, and the grooves on the surface rocks, which descend under their waters, appear to point to glacial action as one of the great causes which have produced these depressions.

This view, it will be noted, is at variance with that of Agassiz (p. 277) and others who preceded him. (See also under Stevens, p. 418.)

Again, in a footnote on the same page, he quoted, with evident approval, the following:

> This hypothesis [*i.e.*, the origin of the lake basins] points to a glacial period when the whole region was elevated far above its present level, and when the Laurentides, the Adirondacks, and the Green Mountains were lofty Alpine ranges, covered with perpetual snow from which great frozen rivers or glaciers extended over the plains below, producing by their movements the glacial drift and scooping out the river valleys and the basins of the lakes.

It was in this same report that Logan first noted the occurrence of a supposed fossil in the Laurentian of Canada, describing under the name of *Stromatopora rugosa* an aggregate of crystalline pyroxene and calcite found by John Mullen in one of the bands of limestone at the Grand Calumet. This was the so-called Eozoön of Dawson, referred to elsewhere.

Logan, for his time, possessed a very profound insight into petrographical problems, though he naturally regarded as traces of an original bedding what is now known to be, in part, at least, foliation due to dynamic causes. This is shown in his reports for 1853-1856, where he wrote concerning the rocks of the Laurentian system:

> They are the most ancient yet known on the continent of America, and are supposed to be equivalent to the iron-bearing series of Scandinavia.

Stretching on the north side of the St. Lawrence from Labrador to Lake Superior, they occupy by far the larger share of Canada, and they have been described in former reports as sedimentary deposits in an altered condition, consisting of gneiss interstratified with important bands of crystalline limestone.

And again: "The Laurentian series are altered sedimentary rocks."

The possibility that the system might not be a single unit was here again recognized in a supplementary note:

If, on exploration to the eastward of the Trembling Mountain, it should be further ascertained that the two inferior limestone bands of the Greenville series disappear on reaching the margin of the anorthosite, it may be considered as conclusive evidence of the existence in the Laurentian system of two immense sedimentary formations, the one superimposed unconformably on the other, with probably a great difference in time between them; and it will be an interesting subject of inquiry, whether the intrusive rocks which have been found intersecting the lower division give any clue to events which may have happened in the interval.

The geology of the islands of Anticosti, Mingan, and the Magdalen River region had been assigned to James Richardson, who made his report in 1857. The fossils collected were worked up by Billings, who considered the rocks of the Anticosti group to consist of beds of passage from the Lower to the Upper Silurian and synchronous with the Oneida conglomerate, the Medina sandstone, and the Clinton group of the New York survey, and with the Caradoc formation of England.

Against this view N. S. Shaler, in a paper before the Boston Society of Natural History, December 18, 1861, had argued that from the base of the level of the Canadian channel to the summit at the southwestern point of the island the beds were entirely Upper Silurian and their deposition synchronous with the Clinton and Niagara, of New York and elsewhere, though the fossils themselves might not be absolutely identical.

The map accompanying the volume of 1863 was beautifully executed in colors, and comprised all of the provinces southeast of the St. Lawrence as well as a narrow belt composed mainly of Laurentian rocks extending from Labrador southwesterly to the Great Lakes, and thence northwesterly to the ninety-sixth meridian. It included also a considerable portion of the United States, data for which were supplied by Hall and other American geologists.

Sketch of Logan.

Logan was born of Scottish parentage at Montreal in 1798, but his father shortly returning to Scotland, he received his early training, which was classical, in the High School and University of Edinburgh. He showed no disposition toward scientific pursuits until chance led him to the keeping of accounts in the establishment of an uncle, who was interested in mining and copper smelting operations in Wales. Here he was attracted by the phenomena of the coal seams and devoted a large share of his spare time to their study.

In 1838 the death of his uncle caused him to give up his position in Wales, and in 1840 he returned to Canada. His first geological paper was on the character of the beds of clay immediately below the coal seams of South Wales. This was communicated to the Geological Society of London in 1840. In this he announced the invariable presence under the coal seams of beds of fire clay carrying Stigmaria. This he regarded as proving the origin of coal through plant growth in place—an opinion which was very generally accepted at that time.

When he began his geological work in Canada a large portion of the country was a wilderness, without roads, and there were no maps. Of the topography of the Gaspé district it is written:

> Little was known of the region beside the coast line; of the geology, practically nothing. Settlements were few, confined almost exclusively to the coast, and made up chiefly of fishermen. There were no roads through the interior, most of which was, and indeed still is, a wilderness, inhabited by bears and other wild beasts, or at best only penetrated in certain regions by a few Indians or lumbermen. The courses of most of the streams were unknown and the mountains untraversed.
>
> I worked like a slave all summer on the Gulf of St. Lawrence, living the life of a savage, inhabiting an open tent, sleeping on the beach in a blanket sack, with my feet to the fire, seldom taking my clothes off, eating salt pork and ship's biscuit, occasionally tormented by mosquitoes. I dialled the whole of the coast surveyed, and counted my paces from morning to night for three months. My field-book is a curiosity. (Letter to De la Beche.)

From early dawn to dusk he paced or paddled, and yet his work was not finished, for while his Indians—often his sole companions—smoked their pipes around the evening fire he wrote his notes and plotted the day's measurements.

Logan is represented to us as strong in body, of active mind, industrious, and doggedly persevering, painstaking, a lover of truth, generous, possessed of the keenest knowledge of human nature, sound of judgment, but always cautious in expressing an opinion.

During his twenty-seven years of office, sixteen reports were submitted, the first, that for the year 1843, appearing in the form of a pamphlet of 159 octavo pages. It contained remarks on the mode of making a geological survey and a short preliminary report containing general observations on the geology of the provinces, and adjacent portions of the United States, together with the Joggins section already mentioned.

Assisted mainly by Alexander Murray, and T. Sterry Hunt, Logan continued his work until 1869, when he resigned to be succeeded by Mr. Selwyn. His reports cover the geology of Ottawa, the Gaspé peninsula, the economic geology of the Lake Superior region, the geology of lower Canada with especial reference to the eastern townships, the region along the north coast of Lake Huron, the gold-bearing fields of the Chaudière region, and the western peninsula, also the region between the Ottawa, the St. Lawrence, and the Rideau. His most important publication was his *Geology of Canada,* noticed above. His geological map, bearing date of 1866 and measuring 8 by 3½ feet, is said to have been the largest and most comprehensive that had appeared up to that time.

Sketch of Billings.

Billings, Logan's paleontologist, was trained as a lawyer and gave his first indications of scientific tendencies in the period 1852-1855, when, in the capacity of editor of *The Citizen,* a Toronto newspaper, he wrote popular articles on related subjects. He began his paleontological researches merely as a matter of recreation and made large collections now in the museum of the Geological Survey in Ottawa. In 1856 he founded the *Canadian Naturalist and Geologist* and the same year was appointed paleontologist to the Canadian survey under Logan, in which capacity he served for twenty years. He was necessarily prominent in the discussions relative to the validity of the Quebec group and many and sharp were the word warfares with Hall across the border. It is stated that he described in the course of his professional work sixty-one genera of fossils new to science and 1,065 new species.

Work of Murray and Howley in Newfoundland, 1864.

Under the directorship of Logan, Murray, assisted by James J. Howley, began work on the geology of Newfoundland in 1864, making a first brief report in May, 1866. The survey was continued until

PLATE XX

SIR WILLIAM LOGAN

1880, and a reprint of all the reports published in book form—an octavo volume of 536 pages—in 1881.

The work as a whole consisted mainly of details of structure of the regions immediately along the coast, with notes on the mines. The various subdivisions of the formations adopted were naturally those of the Canadian survey, which were based to a considerable extent upon those of the New York survey.

Murray's observations were limited mainly to a comparatively narrow belt along the coast. With the exception of the Glacial and post-Glacial material, no formations were found of later date than the Carboniferous, as was the case in Canada. The succession of the Lower Silurian formations of the island from above downward he gave as follows: Sillery, Lauzon, Levis, Upper Calciferous, Lower Calciferous, Upper Potsdam, Lower Potsdam, and St. John's Group. The St. John's Group is now recognized as Middle Cambrian (?), while his Lower Potsdam is Lower Cambrian.

Before the beginning of this work it had already been shown by Richardson, working under direction of Logan, that a trough of Lower Silurian rocks must underlie the northern part of the Gulf of St. Lawrence, gradually narrowing toward the Strait of Belle Isle, one side of the trough rising on the coast of Labrador, while the other formed the western shore of Newfoundland from Bonne Bay to Cape Norman. On each side of the strait these rocks were found to rest on Laurentian gneiss, which was ascertained to extend from the neighborhood of Bonne Bay to within twelve or fifteen miles of Hare Bay.

Murray's investigations proved that the Laurentian rocks spread in breadth to the Atlantic coast of the great northern peninsula of the island, and that the base of the Lower Silurian strata, sweeping around the northern extremity of the gneiss, comes upon the coast near Canada Bay, and again strikes into the land at Coney Arm in White Bay, where the Lower Silurian are overlaid by Upper Silurian, followed by rocks of Devonian age. Farther to the southeast the Laurentian and Silurian series were found to be partially and unconformably covered by rocks of Carboniferous age.

The report was accompanied by a colored geological map on ten sheets, on a scale of twenty-five miles to the inch.

Brief reference may here be made to Thomas Macfarlane, a mining geologist, briefly connected with the Canadian survey, and who wrote somewhat profusely in the *Canadian Naturalist* and *Transactions of the American Institute of Mining Engineers* on the origin and classification of the eruptive and primary rocks. His views do not seem to have been very generally accepted, and his plan of classification is not even noticed by the quartet who have dared seek their immortality through the introduction of a "quantitative" system.

Stevens's Criticisms of Logan's Views.

Logan's views on the origin of the Great Lake basins referred to elsewhere, did not meet with universal acceptance. R. P. Stevens, writing in 1865 on denudation of the North American continent, and with particular reference to the region west of the Appalachians, denied emphatically the efficiency of the glacial erosion in accomplishing all that came under his observation. "H. D. Rogers and his disciple, J. P. Lesley, have written very learnedly upon the valleys of the Appalachian system, or the disturbed portions and have called to their aid upthrows and downthrows, faults, fractures, synclinals, and anticlinals, and all the mathematical nomenclature pertaining to phenomena which are liable to have rigid laws of this science applied rigidly to them, but these fail west of the mountains; we cannot lean upon them; they serve us not." Speaking of the valleys of the Geneseo, Cayuga, Conewango, and others, he says: "Are these valleys of subsidence? I answer, No. . . . What has been the agent or agencies which has accomplished this mighty work—to shave off thousands of feet of our mountains, to scoop out wide valleys—to excavate, the whole width and breadth of the country? The great American expounder of the Swiss system says: 'Glaciers, nay, one glacier, over a mile in thickness covering perhaps the entire northern portion of the North American continent.' Another philosopher of the Pennsylvania school says: 'Some unknown agent accomplished the work with *one fell swoop* of its most tremendous power.' . . . I consider then, that the valleys of the disturbed portions of our country are as old in their origin as the rock system to which they belong and have suffered denudation from that time to this by the same agencies and *no other* which now acts through the whole year upon the North American continent. Through each geologic change the older valleys suffered from widening and deepening, while the hill tops were lowered." This paper of Stevens is of further interest in that he recognized the fact that eminence above the general level does not necessarily mean upheaval. Speaking of the Catskills, he wrote: "They are mountains not because of upheaval but because the basset edges of the strata composing their bulk have been cut away upon their sides fronting the horizon from which they appear as mountains."

Dana's Manual of Geology.

The first edition of Dana's *Manual*, already mentioned, would seem to have been issued in 1862, though the copies commonly found in libraries bear the date of 1863.[14]

This work gave for the first time an authoritative summary of the geology of North America and held its own through all the years following down to the fourth and last edition, that of 1895. In view of its now universally acknowledged standing, it is a little amusing to read in the columns of *The Reader*, London, January, 1863, the review signed C. C. B. The chief plaint of the reviewer seems to have been that the work was not English in viewpoint and nomenclature though he acknowledged "that the work which Mr. Dana has recently published bids fair to attain a place among the few sound educational works of the present day." And again:

> Admiring, as we do, the general tone and scope of the work, we believe that we shall often have occasion to refer to it. But we shall not need to use it so frequently as the works of Lyell or Jukes, for the reason which we have stated above, that the peculiarities of style, diction, classification, and habitual mode of thought will preclude many English students from feeling that unmixed pleasure in its perusal to which the skill, labour, and unquestionable erudition of its author would otherwise be entitled.

Too bad!

Pumpelly's Work in China, 1863-1864.

In the autumn and winter of 1863-1864 Raphael Pumpelly, of Rhode Island, was engaged by the Chinese Government to examine the coal fields west of Peking. Incidentally he made journeys in northern China and Mongolia. Subsequently, in 1864-1865, he crossed into Siberia and journeyed overland to St. Petersburg. The results of his observations on the geology of the region were pub-

[14] The writer possesses the original inscribed copy of the Manual given by Dana to Sir Richard Owen together with the letter which accompanied it. This letter is dated January 24, 1863, and reads, "I send you herewith for your acceptance a copy of the new edition of the Geology in which the corrections I alluded to in my last letter have been made." The title-page of the volume bears the date of 1863 and the preface that of November 1, 1862. Gilman, in his bibliography of Dana (Life and Letters), mentioned the Manual under date of 1862 and says nothing of an edition of 1863, although this is the date given by Dana in his preface to the edition of 1894. It seems most probable that the 1863 edition was but a corrected reprint.

lished in the Smithsonian Contributions to Knowledge, 1866, forming a quarto pamphlet of 144 pages with 8 plates of sections, and a colored geological map, the latter confessedly largely hypothetical. This memoir, which antedated Richthofen's great work on China by some years, gave to the world the first authentic account of the geology of that country.

He showed that in the region extending from the twentieth to a little beyond the fortieth parallel and from near the one hundredth to about the one hundred and twenty-second meridian the oldest sedimentary rocks were Devonian limestones, which prevailed in some cases to the enormous thickness of 11,600 feet.[15] Overlying this, through the greater part of the area, were the Chinese Coal Measures (Mesozoic), interrupted by bands of granitic and metamorphic rock of undetermined age. In the extreme northern part of this region was a comparatively small area of basaltic and trachytic rocks. The region immediately south of Peking, comprised principally within the provinces of Chihli, Nganhwui, Kiangsu, and Shantung, was colored as occupied by post-Tertiary materials, with smaller areas of the same age along the Yangtse-Kiang and Hoangho rivers in the provinces of Hupeh, Sz'chuen, and Shensi.

Fig. 76. *R. Pumpelly.*

Considerable attention was given to the post-Tertiary "Terrace" deposit, or loam, which he found in the valley of every tributary of the Yang Ho, and probably also of the Sankang Ho. This, which has since become more generally known as the Chinese loess, was described in considerable detail as to modes of occurrence, physical properties, and geological distribution. The material he regarded as having been deposited in a chain of lakes extending from Yenkingchau north-northwest of Peking to near Ninghia, in Kansuh, a distance of nearly 500 miles, the lake basins themselves being formed by the dislocations which gave rise to the plateau wall to the north, and being filled by sediments brought by the Yellow River.

The fossil plants from the coal-bearing rocks were studied by Newberry and identified as Mesozoic.

[15] According to Willis's later observations these beds are Cambro-Ordovician, Pumpelly having been misled by certain fossils which were *supposed* to have come from them.

Cook's Survey of New Jersey, 1863-1889.

Professor George H. Cook, who was assistant geologist of New Jersey under the Kitchell survey (suspended in 1856), was appointed state geologist with the reorganization of the same survey in 1863, and continued to serve in this capacity until the time of his death in 1889. His first annual report, that for 1864, bearing the date of 1865, was a small pamphlet of but twenty pages, and contained a single-page colored geological map of the state, the second of its kind to be issued, the first having been by H. D. Rogers.

One of the first tasks which Cook imposed upon himself after his appointment was the preparation of a large octavo volume, accompanied by a portfolio of maps, setting forth the condition of the knowledge of geology of the state up to the date of publication (1868). In this work he gave a general summary of all that had previously been accomplished.

Naturally the question of the age and stratigraphic position of the white limestone came up for discussion. He wrote:

> In regard to the crystalline limestones he [Rogers] was mistaken, They are everywhere conformable to the gneiss and interstratified with it. His mistake is acknowledged by his former assistant, J. P. Lesley, in the American Journal of Science, LXXXIX, p. 221. The true position and identity in age of the crystalline limestone and gneiss was proved by Vanuxem and Keating, in the Journal of the Academy of Natural Sciences in 1822, and this view has been sustained by all the observations of Doctor Kitchell and his assistants and can be easily verified by anyone who will visit the localities cited in this report.[16]

He was inclined to consider the magnetic iron ore of sedimentary origin, deposited in beds just as were the gneiss and crystalline limestone, in this respect agreeing with Kitchell and again disagreeing with Rogers.

As early as 1854 he had called attention to the gradual subsidence of the coast of New Jersey, and before his death was able to give apparently absolute figures on the rate of the depression.

Sketch of Cook.

Cook was born in Hanover, New Jersey, in 1818, and educated in the public schools of the state and the Rensselaer Polytechnic Institute, whence he graduated in 1839 with the degree of C.E. After

[16] Studies by Mr. A. C. Spencer, of the United States Geological Survey, made during the season of 1904, point to an igneous origin for the gneiss. This view, if correct of course, effectively disposes of the idea that the limestone is conformable with the gneiss and interstratified with it.

graduation he remained at the institute as tutor, adjunct professor, and finally full professor, until 1846, when he removed to Albany, New York, where he was engaged at first in business and latterly as professor of mathematics and natural philosophy, and finally principal of the Albany Academy.

In 1852 he was sent to Europe by the state authorities of New York to study the salt deposits, with a view of developing those of Onondaga County. In 1853 he accepted a call to the chair of chemistry and natural sciences in Rutgers College, New Jersey, retaining his connection with the institution during the remainder of his life, though after 1854 being actively connected with the state geological survey as well. In 1880, moreover, he was made director of the state agricultural experiment station, which, indeed, had been established largely through his efforts.

A noble and unselfish man, who, as someone has expressed it, "went in and out of the houses of this State, making friends of every man, woman, and child he met." Far-seeing, persistent, ever calm and judicious in his work, yet light-hearted and cheerful among his friends—

> his broad expanse of face, full of light, his eyes gleaming with kindliness, as well as with shrewdness, and often with a right-merry twinkle; his genial smile, his frank greeting, never marred by any hollow and flippant phrase of mere etiquette, but as honest as it is cordial; his sympathy, so responsive yet so genuine; his massive though quiet strength of purpose; and his self-contained, self-poised nature, all crowned with boundless hopefulness, united to make his very presence an inspiration and benediction.

In the author's memory there are two men among American geologists who stand out as devotees of science, yet entirely free from the narrowness of the specialist or the personal idiosyncrasies that so frequently mar the character of men of their class. These two men are George H. Cook and Edward Orton. They loved science for science' sake, yet did not close their eyes to its economic bearing, nor call upon an overtaxed public to support them in the work they loved, regardless of its outcome. Never a minister of the gospel had the interests of his parishioners more at heart than these two men that of the public they served. For themselves they asked simply the privilege of doing the work and doing it to the best of their ability.

Work of B. F. Mudge in Kansas in 1864.

B. F. Mudge, in the capacity of state geologist, submitted his *First Annual Report on the Geology of Kansas* for the year 1864, in

PLATE XXI

GEORGE H. COOK

form of an octavo volume of fifty-six pages, in 1866. He announced the lowest geological formation of the state to be the upper portion of the Coal Measures, of which he gave a section in Leavenworth County. He accepted the Permian identification of the fossils as made by Meek, Hayden, and Swallow, and noted the occurrence of the Triassic and probably the Jurassic formations also, in a belt of territory crossing the Republican and Smoky Hill valleys; also Cretaceous, the geographic limits of which had not been worked out. The drift and erratic bowlders he thought due to icebergs.

Sketch of Mudge.

Fig. 77. *B. F. Mudge.*

Mudge was born in Maine in 1817, and graduated at Wesleyan University, Connecticut, in 1840. After graduation he studied law and, being admitted to the bar, practiced his profession in Lynn, Massachusetts, until 1859. During 1859 and 1860 he was employed as chemist in the Chelsea, Massachusetts, and Breckinridge, Kentucky, oil refineries. In 1861 he removed to what is now Kansas City, Kansas, where he engaged in teaching. Such a life would now be considered as little fitting a man for the profession he was subsequently to follow, yet in 1864, having by invitation delivered a course of lectures before the Kansas legislature upon the geological resources of the state, he was unanimously elected state geologist, a position which he filled, however, for but a single year, resigning to accept the professorship of natural history in the Agricultural College at Manhattan, Kansas, where he remained until 1873. His resignation from this last position is stated to have been caused by disgust aroused at the political conditions in which the institution became involved and the assumption of its presidency by a well-known politician, with no qualifications for it.

Although educated as a lawyer, Mudge is stated to have been throughout his whole life deeply interested in natural sciences, and while in Lynn to have taken an active part in the organization of the Lynn Natural History Society. In Kansas his scientific work was largely in the line of exploration.

Arduous, intrepid, willingly undergoing hardships and dangers for the sake of science, he explored a very large part of Kansas when explorations meant real dangers and hardships of the most pronounced kind. As early as 1870 he made explorations in the extreme western part of the State in the study of its geology and paleontology, and for years afterwards nearly every summer found him in the midst of the Indian country, usually wholly without protection from the danger of hostile Indians, save such as his own rifle and revolver afforded. In the summer of 1874 he explored the whole length of the Smoky Hill River, an utterly trackless wild, infested by Indians, whose murderous depredations were visible on every side.[17]

Mudge made the first geological map of the state (Kansas), which is fairly correct in its main features, save for the Lower Cretaceous, which he failed to recognize.

He mapped and described with tolerable accuracy and fullness the physical structures of the different Cretaceous and Tertiary horizons. Much, if not most, of the information thus given was based upon his patient researches in wagon or on foot. In general it may truthfully be said that his pioneer work in Kansas geology was important and extensive, though now largely superseded by more detailed and accurate studies. His work in life, however, has chiefly borne fruit as a teacher. He was widely known as an enthusiastic and able lecturer, and his courses were always in demand by the teachers and scientific men of the State. His quiet modesty and unselfishness disarmed all envy and jealousy. Of most charming personality, of wide culture, and unbounded enthusiasm, his teachings made an unusual impression upon all with whom he came in contact.

His bibliography is brief and the papers generally limited to but a few pages. His material he willingly put in the hands of others for publication, and Marsh, Cope, White, and Lesquereux profited thereby. It was during one of these earlier trips that he discovered the first specimen of Ichthyornis, which, coming into the hands of Marsh, did so much toward making the latter famous.

Swallow's Geological Survey of Kansas, 1866.

In 1866, G. C. Swallow, who succeeded Mudge, issued a *Preliminary Report of the Geological Survey of Kansas*, in form of an octavo volume of 198 pages, including a report by Dr. Tiffin Sinks on the climatology, and one by Dr. C. A. Logan on the sanitary relations of the state. Major F. Hawn was assistant geologist.

Special attention was given to the eastern and central part of the

[17] Williston, American Geologist, Vol. XXIII, 1899, p. 342.

state. He found rocks belonging to Quaternary, Tertiary, Cretaceous, Triassic (?), Permian, Lower Permian, and Carboniferous formations, the lower-most division being the Lower Carboniferous. Mudge, it will be remembered, found nothing older than the Coal Measures. The buff, mottled, and red sandstones underlying the Cretaceous were doubtfully referred to the Triassic from their resemblance to the foreign Triassic and the presence of a *Nucula* resembling the *speciosa* of Munster from the Muschelkalk of Bindlock. The presence of Permian beds he had previously announced (see p. 368). The coal-bearing rocks he estimated at 2,000 feet in thickness and underlying an area of over 17,000 square miles. In these he found twenty-two distinct and separate beds of coal, ranging in thickness from one to seven feet.

This work of Swallow in Kansas has been largely overlooked by recent workers. According to Keyes,[18] a large portion of it was not only good but marvelously well done for its day and the conditions under which it was accomplished. The historical importance of it "lies in the fact that some of his geographic names applied to geologic terranes will have to stand as valid terms, although his correlations were often very bad."

Sketch of Swallow.

Swallow, like Mudge, was born in Maine, but was of Norman-French descent. He studied the natural sciences under Parker Cleaveland at Bowdoin College, graduating in 1843, and, after several years in educational work, accepted the chair of chemistry, geology, and mineralogy in the University of Missouri in Columbia. From 1856 to 1861, the date of the discontinuance of the survey, he served as geologist of Missouri, and in 1865 was appointed state geologist of Kansas, as above noted.

It was about the time of his first appointment that he wrote Hall, "I have no ambition to gain a reputation as a geologist, but to do up our Survey as well as may be and then return to my profession, and aid in my humble way those who may be making geological pursuits a profession."

He is represented as a large, fine-looking man, over six feet in height, and a very close observer of all natural phenomena. According to Professor Broadhead, "No other man during the same length of time has ever gone into a strange field, traversed the country, and published a volume all in a year and a half, as he did." He was

[18] American Geologist, June, 1900, p. 347.

connected for a time (1867-1870) with mining operations in Montana, but his scientific field was limited wholly to Missouri and Kansas.[19]

Hawn was a civil engineer, first engaged in railroad work and afterward in a linear survey of the state. It was during this period that he made the observations and collections that brought him to Swallow's notice and that resulted in his appointment as assistant geologist. In 1850 he had accompanied, in the capacity of geologist, a party under the command of Lieutenant E. H. Ruffner, Corps of Engineers, on a reconnaissance into the Ute country. This and his Kansas experience seem to have comprised all of his geological work. It is stated that the first prospecting for coal at Leavenworth, Kansas, was at his instigation.

Fig. 78. *G. C. Swallow.*

N. S. Shaler's Views on Continental Uplift, 1865.

At the December (1865) meeting of the Boston Society of Natural History, N. S. Shaler, then but twenty-four years of age and a graduate of the Lawrence Scientific School in Cambridge, made some interesting remarks on the elevation of continental masses. Referring to the assumptions of Charles Babbage and Sir John Herschel relative to the shifting of isothermal lines and consequent expansion and local uplift along lines of deposition, he argued that for the same reason sea bottoms on which sedimentation was taking place would be areas of depression, since the curving must take place in the direction of greatest expansion. In like manner, uplifting would take place

[19] The path of the state geologist was, and is, almost invariably thorny. Swallow seems to have met some trouble of an unusual type to judge from the following:

St. Louis, Feb. 27, 1862.

My dear Meek:

. . . Our mutual friend Prof. Swallow, is now in this city confined in the Military Prison. Some of his "good friends" at Columbia lodged information with the military authorities here to the effect that he was a Secessionist. Ulffers stole, it is thought, all of the County Geological Maps and wrote a letter which aided in his arrest. His friends here are doing all they can to procure his release, and I presume he will be out in a couple of days. Swallow is not a whit more a "Secesh" than you are. . . .

(*Signed*) B. F. Shumard.

along lines of denudation. The intermediate point between the two zones of movement would naturally be the sea border, and hence here would occur the fracturing of the superincumbent strata and resultant volcanic phenomena. In this way, assuming the original nuclei of the continents, or points first elevated above sea level, to have been in the northern portion of the sphere, he thought it probable they would continue to grow by uplift southward in a succession of southwardly pointed triangles.

Some six months later, in June, 1866, he read before the same society a paper on the formation of mountain chains, which is also of interest in this connection, though in part simply a reiteration of the ideas expressed above. Accepting the theory that the earth's mass consists of a solid nucleus, a hardened outer crust, and an intermediate zone of slight depth in a condition of imperfect igneous fusion, he argued, as in a previous paper, that while the continental folds were probably corrugations of the whole thickness of the crust, the mountain chains were but folds of the outer portion caused by the contraction of the lower portions of this outer shell, the contraction in both cases being due to loss of heat. Further, the subsidence of the ocean's floors would, through producing fractures and dislocations along those lines, tend to promote the formation of mountain chains along and parallel with the sea borders. These ideas should be considered in connection with those of Dana and Le Conte (see pp. 259 and 464).

Still again, in 1868, Shaler, having in the meantime been elected professor of geology in Harvard University, brought up before the Society the matter of the nature of the movements involved in the changes of level of shore lines, and this time with particular reference to changes coincident with or subsequent to the glacial period. He showed that local phenomena of continental uplift or depression, as measured by the level of the sea at the shore line, might be variously modified by the position of the points of rotation, whether immediately at the shore line or at a greater or less distance, either seaward or inland. Of greater interest, however, were his remarks relative to the changes in level at the time of glaciation. Referring to a previous paper, where he showed that a compound bar would, when heated, bend toward the side composed of the most expansive material, he compared such a bar to a portion of the earth's crust covered with several thousand feet of ice and snow. The effect of this blanket would be to cause the isothermal lines to move outward toward the surface, causing thus an expansion of that portion of the crust immediately beneath the ice. But the ice itself would partake very slightly, if at all, of this increased temperature, and, as in the case of the compound bar, the bending would take place in the direction of maximum expansion, *i.e.*, in this particular case, downward. In this way, he suggested, the depression accompanying the period of maximum glaciation might be accounted for.

A. S. Packard on Glacial Phenomena in Labrador and Maine, 1865.

Although A. S. Packard is best known to the scientific world through his researches in entomology, he nevertheless wrote on geological subjects, particularly upon glaciation and incidental phenomena. The most important paper along these lines related to glacial phenomena in Labrador and Maine. This comprised some ninety quarto pages of the memoirs of the Boston Society of Natural History. The entire Labradorian plateau was described as molded by ice to a height of at least 2,500 feet above the present level of the sea, and the surface thickly strewn with bowlders. His conclusions, based upon the observed phenomena, were that, first, the coast line of the northern part of the continent once stood at an elevation some 600 feet greater than now. The physical effects of such an elevation would produce the increased cold and precipitations essential to the formation of glaciers which he conceived to have rivaled in thickness and extent those described by Ross in the Antarctic. Following this there ensued a "slow and gentle" submergence, during which the crude moraine material was reassorted into beds of regularly stratified clay (the Leda clay) 100 to 300 feet in thickness. During this period the adjacent sea was filled with floating ice and there was uniformity in the temperature and character of animal life throughout the region. This period was followed by that of the raised beaches, during which the continent was raised some 400 feet, or approximately to its present level, and which left at its close "the surface of New England covered by broad lakes and ponds with vast rivers and extensive estuaries with deep fiords cutting the coast line." The Terrace epoch followed, during which the deposits of sand and shingle resulting from the drainage of the slowly rising continents were formed in the bays and estuaries.

The paper owed much of its value to the remarks on the distribution of animal and plant life before, during, and after glaciation. He found no proof of an Atlantis as advocated by Heer and others to account for certain peculiarities of faunal distribution, nor indeed any Tertiary or post-Tertiary connection between America and Europe.

Work of A. H. Hanchett, H. H. and R. M. Eames in Minnesota, 1865-1866.

Under authorization of an act passed on March 4, 1864, the governor of Minnesota appointed Dr. A. H. Hanchett state geologist. With him was associated a Mr. Thomas Clark. Each of the gentlemen made one brief report which was considered a creditable production, "but it became apparent that Dr. Hanchett was not intelligently and wholly devoted to the work, and on the passage of a more general act by the legislature of 1865 the Governor conferred the position of State Geologist upon Mr. Henry H. Eames." During his term of office Eames made two reports, of 23 and 58 pages respectively, both in the title-pages bearing the date of 1866. They were devoted almost wholly to matters of an economic nature, although a few preliminary pages were devoted to a somewhat popular disquisition on general geology.

He noted the "immense bodies of the ores of iron, both magnetic and hematitic," in the northern part of the state, and demonstrated that in so far as the counties examined were concerned there were no workable beds of coal within its limits. The fuel possibilities of the peat deposits were discussed, and a few pages devoted to agricultural chemistry. Some attention was also given to the possible occurrence of the precious metals and it is stated that it was owing to his discoveries that the gold fever centering about Vermillion Lake rose in speculative mining circles. At the end of two years the legislature refused further appropriations and the survey came to an end. Richard M. Eames served as an assistant on this survey. His report on the Mesaba iron range was never published, the Governor remarking, on its presentation, that the region was so remote from transportation it could well wait for the next generation!

C. F. Hartt's Work in Brazil, 1865-1867.

C. F. Hartt and Orestes St. John had accompanied Agassiz in the capacity of geologists on the Thayer Expedition to Brazil during the years 1865-1866. In 1867 Hartt made a second journey, spending several months on the coast between Pernambuco and Rio, exploring more particularly the vicinity of Bahia and the islands and coral reefs of the Abrolhos.

The results of this and the previous expedition were published in book form in 1870, under the title of *Geology and Physical Geography of Brazil.* In this work the gneisses of the province of Rio de

Janeiro were considered metamorphosed sedimentary deposits and of Azoic age. Their thickness he did not even estimate, recognizing the fact that their apparent enormous thickness was due to numerous reversed folds, so that one might travel for miles over their upturned edges, finding them always highly inclined and dipping in the same general direction.

Fig. 79. *C. F. Hartt.*

The age of the metamorphic rocks succeeding the gneisses he was unable to determine, though it was suggested they might be Silurian or Devonian. South of Rio he found unmistakable Carboniferous rocks, including beds of bituminous coal, and in the province of Sergipe, underlying the Cretaceous, a thick series of red sandstones, referred to the Triassic.

Marine Cretaceous beds of undetermined extent were found north of the Abrolhos Islands, which were conformably overlaid by clays and ferruginous sandstone, referred to the Tertiary. Overlying this along the whole coast he found an immense sheet of structureless clays, gravels, and bowlder deposits, which he believed, with Agassiz, to have been the work of glacial ice, though he noted that nowhere had there been seen either polished or striated rocks, such as are almost constant accompaniments of glaciation elsewhere. It is almost needless to add that this view is no longer held by anyone, the bowlders supposed to have been erratics being products of decomposition and their distribution the work of gravity and water.

In 1870 Hartt went again to Brazil, and in 1875, while professor of geology at Cornell University, was appointed chief of the geological commission of that country, with Richard Rathbun as assistant. He died in 1878.

Hilgard's Views on the Drift, 1866.

Eugene W. Hilgard, in an article in the *American Journal of Science* for 1866, pointed out the great difference in the character of the drift in the North and Northeast and that of the West or Mississippi Valley. He felt that the glacial theory alone, as then understood, could not account for the deposits north of the Ohio any more than for the Osage sand delta south of it. Though referring to

Agassiz's observation regarding "the melting snow of the declining glacial epoch" and its instrumentality in forming river terraces, he adopted the, to him, more plausible idea, first announced by Toumey, to the effect that the southern drift may have been formed in consequence of the sudden melting of the northern glaciers, "such as would have resulted from a first rapid depression of so large a mass of ice below the snow line." In the beginning the flood action would be violent, producing deep erosion of the underlying formations and the transportation and redeposition in mass of their materials. After the first rush, the stratified deposits would be formed, mingled with more or less bowlder material from floating ice. The influx of cold water from the north would also account for the absence of signs of life in the deposits. The "grandly simple means of a single elevation and redepression in the northern latitudes . . . will equally satisfy the conditions required for the formation of the western and southern drift."

Kerr's Geological Work in North Carolina, 1866-1869.

W. C. Kerr, who succeeded Ebenezer Emmons in the office of state geologist of North Carolina, received his commission on April 4, 1866, and continued in service until the time of his death, in 1885. His first *Report of Progress*, submitted in January, 1867, was an octavo volume of fifty-six pages, in which the purposes of the survey were set forth and a summary given of the geology of the western part of the state so far as known. The rocks of this western area were thought to belong "to the most ancient of the Azoic series," and to have been above sea level since very ancient times. As with his predecessors, Kerr was troubled to account for the drift, noting that while it occurred far beyond the limits usually ascribed to glacial action, yet there were "numerous phenomena which have no other plausible explanation."

Kerr's second report, submitted in 1869, was of equal brevity, but naturally contained more of the author's personal observations. He noted that the mountains, plateaus, and valleys of the French Broad and Lower Catawba areas owed "their existence and all the details of their form and position to the action of water, the basins . . . being . . . without exception, valleys of erosion, having in no case an anticlinal or synclinal origin."

The entire western portion of the state he thought to consist of four groups or formations, first, the

Cherokee slates along the Smoky Mountains, on the northwest border, consisting of clay slates and shales, sandstones, grits, conglomerates, and limestones; second, the Buncombe group, occupying the larger portion of the great transmontane plateau between the Blue Ridge and Smoky Mountains, and consisting of gneissic and granitoid rocks; third, the Linville slates, a narrow belt stretching for the most part along the Blue Ridge and composed, like the first group, of semimetamorphic argillaceous slates and shales, sandstones, limestone, and gneissoid grits; fourth, the Piedmont group, gneissic and granitoid.

He noted further that these four groups constituted two recurrences of the same rocks, in the same order, recalling Rogers's theory of reduplication by folding and overturns, as worked out in Pennsylvania.

John L. Le Conte's Union Pacific Railroad Survey, 1867.

John L. Le Conte, a cousin of the Joseph Le Conte, elsewhere noted, is known to science rather through his entomological than geological writings. Five papers are credited to his pen by Darton in his *Catalogue and Index of North American Geology.* Of these, the most important and the only one that need here be considered is one on the geology of the survey for the extension of the Union Pacific Railroad from the Smoky Hill River, Kansas, to the Rio Grande. He made a detailed study of the coal beds, and on the basis of their molluscan remains maintained that they were of Cretaceous age rather than Tertiary, as claimed by Lesquereux. His reasoning as to the relative value of botanical and molluscan remains for determining the age of beds is worthy of note:

The difference between the plants of our early Cretaceous and those of the European middle Tertiary could be ascertained only after much discussion and by the stratigraphy of the region, and we have no right from a few resemblances in vegetables to infer the synchronism either of the western lignite beds with each other, or any of them with the European Eocene and Miocene, except when supported by paleontological evidence derived from animal remains.

In this most geologists will now agree with him.

Le Conte's views on the general development of the western portion of the continent indicated an ability to deal with the larger problems of geology in a philosophical and highly satisfactory manner, and it is perhaps to be regretted that he should have allowed himself to be drawn off into other pursuits. As noted, this paper was the most important of his geological writings, as it was also the last.

White's Survey of Iowa, 1866-1869.

The second geological survey of Iowa was inaugurated in April, 1866, with Charles A. White director, and Orestes St. John, principal geological assistant. The survey continued to the end of 1869, results being published in the form of two royal octavo volumes, comprising all told some 443 pages, with a colored geological map of the entire state.

It appears from correspondence[20] that Hall was desirous of reappointment at the head of the resuscitated survey, in order that he might complete the work begun in 1855. "I never understood that I was appointed . . . for any specific time, but to complete the survey. And, your instructions to me were that I should make Vol. I of such a character that another volume of similar size would contain the entire result. This plan was adopted and I felt that should there be a disposition to resume the survey an opportunity would be given me to make another volume as promised." Finding that White was likely to receive the appointment, he was at first very much aggrieved, but wrote later that he had no objection could he—Hall—be reimbursed for previous personal outlays.

As Hall and Whitney had devoted a large portion of their attention during the previous survey to the eastern part of the state, so White devoted himself mainly to the western. He found reasons to discourage all explorations for mineral oils or precious metals in the state, and also pointed out the hazard of exploring for coal beyond the northern and eastern boundaries of the coal field as designated in his geological map. He also showed that, though iron ore of a good quality had frequently been found within the area, the deposits were always limited. In these matters he was correct.

Considerable attention was given to the peat deposits and an estimate made of the amount of material which could be utilized for fuel purposes should occasion demand.

Among the phenomena of lesser importance to which he called attention were the moving of the bowlders on the shores of lakes and the piling of them into wall-like masses through the expansive action of the freezing water. The so-called Bluff deposit he considered to be of more recent origin than the drift, and referred it to the earliest part of the so-called Terrace epoch, the material composing it having originated by fluviatile erosion immediately upon the close of the glacial epoch, being afterward deposited as a la-

20 Letters to Hon. R. P. Lowe dated May 27, 1865, and to Gov. Grimes, May 28, 1865.

custral sediment in the broad depression in the surface of the drift left by the retreating glaciers. He differed entirely with Whitney as to the cause of the absence of trees in the prairie region, and felt no hesitation in declaring that the real cause was the recurrence of annual fires.

He divided the formations of the state into Azoic, Lower Silurian, Upper Silurian and Devonian, Carboniferous, and Cretaceous systems, and regarded the Sioux Falls quartzite, with its associated pipestone, as belonging to the Azoic. The Potsdam sandstone, which he found reaching a thickness in the state of about 300 feet, he thought to be probably overlying this Sioux quartzite. This view is generally held today.

White agreed with Whitney that there was no hope of profitable lead mining in the Lower Magnesian limestone. For the so-called Hudson River shales of Hall he substituted the name of Maquoketa shales. All the Devonian rocks of the state he referred to the Hamilton period.

A strict conformability was found in all the rocks from the Potsdam sandstones to the Keokuk limestone, inclusive, but between this last and the rocks of the Coal Measures an unconformability and also one between the St. Louis limestone and the older formations of the sub-Carboniferous group. Instead of there being only one formation of Carboniferous limestone, as had been generally supposed, White claimed to have found two, each possessing similar lithological but different paleontological characteristics, the one overlying and the other underlying the coal-producing strata.

The various foldings in the strata he regarded as having taken place subsequent to the deposition of the latest of the Carboniferous beds and earlier than those of Cretaceous age.

The gypsum deposits were thought to be Mesozoic and to have originated through chemical precipitation in comparatively still waters which were saturated with sulphate of lime; hence the exact determination of their geological age was rendered a matter of some difficulty. It is therefore well to note that Keyes in his report in 1895 referred them to the upper part of the Mesozoic—the Cretaceous.

White, it should be noted, had in 1860[21] described in considerable detail the rocks and their included fossils in the vicinity of Burlington. He identified here eight beds, the lower six of which he thought the equivalent of, though not necessarily contemporaneous with, the Chemung of New York. The two upper beds, which were of lime-

[21] Boston Jour. of Nat. His., VIII, 1859-1863, pp. 205-235.

PLATE XXII

RUSH EMERY
C. A. WHITE ORESTES ST. JOHN

N. H. WINCHELL

stone, he thought to be Carboniferous, though he remarked that the line drawn between the two formations was largely imaginary, indicating merely the limit where the Devonian species ceased to predominate and upward from which the Carboniferous species flourished in full force.

It was suggested that the Devonian species might have originated at the east and migrated westward during the time that the bottom of the Chemung sea was gradually sinking and receiving the deposits forming the Old Red sandstone, thus making the Devonian rocks equivalent to the New York Chemung and contemporaneous, in part at least, with the Old Red sandstone of the Catskill Mountains. It may be added that all of the six beds then supposed to be Devonian are now commonly regarded as belonging to the basal Carboniferous (Kinderhook).

Later (in 1866) Niles and Wachsmuth studied the upper beds, which had become known as the Burlington limestone, and were led by the crinoidal remains to regard the two divisions as two independent formations, which they designated as the Lower and Upper Burlington, a subdivision which still holds at time of writing.

Sketch of C. A. White.

The influence of environment in molding a professional career is rarely better shown than in the Ohio and Mississippi valleys, where the richly fossiliferous and but slightly disturbed strata have, since their earliest discovery, aroused the interest of many an embryo naturalist, several of whom achieved a national reputation. The father of C. A. White moved from Massachusetts to Burlington, Iowa, when the latter was twelve years of age. Here he was educated as a physician and began practice in 1864. The allurements of so rich a paleontological field were, however, too great for one of his tastes, and in 1866, when appointed state geologist, he abandoned medicine altogether. In 1867, while still connected with the survey, he was appointed to the professorship of Natural History in the State University. In 1873 he accepted a professorship in Bowdoin College, Maine, but two years later (1875) removed to Washington and entered upon his life's work in connection with the governmental surveys and the United States National Museum. He was connected successively, as elsewhere noted, with the Wheeler, Powell, and Hayden surveys and with the National Museum until his final retirement from active life. The "results of his official work are characterized by a clear simple style which never permits any doubt of the author's meaning or of his honesty of purpose." The writer recalls him as a genial, kindly man, always ready to help and encourage others, and one whose interests were by no means limited to his special field of work.

Safford's Final Report on the Geology of Tennessee, 1869.

Safford's final report on the geology of Tennessee did not appear until 1869, having been delayed by the Civil War. It was accompanied by a colored geological map of the state, and a geological section, uncolored, extending from the Unaka chain on the east to the Mississippi, and giving, on the whole, a very comprehensive and easily understood idea of the physical geography and geology of the area, as well as its economic resources. He here called attention to the frequent recurrence of the same formation, or series of formations, met with in crossing East Tennessee, and accounted for the phenomena on the supposition that the beds had been thrown into a series of parallel and closely compressed and overturned folds, the crests of which had been subsequently denuded. This, for its time, was an important deduction.

Although on his map a section of the Ocoee conglomerate was put down as belonging at the top of the Azoic series, in his chapter on the Potsdam group it was stated that this and the slates, Chilhowee sandstones, and Knox group of shales, dolomites, and limestones might be regarded as a formation which corresponded to Dana's Potsdam period, and that it was not easy to separate, lithologically, the Ocoee subgroup from the Chilhowee, as they often ran into each other.

The main bulk of the report was given up to a discussion of the distribution, lithological nature, and characteristic fossils of the various formations. He was disposed to regard the Porter's Creek group of the Tertiary as distinct from the Orange sand, this latter name having been provisionally applied by him to a series of strata supposed to be, for the most part, equivalent to Hilgard's northern lignitic. The general grouping, from the Cretaceous upward, was essentially the same as it had been in a previous paper in the *American Journal of Science* (1864), which was as follows, beginning at the bottom:

1. Coffee sand .... Cretaceous.
2. Green sand or the shell bed .... Cretaceous.
3. Ripley group (provisional) .... Cretaceous.
4. Porter's Creek group (provisional) .... Tertiary (?).
5. Orange sand or Lagrange group .... Tertiary.
6. Bluff lignite (provisional) .... Tertiary (?).
7. Bluff gravel .... Post-Tertiary.
8. Bluff loam .... Post-Tertiary.
9. Bottom alluvium .... Modern.

Fifteen new species of invertebrate fossils were described. The work does not seem to have attracted much attention at the time, perhaps not so much as it deserved, and was given but a half-page review in the *American Journal of Science* for that year. In this, attention was merely called to the fact that Safford differed with Hilgard on the question of the age of the Orange sand.

Sketch of Safford.

Personally Safford was a man whom all who knew were bound to respect. It was his misfortune to be isolated, to come so rarely in contact with his fellow workers, and to be so lacking in self-advertising qualities that his value was not fully appreciated by the country at large until through old age and failing health he had retired from the field. Concerning him and his work Stevenson has written:[22]

There were no maps, for much of the state was in wilderness; mines were few; the roads were on natural grades; there was no network of railroads to give sections in critical places. Prof. Safford had no instruments except a compass and a pocket level, and the appropriation was so small that he was without means to procure those additions without which a modern geologist would think himself almost helpless. Over much of the area he could not ride, and a great part of the eleven thousand miles travelled during the prosecution of the work was done on foot. He was compelled to live off the country and to endure the more than inconveniences from lodging in the uncomfortable homes of mountaineers. Yet in this modest volume he gave a complete conspectus of Tennessee geology which has borne the most exacting reviews. . . . For compactness and clearness this work is not excelled, perhaps not equaled by any other official report published in this country.

22 Memoir of James Merrill Safford, Bull. Geol. Soc. America, Vol. XIX, 1908.

In addition to the difficulties already enumerated Professor Safford labored under the altogether too common condition incidental to an unappreciative legislature. His salary, aside from being altogether too small, was, under the law of 1871, reduced to $300 a year, and during a part of the time even this was not provided for in the annual appropriations.

Hilgard's Work in Louisiana, 1869.

In 1869 E. W. Hilgard, acting under the auspices of the New Orleans Academy of Sciences, made a reconnaissance of Louisiana, a summary of the results of which was published in the *American Journal of Science* for 1869. The expense of the trip was paid partly by subscription and partly by an appropriation by the State Board of Immigration, and the time limited to thirty days. The journey (some 625 miles) was made mainly on horseback, passing Petite Anse and New Iberia on the Tèche by way of Opelousas to Bayou Chicot; thence to the Calcasieu River, down that stream to Lake Charles and the sulphur and petroleum wells on the West Fork of the Calcasieu River; thence north to Sabine Town, Texas; thence by way of Many to Mansfield, Louisiana; thence, crossing Red River at Coushatta Chute landing, to the salines on Saline Bayou, and thence, by way of Winnfield and Harrisonburg on the Ouachita River, where the expedition terminated. Among the more striking results announced was the fact that the Gulf coast had in late Quaternary times suffered a depression to the extent of at least 900 feet, and during the Terrace epoch reelevation to the extent of about half that amount. The occurrence of sulphur and gypsum beds was also noted. The various formations were described as: The *Port Hudson Group*, the *Orange Sand Formation*, the *Grand Gulf Formation*, the *Vicksburg Group*, and the *Mansfield Group.*

Winchell's Survey of Michigan, 1869.

In 1869 the geological survey of Michigan, which had been brought to a close in 1861 by the outbreak of the Civil War, was resuscitated through an act of the legislature, establishing a board of survey, consisting of the governor of the state, the president of the board of education, and the superintendent of public instruction, with power to select geologists, disburse the money appropriated, and perform other necessary acts. Under this law Alexander Winchell was again made director, and undertook personally the investigation of the Lower Peninsula, with the assistance of his brother, N. H. Winchell, W. M. Harrington, E. A. Strong, A. M. Wadsworth, C. B. Headley, A. O. Currier, and J. H. Emerson. Later (1873-1876), after Winchell's retirement, C. Rominger was appointed by the board to work on the Lower Peninsula also.

To T. B. Brooks was assigned the survey of the iron regions; to Raphael Pumpelly that of the copper regions of the Upper Peninsula, and to Carl Rominger a study of the Paleozoic rocks and their associated fossils. Brooks's report, submitted in 1873 and forming Part I of the first volume of the reports of this survey, was written with the idea of making it "as complete a manual as possible of information relating to the finding, extracting, transporting, and smelting of the iron ores of the Lake Superior region." With this in view, he presented, first, an historical sketch of the discovery and development of the iron mines; second, the geology of the Upper Peninsula, including the lithology; third, the geology of the Marquette iron region; fourth, the geology of the Menominee iron region; fifth, the Lake Gogebic and Montreal River iron ridge; sixth, a chapter on exploration and prospecting for ore; and seventh, the magnetism of rocks and use of the magnetic needle in exploring, concluding with chapters on the method and cost of mining specular and magnetic ores and the chemical composition of the ores.

The lithological work on the rocks of the region was performed by A. A. Julien, of New York, his report forming the second volume (298 pages) of the survey.

Brooks's Work.

Brooks's work contained little of a speculative nature. With reference to the association of magnetic and specular ores, he wrote:

> If we suppose all our ores to have been once magnetic, and that the red specular was first derived from the magnetite and the hydrated oxide (soft hematites) in turn from it, we have an hypothesis which best explains many facts and will be of use to the explorer.

Again, with reference to the ore of the Negaunee district:

> If we suppose tepid alkaline waters to have permeated this formation and to have dissolved out the greater portion of the siliceous matter, leaving the iron oxide in an hydrated earthy condition, we would have the essential character exhibited by this formation as developed on the New England, Saginaw range, and, as will be seen afterward, at the Lake Superior mine. This is offered not so much as an hypothesis to account for the difference as to illustrate the facts observed.

This view, so modestly put, contains in it, however, the germ of the conclusions arrived at by Van Hise some twenty-five years later.

He noted the monoclinal character of the deposit at the Washington mine property in the Marquette region, and described the ore of the Lake Superior specular and hematite workings and the Barnum mine as occupying the position of "the frustum of a hollow cone lying with its axis horizontal and its small end toward the east," which had been cut in two by a horizontal plane representing the surface of the ground.

Other points, which it is well to note, since Rominger in his later report had occasion to disagree with him, are, first, his regarding the ores of the Cascade Range as the equivalent of the Michigan and magnetic ores of the Mishigami district and as older than any of the iron beds in the Republic Mountain series; and, second, the Felch Mountain ore deposit as belonging to the lower quartzite, the ore itself resting immediately upon and being bounded on the south by hornblendic, micaceous, and gneissic rocks which are undoubtedly Laurentian. Subsequent studies by Wadsworth, Van Hise, and others have shown him, however, to have been substantially correct in both of these conclusions.

Ill health prevented him from carrying out his work in as thorough a manner as he wished, and his letter of transmittal was written from London, where he had gone to recuperate.

Sketch of Brooks.

Brooks, as may readily be inferred, was an eminently practical man. Indeed, his entire training, consisting of two years at the School of Engineering of Union College and a single course of lectures on geology under Lesley at the University of Pennsylvania, was of a practical nature. His early work was in connection with land surveys, but after his retirement from the army in the fall of 1864 he served a year on the geological survey of New Jersey under Cook, and then in 1865 became vice-president and general manager of the Iron Cliff mine, near Negaunee, in the Marquette district of Michigan. Here he began that geological work upon which is mainly based his reputation. The difficulties which he encountered were such as can be scarcely comprehended by those who have not visited the region. The country was, much of it, heavily forested and swampy as well. Outcrops were few and perhaps wholly obscured by the drift or by undergrowth. There were no maps, or, at best, the very poor ones furnished by the Land Office, no railroads, and transportation was limited to canoes and pack animals. There were few prospect holes and fewer developed mines. To these difficulties were added the complications due to the repeated folding and squeezing which the beds had undergone. Yet Brooks, by his persistency and originality in methods, succeeded in producing a work of value as a scientific production as well as of the greatest use to the prospector—a rare combination, indeed—and a work which has been superseded only by one that it took twenty years of study by an able corps of geologists and a hundred-fold better facilities to produce.

Fig. 80. *T. B. Brooks.*

He devised the dial compass and adapted the dip needle to the purposes of the prospector. Persistent and determined to succeed in spite of the poverty of appropriations, he expended over $2,000 of his own means and, worst of all, sapped his own vitality in the work to such an extent that he became a confirmed invalid before reaching middle age.

As already intimated, his health gave out in 1873 and he sought relief abroad, residing in London and Dresden, where his reports were completed. After his return to this country in 1876, he resided at Monroe and Newburgh, New York, and after 1883, during the winters, at Bainbridge, Georgia, living the life of a country gentleman and farmer.

Pumpelly's Work in Michigan, 1869-1873.

Pumpelly's work in the copper district is of interest on account of his theories regarding the age and lithological nature of the copper-bearing rocks and the origin of the metal itself. The conclusions at which he arrived were that, first, the cupriferous series was formed before the tilting of the Huronian beds upon which it rests conformably, and consequently before the elevation of the great Azoic area, the existence of which during the Potsdam period predetermined the Silurian basins of Michigan and Lake Superior. Second, after the elevation of these rocks and after they had assumed their essential lithological characteristics, came the deposition of the sandstone and its accompanying shales, as products of the erosion of these older rocks. These contained fossils which showed them to belong to the Lower Silurian, though it was felt to be still uncertain whether they should be referred to the Potsdam, Calciferous, or Chazy.

In a chapter on the paragenesis of copper and its associates he wrote:

> It is still an open question whether the trap which formed the parent rock of the melaphyr was an eruptive or a purely metamorphic rock. If it was eruptive, it was spread over the bottom of the sea in beds of great regularity and with intervals which were occupied by the deposition of the beds of conglomerate and sandstones. It should seem probable that the copper in the melaphyrs was derived by concentration from the whole thickness of the sedimentary members of the group, including the thousands of feet of sandstones, conglomerates, and shales which overlie the melaphyrs, and including melaphyrs also.

The translocation of the copper he thought to have been initiated by the sulphate resulting from the oxidation of the sulphide, but as this salt must have been soon decomposed by the abundant acid carbonate of lime, he could not suppose it to have been effectual in the final concentration of the large deposits, and it was probable that this last was accomplished by the more permanent solution of carbonate and silicate of copper.

Pumpelly was assisted by A. R. Marvine, L. P. Emerson, and S. B. Ladd.

C. Rominger's Work in Michigan, 1873-1876.

Rominger's report on the Paleozoic rocks of the Upper Peninsula formed Part III of the third volume of the survey reports (1873). His second appeared in 1876, forming one of the four large octavo volumes comprising altogether some 386 pages, with 55 plates of fossils, and a colored geological map of the area surveyed. The geological portion contains a record of the characteristic rocks, their geographical distribution, and their fossils. It should be noted that Rominger showed a disposition to disagree in many of his conclusions with Winchell, Brooks, and others who had preceded him, though not in all cases with good reason.

The presence of large bowlders in the midst or on top of well-stratified drift layers he conceived to be due to transportation by swimming icebergs during periods of flood. In this he agreed, substantially, with Dawson, of the Canadian survey. The dolomites of the Ida quarries, which Winchell identified as belonging with the

Onondaga salt group (Upper Silurian), he considered as Upper Helderberg (Middle Devonian), and he stated that the mapping by Winchell of Upper Helderberg rocks throughout a great portion of Cass, Van Buren, and all over Berrien County, was an error. He also differed with Winchell on the stratigraphic position of the Hamilton rocks of Big Traverse Bay. Winchell's Huron shales he considered, from paleontological evidence, to be identical with the Cuyahoga shales of the Ohio geologists, and therefore belonging to the lower part of the Carboniferous rather than the Upper Devonian. He also accused Winchell of a peculiar stratigraphic blunder in his section west of Flat Rock:

Unfortunately this section is laid across a synclinal undulation of the formation, and begins at one end with the same rock beds (Marshall sandstone), which on the other end are found very near the base. Under the impression that he was all the while descending, he stands again on the horizon from which he started.

The salt brines of the state, according to Rominger, were derived from rocks of the Waverly group, and not from those of a higher horizon, which Winchell had designated as the Michigan salt group. The most valuable part of the report is that relating to the fossil corals, which was, for its time, unsurpassed.

Subsequently (1881) Rominger, still acting as a state geologist (1878-1880), issued a report on the Upper Peninsula, which dealt almost entirely with the economic problems of the iron region.

He regarded the region about Marquette as

a synclinal trough of granite which, by the upheaval of its northern and southern margins, caused the inclosure of the incumbent sedimentary strata between its walls and their simultaneous uplift and corrugations into parallel folds by the lateral pressure from its rising and approaching edges.

Concerning the origin of the iron ores of the Upper Peninsula, he wrote:

These ore deposits are not regular sedimentary layers originally formed of iron oxide in a state of purity, but are evidently the product of decomposition of the impurer mixed ferruginous ledges by percolating water, leaching out the siliceous matter and replacing it by the deposition of oxide of iron held in solution.

This view is not greatly different from the conclusion reached by Brooks, as already noted, and subsequently by Van Hise.

His views as to the origin of serpentinous rocks were naturally not those commonly accepted today. Writing with particular reference to those of Presque Isle, he stated that they occur "generally in bulky, nonstratified masses which, if they ever originated from mechanical sedimentary deposits, are by chemical action so completely transformed as to efface all traces of their former detrital structure. They resemble volcanic eruptive rocks forced to the surface in a soft, plastic condition, and most likely heat was one of the prime agents in their formation, or else transformation, in combination with aqueous vapors." The explanation is certainly suggestive of the present view, which is that the serpentines—in the majority of cases—are eruptive rocks highly altered through a process of hydrometamorphism.

In his first report, that on the Paleozoic rocks of the copper district, Rominger took the ground that the Silurian age of the Lake Superior sandstone was unequivocally proven by its stratigraphical position. This is the view now generally held, though the Potsdam period, to which the beds are referred, is now considered as the upper part of the Cambrian instead of Lower Silurian, as at that date.

Sketch of Rominger.

Fig. 81. *Carl Rominger.*

Rominger's career, like that of Lesquereux and others that might be mentioned, offers an interesting illustration of the difficulties with which the early naturalists had to contend, particularly when foreigners but little acquainted with the language of their adopted country.

Born in Wurtemberg, he came to America in 1848 on account of revolutionary disturbances, and without previous preparation. Though trained as a physician and geologist, yet his poor command of English he felt excluded him from associating with scientific men, and, being without financial resources, he established himself as a physician in Cincinnati, a city containing a large German population. Here he remained for twenty-five years, improving himself in the language and devoting what time could be spared from his professional duties to the study of paleontology and geology. In 1870, through the influence of James Hall and others, he was appointed one of the geologists of the Michigan survey under Winchell, ultimately himself becoming director, in which position he remained until 1883, when, under a new administration, he was removed to make room for another.

His bibliography, although not numerous, is important, particularly that relating to paleontological matters. It had been his intention to continue his work on the fossil corals, but the political changes above mentioned prevented. The edition of his work issued by the state being insufficient to supply the demand, Rominger had printed 250 copies at a personal expense of $4.75 each, hoping to be able to sell them at least for the same figure. But this proved impossible, and he suffered a considerable loss thereby.[23]

Survey of Canada under Selwyn, 1869-1894.

Upon the recommendation of Logan, Dr. A. R. C. Selwyn, an Englishman, for several years connected with the geological survey of Great Britain under De la Beche, and from 1852 until 1869 director of the geological survey of Victoria, Australia, was, upon the retirement of the first-named, made director of the geological survey of Canada. In this capacity Selwyn served for twenty-five years, or until 1894. He was assisted by H. H. Ami, Elkanah Billings, and J. F. Whiteaves, paleontologists; L. W. Bailey, Scott Barlow, Robert Bell, Gordon Broome, G. M. Dawson, R. W. Ells, G. F. Matthew, Charles Robb, and H. G. Vennor, geologists. T. Sterry Hunt continued as chemist until 1873, when he was succeeded by B. J. Harrington. With this efficient corps the work of the survey was pushed vigorously, but, extending as it did beyond the

[23] Excerpt from letter of Carl Rominger to Charles Schuchert, dated Ann Arbor, November 30, 1903:

"My original intention was to continue the work on corals I had begun under the auspices of the Michigan Geol. Survey, but the installation of Gouvernor Alger made a sudden end to my position which I had filled for 14 years, as it seems tō the satisfaction of all the people concerned.

"To continue this work on my own expense I became totally discouraged after I had made the experience with the extra copies I had printed of the 3 volume on my own expense, urged to it by more than a hundred letters of persons wishing to obtain it from me after the state had no more of this volume to give away. I ordered two hundred and fifty copies printed and paid for each volume $4.75, wanted to sell them for the same amount, but to my surprise most of the persons ordering the volume were expecting it as a donation. With difficulty I could sell at the rate of $3.00 about 50 volumes, about 50 I gave away, and about 100 volumes are left in my hands unsold. This experience cost me about $800 direct loss and cured me of every attempt to edit a book at my own expense."

time limits laid down for this sketch, it can be touched upon but briefly.

During the period of Selwyn's administration twenty large annual reports were issued and nine volumes on paleontology and paleobotany. The work of the survey was pushed westward as far as British Columbia, and though ever with economic ends in view, much was accomplished in the way of pure science. But little space was devoted to a discussion of purely theoretical matters. He considered the gold veins of Nova Scotia to be in part of Carboniferous and in part of pre-Carboniferous age, and to have been filled through infiltrated thermal waters and gases, and he felt that there were no *a priori* reasons for supposing they should not continue sufficiently rich for commercial possibilities down to any depth to which mining operations were practicable. The silver deposits of Thunder Bay, on Lake Superior, were also investigated, and the post-Carboniferous age of the anthracite beds of Queen Charlotte Islands for the first time recognized. The stratigraphic problems involved in Logan's "Quebec group" received attention. His aim from the start, as stated by one of his biographers, was to make the survey an eminently practical department in which the records of the mines and mineral statistics should be kept for the use of both the Parliament and the public.

Fig. 82. *A. R. C. Selwyn.*

Selwyn is pictured to us as a scholar of rare ability, social, amiable, and chivalrous in private life, but a strict disciplinarian; tall, graceful, quick, and alert, of a rather highstrung and nervous disposition, and with a keen and observant eye. His bibliography consists mainly of short papers and summaries published in connection with his official reports.

T. Sterry Hunt, chemist to the survey until 1873, began first to make himself prominent in the reports for 1853-1856. He examined and wrote on the mineral waters, the chemistry and origin of the calcareous and siliceous rocks and the general subject of economic geology, including the metallurgy of iron, the saliferous possibilities of sea waters, and the utilization of peat as a source of paraffine, gas, and oils, in all and each of which subjects he seemed to be equally proficient and fluent. Keen enough to foresee the coming necessity of increasing the supply of potash to meet the requirements of the arts and agriculture, and recognizing the apparent insufficiency of the supply to be obtained from wood ashes, and the impossibility of its economical extraction from feldspathic rocks, he felt that what was known as the Ballard process of obtaining it from the bittern, or mother liquor formed in the manufacture of salt from sea water, might fulfill the required conditions. It was about this time, too, that he showed the possible origin of beds of rock phosphate through the gradual elimination of lime-carbonate from beds consisting largely of the remains of fossil lingulæ.

Fig. 83. *T. Sterry Hunt.*

In the report for 1856 Hunt for the first time put out clearly and distinctly his views ever after so strenuously (and vainly) defended relative to the origin of serpentinous rocks, which he regarded as either products of direct chemical precipitation or of a complex form of hydrothermal metamorphism. Even the very evident igneous nature of the serpentine of Syracuse, New York, as described by Emmons, Hall, and Vanuxem (pp. 229, 335, 336) failed to influence him. The dolomitic rocks and magnesites were likewise attributed to direct chemical precipitation from sea water, the possibility of which he proved by laboratory experiment, though unfortunately without reference to a possible identity of conditions.

Hunt's influence on American geology can, as a whole, scarcely be considered as beneficial. An exceptionally brilliant man, he began his career at a period when the possibilities in chemical geology were almost infinite and conditions most favorable for the founding of a lasting reputation. His early papers, as summarized in his volume, *Chemical and Geological Essays*, were inspiring and full of suggestive matter, and had he but held himself in check he might have passed into history as an honored leader.

Unfortunately he early developed an erratic tendency, and a disregard for facts that in any manner conflicted with or failed to substantiate his views. His *Crenitic Hypothesis* relating to the origin of crystalline rocks, in spite of its eloquent appeals to imaginary conditions, was never even provisionally accepted by the majority of workers, while his ideas on the origin of serpentinous rocks were shown to be for the most part utterly fallacious, though he adhered to them to the end in utter defiance of all proof. His *Natural System of Mineralogy*, one of the latest of his productions, made no impression on the scientific world at the time it was put forward (1885) and was speedily forgotten.[24] His influence waned long before his somewhat untimely death in 1892, while his private life led to his being socially ostracised by the majority of his clean-thinking acquaintances.

[24] "Hunt of the Canadian Survey is here [in Washington] lecturing on geology at the Smithsonian. He is a very poor speaker, but when he confines himself to his proper dept., chemistry, gives exceedingly interesting lectures. He is disposed to wander a little from his proper sphere and speculate much on the dynamics of geology without having seen enough of the subject to discuss it intelligently. He adopts Hall's untenable theory of the origin of mountains." (Newberry to Hayden, Feb. 27, 1860.)

CHAPTER VII.

The Era of State Surveys, Fifth Decade, 1870-1880.

ALTHOUGH matters were rapidly shaping themselves in favor of a more comprehensive system of surveys than could be carried on unassisted by the individual states and territories, and though, too, as noted in the last chapter, the National Government had sponsored already a number of exploring expeditions to which geologists or naturalists were attached, state and provincial governments remained by no means inactive. Important work was being done by T. C. Chamberlin in Wisconsin, J. W. Dawson in Nova Scotia, W. C. Kerr in North Carolina, E. A. Smith in Alabama, and J. D. Whitney in California. There was also organized the second geological survey of Pennsylvania, with J. P. Lesley at its head, and a geological and natural history survey of Minnesota, under N. H. Winchell. Both of these last-named organizations continued their work beyond the period of the limit set for this history. An attempt at the establishment of a geological survey in Georgia resulted in the appointment of George Little state geologist. Nothing of value was accomplished, however.

Newberry's Survey of Ohio, 1869-1878.

A second geological survey of Ohio was inaugurated in 1869, and J. S. Newberry appointed chief geologist. This survey was continued in operation until 1878, when, owing to a disagreement between Newberry and the legislature, appropriations were withheld and it came to an end. During the period of its existence two annual reports and four volumes of a final report were published, the latest bearing on the title-page the date of 1882.[1] These volumes were issued each in two parts, Part I of each of the first three being devoted to the physical aspects of the question, and Part II to paleontology. The second part of the third volume, intended as a monograph of the Carboniferous flora, never appeared, owing to failure of the legisla-

[1] An edition of 2,000 copies of this report was printed in the German language, as were also the annual reports of Rogers's survey of Pennsylvania and New Jersey.

ture to make the necessary appropriations. The fourth volume was devoted to botany and zoölogy.

Newberry's principal assistants in 1869 were E. B. Andrews, Edward Orton, who afterwards became state geologist, and John H. Klippart. In 1870 the force was increased by the employment of T. G. Wormley as chemist, and G. K. Gilbert, M. C. Read, Henry Newton, and W. B. Potter as local assistants. Still later others were added, among whom F. B. Meek, E. D. Cope, James Hall, O. C. Marsh, R. P. Whitfield, J. J. Stevenson, and N. H. Winchell, are worthy of mention. The paleontological work of the survey, as may be readily understood, fell to Newberry, Hall, Whitfield, Meek, and Cope, mainly, though Alleyne Nicholson described the Devonian corals and E. B. Andrews the fossil plants. As one of the results of their individual and combined investigations, it was announced that the so-called Cliff limestone had been resolved into seven distinct formations belonging to the two great geological systems—Devonian and Upper Silurian. The discovery of Oriskany sandstone in the northwest quarter of the state was announced and the Carboniferous age of the rocks of the Waverly Group thought to have been established on paleontological evidence. This has, however, been since shown to be erroneous.

The following sequence of events during later Tertiary times was established:

(a) In the Miocene and Pliocene epochs a continent several hundred feet lower than now, the ocean reaching to Louisville and Iowa, with a subtropical climate prevailing over the lake region, the climate of Greenland and Alaska being as mild as that of southern Ohio is now, while herds of gigantic mammals ranged over the plains.

(b) A preglacial epoch of gradual continental elevation which culminated in the glacial epoch, when the climate of Ohio was similar to that of Greenland at present, and glaciers covered a large part of the surface down to the parallel of forty degrees.

(c) This period was followed by another interval of continental subsidence characterized by a warmer climate and melting of the glaciers and by inland fresh-water seas filling the lake basin, in which were deposited the Erie and Champlain clays, sands, and bowlders.

(d) Another epoch of elevation which is still in progress.

Some attention was given to economic geology and the study of the coal beds.[2] From analyses it was shown that the change from woody tissues to peat or lignite, and thence to bituminous and an-

[2] Though apparently not enough to satisfy the legislature.

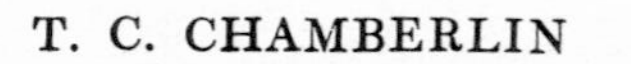

ALEXANDER WINCHELL

T. C. CHAMBERLIN

thracite coal and plumbago, consisted in the evolution of a portion of the carbon, hydrogen, and oxygen, leaving a constantly increasing percentage of carbon behind. This evolution Newberry conceived to be due to the disturbances which resulted in the uplifting of the mountain chains and metamorphosed the included rocks.

The coal beds of Ohio, it should be noted, were considered as always having been separated from those of Illinois by the Cincinnati anticline, which was thought in the report for 1869 to have formed a land surface over a considerable portion of its length, at least during the earlier and probably throughout all the Devonian ages. Later, in discussing the work of Orton in Adams County, Newberry wrote:

> Here we have an indubitable record of the elevation of the Cincinnati arch between the Upper and Lower Silurian ages, and proof that it is far older than the Appalachian system, with which it has been commonly associated.[3]

The carbonaceous matter of the Huron shales was suggested as probably due to an abundance of seaweeds which flourished in a kind of Saragossa sea which occupied the region during the period of deposition.

The petroleum and gas filling the cavities and interstices in the sandstones and conglomerates in the Oil Creek region were thought to have originated by a process of distillation at a low temperature, in the lower-lying Huronian shales, whence they had been forced upward by hydrostatic pressure.

The dolomitic character of the rocks of the Clinton, Niagara, and Water-lime series was ascribed to "a vital rather than a chemical or physical cause," and the occurrence of magnesia in the hard parts of some groups of marine invertebrates as shown by Dana, cited as a possible explanation of their origin. His views on the glacial drift are given elsewhere (p. 635).

For the rocks of the so-called *Blue Limestone* series of the early geological surveys he adopted the name Cincinnati Group, as first applied by Meek and Worthen. The gypsum of the Salina Group was rightly regarded as precipitated in continuous sheets and not to have resulted from a change in the ordinary limestones by sulphuric acid, as had been claimed for the gypsum beds of New York. This is the view now commonly accepted.

The Newberry-Whittlesey Controversy.

It appears that both E. B. Andrews and Colonel Whittlesey were aspirants for the position of state geologist at the time of New-

[3] See also Hall, p. 346.

berry's appointment, and to judge from the tone of an article by Newberry, published in the *Cincinnati Commercial,*[4] Whittlesey adopted rather unfair means to throw discredit upon his work. In his reply Newberry was very bitter, stating that whatever may have been his own qualifications for the work, Whittlesey was too old and in too poor health to do good work, and also that he was not a good geologist; that, further, he held to certain geological heresies which would impair his work; that he believed in the mineral origin of coal, and that the brown hematite ores of the Alleghanies were interstratified with the limestones instead of being mere pockets. He further claimed that Whittlesey was no paleontologist, and without paleontology no man could be a good geologist. As an illustration of his deficiency in this respect, Newberry referred to his (Whittlesey's) paper on the equivalency of the rocks of northeastern Ohio, in which he identified certain beds as equivalent with the Chemung, Portage, and Hamilton groups of New York, on paleontological grounds, whereas in fact every one of the twelve species of fossils on which this identification was based was wrongly named, the fossils actually being wholly of Carboniferous age.[5]

Fig. 84. *Charles Whittlesey.*

From a man of Whittlesey's character the return charge was naturally vigorous. In a pamphlet of ten pages, issued in 1873, apparently privately printed, and entitled *Paleontology and the Moral Sense,* he discussed his relations both with Newberry in Ohio and Hall in Wisconsin, beginning with the statement that "There is no apparent reason why the study of fossil remains should have a demoralizing effect, but the statements I am about to give concerning two leading paleontologists are decidedly suggestive."

And so indeed they were, but as we are trying to write a history of American geology, rather than the bickerings of individuals, they may well be passed over here.

[4] March 28, 1870.

[5] Nevertheless Stevenson credits him with having correctly predicted the existence of iron ores in the Vermillion Range of Minnesota, as subsequently verified by Eames. (See also p. 356.)

PLATE XXIV

EDWARD ORTON

Orton's Appointment as State Geologist.

Newberry was succeeded in office by Edward Orton, who had previously acted as chief assistant.[6] During Orton's administration certain "pardonable errors in identification," which left the stratigraphy of the coal series of Ohio in an almost hopeless tangle, were corrected. He showed the stratigraphical order of the lower Coal Measures of Ohio to be completely in harmony with that of Pennsylvania and that the entire series could be traced from the eastern margin of the state across the same to Kentucky. This is considered by I. C. White to be the masterpiece of Orton's purely geological work, although his contributions to the geology of petroleum and natural gas in the sixth volume, 1888, are of almost equal importance.

Orton's work during his whole life was largely of an economic character, the more important and comprehensive publications being Volumes V (1884) and VI (1888) on the *Economic Geology of Ohio*, his *Report on Petroleum and Gas in Ohio* (1890), and the *Report on the Occurrence of Petroleum, Natural Gas, and Asphalt Rock in Western Kentucky* (1891). He also had an important paper on the Trenton limestone as a source of petroleum and inflammable gas in Ohio and Indiana, in the *Eighth Annual Report of the United States Geological Survey* (1886-1887). Aside from his record as president of the university and a teacher, Orton will be best remembered for his work on the subjects of gas and petroleum, although in his report on the third geological district he makes important observations on the Cincinnati uplift or axis, showing it to have been a slow and gradual formation resulting in a gentle flexure in the earth's crust involving the Lower and Upper Silurian and, to some extent, the Devonian formations of the state. In his own words, his conclusions were as follows:

> First, the Cincinnati axis in southern Ohio was raised above the sea at the end of the blue limestone period, or certainly early in the history of the Clinton epoch. Second, it underwent various oscillations, but the elevatory movements succeeded those of depression. Third, the rate of movement was exceedingly slow.

Sketch of Orton.

Orton was born in Delaware County, New York, and educated at Hamilton College, graduating in 1850. He subsequently studied at

[6] S. A. Miller, a Cincinnati lawyer, with inclinations toward paleontological studies, is stated to have been a candidate for the position.

Harvard and then entered the Andover Theological Seminary, being licensed to preach in 1855, and soon after ordained as pastor of the Presbyterian Church at Downsville, Delaware County, New York. He resigned this position in order that he might accept that of professor of natural sciences in the New York State Normal School at Albany. Becoming convinced, however, that his gradually changing views on religious matters were such that he could not conscientiously continue to hold this position, he resigned it, and accepted the principalship of an academy at Chester, in Orange County, the same state.

In 1865 he became principal of the preparatory department of Antioch College in Ohio, then professor of natural sciences, and afterward president of the same institution.

His active work as a geologist received its first recognition in 1869, when he was appointed Newberry's assistant. In 1873 he was made president of the new Agricultural and Mechanical College, founded under the Morrill act, and also took charge of the chair of geology in the same institution. Under his efficient administration, which lasted until 1881, the institution prospered and finally developed into the Ohio State University.

In 1882, after his voluntary retirement from the State University, he was made state geologist, as already noted.

Orton belonged to the generation beginning work immediately after the Civil War and, according to his biographer, always leaned toward the application of the science to the benefit of his fellow men. "He was painstaking and exact in observation, scrupulous in statement, cautious in speculation." He was one of the first to recognize the possibility of the exhaustion of the supply of petroleum and natural gas, and to issue appeals to the people of Ohio, urging care in husbanding their resources. But these were not received in the spirit in which they were offered, and he had the melancholy satisfaction of seeing his forebodings justified by the event.

By those who knew him he will be remembered as a man of perfect courtesy, dignified, a little stately, and never effusive. His life was full of considerate and helpful kindness—modest and retiring to an unusual degree, and yet one of the few men to whom honors come notwithstanding. In 1891 he suffered from a paralytic stroke which cost him the entire loss of the use of the left hand, yet he continued to teach and to work until almost the last hour of his life. For eight years he had looked upon death as a thing momentarily to be expected. He met it bravely, cheerfully, and fearlessly on October 10, 1899.

Cox's Survey of Indiana, 1869-1879.

In 1869 the fourth attempt at a systematic geological survey of Indiana was made, the appointment as state geologist going this time to E. T. Cox, heretofore known to geological fame only through his work while assistant to Owen in Kentucky and Arkansas in 1856-1860.

Annual reports were issued for each of the ten years which marked the life of this survey. Those of 1869 and 1872 were accompanied by county maps, though no geological map of the state in its entirety was furnished. A colored section across the state from Greencastle to Terre Haute accompanied the report for 1869.

Cox was assisted during the entire or a part of the time by Frank H. Bradley, Rufus Haymond, G. M. Levette, B. C. Hobbs, R. B. Warder, W. W. Borden, M. N. Elrod, John Collett, and E. S. McIntire, the fossil flora being described by Leo Lesquereux and the fauna of Wyandotte Cave by E. D. Cope. Other zoölogical and botanical subjects were treated by D. S. Jordan, J. M. Coulter, and J. Schenk.

These reports as a whole contained little new or impressive. In the eighth, which was the most comprehensive thus far issued, Cox himself called attention to the fact that the geological history of the state "appears tame and devoid of the marvelous interest which attaches to many other regions, and that there is not a single true fault or upward or downward break or displacement of the strata thus far discovered." The oldest rocks of the state were found in the southern portion, extending from the Ohio River near the mouth of Fourteen Mile Creek to the eastern boundary line. These were the so-called *Hudson River* rocks of Hall, which Cox correlated with Safford's Nashville Group, and which Worthen and Meek had included under the name of *Cincinnati Group.* He agreed with others in thinking that the Silurian strata were uplifted, not by a local disturbance, but "by an elevating force that acted very slowly and extending over the entire central area of the United States." The seat of greatest force, he thought, however, was not limited to southwestern Ohio, but was to be looked for in Kentucky.

Cox accepted the general theory of glacial drift as at present understood, and conceived that the climatic changes might be due to the relative position of land and water, possibly a change in the course of the Gulf Stream (see p. 252).

Sketch of Cox.

Cox was a Virginian by birth, but while yet a young boy his family moved to New Harmony, Indiana, and joined the communistic colony founded by Robert Owen. Here he was educated, pursuing his geological studies under David Dale Owen, whose assistant he subsequently became on the geological surveys of Arkansas and Kentucky. After the death of Owen in 1860, he was engaged in commercial work, and in 1865, in connection with the state survey of Illinois under Worthen, reported on the coal beds of Gallatin County and later others in the southern portion of the state. During the time that he was geologist he also occupied the chair of geology in the State University. On retiring from both positions in 1880 he went once more into private work, and for a time was chemist to the Portland Phosphate Company, in Albion, Florida.

C. H. Hitchcock's Work in New Hampshire, 1868-1878.

In 1868 in response to a renewed demand a geological survey of New Hampshire was inaugurated, of which C. H. Hitchcock, son of Dr. Edward Hitchcock, and already mentioned in connection with the work in Maine and Vermont, was made director.

The survey continued until 1878, and from time to time the following assistants were employed: J. H. Huntington, Warren Upham, George W. Hawes, George L. Vose, C. A. Seeley, and A. M. Edwards. Professor Thomas Eggleston of Columbia College was first employed to do the petrographic work, but was obliged to resign on account of ill health and the work was completed by the Dr. Hawes above mentioned. It was in connection with this survey that Hawes prepared the work on microscopic petrography which is elsewhere mentioned. A. M. Edwards's work was limited to the study of diatoms.

The total cost of the survey was upward of $32,000, and the results were given in the form of three quarto volumes published in 1874, 1877, and 1878. From the study of the rocks a triple succession was thought to have been discovered; first, gneiss; second, feldspathic mica-schists; third, hydromica and chlorite schists; and the formations were correlated, as far as possible, with their extensions into Canada. The first and third divisions were considered as Laurentian and Huronian, respectively, while the middle division, that of the feldspathic mica-schists, was given the local name Montalban. All of these were grouped under the general name of Eozoic.

Mineral characters were used chiefly in distinguishing the various

PLATE XXV

RICHARD OWEN

E. T. COX

JOHN COLLETT

divisions, and the foliated igneous rocks were not at the time separated from the related gneisses, supposedly metamorphosed sedimentaries. Many of the quartzites, mica-schists, and slates were, on stratigraphic grounds, referred to the Paleozoic column, although a few Silurian fossils determined satisfactorily the age of certain limestones, slates, and sandstones in the Connecticut Valley.

Murrish's Work in Wisconsin, 1870.

In 1870 John Murrish was appointed by Governor Lucius Fairchild, of Wisconsin, commissioner for the survey of the lead district of that state. Murrish was, according to his own statement, "a practical man" and had served an apprenticeship in the mines of Cornwall, going through the regular course of training for a miner's occupation and a miner's life. Under these conditions, it is perhaps scarcely just to compare his writings with those of men who have had better opportunities. Nevertheless, as he had to do with the survey of an important mining region, it is impossible to ignore his work.

Fig. 85. *John Murrish.*

Under the title *Report of the Geological Survey of the Lead Regions*, Murrish published his observations in the form of a pamphlet of sixty-five pages, in which he set forth his views of the various geological phenomena. He thought to have discovered that the lead-bearing fissures were grouped into ranges with a general east and west trend, the various ranges forming themselves into four well-defined belts of mineral lands running parallel to each other. These he thought to lie at right angles with an axis of elevation running in a generally north-and-south direction, the elevation itself being "a line of physical disturbance." Concerning the character of this physical disturbance he was not perfectly clear. He did not consider it due to an active volcanic disturbance nor earthquakes in the ordinary meaning of the terms, but rather to groups of fissures or faults in the Plutonic and Azoic rocks beneath, which were themselves produced by mechanical forces generated by internal heat. Water entering between the beds would percolate downward through these lines of fracture, where it would come in contact with intensely heated matter under a pressure of several hundred feet of overlying rock. If the temperature was sufficient the water would be converted into steam or elastic vapor, which might possess sufficient mechanical power to bring about the elevation. During the early formation of the stratified rocks, particularly the Potsdam sandstone, the resistance to this expansive force would be comparatively little, since vent for the steam would be easily found through the loosely accumulated sand; but as layer after layer was added to the strata and the more compact limestone began to form and harden above it, resistance would increase until to overcome it a general lifting of the strata would take place, by which escape would be effected through fissures along the line of the original faults in the Plutonic rocks below.

A microscopic examination of the sand grains from the disintegrated Potsdam sandstone having revealed the crystalline nature of the granules, due, as is now known, to the deposition of interstitial silica, he conceived, as did Whitney (p. 349), that the entire deposit was of chemical origin.

> Supposing Iceland should be submerged to a considerable depth beneath the ocean, and those plains situated about 30 miles from that noted volcano Hecla, known now to be full of heated springs, steaming fissures, and boiling geysers, whose waters hold a large amount of silica in solution that is now being deposited on the surface around those places, were pouring their waters into the ocean above, should we not have there on a small scale what perhaps existed on a very large scale during our sandstone formation?

Perhaps so. Who shall say?

Like many men of slight training, or self-trained, Murrish failed to give discrimination and weight to the evidence gathered and was led into many errors, the most serious of which was that of assuming that the Lower Magnesian limestone might prove to be an ore-bearing stratum—this in spite of the opinion to the contrary held by Hall, Whitney, and others.

Marsh's Scientific Expedition, 1870.

During the summer of 1870 O. C. Marsh, professor of paleontology in Yale College, began a series of expeditions into the western part of the United States, having for his primary object the collection of vertebrate fossil remains. The results of these expeditions soon placed him among the leading vertebrate paleontologists of the world.

The Middle West, it should be remembered, was at that time an almost unknown territory, traversed by but a single railroad (the

Union Pacific), and much of it rendered unsafe for the white man through roving bands of Indians, which necessitated military escort in many instances. The expeditions were supported largely at Professor Marsh's private expense, until the organization in 1879 of the United States Geological Survey under Powell, shortly after which Marsh was appointed a United States paleontologist, though still drawing upon his own resources when necessary or when, in his opinion, it became advisable in order that new discoveries might become immediately available.

Under these joint agencies there was brought together the mass of material now forming the nucleus of the vertebrate collections at Yale University and the extensive "Marsh collection" in the National Museum in Washington, which have formed the basis for the numerous monographs and papers included in Marsh's bibliography.

The most noted of his early discoveries was that of the toothed birds, Odontornithes (Hesperornis and Ichthyornis), which furnished the material for monograph VII of the 40th Parallel Survey, published in 1880. Based on later discoveries were monographs on the Dinocerata and the extraordinary dinosauria, of which the Anchisaurus, Brontosaurus, Laosaurus, Ceratosaurus, Camptosaurus, Stegosaurus, Claosaurus, and Triceratops are the best-known representatives.

Marsh's connection with the United States Survey continued up to the time of his death, which took place in March, 1899, and much of his work lay beyond the time limits of this history.

Hager's Work in Missouri, 1870.

In 1870 the legislature of Missouri again passed an act for the establishment of a mining bureau, the board of control of which was to be composed of the Governor and nine members, one from each congressional district. Upon this board was conferred the power of appointing a state geologist, the choice for which fell upon Albert Hager, of Vermont.

Hager's term of service was, however, short, and but one report of progress was published, owing to a disagreement with the board. This report dealt largely with the condition of the survey at the time he took charge. It would appear that, since the suspension of Swallow's survey in 1861, its collections and other property had been stored in the university building at Columbia, which was at one time in the possession of armed troops, and the materials suffered, as would naturally be expected. He found, however, reports of B. F. Shumard in a fairly complete condition. These were on the counties of Crawford, Clark, Cape Girardeau, Phelps, Ste. Genevieve, Ozark, Douglas, Perry, Jefferson, Laclede, Pulaski, and Wright. The reports of Swallow and Meek were also in a satisfactory condition. It is stated, too, that he obtained much material relating to the former surveys from Shumard and Swallow, with apparently the intention of editing their notes and publishing them. So far as known, however, nothing was done, and on March 18 Hager was removed and the law governing the bureau amended. Dr. J. G. Norwood, who had been one of Hager's assistants, remained temporarily in charge from September 1, 1871, until November of the same year, when R. Pumpelly was placed at the head of the organization. Norwood, during his brief period of authority, was assisted by G. C. Broadhead and C. M. Litton.

It appears from correspondence to which the writer has had access that the names of T. Sterry Hunt and James Hall were considered in connection with the appointment. Swallow was opposed to Hager and expressed himself favorable to Hall in case he failed to secure the position for himself. Naturally he was greatly disappointed at the final outcome. "I felt the labor of my life was lost, and I may as well fall back into nonentity."

There is little to show what may have been Hager's qualifications as a geologist or administrator. He appears only in connection with the surveys of Vermont and Missouri, and does not seem to have gained the respect of his associates. Indeed, he was very severely criticised by Swallow, though perhaps somewhat unjustly. After his failure in Missouri he disappeared wholly from the geological field.[7]

Pumpelly's Survey of Missouri, 1871.

Under the management of Pumpelly there were issued in 1873 two volumes of reports, the first a royal octavo of 325 pages which contained all of the previously unpublished material that had been transmitted to him by earlier workers. This comprised the reports of Broadhead, Meek, and Shumard, and contained eight geological maps of counties and numerous sections.

In this same year there was published, also under the supervision of Pumpelly, a preliminary report on the iron ores and coal fields, forming a volume of 440 pages and 190 illustrations in the text, with a large folio atlas of 15 sheets. Pumpelly was assisted in the preparation of this work by G. C. Broadhead, W. B. Potter, Adolph Schmidt, and C. J. Norwood as geologists; and Regis Chauvenet, chemist.

[7] Appendix, pp. 689-692.

In the report on the iron and coal, Pumpelly seems directly responsible only for the ideas advanced in the first twenty-eight pages. It is worthy of note that even at this late date he seems to have considered the porphyries of Pilot Knob and adjacent regions sedimentary Azoic rocks, the exposed portions of the skeleton of the eastern part of the Ozark Range, rising from 300 to 1,800 feet above the level of the surrounding country. "They form an archipelago of islands in the Lower Silurian strata, which surround them as a whole and separate them from each other." The rocks overlying these he thought to belong to the oldest members of the Silurian, and perhaps the deep-seated equivalents of the Potsdam sandstone, or even older.

He recognized the fact that the surface ore at Iron Mountain was a residuary product resulting from the disintegration and gradual removal of the siliceous rocks in which the ore was originally embedded, the instance offering an extreme case where decomposition of the porphyry in mass facilitated the separation of the ore from the rock and the mechanical removal of the latter. Concerning the ultimate origin of the iron, Pumpelly was silent. Regarding the origin of the manganese ore, as at Cuthbertsons Hill, in township 33, he wrote:

It would seem that we have in these occurrences instances of replacement, but it is difficult to imagine a direct substitution of manganese oxides for the decomposition products of a porphyry, and all the more so in this case from the fact that the analyses show the remaining porphyry, which is intimately associated with the ore, to have its normal constitution.

He further described what he called a metamorphic limestone at Huffs, near Ackhursts, as "nearly wholly changed into a porphyry or jasper rock, it having here a schistoid structure in which the alternate laminæ are impure, compact carbonate of lime. . . . Here is a member of the porphyry series which was originally, unquestionably, a limestone, but in which the original physical and chemical characteristics have almost wholly disappeared. It should not seem impossible that the manganiferous rocks which have been described may have had a similar origin, and that the manganese and iron oxides owe their present existence to a former replacement of the lime-carbonate by iron and manganese salts. The porphyry which now surrounds these ores may be due to a previous, contemporaneous, or subsequent replacement of the lime-carbonate by silica and silicates." A "may be" to which few would assent at this date.

A general discussion of the iron ores and their distribution was comprised in the report of Chauvenet, which occupied 192 pages of the volume. According to this writer the ores of Iron Mountain were due to deposition from solution in water which filtered into fissures. Adolph Schmidt regarded them as replacement products and attempted to explain their formation on the following somewhat labored hypothesis.

When a solution of sulphate or chloride of iron containing also carbonic acid remains for any length of time in contact with a porphyry the latter will become decomposed, the alkalies being set free in the form of carbonates. The carbonates, reacting on the iron-bearing solutions, will precipitate the iron as oxides, which will fill the pores in the porphyry made by the removal of the alkalies. The silica set free in the process would be to some extent removed, but most of it would remain. The silicate of alumina would be decomposed by bicarbonate of iron in solution and removed in the form of a soluble bisilicate, as explained by Bischoff.

It is needless to say that these views were not generally accepted.

George Catlin's Views on Geology, 1870.

The annals of American literature present—with the possible exception of Owen's *Key to the Geology of the Globe*—no more extraordinary publication relating to American geology than that of the artist, George Catlin, entitled *The Lifted and Subsided Rocks of America, with Their Influences on the Oceanic, Atmospheric, and Land Currents.* This was published in London in 1870.[8]

One can forgive any amount of ignorance relating to the subject of geology in a man of Catlin's profession, but it is not so easy to forgive him for putting before an indiscriminating public opinions which are founded on wholly insufficient, and in many cases visionary, data.

Catlin had traveled extensively throughout the western portions of North America and naturally had been attracted by the enormous amount of erosion and uplift manifest in a treeless area. That his thoughts should have been turned in this direction is not at all strange, but his conclusions are such as can be accounted for only on the grounds that he had received absolutely no training, had not learned how to observe, nor how to reason from that which he saw.

The most striking features of the work are those relating to the cause of the Gulf Stream and the origin of the Gulf of Mexico. He conceived that from both North and South America there issued

[8] A consideration of the first half hundred pages of the Preliminary Report on the Structural and Economic Geology of Missouri, Bureau of Mines and Geology, 1900, might lead to the making of another exception!

two large subterranean streams of water, one flowing from the north to the south under the main axis of the Rocky Mountains, and the other in an opposite direction along the main axis of the Andes. These becoming heated through proximity to the volcanic fires in the region of the equator, finally issued, giving the necessary volume and temperature to account for the Gulf Stream.

The origin of the gulf itself he imagined to be due to an undermining of the crust of the earth and its subsequent falling in, through the solvent action of the heated water, and he figures the continent both before and after the catastrophe.[9]

F. V. Hopkins's Geology of Louisiana, 1871.

In 1871 F. V. Hopkins, professor of geology, chemistry, and mineralogy in the State University of Louisiana, issued a report on the geology of the state, relating particularly to the post-Tertiary deposits, and containing a colored geological map.

These deposits he divided into (1) Drift, (2) Port Hudson (so-called by Hilgard), (3) the Loess, and (4) the Yellow Loam, the last three being included also under the general name of Bluff Formation.

Hopkins held (and Dana approved) that the Port Hudson and overlying beds were deposited when the land was at a lower level than now, and that the loess was an accumulation over an old flood plain of the Mississippi, as suggested by Lyell. The drift itself, however, he conceived to be due to the agency of icebergs in a sea at least 1,159 feet deep in the Ohio Valley. To this last view Dana, of course, objected.

J. D. Whitney's Views on Mountain Making, 1871.

In 1871 J. D. Whitney, in a series of popular articles on earthquakes, vulcanism, and mountain building,[10] expressed sundry opinions on the last-named subject which are worthy of reference. He took exception to Hall's theories (*ante*, p. 384), both in the original and as defended by T. Sterry Hunt and G. L. Vose,[11] on the ground that he (Hall) had mistaken an effect for a cause, and claimed that

9 Catlin's book was noticed in the American Journal of Science, Vol. L, 1870, and dismissed with the following curt remark: "The writer of this work, well known for his travels among the American Indians, here treats of mountain drainage, upheavals, metamorphism, making of mountain chains, sinking of mountains, and of the Indian races of America. He presents his geological views and criticisms with great positiveness, which is consistent with the fact of his limited knowledge of the subject."

10 North American Review, Vol. CXIII, 1871.

11 Whitney seems to have resented Vose's venture into the field, and refers to him as "one who prints Civil Engineer after his name on the title page, as if he

the sedimentary beds of the Appalachians are thick because mountains preexisted from the destruction of which they could be formed; not that having been already formed they were afterwards made into mountains. As he viewed the problem the great thickness of detritus forming a range came from a more elevated region which has since disappeared, and this disappearance was in the nature of a subsidence which necessarily brought a tangential pressure to bear on edges of the adjacent sedimentary beds, throwing them into a series of folds and producing other evident disturbances. In view of what is to follow, attention may be called to the weak point in Whitney's argument, in that it necessitated a preexisting range, neither the origin nor cause of subsidence of which was accounted for. That the uplift was recognized as due to lateral thrust is, however, important.

Le Conte's Ideas on Mountain Making, 1872.

The year following the subject of the origin of the great features of the earth's surface, including the formation of mountains, was brought up once more by Joseph Le Conte, then professor of geology and natural history in the University of California. In a series of articles in the *American Journal of Science* for this year he reviewed the ideas of Humboldt and others; discussed the probable condition of the earth's interior, whether fluidal or otherwise; and announced himself as convinced that the whole theory of igneous agencies, which formed the foundation of theoretic geology, should be reconstructed on the basis of a solid earth.

On the assumption that such an earth would be not homogeneous, but that some areas would possess greater conductivity than others and would, therefore, cool and contract more rapidly in a radial direction, he affirmed his belief that the present sea bottoms represented the areas of most rapid cooling and contraction and the continents and mountain ranges those of the least. This he felt was borne out by the researches of physicists who had shown that the continental masses were less dense than suboceanic matter, and was, moreover, largely in accordance with the views earlier expressed by Dana.

To him the earth might be regarded as composed of concentric isothermal shells, each cooling by conduction. The exterior being the first to solidify, would, through the shrinking away of the still cooling interior, become subjected to powerful horizontal pressure

feared, by some possibility, he should be taken for a geologist." Vose's book, Orographic Geology, 1866, was merely a review and discussion of the works of others and added little to the knowledge of the subject.

or thrust, which as time went on would find relief in the direction of least resistance (*i.e.*, upward) and along lines of weakness. It was, however, his idea that this yielding was not by upbending into an arch, but by a mashing or crushing together horizontally like dough or plastic clay, with more or less folding of the strata and an upswelling or thickening of the whole squeezed mass.

He showed that, were a mass of sediments 10,000 feet in thickness subjected to horizontal pressure and crushing sufficient to develop well-marked cleavage structure, a breadth of 2½ miles would be crushed into 1 mile, and 10,000 feet thickness would be swelled to 25,000, making an actual elevation of the surface of 15,000 feet. He therefore felt justified in asserting that the phenomena of plication and of slaty cleavage demonstrated a crushing together horizontally and an upswelling of the whole mass of sediments in this case alone sufficient to account for the elevation of the greatest mountain chains.

To Hall's theory, previously noted (p. 384), he also objected, since it wholly failed to explain the actual process by which the chains had been formed; and, taking into account the breadth of the Appalachians (at least 100 miles), he showed that the gentleness of the supposed convex curve of Hall would not produce the amount of crushing necessary for the formation of the immense plication.

He pointed out the fact, moreover, that sedimentation and subsidence were going on together, and therefore that the upper surface was probably never convex at all, but nearly or quite horizontal all the time. Subsidence under such conditions might produce horizontal tension or stretching of the lower strata, but could not produce the crushing and plication of the upper.

To Whitney's idea that plication was the result of the subsidence of a mountain axis he likewise took exception, and contended that chains and ranges were, beyond question, produced by a horizontal thrust crushing together the whole rock mass and swelling it up vertically, the thrust itself being the necessary result of secular contraction of the interior of the earth. In other words, "mountain chains are formed by the mashing together and upswelling of sea bottoms where immense thicknesses of sediments have accumulated, and as the greatest accumulations usually take place off the shores of continents, mountains are usually formed by the uppressing of marginal sea bottoms." He felt that the submarine ridges and hollows shown by the soundings of the Coast Survey to exist along the course of the Gulf Stream, and extending from the point of Florida to the coast of New England, might be true submarine mountain ranges now in course of formation by the processes already described.

Referring to the subject of metamorphism of rocks, as shown in the mountain chains, he argued as follows: Supposing sediments accumulating along the shores of a continent. The first effect is lithification, and therefore increasing density, causing contraction and subsidence *pari passu* with the deposit. Next, if the sedimentation continue, aqueo-igneous softening or even melting would follow, not only of the lower portions of the sediments themselves, but of the underlying strata upon which they were deposited. Finally, this softening would be sufficient to cause a yielding to horizontal pressure along the line, and a consequent upswelling of the line into a chain. Even the granitic axes of mountain ranges he thought might be, in most cases, but the lowermost, and therefore most highly metamorphosed portions of the squeezed mass, exposed by erosion.

He agreed with Richthofen and Whitney in that the great masses of lava, often constituting the chief bulk of mountain chains, had not come from craters but from fissure eruptions, and that volcanoes were themselves only secondary phenomena, produced by the access of meteoric water to the still hot interior portions. He did not, however, agree with them in assuming that this fluidal mass was a part of a universally incandescent liquid interior, but rather a submountain reservoir locally formed. He felt that this theory was powerfully supported by the views of Rose, Bischoff, and others, who regarded the surface materials as having passed by perpetual cycles through all the stages of rocks and soils carried away and deposited as sediments; consolidated into stratified rocks; metamorphosed into gneiss, granite, or even lavas; to be again reconverted into soils, and recommence the same eternal round.

In furtherance of these same views, Le Conte in 1876 gave, in the *American Journal of Science*, the results of observations made in the Coast Range Mountains of California along the line of the Central Pacific Railroad. He found the largely unaltered rocks here thrown into a series of anticlines and synclines with angles of dip varying from sixty-five degrees to seventy degrees. From this he estimated that the original matter as deposited on a sea bottom must, in the building of the range, have been crushed from a breadth or width of fifteen to eighteen miles into six miles, with a corresponding upswelling of the whole mass.[12] Again in 1878, in the same journal and

[12] Some ten years later (*i.e.,* in 1885) E. W. Claypole read a paper before the British Association at Montreal in which he reviewed the work of previous surveys so far as it related to the Pennsylvania Appalachians, and reached the conclusion that in this process of crumpling by lateral thrust the superficial

on the same subject, he argued that mountain ranges were always formed by horizontal pressure, and that this pressure on a large scale could be produced only by the interior contraction of the earth. To overcome the objections raised to this theory, to the effect that interior contraction could not concentrate its manifest results along certain lines without such a slipping and shearing as is impossible in a solid earth, he conceived the presence of a solid nucleus and a solid crust separated by a zone of fused or semifused matter. To an objection that contraction based upon shrinkage alone was inadequate to produce the evident results, he argued that there might be other causes, such as loss of water in form of vapor by volcanic action, and still others concerning which as yet nothing was known. His theory of origin, as he summed it up, consisted of: (1) A stage of preparation by sedimentation; (2) a stage of yielding by horizontal pressure; and (3) a stage of erosion or mountain decay.

Reference should here be made to a paper by Le Conte in the same journal for August, 1877, in which he questioned the correctness of the uniformitarian view, and argued rather in favor of long continued periods of crustal quiescence or repose followed by briefer periods of rapid movement. Only on such a supposition could he account for the almost complete blotting out of certain forms of life and the appearance at a later period of wholly new and more highly developed forms, without intermediate stages. "Geological history," he wrote, "like all other history, has its period of comparative quiet, during which the forces of change are gathering strength, and periods of revolution, during which the accumulated forces manifest themselves in conspicuous changes in physical geography and climate, and therefore in rapid movements in the march of evolution of organic forms—periods when the forces of change are potential and periods when they become active." The latter he referred to a "critical period" in the earth's history. The gaps in the geological record, due to uplift and erosion, and of which there is found no paleontological record, he compared to lost pages in a book. Of these, the first and greatest was that occurring between the Archæan and Paleozoic (as the term was then used); the second that between the Paleozoic and Mesozoic, and the third and least that between the Mesozoic and Cenozoic. The Quaternary he included also as a transition, or critical period.

layer of crust had been reduced from 153 to 65 miles or, otherwise expressed, 88 miles of surface had disappeared. This surface reduction he argued would result in a radial reduction of 13 miles. These phenomena he thought due to the contraction of a gradually cooling and shrinking globe (Am. Nat., March, 1885, p. 257).

Sketch of Le Conte.

Le Conte was born in Liberty County, Georgia, February 26, 1823, and grew to manhood under influences of ease and enjoyment such as have fallen to the lot of few American geologists. Educated as a physician, he early gave up practice and in 1850 removed to Cambridge, in Massachusetts, to become a student of Agassiz. In less than two years, however, he returned to Georgia, where he became professor of natural science at Oglethorpe University. His stay here was brief. In December, 1852, he became attached to the University of Georgia, at Athens, and, in 1856, professor of chemistry and geology in South Carolina College, at Columbia. Here he remained, enduring the vicissitudes of the Civil War, but abandoned his beloved South in 1869 to become professor of geology, zoölogy, and botany in the University of California, at Berkeley. Here he remained until his death, in 1902.

Le Conte was in no sense merely a geologist, nor were his mental activities limited to the geological field. With the possible exceptions of Lesley and Shaler none of his contemporaries ventured to think and write on such a variety of topics. Yet, singularly enough, though he covered in addition to geology, such subjects as "Morphology and Its Connection with the Fine Arts"; "Phenomena of Binocular Vision"; "Genesis of Sex"; "Immortality in Modern Thought"; "Glycogenic Functions of the Liver"; and "Evolution and Social Progress," his expressed ideas were always readable and worthy of consideration.

As a teacher and educator he was one of the most remarkable of men in the ranks of American geologists, as well as one of the most loved and admired for his personal qualities.

Hilgard's Work at Petite Anse, Louisiana, 1872.

In 1866 E. W. Hilgard, at the suggestion of Professor Henry, of the Smithsonian Institution, visited the salt deposits of Petite Anse and published the results of his observations in the Smithsonian Contributions to Knowledge in 1872, under the title *On the Geology of Lower Louisiana and the Salt Deposits of Petite Anse.*

As was to be expected, Hilgard differed completely with Thomassy, whose work has been noted, and who, it will be remembered, regarded the island as of volcanic origin. On the contrary, the island, and others of the group, were considered Cretaceous outliers, with cappings of drift and other alluvial matter, the salt beds themselves being of Cretaceous age.

Both Hilgard and Thomassy believed that the sand and shingle detritus covering a large portion of the states bordering on the lower Mississippi was due to a great flood, which might have resulted from the rapid melting of the northern glaciers. Referring to and apparently accepting the expressed opinion of the western geologists to the effect that the main body of the drift was due to floating icebergs on an inland sea, Hilgard conceived the northern limits of this sea to have been fixed by the ice barriers. Toward the east, southeast, and southwest it would be defined by the Alleghany, Cumberland, and Ozark ranges, the main outlet lying, doubtless, in the axis of the Mississippi Valley, the gap between the Ozark and Cumberland highlands not having been eroded to its present level.

A sudden break in the lower portion of the barriers would, he argued, produce the phenomena now observed, namely, the action of violent currents plowing up and redepositing the material of the more ancient formations, carrying down in the main channel rocks of high northern derivation, restratifying and otherwise modifying toward the end a good portion of the iceberg drift and, as the current diminished, covering over the coarse material with the Orange sand.

Geological Survey of Minnesota under N. H. Winchell, 1872-1888.

Under an act dated March 1, 1872, the Governor of Minnesota appointed N. H. Winchell state geologist; the permanent assistants were Warren Upham, C. M. Terry, and O. W. Westland. A considerable number of others, including C. W. Hall, Leo Lesquereux, and E. O. Ulrich, were employed temporarily. A large number of bulletins and four quarto volumes of annual reports were issued containing the results of this organization, the life of which extended beyond the time limits of this history. The most important of these results were, (1) the discovery of the value of the salt resources and the saving of them for the benefit of the state; (2) the definite settlement of the question of workable coal beds in the state; and (3) a thorough investigation of the water and building stone resources.

Stevenson's Work on Coal, 1872.

The record of the year should not be closed without reference to the work of J. J. Stevenson on the Upper Coal Measures of the Alleghany Mountains. Stevenson was at this time professor of geology in the University of the City of New York, and the paper, which was his third independent contribution to geological literature, was based in part upon observations made while employed on the geological survey of Ohio under Newberry. As a result of his comparisons of the Ohio coal basins with those of adjacent states he concluded that:

(1) The great bituminous trough west of the Alleghanies does not owe its basin shape primarily to the Appalachian revolution.

(2) The Coal Measures of this basin were not united to those of Indiana and Illinois at any time posterior to the Lower Coal Measure epoch, and probably were always distinct.

(3) The Upper Coal Measures originally extended as far west as the Muskingum River in Ohio.

(4) Throughout the Upper Coal Measure epoch the general condition was one of subsidence interrupted by longer or shorter intervals of repose. During the subsidence the Pittsburgh marsh crept up the shore, and at each of the longer intervals of repose pushed out seaward upon the advancing land, thus giving rise to the successive coal beds of these measures.

(5) The Pittsburgh marsh had its origin in the east.

In a subsequent paper in the *Proceedings of the American Philosophical Society* he showed that the commonly alleged parallelism of the coal beds was a fallacy. On the contrary, many of them, as the Mammoth seam, bifurcate and rapidly diverge. From a study in the field of the barren and upper coal groups throughout the northern portion of the bituminous trough, he became convinced that as a whole the subsidence during their formation was fairly regular, but that there were local bulgings, and conditions of deposition that rendered parallelism impossible. He believed that "all the coals of the upper coal group were off-shoots from one continuous marsh, which existed from the beginning of the era to its close and which, in its full extent, is now known as the Pittsburg Coal Seam."

J. S. Newberry on Cycles of Deposition, 1873.

At the Portland meeting of the American Association in August, 1873, J. S. Newberry for the first time brought prominently forward in America an idea, founded upon personal observation and justifiable hypotheses, which had in a crude and unsound manner been suggested much earlier by Amos Eaton (see p. 132). The idea, in brief, was "that each of the great Paleozoic systems represented on the eastern half of the continent may be resolved into a cycle of deposits similar in general character." In other words, he thought to recognize a distinct threefold arrangement or succession of deposits, each cycle beginning with beds of sand and conglomerate, passing to shale, this followed by limestone, and this again by shaly material. The idea at first presented itself to his mind while studying Cretaceous deposits in the West, prior to 1860, and was very briefly alluded to in his first report of the Ohio geological survey. He now brought it out in detail. In support of it he alluded to the Lower Silurian formations, beginning at the bottom with the Potsdam sandstone, a coarse mechanical shore deposit, followed by a calciferous sandrock, a mixed mechanical and organic sediment, an offshore deposit, this by the Trenton limestone, plainly an open sea deposit, and this in turn by the shales and impure limestone constituting the Hudson group and conceded to be deposits of shallow water of retreating seas. The same order of deposits he thought to be able to trace in each of the great formations of the Paleozoic.

The idea was not wholly original with Newberry, nor did he so claim, though he was unquestionably the first in America to set it out in detail. Murchison, Phillips, and Hull had recognized a like trinity in some of the English formations, while James Hall, T. Sterry Hunt, and J. W. Dawson had referred to it incidentally in their American work.

While the generalization was a striking one and at the time it was put forward the conclusions seemed legitimate, recent studies have shown that, as Ulrich[13] has expressed it, "they are not altogether sound from a theoretical standpoint. The cycles are certainly not co-ordinate and none fits a major unit of the stratigraphic column accurately." That there is, however, among the stratified rocks in many localities a more or less distinct threefold arrangement in the order outlined above is unquestioned.

Dana's Views on the Subject of the Earth's Contraction, 1873.

It will be recalled that in 1846, 1847, and again in 1856, Professor J. D. Dana had written somewhat fully on the resultant effects of the earth's cooling and contraction and incidentally the formation of the larger features of the earth's relief. In 1873 he returned once more to the subject. His previously announced (p. 259) views on mountain making he so far amended as to accept those of Le Conte relative to the existence, throughout a large portion of geological time, of a thin crust and of liquid rock beneath that crust, whereby oscillation would be made possible, and a solid nucleus. He summed up his conclusions as follows:

13 Bull. Geol. Soc. Amer., Vol. XXII, 1911.

(1) That in mountain making on the continental borders, the oceanic crust had the advantage through its lower position of *leverage*, or, more strictly speaking, of obliquely upward thrust, against the borders of the continents.

(2) That, among mountain elevations, there are those which, like the Alleghanies, are the result of one process of making, or *monogenetic*, and those that are a final result of two or more processes at different epochs, or are *polygenetic*.

(3) That there are two kinds of monogenetic ranges—those that are geanticlinals, or *anticlinoria*, like the region of the Cincinnati uplift; and those that were the result of a slowly progressing geosynclinal, with consequently a very thick accumulation in the trough of sedimentary beds, ending in an epoch of displacements and solidification, and often of metamorphism of the sedimentary beds, as in the case of the Alleghanies and other *synclinoria*.

(4) That great mountain chains are combinations of synclinorian and of anticlinorian elevations.

(5) The principle advocated by Le Conte (restricted as indicated) that plication, shoving along fractures, and crushing are the true sources of the elevation that takes place *during the making* of the second of the two kinds just mentioned of monogenetic mountain ranges or synclinoria.

(6) That, on the oceanic side of the progressing geosynclinal referred to, there has been generally, as the first effect of the thrust against the continental border, a progressing geanticlinal which usually disappeared in the later history of the region, gravity, and the yielding and plication in the region of the geosynclinal favoring this disappearance.

(7) That the locus of the region of subsidence on a continental border was in general alongside of a region of thickly stiffened unyielding continental crust, and that pressure against the stable area beyond was one source of the catastrophe of mountain making.

(8) That each epoch of plication and mountain making ended in annexing the region upturned, thickened and solidified, to the stiffer part of the continental crust, and that consequently the geosynclinal that was afterward in progress occupied a parallel region more or less outside of the former, either landward or seaward, and commonly the latter.

(9) The principle adopted from Le Conte, that the bottom of a geosynclinal becomes weakened, as subsidence and surface sedimentary accumulations go forward, through the access of heat from below or the rise of the isogeotherms (the change of level in a given

isothermal plane having been seven miles in the Appalachian region), and that this in an important degree has made possible the catastrophe in which synclinoria have resulted.

(10) That, while igneous eruptions and metamorphisms have each attended the formation of synclinoria, still in places where the plication was greatest the igneous eruptions have been least in amount or absent; and that the most extensive igneous eruptions have taken place on continental borders after the crust had become too much stiffened to bend freely before the lateral pressure.

(11) That in the upturning and plication attending mountain making, the heat from the transformation of the motion was sufficient (in connection with other heat from a rise of the isogeotherms due to previous surface accumulations) to cause metamorphism; and also the pasty fusion which obliterates all stratification and gives origin to granite, and which may fill cavities or fissures, and so make veins that have all the aspects of true igneous ejections; and, as a more extreme effect, it may produce, as Mallet says, the degree of fusion which belongs to plastic trachyte, and give rise to trachytic and other ejections through fissures or volcanic vents. But—

(12) That the chief source of igneous rock is the plastic layer situated beneath the true crust, or the local fire seas derived from that layer.

W. C. Kerr on Effects of Earth's Rotation, 1873.

In this same year W. C. Kerr, state geologist of North Carolina, published in the *Proceedings of the American Philosophical Society* a brief but suggestive paper on the possible influence of the earth's rotation on the deflection of rivers. He called attention to a fact so obvious as to have been evident to the common people of that state, that in both the Carolinas the eastward flowing rivers always presented high banks and bluffs on the south side and low plains and swamps on the north; also to the fact that owing to these topographical conditions, the large towns and main roads were always on the south side also.

In seeking for a cause for this Kerr rejected the slow subsidence theory of Tuomey, and suggested that it was due rather to the rotation of the earth, coacting with the force of the river currents, and called in to sustain his opinions the law of motions as developed by Professor W. Ferrel:

> In whatever direction a body moves on the surface of the earth there is a force arising from the earth's rotation which deflects it to the right in the northern, but to the left in the southern hemisphere.

Whether or not the application of this principle was original with Kerr, it is of course impossible to say. This, however, seems to have been its first application in America, though von Baer had suggested as early as 1860 like results in the case of north and south flowing rivers, as the Irtisch and the Volga of Russia.[14]

Kerr's Work in North Carolina, 1870.

Kerr's final report on the geology of North Carolina, although presented in 1870, was not printed until 1875, owing to the parsimonious action of the state legislature, the public printer being allowed finally to set it up only under condition that the work be done in intervals of other work on the laws, documents, supreme court reports, etc.

The report was accompanied by a colored geological map and three colored sections. The classification of the formations adopted was as shown below.[15]

PROGRESSIVE THEORETICAL VIEW.

The following scheme exhibits the progress of theoretical notions which have obtained at different times with regard to the classification of the formations—the age and horizon of the rocks in the state:

Mitchell, 1842.	*Emmons, 1856.*		*Present, 1875.*	
Tertiary	Tertiary	Post-Pliocene ...	Quaternary.	
		Pliocene	Miocene.	Tertiary.
		Miocene		
		Eocene*	Eocene.	
Secondary	Cretaceous		Cretaceous.	
	Triassic		Triassic.	
	Permian			
Transition	Taconic		Silurian?	
			Huronian.	
Primitive	Gneiss		Laurentian.	
	Granite			
	Syenite			

*Also partly Quaternary and partly Cretaceous.

[14] Sir Archibald Geikie in his Textbook of Geology, 4th ed., 1903, expresses doubt if this cause can have much effect, though G. K. Gilbert, in a paper before the National Academy of Science in 1884, argued in its favor.

[15] There were two editions of this report, both bearing the same date and place of publication. They differed in that one carried an appendix of 120 pp. by T. A. Conrad, with eight full-page plates of fossils. There was also a slight difference in the alignment of the matter on the first seven pages of the volumes but none in the subject matter.

This report gave a very good summary of the knowledge of the geology of North Carolina up to date. The thin gravels overlying the eroded surfaces of the Eocene, Miocene, and Cretaceous in the eastern part of the state were still thought to be of glacial origin, the underlying rock having been planed down by the currents and drifting ice which brought the bowlders from the Archean hills of Chatham. This idea is not now accepted, North Carolina being universally recognized as far south of the glacial limit.

The gold-bearing gravels were looked upon as beds of till which had crept down the declivities of the hills and mountains, as glaciers descend Alpine valleys by successive freezing and thawing of the whole water-saturated mass, both gravitation and the expansion of freezing contributing to the downward movement. This is evidently the same idea as that later advanced by I. C. Russell in his description of the débris streams of Alaska.

Appendices attached contained a list and descriptions of the new genera and species of fossil shells, by T. A. Conrad, and the minerals of the state, by F. A. Genth. A chapter on corundum and its associated rocks, by C. D. Smith, is mainly of interest from the fact that Smith considered the chrysolitic rocks of igneous origin, a conclusion which for some time was disputed, but to which later workers have returned.

Sketch of Kerr.

Kerr was of Scotch-Irish descent, though born in North Carolina. His parents dying while he was quite young, he was adopted by a Reverend Dr. Caruthers, under whose instruction he was fitted for the State University at Chapel Hill, where he graduated in 1850 with highest honors. After graduation he taught school for a while at Williamstown in his native state, until elected to a professorship in Marshall University, Texas. This position he resigned in 1852 to accept an appointment as computer in the Nautical Almanac Office, then located at Cambridge, Massachusetts. While here he entered the Lawrence Scientific School and studied under Agassiz, Gray, and the chemist Horsford.

He became nominally state geologist in 1864, though active work, owing to the confusion and disorganization of war, did not begin until 1866.

It is safe to say that few men ever entered upon the work of a geological survey under more unfavorable auspices. He had had no extended experience, nor was he a trained specialist. Little money was available, the industries of the state paralyzed, and social conditions disorganized, if not demoralized. There was never a more genuine and sagaciously public-spirited citizen, but the times were evil, and for several years he shared the fate of all southern public officials who were endeavoring to adjust, to reconcile, and to go forward. His motives were misrepresented, his character assailed, his abilities questioned, his work maligned. Yet in the face of all this, with poor maps and with few roads, good work was done.

Fig. 86. *W. C. Kerr.*

During one of the first years of the survey he traveled, mainly on horseback, 1,700 miles over a mountainous country, with and without roads, and during the season of 1866 and 1867 not less than 4,000 miles. Work thus performed was from necessity largely in the nature of reconnaissance. The subjects of drainage and topography received a large share of attention, and indeed a large part of his work was along physiographic lines. Agricultural and mineral resources were investigated and, so far as possible with the limited means at his command, advertised. His papers which attracted the most widespread attention were those relating to the action of frost on superficial materials and on the unequal erosion of the banks of streams noted above.

Smith's Work in Alabama, 1873.

In 1873 the legislature of Alabama passed an act reviving the survey, which had been discontinued after the death of Tuomey, in 1857, and named Eugene A. Smith, of the University of Alabama, state geologist. This office Professor Smith has continued to hold down to the present time. Up to and including 1880 he had made five reports. These were given up mainly to a discussion of the geographic distribution of the various formations comprised within the state limits, with extensive notes regarding the economic possibilities of coal, iron, and other minerals of less importance.

In his report for 1875-1876 he treated of the geology of Jones Valley and the Coosa Valley region and announced the practical identity of the formations there found with those described by Professor Safford in Tennessee, recognizing the Ocoee, Chilhowee, Knox sandstone, shale and dolomite, and Lower and Upper Carboniferous. In the reports of 1877, 1878, and 1879 special reference was made to the Warrior coal fields and their probable value. Several colored county maps were given, but no geological map of the entire state.

Second Geological Survey of Kentucky, 1873-1891.

In the spring of this same year there was organized a second geological survey of Kentucky, with N. S. Shaler, then professor of paleontology at Harvard, chief geologist and director, and A. R. Crandall, P. N. Moore, C. J. Norwood, and J. R. Proctor, assistants. Robert Peter served as chemist. The organization continued as above until 1880, when J. R. Proctor succeeded to the directorship through the resignation and recommendation of Professor Shaler, the survey coming finally to a close in 1891. Up to 1880 five reports of progress were issued, dealing mainly with economic problems, including agriculture and forestry. In the discussion of the lead deposits in the Upper Cambrian limestone, Shaler thought to have shown that the fissures, with the exception of some of those in the western part of the state, were formed by shrinkage, and not by faulting and that therefore they could not be expected to continue downward indefinitely, probably not more than 1,000 feet. Further, that the limitation of the metalliferous veins to the limestones was strong proof of the efficacy of organic matter in the precipitation of the ore. The chemistry of the process was largely in accord with the views of Whitney.

N. S. Shaler on Recent Changes of Level, 1874.

A paper by Shaler in the *Memoirs of the Boston Society of Natural History*, about this time, is of interest in connection with that of A. S. Packard, referred to on p. 428. Shaler also found evidences of a considerable depression of the land along the New England coast at the close of the first division of the glacial period, which he believed to be due purely to the weight of the accumulated ice upon the yielding crust. The amount of this depression he found to increase from the vicinity of Boston toward the northern regions. Summarizing the probable causes of the glacial period he argued that they were due to intensification of evaporation over the oceanic areas between forty degrees north and south of the equator, such as could be brought about only through an increase in the heat of the earth's surface. The causes of this increased heat he thought to lie in the sun, which he regarded as a variable star. A sudden increase in heat of one-half would make an intertropical region one of intense evaporation, and the cloud-wrapped poles one of correspondingly excessive precipitation. The glacial period he thought must have been simultaneous throughout at least one and probably both hemispheres. From an inspection of the Maine coast alone he was unable to form a satisfactory opinion as to the time which had elapsed since the period of second glaciation, but from evidences on other parts of the coast and Europe he was led to conclude that it might have been 10,000 years. Believing it reasonable to suppose that the basins of the Great Lakes were dug out by glacial action, he thought that we must regard the Gulf of St. Lawrence as a product of the same forces.

Sketch of Shaler.

Shaler, like Le Conte, was born in the South, and studied under Agassiz at Harvard, later becoming instructor in the Lawrence Scientific School, and then professor of paleontology, and dean of the Scientific School faculty. Like Le Conte, too, he is not to be judged from the standpoint of geology only. As a writer and teacher he was unexcelled. "No instructor," wrote one of his biographers, "has held the position which Shaler did in the eyes and hearts of his students." In view of the number at one time engaged in geological teaching at Harvard, the present writer once asked one of the students the particular field or province Shaler occupied. After a moment's hesitation he replied, "Well, Shaler arouses the enthusiasm." And so he did, and kept it aroused as well. His influence was great in all matters of public welfare, and particularly in the matter of good roads, in which he and his state became pioneers. Nor were his writings limited to geological subjects. Less versatile than Le Conte, perhaps, he nevertheless covered a wide field and gained a higher place in what is commonly considered literature. *The Story of the Continent; The United States of America; The Individual, a Story of Life and Death; Elizabeth of England, a Dramatic Romance; Man and the Earth; Aspects of the Earth,* and a volume of poems, are some of the more prominent of his works.

Powell's Exploration of the Grand Canyon, 1869-1874.

The plateau country drained by the Colorado River offers facilities for the study of geological problems of a certain type such as

PLATE XXVI

N. S. SHALER

JOSEPH LE CONTE

are, perhaps, equaled nowhere else in the world. This is especially true with reference to those relating to stratigraphic succession, of uplift with a minimum amount of contortion of the beds, and of erosion. The plateau nature of the country was recognized by Newberry while with the Ives expedition in 1860, and by Blake in connection with the Pacific Railroad surveys in 1856. It remained, however, for later workers to bring out the salient features of the geology of the region and to work out the problems in a way that, in the words of Emmons,[16] has formed the starting point of modern physical geography.

In the summer of 1869, J. W. Powell, a retired officer of the Federal Army, made a boat trip down the Colorado, starting from Green River City on the Union Pacific Railroad May 24, and emerging from the mouth of the Black Canyon, nearly 900 miles below, on August 30 following. It was a journey which, to quote Emmons again, was "unequalled in the annals of geographical exploration for the courage and daring displayed in its execution," but which nevertheless offers an interesting illustration of an undertaking considered foolhardy if resulting in failure, or heroic if successful.

In his reports on these explorations,[17] which were made and published under direction of the Smithsonian Institution, with a congressional appropriation of $12,000, Powell called attention to the fact that the canyons are gorges of corrosion and due to the action of the river upon the rocks, which were undergoing a gradual elevation. As he expressed it, the river preserved its level, but the mountains were lifted up; as the saw revolves on a fixed pivot while the log through which it cuts is moved along. The river was the saw which cut the mountain in two. This view is essentially identical with one advanced by Hayden, in his discussion of the gorges and canyons of the rivers of Montana in 1872 (see p. 516). The more striking illustration by Powell, and its still more striking elaboration by Dutton has caused the first suggestion to be largely overlooked. With whom the idea actually originated, it is impossible to say. Although Powell's final report was not published until 1875, it would seem that the subject must have become one for general discussion shortly after his return from the trip of exploration. We may, however, well dismiss the matter with the fitting quotation from Lowell, "And we call a thing his, in the long run, who utters it clearest and best."

In this same report Powell first made use of the expressions *ante-*

[16] Presidential address, Geological Society of Washington, 1896.
[17] Letter of transmittal dated June 16, 1874; publication in 1875.

cedent and *consequent* valleys, meaning in the first instance that the drainage was established prior or antecedent to the corrugation of the beds by faulting and folding; and in the second case, that the valleys have directions which are dependent upon the corrugations. Valleys which were formed by streams, the present courses of which were determined by conditions not found in the rocks through which the channels are now carved, but which were in existence when the district last appeared above sea level, he called *superimposed* valleys.

In this report, too, use was made for the first time of the term *base level of erosion*, which was defined as an imaginary surface inclining slightly in all its parts toward the end of the principal stream draining the area through which the level is supposed to extend, or having the inclination of its parts varied in direction as determined by tributary streams.

It was pointed out, further, that the region of the Grand Canyon was after all a region of lesser rather than greater erosion; that had the country been favored with a rainfall equal to that of the Appalachian country, the entire area might have been reduced to a base level which would be the level of the sea, though the evidence of such erosion might be almost wholly obliterated.

The daring of Powell in making the trip down the canyon threw a glamour over his subsequent geological work in the region which has resulted in Newberry's previous observations being almost wholly forgotten. In geology as elsewhere "the interest's on the dangerous edge of things," as Browning's Bishop Blougram remarks.

The W. A. Jones Expedition of 1873.

During the summer of 1873 Captain William A. Jones, of the Engineer Department, made, under direction of the Secretary of War, a reconnaissance of northwestern Wyoming, the purport of which was to open up a wagon road from the line of the Union Pacific Railroad to the Yellowstone National Park and Fort Ellis, Montana. Theodore B. Comstock accompanied the party as geologist, rendering a report of 100 pages accompanied by numerous diagrams and a colored geological map. The region was afterward covered more in detail by the Hayden surveys and Comstock's report largely lost sight of as a result. Peale, in his report on the Green River District in 1877, states that so far as the region covered by him is concerned the Comstock map is incorrect. St. John, who reported on the Teton division, merely refers to the Comstock survey, without comment.

T. L. Clingman on Earthquakes and Volcanoes in North Carolina in 1875.

In July, 1874, there appeared in the columns of the *Western Expositor* of Asheville, North Carolina, a long article by T. L. Clingman entitled "Volcanic Action in North Carolina." Mr. Clingman's name appears not elsewhere in geological literature, but as his paper contained numerous references to what were obviously ordinary earthquake phenomena they were of interest and considered of sufficient importance to merit a three-page review in the *American Journal of Science* for the year following. Numerous shocks and evidences of disturbance in the western and mountainous portion of the state were noted, the most recent in 1869 and the earliest of which he found record in 1812. That Clingman, who was not a geologist, should have failed to realize that there was no necessary connection between earthquakes and volcanoes is not strange, but that the *Journal* should have allowed his opinions to pass unchallenged is not easily accounted for. The report that smoke issued for several weeks from a small rock crevice in Madison County, that elsewhere leaves and branches of timber immediately above the chasm in places present the appearance of having been scorched, was accepted as indicative of volcanic activity and that a portion of the globe which ". . . ought to be regarded as being stable as any part of our planet, is nevertheless not free from change." "Whether this is to be considered as evidence of a gradual return of that volcanic action which manifests itself still elsewhere . . . it is perhaps difficult to decide." And it was suggested that the problem might be taken up by the United States Coast Survey. It may be remarked that the geologist, F. H. Bradley, subsequently visited the locality and reported that there was no foundation for the prevalent stories of yawning crevices and smoking pits. The noises heard were the usual rumblings characteristic of earthquakes and "there was nothing properly volcanic about them."

Winchell's Work in the Black Hills, 1874.

In 1874 N. H. Winchell accompanied, as geologist, the military reconnaissance party headed by Captain W. Ludlow to the Black Hills of Dakota. His report, appearing in 1875, occupied pages 21-66 of the quarto volume issued, and was accompanied by a geological map. Winchell recognized the occurrence of Cretaceous, Jurassic, Triassic, Carboniferous, and Silurian (Potsdam) rocks, overlying schists and slates, into and through which granite had been intruded.

He agreed in the main with Hayden's observations, as given in his Second Annual Report (1868), though he failed to find evidence of the unbroken conformability of the fossiliferous formations. On the contrary, he mentioned the Red Beds as lying unconformably on the Carboniferous limestone.

George Bird Grinnell accompanied the expedition in the capacity of paleontologist, though the new species of invertebrate fossils found were described and figured by R. P. Whitfield.

Work of Jenny and Newton in the Black Hills, 1875.

The presence of numerous bands of prospectors on the Indian reservations of the Hills, led there by the reported finding of gold by Custer's and other expeditions, caused the National Government, in 1875, to send "trusty persons" to examine the region and report to the Secretary of the Interior, in order that a proper basis might be secured for future negotiations with the Indians. The locality being then comprised in the Sioux reservation, immediate direction of affairs was put in the hands of the Indian Bureau. Under this authorization W. P. Jenny was appointed to undertake the work, with Henry Newton assistant. The party entered the region on the third of June, some 400 strong, a large military guard being esteemed necessary on account of the manifest discontent of the Indians, and returned to Fort Laramie on October 14, after an absence of four months and twenty days. A preliminary report on the mineral resources of the Hills, accompanied by a map by V. T. McGillycuddy, the topographer, was submitted by Mr. Jenny and published in the report of the Commissioner of Indian Affairs for 1875. The complete report on the mineral resources, climate, etc., with a preliminary map, was published in the form of an octavo pamphlet of 71 pages in 1876, and the final *Report on the Geology and Resources of the Black Hills of Dakota*, in form of a quarto volume of 566 pages, with a large folio atlas, in 1880. This included, also, the previous reports on the mineral resources, noted above. Unfortunately, Newton, to whom was left the general geology of the region, died before his report was fully prepared for the press, the work being ably edited by G. K. Gilbert, and the volume issued as one of the monographs of the Powell survey.[18]

The region, it will be remembered, had already been touched upon and partial surveys made by numerous parties, including John Evans in 1849 and 1853, Thaddeus Culbertson in 1850, Meek and

[18] Appendix, p. 692.

Hayden in 1853, and Hayden, with the military expeditions of Harvey and Warren in 1855 and 1857, and with Captain Raynolds in 1859. Hayden again visited the region under the auspices of the Philadelphia Academy in 1866, and N. H. Winchell with Custer's expedition in 1876, as noted elsewhere. Under all these conditions it is difficult to estimate the value of the work of Newton. Much that he stated in his report had already been made known by the writers mentioned. On the other hand, much that he might have had in mind to say has never appeared, owing to his untimely death.

Fig. 87. *Henry Newton.*

The summary of Black Hills history, as given, shows an older Archean series consisting of shales and sandstones over which, after an interval of erosion and metamorphism, a newer Archean, consisting mainly of sandstones, was deposited, accompanied by abundant intrusions of granitic rock. Then followed the period in which these rocks were raised to a nearly vertical position, metamorphosed, and deeply eroded, after which the various Potsdam, Carboniferous, Red Bed, Jura, and Cretaceous deposits were laid down conformably among themselves. Finally came the uplift and subsequently the unconformable depositions of the White River Tertiaries upon their flanks. The date of the uplift was therefore set as the interval of time between the Cretaceous formation and the beginning of the Miocene Tertiary. (See under Hayden, p. 507.)

It was shown that the drainage from the Black Hills was conse-

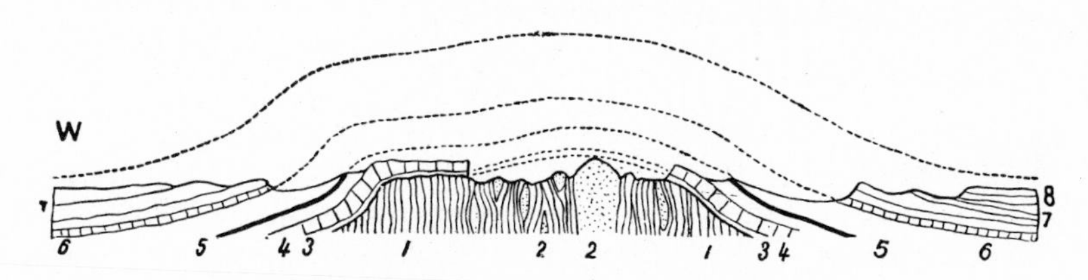

Fig. 88. *Ideal Cross Section of the Black Hills.*

quent—that is, it conformed to the dip of the strata. This was conceived to be caused by the streams having acquired their direction during the uplift of the hills, or while they were being laid bare by the drainage of a lake. In either case the drainage was consequent upon the uplift. With the Belle Fourche and the South Fork of the Cheyenne rivers the case was different, however. These were found to cut across the uplifted strata in a manner showing that their drainage was superimposed. If the streams existed prior to the deposition of the Tertiary sediments they were completely blotted out.

J. H. Caswell described the various metamorphic and eruptive rocks collected by the survey, his paper forming the second on the subject of micropetrography to appear from the government press, Zirkel's fortieth parallel report (1876) being the first. The report was accompanied by two plates with eight colored figures illustrating microstructure and composition, and is of particular interest as giving the first authoritative account of the occurrence of the rock phonolite in the United States. Caswell, although one of the first in a new and fascinating field, seems to have rested on his laurels, content with the one effort, his name in geologic literature appearing here for the first and only time.

W. P. Jenny, to whom fell the economic work, reported gold occurring (1) in veins of quartz traversing Archean schists and slates; (2) in slate mineralized by waters depositing silica and iron pyrites; (3) in the conglomerate forming the lower layer of the Potsdam sandstone and derived from the Archean rocks; (4) in trachyte and porphyry intruded at the time of the elevation of the hills; (5) "in deposits in the slates and sedimentary rocks produced by the intrusion of the trachyte and porphyry," and (6) "in placer gravels resulting from the decomposition and erosion of the above formations in Tertiary times."

The quartz veins he did not consider true fissure veins, but "interlaminated fissures," since they occurred filling crevices between the lamellæ of the schists formed in the process of the folding.

The fossils collected by this expedition were worked up also by Whitfield while at Albany in 1876.

Lapham's Survey of Wisconsin, 1873.

Under an act approved March 18, 1873, a geological survey of Wisconsin was authorized, and I. A. Lapham appointed state geologist by Governor C. C. Washburn. Owing to an apparent oversight Lapham's name was not, however, sent to the senate for approval, and an opportunity thereby offered for Washburn's successor, W. R. Taylor, to supersede him by Dr. O. W. Wight. So far as shown by public records the transaction was a purely political one.[19]

[19] "Our geological survey has gone the fate of its predecessors—or rather a worse one. The governor has appointed a disreputable politician to Dr. Lapham's

Lapham therefore served but two years, during this time rendering two reports. These were not published independently, but subsequently (1877) formed the first sixty-five pages of the second volume of the reports of T. C. Chamberlin, who became state geologist in 1876. Lapham was assisted by R. D. Irving, T. C. Chamberlin, and Moses Strong.

Fig. 89. *I. A. Lapham.*

The two reports referred to were naturally of a preliminary nature, designed to form the basis for future work.

Lapham, perhaps even more than others of his time, was an all-round naturalist—a type not possible in this day and generation. Beginning life as a stonecutter and afterward a civil engineer, he yet found time to study and observe in nearly all branches of the sciences, and this, too, with remarkable accuracy. His first paper of a geological nature is said to have been prepared when he was but sixteen years of age, and was published in the *American Journal of Science* for 1828. This gave an account of the geology in the vicinity of the Louisville and Shippingport Canal, and was illustrated by a map and geological sections of a nature very creditable to a youth of his years.

His bibliography includes upward of fifty titles, embracing, besides geology, articles on climatology, archeology, botany, and cartography. He is represented as a modest, patient, and industrious man, living more for the service of others than for his own aggrandizement, and taking an active part in all movements tending toward the advancement of science and education or the legitimate development of his state. He died of heart trouble while alone on a lake near Oconomowoc, Wisconsin.

As a bit of unwritten history it may here be stated that on March 1, 1850, an agreement was entered into between Lapham and James Hall, whereby the latter was to prepare a work called *American Paleontology,* to be based on manuscript placed in his hands by

position, leaving the survey still unorganized. We had accomplished an immense amount of work, having produced as many as one hundred detailed colored geological and topographical maps, the whole lead region having been covered with contour lines at 50 foot verticals according to Whitney's recommendations in your volume. It is probable that none of it will ever see light. One reason of the trouble was my refusal to call the Penokee ores so rich as Col. Whittlesey makes them to be. Wisconsin has most certainly had ill luck with her surveys." Irving to Hall, February, 1875.

Lapham and stated to embrace descriptions of about 2,000 species. The agreement stipulated that the work should be completed within one year from date of signature, and if no publisher could be found to take it on favorable terms it should be issued at their joint expense. Nothing seems to have come from it, however. In a letter to Hall dated May 19, 1856, pleasure is expressed that the matter "was not forgotten but only delayed for good cause." In 1857 Lapham wrote suggesting that the expense of publication might be reduced by omitting descriptions and figures, and on January 31, 1860, he asks for the return of his "tin box of 'American Paleontology' that has been cumbering your premises so long."[20]

Lapham was succeeded, as above noted, by O. W. Wight, who served but one year (1875). His report of progress was not published at the time, but was likewise included by Chamberlin in the second volume of the final reports. It contained matter of little other than historical interest.

Chamberlin's Survey of Wisconsin, 1876.

In February, 1876, T. C. Chamberlin was placed in charge of the survey. The organization continued in existence until 1879, the final reports appearing in the form of four royal octavo volumes, dated respectively, Volume I, 1883; Volume II, 1877; Volume III, 1880; and Volume IV, 1882, comprising altogether 3,035 pages.

The principal assistants were R. D. Irving, F. H. King, Moses Strong, E. T. Sweet, J. D. Whitney, and L. C. Wooster. R. P. Whitfield served as paleontologist. Aside from acting as director of the survey, Chamberlin himself took charge of the geology of eastern Wisconsin, of which he described in considerable detail the hydrology, soils, and the glacial drift. The so-called "Kettle Range" he believed to be in part moraines, and the kettle holes to be due mainly to the melting of masses of ice buried in the gravels. He discussed the economical value of the clays and shell marls of the region, and for the first time in America demonstrated the usefulness of the microscope in examinations of detrital rocks. In this connection he noted that a microscopic examination of the sand grains of the Potsdam sandstone was entirely fatal to the view still occasionally advanced to the effect that such were produced by crystallization from solution. He, in his turn, however, failed to recognize the possibility of the crystalline form of some of the granules being due to secondary enlargement (see pp. 350 and 458).

20 Appendix, p. 693.

In view of later developments it is perhaps permissible to refer here, briefly, to a system of rock nomenclature proposed by Chamberlin in Volume I of his reports (1883). Recognizing the desirability of conveying in the name an indication or suggestion of the mineral nature of a rock, and realizing that to attempt this by compounding the names of the component minerals in full would be productive of a nomenclature too cumbersome, he proposed a system in which these names should be abbreviated, as *qua* for quartz, *fel* for feldspar, and *mi* for mica. A mica granite would then be known under the expressive term *fel-qua-mi, qua-mi-fel,* or *mi-fel-qua,* according as one or the other of these essential constituents prevailed. For biotite schist he would substitute the compound *schistose bi-ortho-qua,* and for diorite *plagi-amph.* The suggestion seems to have fallen flat, as the saying is, and one can but feel that the uncouth character of the names had much to do with its failure to gain recognition. But what can one say when the latest and most highly authenticated proposed nomenclature contains such linguistic monstrosities as *phyrowyomingose, umptekose, Dugannonose,* and *hornblende-trach-phyro-monzonose?*

The oölitic iron ore lying between the Cincinnati shales and the Niagara limestone was thought to be undoubtedly a marine deposit laid down in detached basins, and it was shown that the three well-marked classes of limestone occurring in the southeastern counties of Wisconsin graded into each other and were doubtless formed contemporaneously, the residual mounds or ridges being ancient coral reefs, while the granular sandrock was formed from calcareous sands derived by wave action from the reef, the compact strata being deeper quiet-water deposits of the same material.

Chamberlin also investigated the zinc and lead deposits of the southeastern part of the state, and discussed thoroughly the problems relating to their origin, disagreeing in many minor points with both Owen and Whitney. The crevices in which the ores occur he believed to have originated primarily in the folding of the rocks by lateral pressure. He pointed out that the ore beds lay mainly in basinlike depressions and argued that the ore material, originally in solution in sea water, was deposited contemporaneously with the limestone through the action of sulphureted vapors given off by the decomposition of organic matter, and that, after the elevation of the beds above sea level, percolating meteoric waters oxidized and dissolved the material once more and carried it into the fissures, where it was the second time precipitated as sulphides. Fourteen-hundredths of 1 per cent of ore material disseminated throughout 100 feet in depth of the limestone would, he showed, be sufficient to have formed all the deposits as they now exist. He did not agree with Whitney in assuming that the oceanic waters of early geologic periods were richer in metallic salts than those of later date. With his predecessors he doubted the probability of the existence of workable deposits of ore in the Lower Magnesian limestone.

Fig. 90. *R. D. Irving.*

Volume I of the reports of this survey was designed as a general treatise on the geology of the state, suited to the wants of explorers, miners, landowners, and manufacturers. In this Chamberlin stated his belief that the original source of the Keweenawan copper ore was the igneous rocks, from whence it has been concentrated through the agency of permeating atmospheric waters. To quote his own words, he believed—

First, that the metals, copper and silver, were primitively constituents of the rocks that were melted to produce the lavas which formed the trappean sheets of the formation; second, that they were brought up and spread out, commingled with the molten rock material; third, that they were chemically extracted thence by percolating waters and concentrated in the porous belts or fissures of the formation, giving rise to the exceptionally rich deposits for which the formation is famous; and fourth, that the surface disintegration of portions exposed in the latter part of the period yielded metallic ingredients to the adjacent sea, from whence they were extracted by organic agencies, giving rise to impregnated sediments, which, in turn, through subsequent concentrations, gave rise to other copper and silver deposits, among which are to be reckoned the later metallic horizons of the Lake Superior region and possibly elsewhere.

Naturally, the subject of the drift was treated in considerable detail. The glacial period was divided into the Terrace or Fluviatile epoch, Champlain or Lacustrine epoch, the Second Glacial epoch, inter-Glacial epoch, and first Glacial epoch. This formal announcement of the fact that there had been two distinct periods of glaciation was here made for the first time, although as noted on page 633, Professor Edward Hitchcock had suggested its possibility as early as 1856. The law of flowage Chamberlin regarded as essentially similar to that of viscous fluids, in accordance with the observations of Agassiz, Forbes, Tyndall, and others. A later study of Greenland glaciers caused him to change his views on this point.

Irving, one of Chamberlin's assistants, dwelt with great detail on the lithological character of the rocks in the regions surveyed by him, and his reports are notable for the beauty of the colored plates of thin sections, which were by far the best that had been prepared and published by an American up to that date. He was assisted in this part of the work by C. R. Van Hise. He also described in some detail the glacial drift, and was the first to announce that the Kettle Range of Central Wisconsin was a continuous terminal moraine.

Third Edition of Dawson's Acadian Geology, 1878.

In the third edition of his work on Acadian geology, which appeared in 1878, Dawson returned once more to a vigorous discussion of the problems of the ice age, and to register again his opposition to the views generally held by American geologists.[21] Many of the arguments used closely resembled those of his former papers and are reviewed in detail on p. 640.

In the first edition (1855) of the work the occurrence of the albertite deposits in the Lower Carboniferous rocks of New Brunswick received considerable attention. The material belonged, he thought, to the bituminous coal series, the original source of which was fresh-water deposits of organic matter altered by mechanical pressure. He, therefore, classed it with the pitch coals or jets, although recognizing the possibility of its being a distinct mineral species intermediate between coal and asphalt.

In the second (1868) edition he, however, modified his opinion, accepting the views of more recent explorers like Hitchcock, Bailey, and Hind, who had described the deposit as a vein filled with bituminous matter, the source of which he now supposed to be the Lower Carboniferous shales. In his own words:

> The deposit of the Albert mine would thus be a vein or fissure constituting an ancient reservoir of petroleum which, by the loss of its more volatile parts and partial oxidation, has been hardened into a coaly substance.

The Nova Scotia gypsum beds he considered, one might well say *imagined*, as due to the alteration of beds of limestone by free sulphuric acid poured into the sea by springs or streams issuing from the volcanic rocks. In this connection he gave a highly interesting

21 The same views, practically unchanged, are given in the edition of 1891.

verbal picture of conditions at what is now the Southern Hants and Colchester at the time when the marine limestones and gypsums were being formed.

At this period, then, all the space between the Cobequids and the Rawdon Hills was an open arm of the sea, communicating with the ocean both on the east and west. Along the margin of this sea there were in some places stony beaches, in others low alluvial flats covered with the vegetation characteristic of the Carboniferous period. In other places there were creeks and lagoons swarming with fish. In the bottom, at a moderate distance from the shore, began wide banks of shells and corals, and in the central or deeper parts of the area there were beds of calcareous mud with comparatively few of these living creatures. In the hills around, volcanoes of far greater antiquity than those whose products we considered in a former chapter, were altering and calcining the slaty and quartzose rocks; and from their sides every land-flood poured down streams of red sand and mud, while in many places rills and springs, strongly impregnated with sulphuric acid, were flowing or rising, and entering the sea, decomposed vast quantities of the carbonate of lime accumulated by shells and corals and converted it into snowy gypsum.

The fauna of the seas of the Lower Carboniferous and Permian periods, both in Europe and America, presented so great similarities that, in a broad view, he felt they were identical. The changes and the subdivisions of this fauna were related not merely to lapse of time but to vicissitudes of physical conditions. It followed that, according to his reasoning, if the peculiar Permian conditions indicated by the rocks came earlier in Nova Scotia than in Europe the character of the fauna might also be changed earlier. In other words, "We have both rocks and shells with Permian aspect in the earlier Carboniferous period." The fact that the marine fauna of the Lower Carboniferous of Nova Scotia more nearly resembled that of Europe than the western states indicated to his mind that the Atlantic was at that time probably an unobstructed sea basin as now, while the Appalachians already, in part, separated the deep-sea fauna of the Carboniferous seas east and west of them.

Concerning the origin of coal he wrote: "Mineral charcoal results from subaerial decay; the compact coal, from subaqueous putrefaction more or less modified by heat and exposure to air." Further, he regarded mineral charcoal as the woody débris of trees, while the compact coal was produced from the bark of these same trees along with such woody and herbaceous matter as might be embedded or submerged before decay had time to take place.

He referred again to the observations and conclusions made by

PLATE XXVII

SIR WILLIAM DAWSON

him in connection with his Joggins work, several years previous, to the effect that the layers of clear, shining coal (pitch or cherry coal) are composed of flattened trunks of trees, and that of these usually the bark alone remains, the lamination of the coal being due to the superposition of layers of such flattened trunks alternating with the accumulations of vegetable matter of successive years, and occasionally with fine vegetable muck or mud spread over the surface by rains or by inundations.

The stigmaria found in the clay underlying nearly every bed of coal he felt proved beyond question the accumulation of the coal-forming materials through growth *in situ*, following in this respect the teaching of Logan. The under or fire clay was hence looked upon as a fossil soil robbed of its alkali and lime through growths of terrestrial vegetation.

The rocks of the Arisaig series, which in 1855 he had considered Silurian (Devonian?), in the edition of 1868 he thought "must be regarded as representing the middle and upper parts of the Upper Silurian, a position somewhat lower than that assigned to it in the first edition."

The provincial predictions of Dawson in this work are somewhat amusing. Thus, on page 4:

> Further, since by those unchanging laws of geological structure and geographical position which the Creator himself has established, this region must always, notwithstanding any artificial arrangements that man may make, remain distinct from Canada on the one hand and New England on the other, the name Acadia must live; and I venture to predict that it will yet figure honorably in the history of this western world. The resources of the Acadian provinces must necessarily render them more wealthy and populous than any area of the same extent on the Atlantic coast from the Bay of Fundy to the Gulf of Mexico, or in the St. Lawrence Valley from the sea to the head of the Great Lakes. Their maritime and mineral resources constitute them the Great Britain of eastern America, and though merely agricultural capabilities may give some inland and more southern regions a temporary advantage, Acadia will in the end assert its natural preeminence.

A prediction of which there are as yet no signs of fulfilment.

Sketch of Dawson.

Dawson was a Nova Scotian by birth, having first seen the light of day at Pictou in 1820. He was educated at Pictou Academy and subsequently at the University of Edinburgh, Scotland, where he came under the influence of Jamieson, Forbes, Balfour, and Alexander Rose. In 1847 he returned to Nova Scotia, and in 1855 assumed the principalship and chair of natural history in McGill University, Montreal, at the head of which institution he remained until 1893, when he was forced by ill health to resign. His first original paper was on a species of field mouse found in Nova Scotia, and was read before the Wernerian Society of Edinburgh. From so insignificant a beginning he developed into one of the most prolific and voluminous writers. *The Geological History of Plants, The Air Breathers of the Coal Period, The Canadian Ice Age, The Dawn of Life, Story of the Earth and Man, Fossil Men and their Modern Representatives, The Meeting Place of Geology and History,* and *Modern Science in Bible Lands* are among the best known of his writings, in addition to the work above reviewed. In educational matters he was always prominent, and the present standing of McGill University is largely due to his industry and ability as an administrator. He is represented by his biographer as a man of quiet geniality, gentle and courtly in manner, but, as may be readily surmised from his record in the Eozoön controversy, decided in opinion and firm in action. Like Alexander Winchell he took a prominent part in overcoming the popular prejudice concerning the supposed antagonism between religion and the sciences, particularly geology. An advanced and liberal thinker, he was, nevertheless, not an evolutionist in the ordinary acceptance of the term. He believed that the introduction of new species of animals and plants had been a continuous process, not necessarily in the sense of deviation of one species from another, but in the higher sense of the continued operation of the cause or causes which introduced life at first. The history of a life, he argued, presents a progress from the lower to the higher, from the simpler to the more complex, and from the more generalized to the more specialized. In this progress new types are introduced and take the place of the older ones, which sink to a relatively subordinate place and become thus degraded. To him paleontology furnished no direct evidence as to the actual transformation of one species into another or as to the actual circumstances of creation of a species; but the drift of its testimony to him showed that species come in *per saltum* rather than by any slow and gradual process.

Whitney's Auriferous Gravels, 1879-1880.

In 1879 there was published, as already noted (p. 407), Part I of Whitney's *Auriferous Gravels of the Sierra Nevada of California*, and in 1880 Part II of the same work, the combined papers comprising a volume of 569 pages, in which the distribution, origin, and characteristics of these gravels were fully discussed and much information was given regarding their method of mining and economic value.

Concerning the origin of the conglomerates and the prevailing theories regarding them, he wrote:

> Again, these detrital deposits are not distributed over the flanks of the Sierra in any such way as they would have been if they were the result of the action of the sea. On the contrary, there is every reason to believe that they consist of materials which have been brought down from the mountain heights above and deposited in preexisting valleys; sometimes very narrow accumulations, simply beds of ancient rivers, and at other times in wide lake-like expanses of former water courses.

Subsequent work has apparently fully confirmed this view. The time of deposition was during the later Pliocene epoch, and not as late as the drift of the Diluvial period.

Many pages were devoted to the vertebrate fossils found in the auriferous gravels, and particularly to the evidences of man, and the Calaveras skull. The question of the contemporaneous deposition of this skull with the gravels has been too frequently discussed by ethnologists to need more than a brief mention here. Suffice for the present to state that Whitney himself seems to have been fully convinced of the genuineness of the find, and as establishing beyond doubt the existence of Tertiary man in California.

In discussing the source of the gold and its distribution in the gravel he expressed the opinion that, as a whole, the occurrence of metalliferous ores is rather a surface phenomenon than a deep-seated one, and that this is due to the favorable conditions for fissure formation and deposition from solution controlled by temperature and pressure.

The source of the gold he thought to be undoubtedly the quartz veins which traverse the Jurassic slates, a fact to his mind refuting Murchison's theory of the occurrence of gold exclusively in Paleozoic rocks. He found no evidence to support the prevailing opinion that the large size of the nuggets in the gravel was due to a gradual growth through chemical precipitation, and regarded such as more or less dendritic and branching masses which have been liberated from the gangue and reduced to pebble form by the pounding they received in the moving gravels.[22]

[22] Recent work of A. Liversidge, the Australian mineralogist, confirms this view.

Whitney's Climatic Changes, 1880-1882.

The work on the auriferous gravels was followed in 1880 and 1882 by an equally comprehensive work on *Climatic Changes of Later Geological Times*, the discussion being as in the last-mentioned case, based largely on observations made during the work of the California survey. Struck by the appearance of recent desiccation in the West, as illustrated by the lake regions of the Great Basin, and from a study of the phenomena there met with as compared with those in other regions, Whitney was led to conclusions radically at variance with those commonly accepted by his coworkers. It will be remembered that in the work on the auriferous gravels he took the ground that there had been no appreciable change in elevation of the Sierras since the close of Cretaceous times, and he argued that the Tertiary auriferous gravels were laid down by the rivers of that period which flowed through broad channels, the present deep-cut V-shaped forms being due to the smaller volume of water which was the result of a decrease in annual precipitation.

Concerning such, he wrote:

> It is, as a general rule, safe to assume that where U-shaped valleys exist the perpendicular walls have an orographic origin and that those of V form have had that shape given to them by the débris piles which have accumulated against their sides. The farther we descend the mountain slope the less the grade and consequently the less the carrying power of the stream. Hence the valley which is U-shaped in the upper part of its course acquires more and more of a V form as it approaches the plain at the base of the range from which it heads.

He thought to be able to trace, not to go too much into detail, a period of warmth and heavy precipitation followed by one of desiccation, but anticipated by one of cold and glaciation, the glaciers, however, being limited to the most elevated ranges of the Cordilleras.

In the discussion of the question he called attention to the fact that the Great Lakes of North America and most of those of other countries as well are included in areas underlain by Paleozoic rocks or those partly Paleozoic and partly Archean, and are due not to glacial erosion, but to orographic movements—Lake Superior, for instance, occupying a synclinal depression in Paleozoic rocks just along the edge of the Azoic series. These views, it will be noted, are quite at variance with those of Agassiz and Logan, elsewhere expressed.

The lake region of the Great Basin while likewise orographic in origin had become desiccated through climatic changes, no evidence being found that there had been any essential alteration in the configuration or topography of the western side of the continent since the glacial epoch—that is, since the time when the crests of the highest ranges were to some extent covered with snow and ice. Therefore, no part of the desiccation which appears to have taken place since that time could be due to orographic changes; the phenomenon must have been a climatological one. (See further, on p. 641.)

Second Geological Survey of Pennsylvania, 1874-1887.

The rapid development of the economic resources of Pennsylvania, particularly coal, iron, and petroleum, during the years intervening between the publication of Rogers's final report (1858) and 1874, aroused a great public demand for more detailed information. An appeal was, therefore, made to the state legislature in 1873 for the establishment of a second geological survey. This culminated in 1874 in an enactment providing for the appointment of ten commissioners, having authority to appoint a state geologist "of ability and experience" who should, with ten competent assistants, make such investigations as might be required to elucidate the geology of the state and put the results of this and previous work, of either individuals or surveys, into a convenient form for reference.

Acting under this authority the commissioners in 1874 appointed to the office J. P. Lesley, an assistant on the first survey under Rogers, and subsequently connected, in the capacity of topographer and geologist, with various private surveys, and at that time professor of geology in the University of Pennsylvania. Annual appropriations were made, amounting, during the thirteen years in which the survey existed, to $545,000. Under these seemingly favorable auspices the second survey was inaugurated. With Lesley were associated from time to time, in one capacity or another, a considerable number of individuals, among whom mention may be made of C. A. Ashburner, C. E. Beecher, J. C. Branner, J. F. Carll, H. M. Chance, E. W. Claypole, E. V. d'Invilliers, L. G. Eakins, Persifor Frazer, F. A. Genth, C. E. Hall, T. S. Hunt, A. E. Lehman, Leo Lesquereux, A. S. McCreath, F. Prime, Jr., J. J. Stevenson, I. C. White, Arthur Winslow, and G. F. Wright.

From the work of this organization has sprung the most remarkable series of reports ever issued by any survey. Up to and including 1887, when field work was practically discontinued, there had appeared some seventy-seven octavo volumes of text, with thirty-five atlases, and a "Grand Atlas." These were followed in 1892 and 1895 by the three octavo volumes constituting the final report.

A very large portion of the work of this survey fell beyond the time limits mentally set for the present history. It may, however, be stated that the energies of the organization were directed toward the solving of economic problems. "I have been obliged," Lesley wrote, "for the last fourteen years to direct the State survey almost exclusively in an economic direction, so as to make the whole of every appropriation bring as much fruit to the business community as possible, neglecting, in what systematic geologists may possibly or probably consider a shameful manner, strictly scientific researches. Even when I have ordered long and extensive scientific researches (as in the case of the analyses of the Lower Silurian limestone series opposite Harrisburg) it has been, not in the spirit of transcendental science, but with the express intention to use the results directly as applied science to the economical demands of the State. It can hardly be understood by outsiders how completely a State bureau is shut up to this necessity." And again, in one of his letters to Hall: "Our paleontological work was always hap-hazard, subordinate to structural field work and at the discretion of the assistants. I did not dare to turn them aside from stratigraphy by ordering collections." Lesquereux, who had been responsible for what paleobotanical work was done under Rogers, was commissioned to extend his investigations, and in 1880 and 1884 issued three reports of text and one atlas on the coal flora.

In 1881 and 1883 I. C. White studied the fossils of the middle belt of counties on the Delaware, Susquehanna, and the upper Juniata rivers. J. J. Stevenson did the same for the Maryland border, while E. W. Claypole was commissioned to prepare a special report on all forms discoverable in the district of the lower Juniata.

The main energies of the survey, as above noted, were devoted to economic problems, with particular reference to the extent, both geographic and geologic, of coal and petroleum. A great amount of good was undoubtedly accomplished, perhaps as much as one has a right to expect. Work in later years has, however, showed that the fundamental defect of the survey was lack of accurate topographical maps. This seems the more remarkable in view of the fact that the same defect became obvious during the progress of the Rogers survey.

To quote Lesley's own words (manuscript), the facts of dominant importance which became impressed upon the Pennsylvania survey were as follows:

PLATE XXVIII

J. PETER LESLEY

From a portrait by Mrs. L. Bush-Brown.

(1) The Paleozoic formations reach their maximum thickness in this State; and consequently admit of a greater differentiation than elsewhere into special groups of beds.

(2) The middle region of the State is magnificently plicated and eroded, exposing innumerable outcrops, connected in zigzags, and of immense length.

(3) No unconformable later deposits cover and conceal these outcrops, so that there is an unexampled opportunity for the study of variable thickness and changes of type.

(4) The topographical features are so dependent upon the lithology and structure that any geological survey of the region must be virtually a topographical survey.

(5) The geological areas are of great size and so clearly defined, and so distinct in character, that they naturally claimed and received each one a survey of its own. These areas are: 1, The Bituminous Coal field of the west and north; 2, the Anthracite Coal fields in the east; 3, the middle belt of Devonian and Silurian formations; 4, the Mesozoic belt of the south and east; 5, the South Mountain Azoic; 6, the Philadelphia belt of Azoic rock; and 7, the region of Glacial Drift.

(6) The natural section of the Bituminous Coal Measures, down the Monongahela and up the Allegheny rivers, relieved the study of that part of the Paleozoic system of all ambiguity.

(7) The great amount of mining done in the anthracite fields made that part of the survey peculiarly exact and correct.

(8) The great size and number of the brown hematite mines furnished unusual opportunities for the study of that kind of mineral.

(9) The great size and number of limestone quarries, exploited for the manufacture of iron and for fertilizing farms, opened to view every part of the great Siluro-Cambrian formation, the whole of the Lower Helderberg, all the Devonian, and most of the Carboniferous limestone beds.

(10) On the other hand, Pennsylvania is singularly destitute of workable veins of the precious metals. Its poverty in gold, silver, copper, and lead is extreme. It has but one important zinc deposit; and but one nickel mine.[23] In fact its Azoic regions as a whole are barren country, containing but a few small magnetic iron ore beds, in strong contrast to the adjoining Azoic region of northern New Jersey. What little white marble it possesses makes a narrow outcrop for a few miles along a single line. Some serpentine rock, a little chrome iron, one large soapstone quarry, and some kaolin deposits conclude the list of its Azoic minerals.

Practically viewed, the geology of Pennsylvania is wholly Paleozoic, on the most magnificent scale, with an unexampled wealth of anthracite and bituminous coal, brown hematite iron ore, limestone, rock oil, and rock gas; and to the study and description of these its geological survey has from first to last been devoted.

[23] Not worked since 1891.

Little attention has been paid to the lithological study of the building stones of the State, or to their economic description. The entire State is a rock quarry. Every known building stone from the granites, gneisses, quartzites, and traps, to hearthstones, flagstones, brownstone, and limestone can be got with ease and with infinite abundance on lines of transportation. All the principal outcrops of these building stone formations have been located and their places in the Paleozoic series defined in the reports, with sufficiently precise descriptions of their qualities and uses; but beyond this the survey could not go.

Sketch of Lesley.

Lesley was born in Philadelphia in 1819, and graduated at the University of Pennsylvania in 1838, becoming almost immediately connected with the state geological survey under Rogers, in the capacity of topographer. In 1841, during the temporary suspension of this work, he entered upon the study of theology at Princeton, and was licensed to preach in 1844. He then went to Europe, where he studied in the University of Halle during the winter of 1844-1845.

Returning to America, he assumed the pastorate of a Congregational church in Milford, Massachusetts, but resigned in 1851 and owing, it is stated, to a change in his religious views, gave up the ministry altogether. Returning to Philadelphia, he soon became secretary of the Iron and Steel Association and of the American Philosophical Society, and prominent in geological matters, particularly those relating to iron and coal.

He was a man of tall, lank, but commanding figure, and, according to his biographer, of impressionable and emotional nature, an enthusiast and optimist, but often lamentably melancholy, undemonstrative, and even cynical. A man of tremendous nervous energy, aggressive and outspoken, his writings are full of expressions which, for terseness and unpolished emphasis, are unequaled.[24] Thus, in a letter to Powell relative to the coloring of maps and the names of the various formations, he wrote:

> The fact is these English names are good for nothing in America and ought to be ignored. If I were an expert in profanity, I should say damn the text-books, Dana's, Le Conte's, and all of them. They are mere museums of embarrassments, so far as classification is in question. . . . Fortunately Pennsylvania is a very small corner of the United States, and I suppose it matters very little whether its structure appears on your map or not. But in heaven's name (I mean the heaven of geology), what do you gain by distinguishing a miserable subformation like the Permian

[24] Appendix, p. 695.

(one or two patches) and not distinguishing an enormous subformation 5,000 feet thick like the Subcarboniferous?[25]

And again, in a letter to Hall, toward the latter years of his life:

My children beg me to write my autobiography. How would that be possible? I have lived many lives in one. I have made forty voyages by sea, and twice forty by land. I have been all sorts of man by turn except speculator and politician. I have made many friends of worth and note, and lost them all except two or three. I have lived in palaces and hovels, in cities and in forests, in workshops and in garrets and cellars, underground and on mountain tops. But, why should it be told over again? Who reads the adventures of the most adventurous—even heroes, poets, saints! The XIX Century is dyspeptic with book diet. Better write fairy tales and boys' stories, I think.

[25] Some of Lesleys "digs" at his fellow workers are masterpieces of their kind. Thus in Volume I of his final report, where attempting to describe the chaotic conditions existing on the earth during the earliest Archean times, and the intense chemical activity incidental to the deluges of "sour rain" falling upon the hot surface, he says (p. 53): "All this had taken place before the first age of which we have any geological monuments and is only known to God and Dr. Sterry Hunt, who has described it magnificently in his Chemical Researches."

CHAPTER VIII.

The Era of National Surveys.

THE period of the Civil War had brought to light a considerable number of men for whom the piping times of peace, even when varied by Indian outbreaks in the West, afforded insufficient opportunities. They were men in whom the times had developed a power of organization and command. They were, moreover, men of great physical and moral courage. It was but natural, therefore, particularly when the necessity for military routes in the West and public land questions were taken into consideration, that such should turn their attention toward western exploration. Further, the surveys made in the third decade, in connection with routes for the Pacific railroads, and the work done by Evans, Hayden, and Meek in the Bad Lands of the Missouri, had whetted the desires of numerous investigators. Willing workers were abundant and Congress not difficult to persuade into granting the necessary funds. Hence expedition after expedition was organized and sent out, some purely military, some military and geographic, with geology only incidental, and others for the avowed purpose of geological and natural history research.

Under such conditions was inaugurated the work which culminated, in 1879, in the organization of the present United States Geological Survey, which, for breadth of scope and financial resources, is without counterpart in the world's history of science.

The more important of the expeditions above referred to were Hayden's Geological Surveys of the Territories; King's Geological Survey of the Fortieth Parallel; Powell's United States Geological and Geographical Survey of the Rocky Mountain Region; and Wheeler's Geographical Surveys West of the One Hundredth Meridian.

These expeditions demanded men in the prime of life and bodily vigor—men who could endure exposure and fatigue and, if necessary, face danger. It is a natural consequence that there should be found among the workers many names and faces which have not heretofore appeared in our chronicles. Among those who appear now, if not for the first time, at least for the first time prominently, mention may be

PLATE XXIX

FERDINAND VANDEVEER HAYDEN

made of C. E. Dutton, S. F. Emmons, G. K. Gilbert, Arnold Hague, F. V. Hayden, W. H. Holmes, Clarence King, Joseph Le Conte, O. C. Marsh, A. R. Marvine, A. C. Peale, J. W. Powell, I. C. Russell, Orestes St. John, J. J. Stevenson, and R. P. Whitfield.

The impracticability of strict chronology in this history has been made repeatedly apparent, and it now seems advisable to turn back and review briefly Hayden's career with various exploring expeditions, before entering upon his career as a U. S. geologist in 1867.

Hayden and Meek in the Bad Lands, 1853.

The reports of D. D. Owen and Dr. John Evans on the collections made by the latter in the *mauvaises terres* of the White River in 1849, as published in 1852, had attracted a great deal of attention, and in the spring of 1853 F. V. Hayden and F. B. Meek were employed by James Hall to visit the "Bad Lands" to make collections. In view of the subsequent relation of the first two mentioned, it is of interest to note here that at Hall's suggestion Meek was placed in charge of the party, Hayden acquiescing, "He [Meek] has had experience in such matters, and I have had none. He will be more cool, and calm."

The country being wholly without railroads, the journey was made by boat up the Missouri. The party met at St. Louis on Saturday, May 14, 1853, where they found Dr. Evans bent upon a similar errand under the direction of B. F. Shumard.[1]

Meek's letters to Hall, written on the way, are full of interest, and it is to be regretted that they cannot be reproduced entire. (See Appendix, pp. 696-709.) Owing to the swiftness of the current and other difficulties and dangers attendant upon river navigation, the boat ran only during the hours of daylight, making even then but some three and one-half to four miles an hour, and tying up to the banks at night. No sooner tied up, however, than the two were on shore with candles or a lantern collecting plants or whatever might prove of interest. They left St. Louis on the twenty-first of May and arrived at Fort Pierre the nineteenth of June, nearly a month accomplishing what is now a journey of but a day. But the distance was then 1,600 miles!

In spite of the difficulty of holding the party together, through fear of hostile Indians, they remained in the field for a period of

[1] The young Prince of Nassau, with a companion, was also there on his way to the Rockies and the Pacific coast. Although there are frequent references in Meek's letters to the Evans party, even up to the date of their leaving the steamer at Fort Pierre, I can learn of nothing that they accomplished.

several weeks, returning to Fort Pierre on July 18. They brought with them a large and valuable collection, including mammalian remains which were investigated by Joseph Leidy.

The Cretaceous invertebrate fossils were studied by Hall and Meek and described by them in a memoir published by the American Society of Arts and Sciences of Boston in 1854. This paper was accompanied by a brief vertical section by Meek, showing the order of superposition of the Cretaceous beds. As this is believed to be the first section of the region, it is here reproduced in full:

Section of the Members of the Cretaceous Formation as Observed on the Missouri River, and thence Westward to the Mauvaises Terres.

Eocene Tertiary formation:

Clays, sandstones, etc., containing remains of mammalia. The entire thickness of this formation in the Bad Lands is from 25 to 250 feet.

Cretaceous formation:

5. Arenaceous clay passing into argillo-calcareous sandstone. 80 feet.
4. Plastic clay with calcareous concretions containing numerous fossils. 250 to 350 feet. This is the principal fossiliferous bed of the Cretaceous formation upon the Upper Missouri.
3. Calcareous marl, containing Ostrea congesta, scales of fishes, etc. 100 to 150 feet.
2. Clay containing few fossils. 80 feet.
1. Sandstone and clay. 90 feet. Buff-colored magnesian limestone of the Carboniferous period.

Hayden's Work on the Upper Missouri, 1854.

In the following spring (1854) Hayden, having severed connections with Hall, again ascended the Missouri River, this time partly under the auspices of the American Fur Company. He spent a part of two years on this expedition, during which period he visited the various portions of the Upper Missouri, being without other means than what he earned or secured in various ways as he went along, and dependent even for subsistence on such friends as he met in the country, among whom were Colonel A. J. Vaughan, the Indian agent, and Mr. Alexander Culbertson, of the American Fur Company.

He traversed the Missouri River to Fort Benton and the Yellowstone to the mouth of the Big Horn, and also considerable portions of other districts not immediately bordering on the Missouri. As the boats of the fur company had to be towed in ascending the river, the progress was necessarily slow. The time thus occupied by the boats was utilized by Hayden on the shores, and as a result he traversed a considerable portion of the journey on foot.

The vertebrate remains collected on this trip were, as in the previous case, described by Leidy, mainly in papers read before the Academy of Natural Sciences of Philadelphia, while the invertebrates were described by Hayden himself in connection with Meek. The collections, which were deposited partly with the Academy of Natural Sciences of Philadelphia, contained a larger number of species than all those previously known from that region, many of them being new and of a remarkable character.

Early in 1856 Hayden returned to St. Louis, and by request prepared a report on the geology of the region which was incorporated in that of Lieutenant G. K. Warren for 1855.[2]

Hayden with Warren in 1856 and 1857.

In May of 1856 Warren appointed Hayden with some highly interesting stipulations[3] an assistant in the exploration of the Yellowstone River and the Missouri River from Fort Pierre to a point sixty miles north of the mouth of the Yellowstone.

The field work of this expedition began on June 28, the party returning to Fort Pierre on October 22, and reaching Washington in November.

In May, 1857, Hayden was again appointed geologist by Lieutenant Warren, this time on an expedition to the Black Hills. The party was organized at Sioux City in June and proceeded up the Loup Fork of the Platte, returning to Fort Leavenworth, Kansas, early in the following December.

Hayden noted the occurrence of—

1, Metamorphic Azoic rocks, including granite; 2, Lower Silurian (Potsdam); 3, Devonian (?); 4, Carboniferous; 5, Permian; 6, Jurassic; 7, Cretaceous; 8, Tertiary; 9, Post-Tertiary and Quaternary.

In a preliminary report given by Warren in the *American Journal of Science* for May, 1859, attention was called to the important physiographic fact that the Niobrara River seemed "to run along a swell or ridge on the surface and to be practically without tribu-

2 Explorations in the Dacota Country, in the year 1855. Sen. Ex. Doc. No. 76, 1856.

3 Appendix, p. 711.

taries." This would seem to be a recognition of the fact, though not the principle, that streams flowing from a mountainous country and laden with silt may, in their lower levels where the current is less rapid and the carrying power less, so deposit their load as to build up both the bottom and banks, and this until the stream actually occupies the crest of a ridge. The Platte River is, however, a better illustration of this than is the Niobrara. That portion of the channel running between steep bluffs he thought "must have originated in a fissure in the rocks which the water basins enlarged and made more uniform in size." The failure at so late a date to realize that a stream may carve out its own channel can be excused on the ground that Warren was not a geologist.

Meek and Hayden in Kansas, 1858.

During the summer of 1858 Hayden and Meek explored a portion of what was then called the Territory of Kansas. The route followed, as given in the report published in the *Proceedings of the Academy of Natural Sciences of Philadelphia,* was as follows: From Leavenworth City on the Missouri, across the country to Indianola near the mouth of Soldier Creek and the Kansas River; thence up the north side of Kansas and Smoky Hill rivers to the mouth of Solomon's Fork. Here they crossed the Smoky Hill, following it up on the south side to a point near the ninety-eighth degree of west longitude, from which point they struck across the country in a southwest direction to the Santa Fé road, which was followed northeastward to the head of Cottonwood Creek. Leaving the road here, they went down the Cottonwood Valley some thirty miles, when they turned due east to Council Grove, whence the Santa Fé road was followed southwest for about twenty-four miles to a watering place known as Lost Springs. Here they struck across the country in a northwest direction to Smoky Hill River again at a point nearly opposite the mouth of Solomon's Fork. Thence the route lay down the south side of Smoky Hill and Kansas rivers to Lawrence, and thence across the Kansas in a northeastward direction to Leavenworth City. The explorations were very successful, and the results embodied in numerous papers in the *Proceedings of the Academy of Natural Sciences of Philadelphia* and elsewhere.

Hayden with Raynolds's Expedition, 1859.

In April, 1859, Captain W. F. Raynolds was instructed to organize an expedition for the exploration of the country from which

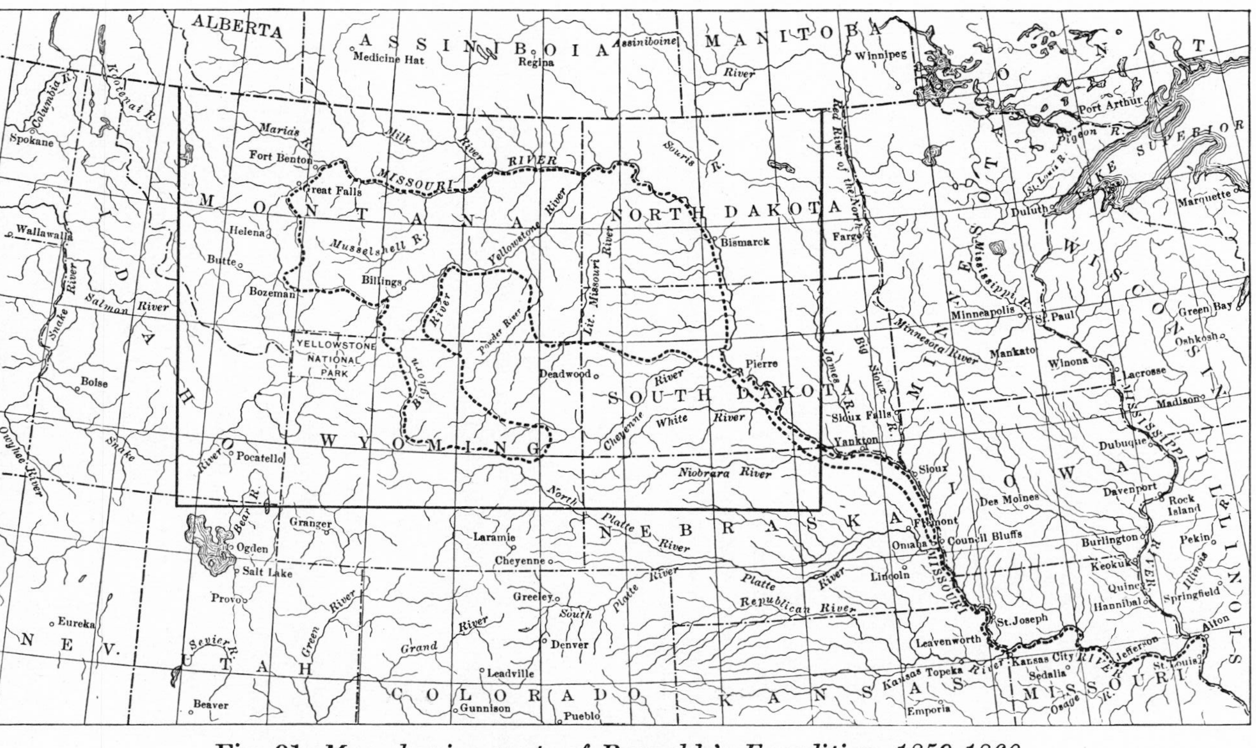

Fig. 91. *Map showing route of Raynolds's Expedition, 1859-1860.*

flow the upper tributaries of the Yellowstone River and of the mountains in which these tributaries and the Gallatin and Madison forks of the Missouri have their source. On April 22 Hayden was appointed surgeon and naturalist of this expedition. (Appendix, p. 714.) The party left St. Louis May 28, 1859, going up the Missouri to Fort Pierre, which point was reached about the middle of June. From here the route was westward and northwestward to the Yellowstone River by way of the Cheyenne, the Belle Fourche, and Powder rivers. The Yellowstone was reached near the mouth of the Big Horn about the middle of August. From the Yellowstone they turned southward early in September and followed up the Big Horn, skirting the eastern edge of the Big Horn Mountains, and going into winter quarters on Deep Creek, near the North Platte River, about the middle of October.

The season following explorations were continued as far north as Fort Benton on the Missouri by way of the North Platte, Wind, Snake, and Madison rivers. Fort Benton was reached on July 14. The return trip was made by the more direct route down the Missouri to Fort Union and thence to Omaha, Nebraska, where they disbanded on October 4.

Hayden's report on this expedition was not published until early in 1869. It comprised some 174 octavo pages, including 30 pages of paleontological notes by J. S. Newberry, and was accompanied by a colored geological map of the region north of the forty-second parallel and lying between the ninety-eighth and one hundred and fourteenth meridians (Fig. 91).

Hayden's Second Expedition to the Bad Lands, 1866.

In the summer of 1866 he undertook a second expedition to the Bad Lands, this time under the auspices of the Academy of Natural Sciences of Philadelphia. In company with James Stevenson he left Fort Randall, South Dakota, August 3. The trip was made with a team of six mules and occupied fifty-two days, during which a circuit of 650 miles was accomplished. The large and valuable collection of mammalian fossils was described by Leidy in his great work of upward of 450 large octavo pages and 30 plates, published under the auspices of the Academy of Natural Sciences of Philadelphia. This began with an introduction of 12 pages by Hayden on the geology of the Tertiary formations of Dakota and Nebraska and was accompanied by a map. In this he pointed out the possibility of bridging over the chasm heretofore existing between the Cretaceous and Tertiary periods by means of Transition beds belonging to the lignite series. He reiterated some of the statements made in previous writings, to the effect that, at the close of the Cretaceous period, the Rocky Mountain area was occupied by the waters of an ocean with perhaps a few peaks projecting. Near the close of the period the surface had reached a sufficient elevation to form long lines of separation between the waters of the Atlantic on the east and those of the Pacific on the west; then this great watershed began to rise above the surrounding country and the period of fresh-water lakes was inaugurated. The elevation during the Cretaceous period he regarded as slow and gradual, but at about the close of the period or in the early part of the Tertiary the limit of tension in the crust was reached and long lines of fracture commenced which form the nucleus of the present mountain ranges, including the lofty continuous ranges with a granitoid nucleus along the eastern portion of the Rocky Mountains, as the Wind River, Big Horn, Laramie, or Black Hills. He showed that the Tertiary deposits were in part laid down before the upheaval, as indicated by the inclination of the lignite beds.

The lower Tertiary fossils included brackish-water forms, and he thought to trace the "history of the growth of the continent step by step from the purely marine waters of the Cretaceous ocean to the period when the mountain ranges were elevated," the ocean waters excluded, and inclosed lakes formed, at first salt, but gradually freshened by influx from fresh-water streams. During the Tertiary period there were at least four and possibly five of these fresh-water lakes in the West, two of which were of great extent. The deposits in them formed the bad lands of the Judith, the great Lignite basin, the Wind River basin, and the White River group.

Summary of Hayden's Early Work.

Some of the more important conclusions arrived at by Hayden as a result of observations on these expeditions are as follows: He announced in 1857 the discovery of Potsdam sandstone in the country about the headwaters of the Yellowstone, and in a preliminary publication in the *American Journal of Science* for 1861 described it as more or less changed by heat from beneath. The other formations noted were Carboniferous (including Permian?), red arenaceous deposits overlying the Carboniferous, but of uncertain age; Jurassic, Cretaceous, and Tertiary deposits. During the long interval that elapsed between the deposition of the earliest part of the Silurian and the commencement of the Carboniferous, he believed dry land to have prevailed over a large portion of the West, and he found no evidence of deep-water deposits until far up into the Cretaceous. Near the close of this epoch the waters of the great Cretaceous seas receded toward the present position of the Atlantic on the one side and the Pacific on the other, leaving large areas in the central West dry land with but a slight elevation above the sea level. He showed that the White River Tertiary deposits were younger than the Lignite, and that the older members of the western Tertiary were clearly separable into four divisions exclusive of the Pliocene of the Niobrara. Further that the estuarian deposits ushered in the dawn of the Tertiary epoch and belonged to the Eocene period. The evidence of the fossils was regarded as indicating a much milder climate throughout the West during the greater part of the Tertiary than now—a climate somewhat similar to that of the Gulf States at the present day.

In an article on the Primordial sandstone of the Rocky Mountains, published in the *American Journal of Science* for 1862, he announced finding undoubted evidence of the existence of the equivalent of the Potsdam sandstone of the New York series in two important outliers of the Rocky Mountain chain. He pointed out the singular uniformity in the nature of the sediments and general lithological resemblance to the eastern type, which he thought due to a similar uniformity on the part of the underlying rocks from which sediments were derived—that is, he believed that the source of all the sediments composing the Primordial rocks in the West could be traced to a source near at hand. He called attention to the gradual thinning out of this Primordial sandstone toward the West, and quoted the observations of D. D. Owen in Minnesota, Whitney in Iowa, Safford in Tennessee, and Shumard in Texas as confirmatory. The lower secondary formations, on the other hand, gradually increased in thickness.

No unconformability was found by him in any of the fossiliferous sedimentary rocks of the Northwest from the Potsdam sandstone to the summits of the true Lignite Tertiary, but there was found proof of two great periods of disturbance, the one prior to the deposition of Potsdam sandstone, when the Azoic or granitic rocks were elevated to a more or less inclined position, and the other, much the more important, at the close of the Lignite Tertiary, when the "massive nuclei of the ranges were raised above the surrounding country."

In the same volume Hayden had an important paper on the period of elevation of the ranges of the Rocky Mountains near the sources of the Missouri River and its tributaries. He thought the evidence clear that the great subterranean forces which elevated the western portion of the continent were called into operation toward the close of the Cretaceous epoch, and that the gradual rising continued without a general "bursting" of the earth's surface until the accumulation of the Tertiary lignite deposits, or at least the greater part of them. Also, that after the fracture of the surface commenced and the great crustal movements began to display themselves, the whole country continued rising, though perhaps with intervening periods of subsidence, up to and even including the present period. Up to this time, it will be noted, Hayden seems to have regarded all the phenomena of uplift as due to forces acting directly from beneath.

During the years of the Civil War western exploration of all kinds was interrupted. Hayden served in the Federal Army as a surgeon from 1862 until 1865, resigning to accept the position of professor of mineralogy and geology in the University of Pennsylvania, a position which he retained until 1872.

The United States Geological Surveys under F. V. Hayden.

In the spring of 1867 F. V. Hayden, acting under the direction of the General Land Office, and with funds representing "the unexpended balance of the appropriations heretofore made for defraying the expenses of the legislative assembly to the territory of Nebraska" amounting to some $5,000, entered upon his work as a United States geologist and in so doing laid the foundations for the United States Geological Survey as it exists today.[4] Hayden was allowed a salary of $2,000 for himself, $1,000 for an assistant geologist, and $700 for collectors and laborers. Three hundred dollars was set aside for chemistry and natural history and $1,000 for the general expenses of the outfit. His first annual report comprised sixty-four octavo pages and dealt largely with the possible occurrence of workable coal beds within the territory, a question decided in the negative. The loess he considered a silt brought by streams and deposited in a fresh-water lake. Consideration was also given to the distribution of the Cretaceous and Tertiary formations. The final report appeared as House Executive Document No. 19, 1871 (1872), and comprised 264 pages, of which 180 pages were devoted to paleontology, mainly by F. B. Meek, and entomology. It was accompanied by a colored geological map of the region west of the Missouri, in-

[4] It appears from correspondence that this opportunity came to Hayden through Professor S. F. Baird, Secretary of the Smithsonian Institution. See Appendix, p. 715.

cluding the Black Hills. This was, however, little more than an enlargement of the map previously published in the *Transactions of the American Philosophical Society* of Philadelphia.

In 1868 the appropriation of $5,000 was renewed and the field of work extended into Wyoming. In his report of 102 pages Hayden called attention to the probable Tertiary age of all the coal of both Wyoming and Colorado.[5] In 1869 $10,000 was appropriated and the field of work transferred to Colorado and New Mexico, the survey being at the same time placed under the direction of the Secretary of the Interior. This may well be considered the fundamental act establishing the national survey on a permanent basis. The assistants for this season were James Stevenson; Henry W. Elliott, artist; Persifor Frazer, mining engineer and metallurgist; Cyrus Thomas, entomologist and botanist; E. C. Carrington, zoölogist, and B. H. Cheever, Jr.

Fig. 92. *Map showing approximate routes traversed by the Hayden Survey in 1869.*

Although what is here written regarding the work of the organization has reference only to the geological results secured, it is well to call attention to the fact, suggested by this personnel, that it was not merely a geological, but a general natural history survey.

In his report for this season Hayden reiterated the statement made by him in the *American Journal of Science* for March, 1868, to the effect that all the lignite Tertiary beds of the West are but frag-

[5] E. L. Berthoud in a letter to Dr. Hayden, dated December 18, 1866, had called attention to the lignite beds near Denver, Colorado, and their probable fuel value, and gave a section showing their inclination with reference to the mountain range. The fossil associations suggested a probable Tertiary age for the beds. The Raton, New Mexico, coal was thought to be a continuation of the same bed. Professor John Le Conte, who examined the coal east of San Antonio and the Rio Grande, concluded it to be of Cretaceous age. He also showed that this coal near Placer Mountain had been converted into anthracite by the outpouring of a stream of lava. (Am. Jour. Sci., Vol. XLV, 1868.)

FIELD PARTY OF HAYDEN SURVEY

Sir Joseph Hooker. Mrs. Strachy. F. V. Hayden.
Asa Gray. Gen. R. Strachy.

ments of one great basin, interrupted here and there by the upheaval of mountain chains or concealed by the deposition of newer formations.

He pointed out for the first time that the main range of the Rocky Mountains "is really a gigantic anticlinal and all the lower ranges and ridges . . . only monoclinals, descending steplike to the plains on each side of the central axis." Also that there were two kinds of ranges in the system—one with a granitoid nucleus, with long lines of fracture, and in the aggregate possessing a specific trend; the other with a basaltic nucleus, composed of a series of volcanic cones or outbursts of igneous rocks, in many cases forming saw-like ridges like those of the Sierras.

No evidences were found of any unconformity between the Cretaceous and lower Tertiary beds and no such changes in the sediments as would account for the sudden and apparently complete destruction of organic life at the close of the Cretaceous period.

He visited the Salt Lake Valley and examined the terraces and old shore lines of Great Salt Lake, describing the beds as of post-Pliocene or Quaternary age and correlating them with the terraces noted by him above the Wasatch Canyon. This series of beds was so widely extended and so largely developed in Weber and Salt Lake valleys that he felt it worthy of a distinct name, and in consequence called it the *Salt Lake* group. Afterward (in 1871), the name was limited to the older beds, which he considered of later Pliocene age, recognizing the more modern character of the terraces in which were found a great abundance of fresh-water shells.

The question of priority in this region having arisen between the King and Hayden surveys, it may be well to state that, as I am informed by A. C. Peale, Hayden's first work in the Salt Lake Valley was done in the years 1868, 1869, and 1870, and the results published in February, 1869, during the latter part of 1869, and the early part of 1871. The report of the field work of 1870 in Wyoming was first printed in 1871, and a second edition issued in 1872.

Work of Hayden in Wyoming, 1870.

In 1870, with appropriations increased to $25,000, Hayden's field of operations was transferred to Wyoming and portions of contiguous territories. Stevenson, Elliott, and Thomas were with him as before, while W. H. Jackson, photographer; John H. Beaman, meteorologist; A. L. Ford, mineralogist; C. P. Carrington, zoölogist; and Henry D. Schmidt, naturalist, were added to the scientific corps.

The party outfitted at Cheyenne, in Wyoming, and proceeded

northward along the eastern base of the Laramie Range, exploring the Platte River as far as the Red Buttes, and thence passing across the divide to the Sweetwater; thence up that stream to its source in the Wind River Mountains, passing down Big and Little Sandy creeks to Green River, and along the northern slope of the Uinta Mountains. From Fort Bridger the route lay southward to Henrys Fork, which was explored down to its junction with Green River proper. From Green River Station the route followed the old stage route up Bitter Creek by way of Bridger Pass and the Medicine Bow Mountains, across the Laramie Plains, and through the Laramie Range by way of Cheyenne Pass, back to the point of departure.

Studies were made also along the line of the Pacific Railroad between Cheyenne and the Salt Lake Valley. No topographer accompanied the party, and the maps used were those constructed by the engineering department of the army, which were so inaccurate that to delineate the geology upon them in any but the most general way was impossible.

During this season Hayden worked out the sequence of the Carboniferous and Cretaceous rocks, and made the subdivisions of the latter into Dakota, Fort Pierre, and Fox Hill groups, which are still recognized. He remarked that some of the fossils found in southern Nebraska seemed to possess Permian affinities, though as they all extended down into the Coal Measures they could not be considered characteristic, and therefore those rocks which he had previously mapped and colored Permian should be relegated to the Permo-Carboniferous.

He noted the occurrence of Potsdam sandstone with the fossils *Obolella nana* and *Lingula* at South Pass on the south side of the Sweetwater. The massive granites as well as the intercalated stratified gneisses extending from South Pass City nearly to Pacific Springs were all thought to be of sedimentary origin. He showed that, near the close of the Cretaceous period, the ocean extended all over the area west of the Mississippi from the Arctic circle to the Isthmus of Darien. A little later the great watershed of the continent was marked out and the marine waters were separated into more or less shallow seas, lakes, and marshes, within which grew the abundant forests that went to form the coal beds.

From a study of the character of the vegetable impressions found in these beds, he argued that coal strata of contemporaneous origin may be purely marine, purely fresh water, or brackish, dependent upon local conditions. He pointed out that the sea had not had access to the Salt Lake Valley since middle Tertiary times, the sediments from 800 to 1,200 feet in thickness, called by him the Salt Lake group, being of Pliocene age and contemporaneous with the Niobrara, Arkansas, and Santa Fé groups, and of fresh-water origin.

This report of Hayden was accompanied by special reports by Meek on the invertebrate paleontology, by Cope and Leidy on the vertebrate paleontology, by Lesquereux on paleobotany, and by Newberry on the ancient lakes of western North America. The volume marks the beginning of Cope's work with the Hayden survey, which resulted later in the production of the two monographs on *The Vertebrata of the Cretaceous, and of the Tertiary Formations of the West*, the latter, popularly known as "Cope's Primer," a pudgy quarto volume of 1,009 pages and 134 full-page plates. With this year, too, began Jackson's work, which resulted in the production of what were at the time the finest landscape photographs ever taken, and which excited the wonder and admiration of geologists the world over.[6]

Work of Hayden Survey in Montana and the Yellowstone Park, 1871.

In 1871, with an appropriation of $40,000, field operations were transferred to Montana and portions of adjacent territories, including what is now the Yellowstone Park. To the party of the year previous was added Anton Schönborn, topographer; G. N. Allen, botanist; and A. C. Peale, mineralogist. The route lay from Ogden, Utah, along the shore of Salt Lake to Willard City; thence through the Wasatch Range to Cash Valley, and up the valley to the divide between the Salt Lake and Snake River basins. From this point they descended Marsh Creek to the Snake River basin and Fort Hall. Following the stage route to Virginia Junction, they crossed Black Tail Deer Creek near its source; thence down the Stinking Water to Virginia City; then, crossing the divide eastward to the Madison River, they descended the valley about thirty miles and crossed over the other divide to Fort Ellis, at the head of the Gallatin River. From Fort Ellis they passed again eastward over the divide between the Yellowstone and the Missouri to Bottler's ranch, where was established a permanent camp. A portion of the party then proceeded up the Yellowstone and entered the park area, surveying the mammoth hot springs on Gardiner River, the Grand Canyon of the Yellowstone, the Upper and Lower geyser basins, and the lake. On returning to Bottler's ranch in August, they passed down the Yel-

[6] This, it must be remembered, was long before the convenient "dry plate" and film were invented.

lowstone to Shields River and Fort Ellis, and thence down the Gallatin to Three Forks, up the Jefferson to the Beaver Head Branch and to Horse Plain Creek, and across the main Rocky Mountain divide to the headwaters of Medicine Lodge Creek, into the Snake River basin to Fort Hall once more, and thence across the mountains to the head of Bear River and up the river to Evanston, on the Union Pacific Railroad, where the party disbanded.

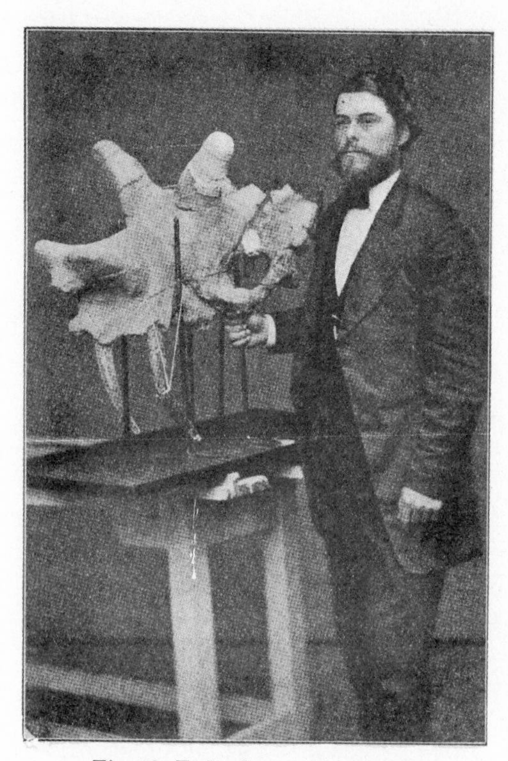

Fig. 93. *E. D. Cope with skull of Loxolophodon Cornutus.*

As in years previous Messrs. Cope, Lesquereux, Leidy, Meek, Newberry, and others collaborated in their especial fields. The hot springs and geysers were described in considerable detail, and the fact that they were but the feeble manifestations of dying volcanic energy recognized. It was shown that the mountain ranges passed over lie along the borders of synclinal valleys, which were originally the basins of fresh-water lakes, and that all the ranges had a general north and south or northwest and southeast trend, and were here and there connected by cross chains; that the three main branches of the Missouri—the Madison, Jefferson, and Gallatin—flowed through valleys now extending to a width of three to five miles and now contracting to narrow canyons, the expansions of which had all been lake basins within late Tertiary and perhaps early Quaternary times. The valleys were considered in part due to erosion, but for the most part as synclinal folds, the intervening mountain ridges being wedge-like masses of Carboniferous limestones.

The work of this year resulted in the setting aside of the Yellowstone region as a national park.[7]

[7] I am informed by Dr. A. C. Peale that the idea of reserving this region as

Work of Hayden in Montana, 1872.

In 1872, with appropriations increased to $75,000, Hayden divided his force into two parties. The first, under his immediate charge, consisting of Adolf Burck, chief topographer; Henry Gannett, astronomer; A. E. Brown, assistant topographer; E. R. Wakefield, meteorologist; A. C. Peale, mineralogist; W. H. Holmes, artist; and W. B. Platt, naturalist. This division left Fort Ellis, Montana, and explored the headwaters of the Yellowstone, Gallatin, and Madison rivers in much more detail than had been done during the previous year.

Fig. 94. *F. H. Bradley.*

The second or Snake River division, under the directorship of James Stevenson, included Frank H. Bradley, chief geologist; W. R. Taggert, assistant geologist; Gustavus R. Bechler, chief topographer; Adolph Herring and Thomas W. Jaycox, assistant topographers; William Nicholson, meteorologist; John M. Coulter, botanist; Josiah Curtis, surgeon and naturalist; and William H. Jackson, photographer. This division visited the Snake River or Lewis Fork of the Columbia in Idaho and Wyoming territories, a region up to that time little known. The Teton Mountains—a prominent range—were ascended by Stevenson on this trip, the first time, it was claimed, that the feat had been accomplished by a white man.

Leidy and Cope spent a large part of the summer in studying the ancient lake basins in the interior, and obtained the materials described in Volumes I and II of the quarto monographs. Lesquereux spent several months in exploring the coal beds to ascertain their geological position, and F. B. Meek and H. M. Bannister studied the invertebrates.[8]

From a preliminary study Lesquereux was inclined to call the lignite beds mostly Eocene. Meek thought them to be Upper Cretaceous, passing through Transition beds to the Eocene, and Cope regarded them Cretaceous. Hayden in this report gives a brief review

a park or pleasure ground for the people originated with Dr. Hayden, and the law setting it apart was written in great part by him, and it was also largely due to his personal efforts that the bill was passed by Congress.

8 Bannister also reported on a geological reconnaissance along the Union Pacific Railroad this same year.

of the opinions held and the evidence on which same were based, and concluded that the deposition of the lignite strata began during the latter portion of the Cretaceous period and continued on into Tertiary time without any marked physical break, so that many of the Cretaceous types, especially of the vertebrates, may have lingered on through the Transition period even into the Tertiary epoch.

He called attention to one feature in the geological structure of the mountains of Montana observed during this season for the first time and not noticed in such a marked degree in any other portion of the West—the inversion of the sedimentary beds, so that the oldest incline at a greater or less angle on the more recent. As illustrative of this, he gave an east and west section across the Flathead Pass in the East Gallatin Range, the central portion of this range being composed of Carboniferous limestone standing nearly vertically. While disclaiming any intention of discussing the origin of such a structure, he nevertheless suggested that there were two forces engaged, one in raising the mass vertically and one acting tangentially, the crowding of the Silurian beds over beyond the vertical being due to the latter. A similar illustration of inversion was given by Peale in his report on the geology of Jackass Creek on the upper Missouri River.

Another point to which Hayden called attention and to which reference has been made in discussing the work of Powell (p. 479) was the fact that the streams seem to have cut their way directly through mountain ranges, instead of following synclinal depressions. This seemed to him to indicate that they began the process of erosion at the time of the elevation of the surface. "We believe that the courses of these streams were marked out at or near the close of the Cretaceous period, and as the ranges of mountains were in process of elevation to their present height, the erosion of the channels continued." It will be remembered that this faculty of a river to carve out its channel across a slowly rising land was first noted by the elder Hitchcock, who was, at that time, quite unable to account for it satisfactorily. The period of intense volcanic activity manifested in the Yellowstone region Hayden thought probably to have commenced somewhere during later Miocene or early Pliocene epochs.

With Cope, Leidy, and Marsh all in the field of vertebrate paleontology at one time, it is not strange that a spirit of rivalry, if not of personal jealousy, should have arisen. This found expression in numerous instances of scrappy descriptions which at this date are only amusing, however serious they may at the time have seemed to those most interested. Thus, Cope in 1868 described and figured remains of a marine saurian from the Cretaceous of Kansas, to which he gave the name of *Elasmosaurus platyurus*. Leidy, ever on the alert, made a reexamination of the materials, and at the meeting of the Philadelphia Academy on March 8, 1870, announced that the remains were, in reality, those of an Enaliosaurian and closely allied to Plesiosaurus, and, further, that Cope's error in identification lay in his having described the animal—the skeleton of which was without a skull and quite incomplete—in a reversed position from the true one, or, as commonly expressed, "wrong end to." Again, through fear of being anticipated by Marsh, Cope in one instance at least, actually telegraphed from the field to the Philadelphia Academy his description of a newly found species, a proceeding which brought forth a hearty protest from Dana. (See Appendix, pp. 716-719.)

Work of Hayden Survey in Colorado, 1873-1876.

In 1873, with appropriations the same as for the previous season, the field of operations was transferred to Colorado, this in part owing to the expense of transporation, subsistence, and labor in regions so remote as those of the upper Missouri, and in part to the hostility of the Indians.[9]

Fig. 95. *F. M. Endlich.*

The party rendezvoused at Denver. The area decided upon to be surveyed comprised the eastern portion of the mountainous part of Colorado, and was divided into three districts known as the North, Middle, and Southern. The personnel and their assignments were as follows:

The first or Middle Park division was directed by A. R. Marvine, assistant geologist, with G. R. Bechler, topographer, and S. B. Ladd, assistant topographer. The second or South Park division had Henry Gannett as topographer in charge; A. C. Peale, geologist; W. R. Taggert, assistant geologist; Henry W. Stuckle, assistant topographer; and J. H.

[9] "The red devils have attacked Lieut. Hayden's exploring party on the La Platte river, Colorado. The Indians passed the surveying party, appearing friendly, and then turned and fired on them from the rear. Each party sought an advantageous position and the fight began. The battle lasted five hours, each firing from cover, Indian style. The enemy drew off at midnight. Hayden's party took advantage of the cessation of hostilities to obtain a better position, but were again attacked at sunrise. The red devils fought at long range, and did but little

Batty, naturalist. The third or San Luis division was in charge of A. D. Wilson, topographer, with George M. Chittenden, assistant topographer, and F. M. Endlich, geologist.

The work of this year extended as far westward in Colorado as Middle Park, the Elk Mountains, and the San Luis Park. It was during this season's work that were described and figured the peculiar examples of subaerial erosion of Monument Park in Colorado which have been so frequently reproduced in the later textbooks. The wonderful instances of complete overturning of immense groups of beds, as illustrated in the Elk Range, were again referred to, attention being called to the fact that for several miles there is a double series from the Silurian up to the Cretaceous, inclusive, which had been thus inverted. In this report, too, were given the examples of inversion in the Snow Mass Range and the view on *Roches Moutonnés* Creek, both of which have served their purpose in the textbooks of Dana and Le Conte.

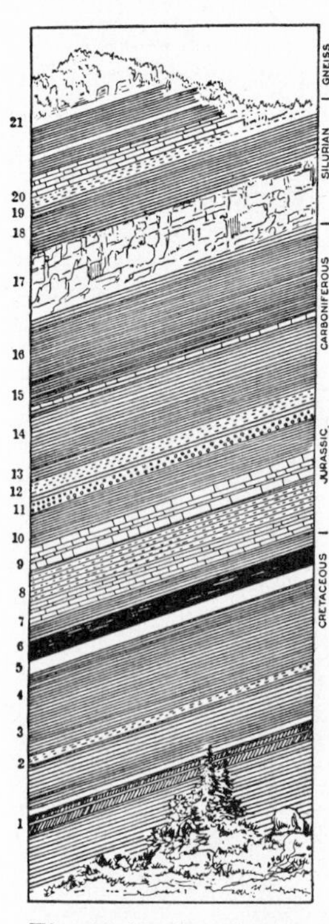

Fig. 96. *Peale's Section of Inverted Beds on Jackass Creek.*

The question of the age of the lignite beds occupied the attention of nearly all the workers in the field (see p. 579). Lesquereux, in his chapter on the lignite flora, argued in favor of the Eocene and Miocene age of the beds. Though not denying the presence of Cretaceous animal remains in the lignite strata, he thought the "presence of some scattered fragments of Cretaceous shells of little moment in comparison with the well-marked characters of the flora." Meek's invertebrate work, moreover, he felt to be rather in favor of the Tertiary hypothesis.[10] To Cope's conclusions "that a Tertiary flora was contemporaneous with a Cretaceous fauna, establishing an uninterrupted succession of life across what

execution. They feared to make a charge on the brave little party under Hayden, who stood their ground and eventually fought the Indians off, when the party returned to camp without loss of human life, but with stock killed and wounded. The fight lasted forty-eight hours." (Newspaper clipping without date. See also Appendix, p. 719.)

[10] See Appendix, p. 720.

is generally regarded one of the greatest breaks in geological time," he took exception, as it did not appear exactly to conform with facts.

During the years 1874 to 1876 under essentially the same conditions work was continued by the survey throughout Colorado, the appropriations being $75,000 annually, with the exception of 1876, when they dropped to $65,000. The individual work of Hayden himself becomes gradually less conspicuous in the reports issued, owing to the increased amount of administrative work.

Fig. 97. *A. C. Peale.*

In 1874 the party under direction of A. R. Marvine was engaged in the southern portion of North Park; that under A. C. Peale in the region bounded on the north by the Eagle and Grand rivers, on the east by the one hundred and seventh meridian, on the west by the state line, and on the south by latitude 38° 20′. The third division, under A. D. Wilson, with F. M. Endlich, geologist, was assigned to what is known as the San Juan district, and the fourth, under the immediate direction of Hayden, with W. H. Holmes as artist and geologist, to the Elk Mountain region.[11]

Fig. 98. *A. R. Marvine.*

In the report for 1874 considerable attention was given to the stratigraphic position of the Lignite group, a discussion of which may be referred to later. Perhaps the most striking feature brought out in the work of this year was that relative to the Elk Mountains. This range was regarded by Hayden as a grand illustration of an eruptive range, "the immense faults, complete overturning of thousands of feet of strata, and the great number of peaks, all composed of eruptive rocks," indicating to him periods of violent and catastrophic action. The great thickness of sedimentary strata which had been carried to the loftiest points of the axial ridge in a nearly hori-

[11] An interesting light is thrown upon Marcou's character in a letter he wrote Hayden on receipt of the report of this year. "It is to be regretted that you

zontal position he thought might be explained on the supposition that at one time the sedimentary mass rested on a floor of pasty or semipasty granite, and that the forces in the interior were struggling to find vent and carried upward the entire overlying mass. This description is of interest when taken in connection with one by Marvine the year previous as showing the gradual inception of the laccolithic idea afterward worked out in detail by Holmes, Peale, and Gilbert, to which I will now refer.

Fig. 1. Section across Northern Group. 7 miles

Fig. 2. Section across Middle Group. 7 miles

Fig. 3. Section across Southern Group. 7 miles

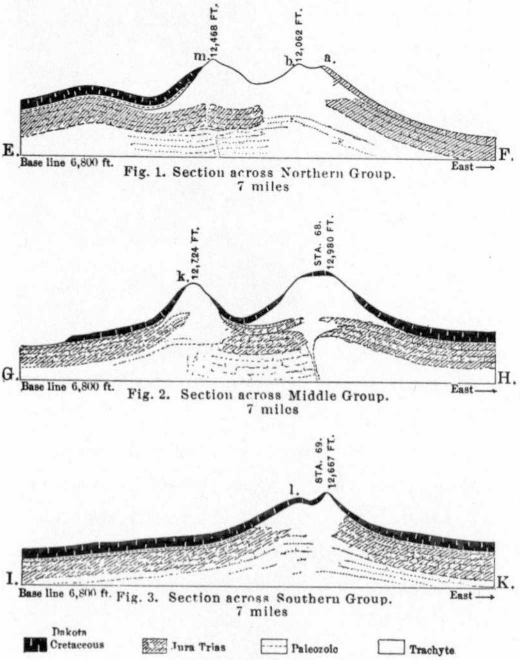

Sections across the SIERRA LA SAL. for lines see map on Plate V.

Fig. 99.

In describing the Sierra La Sal south of the Gunnison in the report for 1875, Peale compared them in structure with the Elk Mountains—that is, as of eruptive origin. "By this," he wrote, "I mean that the sedimentary strata have been lifted up by eruptive rock which has broken through them in some places, and in others is seen only as the result of subsequent erosion." Illustrative of this, he gave the figures reproduced here. (Fig. 99.)

The idea thus advanced seems to have been contagious, for in the same report Holmes described the Sierra El Late as formed of a number of distinct bodies of trachytic rock that had reached their present horizon through closely associated vents, frequently bending up the sedimentary rocks at a high angle around the borders, the upturned strata including the lower part of the Middle Cretaceous

spend so much on map making. Topography is a far more expensive business than geology, and has little to do with it, at least for a reconnaissance, unless you want to publish a geological survey like the English, on a very large scale, but such a survey is neither wanted nor even possible in the U. S." Appendix, p. 721.

shales and portions of the Dakota sandstone. His observations tended to show, to quote his own words, "that there had been a sort of absorption, so to speak, of the shales, and that at least half of the space through which the trachyte is distributed is occupied by the crushed and metamorphosed fragments of shale. As a consequence the height of the arch—such as may once have existed—would not equal the height of the trachytic mass, as only the higher layers of shale extend entirely over it." His idea can be best understood by reference to Fig. 100, copied from Plate 46 in the report for 1875.[12] (See further on p. 546.)

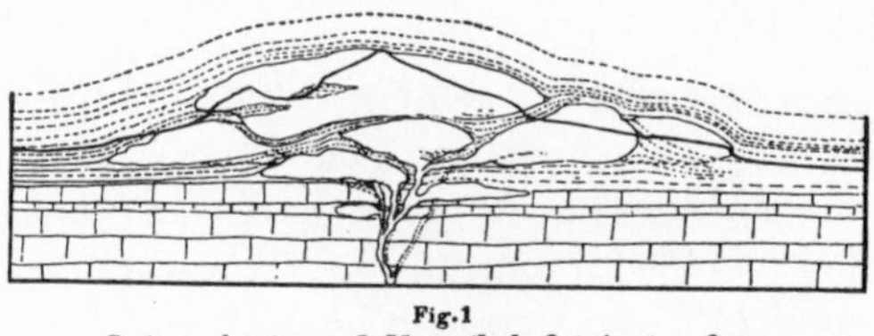

Fig. 1
Section showing probable method of intrusion of masses of trachyte.

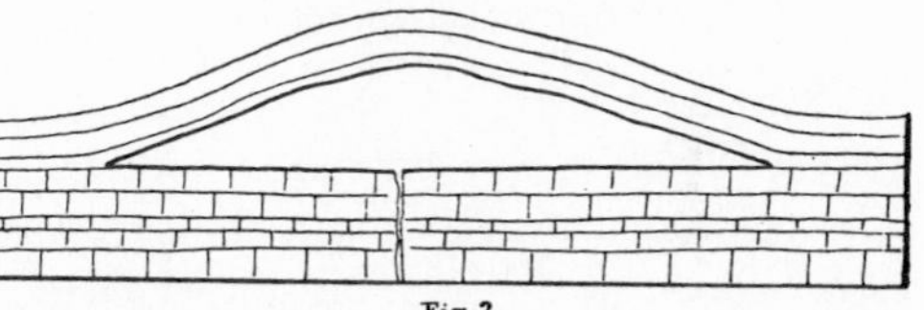

Fig. 2.
Arching of strata produced by intrusion of single mass uniformly destributed.

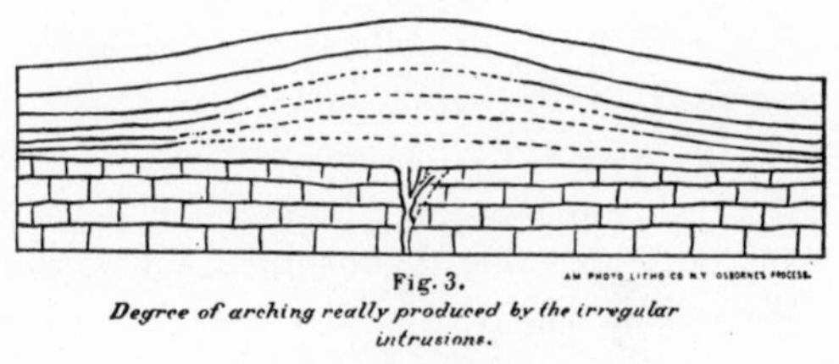

Fig. 3.
Degree of arching really produced by the irregular intrusions.

Intrusion of masses of Trachyte.
Sierra el Late.

Fig. 100.

The topographic party under J. T. Gardner, which established the triangulation station on the Sierra La Sal, was later attacked by hostile Indians but escaped with loss of their stock only. (Appendix, p. 719.)

In the season of 1876 C. A. White was at work in northwest Colorado, including the area lying between the Uinta and Park ranges. F. M. Endlich was engaged in the survey of the White River district, A. C. Peale of the Grand River district, and W. H. Holmes of the Sierra Abajo and West Miguel mountains.

In the interval between the issue of the reports for 1875 and 1876 Hayden, at the suggestion of King, had decided to call the transi-

12 The views regarding this method of mountain formation were subsequently summed up by Peale in an article, "On a Peculiar Type of Eruptive Mountains in Colorado," which was published in No. 3 of the bulletins of the survey, May 15, 1877. Obviously this was published in order that the Hayden survey might claim their full share of credit of discovery of laccolithic mountain structure.

tion group, heretofore referred to by him as *Lignitic* and the exact geological position of which was still in dispute, the *Laramie* group.

The report of this year contained little that is new or striking, the work resulting mainly in an extension of knowledge of the geographic range of various geological formations. White, working in the Uinta region, aptly compared the structure of Junction Mountain to a displacement which might be illustrated by the action of a large punch worked by machinery, the perforated heavy iron plates being somewhat torn in places and nowhere clearly cut through in the process of punching.[13] The work of White this year as a whole confirmed the view held by Hayden that the lignitic or Laramie beds, as they were now called, were of a transitional nature.

Work of Hayden in Idaho and Wyoming, 1877.

The field work for Colorado was completed in 1876. The following year (1877) it was extended northward into Idaho and Wyoming. The geological work was, as before, assigned to F. M. Endlich and A. C. Peale, with the addition of Orestes St. John. Endlich worked in the Sweetwater region, Peale in the Green River district, and St. John in the Teton district.

S. H. Scudder spent two months of the year in Colorado, Wyoming, and Utah in collecting fossil insects, which were subsequently described in the thirteenth monograph of this survey. Sir Joseph Hooker, director of the Kew Gardens, England, and Professor Asa Gray, of Cambridge, Massachusetts, accompanied the party for a time, making valuable botanical collections (see Plate 33).

St. John noted the overturned character of a portion of the Caribou Range and made numerous sections across the Teton Range. Peale noted that in the region of the Blackfoot Basin the structure was that of a series of anticlinal and synclinal folds, the streams sometimes occupying the synclines and sometimes the monoclines. Also that there were at least three parallel anticlinal axes having the general direction northwest and southeast.

Work of Hayden in Wyoming, 1878.

Hayden's twelfth and last annual report, bearing date of 1879 (1883), was issued in the form of two volumes of upward of 1,200

[13] Suess in his work, Face of the Earth (English translation), p. 572, commenting on the above says: "I would propose another. A stake is driven in the edge of a pond, so that the upper end stands just below the level of the water. The water becomes covered by a sheet of ice; the level of the water falls; the covering of ice gives way and the stake is exposed to view."

pages, and included the work of the corps for the field season of 1878 and the office work until the closing up of the survey, which, by law, took effect June 30, 1879.

The headquarters of the survey were at Cheyenne, Wyoming, as in previous years, and but four parties organized. The geological work was under the charge of W. H. Holmes, A. C. Peale, and Orestes St. John, and the paleontological work under C. A. White. Holmes made a general survey of the park, while Peale, assisted by J. E. Mushback, was occupied in making detail studies of the geyser and hot-spring localities.

Fig. 101. *W. H. Holmes.*

The party with St. John surveyed the Wind River Mountains and a portion of the Wyoming and Gros Ventre ranges. The work of the topographic party in the Wind River and Grand Teton regions was hampered by their being robbed of all their animals and a portion of their outfit by hostile bands of Indians.

During the summer of 1877 Professor S. H. Scudder, with a party, visited the Tertiary lake basin at Florissant and made an extensive collection of fossil insects, the published descriptions of which have made this region classic.

The two volumes mentioned are almost monographic as far as the hot springs and geysers are concerned, and were rendered unusually attractive for their time by the sketches and panoramas of Holmes. Peale gave a detailed description of all the springs and geysers of any importance found in the park, describing and tabulating over 2,000 of the former and 71 of the geysers. Holmes's report was accompanied by some brief petrographic descriptions by C. E. Dutton.

Sketch of Hayden.

Dr. F. V. Hayden was born at Westfield, Massachusetts, September 7, 1829. His father dying when he was but ten years of age and his mother marrying again, he went at the age of twelve to live with an uncle in Rochester, Ohio, where he stayed until he was eighteen, beginning when he was sixteen to teach during the winter months in the district schools of the neighborhood. When eighteen, ambitious for an education and without money, he walked to Oberlin and laid his case before President Finney, of Oberlin College, who gave him such encouragement and sympathy that he set about preparing himself for college, working meanwhile at whatever he found to do to pay for his support and tuition.

He entered college and graduated in 1850, paying his own expenses. These facts are mentioned, since they show the character of the man and enable one to understand better the causes of his success in after life. It is important to note, however, that, owing to his shyness and diffidence, his fellow students did not predict for him a prominent career, notwithstanding his acknowledged scholarly habits.

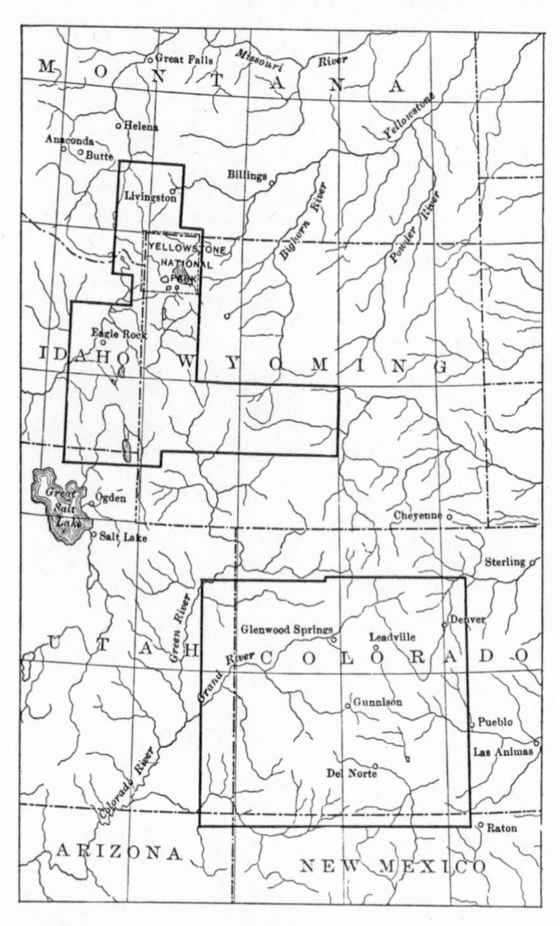

Fig. 102. *Map showing areas covered by Hayden Survey.*

After graduating he studied medicine and received the degree of M.D. in 1853 at Albany, New York. While here he became acquainted with James Hall, and in the spring of 1853, together with F. B. Meek, went to the Bad Lands of the upper Missouri to make collections, as already noted. This trip marked the beginning not only of his scientific career but also that of his association with Meek, which lasted until the latter's death in 1876, and, incidentally, furnished an example of uninterrupted collaboration without a counterpart in the history of American geology.

The scientific career, so promisingly begun, was, however, interrupted by the outbreak of the Civil War, when he promptly volunteered, entering the Union Army as an assistant surgeon, gradually rising to become post-surgeon and surgeon-in-chief to the Twenty-second Division of Cavalry. In June, 1865, he resigned from the army, but was subsequently breveted lieutenant colonel for meritorious conduct. In the same year he was elected professor of geology and mineralogy in the University of Pennsylvania, holding this chair until 1872, when he was forced to resign, owing to his rapidly increasing duties in connection with the Geological Survey.

In 1879, after the consolidation of the various surveys, Hayden was appointed one of the geologists on the new organization. During the next four years his time was devoted mainly to the completion of the business and final publication of the reports of the Geological and Geographical Survey of the Territories. His health began failing soon after his acceptance of this position, and in June, 1882, at his own request, he was relieved from the work of supervision of printing the reports and assigned to duty in the field, spending the summers of 1883-1886 in Montana. His disease, however, grew steadily upon him, and in 1886 he resigned on account of complete incapacity for duty, thus closing a career of nearly thirty years of actual service as naturalist, surgeon, and geologist in connection with the Government.

The apparent diffidence which impressed Doctor Hayden's fellow students at Oberlin, and led them to be doubtful as to his future course in life, characterized his maturer years, and to those not well acquainted with him made it difficult to account for his success. However, enthusiasm, perseverance, and energy were qualities equally characteristic of him all his life, and what seemed to be diffidence was largely the result of his nervous temperament. The secret of his success is to be found in his enthusiastic frankness and his energetic determination to carry through whatever he undertook. He was absorbed in the work of the Geological Survey and bent all his energies to its success. Excitable in temperament and frequently impulsive in action, he was generous to a fault, and, although ever ready to defend what he believed to be right, he was willing upon the presentation of evidence to modify his views.

He was always careful to give due credit to all who had worked in the fields he afterward explored. In one of his reports, speaking of those who had preceded him, he says:

> Any man who regards the permanency or endurance of his own reputation will not ignore any of those frontiersmen who made their early explorations under circumstances of great danger and hardship.

The same spirit actuated him in his treatment of his subordinates and coworkers. His honesty and integrity were undoubted and his work for the Government and for science was a labor of love.

The following extract is from an article by Dr. Archibald Geikie, then director of the Geological Survey of Great Britain:

There can be no doubt that among the names of those who have pioneered in the marvelous geology of western North America that of F. V. Hayden will always hold a high and honored place. This place will be his due not only because of his own personal achievements in original exploration. His earlier work exhibits much of that instinctive capacity for grasping geological structure which is the main requisite for a field geologist. He had a keen "eye for country." But he likewise possessed the art of choosing the best men for his assistants and the tact of attracting them to himself and his corps. In this way he accomplished much excellent work, keeping himself latterly in the background so far as actual personal geological investigations were concerned, and contenting himself with the laborious task of organization and supervision while he encouraged and pushed forward his coadjutors.[14]

In an obituary notice of Hayden, read before the American Philosophical Society, J. P. Lesley, who had known him for many years, pays him the following tribute:

He represented in science the curiosity, the intelligence, the energy, the practical business talent of the western people. In a few years they came to adopt him as their favorite son of science. He exactly met the wants of the Great West. There was a vehemence and a sort of wildness in his nature as a man which won him success, cooperation, and enthusiastic reputation among all classes, high and low, wherever he went. In the wigwam, in the cabin, and in the court-house he was equally at home, and entirely one with the people. He popularized geology on the grandest scale in the new States and Territories. He easily and naturally affiliated with every kind of explorer, acting with such friendliness and manly justice toward those whom he employed as his coworkers that they pursued with hearty zeal the development of his plans.

In dealing with the public men of the country he was so frank, forcible, and direct that it was impossible to suppress or resist him. He had the western people at his back so heartily and unanimously that he was for a long time master of the scientific situation at Washington. He was a warm personal friend of some of the highest officials of the Government, who never failed to support strenuously and successfully his surveys.

I think that no one who knows the history of geology in the United States can fail to recognize the fact that the present magnificent United States Geological Survey, now under the direction of Major Powell, is the legitimate child of Doctor Hayden's Territorial surveys.

[14] Nature, XXXVII, February, 1888, pp. 326-327.

PLATE XXXI

FIELDING BRADFORD MEEK

According to Cope,[15] the Sioux Indians gave to Hayden the name "The-man-who-picks-up-rocks-running," a name which was certainly descriptive of the manner in which much of his work was necessarily done. The same writer states that at one time, when engaged in the exploration of the Laramie beds of the upper Missouri, he was pursued and finally overtaken by a band of hostile Indians. Finding him armed only with a hammer and carrying a bag of rocks and fossils, which they emptied out and examined with much surprise and curiosity, they concluded he was insane and let him alone.

Sketch of Meek.

F. B. Meek, Hayden's collaborator, was born of Irish parentage in the city of Madison, Indiana, in 1817, and during 1848-1849 served as assistant to D. D. Owen on the surveys of Iowa, Wisconsin, and Minnesota. Afterward (from 1852-1858) he was assistant to James Hall at Albany, New York,[16] and for a portion of the time connected with the geological survey of Missouri with Professor G. C. Swallow. As noted elsewhere, he became early associated with Hayden, and, though refusing to become officially connected with the survey, was tacitly associated with him until the time of his death. Indeed, all the invertebrate fossils collected by Hayden in his western explorations were studied and the results prepared for publication by Meek, although appearing mainly under the joint name of Meek and Hayden.

On leaving Albany in 1858 Meek came to Washington, where he

15 American Geologist, I, 1888, p. 110.

16 Meek's correspondence with Hall began in 1849, on the twelfth of November of that year, he having written to Hall stating that he understood he was looking for a man to collect fossils for him in the West, and offering his services. "Not knowing exactly the nature of the arrangements you wish to make, I would merely remark that I am a young man with no one but myself to support, and being passionately fond of geology I think I would probably be able [to] engage on as liberal terms as almost any other person." Later, in response to other letters from Hall, he wrote, relative to salary: "I hardly know what to say. . . . While connected with the U. S. Geol. Corps in the N. W. my pay was two dollars per day, my travelling expenses being paid to and from the place of rendezvous. In the field I incurred no travelling expenses as we generally followed the course of the streams, in bark canoes manned by half-breeds or Canadian voyagers who were paid by the government. We had, however, to bear our own portion of camp expenses which amounted generally to about 20 cents per day for each man. Although I can generally do better than this (in a pecuniary point of view) at home, I am always ready to make a sacrifice in order to be engaged in a pursuit congenial to my taste." Later (December 29, 1851) he expressed a willingness to engage for a year, at the rate of "fourteen dollars per month in money in addition to the other considerations mentioned." (See also Appendix, pp. 722-724.)

principally resided during the remainder of his life, having a room in the Smithsonian building. He was moderately tall and rather slender in build, with a dignified bearing, though quite diffident. During his later years he was also deaf, which caused him to avoid social gatherings almost altogether. He was never a man of robust health; indeed, for a large part of his life he was more or less an invalid. Genial, sincere, pure-minded, and honorable, such are the adjectives applied to him by his biographer. With the exception of James Hall, he is perhaps the most widely known of American paleontologists. Indeed, had Meek possessed the tremendous physique of Hall or the nervous energy of Hayden, he might have stood alone head and shoulders above most, if not all, of his contemporaries. As it was, he did his best work only in the service of, or in collaboration with, others, never as an organizer or leader. It is true that in his first expedition (under the patronage of Hall) to the Bad Lands he outranked Hayden (see Appendix, p. 696), and that in numerous instances his work was of a higher order. Yet his mildness of character and lack of disposition to assert himself, until perhaps too late, caused him to almost invariably occupy a second place in any organization with which he was connected.[17]

Beginnings of Cope and Marsh in Vertebrate Paleontology.

Prior to the advent of Cope and Marsh a very large proportion of the work in vertebrate paleontology in America had been performed by Leidy, and that, too, on fragmental material that had weathered out of the matrix and been gathered in many cases without an exact knowledge of the beds from which they were derived, during the haste and hurry of reconnaissance surveys. It remained for these men to take to the field for themselves and for Marsh in particular to adopt new methods, train collectors, and, in short, to change entirely the mode of procedure. The results became shortly the wonder of the scientific world. The material was no longer collected haphazard and

[17] "My dear Doctor:

"I am working very hard with the hope of being able to join you by the 15th of next month, or possibly a little sooner. Is there any little nook or corner about your museum rooms where I could have a little cot to sleep on while I am with you? I can bring my blankets and sheets with me. If there is and I could find an eating house in the neighborhood, that would suit me exactly. You know it does not require much room to hold me, and I prefer to live in that way, because I do not like hotels, boarding-houses or private families, and prefer two meals per day—breakfast at 8 o'clock a.m. and dinner and supper together at 4 p.m. I also prefer to spend my evenings with the books and specimens. I can hear and understand what they say, and they require neither small talk nor formalities." (Meek to Newberry, November, 1869.)

PLATE XXXII

E. D. COPE

O. C. MARSH

JOSEPH LEIDY

in form of weathered fragments, but actually shipped in the matrix in which it was embedded, to the laboratories in the east where proper time and facilities could be devoted to it. In this way it became possible to restore entire skeletons and gain an idea of external form, before approximated only by guessing.

Cope, although he had served temporarily with the Indiana and other state surveys, really began his public career with Hayden in 1870. Marsh, as already noted (p. 458), began at his own expense, but after 1882 was associated with the United States Geological Survey under Powell. In his early youth Cope was precocious almost to the danger point, but nevertheless developed into a combination of nerve and intellect the like of which is rarely met with. He erred often through too hasty judgment, impressed as it were with the immensity of the work to be done and the brief time to be allotted to it. He was a man of whom it has been said that his intuition was better than his logic. According to Scott, Cope's most valuable services to geology were rendered in unraveling the complexities of the fresh-water Tertiaries which cover so much of the West. His discovery and identification of the Puerco or oldest Eocene formation brought to light an entirely new and highly significant fauna and demonstrated a long-time interval between the periods of deposition of the Laramie and the Wasatch beds. The last-named had been discovered by Hayden and its correlation with the Suessonian beds of Europe had been pointed out by Marsh, but to Cope is due much the greater part of our knowledge regarding its distribution, its relation, and its place in the geological column. Personally, or through his collectors, he thoroughly explored the Wasatch of New Mexico and Wyoming, elucidating its fauna with wonderful skill and insight. He added much to what was known regarding the fauna formation and made a classical series of investigations on the fossil fishes of the Green River shales and Florissant beds of Colorado.

Marsh, like Cope, was at the beginning of his career a man of independent means. Through more careful and judicious investments he was able to so remain throughout his life and to devote funds toward collecting such as his less careful collaborator had to beg from an—at times—not overgenerous Government. It was through him, ably assisted by Hatcher, more than any other one man, that was brought about the enormous improvement in the manner of collecting and preparing fossils above referred to. "He not only had the means and the inclination, but entered every field of acquisition with the dominating ambition to obtain everything there was in it, and leave not a single scrap behind." This, and a natural disposition to resent the intrusion of others into a field which he felt he had created, to a considerable extent alienated him from coworkers in his particular department.

His first western collecting was done in 1870, when he organized a Yale Scientific Expedition, consisting of thirteen persons in addition to a military escort, and explored the Pliocene deposits of Nebraska and the Miocene of northern Colorado, then crossing into Wyoming and southward into Utah, then west into California, and on the return route making collections in Kansas. In each of the three following years similar expeditions were carried on. In 1875 he visited the Bad Lands of Nebraska and South Dakota, and later (1882) became paleontologist to the survey as noted. Under these conditions the materials accumulated more rapidly than they could be studied thoroughly, and it is not surprising that at the time of his death in 1899 several of the projected monographs should have been scarcely begun. Aside from the short papers in the *American Journal of Science* his principal monographs are those on the toothed birds, *Odontornithes*, which appeared in 1880, and on the *Dinocerata* in 1885. According to Beecher,[18] from whom we are quoting, Marsh's greatest discovery in the domain of geology was the Eocene deposit of the Uinta Basin. He believed vertebrate fossils were more accurate time markers than invertebrate or botanical, and made his determinations accordingly. The results have not been universally accepted. Personally he was inclined to be reticent and self-reliant. "As a friend he was kind, loyal and generous. As a patron of science he has seldom been equalled."

The United States Geological Survey of the Fortieth Parallel, 1867-1877.

In 1867 there was established by congressional action, and almost wholly through the personal efforts of Clarence King, what has since been known as the Geological Survey of the Fortieth Parallel. This, though subject to the administrative control of General A. A. Humphrey of the United States Engineers, was under the immediate direction of King, to whom must be given almost the entire credit of its inception and successful execution.

The immediate excuse for the survey was the desirability of ascertaining the character of the mineral resources of the country to be traversed by the Pacific Railroad. The region explored, however, was a very extended one, reaching from the eastern Colorado range

[18] Amer. Jour. Sci., Vol. VII, 1899.

PLATE XXXIII

ARNOLD HAGUE

J. D. HAGUE

S. F. EMMONS

to the Sierra Nevadas, with an average width of about 100 miles along the fortieth parallel.

The plan of the work, as outlined by Emmons,[19] contemplated making a topographical map of the region on the general plan of those made at the present time, *i.e.*, one on which the topography was to be indicated by contour lines rather than hachures on the hill-sides, the then prevailing custom. The scale adopted was four miles to the inch, and the original area divided into three rectangular blocks or atlas sheets, each about 165 miles in length by 100 miles in width. Subsequently two more blocks were surveyed, making the total area surveyed and mapped some 82,500 square miles.

The party, according to the writer above quoted, rendezvoused in California in the early part of the summer of 1867, and began their work at the east base of the Sierras in August, with J. D. Hague, Arnold Hague, and S. F. Emmons geological assistants. Though few in number, the force was beyond question the best equipped by training of any that had thus far entered the field of American geology.

The winter of 1867-1868 was spent at Virginia City, Nevada, in the study of the Comstock Lode, the mines of which, then but 1,000 feet in depth, had already produced some one hundred millions of dollars. The results of this work appeared in J. D. Hague's mono-graph, *The Mining Industry*, published in 1870.

During the season of 1868 the work of the survey was pushed eastward entirely across the Great Basin to the western shore of Great Salt Lake. In that of 1869 the desert ranges of Utah, the Wasatch, and the western end of the Uinta ranges were surveyed. This completed the work as originally planned, and with headquar-ters at New Haven, the task of working up the collections and plat-ting the topographic and geologic notes was undertaken.

This work was, however, abruptly interrupted in the summer of 1870 by telegraphic orders from General Humphrey, directing the party to once more take the field, Congress, without solicitation, having appropriated money for the continuation of the survey. It being then too late in the season to prepare the necessary outfit for work in the high mountain regions east of the Wasatch, the season was devoted to a study of the extinct volcanoes, Mount Shasta, Mount Rainier, and Mount Hood. Among the results of this study was the discovery of the first-known active glaciers within the limits of the United States.

During the summers of 1871-1872 the survey was carried eastward

19 Presidential address, Geological Society of Washington, 1896.

to the Great Plains and included an examination of the Eocene beds of the Green River Basin, the Uinta and Elk mountains and the intervening Mesozoic and Tertiary valleys of the North Park and the Laramie Plains, the Medicine Bow Range, and the northern extension of the Front Range.

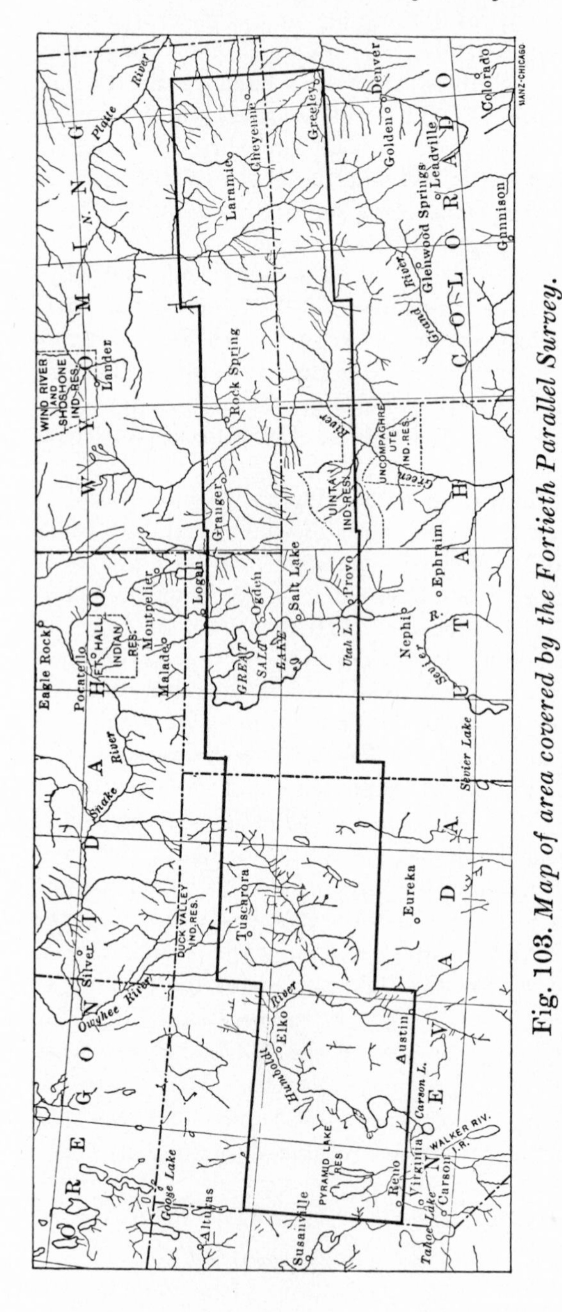

Fig. 103. *Map of area covered by the Fortieth Parallel Survey.*

The survey made no annual or preliminary reports, and as the thorough study of the data collected consumed several years of time, some of the results obtained were anticipated by other organizations through priority of publication.

The final reports, which appeared in the form of quarto volumes, named in the order of their appearance, were: Volume III, *Mining Industry*, by James D. Hague, 1870; Volume V, *Botany*, by Sereno Watson, 1871; Volume VI, *Microscopical Petrography*, by Ferdinand Zirkel, 1876; Volume II, *Descriptive Geology*, by Arnold Hague and S. F. Emmons, 1877; Volume IV, consisting of Parts I and

II, *Paleontology*, by F. B. Meek, James Hall, and R. P. Whitfield, and Part III, *Ornithology*, by Robert Ridgeway, 1877; Volume I, *Systematic Geology*, by Clarence King, 1878; and Volume VII, *Odontornithes*, by O. C. Marsh, in 1880. Volume I was accompanied by a geological atlas containing ten large double maps.

To properly summarize the work of the survey, as made known in these volumes, is a practical impossibility in the space that can be devoted to it. It was noted that, in the grand total of 120,000 feet of sedimentary accumulations found, the main divisions of Archean, Paleozoic, Mesozoic, and Cenozoic were all distinctly outlined by divisional periods of marked unconformity. Considered as a whole, there was a noteworthy fullness in the geological column, none of the important stratigraphical time divisions being wholly wanting, except some of the obscure intermediate deposits which in other localities have been found lying between the base of the Cambrian and the summit of the Archean series.

Fig. 104. *R. P. Whitfield.*

From the data furnished by Emmons and Hague, and his own observations, King felt himself able to reconstruct with a considerable degree of accuracy the topographical configurations of the Archean surface, and pictured with great clearness the growth of that portion of the American continent included within the area surveyed. He conceived that, at the close of the Archean age, there was a great mountain system built up of at least two sets of nonconformable strata, referred to Laurentian and Huronian, which was coextensive with the greater part of the Cordilleras.[20] This, west of 117° 20′, formed a land area and to the eastward a sea bottom, upon which, throughout the entire Paleozoic period, were conformably deposited the gradually accumulating detritus from the land, brought down by eastward-flowing rivers. These Paleozoic sediments he found in the region of the Wasatch to be 32,000 feet in thickness, and in the extreme western limit, upward of 40,000 feet.

He divided the series into four great groups: The first, which is purely detrital, being wholly of Cambrian age; the second, a lime-

[20] King used the word *Cordilleras* to designate the entire series of mountain chains bordering the Pacific front of North America, limiting the term *Rocky Mountains* to the eastern front only.

stone series of 11,000 feet in thickness, extending from the Cambrian to the top of the lower Coal Measures, and indicating a deep-sea deposit. Succeeding this came the Weber quartzite, a purely siliceous deposit of from 6,000 to 10,000 feet in thickness—followed by the fourth group of upper Coal Measure limestone, about 2,000 feet in thickness. The entire Paleozoic series he summed up as composed of materials of two periods of mechanically accumulated detritus, interrupted by one and followed by another period of deep-sea limestone formation.

At the close of this great conformable period of Paleozoic deposition there were widespread mechanical disturbances. All the thickest part of the sediments, from the western shore line eastward to and including what is now the Wasatch, were raised above the ocean level to become a land area. East of the Wasatch the ocean bottom, with its Upper Carboniferous sediments, remained practically undisturbed.

Contemporaneous with or immediately succeeding this uplift, the old land mass to the westward went down. What was sea bottom had become land, and what was land became sea bottom. But the new land area extending from the Wasatch westward to the Havallah Range (longitude 117° 30′), under the combined action of heat, cold, and chemical action of the atmosphere, began at once to yield the materials which, in the form of sand and silt, were carried west and east to be laid down, in the first-named instance, on a gradually sinking Archean bottom, until a thickness of 20,000 feet was reached, and in the last-named, conformably upon a bottom of Upper Carboniferous rocks, until some 3,800 feet had accumulated.

At the close of this period of sedimentation, which includes the Triassic and Jurassic, the western ocean with its twenty-odd thousand feet of sediments underwent a sharp folding and uplifting, whereby the shore line was pushed outward as far as the western base of the present Sierra Nevadas. The force causing this uplifting acted tangentially and was most severe in the extreme western portion, *i.e.*, the Sierras, where, in a belt of about fifty miles' width, the Triassic and Jurassic sediments were crumpled and crowded together and crushed into a mass of almost undistinguishable folds. During all this powerful disturbance in the western sea, the region east of the Wasatch remained practically undisturbed, as before.

King's views regarding the geographic distribution of land and water during the subsequent period of geological history were largely in harmony with those of Hayden. During the Cretaceous times he believed a Mediterranean ocean to have stretched from the eastern base of the Wasatch into Kansas. Over the bottom of that body of water an almost continuous conformable sheet of Cretaceous sediments was laid down, the greatest thickness of which was against the western shore of the ocean—that is, along the base of the Wasatch, where were found, conformably over the Jurassic shales, about 12,000 feet of Cretaceous beds.

With the close of the Cretaceous period of sedimentation the entire area from and including the Wasatch, eastward as far as the Mississippi Valley was uplifted, and in its western portion faulted or thrown into sharp or gently undulating folds. The immediate effect of this uplifting was, first, the development of the broad level region now occupied by the Great Plains; secondly, the outlining of the basin of the Vermilion Creek (Wasatch) Eocene lake; thirdly, the formation of the distinct folds of which the Wasatch and Uintas are the most prominent examples; and fourthly, the relative upheaval of the old Archean ranges, with all their superincumbent load. The folds of the Wasatch involved a conformable series of strata extending from the base of the Cambrian to the top of the Cretaceous—in all, 44,000 feet in thickness. The astonishing and almost incredible feature of the case lies in the fact that, if King's ideas are to be accepted, this stupendous fold, together with the one of 30,000 feet forming the Uintas, was not a gradual uplift, but of sudden and necessarily catastrophic origin, and that, before the forces of erosion had accomplished their work, there actually here existed mountain ranges from five to eight miles in height.[21] From the date of this

[21] King's exact words are: "So that, since the ocean level was banished to somewhere near its present position, the fold itself [*i.e.*, the Wasatch] was not less than 44,000 feet in altitude. The Uinta was not so imposing a body, but its summit, before erosion began, was certainly 30,000 feet above the sea level." (Systematic Geology, p. 748.)

Extreme as these views may now seem, they had been equaled by at least one previous writer. In describing the occurrence of the iron ores of Spruce Creek, Warriors Mark Run, and Half Moon River, in Huntingdon and Center counties, Pennsylvania, in 1874, J. P. Lesley had argued that they were residual deposits from the decomposition of ferriferous Lower Silurian beds, which were at one time buried beneath more than 16,000 feet of Upper Silurian, Devonian, and Carboniferous rocks, and had been subsequently exposed through decomposition and erosion.

"At the end of the coal era the Middle States rose from the waves. . . . The edges of the Bellefonte fault stood as a mountain range as high as the Alps, and the backs of some of the great anticlinals of Pennsylvania must have formed plateaus then as high as Tibet and Bootan are now.

"Erosion commenced and has continued through the Permian, Jurassic, Cretaceous, and Tertiary ages to the present day, and still goes on. The high plateau was gradually worn down to the present surface. Mountains once 30,000 or 40,000 feet high are now but 2,000 or 3,000 above sea level."

elevation no marine waters have invaded the middle Cordilleras, the subsequent strata being, as noted by Hayden, of lacustrine origin.

Studying in detail the underlying geology, in connection with these folds, King announced the principle that "wherever an Archean mountain range underlay the subsequent sheets of sediment, there a true fold has taken place"; and further, that when one observes "the continuity of the strata across such a valley as that of the Laramie Plains, and then sees them sharply and suddenly rise against the foothills of the Archean, it becomes evident that the entire area of the Rocky Mountains has suffered actual lateral compression, and that the diminution of surface amounts to from 6 to 10 per cent." When he further considered that the post-Archean sedimentaries were a mere thin covering over the subjacent crust, he added that "this diminution of area of actual surface means an actual compression of the solid Archean shell of the earth."

Pursuing the same line of thought, King noted that the configuration of America today is due to the configuration or topography of the pre-Cambrian continent. Where Archean faults or mountain chains existed, there were the lines of weakness along which later orographic movements made themselves manifest. A comparatively thin coating of sedimentary beds, for illustration, overlies the generally smooth Archean rocks of the Mississippi Valley, and here no subsequent disturbances have taken place. On the other hand, the high Archean Wasatch ridges, which were covered by 10,000 feet of sediment in post-Archean times, were again and again uplifted during the subsequent periods of disturbance.

It was noted above that, at the close of the post-Cretaceous uplifting, the Wasatch highland stood at an elevation of upward of 40,000 feet. Under such extraordinary conditions rapid and intense erosion was inevitable, and in a comparatively brief period, geologically, the Vermilion Creek Eocene lake was filled with sediments derived therefrom to a depth of 5,000 feet.

Then ensued another period of disturbance, involving the upturning of the western portion of the Vermilion Creek beds, while the region to the immediate west, from which their sediments had been derived, was as suddenly depressed, allowing the waters of the lake to extend themselves 200 miles westward into Nevada.

Another series of crustal movements was now inaugurated in the east, whereby the Great Plains land area also underwent a subsidence, which was most pronounced along the foothills of the Rocky Mountains and gradually died out to the eastward. This movement marks the dividing line between the Eocene and the Miocene periods.

Contemporaneously with this the entire Great Basin area, including portions of Washington, Oregon, Nevada, and California, lying to the east of the Sierras and the present Cascade Range, became depressed, and, receiving the drainage of the surrounding hills and mountains, was converted into two large lakes which, throughout the Miocene period, were depositories of sediment from the adjacent land. Powerful and profound crustal movements at the close of the Miocene threw the beds into folds, but did not apparently raise them above the surface level of the lake. Contemporaneously with this movement the Miocene lake of the east, through the subsidence of the surrounding country, was so increased as to cover almost the entire province of the Great Plains.

The beginning of the Pliocene period found, then, two enormous fresh-water lakes, the one covering the basin country of Utah, Nevada, Idaho, and eastern Oregon, the other occupying the Plains province. The period was brought to an end by crustal movement which, however, affected the two areas quite differently, the sediments of the Great Basin area being broken through the middle and the halves depressed from 1,000 to 2,000 feet, while those of the Plains were bodily tilted toward the south and east.

The result of the post-Pliocene movement in the department of the Plains was to give thereafter a free drainage to the sea. The result in the area of the Great Basin, on the other hand, was to leave two deep depressions, one at the western base of the Wasatch and one at the base of the Sierra Nevada, which in Quaternary times received the abundant waters of the glacial period and formed the two now nearly extinct lakes known as Lake Lahontan and Lake Bonneville.

According to Emmons,[22] the most important result of the King survey, in its relation to mining, was the publication in 1870 of the volume on *The Mining Industry*, a work which marked a new departure in geological reports. This was devoted mainly to an exhaustive study of the Comstock Lode, and formed "a scientific manual of American precious metal mining and metallurgy, and an invaluable work of reference concerning the mode of occurrence of gold and silver."

Sketch of King.

King was born at Newport, Rhode Island, in 1842, and graduated from the Sheffield Scientific School in 1862, being a member of the first class to receive degrees from that institution. The year follow-

[22] Eng. and Mining Jour., 1902.

ing his graduation he, in company with James T. Gardner, joined an emigrant train starting from St. Joseph, Missouri, on an overland trip to California. It is stated that it was during this trip, at that time a slow and eventful one, that he conceived the idea, afterward carried into execution, of a geological section across the entire Cordilleran system. Reaching California, he attached himself as a volunteer assistant to the state survey under J. D. Whitney. Later he was connected with parties under General McDowell in the examination of the mineral resources of the Mariposa grant. It was during this expedition that he and his companion were captured by Apaches, but were fortunately rescued just as the fires were being prepared for their torture.

After the Civil War and the passage of the bill subsidizing the Pacific Railroad, King recognized that the time had come for carrying out his scheme for connecting the geology of the East with that of the West and making the cross section above referred to. With this project in view he went to Washington in the winter of 1866-1867, and in spite of the disadvantage of his youth—being then scarcely twenty-five years of age—and still more youthful in appearance, he was so successful in impressing Congress with the importance of ascertaining the character of the mineral resources of the country about to be opened that not only was a generous annual appropriation voted, but King was himself placed in charge, subject only to the administrative control of General A. A. Humphrey, as already mentioned.

The published results of this survey we have already reviewed. As described by Emmons, from whose sketch most of what is here given of the personal history of King is quoted, "probably no more masterly summary of the great truths of geology had been made since the publication of Lyell's *Principles*." Making due allowance for the enthusiasm of one who was an associate and warm personal friend, attention need only be called to the fact that the entire work was consummated before its author saw his fortieth birthday, to establish once and for all King's fame as an organizer and geologist.

Aside from the publications under this survey, King's bibliography contains few titles, but this may mean simply that neither ink nor words were wasted. In 1877 he delivered an address at the thirty-first anniversary of the Sheffield Scientific School on "Catastrophism and the Evolution of Environment," which Emmons characterized as a "protest against the extreme uniformitarianism of that day." It was naturally based largely on his Fortieth Parallel work. This uniformitarianism he described as "the harmless un-

PLATE XXXIV

CLARENCE KING

destructive rate [of geological changes] of to-day prolonged backward into the deep past." He contended that while the old belief in catastrophic changes had properly disappeared, yet geological history, as he read it, showed that the rate of change had not been uniform, as was claimed by the later school. He believed rather, as a result of his own observations, that at certain periods in geological history the rate of change had been accelerated to such a degree that the effect produced upon life was somewhat catastrophic in its nature.[23]

One act in King's professional career should be here referred to, although the story has often been told. It will be remembered that in 1872 there was made a reported discovery of a diamond field in southern Arizona within an area that had been gone over in the course of the work of the Fortieth Parallel Survey. King, for purely scientific purposes, undertook a study of the region, with the purpose of discovering something regarding the matrix and the origin of the diamond. He discovered, rather, that the whole matter was a stupendous fraud; that, so far from there being a diamond mine, the ground had been "salted." So soon as this discovery was made he started for San Francisco, traveling night and day, that he might outstrip all other possible sources of information. On his arrival he at once visited the offices of the directors of the company organized for the selling of stock, and demanded peremptorily that all issues should be stopped. To a suggestion that his announcement of the fraudulent nature of the claim be delayed temporarily, he replied: "There is not money enough in the Bank of California to induce me to delay this announcement a single hour." And it was not delayed.

By those who knew him, King is described as a man of rare charm of manner, of cheerful and courteous disposition, and unrivaled as a raconteur. A man of remarkably robust physique, he yet broke down almost in the prime of life, and died in 1901.

Emmons was not only one of the best-trained but one of the best-poised, mentally, of the American geologists who have devoted themselves to economic problems, but with the exception of that already mentioned his work lies beyond the time limit of our history. He was a man whose opinions were everywhere respected; quiet, unassuming, yet ready enough when occasion required to discuss or defend a principle. His greatest work, and the one which gave him an international reputation, was his monograph, *The Geology and Mining Industry*

[23] The writer was informed by the late Professor W. H. Brewer that King's views on these subjects later underwent considerable modification and that he would have been glad to suppress this pamphlet, had it been possible.

of Leadville, Colorado, issued in 1886. He was not a profuse writer, but all that he wrote was good.

United States Geological Surveys West of the One Hundredth Meridian under Lieutenant G. M. Wheeler.

In 1869 Lieutenant G. M. Wheeler, of the United States Engineers, was authorized to undertake a military reconnaissance for topographical purposes in southwest Nevada and western Utah. No geologist accompanied the party until 1871, the work being purely topographical. In the last-named year G. K. Gilbert was appointed chief geologist, serving through three field seasons, with A. R. Marvine, assistant in 1871, and E. E. Howell, in 1872-1873. J. J. Stevenson was geologist with one of the parties in 1873, Jules Marcou in 1875, A. R. Conkling in 1875-1877, and J. A. Church in 1877. In 1878-1879 Stevenson was again attached to the survey, with I. C. Russell assistant. Oscar Loew served as mineralogical assistant during 1871-1875, and E. D. Cope as vertebrate paleontologist in 1874.

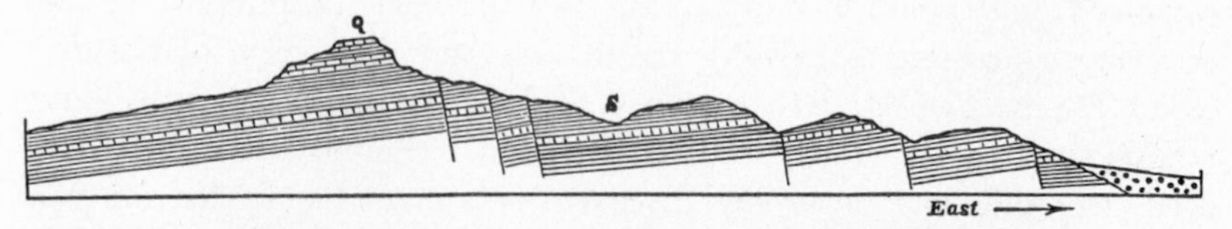

Fig. 105. *Gilbert's Section of the Pahranagat Range.*

The invertebrate paleontological collections made by all of the parties were worked up by C. A. White, then connected with Bowdoin College, after a preliminary examination by F. B. Meek. The results of the geological work appeared in the form of two quarto volumes, the first, of 681 pages, in 1875; and the second, of 420 pages, in 1881—the latter with an appendix by Doctor White, comprising 36 pages of text and 2 plates of fossils. The paleontological report proper, comprising the work of both Cope and White, appeared under date of 1877, a quarto volume of 599 pages and 83 plates of fossils.

The first vertebrate fossils from the western Triassic were described in this report, the materials having been obtained by Newberry when attached to the Macomb expedition in 1855.

Gilbert applied the names *Basin Range system* and *Basin Ranges* to all that system of short mountain ridges separated by troughlike valleys which lie west of the plateau region, though not quite coincident with the Great Basin itself. This Basin system of mountain uplift he considered due mainly to faulting and tilting, in this respect differing with the geologists of the Fortieth Parallel Survey, who thought the primary features due to folding, the now evident faulting being a phase of late Tertiary or post-Tertiary time.[24] He noted that the ranges were parallel, recurring at regular intervals and of only moderate dimensions; further, that the ridges of the system occupied loci of upheaval, and were not residua of denudation; and that the valleys were not valleys of erosion, but mere intervals between the lines of maximum uplift.

He dwelt in considerable detail upon the phenomena of erosion by wind-blown sand and silt-laden streams, and discussed the glacial phenomena and the conditions attending the drying up of the great inland lakes, applying the name Lake Bonneville, in honor of Captain Bonneville, the explorer, to the great body of fresh water that once occupied Sevier and Salt Lake valleys and of which the present bodies of salt water bearing these latter names are but the puny residuals. Those great bodies of water, which obviously could have existed only under conditions of climate quite different from those of today, he believed to be coeval with the glacial period of the northeastern states. He found, however, no counterpart in this region of the general glaciation of the eastern states, though there were local glaciers high up on the flanks of the mountains.

The abundant volcanic phenomena presented by the region were discussed in considerable detail and the recency of many of the lava flows noted, the conclusions arrived at being that while "we are not merely permitted to think of a renewal of that activity as possible . . . we are logically compelled to regard it as probable."

The geological history of the basin region as read by Gilbert was to the effect that the area was depressed below sea level from the close of the Archean period until late Carboniferous. From the close of the Carboniferous to the beginning of the Cretaceous a great area, including the entire plateau country, was covered by the waters

[24] In Chapter VII of the volume on Mining Industry (Vol. III, Report, 40th Parallel Survey, 1873) King refers incidentally to these mountains as the tops of folds whose deep synclinal valleys are filled with Tertiary and Quaternary detritus. Just how fully the problem may have been worked out in his own mind at that time is not apparent, but in 1878, after the appearance of Gilbert's monograph, he expressed the opinion noted above. The entire subject, it may be stated, was gone over by J. E. Spurr in a paper read before the Geological Society of America in 1900 (see Bull. Geol. Soc. Amer., XII, pp. 217-270), and the matter rediscussed at the winter meeting of the same society in Washington in 1902. The prevailing opinions there expressed were in accordance with the views put forward by Gilbert in 1875.

of an inland sea entirely cut off from the main ocean. Only once did the sea regain a temporary sway, bringing with it a Jurassic fauna and then retreating. Throughout the Cretaceous age the plateau country was the scene of a shallow ocean, the shores of which were ever advancing and receding. Through the upraising of some remote barrier the ocean was permanently shut out and the inland sea gradually converted into an immense fresh-water lake, and finally drained till the whole region became terrestrial. Since the expulsion of this sea the elevation of the continent which caused it has continued, and the plateau country, which from early Silurian to late Cretaceous times was slowly sinking to an extent of not less than 8,000 feet, has been bodily uplifted to its former altitude.

Gilbert called attention to the almost entire absence of Upper Silurian and Devonian fossils in the region, and described the volcanic necks or plugs as vestiges of the flues through which the eruptions reached the surface, the last contents congealing in the flue to be subsequently exposed by erosion.

A. R. Marvine, who was attached to the party in 1871 as an astronomical assistant, was later detailed for geological work, his report on the region between Fort Whipple, New Mexico, and Tucson, Arizona, occupying pages 191-225 of the third volume.

E. E. Howell, as a member of the survey during 1872-1873, worked throughout the first season in western Utah and eastern Nevada and the plateau region of central Nevada. In 1873 he once more entered the plateau country and continued upon it to Arizona and New Mexico. His report is included in pages 227-300 of the same volume.

The party to which J. J. Stevenson was attached as a geologist was assigned in 1873 to work in southern Colorado, the region including portions of the drainage areas of the South Platte, Arkansas, Rio Grande, San Juan, Grand, and Gunnison rivers.

Stevenson devoted considerable attention to a discussion of the geological age of the Colorado lignites, which he considered (Lesquereux to the contrary, notwithstanding) Cretaceous. He agreed with J. L. Le Conte in regarding the Rocky Mountains as not the product of a single upheaval, nor the several axes wholly synchronous in origin. The first great epoch of accelerated disturbance in the region, resulting in the permanent elevation of the surface, he thought to have been synchronous with that during which the Appalachian chain was completed. Further, that the second epoch of elevation began toward the close of the Triassic. This was followed by a period of subsidence during which the Cretaceous beds were laid down, a third period of uplift marking the close of the Cretaceous. During this latter period volcanic agencies were in a state of intense activity, and a vast sheet of lava, two to three thousand feet in thickness, flowed out over the whole region of the Grand and Gunnison rivers. During the Tertiary period still another elevation took place, sufficient to give the rocks of that age a dip of some five degrees.

In his report of the work for 1879 were discussed in considerable detail the relations of the Laramie group, which he considered but the upper part of Hayden's Fox Hill group—that is, of very late Cretaceous age. He felt that there was no doubt but that the coal fields of the Galisteo area and of southern Colorado were of the same age as those of northern Colorado and Wyoming.

The United States Geological and Geographical Surveys under J. W. Powell.

Powell's Geology of the Uinta Mountains, 1874-1875.

J. W. Powell's first observations on the geology of the Uinta Mountains were made in 1869, when engaged in his famous exploration of the Grand Canyon, already noted. In 1871, 1874, and 1875 he again visited the plateau region, the last year being accompanied by C. A. White. The results of these later years of exploration are given in the quarto monograph on the *Geology of the Eastern Portion of the Uinta Mountains*, published in 1876. This comprised, all told, 218 pages, with a large folio atlas.

The expedition of 1875 and those of the intervening years until 1880 were made under the authority of the Department of the Interior, the organization, with Powell as director, being known as the second division of the United States Geological and Geographical Survey of the Territories. In 1877 the name was changed to the United States Geographical and Geological Survey of the Rocky Mountain Region.

In none of the early Powell surveys was an attempt made at systematic and detailed areal work. Certain striking and well-exemplified features were selected and made the subject of special monographs. In the work of 1876, above noted, Powell had divided the region west of the Great Plains, east of the Sierra Nevada, and south of the North Platte, Shoshone, and Sweetwater rivers into what he designated as geological provinces—the Park Province, the Plateau Province, and the Basin Province.

The first-named he described as characterized by broad, massive ranges, sometimes distinct and sometimes coalescing, so as to include the great parks. The mountains comprise high, lofty, snow-clad peaks which form perennial reservoirs for the multitude of streams discharging in part into the Colorado River and thence into the Gulf of California, and in part into the Rio Grande and thence into the Gulf of Mexico.

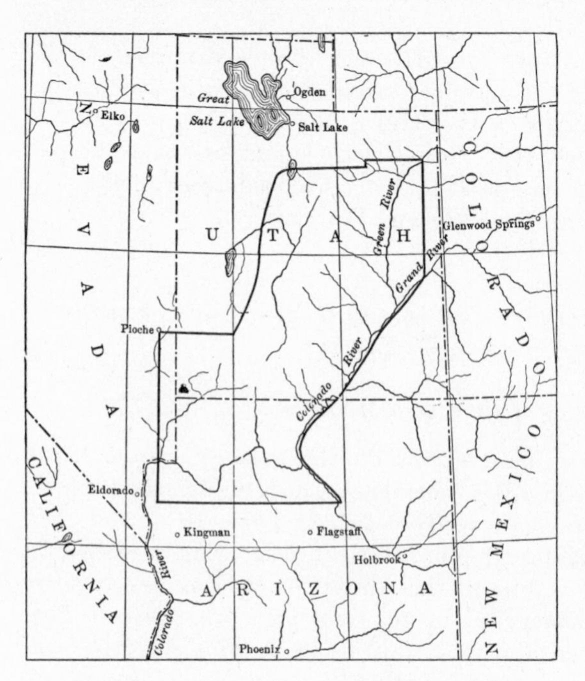

Fig. 106. *Map of Area Covered by Powell's Surveys.*

His Plateau Province was characterized by many tables bounded by canyon and cliff escarpments, on which stand lone mountains and irregular groups of mountains and short ranges. This region drains almost wholly into the Colorado River.

The Basin Province was characterized by short north-and-south mountain ranges and ridges separated by desert valleys, with a drainage which is almost wholly into the interior salt lakes and sinks.

Devoting himself mainly to the Uintas, he showed that they owed their present configuration to the degradation of a great upheaved block having its longer axis in an east-and-west direction. This upheaval, which he thought took place very slowly and gradually, began at the close of Mesozoic time and continued with slight intermissions until late Cenozoic. The total amount of upheaval in the axial region was more than 30,000 feet. Contemporaneously with upheaval the forces of degradation were at work, though not at the same rate of progress, along the axial line of uplift, the degradation amounting to more than 25,000 feet ($4\frac{3}{4}$ miles), and the mean degradation to $3\frac{1}{2}$ miles; so that over the entire area of about 2,000 square miles some 7,100 cubic miles of rock material had been removed.

PLATE XXXV

JOHN WESLEY POWELL

Dutton's Work on the High Plateaus, 1874-1876.

During the years 1874, 1875, and 1876 Captain Clarence E. Dutton, of the Ordnance Department, United States Army, made, under the Powell survey, further studies in the plateau region. Dutton's monograph—a *Report on the Geology of the High Plateaus of Utah,* a quarto volume of 307 pages, with an atlas—appeared under date of 1880. The particular territory studied lay in Utah, occupying a belt of country extending from a point about 15 miles east of Mount Nebo, in the Wasatch, south-southwest for about 175 miles, and having a breadth of from 25 to 80 miles—a total area of some 9,000 square miles.

As was the case with Powell, as indeed must have been the case with every observing man, Dutton was impressed with the vast amount of erosion which the country had undergone during the Miocene and more recent geological times. Noting that from an area of 10,000 square miles from 6,000 to 10,000 vertical feet of strata have been removed, he fell to speculating on the probable effect upon the earth's equilibrium of such a transference of materials. If the slow accumulation of great masses of sediment on sea bottoms brings about a gradual subsidence, why should not, he argued, the removal of a like load from any land area result in a corresponding uplift or elevation? Thus, for the second time in the history of American geology, was broached the now well-known subject of isostacy. (See p. 385.)

He noted that the great structural features of the high plateaus were due to faults and monoclinal flexures; also that the one form of displacement passed continually into the other—that what is here a simple fault passed into what he designated as step faults, and still farther on into unbroken anticlinals. All of these grander displacements belonging to the same system he felt had their commencement in the latter part of Pliocene times.

The great amount of volcanic activity evident received attention, the most ancient dating back to Eocene times. The character of the products was studied and the various lavas classified as (1) propylite, (2) andesite, (3) trachyte, (4) rhyolite, and (5) basalt, named here in order of their extrusion. The conditions in this region he believed to be as a whole confirmatory of Richthofen's law of the sequence of volcanic rocks.[25] He noted his objections to the German method of rock classification based upon geological age.

[25] Richthofen divided the Tertiary and post-Tertiary volcanic rocks into the five classes given above, and regarded them as having been the product of a regular sequence of eruptions, propylite being the most ancient and basalt the most recent. This has since been shown to be incorrect.

Gilbert's Work in the Henry Mountains, 1875-1877.

In 1876 G. K. Gilbert, then a member of Powell's corps, made a study of the Henry Mountains of Utah. His report appeared in 1877 in the form of a quarto volume of 160 pages, with 5 folding plates, and numerous figures in the text. In this he showed that these mountains were due to the intrusion from below of igneous matter through Carboniferous and Triassic strata, causing the overlying Cretaceous and Tertiary beds to arch or bulge upward, the present aspect of the mountains having resulted from erosion, whereby the overlying bulged beds and a portion of the igneous rocks as well were cut away. Phenomena of this type had been previously noted by Holmes, Peale, and Marvine, of the Hayden survey (pp. 520, 521), but it remained for Gilbert to fully elaborate the idea, and that in a manner that must ever connect it with his name. Here again we have an illustration of the saying that an idea belongs to him who puts it to the best use. To intrusions of this type he gave the name of *laccolites.*[26]

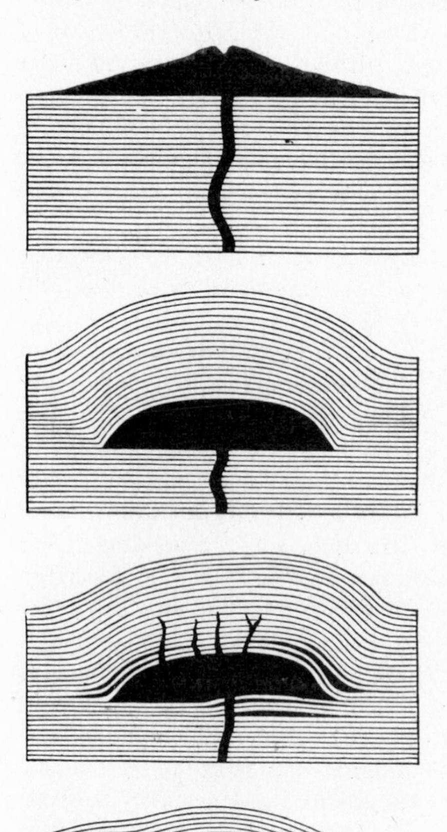

Fig. 107. *Laccoliths.* (*After Gilbert.*)

Gilbert showed that, in the uplifting of the sandstone to form these domes, the beds, as in the case of that of the Vermilion Cliff sandstone, might be elongated as much as 300 feet in a distance of three miles. That this could be done, and that suddenly, he believed to be

26 Later changed to lacco*liths.*

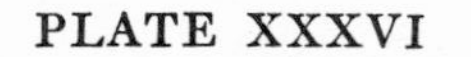

PLATE XXXVI

CLARENCE DUTTON

J. J. STEVENSON

G. K. GILBERT

due to the pressure of the overlying beds. The material was in a quasi-plastic condition, and no fissures could be opened unless coincidentally filled by some material which would resist the tendency of the walls to flow together. This consideration led him to the conclusion that just as for each rock there is a crushing weight, so there is for each rock a certain depth at which it cannot be fissured. Applying this principle to faults and fissure veins he concluded that

> if the fault extend to a great depth it will finally reach a region where the hardest rocks which it separates are coerced by so great a pressure that they can not hold themselves asunder, but are forced together before the fissure can be filled by mineral deposits. Thus, there is a definable inferior limit to the region of vein formation.

This the present writer believes to be the first time the idea of rock flowage was put forth in America and by an American.

The matters of instability of drainage lines, planation, and formation of river terraces were discussed, and it was shown that Hitchcock's idea of the formation of river terraces by successive periods of sedimentation (p. 344) was erroneous, such in fact being but the recorded stages of progressive erosion. In this work Gilbert used the terms *consequent*, *antecedent*, and *superimposed* as introduced by Powell, and showed that the drainage of the Henry Mountains, while as a whole consequent, was not so in all of its details.

The igneous rocks of the mountains were studied microscopically by C. E. Dutton, who classed them as trachytes.

Gilbert continued his work under King and Powell after the reorganization incidental to the formation of the United States Geological Survey in 1879, and remained one of its prominent and respected members down to the time of his death. His work was mainly along the broader lines, as is shown in the publications reviewed and in his later monograph on Lake Bonneville (1890). A quiet, studious, and dignified writer, rarely entering upon the sensational or popular field, yet invariably so stating his views as to render them easily understood by the popular mind, had it been so inclined, probably none of our geologists attacked their problems in a more calm and dispassionate tone, and none have rendered reports more nearly free from personal bias. "It is doubtful," wrote Professor Chamberlin,[27] "whether the products of any other geologist of our day will escape revision at the hands of future research to a degree equal to the writings of Grove Karl Gilbert."

27 Jour. of Geology, Vol. XXVI, 1918.

Sketch of Powell.

Powell was born of English parentage in Mount Morris, New York, March 24, 1834, and died September 23, 1902. From childhood he manifested a deep interest in all natural phenomena and early gave evidence of that bold and self-reliant spirit which in later years found vent in his hazardous exploration of the Grand Canyon of the Colorado. Vigorous, impetuous, and sometimes intolerant of the opinions of others, he made warm friends and strong enemies. Rising from obscurity, without university training, indeed almost wholly self-taught, to become a bold and aggressive thinker, and finally the head of the United States Geological Survey in 1881, it is little to be wondered at that he became for many a target for sneers as well as an object of envy. But however much men may differ as to the value of his individual work in the geological as well as ethnological field, no one will for a moment question that it was through his efforts that others were given the opportunity to carve out their own immortality. Upon his success as an administrator his fame may safely rest.

After his retirement from the Survey in 1894 Powell limited himself mainly to abstruse psychological problems and the directorship of the Bureau of American Ethnology, which, it should be said, he had been largely instrumental in founding in 1880.

He served with gallantry during the Civil War, rising to the rank of major of artillery, and lost his right arm at the battle of Pittsburg Landing.

Consolidation of All Geological Surveys, 1879.

With four separate parties operating in the western fields and with the little or no attempt at mutual collaboration more or less overlapping, with attendant wasteful expenditures of time and public funds, was inevitable. The accompanying map, from a report made by Powell to Congress in 1878, shows the amount of this overlapping. There were other reasons which rendered an attempt at coordination desirable. The Wheeler surveys as carried on during the earlier days, at least, were for geographic and military purposes. The topography was indicated by hachures and the maps drawn on a scale of eight miles to the inch, a scale altogether too small for any future geological purposes. The King survey, on the other hand, organized purely for geological work, indicated topography by contour lines and on a scale of four miles to the inch. The Powell surveys, organized for geological and geographic purposes, drew its

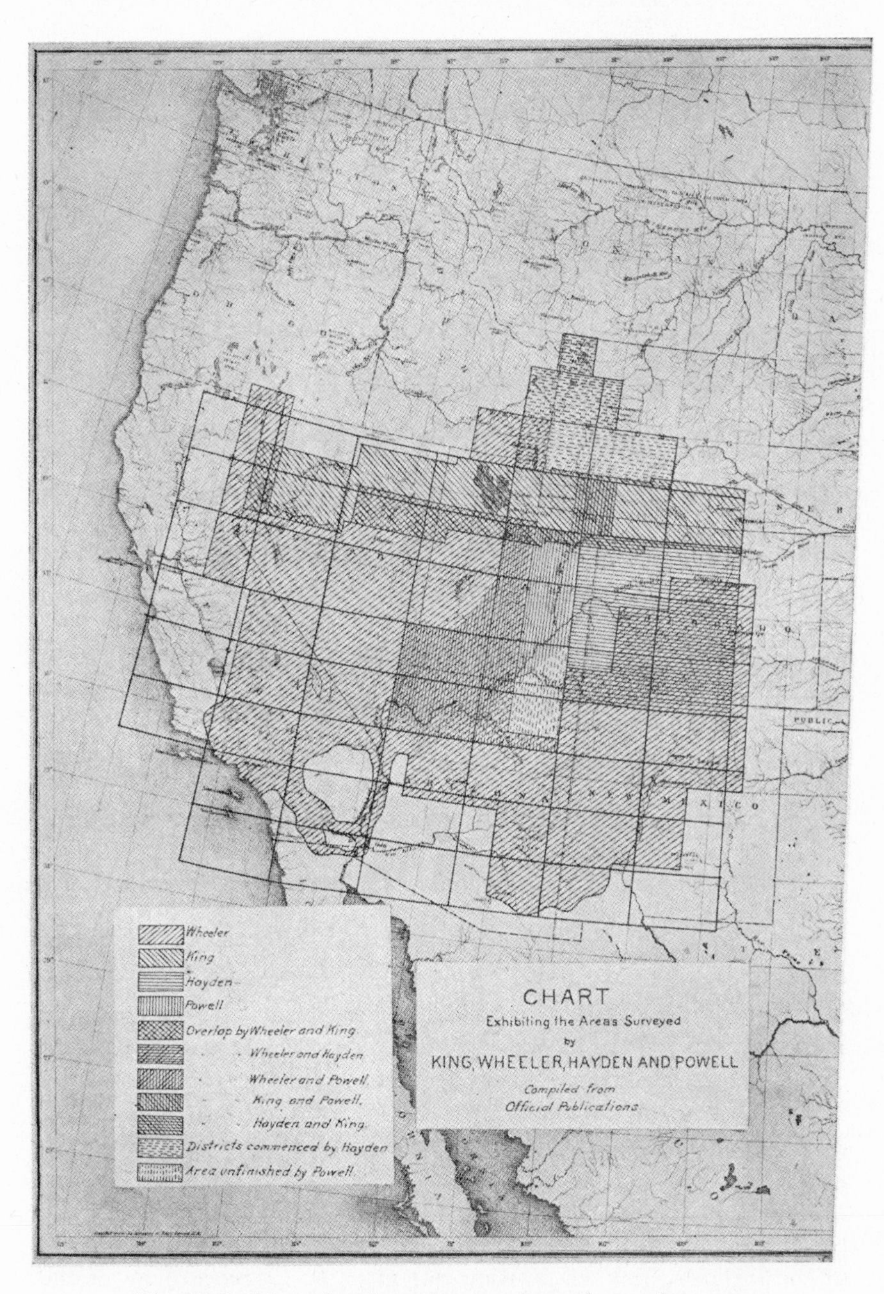

Fig. 108. *Map showing areas and overlaps of surveys.*

maps on a scale of four miles to the inch and indicated topography by contour lines with accessory hachures. The Hayden surveys in the early days were of a purely reconnaissance type and it was not until 1873 that the contour system and a scale of four miles to the inch was adopted.[28]

As early as 1874, the possibility of interference by the various surveys seems to have been foreseen (see Appendix, p. 727), and the question of the advisability of their consolidation was considered by the Committee on the Public Lands to which Congress had referred that portion of the message of the President relating to the subject. This committee made exhaustive investigations, receiving testimony from many interested parties. Undoubtedly, a strong effort was made to have the surveys all conducted under the direction of the United States Army. In opposition to this was successfully arrayed the almost unanimous opinion of the scientific and professional men of the country.

In 1878 the matter came up once more in Congress, in connection with the annual appropriation bills, and was referred to a committee of the National Academy of Science for consideration. This committee consisted of O. C. Marsh, J. D. Dana, William B. Rogers, J. S. Newberry, W. P. Trowbridge, Simon Newcomb, and Alexander Agassiz. It recommended "that Congress establish under the Department of the Interior an independent organization to be known as the United States Geological Survey, to be charged with the study of the geological structure and economical resources of the public domain, such survey to be placed under a director who shall be appointed by the president, and who shall report directly to the Secretary of the Interior."

This report, which was adopted, did away with the then existing Geographical and Geological Survey west of the One Hundredth Meridian, under Lieutenant Wheeler, excepting such as might be necessary for military purposes; the Geographical and Geological Survey of the Territories under Dr. F. V. Hayden; and the United States Geological Survey of the Rocky Mountain Region under J. W. Powell. Naturally the head of each of these discontinued organizations was ambitious to become the director of the new. Hayden as longest in the field would seem to have been the logical candidate, but his appointment was opposed by Carl Schurz, then Secretary of the Interior, and the position was given to Clarence King of the late Fortieth Parallel Survey. King, however, served but a single year, resigning in favor of Major Powell.

28 See Executive Document, No. 80, 2d Session, 45th Congress, 1878.

The total expenses of these various surveys as given in the report of the investigating committees and including the appropriations for the fiscal year ending June 30, 1879, were as follows:[29]

U. S. Geological Exploration of the 40th Parallel, under Clarence King:	
Amount expended up to June 30, 1878 (1868-1872)	$386,711
U. S. Surveys and Explorations west of the 100th Meridian, under Lieutenant George M. Wheeler:	
Amount expended up to June 30, 1878 (1869-1878)	$499,316
Appropriation of fiscal year ending June 30, 1879	50,000
U. S. Geographical and Geological Survey of the Territories, under Professor F. V. Hayden:	
Amount expended up to June 30, 1878 (1867-1878)	$615,000
Appropriation for the fiscal year ending June 30, 1879	75,000
U. S. Geographical and Geological Survey of the Rocky Mt. Region, under J. W. Powell:	
Amount expended up to June 30, 1878 (1871-1878)	$209,000
Appropriation for fiscal year ending June 30, 1879	50,000

Below is given a transcript of the Act of March 6, 1879, establishing the new organization and giving its scope as well as limitations. This, with an additional clause in the act of appropriations for 1882-1883, remained in force up to the time set as a limit to this history. By this additional clause it should be stated the Survey was called upon to prepare a geological map of the United States, a work the magnitude of which it is safe to say the responsible legislators had no adequate conception.

For the salary of the Director of the Geological Survey, which office is hereby established under the Interior Department, who shall be appointed by the President, by and with the advice and consent of the Senate, six thousand dollars: *Provided,* That this officer shall have the direction of the Geological Survey, and the classification of the public lands, and examination of the geological structure, mineral resources, and products of the national domain. And that the Director and members of the Geological Survey shall have no personal or private interests in the lands or mineral wealth of the region under survey, and shall execute no

[29] House Miscellaneous, 34th Session, 45th Congress, 1878-1879. Vol. I, No. 5, p. 22.

surveys or examinations for private parties or corporations; and the Geological and Geographical Survey of the Territories, and the Geographical and Geological Survey of the Rocky Mountain Region, under the Department of the Interior, and the Geographical Surveys west of the one hundredth meridian, under the War Department, are hereby discontinued, to take effect on the thirtieth day of June, eighteen hundred and seventy-nine. And all collections of rocks, minerals, soils, fossils, and objects of natural history, archæology, and ethnology, made by the Coast and Interior Survey, the Geological Survey, or by any other parties for the Government of the United States, when no longer needed for investigations in progress, shall be deposited in the National Museum.

Fig. 109. *Hayden party in field, 1871.*

CHAPTER IX.

The Fossil Footprints of the Connecticut Valley.

IN the spring of 1835 Dr. James Deane, a practicing physician of Greenfield, Massachusetts, called to the attention of Dr. Edward Hitchcock of Amherst College some footlike markings thought by him to be turkey tracks, which were found in slabs of flagstones being laid in the streets of that village. The animal origin of these markings Dr. Hitchcock was at first slow to admit.

"I am not without strong suspicions," he wrote, "that the case you mention may be a very peculiar structure of certain spots in the sandstone. . . . The layers of rock having this structure sometimes present an appearance resembling the foot of a bird. But I am satisfied that it is not the result of organization, though I confess myself unable to say precisely from what principle it resulted; though perhaps the case you mention is not of this sort." Dr. Deane was not satisfied with this answer and made further investigation, writing again to Dr. Hitchcock and declaring that "in my mind there is not a doubt but what they [the tracks] are really impressions of the feet of some bird."

To this letter and one subsequently written, no reply appears to have been sent. Apparently under the impression, however, that the matter had been left wholly in his hands, Hitchcock seems to have entered upon a systematic study of the markings, which resulted in an entire change of view from that first expressed, and the year following he began a series of papers on the subject, the publication of which extended over a period of more than twenty years, opening up an entirely new line of paleontological research to which he gave the name *ichnolithology.* His first paper appeared in the *American Journal of Science* for 1836. In this he described and figured tracks which he considered to have been made by as many as seven different species of animals. His conclusions were, first, that they were all the impressions of biped animals; second, that they could not have been made by any known biped except birds, and third that they well corresponded with the tracks of birds in having the same ternary division of the feet, with frequently, and perhaps always, the toes terminating with claws, as do birds. To these he gave the name *Or-*

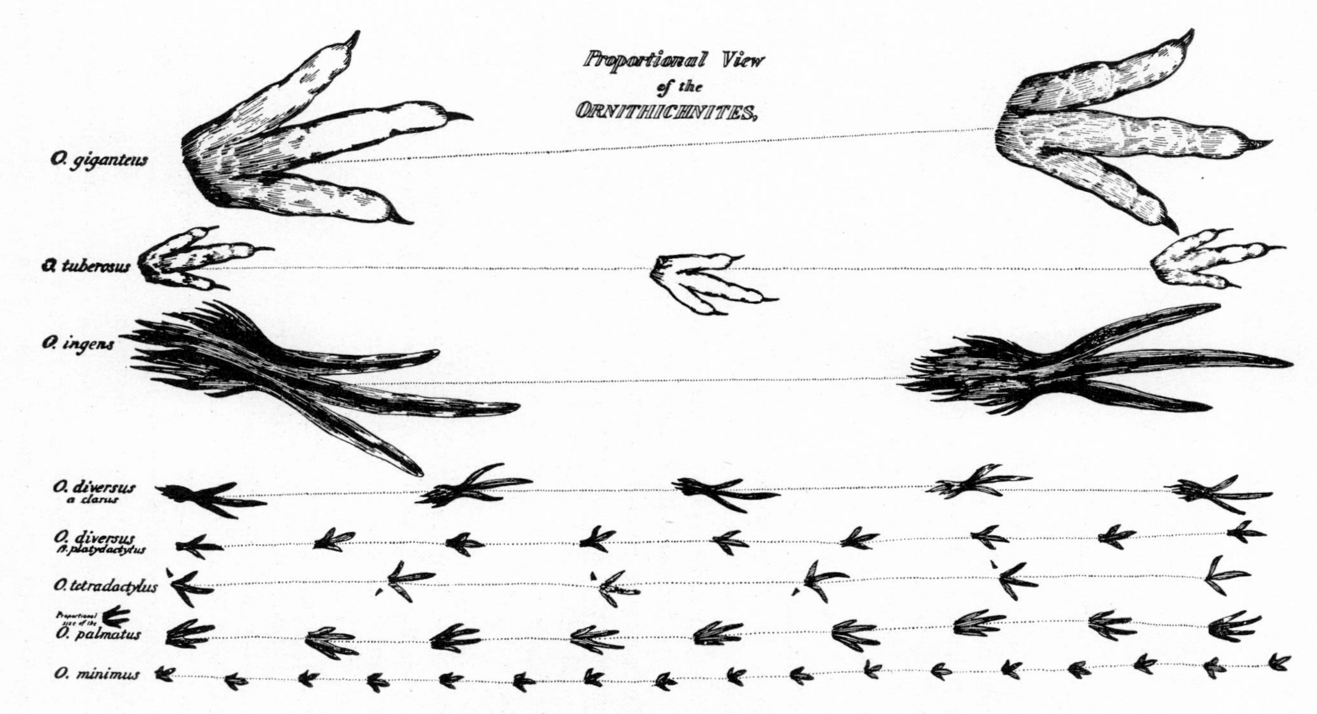

Fig. 110. *Fossil footprints.* (*After Hitchcock.*)

nithichnites (from ορνίς and τίχνοσ), signifying stony bird tracks. Five years of further examination enabled him to swell the list of species to twenty-seven, which were described and figured, natural size, in 1841, in his final report on the geology of Massachusetts.

Up to this last date he had found no certain evidence that any of the tracks were made by quadrupeds, yet a considerable proportion of them bore so strong a resemblance to those of saurian reptiles that he denominated them *sauroidichnites*, or tracks *resembling* those of saurians, intending, however, by the term merely to convey an intimation that they might prove reptilian. To the others he now applied the name *ornithoidichnites*, or tracks *resembling* those of birds. To animal tracks of all kinds, and of whatever nature, he proposed the general name *ichnolites*.

At the April meeting, 1842, of the American Association of Geologists, Professor Hitchcock brought the matter up once more, describing five new species, one of which, *O. tuberosa*, from Pompton, New Jersey, was the first thus far found outside of the Connecticut Valley.

In this same paper he also announced the finding of tracks which afforded the first "certain evidence that any of the numerous tracks upon the sandstone of the Connecticut Valley were made by a quadruped," though he had suspected that such might be the case. These quadrupedal tracks he described under the name *Sauroidichnites deweyi*, and it was suggested that they were comparable with those of the *Cheiotherium*, an amphibian found in the European Keuper beds (Triassic). This discovery, naturally, raised again the question as to the possible quadrupedal nature of the others described, but, with characteristic conservatism, he still refrained from committing himself, simply noting that it would be "contrary to the cautious spirit of science" to decide on the evidence then at hand as to their exact nature, and he therefore adhered to the course adopted in his previous report, and grouped all under the names *ornithoidichnites* and *sauroidichnites*.

The discovery of these tracks naturally excited a very lively interest among geologists and paleontologists both at home and abroad, and references and suggestions concerning them multiplied in the literature of the day. In the *American Journal of Science*, of 1843, under the title "Ornithichnites of the Connecticut River Sandstones and the Dinornis of New Zealand" are given the opinions of Deane, Hitchcock, Mantell, and Owen regarding their nature. Deane, in a letter to Mantell, of London, written in 1842, had said:

These beautiful fossils, indicating a high grade of animal existence in a period of the earth so immensely remote, may well be regarded among the wonders of paleontological science. . . . That the footsteps of the Connecticut River are, however, the authentic traces of extinct birds is confirmed by the undeviating comparisons they bear to living nature.

He wrote further of referring the footprints to Professors Silliman and Hitchcock, both of whom admitted the plausibility of his statements, yet remained incredulous as to the inferences drawn until accurate models were submitted, when Professor Hitchcock had pronounced his unqualified conviction that they were those of birds. Mantell himself in his reply to Deane seems to have accepted the opinion of their birdlike nature, and stated further that at first both Professors Owen and Murchison were in doubt as to whether they were made by birds or reptiles. Professor Lyell, however, stated his conviction that they were genuine *ornithichnites*. Later, Murchison "confessed that the gigantic bones from New Zealand, evincing as they did most unequivocally the existence even in our own times of birds as large as any required by the American footmarks, had removed his skepticism, and that he had no hesitation in declaring his belief that the *ornithichnites* had been produced by the imprints of the feet of birds which had walked over the rocks when in a soft and impressible state." "An opinion," added Mantell, "in which I entirely concur."

Professor Owen,[1] too, in a letter to Professor Silliman, under date of March 16, 1843, after calling attention to the need of caution in assuming the existence of highly organized birds at so early a period, particularly when there were known to be large reptiles which might make very similar tracks, went on to speak of the then recently described Dinornis[2] of New Zealand, and wrote as follows:

It seems most reasonable, therefore, to conclude that the *ornithichnites* are the impressions of feet of birds which had the same low grade of organization as the Apteryx and Dinornis of New Zealand, and this latter may be regarded as the last remnants of an apterous race of birds which seems to have flourished in the epoch of the New Red sandstone of Connecticut and Massachusetts.

The concluding paragraph in the article from which these abstracts have been made is interesting as showing the unqualified

[1] Richard Owen, the paleontologist, of London, England.

[2] Bones of the Dinornis, the giant bird of New Zealand, it should be mentioned, first came to the notice of the scientific world through the publications of Professor Owen, in 1839-1843.

acceptation by Murchison of the opinions by the various authorities. He is quoted as having said:

From this moment, then, I am prepared to admit the value of the reasoning of Doctor Hitchcock and of the original discoverer, Dr. James Deane, who, it appears by the clear and modest paper laid before us (by Doctor Mantell), was the first person to call the attention of the professor to the phenomenon, expressing then his own belief, from what he saw in existing nature, that the footprints were made by birds. Let us now hope, therefore, that the last vestiges of doubt may be removed by the discovery of the bones of some fossil Dinornis. In the meantime let us honor the great moral courage of Professor Hitchcock in throwing down his opinions before an incredulous public.

In a paper read before the American Association of Geologists and Naturalists at Washington in May, 1844, Hitchcock took the subject up once more and gave a brief history of the discovery of footprints in other countries, and also certain data regarding an earlier discovery than that of Dr. Deane, referring to the fact that in 1802 Mr. Pliny Moody, of South Hadley, in Massachusetts, had turned up with the plow upon his father's farm a stone containing in relief five tracks of what he, Hitchcock, now referred to as *Ornithichnites fulicoides*.

Hitchcock reviewed the entire subject up to date and described a fine large specimen found in the impure limestone of Chicopee Falls, in Springfield, under the name of *Ornithoidichnites redfieldii*, and also several other new species. The resemblance of these to saurian remains was still recognized, but the proof was not regarded as sufficient to refer them unequivocally to quadrupedal forms.

In this paper the following classification was adopted for tracks or markings of various kinds:

Order 1. *Polypodichnites,* or many-footed tracks.
Order 2. *Tetrapodichnites,* or four-footed tracks.
Order 3. *Dipodichnites,* or two-footed tracks.
Order 4. *Apodichnites,* or footless tracks.

Under the last term were included certain markings, made by fishes, mollusks, and annelid worms.

There appears to have been gradually developing in the mind of Dr. Deane the feeling that he had "too inconsiderately surrendered" a promising field, and that he was not receiving all the credit for the discovery he deserved, and perhaps also that Hitchcock's conclusions were not in all cases correct. It was this feeling evidently that led him to correspond with Dr. Mantell, as already mentioned, and to take up for himself the description and discussion of his subsequent finds. Dr. Hitchcock replied to the complaints of Dr. Deane, as published in the *American Journal of Science* for 1844, in a clear and dignified manner, but reading both versions in the cold light of history one can but feel that here, as in the case of Peter Dobson noted elsewhere, he showed a disposition to undervalue the opinions of the layman and perhaps overvalue his own mental superiority. Dr. Deane's first paper, in his attempt to recover the ground, was published in the *American Journal of Science* for 1844. Here he figured a slab of sandstone some six by eight feet in dimensions, containing over seventy-five impressions. He did not, however, attempt any scientific descriptions of the same, and referred to them as *Ornithichnites fulicoides* of Hitchcock. In the same *Journal* for the following year he had two papers descriptive of footprints found by him, some of which he thought to be undoubtedly those of a batrachian reptile, the method of progression of which, as indicated by the tracks, was by kangaroo-like jumps, the fore limbs not reaching the ground. Again in 1847 he had a paper describing not merely the tracks but discussing as well the possible conditions under which they might have been formed and preserved. The forms described were considered to have been made in part by birds, and in part by quadrupeds. Their preservation was thought to have been due to a submergence of the mud flats upon which they were made during seasons of flood.

Fig. 111. *James Deane.*

Through the same medium and this same year Hitchcock reverted once more to the subject and described two new species of thick-toed bipeds, the one renamed *Brontozoum* (from *Ornithoidichnites*) *sillimanium*, and the other *B. parallelum*. In connection with these he described a large and extraordinary track thought to be that of a batrachian to which he gave the generic name *Otozoum*.

In 1848, Dr. Deane came forward with a brief paper and figures of tracks which he thought might with propriety be referred to some member of the tailed or salamandrian family of batrachians, since he discovered markings which seemed to him probably due to the trailing of the caudal appendage. In this same year Dexter Marsh, the man who first called Dr. Deane's attention to the footprints in 1836, in a long letter to the editor of Silliman's *Journal*, described and figured footprints which he thought to be those of a quadruped and one that walked step by step and not by leaps.

Previous to the forties Hitchcock had given names only to the tracks, but in 1845, acting under the suggestion made to him by Professor Dana, he presented, at the meeting of the Association of American Geologists, a catalogue of the animals themselves, so far as known, a plan which he adhered to and defended in a paper in the *Memoirs of the American Academy of Arts and Sciences*, presented in 1848. In this he discussed with his usual caution the possibilities of identification and classification from footprints alone, and while he acknowledged that he had no great confidence in the arrangement adopted except in a few instances, he submitted the following list of genera:

1. *Brontozoum.*	9. *Triænopus.*	17. *Hoplichnus.*
2. *Æthiopus.*	10. *Harpedactylus.*	18. *Macropterna.*
3. *Steropezoum.*	11. *Typopus.*	19. *Xiphopeza.*
4. *Argozoum.*	12. *Otozoum.*	20. *Ancyropus.*
5. *Platypterna.*	13. *Palamopus.*	21. *Helcura.*
6. *Ornithopus.*	14. *Thenaropus.*	22. *Herpystezoum.*
7. *Polemarchus.*	15. *Anomœpus.*	23. *Harpagopus.*
8. *Plectropus.*	16. *Anisopus.*	

Under these names he described 49 species, of which he thought 12 were certainly quadrupeds: 4 probably lizards, 2 chelonians or turtles, and 6 batrachians. Two were annelids or mollusks; 3 of doubtful character; and the remaining 32 bipeds. Eight of these 32 he regarded as being thick-toed tridactylous birds; 14 others were probably narrow-toed tridactylous or tetradactylous birds; 2 were perhaps bipedal batrachians, and the remaining 8 might be birds, but would more probably turn out to be either lizards or batrachians.

It may be noted in this connection that, in 1845, Dr. Alton King found fossil footprints in Carboniferous sandstone in Westmoreland County, Pennsylvania. These, though now known to have been tracks of amphibians, were described under the names of *Ornithichnites gallinuloides* and *O. culbertsoni*, King believing in their birdlike nature, though assuring the reader that the "popular error that these are the tracks of wild turkeys needs no discussion." Certain other tracks found at the same time were recognized as those of quadrupeds, and were described under the generic name of *Spheropezium.*

In the *Memoirs of the American Academy of Arts and Sciences* for 1849, Deane had again a paper on fossil footprints of the Connecticut Valley, which he thought to be in part those of birds and in part those of reptiles. These were described and figured, but no attempt at classification was made.

In 1858 Hitchcock published his *Ichnology of New England,*[3] a quarto volume of some 200 pages, with 60 plates of footprints and a map of the Connecticut Valley, showing the distribution of the Triassic sandstone with its included tracks, together with a section across the valley at Springfield. This work is of importance, not merely on account of the detailed description of the tracks, but also as summing up the knowledge of the subject and the prevailing opinion regarding the origin and age of the sandstone itself. The sandstone, it should be stated, from its inclined position had been regarded by many geologists as having been laid down upon a sloping floor. To this view Hitchcock now took exception and could find no way of escaping from the opinion that it had been upheaved since its deposition. The age he considered undoubtedly Jurassic. He discussed the tracks in great detail and announced some thirty characteristics which he considered based on the principles of comparative anatomy and zoölogy, and which he thought afforded him reliable grounds from which to judge of the nature of an animal from its tracks. He felt that he could now decide with a good degree of confidence, first, whether the animal making the footprints was vertebral or invertebral; second, whether a biped, quadruped, or multiped; third, to which of the four great classes of vertebrates it belonged, and, with less certainty, to what order, genus, or species the animal might be referred. It was in this work that he first proposed the term *Lithichnozoa* (stony-track animals), to embrace animals the characters of which he described from their tracks. The classification adopted divided the animals into ten groups; First, *marsupialoid;* second, *pachydactylous,* or thick-toed birds; third, *leptodactylous,* or narrow-toed birds; fourth, *ornithoid* lizards or batrachians; fifth, *lizards;* sixth, *batrachians;* seventh, *chelonians;* eighth, *fishes;* ninth, *crustaceans, myriapods,* and *insects;* and tenth, *annelidans.* It will be noted that the *Brontozoum* was still included under the group of pachydactylous or thick-toed birds.

At the Newport, Rhode Island, meeting of the American Associa-

[3] The cost of this publication was borne by the state. (Reminiscences of Amherst College, 1863.)

tion for the Advancement of Science in August, 1860, he came forward with a further discussion of the position of the trackmakers in the zoölogical scale and, while recognizing their reptilian (Saurian) nature in part, still felt that there were some fourteen species that could not be so considered, since no traces of forefeet had been discovered among the several thousand tracks he had examined. Inasmuch, moreover, as Owen had described a fossil bird from the English Upper Oölite he felt there was no longer any presumption against the existence of these higher types during Triassic times. "While, therefore, in some respects the argument for the ornithic origin of our foot-marks has been weakened by subsequent discoveries, in other respects it has been strengthened, and while I would not be dogmatic on such a subject it seems to me quite premature to abandon my opinion which, however, I am quite ready to do when the analogies of science demand it." Referring to his describing in his previous paper certain of the tracks as marsupialoid he announced that further reflection had convinced him of the correctness of this identification, and he felt that here also he was supported by the finding of marsupial remains in the European Oölite, and even older rocks in North Carolina and Germany.

In 1863 Hitchcock returned to the subject in a brief paper in the *American Journal of Science*, in which he announced that, though having been compelled to give up ten or a dozen of his old species of footmarks, he had described over thirty new ones. Further, finding that an error had been made as to the number of phalanges in the tracks of some of the animals classed as *Lithichnozoa*, he confessed to a doubt if the animals heretofore grouped under this name should be classed as birds; and, as a general conclusion, announced that in fossil footmarks birds could not be distinguished from quadrupeds by the number of phalanges. In this same paper he announced the finding of tracks which were accompanied by markings evidently made by the tail of the animal, such being particularly characteristic of his *Anisopus gracilis*, an animal the quadrupedal nature of which had already been recognized. The presence of such tail-like markings in the case of the *Anomœpus* led him into a somewhat lengthy discussion of the affinities of this animal, whether bird or reptile, though he was evidently inclined to regard it as most nearly related to a lower order of birds like the *Archæopteryx*. In support of this, he appended a letter from Professor Dana, in which he referred to the generalized characters of the early birds.

Here the matter would appear to have rested until Mr. Roswell Field, a farmer of Greenfield, Massachusetts, on whose property the first slabs were found, for the first time in print voiced his own opinions. In the *American Journal of Science*, XXIX, 1860, he gave briefly and modestly the results of his experience in collecting and observing, and announced it as his opinion that the vertebral tracks should all be classed as reptilian. That the animals that made them usually walked on two feet he admitted, but contended that they could as well have walked on four had they chosen. In proof of this he added:

> We find tracks as perfect as if made in plaster or wax which, to all appearances, as to the number of toes and the phalangial or lateral expansions in the toes, agree perfectly well with those of living birds, and still we know, by the impressions made by their forward feet, that these fossil tracks were made by quadrupeds.

In still other cases he noted traces of the tails trailing in the mud. Enumerating some of the cases in which the so-called bird tracks—as, for instance, the *Otozoum*, had been proven to be reptilian—he added that this he verily believed is the place for them all. And in this he was right, but it was not until 1893 that the actual bones of one of the reptiles were found in a sufficiently good state of preservation to allow Professor O. C. Marsh to make the skeletal restoration on which our Fig. 112 is based. This particular form was about six feet in height, when standing, as in the figure. "One of the most slender and delicate dinosaurs yet discovered, being only surpassed in this respect by some of the smaller, bird-like forms of the Jurassic." The creature is shown in the position it is thought to have habitually assumed. On a firm but moist beach only three-toed impressions would have been left by the hind feet and the tail would have been kept free from the ground. On a soft muddy shore the claw of the fifth digit would have also left its mark, and perhaps the tail would have dragged. When, perhaps momentarily, he rested on his forefeet, tracks of a quite dissimilar nature would have resulted, such as at first were assumed to be those of an animal of another species.

And so the matter rests.

Sketch of Doctor Deane.

Dr. James Deane was born at Coleraine, Massachusetts, February 24, 1801, and became by profession a physician, practicing at Greenfield, in the same state. Eminently successful in his practice, he nevertheless found time for study and work in other lines. "Not a moment was lost that he could spare from the great labors of his profession. Late into the night was his lamp seen glittering from his casement

. . . while he was copying, with his masterly touch, these relics of an ancient era" (Bowditch). He is described as a man of few words, though of a genial and social character; of a tall and commanding form, and a well-knit, compact frame; a man whose very walk conveyed an idea of strength. After his death his work on the footprints was summarized in a quarto volume of sixty-one pages of text and forty-six full-page plates, entitled *Ichnographs of the Sandstone of the Connecticut River.*

Fig. 112. *A Maker of "Bird Tracks."*

CHAPTER X.

The Eozoon Question.

"The attainment of scientific truth has been effected to a great extent by the help of scientific error." *Huxley.*

IN his work on the geology of Canada, 1863, W. E. Logan described somewhat briefly certain forms strongly resembling fossils which were discovered three years earlier in the Laurentian rocks belonging to the Grand Calumet series.

The specimens represented parallel or concentric layers of pyroxene, the interstices of which were filled with crystalline carbonate of lime, the arrangement of the two constituents being such as to resemble somewhat the structure of the fossil *Stromatopora rugosa* found in the Bird's-eye and Black River limestone (see Fig. 113). Logan realized the fact that organic remains entombed in these limestones would, if they retained their calcareous nature, have been almost certainly obliterated by crystallization, and only through the replacement of the original lime carbonate by some different mineral substance would there be any chance of the forms being preserved. Nevertheless, their resemblance to fossil remains was so great that, had the specimens been obtained from the altered rocks of the Lower Silurian series instead of the Laurentian, he thought there would have been little hesitancy in pronouncing them to be true fossils.

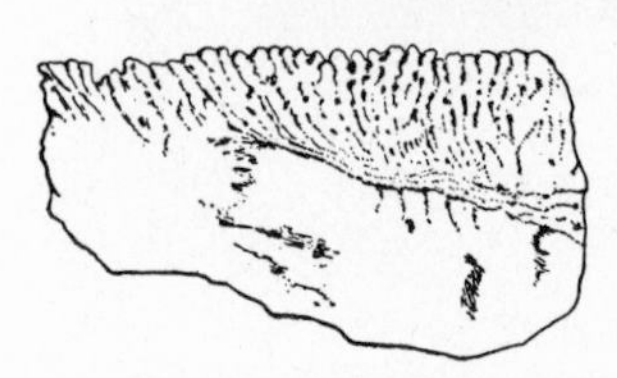

Fig. 113. *Supposed Fossil from Laurentian Limestone. a, Weathered surface; b, vertical transverse section (reduced about one-half).*

In the *Quarterly Journal of the Geological Society of London* for 1865 Logan returned to the subject, bringing much new evidence. He described as the oldest rocks in North America those composing the Laurentide mountains of Canada and the Adirondacks of New

York, dividing them into the Lower and Upper Laurentian series, the united thickness of which was estimated to probably exceed 30,000 feet, this overlaid by a third group, the Huronian, which had been estimated to be some 18,000 feet in thickness. In both the Upper and Lower Laurentian series he found zones of limestone which had sufficient volume to constitute independent formations.

While studying these rocks he had naturally fallen to speculating upon the possible occurrence of life during the period in which they were being laid down, and his mind was doubtless in a condition to readily accept any promising discoveries. He recognized the fact that the mere absence of fossils did not offer any proof in negation, since such might have been obliterated by metamorphic action, and he referred to arguments of T. Sterry Hunt in favor of the possible organic origin of the beds of iron ore and metallic sulphurets in the older rocks. When, therefore, these imitative forms were brought to his attention, he candidly acknowledged himself as being disposed to look upon them as fossils, and, indeed, he exhibited them as such at the meeting of the American Association for the Advancement of Science at Springfield in 1859.

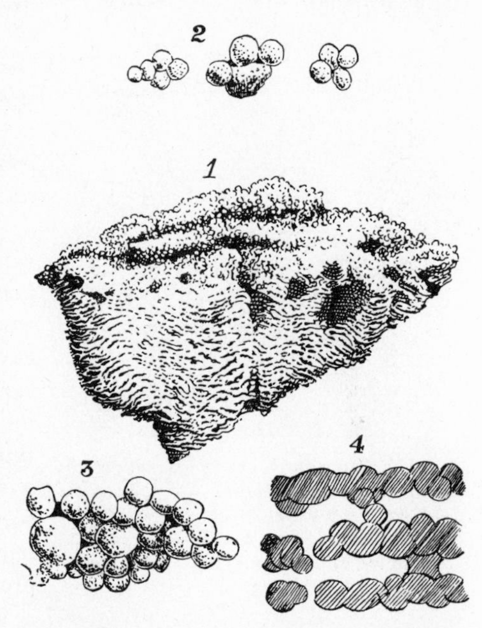

Fig. 114.—(*nos. 1 to 4*)—*Small weathered specimens of Eozoon canadense. From Petite Nation. 1, Natural size, showing general form and acervuline portion above and laminated portion below; 2, enlarged casts of cells from upper part; 3, enlarged casts of cells from the lower part of the acervuline portion; 4, enlarged casts of sarcode layers from the laminated part.* (*After J. W. Dawson.*)

Subsequently thin sections prepared from these forms were referred to Dr. J. W. Dawson, of Montreal, who was regarded as the "most practiced observer with the microscope." To Doctor Dawson, then, may be given a large share of whatever credit is due for the recognition of the supposed animal nature of this much disputed body.

The specimens first examined were from the base of the so-called Grenville limestone, belonging to the highest zones of the Laurentian, the mass of the rock being composed of great and small irregular aggregates of white crystalline pyroxene, interspersed with a multitude of small spaces, consisting mainly of carbonate of lime, many of which showed minute structures similar to that of the supposed fossil. The general character of the rock he thought conveyed the impression that it was a great foraminiferal reef in which the pyroxenic masses represented a more ancient portion, which had become much broken up and worn into cavities and deep recesses after the death of the animal, affording a seat for a new growth, represented by the calcareous and serpentinous part. This in turn became again broken, leaving in some places uninjured portions of the general form.

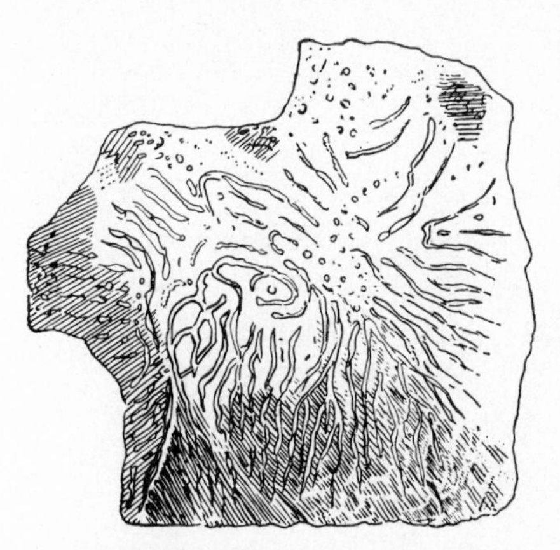

Fig. 115. *Magnified group of canals in supplemental skeleton of Eozoon. Taken from the specimen in which they were first recognized (Life's Dawn on Earth). (After J. W. Dawson.)*

The main difference between this foraminiferal reef and more recent coral reefs, he thought to be that, while with the latter there are usually associated many shells and other organic remains, in the more ancient one the only remains yet found were those of the animal which built the reef.

Dawson attacked the problem from the zoölogist's standpoint. To properly understand his position and also that of those who combated his arguments. it must be remembered that the science of micro-petrography was then in its infancy; in fact, the possibilities of metasomatosis or alteration by indefinite substitution and replacement were only beginning to be realized by even the most advanced of mineralogists and geologists. Hence it was possible for two men, attacking the same problem from opposite standpoints—the one as a chemist and physicist and the other as a zoölogist—to arrive at conclusions diametrically opposed to one another. It will become evident as we proceed, that full advantage was taken of the opportunity.

The specimens examined, as already indicated, were masses often several inches in diameter,. presenting to the naked eye alternate laminæ of serpentine, or pyroxene, and carbonate of lime, in general aspect simulating the Silurian hydroids of the genus *Stromatopora* (see Fig. 114). Under the microscope Dawson found the laminæ of serpentine and pyroxene to present no organic structure, the pyroxene being highly crystalline. The laminæ of carbonate of lime, on the other hand, retained distinct traces of structure which could be considered organic constituting parallel or concentric partitions of variable thickness inclosing flattened spaces or chambers, frequently crossed by transverse plates or septa. The laminæ themselves were frequently excavated on their sides into rounded pits, and in some places traversed by canals and penetrated by numerous minute tubuli, a structure which can be best understood by a reference to Fig. 115.

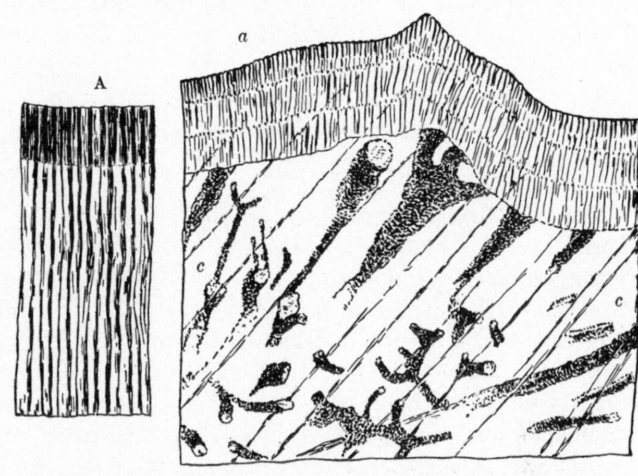

Fig. 116. *Portion of Eozoon magnified 100 diameters, showing the original cell-wall with tubulation and the supplemental skeleton with canals. a, Original tubulated wall or "Nummuline layer," more magnified in Fig. A; b, c, intermediate skeleton with canals.* (*After W. B. Carpenter.*)

According to the conclusions of Dawson, the calcareous portion represented the original shell, the serpentinous matter, on the other hand, being material infiltrated into the cavities which had been occupied by sarcode during the life of the animal. He referred the supposed organism to the group of rhyzopods of the order of foraminifera, and conceived that during life they were sessile by a broad base and "grew by the addition of successive layers of chambers separated by calcareous laminæ but communicating with each other by canals or septal orifices sparsely and irregularly distributed." He imagined, further, that the organisms grew in groups which ultimately coalesced and formed large masses penetrated by deep, irregular canals and that they continued to grow at the surface while the lower parts became dead and filled up with the infiltrated matter or sediment, assuming the aspect of a coral reef.

On account of their geological position, he proposed to designate the animals by the name of *Eozoon*,[1] and these particular ones by the specific name of *canadense*, and in his annual address before the Natural History Society of Canada for that year, he referred to the discovery as "one of the brightest gems in the scientific crown of the Geological Survey of Canada."

The paper by Dawson was immediately followed by one by the well-known microscopist, W. B. Carpenter, whose conclusions were in every way confirmatory of those of Dawson. To quote his own words:

> That the Eozoon finds its proper place in the foraminiferal series, I conceive to be conclusively proved by its accordance with the great types of that series in all the essential characters of organization, namely, the structure of the shell forming the proper wall of the chambers, in which it agrees precisely with Nummulina and its allies; the presence of an intermediate skeleton and an elaborate canal system, the disposition of which reminds us most of *Calcarina;* a mode of communication of the chambers when they are most completely separated, which has its exact parallel in *Cyclocypeus;* and an ordinary want of completeness of separation between the chambers, corresponding with that which is characteristic of *Carpentaria.*

It is not surprising that the finding of supposed fossil remains in rocks so old as those of the Laurentian should have excited the widest attention, and it would have been strange indeed had such conclusions been allowed to go unchallenged. In 1866[2] there was read before the same society a detailed account of the investigations made by Professors William King and T. H. Rowney, of Queen's University in Ireland. These gentlemen, although claiming to have had no misgivings at the commencement of their investigations as to the eozoonal (*i.e.*, organic) origin of the material, nevertheless attacked the problem from the physical and chemical rather than the biological standpoint. They examined in great detail the structures described by Dawson, though not having access to all his materials. Their conclusions were to the effect that (1) the so-called chamber casts or granules of serpentine were more or less simulated by minerals like chondrodite, coccolite, pargasite, etc.; (2) that the "intermediate skeleton" was closely represented both in chemical composition and other conditions by the matrix of these minerals; (3) that the "proper wall" of Dawson was structurally identical with an asbestic-form layer which was frequently found investing the grains of chondrodite, and that instead of belonging to the skeleton

[1] ηως, Dawn, and ζωον, animal.

[2] Quarterly Journal of the Geological Society of London, XXII, 1866, p. 185.

it was altogether independent of that part and formed an integral portion of the serpentine constituting a chamber cast; (4) that the canal system was analogous to the embedded crystallizations of native silver and other similarly conditioned minerals; and (5) that the type examples of casts of stolon passages were isolated crystals apparently of pyrosclerite. From these considerations and the perhaps even more important one that the eozoonal structure was found only in metamorphic rocks and never in unaltered sedimentary deposits, they concluded that every one of the specialties which had been diagnosed for the *Eozoon canadense* was of purely crystalline origin.

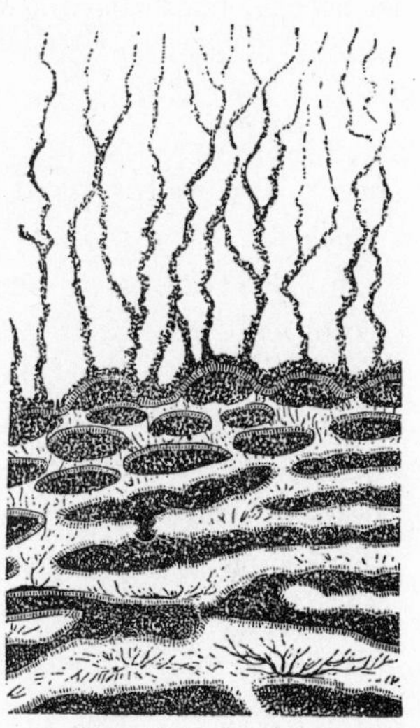

Fig. 117. *Magnified and restored section of a portion of Eozoon canadense. The shaded portions show the animal matter of the chambers, tubuli, canals, and pseudopodia; the unshaded portions the calcareous skeleton.* (*After J. W. Dawson.*)

Such conclusions naturally brought forth prompt rejoinders from Carpenter and others, the controversy becoming in some cases almost viciously personal. Thus Carpenter accused Rowney of never having seen any sections of Eozoon thin enough to give a good view of its tubuliferous layer until he himself had shown it to him; and, further than that, he had no practical knowledge of the structure of nummuline shells, at best. His paper in reply to the statements of Rowney concluded with the remark that he had recently had his attention called to an occurrence of Eozoon, which was preserved simply in carbonate of lime without any serpentine or other foreign mineral, and showed the canal system very perfectly. This he felt was a conclusive answer to King and Rowney's objection No. 3, to the effect that the canal system of Eozoon was no other than crystallization of metaxite, an allomorphic variety of serpentine.

In this same year Ferdinand von Hochstetter found in the crystalline Azoic limestone of Krumman, Bavaria, structures which, when examined by Carpenter, were pronounced by him to be unmistakably those of Eozoon.

The forms thus far found were limited to the Laurentian rocks, and were regarded by both Logan and Dawson as important "horizon markers," *i.e.*, as affording presumptive evidence in favor of the

Laurentian age of all rocks in which they might occur. The announcement by C. W. Gümbel of finding Eozoonal structures in the Hercynian gneisses of Bavaria, of Huronian or Cambrian age, was, therefore, both disconcerting and encouraging to the "Eozoonists." Gümbel himself adopted enthusiastically the conclusions of Carpenter and Dawson as to the organic nature of the forms. "This discovery," he wrote, "at once overturns the notions hitherto commonly entertained with regard to the origin of the stratified primary limestones. In this discovery we hail with joy the dawn of a new epoch."

In 1867 Logan again brought the matter to the attention of the Geological Society of London, submitting a specimen obtained in the township of Tudor, Hastings County, Canada, in which the serpentine minerals were quite lacking, it being, in fact, the specimen above referred to by Carpenter. This was examined and described by Dawson, who thought to be able to detect all the essential characteristics of the true Eozoon. In presenting his description of this new find, Dawson took occasion once more to challenge the work of Messrs. King and Rowney, and accused them of making the fundamental error of defective observation in failure to distinguish between organic and crystalline forms, an accusation which he substantiated (?) by a long argument not necessary to reproduce here.

From this time on, literature on both sides of the Atlantic contains many references and matter more or less relevant to the discussion. By many of the best-informed the organic nature of the fossil was considered to be an open question, and but few American geologists, so far as I am aware, ever openly advocated it, or at least committed themselves in public print either for or against it. James Hall, in a letter to T. Sterry Hunt, dated November 16, 1869, did however say: "I am glad to hear of the preeminent success of the Eozoon. It is the greatest discovery in geology for half a century at least, and its influence of vastly more importance than anything I can think of as a single discovery affecting our views of the early history of the earth." Hunt himself, from his well-known vagaries regarding the origin of serpentine and the older crystalline rocks in general, might have been expected with equal reason on either side of the question, and apparently favored Dawson's views. In a letter evidently in response to the one mentioned above he wrote: "You will be glad to hear that we have beautiful Eozoon from the crystalline limestones of Chelmsford, near Lowell, Mass. The rocks there are granitoid gneiss, very like the Lawrentian elsewhere. . . . Bicknell of Salem has Eozoon from Newburryport, so Hyatt writes." It should be stated that Hunt had previous to this made some chemical analyses of Dawson's materials and found that the substance filling the chambers of the supposed organism was composed of serpentine and pyroxene. The genetic relationship of the two minerals was not, however, understood.

In 1871 Messrs. King and Rowney read before the Royal Irish Academy in Dublin a paper on the geological age and structure of the serpentine marble of Skye, in which reference was made to structures closely simulating those of the Eozoon, the rock in this case being of Liassic age, and the imitative forms the result of structural and chemical changes to which such siliceous minerals as malacolite and other varieties of lime magnesian pyroxenes were characteristically liable.

This paper was followed the same year by another on the mineral origin of the *Eozoon canadense,* which passed in review all of the accumulated evidence both for and against, reiterating and emphasizing many of the statements made in previous papers. They held that the intermediate skeleton and the chamber casts of the Eozoon were completely paralleled in various crystalline rocks containing pyroxene, the chamber casts being, in fact, composed occasionally of loganite and malacolite besides serpentine, a fact which, instead of favoring the organic origin, as claimed, was to be held as proof of their having been produced by mineral agencies.

With reference to Gümbel's observations regarding the rounded cylindrical or tuberculated grains of coccolite and pargasite which he found in various crystalline marbles and supposed to be chamber casts, they claimed to have found upon such grains crystalline planes, angles, and edges, "a fact clearly proving that they were originally simple or compound crystals that have undergone external decretion by chemical or solvent action."

The so-called nummulitic layer they contended had originated directly from closely packed fibrous or asbestiform serpentine, and that it occurred in cracks or fissures in both the Canadian and Connemara (Ireland) ophite. The fact that the two superposed asbestiform layers forming the upper and under proper wall and their component aciculi often passed continuously and without interruption from one chamber cast to another, they argued was totally incompatible with the idea of this nummuline layer having resulted from pseudopodial tubulation. The canal system they found to be composed of serpentine or malacolite, and to be completely paralleled by crystalline configurations in the coccolite marble of Aker in

Sweden, and in the crevices of a crystal of spinel embedded in a calcite matrix from Amity, New York.

Finally, they argued that the occurrence of the Eozoon solely in crystalline or metamorphosed rocks, and never in ordinary unaltered deposits, must be assumed as completely demonstrating its purely mineral origin. The paper was aggressively argumentative from the start and contains a very scathing review of Dawson's work, further expressing a willingness to renounce the controversy altogether, "fully believing that Doctor Dawson can employ his time more usefully on other subjects than that of Eozoon."

The testimony, however, was not altogether in one direction, Professor Max Schultze, of Wurzburg, after an examination of a specimen transmitted him by Dawson, announcing that "there can be no serious doubt as to the foraminiferal nature of *Eozoon canadense.*" Further evidence against its organic nature was, however, furnished by Mr. H. J. Carter in 1874, in a letter to Professor King, and published in the *Annals and Magazine of Natural History* of that year. After a somewhat elaborate comparison of the structure with that of known foraminifera, Mr. Carter wrote:

> But in vain do we seek in the so-called *Eozoon canadense* for an unvarying perpendicular tubuli, the *sine qua non* of foraminiferal structure. . . . In short, in vain do we look for the casts of true foraminiferous chambers at all in the grains of serpentine. They, for the most part, are not subglobular but subprismatic. With such deficiencies I am at a loss to conceive how the so-called *Eozoon canadense* can be identified with foraminiferous structure except by the wildest conjecture.

To these conclusions Carpenter took violent exception and accused Mr. Carter of not having read anything that had been written bearing upon the other side of the question; of having no comprehensive knowledge of foraminiferal structure, and of having seen only those samples submitted by Professor King; and hence his imputation to the effect that the organic nature of the Eozoon had no other basis than "the wildest conjecture" was to be regarded "simply as specimens of a new method of language which might be termed *Carterese.*" He then went over once more certain disputed points and concluded by saying:

> An experience of thirty-five years . . . has given me, I venture to think, some special aptitude for recognizing organic structure when I see it; and I never saw, in any fossil whatever, more distinct evidences of organic structure than are to be seen in these finer ramifications of the canal system of Eozoon.

As Chaucer wrote:

> "For treweley there is noon of us all,
> If any wight wol claw us on the galle,
> That we nyl kike."

Perhaps the most important paper on the subject after those mentioned was that of Dr. Carl Moebius, of Kiel, in 1879. Moebius, it should be noted, was a zoölogist and microscopist. He claimed to have been first led to the study of the eozoon through observations of the structure of a rhyzopod found by him in 1874 on the coral reefs near Mauritius. Sections of these growths so closely resembled the representations of eozoon published that he resolved to make a careful investigation of the latter for purposes of comparison with his *Carpentaria raphidodendron* and other foraminifera. With this object in view, he investigated upward of ninety Eozoon sections, which were placed at his disposal through the kindness of Carpenter, and many of which originally belonged to Professor Dawson.

The result of Moebius's work was not at all, presumably, what Carpenter was led to expect or Dawson to hope. He wrote:

> My task was to examine the Eozoon from a biological point of view. I commenced it with the expectation that I should succeed in establishing its organic origin beyond all doubt, but facts led me to the contrary. When I saw first the beautiful stem systems in Professor Carpenter's sections, I became at once a partisan of the view of Professor Dawson and Professor Carpenter, but the more good sections and isolated stems I examined the more doubtful became to my mind the organic origin of Eozoon, until at last the most magnificent canal systems taken altogether and closely compared with foraminiferal sections preached to me nothing but the inorganic character of Eozoon over and over again.

Dawson thereupon became particularly indignant, and characterized Moebius's work as furnishing "only another illustration of partial and imperfect investigation, quite unreliable as a verdict on the question in hand." He claimed that Moebius should have studied the fossil *in situ* and in its various stages of preservation; that he confounded the "proper wall" with the chrysotile veins traversing many of the specimens; and that, further, in his criticisms he regarded each structure separately, and did not "consider their cumulative force when taken together." This cumulative force he presented as follows:

> 1. It [*i.e., Eozoon*] occurs in certain layers of widely distributed limestones, evidently of aqueous origin, and on other grounds presumably organic.

2. Its general form, lamination, and chambers resemble those of the Silurian *Stromatopora* and its allies, and of such modern sessile foraminifera as *Carpentaria* and *Polytrema.*

3. It shows under the microscope a tubulated proper wall similar to that of the nummulites, though of even finer texture.

4. It shows also in the thicker layers a secondary or supplemental skeleton with canals.

5. These forms appear more or less perfectly in specimens mineralized with very different substances.

6. The structures of Eozoon are of such generalized character as might be expected in a very early Protozoan.

7. It has been found in various parts of the world under very similar forms, and in beds approximately of the same geological horizon.

8. It may be added, though perhaps not as an argument, that the discovery of Eozoon affords a rational mode of explaining the immense development of limestones in the Laurentian age; and on the other hand that the various attempts which have been made to account for the structures of Eozoon on other hypotheses than that of organic origin have not been satisfactory to chemists or mineralogists, as Doctor Hunt has very well shown.

Singularly enough, although first found on the western continent, active work regarding the nature of the Eozoon was confined largely to the Canadians and English, with an occasional European collaborator, workers in the United States taking little part in the dispute so far as indicated by literature, although watching the contest with interest and becoming more or less partisans, according to the extent of their own observations or the character of the evidence offered. In 1871 Messrs. Burbank and Perry made a study of the eozoonal limestone occurring at Chelmsford, Bolton, and Boxboro, in Massachusetts, and arrived at the conclusion that the limestones were not themselves of a sedimentary, but of a veinlike character, and that consequently the Eozoon itself, as there occurring, must be of mineral and not organic origin.

Meantime studies in micro-petrography were steadily advancing, and it was noticeable that none of the workers along these newly developed lines of research were disposed to regard the Eozoon as of other than inorganic origin. In 1884 Professor T. G. Bonney, of St. John's College, Cambridge, England, visited one of the most noted of the Canadian localities in company with Doctor Dawson, and published the results of his observations in the *Geological Magazine* for 1895, but contented himself with describing the mode of occurrence and structure of the mass without committing himself definitely as to its origin. From this paper was taken the accompanying figure. His conclusions so far as given were to the effect that the Eozoon often occurs in close relation, on the one hand, with a fundamental mass of almost pure pyroxene (or serpentine resulting from its alteration); or, on the other hand, with a fairly large mass of crystalline limestone containing more or less numerous grains of pyroxene or serpentine. He found nothing to lead him to think the eozoonal specimens were blocks of foreign material metamorphosed by becoming included in either a volcanic or plutonic mass as has

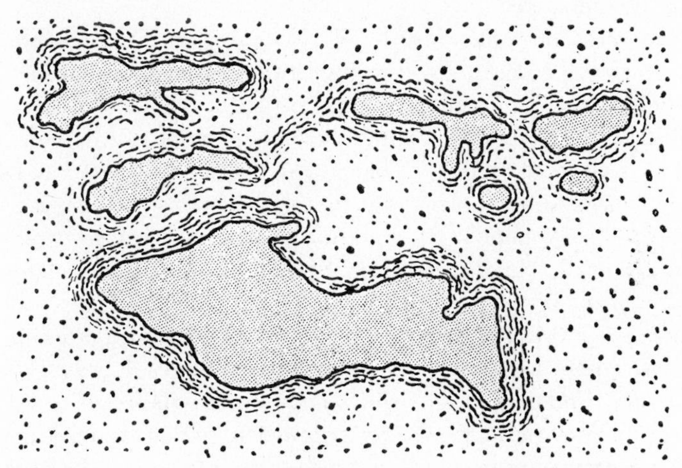

Fig. 118. *Diagram of Eozoonal rock at Cote St. Pierre. (After Bonney.)*

been suggested (see p. 577). The structure, to his mind, offered a choice between two interpretations only. It "is either a record of an organism or a very peculiar and exceptional condition of a pyroxene marble of Laurentian age."[3]

The fact that the lime-magnesian pyroxenes, through a process of chemical metamorphism, passed over into serpentine with the

[3] "I remember that, till 1884, I always upheld, in lecturing to my students, the organic origin of Eozoon. But in the later years prior to my visit to Canada, I began, owing to increasing knowledge of the Archaean rocks and metamorphism generally, to feel more and more strongly the difficulties involved in the organic origin view, and when I went to Cote St. Pierre I expected to find the limestone later in date than the gneisses. That was not so, and I returned from Cote St. Pierre a sceptic. You may be surprised that I did not say so in 1895 (or before). I was in a rather delicate position. I had been Dawson's guest at Montreal and Cote St. Pierre; I had the greatest regard for him, and knew that he really felt strongly about this point—'Eozoon' was a favorite child! So I refrained as long as I could and then restricted myself to stating the facts, without expressing an opinion, what, however, I thought few would fail to infer." (Letter from T. G. B. to G. P. M., May 22, 1906.)

formation of secondary calcite, and that many of the supposed sarcode chambers were filled with granules of this material in all stages of this alteration was from time to time noted by Messrs. King and Rowney, Julien, Williams, Wadsworth, Merrill, and others. The evidence thus presented was, however, wholly without effect on Messrs. Carpenter and Dawson, who to the very last remained true to their first convictions, and as late as 1895 Dawson reviewed the subject with seeming thoroughness and announced his original opinion as unchanged.[4]

It remained, however, for J. W. Gregory and H. J. Johnston-Lavis, to give the death blow to the organic theory as late as 1891 and 1894. Gregory, from an exhaustive study of Dawson's original specimen of the so-called Tudor Eozoon, arrived at the conclusion that the same was wholly of an inorganic nature. After a careful examination of all the slides and figures, he wrote: "I must confess myself absolutely unable to recognize in the specimen any trace of the proper wall, canals, or stolon passages which are claimed to occur in Eozoon."

The case against the organic nature of the specimen did not, however, rest upon negative evidence alone. The rock was intensely cleaved and crumpled. The twin laminæ in the planes of crystalline cleavage in the calcite bands were, however, not bent. Further than this, the bedding plane could be traced directly across the specimen, traversing the limestone in the supposed body cavities. These facts would seemingly prove conclusively that the supposed organic forms were not original, but wholly secondary and due to metamorphism.

Johnston-Lavis's later work was perhaps even more decisive, since he showed that structures in every way similar could be produced by the action of heat upon limestone. His conclusions were based upon an examination of microscopic slides from blocks of limestone in the volcanic tuffs of Monte Somma. To appreciate his evidence it must be remembered that these blocks are ejected from the volcano and are found embedded in a tuff consisting largely of pumiceous lava. They occur as irregular, angular, or subangular masses ranging to more than a cubic meter, though commonly less than a quarter of that size. They have been acted upon by the heat and vapors within the volcano and more or less completely metamorphosed, with the production of various silicate minerals, such as

[4] "The month of September I spent . . . in England. I met there Mr. Dawson. He still keeps up his unfortunate Eozoon although no one else believes in it since Prof. Möbius of Kiel refuted completely its organic origin." (Letter from F. Roemer to James Hall, November 8, 1883.)

pyroxene, olivine, epidote, and mica. The structure of the altered limestone is entirely different from that of the unaltered material, and corresponds in all details with that of the original Canadian specimens, indeed, in many cases exhibiting some of the pseudo-organic structure details—such as stolon passages—in far greater perfection than does the true Eozoon.

It has been remarked that American workers took comparatively little part in active research, though it does not necessarily follow that examinations of the problematic bodies were not made sufficient to enable them to hold opinions of their own. It is interesting, therefore, to note that few accepted unhesitatingly the organic theory, and many of them rejected it entirely. The consensus of opinion today is so decidedly against the organic nature of the body that it may be considered as practically settled, although Dana, in the latest issue of his *Manual* (1895), includes a copy of one of Dawson's original figures, and sums up the evidence pro and con without prejudice.[5]

There is apparently no question but this simulative form is due to a process of chemical metamorphism, a process of indefinite substitution and replacement, technically *metasomatosis*, acting upon rocks which were originally granular aggregates of lime-magnesian pyroxenes, with more or less calcareous matter, the serpentine being, in all cases, secondary. Such an origin is suggested at once, even to the uninitiated, by reference to figures like that of Professor Bonney, on page 575. Similar structures have, moreover, been noted by various observers in rocks which were unmistakably of igneous origin.

Specialization is, undoubtedly, essential to the rapid advancement of knowledge, but there is danger of specialization being carried too far—danger that the individual, through insufficient breadth of training or through too close application to his own particular hobby, may ignore the work of his neighbor along other lines, and perhaps in time become so immune as to be unable to appreciate that work, even when its details are laid before him.

In the review of the Eozoon question, much must be allowed for the growth of science—the gradual increase in knowledge regarding both mineral structure and mineral alteration. Still, one who peruses these papers can but feel that had Messrs. Dawson and Carpenter

[5] In 1888 an attempt was made by Professor Persifor Frazer of Philadelphia to ascertain the opinions of leading American geologists relative to the nature of the Eozoon. Of the fourteen whose answers were published four—three of whom were paleontologists,—pronounced in favor of an organic origin. The remaining ten, stratigraphers and men whose work lay largely along the lines of physical and chemical geology, considered it inorganic. (Am. Geologist, 1888.)

had a little more knowledge of mineralogy they would have been less dogmatic in their assertions, and it is possible that had Messrs. King and Rowney had more knowledge of foraminiferal structure they might have been less harsh in their criticisms.

CHAPTER XI.

The Laramie Question.

"What's in a name?" *Shakspeare.*

IT will be recalled that Hayden, in his annual report for 1872, referred to the fact that during his explorations of the Tertiary formations along the upper Missouri River in 1854-1855 he made large collections of shells and plants, many of which were quite new to science. During the succeeding years up to the autumn of 1860 these explorations were extended and additions made to the collections, which were described from time to time in the current literature.

The shells were of extinct fresh-water species and, while they did not appear to be positively characteristic of any age, were regarded by Meek as most nearly resembling Tertiary types. The fossil plants were mainly of extinct species and were regarded by Newberry as also of Tertiary age, probably Miocene.

From evidence of this kind, accumulated during the various expeditions, Hayden had announced the conviction that these *Lignitic* strata, as he called them, which had been found to occupy such vast areas in the upper Missouri Valley—extending southward, with very little interruption, to New Mexico, and westward into the interior of the continent—were probably all portions of one great group, interrupted here and there by mountain chains or concealed by more modern deposits, and, from the identity of their fossil flora, all of Tertiary age.

He then went on to state that his studies of the lower coal beds at Bear River City, Wyoming, and Coalville, Utah, in 1868, had convinced him that these particular beds were of Cretaceous age, but admitting this, he felt, would be to admit the Cretaceous age of all the coal beds of the Northwest, and in so doing to ignore the evidence of the fossil flora altogether. The facts then at hand, he thought, seemed rather to point to the conclusion that the deposition of all the Lignitic strata began during the latter portion of the Cretaceous period and continued on into the Tertiary without any marked physical break, so that many of the Cretaceous types, especially of the vertebrata, may have lingered on through the Transition period, even into the Tertiary era. Inasmuch as this statement contains the first recognition of the full importance of what later became known as the Laramie question, and inasmuch, further, as the discussion brought out many interesting facts and opinions as to the relative value of the various kinds of fossil remains as horizon markers, the subject may be dwelt upon in some detail.

It is well to anticipate, however, in order more readily to understand what is to follow, that Hayden made a fundamental error in assuming an identity of age for all the lignite strata. Had he realized the fact, afterward abundantly proven, that essentially similar conditions existed at various periods in localities not widely remote from one another, which were productive of very similar results, the so-called "Laramie question," as it is known today, would never have arisen. The apparently conflicting character of the fossil remains of plants and animals was, nevertheless, extremely confusing, and some discussion and verbal warfare during the gradual accumulation of the necessary data for the final settlement of the problem was bound to arise. It will be well to note, further, that the localities whence was derived the major portion of the evidence brought to bear were: Fort Union, Nebraska; the Judith River, on the upper Missouri, Montana; Coalville, Utah; and Bear River and Bitter Creek, in Wyoming. Incidentally, other localities in Colorado and New Mexico came in for discussion.

During the season of 1872 Lesquereux, Meek, and Cope were assigned by Hayden to work in areas which seemed to afford the most promising opportunities for deciding the question as to the precise position of the beds in the geological scale. Their conclusions as rendered in the annual reports for 1872-1873 were widely divergent and did little more than emphasize the existing confusion.

Lesquereux naturally worked wholly from a paleobotanical standpoint. He explored the plant-bearing Cretaceous strata of the Dakota group and the valley of the Saline River, as well as the Smoky Hill Fork of the Kansas River and the Lignite formations of the Rocky Mountains from Trinidad to Cheyenne and along the Union Pacific Railroad to Evanston. He made extensive collections and studied the materials in great detail, comparing the forms found with those from known geological horizons in Europe. The summary of his conclusions, as given in his own words, was—"that the great Lignitic group must be considered as a whole and well-characterized formation, limited at its base by the fucoidal sandstone, at its top by the conglomerate beds; that independent from the Cretaceous under it and from the Miocene above it our Lignitic formations represent the American Eocene."

Meek regarded the Coalville and Bear River beds as Cretaceous, but argued for the Bitter Creek beds that the entire absence among the invertebrate fossils yet found of *Baculites*, *Scaphites*, *Ancyloceras*, *Ptychoceras*, *Ammonites*, *Gyrodes*, *Anchura*, *Inoceramus*, and all of the other long list of genera characteristic of the Cretaceous certainly left its molluscan fauna with strong Tertiary facies. But when he came to consider these fossils in their specific relation he found that, with possibly two or three exceptions, all were new to science and quite different from those yet found either at Bear River or Coalville, or, indeed, elsewhere in any of the established horizons. He felt, therefore, that he could scarcely more than conjecture from their specific affinities to known forms what the probable age of the rocks might be in which they occurred. He, however, called attention to the following facts relative to the age of the formation as found at Bitter Creek:

First, that it was conformable to an extensive fresh-water Tertiary formation above, from which it did not differ materially in lithological characters except in its containing numerous beds of coal.

Second, that it seemed also to be conformable to a somewhat differently composed group of strata below, apparently containing little, if any, coal, and believed to be of Cretaceous age.

Third, that it showed no essential difference of lithological characters from the Cretaceous coal-bearing rocks at Bear River and Coalville.

Fourth, that its entire group of vegetable remains as determined by Lesquereux presented exclusively and conclusively Tertiary affinities with the one exception of a marine plant (*Halymenites*) which also occurred under thousands of feet of undoubted Cretaceous strata.

Fifth, that all its animal remains thus far known were specifically different from any of those found in any other formation of this region, with possibly two or three exceptions.

Sixth, that all its known invertebrate remains were mollusks, consisting of about thirteen species and varieties of marine, brackish, and fresh-water types, none of which belonged to genera peculiar to the Cretaceous or any older rocks, but all to such as are alike common to the Cretaceous, Tertiary, and present epochs, with possibly one exception.

Seventh, that, on the other hand, two or three of its species belong to sections or subgenera apparently characteristic of the Eocene-Tertiary of Europe, and even very closely allied to species of that age found in the Paris basin, while one species seemed to be conspecific with and two congeneric with forms found in brackish-water beds on the upper Missouri containing vertebrate remains most nearly allied to types hitherto deemed characteristic of the Cretaceous.

Eighth, that one species of *Anomia* found in it is very similar to and perhaps identical with a Texas Cretaceous shell, while a *Viviparus,* found in one of the upper beds, is almost certainly identical with the *V. trochiformis* of the fresh-water Lignite formations of the upper Missouri, a formation which has always been considered as Tertiary.

He summed up his conclusion as follows: It thus becomes manifest that the paleontological evidence bearing on the question of the age of this formation, so far as yet known, is of a very conflicting nature; though aside from the supposed Cretaceous evidence furnished by the dinosaur bones found by him at Black Butte in 1872, the organic remains favor the conclusion that it is Tertiary. The testimony of the plants, however, on this point, although they doubtless represent what would be in Europe considered as clearly a Tertiary flora, is weakened somewhat by the fact that there is in Nebraska, in clearly Cretaceous rocks, a flora that was then referred by the highest European authority to the Miocene.

In the report for 1873 (printed in 1874) Lesquereux returned once more to the subject and answered more or less satisfactorily various objections which had been made to his previous conclusions. He referred to the Lower American Eocene all the coal strata of the Raton Mountains; those of the Canon City coal basin; those of Colorado Springs; those of the whole basin of central and north Colorado extending from Platte River or from the Pinery divide to south of Cheyenne, including Golden, Marshall, Bowlder Valley, Sand Creek, etc.; and in Wyoming, the Black Butte, Hallville, and Rock Spring coal. He considered as American Upper Eocene or Lower Miocene the coal strata of Evanston, and from identity of the characters of the flora, those six miles above Spring Canyon near Fort Ellis, those of the locality marked near Yellowstone Lake among basaltic rocks, and those of Troublesome Creek, Mount Brosse, and Elk Creek, Colorado. The coal from Bellingham Bay, in Washington, he also referred to the same horizon. To the Middle Miocene he referred the coal basin of Carbon and those of Medicine

Bow, Point of Rocks, and Rock Creek; to the Upper Miocene, the coal of Elko Station, Nevada.

Concerning the evidence furnished by the invertebrates he simply remarked, "I regarded and still regard the presence of some scattered fragments of Cretaceous shells as of little moment in comparison with the well-marked characters of the flora, characters which have been wholly established by a large number of specimens obtained from all the localities referred to the Lignitic."

Cope, in his report for this same year, was inclined to agree with Hayden in thinking that the period of the deposition of the sediments of this Lignite or Fort Union group, as it was also called, was one of transition from marine to lacustrine conditions and added: "It appears impossible, therefore, to draw the line satisfactorily without the aid of paleontology, but here, while evidence of interruption is clear from the relations of the plants and vertebrate animals, it is not identical in the two cases, but discrepant."

He then went on to discuss the evidence as given by the various workers, and, referring to Lesquereux and Newberry's opinions, based upon the flora, to the effect that the whole series of formations is of Tertiary age, summed up his results as follows:

> I regard the evidence derived from the mollusks in the lower beds and the vertebrates in the higher as equally conclusive that the beds are of Cretaceous age. There is, then, no alternative but to accept the results that a Tertiary flora was contemporaneous with a Cretaceous fauna, establishing an uninterrupted succession of life across what is generally regarded as one of the greatest breaks in geological time.

Practically the same conclusion was independently arrived at by King in his study of the Green River coal basin in connection with the surveys of the fortieth parallel.[1] He wrote:

> We have, then, here the uppermost members of the Cretaceous series laid down in the period of oceanic sway and quite freely charged with fossil relics of marine life; then an uninterrupted passage of conformable beds through the brackish period up until the whole Green River basin became a single sheet of fresh water.

It will be seen then that the transitional character of the beds was very generally recognized, the main point in dispute from now on being that of their Upper Cretaceous or Lower Tertiary age.

A. R. Marvine, in his report for this year (1873), discussed the problem, but somewhat guardedly, since his opportunities for ob-

[1] The third volume, 1870, p. 453.

servation had confessedly been somewhat limited. He wrote, after summarizing the opinions of others:

It must be supposed, then, that either a Cretaceous fauna extended forward into the Eocene period and existed contemporaneously with an Eocene flora, or else that a flora wonderfully prophetic of Eocene times anticipated its age and flourished in the Cretaceous period to the exclusion of all Cretaceous plant forms.

Again, and much more to the point, he wrote:

Much of the confusion and discrepancy has, in my opinion, arisen from regarding different horizons as one and the same thing. It must be distinctly understood that this group as it exists east of the mountains in Colorado is very different from, and must not be confounded with, the horizon in which much of the Utah and New Mexican lignite occurs, and which belongs undoubtedly to the Lower Cretaceous; and, further, that the extended explorations of Hayden and others would seem to prove almost conclusively that the Colorado lignite group is the direct southern stratigraphical equivalent of the Fort Union group of the upper Missouri, which is considered generally to be no older than the Eocene, while Newberry asserts that it is Miocene.

To Lesquereux's conclusions in the report for 1872, Newberry, in an article in the *American Journal of Science,* 1874, took exception, calling in question the accuracy of many of his statements, and affirming that, to his "certain knowledge," a considerable portion of the flora he called Eocene was really Cretaceous and another portion Miocene. Further than that, having spent nearly two years in New Mexican explorations, he felt authorized to state that all the lignite beds yet known in New Mexico were unmistakably of Cretaceous age. While, through lack of acquaintance with the Colorado localities cited by Lesquereux, he would not venture to doubt the truth of his assertions regarding them, he nevertheless reminded him that: (1) The flora of the Colorado lignite beds had almost nothing in common with that of the European Eocene, in his judgment not a single species and scarcely any genera being common to both; (2) that the fucoid *Halymenites,* considered by Lesquereux as diagnostic of the Eocene, was really in New Mexico the most characteristic fossil of the Cretaceous; and (3) that, guided by their animal remains, Professors Marsh, Meek, Cope, and Stevenson had all regarded the Colorado lignites as Upper Cretaceous.

Newberry further contended that in the plant beds which he had himself designated Miocene the entire aspect of the flora was identical with the Miocene of Europe, and contained a very considerable number of well-marked Miocene species, not one of which deserved to be called Eocene. The lignite plant beds of New Mexico, which he called Cretaceous but which Lesquereux referred to the Tertiary, were for the most part derived from the lower portions of our Cretaceous series, and were overlaid by many hundreds of feet of unquestionable Cretaceous strata in which all the typical Cretaceous forms were represented.

He further announced the principle that:

> In the absence of any distinctive or unmistakably Eocene plants, if the strata which contain them (the lignite deposits) shall be found to include vertebrates or mollusks which have a decidedly Mesozoic aspect, we shall have to include them in the Cretaceous system.

To Lesquereux's claim that the testimony of his 250 species of fossil plants far outweighs that of the Cretaceous mollusks, he rejoined that these plants were probably all distinct from European Cretaceous and Eocene species and had little or no bearing on the question in hand. He acknowledged it was not impossible that the physical condition of the continent may have been such that the Cretaceous age faded gradually into the Tertiary, and that consequently some forms of Cretaceous life might be found interlocking with those of Tertiary age, but of this he demanded proof, and asserted that as yet none such had been found.

J. J. Stevenson, meeting with these same beds in his work in connection with the Wheeler survey (1874), pronounced in favor of their Upper Cretaceous age. As for the fucoid *Halymenites*, he agreed with Newberry that it was not indicative of an Eocene age, since it was never found with any but a Cretaceous fauna. Neither was the evidence of the land plants acceptable to him, the materials being mostly single leaves in a state of preservation showing that they had been blown from trees growing near streams or on the shore, where they were washed into the sea and became associated with at least one fossil identified with the Cretaceous, and a few of which were identical with European species.

It was more reasonable, he thought, to suppose that in the later portion of the Cretaceous period the climate in our northwestern region was like that of the European Eocene than to imagine that our Cretaceous fauna is useless for determination of horizon in the narrow strip east of the mountains in Colorado, while acknowledging it to be decisively of Cretaceous age in New Mexico, the rocks being the same, but the leaves being absent.

To Newberry's criticisms Lesquereux replied[2] that the Cretaceous age of the so-called halymenites sandstone had not yet been proven and could not be decided on mere affirmation; but that when Dr. Newberry had furnished sufficient proof or evidence on the geological age of the lignites of New Mexico he was prepared to accept his decision. The quoted opinions of Marsh, Cope, and Stevenson he claimed were based on insufficient evidence. All the repeated assertions of the finding of fossil shells and bones of Cretaceous age in the lignites of Colorado, when carefully sifted down, reduced themselves in his opinion to the finding of a single badly preserved specimen of *Inoceramus.*

He asserted that the lignitic formation being composed wholly of detritus from the land or a *land formation,* the evidence presented by the fossil plants should outweigh in importance that of some Cretaceous animal remains, the presence of which could be considered as of casual occurrence. Cope's conclusions, he argued, did not in the least interfere with his own, simply proving the noncoincidence of animal and vegetable types in certain formations, but if, he added, "Tertiary and Cretaceous faunas are regarded as contemporaneous, even inhabiting the same repositories, we may more easily admit that a Cretaceous fauna and Tertiary flora have sometimes succeeded each other in alternating strata."

Lesquereux further contended that the specimens on which Dr. Newberry had relied to substantiate the sum of the opinions he advanced had become mixed and had, in reality, come from different localities and represented different horizons.

In his report for 1874 (letter of transmittal written in October, 1875) Hayden took up the matter once more, with particular reference to the results of investigations in Colorado. These he felt warranted him in drawing the following conclusions:

First, that through the upper portion of the Fox Hill group (Cretaceous) there are clear proofs of a radical physical change, usually with no break in the sequence of time. In this portion of the group are well-marked Cretaceous fossils of purely marine types and no others.

Second, that above the Fox Hill group there are about 200 feet of barren beds, which might be regarded as beds of passage to the lignitic group and which more properly belong with the Fox Hill group below. In this group of transition beds all trace of the abundant invertebrate life of the great Cretaceous series below has disappeared.

2 American Journal of Science, 1874.

Third, in almost all cases he found, at the base of the true lignitic group, a bed of sandstone in which the first deciduous leaves peculiar to the group occur. No purely marine mollusks pass above this horizon. Estuary or brackish-water shells are found in many localities in great abundance, but these soon disappear and are succeeded farther north by fossils of purely fresh-water origin. He added:

Whatever view we may take in regard to the age of the Lignitic group, we may certainly claim that it forms one of the time boundaries in the geological history of our western continent. It may matter little whether we call it Upper Cretaceous or Lower Eocene, so far as the physical result is concerned. We know that it plays an important and, to a certain extent, an independent part in the physical history of the growth of the continent. Even the vertebrate paleontologists, who pronounce with great positiveness the Cretaceous age of the Lignite group, do not claim that a single species of vertebrate animals passes above the horizon I have defined from the well-marked Cretaceous group below.

Peale, in his report for that year, threw a ray of light upon the subject by suggesting that the reason of the difference of opinion as to the age of the disputed beds might be the existence of two sets of lignite-bearing beds close together, one belonging to the horizon of the Fox Hill beds of the Cretaceous, or possibly a little above it, and the other belonging to the horizon of the Fort Union group (Lower Eocene). He incidentally called up the question relative to the value of different types of fossils as criteria in determining the precise geological horizon.

He summed up his own conclusions as follows:

First. The lignite-bearing beds east of the mountains in Colorado are the equivalent of the Fort Union group of the upper Missouri, and are Eocene-Tertiary; also, that the lower part of the group, at least at the locality two hundred miles east of the mountains, is the equivalent of a part of the Lignitic strata of Wyoming.

Second. The Judith River beds have their equivalent along the eastern edge of the mountains below the Lignite or Fort Union group, and also in Wyoming, and are Cretaceous, although of a higher horizon than the coal-bearing strata of Coalville and Bear River, Utah. They form either the upper part of the Fox Hill group or a group called "Number Six."

In his annual report to Hayden for the same year (printed in 1875), Lesquereux went over all the ground once more, showing to his own satisfaction, presumably, the conclusive character of the evidence offered by fossil plant remains, and announced again his conviction to the effect that the authority of animal remains should be unquestioned so far as it relates to marine formations, but when land formations are considered, the plant remains should be given precedence.[3]

Meanwhile, Dr. C. A. White, a paleontologist, became connected with the Hayden survey, and was assigned for his first season (1877) to the area of northwestern Colorado. In this connection he came quickly in contact with the problem under discussion, and early committed himself in favor of the post-Cretaceous age of the beds.

He argued that it was a well-known fact that the evolutional advance of the vegetable kingdom had been greater on this continent than in Europe. Hence, a student of the flora of the American strata, using a series of European standards, would naturally refer those which he found to contain certain vegetable forms to the Tertiary period, while the associated or superimposed remains of animal life might all show them to be of Cretaceous age, according to the same series of European standards.

Taking into consideration the fact that the physical changes which took place in western North America during the Mesozoic and Cenozoic periods were very gradual and without any important break, he would be led to expect to find those animals whose existence was not necessarily affected by a change from a saline to a fresh condition of the waters, to have propagated their respective types beyond the period which those types in their culmination distinctly characterized. For this reason he felt that these perpetuated types being evidently the last of their kind did not necessarily prove the Cretaceous age of the strata, and moreover, all the other known fossil remains of the group indicated a later period. He would, therefore, refer the beds, which had now, according to an agreement between King and Hayden, become known as the *Laramie*, to a post-Cretaceous age.[4]

Referring again to the matter in his report for 1877, White conceded that Cretaceous types (dinosaurs) of vertebrate animals were

[3] Marsh, it may be noted incidentally, in 1877, in discussing the problem of the boundary between the Cretaceous and Tertiary, announced that in his opinion the evidence of the numerous vertebrate remains was decisive in favor of the Cretaceous view.

[4] Conformably over this [the Fox Hill Cretaceous] lies the group which Hayden and I have agreed to call the Laramie, which is his Lignitic group, and is considered by him as a transition member between Cretaceous and Tertiary. There is no difference between us as to the conformity of the Laramie group with the underlying Fox Hill. It is simply a question of determination of age upon which we differ. (King, Vol. I, Reports Fortieth Parallel Survey, 1878, p. 348.)

found in the higher strata of the Laramie group[5] and did not question the correctness of referring the plant remains even of the very lowest beds to the Tertiary; noting also that the invertebrate fossils were indecisive, since the species were new to science and could not be safely compared with those found elsewhere. Without committing himself, he then offered a suggestion in effect as follows: Since none of the American Cretaceous could be considered as equivalent to the Lower Cretaceous of other parts of the world, but must be considered as Upper Cretaceous, these Laramie rocks which, if Cretaceous at all, were certainly at the very top of the Upper Cretaceous, must represent a great and important period wholly unrepresented in any other part of the world.

Further, the finding of evidence of an abundant mammalian life (Tertiary) immediately following the Laramie period, which in itself contained only Cretaceous dinosaur remains, suggested to him a sudden ushering in of the Tertiary types which could be accounted for only on the supposition that they originated elsewhere, and were contemporaneously in existence with the Cretaceous forms, *i.e.*, prior to the close of the Laramie period. Their apparent sudden appearance could then be accounted for on the supposition that the physical barrier was removed by some of the various earth movements connected with the evolution of the continent.

To these same opinions he held in his report for 1878, adding that, if the conclusions of all the leading paleontologists regarding the Eocene-Tertiary age of the Wasatch, Green River, and Bridger groups be accepted, "then is there additional evidence of the correctness of the view that the Laramie is a transitional group between the Cretaceous and Tertiary, partaking of the faunal characteristics of both periods."

Hayden, in his report for this year (1878), argued that the Fort Union beds of the upper Missouri River were the equivalent of the lignitic formation, as it exists along the base of the Rocky Mountains in Colorado, and of the Bitter Creek series west of the mountains; also that it was probable that the brackish-water beds on the upper Missouri must be correlated with the Laramie, and that the Wasatch group, as then defined, and the Fort Union group were identical as a whole, or in part, at least.

In his report on the systematic geology of the Fortieth Parallel Survey (1878), King again attacked the advocates of the Eocene

[5] It should be remembered that White included with the Laramie the Fort Union formation, which was the source later on of interminable error.

age of these beds. He reviewed the evidence and announced his convictions, as before, in favor of their being Cretaceous.

He agreed with Hayden in regarding the Laramie and underlying Fox Hill as strictly conformable, but found a very decided unconformability between the uppermost Laramie containing the dinosaur remains and the immediately overlying rocks of the Vermilion Creek group, which carried mammalian remains. The unconformity here mentioned for the first time is, with the exception of the suggestion made by Peale, perhaps the most important feature yet introduced into the discussion, and it is probable, as elsewhere suggested by King, that had Hayden seen this locality earlier, the question as to the exact position of the Laramie beds might never have arisen.

His conclusions, as summed up in his recapitulation of the Mesozoic, were as follows:

> The Laramie, by its own vertebrate remains, is proved to be unmistakably Cretaceous and the last deposit of that age, and it contains no exclusively fresh-water life. Its plants resemble European Tertiary, but its Dinosaurs are conclusive of Cretaceous age. It was the last of the conformable marine deposits of middle America. Its latest period of sedimentation was immediately followed by an energetic orographic disturbance, which closed the Mesozoic age. In that orographic action the inter-American ocean was obliterated and the Cretaceous locally thrown into great and steep folds. The following deposits over the Green River area were fresh-water lacustrine lowest Eocene strata, lain down nonconformably with the Cretaceous, except in accidental localities.

In 1885 the problem was taken up by L. F. Ward in an exhaustive paper on the flora of the Laramie group. This, although destructive in its criticism rather than decisive, nevertheless contained many important suggestions. After a summary of opinions held by previous workers, he wrote:

> Taking all these facts into consideration, therefore, I do not hesitate to say that the Laramie flora as closely resembles the Senonian (Cretaceous) flora as it does either the Eocene or the Miocene flora, but I would insist that this does not necessarily prove either the Cretaceous age of the Laramie group or its simultaneous deposit with any of the Upper Cretaceous beds. The laws of variation and geographical distribution forbid us to make any such sweeping deductions. With regard to the first point, it is wholly immaterial whether we call the Laramie Cretaceous or Tertiary, so long as we correctly understand its relations to the beds above and below it. We know that the strata immediately beneath are recognized Upper Cretaceous, and we equally know that the strata above are recognized Lower Tertiary. Whether the first intermediate deposit be known as Cretaceous or Tertiary is, therefore, merely a question of name, and

its decision one way or another can not advance our knowledge in the least.

In this, it will be noted, he followed closely the opinions already expressed by Hayden, but as now known Ward's Laramie included Montana (Mesaverde) Arapaho, Denver, Laramie, Lance, and Fort Union. It is therefore not surprising that he could come to no decision. As Newberry said at the time, Ward's *Laramie Flora* was in the main a contribution to the Fort Union flora.

Other statements of Ward's in this connection are worthy of consideration. He pointed out that there was no probability that the conditions existing during the Laramie deposition would be ever exactly reproduced elsewhere, and hence the chances were as infinity to one against the existence of other beds that should contain an invertebrate fauna identical with that of the Laramie group. Further than this, he regarded the law laid down by paleontologists that the same epochs in geologic time produced the same living forms, as quite contrary to the now well-established principles of geographical distribution.

Amidst all this confusion and conflict of opinion, the fact was becoming more and more apparent that under the term *Laramie* had been included beds belonging to various but not widely separated horizons. King and Hayden, it will be recalled, had believed it to contain all the coal beds of the region in which it occurs and to be sharply circumscribed both above and below, the main point in dispute being whether it should be relegated to the Cretaceous or the Tertiary. As time went on, but particularly after the organization of the present survey (*i.e.*, after 1879), more careful and detailed work became possible. The Bear River beds were shown by Drs. C. A. White and T. W. Stanton, in 1891, to belong to the marine Cretaceous; the upper portion in Colorado and Wyoming was found by Cross, Hills, and Weed to be out of harmony with that beneath, and was relegated to the Eocene-Tertiary; the Laramie proper becoming more and more restricted as the work of differentiation went on.

In 1897 Drs. F. H. Knowlton and T. W. Stanton, the one a paleobotanist and the other a paleontologist, together made a personal inspection of many of the important localities, and, after passing the evidence in review, concluded that the so-called Ceratops beds of Converse County, Wyoming, should be referred to the Laramie group, though later, when it was abundantly demonstrated that the Ceratops-bearing beds were separated from the underlying beds by a widespread unconformity, Knowlton changed his views and concluded that they have no relation to the true Laramie but are Tertiary in age, and that the coal-bearing series of the Laramie Plains is older than the "true" Laramie, and, instead of conformably overlying the Fox Hills group, is itself overlain by it.

The Bitter Creek and Black Buttes beds they considered as belonging to the "true" Laramie, and also those of Crow Creek, Colorado, while those of Point of Rocks, in the Bitter Creek Valley, were regarded as Cretaceous (Montana). The base of the Laramie (after a review of the opinions of Hatcher, Hills, King, and Hayden) they would place "immediately above the highest marine Cretaceous beds of the Rocky Mountain region," the top being marked by the Fort Union beds. In other words, the Fort Union beds are now regarded as Eocene and the lower-lying as Laramie Cretaceous.

Both these workers, it is well to note, concurred with the generally received opinion that "marine invertebrates (fossils) are more accurate and definite horizon markers than either plants or nonmarine invertebrates, because they have a less extended vertical and a broader geographical distribution."

O. C. Marsh, the reader will perhaps remember, had in 1891 announced the general principle that all forms of animal life are of value as horizon markers "mainly according to the perfection of their organization or zoölogical rank." Following out this principle, he regarded plants as unsatisfactory witnesses, invertebrates as much better, and vertebrates as the best of all, as offering "reliable evidence of climatic and other geological changes." Dana had expressed the same view, so far as plant remains were concerned, as early as 1876.

Today (1923) the Laramie is restricted to the type locality along the Front Range, especially in the Denver Basin, and is defined in accordance with the original definition of King as the uppermost part of the conformable Cretaceous series above the Fox Hills. The Laramie is also accepted as occurring in Carbon County, Wyoming, where it is 6,000 feet thick, whereas in the Denver Basin it is only about 1,200 feet thick. The Ceratops beds are now known as the Lance formation and most geologists recognize an unconformity between the Lance and underlying formations, this unconformity in many places cutting down even into the Pierre. Sedimentation was continuous and uninterrupted between the Lance and the overlying Fort Union. The Lance formation contains besides the well-known Ceratopsia a few marine and brackish-water Cretaceous shells in the lower portion, but in addition it holds an abundant and widespread Tertiary flora with a majority of its species in common with the Fort

Union beds. Whether the Lance is to be considered as Tertiary or Cretaceous now becomes a matter of principles involved in distinguishing the great periods of time. From the standpoint of diastrophism and on the evidence of paleobotany the Lance formation is certainly of Tertiary age. Ceratopsia and the few fossil shells ally it to the Cretaceous. However, in dealing with fossils many paleontologists believe that the introduction of a new fauna is of more importance in distinguishing the age of the rocks than the long enduring old species. Following this line of evidence the Lance formation (Ceratops beds), in spite of its Cretaceous appearing dinosaurian remains, must be regarded as of Tertiary age.

CHAPTER XII.

The Taconic Question.

"Now, who shall arbitrate?
Ten men love what I hate,
Shun what I follow, slight what I receive;
Ten, who in ears and eyes
Match me: we all surmise,
They, this thing, and I, that: whom shall my soul believe?"

Browning.

IT is presumably scarcely necessary to call attention to the fact that the older rocks of the earth's crust are exposed, in a majority of cases, only where this crust has been disturbed through such folding and faulting as is incidental to mountain making. As a result of such processes these older rocks are, in the main, considerably altered and their origin as well as geological age at times quite indeterminable.

The attempt to fix the base of the Paleozoic strata or, in other words, to find a line of demarkation and division between the non-fossil-bearing and the most ancient members of the overlying fossiliferous strata has, therefore, proven a matter of the greatest difficulty both in America and abroad. In Great Britain and on the Continent manifestation of this is found in the voluminous literature relating to Sedgwick's Cambrian and Murchison's Silurian systems. In America a similar controversy was contemporaneously waged, which has come down to history under the name of the Taconic question.

In his report on the second geological district of New York, published in 1842, Ebenezer Emmons gave his first detailed account of the Taconic system. A review of the subject may well, therefore, begin with this paper, though an occasional earlier reference may be necessary.

In his report for 1838 Emmons had stated that the Potsdam sandstone was the oldest sedimentary rock occurring in the vicinity of Potsdam (New York) and there was nothing intervening between it and the Primary. In this opinion, to which he ever afterward adhered, he was quite correct. Overlying the Potsdam sandstone and

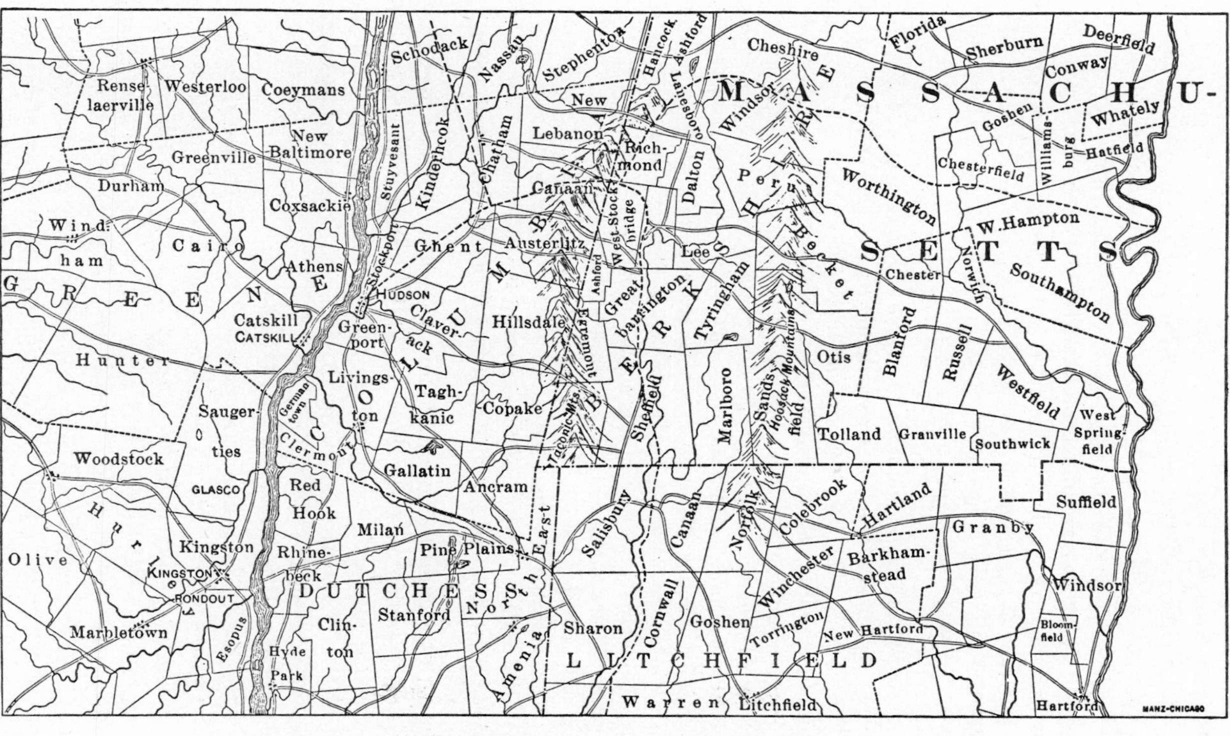

Fig. 119. *Map of the Original Taconic Area.*

always in the same order he found the Calciferous sandrock, the Chazy, Bird's-eye, and Trenton limestones, the Utica and Hudson River slates, etc.

An examination of the country at the foot of the Hoosac Mountains in western Massachusetts showed what appeared to him an entirely different series resting, like the Potsdam, directly upon the gneiss, but in which none of the rocks of the previous series appeared. This series was described in his own words as below, the order of superposition being shown in the section reproduced herewith, which the reader will note is given in a reversed position, the left being east and the right west (Fig. 120):

1. A coarse granular limestone of various colors which I have denominated the *Stockbridge limestone,* taking its name from a well-known locality, one which has furnished the different parts of the Union a large portion of the white and clouded marbles which have been so extensively employed for building and other purposes in construction.

2. *Granular quartz rock,* generally fine grained, in firm, tough, crystalline masses of a brown color, but sometimes white, granular, and friable.

3. Slate, which for distinction I have denominated *Magnesian slate,* from its containing magnesia, a fact which is distinctly indicated by the soft feel peculiar to rocks when this earth forms a constituent part.

4. Sparry limestone, generally known as the *Sparry limerock.*

5. A slate, which I have named *Taconic slate,* and which is found at the western base of the Taconic range. It lies adjacent to the Lorrain or Hudson River shales, some varieties of which it resembles. In composition it contains more alumina and less magnesia than the magnesian slates.

The series occupied an area extending from the Hoosac Mountains westward, passing over the Hoosac Valley, Saddle Mountain, and also over the high ridge of granular quartz known as Oak Hill, just north of the Williamstown Plain, the Taconic Mountains next west of the Massachusetts boundary, and the eastern border of New York west of this boundary to the Hudson.

To this series, which Emmons conceived to be older than the Potsdam, he proposed in 1841 to give the name Taconic, after the Taconic range, elevating it to the dignity of a system. His persistent advocacy of the actuality of this system gave rise to a controversy extending over more than half a century and equaled by none in the annals of American geology, not exceeded even by the Eozoon question noted elsewhere. Unmoved by argument, to the day of his death Emmons adhered faithfully to his "system," although from the very first he noted its most inherent weakness—that in no case had the Potsdam sandstone been found resting upon any of its members.

That this system was actually an older, lower-lying series was indicated only by the fact that neither were any of its members found intercalated with the overlying series, which always occupied the position and relationship given above.

It must be remarked, by way of preliminary explanation, however, that the region, as shown by subsequent studies, is one where faulting, folding, erosion, and metamorphism have prevailed to an extent then undreamed of, and where, as a consequence, all natural criteria had become so obscured as to make a prompt solution of the problem

Fig. 120. *Section of Taconic Rocks.*

impossible. The science of geology has grown through cumulative evidence. The greatest minds and the most acute of observers were bound, by the then existing condition of knowledge, to make incomplete observations and faulty deductions.

The original section, as noted, was some fifteen miles in length. The rocks were all pronounced unfossiliferous and dipping throughout to the eastward, their relative ages being judged wholly by superposition, the Stockbridge limestone, where it rested immediately upon the gneiss, being considered the oldest of the series.

The first announcement of this conclusion appears to have been made at the Philadelphia meeting of the American Association of Geologists and Naturalists, held in April, 1841. Unfortunately neither the paper nor the discussion which followed was printed in detail. It is stated by Dana, however, that the matter was discussed by H. D. Rogers, Edward Hitchcock, William Mather, James Hall, and Lardner Vanuxem, all of whom had worked in the region. It is stated, further, that none of the gentlemen, with the exception of Vanuxem, favored the views put forward by Emmons.

During the summer following both Rogers and Hall studied the section in the field, Rogers rendering a report to the American Philosophical Society at Philadelphia at the meeting in 1842. In this he reiterated the views previously expressed by Hitchcock, Hall, Mather, and himself, to the effect that the rocks were Lower Silurian (as the term was then used) extending from the Potsdam upward, but much flexed and disguised by partial metamorphism. Hall for some reasons failed to make a report at the time, though later claiming to have written out his notes very fully (see p. 608).

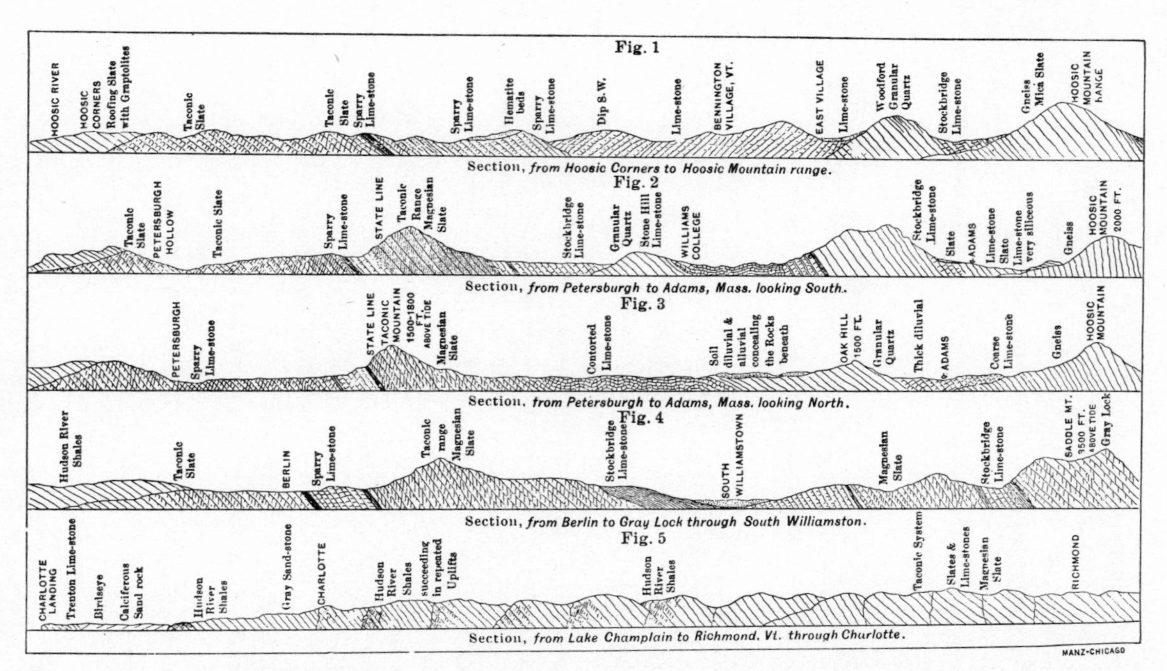

Fig. 121. *Sections Explanatory of the Taconic. (After Emmons.)*

In his paper in the *Report on the Geology of the Second District of New York*, in 1842, where the system is first elaborated, Emmons referred to Rogers's conclusions, and seemingly himself recognized the possibility of the various beds of limestone being but portions of the same bed, brought to the surface by successive uplift. He did not, however, regard this as probable. The system as a whole he thought to be the equivalent of the Lower Cambrian of Sedgwick, the upper portion being the lower part of the Silurian system.

Although the proceedings of the American Association of Geologists were not given in full, as already noted, the opinions of Mather have fortunately been handed down in his *Report on the Geology of the First New York District*, dated 1843. In this he made use of the term in his descriptions, but stated emphatically that the "Taconic rocks are the same in age as those of the Champlain division, but modified by metamorphic agency and the intrusion of plutonic rock."

In December, 1844, Emmons brought out in pamphlet form a revision of the Taconic system, with additions and an extension of its limits. This was published without change in his *Report on the Agriculture of New York*, under date of 1846. The most important feature of this revision related to the finding of fossil crustacean remains in the Black slate of Bald Mountain, in Washington County. These he accounted for on the supposition that the beds, instead of being the lowermost, as he first supposed, belonged in reality to the top of the series, and had come into their present position through a reversion of the strata, adopting thus in part Rogers's ideas as to flexures and overthrust faults. The order of succession of the strata would then stand as below, that given in 1842 being introduced for comparison. It will be at once noted, as later charged by Dana (1888), that "the system is for the most part turned the other side up."

Taconic system, 1842.	*Taconic system, December, 1844.*
6. Stockbridge limestone.	6. Black slate, Bald Mountain.
5. a. Magnesian slate of Greylock—perhaps a repetition of No. 3. b. Granular quartz.	5. Taconic slate.
4. Limestone.	4. Sparry limestone.
3. Magnesian slate of Taconic Mountain.	3. Magnesian slate.
2. Sparry limestone.	2. Stockbridge limestone.
1. Taconic slate.	1. Granular quartz.

As here outlined, Emmons argued that the system occupied a position inferior to the Champlain division of the New York system, or the lower division of the Silurian of Murchison.

The fossils found in the Black slate, it should be mentioned, were trilobites which Emmons described under the names *Atops trilineatus*, a form allied to *Triarthrus beckii* and *Elliptocephala asaphoides*. In addition there were annelid markings, chiefly those of *Nereites* and *Myrianites*. These were subsequently (in 1846) redescribed by Hall, who contended that the *Atops* and *Triarthrus* were identical and referable to the Hudson River group, while the *Elliptocephala* should be referred to the genus Olenus,[1] an Upper Cambrian form.

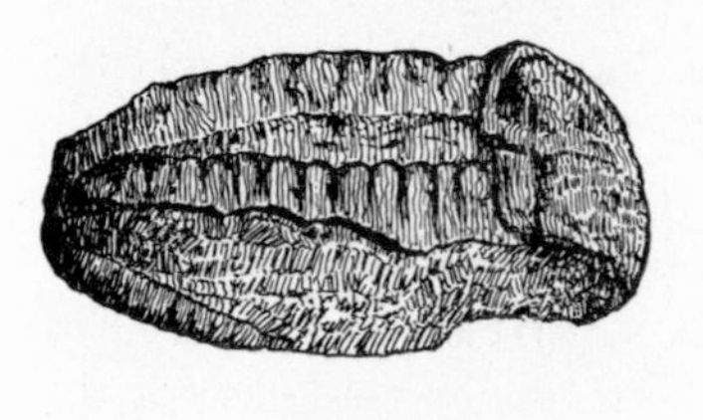

Fig. 122. *Atops Trilineatus.*

At the Boston meeting (1847) of the Association of American Geologists and Naturalists, Professor C. B. Adams discussed the Taconic rocks of the northern part of Addison County, Vermont, a locality judged to be particularly favorable for study, since the rocks pass here from a highly to but a slightly metamorphosed and disturbed condition. He exhibited sections of Snake and Bald mountains and from Lake Champlain to the Green Mountains, and thought to show that the Taconic quartz rock was probably the metamorphic equivalent of the red sandrock which he regarded as belonging to the Champlain division of the New York geologists, and that the Stockbridge limestone was the equivalent of the calcareous rocks overlying the red sandrock, rather than of the lower limestone of the Champlain division.

Inasmuch as the value of the evidence furnished by the fossil remains had become largely a matter of individual opinion, a committee, of which S. S. Haldeman was chairman, was appointed at the same meeting of the association to investigate. Their report, given in the *American Journal of Science* for 1848, was to the effect that so far as could be determined from the fragmental character of the specimens submitted, *Atops* and *Triarthrus* were not identical, but *A. trilineatus* was "a fossil characteristic of the stratum investigated and named by Professor Emmons."

[1] Later, *Olenellus* of the Lower Cambrian.

To this report Hall naturally took exception, and in the *Journal* for the same year reviewed the subject and published figures giving reasons for thinking that the distinctions noted by the committee were not "actual and constant," but merely those of individuals, and reaffirming the statement that the two forms were essentially identical. Hall was in turn replied to by Emmons, but nothing conclusive was brought forward. The last-named, however, in a paper presented

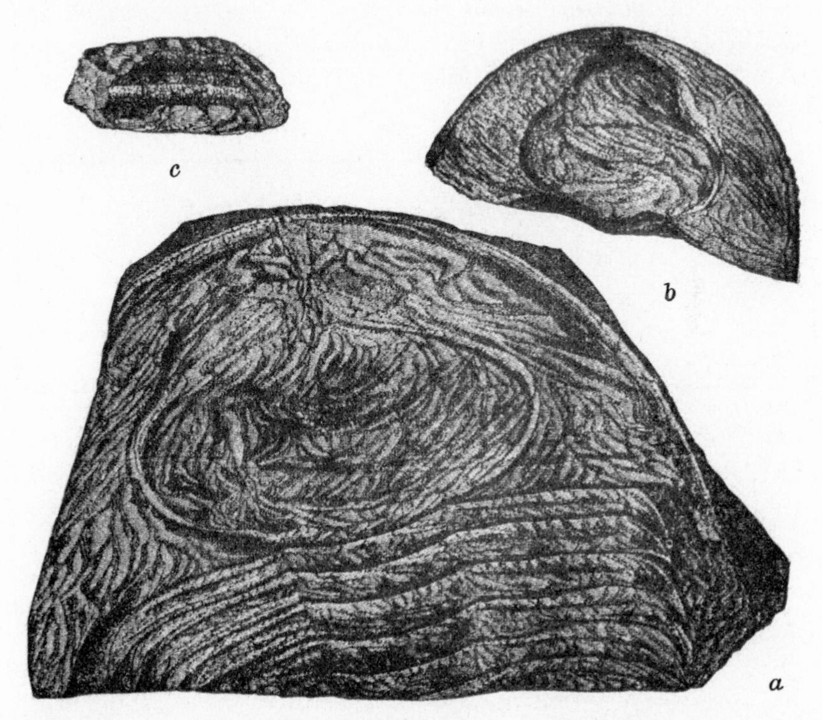

Fig. 123. *Elliptocephala asaphoides.*

before the American Association, this same year, stated: "I wish it to be fully understood that the separation of the Taconic Rocks from the New York, or Silurian, is called for on grounds of far more importance than the presence of certain fossils; that evidence is found in the succession and superposition of the two systems, and in the vastness of the Taconic system compared with the Silurian."

In 1849 the ubiquitous Sterry Hunt entered the field, siding, as it chanced, with the opinions of the opponents of the system, though later he shifted his ground. No new facts were, however, presented, and the matter is mentioned here only on account of the characteristically emphatic and apparently decisive manner in which his opinions were expressed, however fanciful may have been their basis.[2]

Little of consequence now occurred until 1855, when Professor Emmons brought out his volume on American geology, in which he made his third presentation on the subject. In this he extended the system from Maine to Georgia and subdivided it into an upper and lower portion, the fossiliferous portion being called the Upper Taconic and the nonfossiliferous the Lower Taconic. The Sparry and Stockbridge limestones were brought together as one formation, while the synclinal character of Mount Greylock was recognized and figured.

Taconic system in 1855.

Upper Taconic	2. Black slate of Bald Mountain.
	1. Taconic slate.
Lower Taconic	3. Magnesia slate.
	2. Stockbridge limestone.
	1. Granular quartz.

As here given, the system, as stated by Dana with reference to that of 1844, had a top and bottom of Cambrian rocks. The succession in the Lower Taconic was the same as in the publication of 1844.

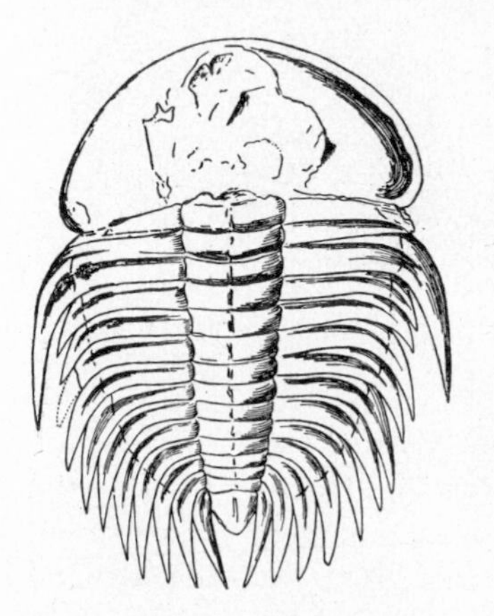

Fig. 124. *Olenus Thompsoni.*

In 1854 or 1855 trilobites related to those of Bald Mountain were found in the black slates of West Georgia, Vermont. Passing into the hands of Zadock Thompson, the assistant in the geological survey of the state, they were sent to Professor Hall, and in 1859 figured and described under the names *Olenus thompsoni*, *O. vermontana*, and *Peltura holopyga*, and the beds made equivalent to the Bald Mountain slates already noted. In this he was in error.

The problem was not, however, limited to the United States. In 1856 there were found in the limestone belonging to the so-called graywacke series at Point Levis,

[2] Hunt's views may be found *in extenso* in his Mineral Physiology and Physiography, 1886, pp. 517-686.

opposite Quebec, a trilobite fauna of a nature sufficient to convert the paleontologist, Elkanah Billings, of the Canadian survey, to the views of Emmons, and cause him to affirm that the graywacke group lay below the Trenton in stratigraphic position.[3] In this view Logan at first acquiesced, but applied to the beds as there developed the name *Quebec* rather than Taconic, subsequently extending the name to the whole belt of Taconic rocks reaching from the St. Lawrence to the Hudson River.

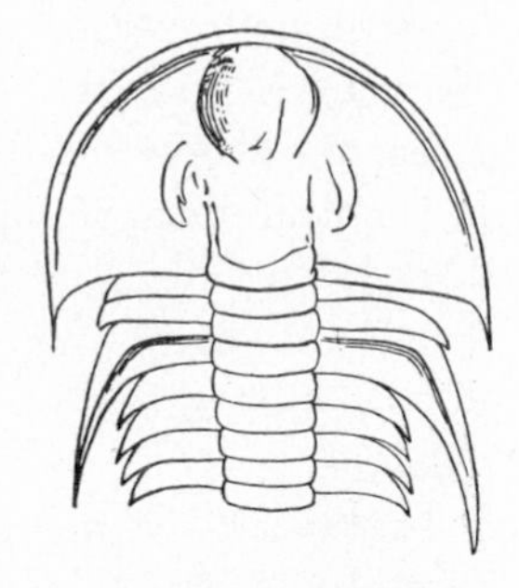

Fig. 125.
Olenus Vermontana.

In 1860 Barrande, the eminent Bohemian paleontologist and authority on the Silurian, read before the Geological Society of France a memoir in which he adopted in full the conclusions of Emmons and pronounced the Georgia, Vermont, trilobites unquestionably of Primordial age and characteristic of a great Taconic system extending far below the *Olenus* or *Paradoxides*[4] zone.

Subsequent discoveries seemed to show that Barrande was misled through the character of the evidence available, as he had not studied the question in the field. Be this as it may, his accession to the ranks of the "Taconists" for a time greatly strengthened their cause, through sheer weight of authority, and did much to complicate the situation, while the use made of his writings and personal letters by

3 Subsequent investigations have shown that the masses in which these fossils were found were but blocks or fragments in a conglomerate, itself a member of the Quebec group. Billings's conversion was, however, thorough, as shown by the following:

MONTREAL, *10th Nov., 1860.*

MY DEAR MR. MEEK:

. . . The discovery of the Point Levi fossils, and also of those published by Hall under the name of *Olenus* (from Vermont), opens out a new field in American geology. I have received several letters from Barrande on the subject. In the last one received, two days ago, he says that Angelin was then in Paris, and that he had shown him both Hall's figures and mine. Angelin agrees with him that Hall's species, and also mine from limestone No. 1, are of the Primordial type. The Quebec rocks are undoubtedly the Taconic rocks of Emmons. It will be rare sport if Emmons, after standing alone for 25 years against the majority of the *élite* (or at least those who consider themselves the *élite*) of American geologists, should after all be right. His unfortunate Taconic system has been annihilated and proved not to exist regularly once a year for the last twenty years, and yet it once more raises its head with a new life. I firmly believe that in the main he is right, although he may be wrong in some minor details. . . .

E. BILLINGS.

4 The Paradoxides is a trilobite found in Sedgwick's Middle Cambrian, of England.

Marcou swelled the literature and confused the question until for a time the correct solution seemed hopeless. And here it may be remarked that, however conclusive and convincing the writings of Marcou may appear, the arguments he advanced were founded almost altogether upon the works of others, or, in some cases, upon no other basis than prejudice in favor of or against certain individuals. From the time of his Pacific Railroad survey work Marcou did little in the field, but contented himself with sitting in judgment upon the work of others. For this reason he was of little actual service to American geology and his writings and opinions are not quoted in detail, although unquestionably many of his suggestions were of value.

In 1861 there was issued the final report on the geology of Vermont by Edward Hitchcock and A. D. Hager. On paleontological evidence it was decided that the Stockbridge limestone was not older than Silurian. Cross sections showed Mounts Anthony, Equinox, Æolus, and others to have a synclinal structure, the limestone beneath and the slates above as Emmons had shown for Mount Greylock. The stratigraphy was, therefore, in favor of Emmons's view, but the paleontology was against it and no decision was reached concerning the age of the quartzite.

C. H. Hitchcock had at the March, 1860, meeting of the Boston Society of Natural History exhibited the geologic map of Vermont and incidentally had stated that he had found no basis for Emmons's theory, the Vermont rocks he had included in his system being in part Silurian and in part Devonian.

In 1863, the year of Emmons's death, there appeared the first edition of Dana's *Geology*. In this the Potsdam sandstone was made to include the Primordial and the equivalent of the era of the *Paradoxides* and the Primordial beds of Scandinavia and Bohemia. The Georgia slates were also recognized as Primordial, in this conforming to the ideas of Barrande and Emmons.

In 1869 J. B. Perry, of Vermont, entered the lists and argued in favor of the system as presented by Emmons. He was subsequently shown, through the discovery of fossils and the existence of numerous unsuspected faults and folds, to have been in error. The following year the Reverend A. Wing, of Vermont, working with the avowed intention of settling the vexed question as to the age of the limestone, slates, and quartzites in the West Rutland region, found fossils in the limestone which Billings identified as probably belonging to the Chazy epoch of the Canadian (Lower Silurian) period. Still others were found, sufficient to show that the beds range from the Calciferous to and including the Trenton, and that consequently the overlying slates must be of Utica or Hudson River age, and not limited to the Quebec group, as Logan had supposed.

Two years later Elkanah Billings published in the *Canadian Naturalist* an article on the age of the black slate and red sandrock of Vermont, in which he took occasion to refer to the Taconic system and announce his views on the subject. He contended that in the consideration of this question "nearly all of the leading geologists of North America" had ranged themselves upon the wrong side; that, while for nearly a quarter of a century Doctor Emmons stood almost alone, during the last thirteen years a great revolution of opinion had come about; and that the idea that the rocks of the Taconic system were really above the Potsdam sandstone (as had been contended) had been exploded. (See letter to Meek quoted on p. 603.)

Fig. 126. *Elkanah Billings.*

As he understood the matter at the time of writing, some of the Taconic rocks were certainly more ancient than the Potsdam, while others might be of the same age, and perhaps some of them more recent. The details, he felt, had not yet been worked out, and on account of the extremely complicated structure of the region, he ventured to say that no man at that time living would ever see a perfect map of the Taconic region. The present indications are favorable to this view of the subject.

The theory that the Taconic rocks belong to the Hudson River group, he went on to say, was an "enormous error" that originated in the geological survey of New York and thence found its way into the Canadian survey. The mistake was doubtless due to the extraordinary arrangement of the rocks, the more ancient strata being elevated and often shoved over the more recent, so that, without the aid of paleontology, it was impossible to assert positively that they were not the age of the Hudson River formation, as they appeared to be. The main object of his article was, as he acknowledged, to show that while the error had originated in New York, it was corrected by the geological survey of Canada.

This article brought out a reply by J. D. Dana in the *American Journal of Science* for June, 1872, in which he called Billings's attention to the fact that, while he, Dana, might differ with Billings about the Taconic, the differences, after all, were not material, since Billings viewed the Taconic as developed by Emmons through successive interpolations year after year, and not as first announced in 1842. He called his attention, further, to the fact that the system was based on a section fifteen miles long, made across the Taconic Range through Williamstown and Greylock, to North Adams on the east and to Petersburg or Berlin on the west; that the dip was originally throughout to the eastward; and that the beds were destitute of fossils and their relative age judged by superposition, according to which the Stockbridge or North Adams limestone—the most eastern rock in the section—would be the most recent.

Referring to Emmons's discovery of a fossiliferous black slate at Bald Mountain, New York, he stated that, according to Emmons's principle adopted in 1842, this slate being to the west of the Taconic, should be older than the Taconic proper, and therefore that his Taconic system was actually newer than the fossiliferous rock. This evidently being to his (Emmons's) mind impossible, he was thence led to think out a way by which rocks might dip eastward and still be newer to the westward and that, "without a fact or even an argument to sustain it, he announced in his agricultural report, published in 1843, this as the true order. He thus, by a stroke of his pen, tipped over the Taconic system and got the black slate to the top with all other Taconic rocks beneath it."

This black slate interpolation in 1843 thus brought mischief to the Taconic system and to much American geology and was called by Dana "a most desperate blunder." Billings's work he regarded as eliminating the black slates from the system. He went on to state that the quartzite, which in the publication of 1842 occurred toward the middle of the section, was, in that of 1843, placed at the bottom of the Taconic series. Hence in this "perfected Taconic" the rocks which Billings had shown to be nearest to the pre-Silurian of all the Taconic beds were placed at the remote ends of the system, the black slate at the top and the quartzite at the bottom, the former being of Primordial age and the only rock series which had yet proven to be pre-Potsdam.

Dana acknowledged that Emmons was deserving of honor for combating the old idea which had prevailed among geologists and paleontologists, to the effect that the Taconic slates belong to the Hudson River period; yet he contended that he "blundered in everything else," determining nothing correctly as to the age or order of succession of the rocks of the system, and "his assumptions after 1842 were so great as to order of stratification and faults, and his way of sweeping distant rocks into his system so unscientific, that his opponents had abundant reasons for their doubts." He went on to say that no one knew, even at that date, what the precise age of the slates of the Taconic Mountains might be, although Logan's view that they belong to the Quebec seemed nearest the truth. The only way, he argued, for geologists to get out of the Taconic perplexity was to go back to Emmons's original report and section of 1842. "The name Taconic," he wrote, "belongs only to the era represented by the rocks of the Taconic Mountain," and nowhere else.

In a series of articles in the *American Journal of Science*, beginning with December of 1872, Dana showed the conformability of the Taconic slates and schists of the Taconic Mountains and the Stockbridge limestone and quartzite, and on the basis of the discoveries of Wing and Billings, pronounced the limestone to be of Trenton and Chazy age and the schists and slates to be of Hudson River age, in this agreeing with Rogers. He also pointed out that the same beds of metamorphic rocks might vary as do their unconsolidated representatives, being quartzite in one part and mica-schist or even gneiss elsewhere, and hence that purely lithological evidence as to the identity of beds was practically worthless. This is a little amusing as coming from Dana, who himself accepted the presence of the mineral chondrodite in the limestones of Berkshire as evidence of their Archean age.

In 1878, T. Nelson Dale found brachiopods belonging to the Hudson River group in the Taconic slates at Poughkeepsie. In this same year W. B. Dwight began work in the "Sparry limestone" of Dutchess County, New York, finding fossils of undoubted Lower Silurian age.[5] These finds and others made by S. W. Ford and I. P. Bishop in adjacent localities were made use of by Professor Dana in his subsequent papers.

Dana continued in the field at intervals until 1889, accepting as his working basis the Chazy fossils found at West Rutland by Dr. Wing, with whom he worked throughout the period. In 1879 he showed on stratigraphic and fossiliferous evidence that the Taconic schists, so called, as developed in Dutchess and adjacent counties, were of the age of the Hudson River group, and the five limestone belts there found but five successive outcroppings of the Lower Silurian limestone brought to the surface by a series of flexures. In this he agreed in the main with Mather.

[5] Dwight later found Cambrian fossils in these same rocks.

In 1884, Hall for the first time put himself fully on record as opposed to the Taconic system on stratigraphic as well as paleontological grounds. In this year he sent Dana copies of two sections of the Taconic area and manuscript notes claimed by him to have been made prior to 1845, which gave the result of his own studies. In these Mounts Anthony and Equinox were shown to have a synclinal structure, the limestone underlying a broad synclinal of slates and schists, the former being put down as Trenton and lower, while the slates and schists were of Hudson River age. Prompt publication would have given Hall priority over the Vermont survey and others, but owing to the long delay the matter is of only historical interest.[6]

The opinions of T. Sterry Hunt, as already noted, were summed up by him in his *Mineral Physiology and Physiography*, published in 1886. In this he devotes a chapter of some 170 pages to a seemingly exhaustive discussion of the question, beginning with the work of Eaton in 1824. He claimed, on apparently good grounds, to have made use of the extensive literature, so far as it seemed important, and to have supplemented the researches of the various investigators by personal observations extending over a wider field and a greater number of years than those of any of his predecessors. With characteristic assurance and verbosity, he states his conclusion that "there exists in eastern North America a great group of stratified rocks consisting of quartzites, limestones, argillites and soft crystalline schists, having all together a thickness of 4000 feet or more and resting unconformably upon various more ancient crystalline rocks from the Lawrentian to the Montalban inclusive." This was the series called Transition by Maclure and includes the primitive quartz-rock, primitive limestone, and Transition argillite of Eaton and the *Itacolumite* group of Lieber. It was the Lower Taconic of Emmons, now called Taconian by Hunt. He worked out the distribution of the various series to his own satisfaction, over North America, the West Indies, South America, Hindostan, Russia, the Alps, Bavaria, Norway, and Spain and left the subject with the safely conditioned statement that should further studies confirm his view, it will appear that the Taconian is a "great and widespread group of strata which cannot henceforth be overlooked in geognostical history."

In Volumes XXIX and XXXIII (1885 and 1887) of the *Ameri-*

[6] "I have never been able to find from yourself or your publications that you have published anything on the age of the Taconic System, although you talked enough on the subject years ago to make a memoir as long as Emmons's." (Dana to Hall, September 16, 1885.) See further, Appendix, p. 728.

can Journal of Science, Dana again took up the subject systematically under the caption of "Taconic Rocks and Stratigraphy," the paper being accompanied by a map of the region and numerous sections. He showed that the flexures throughout the Taconic area were of a prevailing synclinal habit; that the limestone was a continuous formation lying underneath the mountains, overlaid conformably by strata of quartzite and quartzitic and ordinary mica-schist, and underlaid by similar rocks along the eastern border, also. He further showed that within the Taconic region the texture and mineral nature of the limestone beds varied geographically, the crystalline texture being coarser to the southward and eastward. He found no evidence of general overthrust faulting affecting the entire region. As to the age of the rocks he remained conservative, but still regarded the limestone beyond doubt Lower Silurian, though whether Trenton, Chazy, or Calciferous remained an open question.

Fig. 127. *C. D. Walcott.*

In the *Journal* for 1886 he referred again to the matter, questioning the reliability of the lithological evidence put forth by Emmons as to the identity of his Taconic rocks and those of Sedgwick's Cambrian, asserting that "geological investigation with reference to the Cambrian had not advanced so far as to make its application safe." In this paper he reported also the finding in Canaan, New York, of Lower Silurian fossils in the Sparry limestone, this being the oldest stratum of the Taconic system, as announced by Emmons in 1842. These fossils were studied by Professor Dwight, of Poughkeepsie, and S. W. Ford, and identified as belonging probably to the Trenton period.

In 1886, C. D. Walcott, then paleontologist of the United States Geological Survey, took up the subject, giving a summary of his results, with map, in the *American Journal of Science* for April and May, 1888. Walcott began with a systematic study of the slates, limestones, and quartzites of Vermont and the adjoining counties of New York, continuing his work the following season and paying particular attention to areas within the counties of Washington and Rensselaer, New York; Bennington, Vermont, and Berkshire, Massachusetts, since here were found the series of sections described by Emmons and nearly all the localities mentioned by him.

Walcott showed to his own satisfaction and that of most of those having any detailed knowledge of the subject that the quartzite series belonged to the Middle Cambrian, the talcose slates to the Upper Cambrian, the limestones to the Calciferous,[7] Chazy, and Trenton,

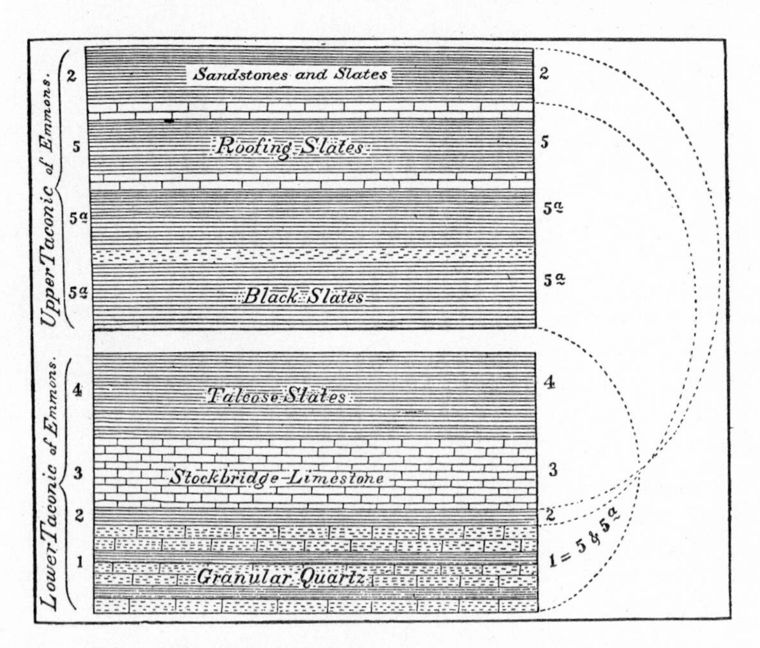

Fig. 128. *Tabular View of the Taconic System.*
(*After Walcott.*)

and the slates and sandstones to the Hudson River group. He agreed mainly with Emmons in his lithological descriptions and the general dip and arrangement of the strata, but disagreed with him in his identification of the geological age of the formations of the Lower Taconic, the stratigraphic relations of both the Lower and Upper Taconic, and also as to the value of the stratigraphic and paleontological identifications of age. He showed that the granular quartz, supposed by Emmons to be unfossiliferous and to lie at the base of his Taconic system, was actually fossiliferous and the equivalent of the greater portion of the Upper Taconic;[8] also that this quartzite was a shore deposit formed at the same time as were the silico-cal-

[7] The supposed Middle Cambrian and a part of the Calciferous here described were subsequently relegated to the Lower Cambrian.

[8] These are now accepted as basal Lower Cambrian.

careous muds in deeper waters, which Emmons had included in his Upper Taconic. The Stockbridge limestone, which Emmons had regarded as a peculiar pre-Silurian deposit, he showed on paleontological evidence to be the equivalent of the Trenton, Chazy, and Calciferous limestones of the Lower Silurian,[9] while the Talcose slate resting conformably upon the Stockbridge limestone was found to contain graptolites of Hudson River age. Emmons's subdivision of Upper Taconic he thought due merely to a repetition of certain beds brought up by an overthrust fault, as shown in Fig. 129.

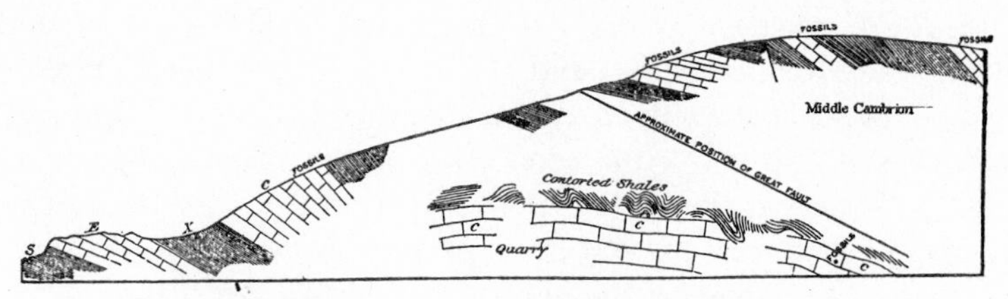

Fig. 129. *Section of Bald Mountain from the South.*

In his summary of the paleontological evidence relative to the Taconic, Walcott stated: (1) The trilobites described by Emmons in 1844-1847 from the black slate were referred then to the highest member of the Taconic system on stratigraphic evidence; but (2) in 1856, on evidence of the same kind, were referred to the lowest member.[10] (3) That in 1859 they were referred to a pre-Potsdam position[11] by comparison with a fauna the position of which had been stratigraphically determined with relation to the Silurian fauna. Further, (4) that the *Nereites* and other trails which Emmons regarded as typical for the Taconic had not yet been stratigraphically located, while (5) the graptolites formed part of the fauna of the Hudson terrane.

Emmons's errors, according to Walcott, were due almost wholly to his trust in lithological characters; his supposed unconformity between the Taconic and Champlain systems was based primarily on the similarity in lithologic characters of the Calciferous sandrock of the Lower Silurian, and the Calciferous sandrock of what is now known, from its fossils, to be a part of his Upper Taconic; also that he confused the dark shales of the Lower Silurian with those of his

[9] But, later, Foerste and Wolff found Lower Cambrian forms near the base of these rocks at Rutland, Vermont.

[10] Subsequent studies seem to show that Emmons was correct in this.

[11] They are today referred to the Lower Cambrian.

Upper Taconic and failed to recognize the obvious fact that the Calciferous terranes were frequently represented in geological sections by a shale undistinguishable from that of the Hudson River; and that in several places the Trenton limestone is replaced by shale. According to T. N. Dale, Emmons was also in error in assuming that the slates along the east foot of the Taconic range dipped to the east, he having confused cleavage with bedding.

At the eighth meeting of the American Committee of the International Geological Congress, which was held in New York in April, 1888, the subject of the subdivision and nomenclature of the American Paleozoic formations was discussed, and incidentally the matter of using the name Taconic, and its limitations, if used at all, was pretty thoroughly gone over. A considerable diversity of opinion was found to exist; although the committee at first reported in favor of retaining the name, they were apparently subsequently led to change their views upon the presentation of new evidence by Walcott. A brief summary of some of the views adopted is given below.

Dana objected to the retention of the name, thinking it would be regarded only as "a reminder of Emmons's blundering work—a succession of unstudied assumptions that brought only evil to the science." S. W. Ford favored the adoption of the name Taconic for the middle portion of the Cambrian, as the term was then used, or that marked by the presence of the fossil *Olenellus*. Hall felt that it might be well to retain the name for those rocks lying below the Potsdam, while C. H. Hitchcock considered the lowest Taconic as of Lower Potsdam age. Newberry, on the other hand, would retain the name Taconic as a group name for one of the minor subdivisions of the Upper Cambrian, while Alexander Winchell, from a perusal of the literature only, would retain the name for those rocks underlying the Cambrian and retaining the primordial fauna (*Paradoxides*). The Canadian geologists, Dawson and Selwyn, both thought the term useless and unnecessary, while H. S. Williams and Joseph Le Conte recommended that it be dropped entirely, owing to the existing conflict of opinion.

The committee finally recommended that all the strata lying between the Devonian and Archean be divided into three great groups, the Silurian, Cambrian, and Taconic, the last named to be subdivided as below:

As above noted, this report of the committee was not adopted, and it is probably due more to the work of Dana and Walcott than to all others that the term today finds no place except historically in American geology.

TACONIC.		
Faunal (systematic) designations.		Rock masses of New York and New England. Stratigraphic designations.
Taconic-Primordial or First Fauna.	Sub-Faunas.	
	St. Croix.	Lower portion of the Calciferous sandrock of New York. The St. Croix beds (so-called western Potsdam) of the Mississippi Valley.
	Taconic.	The Georgia group. The Taconic black slate and granular quartz.
	Acadian.	Paradoxides beds of Braintree, Massachusetts, and St. John's group of New Brunswick.

As to the justice of this decision, there may be some question, particularly when one recalls the fact that Walcott's own studies later showed the existence of his *Olenellus* (Georgian) fauna *below* rather than above the *Paradoxides* (Acadian) beds.

When one considers further the condition of the science at the time Emmons first proposed the system and the conditions under which he labored, without satisfactory maps, it is obviously unfair to hold him to as strict an account as would be justifiable with reference to the later workers. Even were it true, as stated, that there is today "*no known stratum of rock in the Taconic range*" that is of the geological age assigned it by Doctor Emmons,[12] and even though it were also true that "all his reasons for calling the Hudson terrane *Taconic* were based on errors of stratigraphy, and it was only a fortunate happening that any portion of the Upper Taconic rocks occur where he placed them in his stratigraphic scheme," still there would, to the unprejudiced, seem to be abundant room for the recognition of the name, a fact which Walcott has himself recognized by the adoption of the borrowed term Ordovician. Moreover, ruling out the term on the ground of blundering is scarcely just, since, as Winchell pertinently remarks, a similar ruling would take from Columbus the credit of having discovered America, since he blundered upon it, expecting to strike India.

[12] Emmons's granular quartz and the lower part of the Stockbridge limestone are today regarded as older than the Potsdam, as a part of the Lower Cambrian.

The following tables illustrate the three principal stages of the Taconic controversy up to 1903, in the columns to the left the subdivisions of Emmons being given, and in those to the right the equivalents as recognized by Dana and other authorities.

Taconic of 1842.	Equivalents of 1888.	Equivalents of 1903.
6. Stockbridge limestone ...	II. Lower Silurian	Upper part Lower Silurian; lower part Lower Cambrian.
5 a. Magnesian slate of Greylock, perhaps a repetition of No. 3.	III. Hudson slates	Lower Silurian.
5 b. Granular quartz	I. Cambrian	Lower Cambrian.
4. Limestone	II. Lower Silurian	Lower Cambrian.
3. Magnesian slate of Taconic Mountains.	III. Hudson slates	Lower Silurian.
2. Sparry limestone	II. Lower Silurian	Upper part Lower Silurian; lower part Lower Cambrian.
1. Taconic slate	III and I. Hudson slates and Cambrian.	Lower Silurian.
Taconic of 1844.		
5 a. Bald Mountain black slate.	I. Cambrian	Lower Cambrian.
5 b. Taconic slate	III and I. Hudson slates and Cambrian.	Lower Silurian.
4. Sparry limestone	II. Lower Silurian	Upper part Lower Silurian; lower part Lower Cambrian.
3. Magnesian slate	III. Hudson slates	Lower Silurian.
2. Stockbridge limestone ...	II. Lower Silurian	Upper part Lower Silurian; lower part Lower Cambrian.
1. Granular quartz	I. Cambrian	Lower Cambrian.
Taconic of 1855.		
Upper Taconic. 2. Bald Mountain black slate.	I. Cambrian	Lower Cambrian.
Upper Taconic. 1. Taconic slate ..	III and I. Hudson slates and Cambrian.	Lower Silurian.
Lower Taconic. 3. Magnesian slate	III. Hudson slates	Lower Silurian.
Lower Taconic. 2. Stockbridge limestone.	II. Lower Silurian	Upper part Lower Silurian; lower part Lower Cambrian.
Lower Taconic. 1. Granular quartz.	I. Cambrian	Lower Cambrian.

CHAPTER XIII.

The Development of the Glacial Hypothesis.[1]

WITH the coming of Louis Agassiz to America in 1846 the glacial question to which frequent mention is made in previous pages became preeminently an American problem, and here was worked out an hypothesis which in details of cause and effect has withstood the tests of time, but which is yet deficient in that it fails to explain the conditions through which the period itself was brought about. The matter has been given repeated notice in the pages which have gone before, but the subject is of sufficient importance to merit a review, even at the risk of considerable repetition. It may be added that the development of the hypothesis as it exists today furnishes an excellent illustration of the growth of knowledge through cumulative evidence, since of all geological phenomena in America those relating to the drift have been perhaps longest under observation. Further than this, it is a phase of geological investigation which in its earlier stages was least dependent upon the allied sciences.

Fig. 130. *L. Agassiz.*

Out of numerous early observations relating to the phenomena those which now seem most worthy of consideration were by Benjamin De Witt and are to be found recorded in the second volume of the *Transactions* of the Philadelphia Academy for 1793.

De Witt noted the occurrence along the shores of Lake Superior, of bowlders representing sixty-four different rock types. In the discussion of this occurrence, he wrote:

> Now, it is almost impossible to believe that so great a variety of stones should be naturally formed in one place. . . . They must, therefore, have been conveyed there by some extraordinary means. I am inclined to

[1] Adapted from the writer's presidential address delivered before the Geological Society of Washington, December 13, 1905, and published in the Popular Science Monthly the following April.

believe that this may have been effected by some mighty convulsion of nature, such as an earthquake or eruption, and perhaps this vast lake may be considered as one of those great "fountains of the deep" which were broken up when the earth was deluged with water, thereby producing that confusion and disorder in the composition of its surface which evidently seems to exist.

This, so far as I am aware, is the first attempt to account for the wide dispersion of bowlders over the northern part of the country, although the glacial character of the same was not dreamed of.

The paper, so far as existing literature shows, caused little discussion, and I have next to refer to observations by Dr. Samuel Akerly, who, in 1810, published in Bruce's *American Mineralogical Journal* a geological account of Dutchess County, New York. After referring to the Highlands and the country to the northward, he described the southern part—that upon which the then existing city was built—as composed of "an alluvion of sand, stone and rocks." This he looked upon as a recent deposit, "subsequent to the creation and even the deluge." The manner in which this material was deposited he described in the following language:

After the waters of the deluge had retired from this continent, they left a vast chain of lakes, some of which are still confined within their rocky barriers. Others have since broken their bounds and united with the ocean. The Highlands of New York was the southern boundary of a huge collection of water, which was confined on the west by the Shawangunk and Catskill mountains. The hills on the east of the Hudson confined it there. When the hills were cleft and the mountains torn asunder, the water found vent and overflowed the country to the south. The earth, sand, stones and rocks brought down by this torrent were deposited in various places, as on this island, Long Island, Staten Island and the Jerseys.

"This opinion," he added, "is mostly hypothetical, because unsupported by a sufficient number of facts." A candid acknowledgment upon which the author is to be congratulated!

Whose was the master mind that first conceived of this great barrier which held back the flood of waters, so long made responsible for the drift, I have not been able to ascertain. The theory is, however, given in greatest detail by Dr. Samuel L. Mitchill in his *Observations on the Geology of North America*, published in 1818.[2]

It was Mitchill's idea that the Great Lakes are the shrunken representatives of great internal seas of salt water, which ultimately broke through their barriers, the remains of which he thought to be

[2] See also Volney's writings, p. 24.

still evident. One of these barriers, he wrote, seemed to have circumscribed the waters of the original Lake Ontario and to be still traceable as a mountainous ridge beyond the St. Lawrence in upper Canada, passing thence into New York, where it formed the divide between the present lake and the St. Lawrence and continued to the north end of Lake George, apparently crossing the Hudson above Hadley Falls. Thence he believed it to run toward the eastern sources of the Susquehanna, which it crossed to the north of Harrisburg, and continued in a southeasterly direction until it entered Maryland, passing the Potomac at Harpers Ferry into Virginia, where it became confounded with the Alleghany Mountains.

To appreciate Mitchill's views, then, we have to imagine this now broken and gapped ridge as continuous. A time came, however, when at various points it gave way, the pent-up waters rushing through and carrying devastation before them like the waters from cloudburst or bursting reservoirs of today, but on a thousand-fold larger scale. By this bursting all the country on both the Canadian and Fredonian sides must have been drained and left bare, exposing to view the water-worn pebbles, and the whole exhibition of organic remains there formed. Great masses of primitive rocks from the demolished dam, and vast quantities of sand, mud, and gravel were carried down the stream to form the curious admixture of primitive with alluvial materials in the regions below.[3]

A fresh contribution to the subject was rendered this same year, in the publication of Amos Eaton's *Index to the Geology of the Northern States.* In discussing the so-called alluvial class of rocks and attempting to account for the great masses of granite and syenite which he found scattered throughout the Connecticut River region, he wrote:

> What force can have brought these masses from the western hills, across a deep valley seven hundred feet lower than their present situation? Are we not compelled to say that this valley was once filled up so as to make a gradual descent from the Chesterfield range of granite, syenite, etc., to the top of Mount Tom? Then it would be easy to conceive of their being rolled down to the top of the greenstone where we now find them.

It was not easy in all cases for the geologists of these early days to distinguish between the younger and earlier drift, or between the

[3] Somewhat similar ideas were prevalent abroad, at this time. Sir James Hall, writing on the "Revolution of the Earth's Surface," in 1812-1814, expressed the opinion that the glacial markings observed by him in Scotland were due to "Sudden and violent debacles which, set in motion by gigantic earthquakes and laden with mud and stones, swept from the ocean across the face of the country."

material which is now considered glacial drift, and the loosely consolidated alluvial deposits of the Tertiary period. This seems particularly true of H. H. Hayden, a Baltimore dentist and one-time architect, whose volume of geological essays has been given consideration in the earlier pages of this work. In these he dwelt very fully upon the lowlands, or the area at present comprised within the so-called coastal plain. After referring to the geographical limits of this plain and combating the opinions of previous observers, he elaborated his own theories somewhat as follows:

> Viewing the subject in all its bearings, there is no circumstance that affords so strong evidence of the cause of the formation of this plane as that of its having been deposited by a general current which, at some unknown period, flowed impetuously across the whole continent of North America in a northeast and southwest direction, its course being dependent upon that of the general current of the Atlantic Ocean, the waters of which were assumed to have risen to such a height that it overran its limits and spread desolation on its ancient shores.

In seeking the cause of this general current Hayden referred first to the seventh chapter of Genesis: "For yet seven days and I will cause it to rain upon the earth forty days and forty nights, and every living substance that I have made will I destroy from off the face of the earth." He then proceeded to show the inadequacy of rainfall alone, since the water being thus equally distributed over the ocean and the land, there could be no tendency toward a current. Some other cause must be sought, and, fortunately, his imagination proved equal to the task, as has been noted elsewhere (p. 81).

Accepting as a possibility the shifting of the earth's axis of revolution so that the sun would pass immediately over the two poles, he graphically pictured the devastation produced through the floods consequent to the melting of the ice caps, the stripping of the soil and rock débris from the northern latitudes and its transportation.

> At length the floods of the pole, forming a junction with Baffins Bay and the Arctic Sea, defying all bounds, overran their ancient limits and hurled their united forces in dread confusion across the bleak regions of the north to consummate the awful scene. Thus, lakes and seas uniting, formed one common ocean which was propelled with inconceivable rapidity across the continent between the chains of mountains, into the Gulf of Mexico, and probably over the unpeopled wilds of South America into the southern ocean. Fulfilled in this way were the awful denunciations of an offended God, by the sure extermination of every beast of the field and every creeping thing that crawled upon the face of the earth.

To these causes Hayden believed to be due the deposits of the coastal plain as well as the barrenness of Labrador and the northeastern portion of the continent, and the general phenomena of the glacial drift, the bowlders of the latter being conceived as transported by floating ice.

These essays, notwithstanding their seeming crudity today, were favorably reviewed by Silliman in the third volume of his *Journal*, even the idea of the fusion of the polar ice cap being allowed to pass with no more serious criticism than that the flood of waters might have been produced through the expulsion of the same from cavities in the earth. J. E. De Kay, however, writing some years later, ventured to take exception to the views regarding drift bowlders, wisely suggesting that, since the speculative part of geology is but a series of hypotheses, we should in every case admit that which explains the phenomena in the simplest possible manner. To his mind the simplest manner of accounting for them was to suppose that such, as igneous material, had been extruded through the superincumbent strata, forming peaks which have since been destroyed through some convulsion of nature or through the resistless tooth of time, the bowlders thus being fragments which had escaped destruction, though their place of extrusion had become completely obscured.

The observations thus far recorded display a lack of close attention to details and, in some cases, indicate a decided leaning toward cosmogony. Those to which attention is next called were of quite different type and show their author to have been a man of more than ordinary discernment.

While superintending excavations preliminary to the erection of a cotton mill at Vernon, Connecticut, Mr. Peter Dobson observed scarred and scratched bowlders in what is now called the ground moraine or till, and in a letter to Silliman dated November 21, 1825 (soon after published in the *Journal*), described them as worn smooth on their under side as if they had been dragged over rocks and gravelly earth in one steady position. They also showed scratches and furrows on the abraded parts. These appearances he could account for only by assuming that the blocks had been worn by being carried in ice over rocks and earth under water.

These observations seem to have attracted no attention at the time, and even Dr. Edward Hitchcock thirteen years later attached no serious importance to them, although his attention was called to the matter by another letter from Dobson, this time addressed to himself. In this last letter (written in 1838) Dobson described the bowlders as having first been rounded by attrition and then worn flat on one side by a motion that kept them in one relative position, as a plane slides over a board in the act of planing. Some of them were described as worn and scratched so plainly that there was no difficulty in pointing out which side was foremost in the act of wearing, a projecting bit of quartz or feldspar protecting the softer material behind it. In this letter he again announced his inability to account for the appearances except on the supposition that they had been enveloped in ice and moved forward over the sea bottom by currents of water. The drifting icebergs of the Labrador coast he thought might well illustrate the conditions of their production.

Perhaps it may have been because Dobson was a cotton manufacturer and not a member of one of the learned professions, or there may have been other reasons, but Hitchcock allowed the observations to pass unnoticed until 1842, when the subject was brought up by Sir Roderick Murchison in his anniversary address before the Geological Society of London.

> I take leave of the glacial theory in congratulating American science in having possessed the original author of the best glacial theory, though his name has escaped notice, and in recommending to you the terse argument of Peter Dobson, a previous acquaintance with which might have saved volumes of disputation on both sides of the Atlantic.[4]

Supported by this somewhat enthusiastic endorsement, Hitchcock then gave the second letter to the public through the *American Journal of Science*, at the same time remarking that he had himself derived his ideas concerning the joint action of ice and water from the writings of Sir James Hall.

With this much in the way of anticipation, we will turn back to 1825 once more and refer to the writings of William Keating, a mineralogist, who accompanied Major Long's expedition to the sources of the St. Peters River.[5] This observer noted that the entire region of the present headwaters of the Winnipeek (Winnipeg) River had been at a comparatively recent period an immense lake, interspersed with innumerable barren rocky islands, which had been

[4] Nevertheless Murchison, as late as 1846, maintained that the distribution of the erratics and the general phenomena of glaciation had been produced by powerful currents or waves of translation caused by sudden uplifts. In this he was supported by the mathematician, W. Whewell, who looked upon the drift as of paroxysmal origin. "If we suppose," he wrote, "a sea bottom 450 miles long by 100 miles broad which is one tenth of a mile below the surface of the water, to be raised to the surface by paroxysmal action, we shall have the force we require for the distribution of the northern drift." (Quar. Jour. Geol. Soc. of London, Vol. III, 1847, p. 227.)

[5] Now the Minnesota River.

drained by the bursting of the barriers which tided back the waters. This was plainly a recognition of the now extinct glacial Lake Agassiz.

Although the cosmogonist was fast drifting into the obscurity of the past, there were, nevertheless, occasional writers who preferred to ignore facts of observation or the efficiency of simple causes, and to seek for more difficult or more mystical methods of accounting for phenomena than those afforded by the observation of processes now in action. Benjamin Tappan, in discussing in 1828 the bowlders of Primitive and Transition rocks found in Ohio, objected to the commonly accepted idea that such were necessarily foreign to the locality and brought by currents of water or floating ice. He frankly acknowledged, however, his own inability to account for their presence, but ingenuously claimed that "ignorance is preferable to error. It may, therefore, be asked why may not these rocks have been created where they are now found, or, again, why may they not have been thrown up by earthquakes or volcanoes?"

Unsatisfactory as such conclusions may seem to the modern reader, they were certainly not more so than those advanced by a writer in the *American Journal of Science*, who two years previously had accounted for the drift on the supposition that the earth's revolution, amounting to 1,500 feet a second, was suddenly checked. This, he thought, would result in the whole mass of the surface water rushing forward with inconceivable velocity until overcome by opposing obstacles or exhausted by continual friction and the counterbalancing power of gravitation. The Pacific Ocean would thus rush over the Andes and the Alleghanies into the Atlantic, which would, in the meantime, be sweeping over Europe, Asia, and Africa.

> A few hours would cover the entire surface of the earth, excepting, perhaps, the vicinity of the poles, with one rushing torrent in which the fragments of disintegrated rocks, earth, and sand would be carried along with the wreck of animal and vegetable life in one all but liquid mass.

The first geological survey of an entire state carried through at public expense, it will be remembered, was that of Massachusetts, authorized by legislative enactment in 1830. Dr. Edward Hitchcock, then professor of chemistry and natural history in Amherst College, was selected to carry out the work. The report presented early in 1832 was, therefore, a document of unusual importance and, to a certain extent, epoch-making. Much that is of interest within its pages has been noted (p. 142), and we must here limit ourselves to that relating purely to the distribution of the numerous erratics, examples of which are common throughout the state.

It is but natural that this drift should have been attributed to the Noachian deluge, when one considers that Hitchcock's training was that of a clergyman. Writing of that about Cape Ann, he says:

> It can not fail to impress every reasoning mind with the conviction that a deluge of tremendous power must have swept over this cape. Nothing but a substratum of syenite could have stood before its devastating energy.

This observation is of importance, since here, for the first time, Hitchcock put himself on record in a line of investigation in which he became more widely known than in any other, with the possible exception of that relating to the fossil footprints of the Connecticut Valley.

In 1836 there was established a state geological survey of Maine, with C. T. Jackson, of Boston, at its head. Jackson's views on the glacial deposits, as expressed in the annual reports, were perhaps not more crude than those of the average geologist of his day. The "horse-backs" (ridges of glacial gravel) were regarded by him as of diluvial material, transported by a mighty current of water which, as he supposed, rushed over the land during the last grand deluge, accounts of which had been handed down by tradition and preserved in the archives of all people. "Although," he wrote, "it is commonly supposed that the deluge was intended solely for the punishment of the corrupt ante-diluvians, it is not improbable that the descendants of Noah reap many advantages from its influence, since the various soils underwent modifications and admixtures which render them better adapted for the wants of man." "May not the hand of benevolence be seen working even amid the waters of the deluge?" It is, perhaps, doubtful if the hard-fisted occupants of many of Maine's rocky farms would be disposed to take so cheerful a view of the matter.

Substantially the same views were advanced by Jackson in his report on the geology of Rhode Island, which appeared in 1839.

> There can not remain a doubt but that a violent current of water has rushed over the surface of the state since the elevation and consolidation of all the rocks and subsequent to the deposition of the Tertiary clay, and that this current came from the north. . . . Upon the surface of solid ledges, wherever they have been recently uncovered of their soil, scratches are seen running north and south, and the hard rocks are more or less polished by the currents of water which at the diluvial epoch coursed

over their surfaces, carrying along the pebbles and sand which effected this abrasion, leaving striæ, all of which run north and south, deviating a few degrees occasionally with the changes of direction given to the current by the obstacles in its way.

He did not accept the theory of drifting icebergs; "nor," he wrote, "can we allow that any glaciers could have produced them by their loads of sliding rocks, for in that case they should have radiated from the mountains instead of following a uniform course along hillsides and through valleys."

Although primarily a paleontologist, Timothy Conrad was sometimes drawn out of his chosen field by phenomena too obvious to be overlooked and concerning the nature of which little was actually known by the best authorities. The occurrence of enormous bowlders in the drift, resting often upon unconsolidated sand and gravel, fell within this category. That such could not have been brought into their present position through floods was to him obvious, neither could they have been floated by ice floes from the North during a period of terrestrial depression. He assumed, rather, that the country previous to what is now known as the glacial epoch was covered with enormous lakes, and that a change in climate ensued, causing them to become frozen and converted into immense glaciers. At the same time elevations and depressions of the earth's surface were in progress, giving various degrees of inclination to the frozen surfaces of the lakes, down which bowlders, sand, and gravel would be impelled to great distances from the points of their origin. The impelling force, he thought, in some cases might be gravity alone, but during the close of the epoch, when the temperature had risen, vast landslides—avalanches of mud filled with detritus—would be propelled for many miles over these frozen lakes, and when the ice disappeared, the same would be deposited in the form of a promiscuous aggregate of sand, gravels, pebbles, and bowlders.

In 1840 an immense stride in the study of drift deposits was made through the publication in Europe of Louis Agassiz's *Etude sur les Glaciers*, a work comprising the results of his own study and observation combined with those of Jan de Charpentier, E. T. Venetz and F. G. Hugi. The work was published in both French and German, and brought to a focus, as it were, the scattered rays by which the obscure path of the glacial geologist had been heretofore illuminated. It was Agassiz's idea, as is well known, that at a period geologically very recent, the entire hemisphere north of the thirty-fifth and thirty-sixth parallels had been covered by a sheet of ice possessing all the characteristics of existing glaciers in the Swiss Alps. Through this agency he would account for the loose beds of sand and gravel, the bowlder clays, erratics, and all the numerous phenomena within the region described, which had been heretofore variously ascribed to the Noachian deluge, the bursting of dams, the sudden melting of a polar ice cap, or even to cometary collisions with the earth. His ideas were favorably received by the majority of workers, though there was, naturally, a highly commendable feeling of caution against their too hasty acceptation. As a reviewer in the *American Journal of Science* expressed it:

> These very original and ingenious speculations of Professor Agassiz must be held for the present to be under trial. They have been deduced from the limited number of facts observed by himself and others and skillfully generalized, but they can not be considered as fully established until they have been brought to the test of observation in different parts of the world and under a great variety of circumstances.

The effect of the publication, however, soon made itself apparent in the current literature.

In 1843 Professor Charles Dewey, writing on the striæ and furrows on the polished rocks of western New York, argued that, while the bowlders of the drift indicated that a mighty current had swept from north to south, the polishing and grooving *might* be due to glaciers. "Glaciers or icebergs and the strong currents of water—a union of two powerful causes—probably offer the least objectionable solution of those wonderful changes," he wrote. Though thus disposed to accept in part Agassiz's conclusions, Dewey yet failed to realize their full possibilities. He could not conceive, for instance, how it was possible for a glacier to transport sandstone bowlders from the shore of Lake Ontario to the higher level of the hills to the southward. Bowlders of graywacke removed from the hills in the adjoining part of the state of New York and scattered throughout the Housatonic Valley furnished a like difficulty, since between the place of origin and that of deposit lay the Taconic range of mountains. "If," he wrote, "the boulders were once lodged on the glacier, the ice and the boulders must have been transported by a flood of waters over the Taconic Mountains."

In 1842 Edward Hitchcock, already referred to, read an important paper before the Boston meeting of the American Association of Geologists on the phenomena now under discussion, which is particularly interesting as showing the gradual evolution of the present theory from that of the Noachian deluge idea, advocated by the earlier workers. His views evidently had been modified by Agassiz's publication and by the writings of Buckland, Murchison, Dobson, and others, and he expressed at the outset the conviction that nearly all geologists would agree in the principle that the phenomena of drift were the result of joint and alternate action of ice and water. To express this joint and alternate action he made use of the term *aqueo-glacial.*

In this paper Hitchcock devoted some fifty pages to a description and discussion of observed phenomena, after which he proceeded to discuss the theories of the various European authorities and state his agreement or objections to the same. He objected to the theory of Lyell, to the effect that the results observed by him in North America were produced by floating icebergs derived from glaciers formed on mountains as the land gradually emerged from the ocean, because, first, it failed to account for the lower temperature which was necessary; second, because there was no evidence that the glaciers descended from the mountains; and third, because the deposits of vegetable matter derived from land plants showed that the continent must have been above sea level long before the drift period.

The theory of De la Beche, which supposed the contents of the northern ocean to have been precipitated over the countries further south by the elevation of the polar regions, Hitchcock thought possibly applicable to the low countries of Europe, but not to New England, since it would require a rise of the ocean amounting to some 6,000 feet, and he could find no facts to justify such an assumption, although recognizing the fact that the aqueo-glacial agency had operated well over the summits of the White Mountains.

To Agassiz's theory, which supposed an immense accumulation of ice and snow around the poles during the glacial period and a consequent sending out of enormous glaciers in a southern direction, followed by floods of water and transportation of icebergs on return of a warmer period, he likewise took exception, since he was unable to conceive how such effects could be brought about. Nor, indeed, could he understand how such causes could operate when the land was rising from the ocean and the water consequently retreating, as it must have been to account for the various observed phenomena—such phenomena as would necessitate the occurrence of water loaded with ice and detritus floating for centuries at least over a large part of the earth's surface.

His paper showed a very clear insight into what had taken place, but an inability, with the information at that time available, to account for it in a satisfactory manner. Thus, in describing the striæ found by himself on the top of hills and mountains like Monadnock, he wrote:

> Could immense icebergs have been stranded on the northern slope of the hills and afterwards, by the force of currents, have been driven over the summits; or would it be necessary to suppose that, after the stranding, the water must have risen so as to lift the iceberg; or would a vast sheet of ice lying upon the earth's surface, by mere expansion, without the presence of water, have been able to produce the smoothing and furrowing in question?

After considering the phenomena and weighing all the theories advanced from time to time by the authorities quoted, he summed up the matter in the following words:

> Is it not possible that the phenomena of the drift may have resulted from all the causes advanced in the theories under consideration? . . . I feel . . . that the proximate cause of the phenomena of drift has at last been determined, namely, the joint action of water and ice.

It will be recalled that in 1836 there was organized a state geological survey of New York, which was placed in charge of W. W. Mather, Ebenezer Emmons, Lardner Vanuxem, and James Hall—men whose names have since become too thoroughly identified with American geology to ever become effaced. Naturally drift phenomena attracted the attention of these workers, and each expressed opinions, some of which may be referred to in detail.

Seventy-five pages of Mather's report, published in 1843, were given up to a discussion of the subject. He concluded that the transport of the material and the production of scratched surfaces were contemporaneous, the drift itself being transported in part by currents and in part by ice, itself drifted by the currents. The period of the drift and that of the Quaternary deposits were separated by a partial submergence of the land, and, further, the periods of the drift were periods when the currents were stronger than at the present time. He conceived this to be due to a collapse of the crust of the globe upon its nucleus, causing an acceleration of the velocity of rotation, this causing, in its turn, a disturbance of the form of equilibrium of the spheroid of rotation which had been compensated by the flow from the polar regions and an accelerated flow to the equatorial regions. This sudden acceleration of the ocean currents he felt would be sufficient to cause the transportation of vast quantities of detritus-laden ice from the polar regions southward. The large amount of drift scattered over the central and northern Mississippi region he ascribed incidentally to ice-laden currents from Hudson

Bay and the polar seas, which, floating over the northern part of the United States, would be met by the warm waters of the Gulf Stream, damming them back and causing them to deposit their loads. The warm current flowing northward would be superimposed on the cold current, the latter continuing southward beneath it, transporting the finer materials, such as now occupy the lower Mississippi Valley.

Emmons likewise believed the agent of drift transportation to be water and ice. The bowlders he thought to be the work of icebergs, but the striations and polishing he felt could not be due to this agency, since the bottom of the ocean is not bare rock, but covered by débris, and, moreover, icebergs would not move in straight lines, a point which some more recent writers have quite overlooked. The bergs might act as agents of transportation, he argued, but not of erosion. According to his ideas the drift-covered region was, during the drift period, depressed, the country low and connected at the north with an extensive region giving rise to large rivers which flowed in succession over different parts of the region lying between Champlain and the St. Lawrence. These rivers united with the Atlantic on the south through the Champlain, Hudson, and Mohawk valleys. They bore along ice loaded with sand, pebbles, etc., which scratched and grooved the surface of rocks over which they flowed, and were the agents also perforating the rocks in the form of pot holes.

Hall's ideas were somewhat hazy. That he did not accept Agassiz's doctrine of a vast ice sheet is very evident. Thus, he wrote that "Blocks of granite, either enclosed in ice or moved by other means, have been the principal agents effecting the diluvial phenomena; that they have scored and grooved the rocks in their passage and, breaking up the strata and mingling themselves with the mass, have been drifted onward carrying everything before them in one general mêlée. That such may have been the case in some instances or in limited localities, can not be denied; but that it ever has been over any great extent of country will scarcely admit of proof."

In his annual report for 1838 he had expressed the opinion that the "diluvial" phenomena attested "the action of violent currents and not a single uniform current, but of opposite or conflicting ones." The extent of the deposits, the number of erratic blocks, and the evidence of long continued wearing proved, as he thought, that the force which produced them was not sudden and violent, but continued for an indefinite period.

Hall was at this time evidently a catastrophist and thought the drift soils, terraces, and the deep valleys and water courses due to the violent action of water which may have been caused in part by a sudden submergence and the rapid passage of a wave over its surface. His views, indeed, were in many respects little, if any, in advance of those held by Mitchill twenty-five years earlier. Like Mitchill, he conceived of an inland sea bounded and held back by the Canadian highlands on the north, the New England range on the east, and highlands of New York and the Alleghanies on the south, and the Rocky Mountains on the west. These presented barriers of from one thousand to twelve hundred feet above the level of the ocean until broken through by the St. Lawrence, the Susquehanna, the Hudson, partially by the Mohawk at Little Falls, and perhaps also by the Connecticut.

> But to whatever cause we attribute the phenomena of the superficial detritus of the fourth district, the whole surface has been permanently covered by water, for it seems impossible that partial inundations could have produced the uniform character and disposition of the materials which we find spread over the surface. (See Appendix, p. 733.)

It will be recalled (see p. 250) that in 1845 Hitchcock had described a singular chain of bowlders in eastern New York and extending over into western Massachusetts. He recognized the similarity of the southeast trains to the lateral moraines described by Agassiz, but he could not conceive of a glacier traveling directly across the intervening ridges, even were there mountains in the vicinity of sufficient altitude to give rise to the same. Neither did the consideration of river drift or floating ice afford him a satisfactory conclusion.

> In short, I find so many difficulties on any supposition which I may make that I prefer to leave the case unexplained until more analogous facts have been observed.

Neither W. B. nor H. D. Rogers seems to have taken kindly to the idea of a glacial period or even to the possibility of the drift and striations being due to ice action. In his address before the Association of Geologists in 1844, H. D. Rogers had reviewed the prevailing opinions, stating the objections to each as he saw them, and favoring, rather, an origin through the transporting power of currents of water set in motion by undulations of the earth's crust. He conceived a wide expanse of water, somewhat less than 1,000 feet in depth, which had been dislodged from some high northern region by a general uplifting of the crust amounting to perhaps a few hundred feet, and an equal subsidence at the south. "Imagine," he says, "this whole mass converted by earthquake pulsations into a series of stupendous and rapid moving waves of translation, helped on by the still more rapid flexures of the floor over which they move, and then advert to the shattering and loosening power of the tremendous jar of the earthquakes, and we shall have an agent adequate in every way to produce the results we see, to float the northern ice from its moorings, to rip off, assisted with its aid, the outcrops of the hardest strata, to grind up and strew wide their fragments, to scour down the whole rocky floor, and, gathering energy with resistance, to sweep up the slopes and over the highest mountains." It was to similar catastrophic conditions, it will be recalled, that they referred the distribution of the rocks of the Richmond bowlder train mentioned above and more in detail on p. 249.

David Christy, whose letters on geology have been noted on previous pages, conceived a like catastrophic and diluvial origin for the erratics, "They either have been transported by icebergs during the period of submergence or by the force of the receding current at the period of the second elevation." Consideration of the observed facts forced him to doubt the correctness of the iceberg theory and "the glacial theory has no one fact to support it." Nevertheless, Christy's paper (letter to M. de Verneuil, 1847) has the merit of first delineating the southern limit of the drift which is given as from a point thirteen miles south of Mercer, Pennsylvania, extending past Zanesville, Ohio, Madison and Princeton, Nebraska, Pinckney, Illinois, and to within two miles of Chester on the Mississippi River.

Agassiz, it will be remembered, came to America in 1846, and in 1847 was appointed to the professorship of geology and zoölogy in Harvard. Naturally, he was disposed to apply his views on glaciation to the phenomena of the drift in his new field of labor. In the summer of 1848, in company with Jules Marcou and a party of students, he undertook the exploration of the Lake Superior region, the results of which were published in 1850. The views set forth relating to the glacial phenomena of the region are of paramount interest.

He argued that the drift of all northeast America and northwest Europe was contemporaneous and due to a general ice sheet. Through a repetition of many of his former arguments, he showed that a current of water sufficiently powerful to transport the large blocks found would have swept practically over the entire globe and not have stopped abruptly, as did the drift, after reaching latitude 39° north. This limit of distribution of the bowlders to the northern latitudes also indicated to his mind that the matter of climate was an important factor. Water-transported material, he argued, would not cause straight furrows and scratches, and the theory that such might be due to drifting icebergs was rejected on the ground that existing bergs were insufficient, and to produce such as were would necessitate a period of cold sufficient for the formation of his hypothetical polar ice cap. He pointed out that the northern erratics were rounded and widespread; that the highest hills were scratched and polished to their summits, while to the south the mountain tops had protruded above the ice sheet and supplied the glaciers with their load of angular bowlders. He also called attention to the absence of marine or fresh-water shells from the ground moraine deposits, showing that it was not subaqueous.

Referring to the stratified deposits overlying the drift, he wrote:

> The various heights at which these stratified deposits occur above the level of the sea show plainly that since their accumulation the mainland has been lifted above the ocean at different rates in different parts of the country; further, it must be at once obvious that the various kinds of loose material all over the northern hemisphere have been accumulated, not only under different conditions, but during long-continued subsequent periods. To the first, or ice, period belong all phenomena connected with the transportation of erratic boulders, polishing, scratching, etc., during which the land stood at a higher level. To the second period belongs the stratified drift such as indicates a depression of the continent.

In 1849 at the Cambridge meeting of the American Association for the Advancement of Science, Rogers returned again to the subject of glaciation, having evidently become aroused though not convinced by the writings of Agassiz. While apparently recognizing the validity of the arguments of Lyell, Murchison, Chambers, Agassiz, and others, he still felt that the question of the origin of the drift had as yet scarcely received that clear or critical analysis which its complexity required. Such an analysis he then proceeded to give, arriving, however, at essentially the same conclusions as in his previous paper. He still adhered to his diluvial hypothesis. All the features of the glacial drift and all the confused and tumbled intermixture of rocky débris found in the true moraine could be accounted for under the supposition of violent and interrupted wave or current action. "What a glacier can do moving at the rate of 200 feet a year in scratching the surface of the earth, a sheet of gravel consisting of small pieces moving at the rate of a mile a minute can readily effect." It is not recorded that Professor Agassiz made any further reply to this paper than to state that however correct the theory of translation of waves might be for other places, it would not apply to

Switzerland, where it was physically impossible that water action could have had anything to do with the observed phenomenon.

Who it was that first suggested an elevation of the northeastern portion of the American continent as a height essential to a glacial period is not apparent from the printed literature. At the New Haven meeting of the American Association, held in August, 1850, Professor C. B. Adams suggested a cause involving an ice sheet 5,000 feet in thickness, in "a great elevation of the land in the northern regions." Whether or not the elevation did actually account for the glacial sheet, he found direct evidence that it had prevailed in the presence of drift striæ extending beneath sea level. The extent of the postglacial subsidence, amounting to not less than 2,000 feet below the present level, he also found substantiated by the beds of stratified clay and fine silt at the summits of the longitudinal valleys.

The views of Professor J. D. Dana, which have thus far been passed over, were clearly set forth in his address as retiring president of the American Association for the Advancement of Science in 1855.

> The drift was one of the most stupendous events in geological history. In some way, by a cause as wide as the continent—and, I may say, as wide nearly as the world—stones of all sizes, to immense boulders of one to two thousand tons weight, were transported along with gravel and sand, over hills and valleys, deeply scratching the rocks across which they travelled. Although the ocean had full play in the many earlier ages, and an uneasy earth at times must have produced great convulsions, in no rock strata, from the first to the last, do we find imbedded stones or boulders at all comparable in magnitude with the immense blocks that were lifted and borne along for miles in the drift epoch.
>
> Much doubt must remain about the origin of the drift, until the courses of the stones and scratches about the mountain ridges and valleys shall have been exactly ascertained. The general course from the north is admitted, but the special facts proving or disproving a degree of independence on the configuration of the land have not yet been sufficiently studied.
>
> One theory, the most prevalent, supposes a deep submergence over New England and the North and West, even to a depth of four or five thousand feet, and conceives of icebergs as floating along the blocks of stone, and at bottom scratching the rocks. Another, that of the Professors Rogers, objects to such a submergence, and attributes the result to an incursion of the ocean from the north, in consequence of an earthquake movement beneath the Arctic seas.

Professor Dana then went on to state the objections to the theory of submergence, as lack of fossils, beach lines, etc., and concluded;

"This much, then, seems plain, that the evidence, although negative, is very much like positive proof that the land was not beneath the sea to the extent the explanation of the drift phenomena would require." He did find positive proof, however, that, so far as New England is concerned, the land was once at a higher level than now, as witnessed by the deep fjords or drowned rivers along its coasts. As to the cause of the cold necessary to the production of a glacial ice sheet, he acknowledged the "inefficiency of all offered explanations." On the question of the drift, we therefore seem to be forced to conclude, whatever the difficulties we may encounter from the conclusion, that the continent was not submerged, and therefore icebergs could not have been the main drift agents: the period was a cold or glacial epoch and the increase of cold was probably produced by an increase in the extent and elevation of northern lands. Further than this, in the explanation of the drift, known facts hardly warrant our going.

In 1856 Edward Hitchcock came once more to the front, through the medium of the *Smithsonian Contributions to Knowledge,* with a paper of some 150 royal quarto pages and 12 plates, in which he considered the changes which had taken place in the earth's surface since the close of the Tertiary period. The products of these changes he classed as, first, drift unmodified, and second, drift modified, including under the latter such deposits as ancient and modern beaches, submarine ridges, sea bottoms, osars, dunes, terraces, deltas, and moraines. The drift proper he considered, as before, a product of several agencies, including icebergs, glaciers, land slips, and waves of translation, which, though more active in the past than now, are still in operation.

To account for the drift accumulations at various altitudes he conceived that the water must have stood some 2,500 feet above its present level and, further, that all the northern part of the continent—at least all east of the Mississippi—had been covered by the ocean since the drift period.

Concerning the origin of the material of the irregular coarse deposit beneath the modified beaches and terraces (ground moraine), he agreed essentially with the German geologist, Naumann, in supposing that, first, the eroding materials must have been comminuted stone; second, they must have been borne along under heavy pressure; third, the moving force must have operated slowly and with prodigious energy; and fourth, moving in a nearly uniform direction, though liable to local divergence; fifth, the vehicle of the eroded material could not have been water alone; but, sixth, a firm and heavy mass, somewhat plastic. The exact period of operation of the drift agency he naturally found difficulty in determining, and felt that, while the greater part of the work was accomplished before the continent had emerged very considerably from the waters, nevertheless, the work of erosion went on for some time after emergence began.

It was in this connection that was made the first suggestion, so far as I am aware, of a possible recurrence of glacial periods, as fully elaborated later by Chamberlin. Referring to the occurrence of two series of striæ, the direction of which did not coincide, and the possible existence of still a third series, he wrote, "Perhaps there were two periods of glaciers, one before, and one subsequent to the drift."[6]

The facts concerning the dispersion of bowlders Hitchcock thought could be more satisfactorily explained by icebergs than glaciers, since the transportation and scattering continued until after the time when a large part of the beaches and terraces were formed. Glaciers, he thought, would have plowed tracks through stratified deposits. Icebergs such as now traverse the Atlantic might carry bowlders over the beaches and terraces and drop them from time to time, forming thus the intermixture of coarse angular blocks and beach and terrace material, as we now find it. "The supposition that a glacier once existed on this continent wide enough to reach from Newfoundland to the Rocky Mountains is the grand difficulty in the way of the glacial theory."

The writings of the Rogers brothers are, with the exception already noted, singularly lacking in more than casual references to the drift. In the, for its time, magnificent publication of the first geological survey report of Pennsylvania, one would naturally look for an extension of the views of Professor H. D. Rogers, but such are not found. The fact that he considered, if not fully comprehended, Agassiz's views is shown only by a brief paragraph in which he described and figured drift striæ seen on an exposed surface of Umbral sandstone on the south side of the Wyoming Valley. These he described as "pointing up the slope toward the southwest, as if produced by fragmentary débris violently propelled against the slope of the mountain wall of the valley from the south." The presence of

[6] Chas. Whittlesey, in a paper in the Am. Journal of Science of 1848, on drift and alluvium of Ohio, suggested that "perhaps there were so many different periods of diluvial action, or a different action at distant places at the same time that the deposits in question cannot be arranged under the general law of superposition and order as to age." It does not appear, however, that he had the idea of glacial periods clearly in mind.

such ascending striæ, both here and elsewhere, effectually refuted, according to his conception of it, the glacial theory of their origin.

Like Hitchcock, he failed to conceive of other than local mountain glaciers of the Swiss type, and he gave the following, even then antiquated, matter for a general discussion of the distribution of the drift and the various phenomena accompanying it.

> A ready explanation of the origin of this newest Pleistocene deposit (*i.e.,* that of the Hudson and Lake Champlain districts) suggests itself when we consider the nature and energy of the crustal movements which lifted the Laurentian clays and sands to a height, in one locality at least, of not less than five hundred feet, and which drained wide tracts of the upper Laurentian Lakes.
>
> The mere agitation or pulsating movement of the crust, if accompanied by any permanent uplift of the land, would suffice, we would think, by lashing the waters of the tidal estuaries in one quarter and the lakes in the other, to strew a portion of the older drift bordering all those basins in wide dispersion upon the top of the more tranquil sediments. But if such a pulsation of the crust were accompanied by successively paroxysmal liftings of wide tracts of the land, then the inundation would take the form of stupendous currents, the strewing power of which would be adequate to any amount of superficial transportation, even to the remote transportation of the larger erratics.

Of the earlier drift, it should be noted, he offered no explanation whatever, other than that implied in a reference to a period of repose "which separated the convulsed epochs of the earlier general and later local drift."[7]

Compared with this the ideas of Professor O. N. Stoddard of Miami University, Ohio, whose geological bibliography includes but a single title, are quite refreshing. Writing in 1859 on *Diluvial Striæ, or Fragments in Situ,* and describing what was evidently ground moraine material, this writer says: "The agency of running water may be dismissed as utterly inadequate to explain the facts in question. . . . It seems necessary to admit that they [the striated pebbles] were firmly frozen into the clay and thus held in position, while some overlying mass slowly ground off their exposed surfaces." No agency of which he could conceive so satisfactorily accounted for this as glaciers moving southward over the region, holding the beds frost-bound, and under enormous pressure, thus grinding down

[7] "The explanation of Rogers's failure in glacial geology, and conclusions as to Niagara" (see p. 167), probably his blindest blunder, is that he was a better structural than physiological geologist. His judgment was sounder in the interpretation of a complex mountain country than in regard to the process which subsequently moulded its surface. (J. W. Judd, *op. cit.,* p. 32.)

both the rocks in place and those frozen in the glacial ice as well. "Probably these fragments were first embedded in the glacier and received, while in that position, the scratches on their under surfaces, but were subsequently detached from the glacier, embedded and frozen in the clay, where they were reduced to the condition in which they are found."

In 1861 the Natural History Survey of Maine was inaugurated and C. H. Hitchcock placed in charge of that portion relating to geology. Of his work, only that relating to glaciers concerns us here. He noted that the fossiliferous marine clays of the same age as similar deposits along the St. Lawrence and Champlain valleys and referable to the Terrace period, sometimes underlay a coarse deposit referable to the modified drift. Without committing himself definitely on this point, he suggested the possibility, therefore, of a recurrence of the drift agencies, that is, a period of second drift, as had the elder Hitchcock fifteen years earlier.

The period itself, according to Hitchcock's view, was inaugurated by a depression of this portion of the continent amounting to at least 5,000 feet below that of today, and it was during this period of depression and reelevation that the deposits were formed through the joint agency of icebergs and glaciers.

In 1862 J. S. Newberry expressed his views on glaciers in an article on the "Surface Geology of the Basins of the Great Lakes." After reviewing the conditions as he saw them, he came to the conclusion that, at a period corresponding in climate, if not in time, with the glacial epoch of the Old World, the lake region, in common with all the northern portion of the American continent, was raised several thousand feet above the level of the sea. This was to him the glacial period, during which the surface of the country was planed down and the deep fiords along the Atlantic coast formed. This was followed by a period of depression, when all the basin of the Great Lakes was flooded with fresh water, forming a vast inland sea in which the laminated blue clays (the oldest drift deposits) were precipitated. Subsequently an immense quantity of gravel and bowlders was transported from the region north of the Great Lakes and scattered over a wide area south of them. This he thought due to floating ice and icebergs.

It would seem that, if one were looking for original observations on drift phenomena, he might turn with safety to the writings of the Canadian geologists. Singularly enough, the views there expressed are, if anything, less original than those of the workers on the southern side of the line.

The establishment of a geological survey of Canada under Logan led to the publication of the now well-known volume of 1863. The views there given may be accepted as a summary of the knowledge relating to the glaciation on Canadian territory, as it then existed.

Concerning the region of the lake basins of western Canada, Logan wrote: "These great lake basins are depressions, not of geological structure, but of denudation, and the grooves of the surface rocks which descend under their waters appear to point to glacial action as one of the great causes which have produced these depressions. This hypothesis points to a glacial period when the whole region was elevated far above its present level and when the Laurentides, the Adirondacks, and the Green Mountains were lofty Alpine ranges covered with perpetual snow from which great frozen rivers or glaciers extended over the plains below, producing by their movements the glacial drift and scooping out the river valleys and the basins of the great lakes."

In his address before the Natural History Society of Montreal in 1864 J. W. Dawson took occasion to combat vigorously these ideas of Logan, and this on the ground that "it requires a series of suppositions unlikely in themselves and not warranted by facts"; that it seems physically impossible for a sheet of ice to move over an even surface striating it in uniform directions over vast areas; that glaciers could not have transported the large bowlders and left them in the positions found, having no source of supply; that the peat deposits, fossils, etc., show that the sea at that period had much the same temperature as the present arctic currents, and that the land was not covered by ice.

In describing the course of the rock striæ he announced that he had no hesitation in asserting that the force which produced those having a westerly direction was from the ocean into the interior against the slope of the St. Lawrence Valley, and as he could not conceive of a glacier moving from the Atlantic up into the interior, he considered this as at once disposing of the glacial theory. He conceived, rather, that a subsidence took place sufficient to convert all the plains of Canada, New York, and New England into seas. This, he felt, would determine the direction of the Arctic current which moved up the slope. He would account for the excavations of the basins of the Great Lakes by supposing the land so far submerged that an Arctic current from the northeast would pour over the Laurentian rocks on the northern side of Lake Superior and Lake Huron, cutting out the softer strata and at the same time transporting the débris in the form of drift to the southwest. Glaciers, in his argument, were not wholly dispensed with, but limited to regions of mountainous elevation.

J. S. Newberry, while director of the geological survey of Ohio (1869-1878), had frequent occasion to discuss glacial phenomena, and a review of his opinions may be given in some detail. Beginning with the later Tertiary times, he believed the following sequence of events to have been established.

(*a*) During Miocene and Pliocene epochs a continent several hundred feet lower than now, the ocean reaching to Louisville and Iowa, with a subtropical climate prevailing over the lake region, the climate of Greenland and Alaska being as mild as that of southern Ohio at present.

(*b*) A preglacial epoch of gradual continental elevation, which culminated in the glacial epoch, when the climate of Ohio was similar to that of Greenland at present, and glaciers covered a large part of the surface down to the parallel of forty degrees.

(*c*) This period followed by another interval of continental subsidence characterized by a warmer climate and melting glaciers and by inland fresh-water seas filling the lake basins, in which were deposited the Erie and Champlain clays, sands, and bowlders.

(*d*) Another epoch of elevation which is still in progress.

The sheet of clay and bowlders which was found directly overlying the polished surface of the rocks over so large a part of the state, now known under the name of till and bowlder clay, he described under the general name of *glacial drift,* while the loose bowlders which he found scattered indiscriminately over the surface, frequently resting on the fine stratified clays, were known under the name of *iceberg drift.*

"If," he wrote, "we restore in imagination this inland sea, which we have proved once filled the basin of the lakes, gradually displacing the retreating glaciers, we are inevitably led to a time in the history of this region when the southern shore of this sea was formed by the highlands of Ohio, etc., the northern shore a wall of ice resting on the hills of crystalline and trappean rocks about Lake Superior and Lake Huron. From this ice wall masses must from time to time have been detached, just as they are now detached from the Humboldt glacier, and floated off southward with the current, bearing in their grasp sand, gravel and boulders—whatever composed the beach from which they sailed. Five hundred miles south they grounded upon the southern shore—the highlands of now western New York, Pennsylvania and Ohio, or the shallows of the prairie region of Indiana,

Illinois and Iowa, there melting away and depositing their entire loads."

The loess, as one would naturally expect from the foregoing, was looked upon as the finer sediment deposited in the quiet waters of one of these inland seas, to which the icebergs had no access. The lake basins, with the exception of that of Lake Superior, were regarded as excavated by glacial action—thus agreeing with Logan.

E. B. Andrews, an assistant of Newberry's on the Ohio Survey, held, however, opinions somewhat at variance with those just quoted. To his mind the western drift was beyond all question a stratified water deposit. It would seem to be an unavoidable inference, he wrote, that the drift of that region came in a vast sweep of water deep enough to cover gravel hills more than 800 feet high and with velocity enough to throw such coarse material into lofty steeps and summits.

The views of Orton, who succeeded Newberry as state geologist, were not widely different, and, briefly expressed, were as follows:

A threefold division of glacial time may be considered as demonstrated: (1) An age of general elevation of northern land, accompanied by intense cold and the formation of extensive continental glaciers; (2) a general depression of the land, with the return of a milder climate; (3) a partial reelevation of the land and a partial return of the cold climate, producing local glaciers and icebergs.

E. T. Cox, while state geologist of Indiana, encountered phenomena in every way similar to those described by Newberry and Orton, and it is to be expected that his mode of accounting for the same would be somewhat similar. In his report of the conduct of the survey for 1869-1879 he announced his acceptance of the general theory of glacial drift, as at that time understood, and conceived that the necessary climatic changes might be due to the relative position of land and water, and, possibly, a change in the course of the Gulf Stream. He could find, however, no evidence of a subsidence of the land to terminate the glacial period, nor could he find in Ohio, Indiana, or Illinois anything to militate against the commencement of the glacial period in Tertiary times and its continuation until brought to a close by its own erosive force aided by atmospheric and meteorological conditions. By these combined agencies acting through time the mountain home of the glacier was cut down and a general leveling of the land took place. This suggestion that the glacial epoch worked out its own destruction through a process of leveling, whereby the altitudes which gave it birth were so far reduced that glaciers could no longer exist, is unique and, so far as the present writer is aware, original with Cox.

The organization in 1876 of a state geological survey of Wisconsin afforded Professor Chamberlin and his assistants opportunity for investigation of the drift phenomena of that state, and in the pages of his reports his views are distinctly formulated. He divided the glacial period into: (1) The terrace or fluviatile epoch, (2) Champlain or lacustrine epoch, (3) the second glacial epoch, (4) the interglacial epoch, and (5) the first glacial epoch. This formal announcement of the possibility of two distinct periods of glaciation was here made for the first time, although, as before noted, Edward Hitchcock had at an earlier date suggested such a possibility.

Not content with a mere discussion of the glacial phenomena, Chamberlin considered also matters relating to the cause of glacial movement. The law of flowage he announced was, in his opinion, similar to that of viscous fluids—this in accordance with the observations of Agassiz, Forbes, Tyndall, and others. A later study of Greenland glaciers, as is well known, caused him to change his views on this point.

In the third edition of his work on *Acadian Geology*, which appeared in 1878, J. W. Dawson returned once more to a vigorous discussion of the problems of the ice age, and to register again his opposition to the views generally held by American geologists. Many of the arguments used closely resembled those of his former papers and may be reviewed here for the last time.

He believed the phenomena of the bowlder clay and drift in eastern America due to the action of local glaciers, drift ice, and the agency of cold northern currents. Against the theory of a universal glacier he again argued on the ground that such suppositions were not warranted by the facts. "The temperate regions of North America could not be covered with a permanent mantle of ice under existing conditions of solar radiation; for, even if the whole were elevated into a tableland, its breadth would secure a sufficient summer heat to melt away the ice except from high mountain peaks."

For the supposition that such immense mountain chains existed and have disappeared, he found no warrant in geology, and for such an "unexampled astronomical cause of refrigeration" as the earth's passing into a colder portion of space, no evidence in astronomy. He agreed with Lyell in regarding the theory of the varying eccentricity of the earth, as expounded by Croll, insufficient; moreover, it seemed to him physically impossible that a sheet of ice, such as that supposed, could move over an uneven surface, striating it in directions uniform over vast areas and often different from the present inclination of the surface.

He was further influenced in his opinion by the work of the English physicist Hopkins, who showed, apparently, that only the sliding motion of glaciers could polish or erode rock surfaces, and the internal changes in their mass—the result of weight—could have little or no effect. Glaciers, moreover, he argued could not have transported the bowlders great distances and lodged them upon the hilltops, and the universal glacier would, moreover, have no gathering ground for its materials. The huge feldspar bowlders from the Laurentide Hills, stranded at Montreal Mountain at a height of 600 feet above the sea and from 50 to 60 miles further southwest, and which must have come from little, if any, greater elevation and from a direction nearly at right angles to that of the glacial striæ, were against the ice-sheet theory, as were also the large bowlders scattered through the marine stratified clays and sands, and the occurrence of marine fossils in the lower part of the drift in the true till near Portland, Maine, and at various points on the St. Lawrence in Canada.

To substantially these views Dawson held to the very last. In his *Ice Age in Canada* (1893) he is found still combating vigorously the idea that all northern Europe and America were covered by a *mer de glace* moving to the southward and outward to the sea, and which moved not only stones and clay to immense distances, but glaciated and striated the whole surface. The glacial theory of Agassiz and others he described as having grown, until, like the imaginary glaciers themselves, it overspread the whole earth. He adopted, rather, what he called the moderate view of Sir Roderick Murchison and Sir Charles Lyell to the effect that Pleistocene subsidence and refrigeration produced a state of our continents in which the lower levels, and at certain periods even the tops of the higher hills, were submerged under water filled every season with heavy field ice formed on the surface of the sea, as at present in Smith's Sound, and also with abundant icebergs derived from glaciers descending from unsubmerged mountain districts. The later Pliocene, so far as Canada was concerned, he considered to be a period of continental elevation and probably of temperate climate.

Thus far the discussion relating to the ice period has been limited wholly to workers and areas east of the Mississippi River. In 1880 and 1882 J. D. Whitney, one-time state geologist of California, issued his well-known work on climatic changes of later geological time, in which he discussed the occurrence of glaciers in the West, particularly in the region of the Great Basin, and their possible origin. He thought to be able to trace a period of warmth and heavy precipitation, followed by one of desiccation, but anticipated by one of cold and glaciation, the glaciers, however, being limited to the most elevated ranges of the Cordilleras. At the outset he announced himself as opposed to the "wild and absurd ideas" that had prevailed regarding glaciation in the Sierras, and stated it as his belief that here, at least, ice had played but an extremely subordinate part as a glacial agent, though "there is no doubt but that the great California range was once covered with grand glaciers, but little if at all inferior to those which now lend such a charm to the Swiss Alps."

It was Whitney's opinion, further, that the geological importance of the ice sheet had been greatly exaggerated. It seemed to him beyond question that icebergs had played an important part in carrying and distributing the large angular bowlders which in many places rest upon the surface in such a manner as to show that they could not have been placed in their present position by running water or by a general ice sheet.

He thought it evident enough that the climate of northeastern America during the glacial epoch was a period of greater precipitation than now, but that it was a period of intense cold he would not admit. Glaciation or a glacial period was due merely to increased precipitation. In order that such precipitation should take place, an increased evaporation from the land and water was necessary. This could be brought about only by a general increase of temperature. The amount of precipitation being sufficient, the production of glaciers would depend upon temperature, which itself would be dependent upon local conditions, which again might, or might not, be due to elevation of land surfaces. His idea, in brief, was that while during the glacial epoch there might be over the entire globe a period of sufficient warmth to produce the desired evaporation, the precipitation would fall as rain or snow, according to the local uplift or depression. That the glaciers are now retreating in nearly every instance, he thought due, not so much to a change in climate, at least not to a gradual increase of temperature, but rather to a gradual decrease in the amount of annual precipitation.

In this connection, it may be mentioned that Whitney considered the movement of glacial ice due to water.

Glacier ice is not simply ice, but a mixture of ice and water, and it is to the presence of the latter that the whole mass owes its flexibility. The larger the amount of water, other things being equal, the more easily the glacial mass moves. When the water increases so as to get the upper hand, the ice gives way with a rush and becomes an avalanche. . . . The extreme variation of the rate of motion of different glaciers coming down from the inland ice of Greenland is due to the different amounts of water which they have imbibed.

More recent observations than those quoted are familiar, and one may well stop here. That, at a period geologically not very remote, a vast sheet of ice and snow, with all the attributes of a modern glacier, or series of glaciers, covered the northeastern United States and eastern Canada, that this sheet advanced, retreated, and again advanced, and finally utterly disappeared, is the commonly, though not universally, accepted view. The causes which led up to this condition are still problematical. Whether due to cold from increased elevation, as taught by Dana, to astronomical causes, as taught by Croll, or merely to an increase in precipitation, as argued by Whitney, to interception of the sun's heat by volcanic dust, or to a combination of any or all of these causes, is the great problem awaiting solution, if solution is possible on other than a theoretical basis. Chamberlin, the Dawsons, Gilbert, Hall, the Hitchcocks, Lewis, Mather, Newberry, Salisbury, Upham, Winchell, Wright, and a score of others have made us acquainted with the physical characteristics of drift deposits and their geographic distribution, but the first-named alone, among Americans, has put forward a satisfactory working hypothesis as to the cause of glacial motion.

Leaving out of consideration Peter Dobson, whose views were not pushed to their legitimate conclusions, the world at large must credit Louis Agassiz, born in Switzerland, but an American by adoption, with being the great promoter and, perhaps, originator of the glacial hypothesis as it exists today. His method of procedure, it is interesting to note, consisted in applying what one of our prominent geologists has slightingly referred to as the principle of prolonging the harmless and undestructive rate of geological change of today backward into the deep past.

CHAPTER XIV.

The Development of Micro-Petrology.

PRIOR to 1870 the microscope had been of little use in the solution of problems in mineralogy and petrology, and its possibilities were not realized. Naturally there had been experienced much difficulty in the determination of the mineral nature and structure of the finer grained rocks; in fact, it had in many instances proven to be an impossibility.[1]

The first steps leading directly toward the study of rocks by means of a microscope and thin sections, or micro-petrology as it is now called, were taken when Sir David Brewster began the use of polarized light in the examination of inclusions in crystals. A second and most important step was taken when Clifton Sorby, also an Englishman, taking advantage of the newly invented nicol prism and William Talbot's adaptation of the microscope, began, about 1850, the examination of sections of rocks cut sufficiently thin to be transparent, or at least translucent. Singularly enough, although his results were promptly made public,[2] little notice was taken of them

1 The following review from the American Journal of Science for 1831 illustrates well the now seeming crudity of some of the early attempts:

"In connection with Geology, we shall advert to C. G. Gmelin's elegant examination of Clinkstone. He has found that this volcanic rock is an aggregate of mesotype and feldspar. He shows this in a very interesting way. He treats the mineral with muriatic acid, and separates the dissolved portion, after which the silica of the decomposed part is dissolved out by boiling with carbonate of potash. In the mesotype, a portion of the soda is compensated by potash and lime and a part of the alumina by peroxide of iron and manganese, and in like manner, in the feldspar, a part of the potash is compensated by soda and lime, and of the alumina by oxides of iron and manganese. In this investigation the water in the mesotype was found to be less than in the same substance when crystallized. The water might possibly be merely hydroscopic, and its quantity in the mineral varied from 0.633 to 3.19 per cent. It is probable that the application of this principle to other rocks might be productive of very interesting results, and might throw light upon geological formations, which we shall seek in vain from the analysis of specimens of rocks in the aggregated state."

Incidentally it may be remarked that the rock *clinkstone* became known later under the name *phonolite,* and was found to be composed essentially of the minerals sanidin and nepheline.

2 His first paper giving results of studies on thin sections bears the title "On the Microscopical Structure of the Calcareous Grits of the Yorkshire Coast,"

either at home or abroad, and it was not until, through his personal influence and conversations with a foreigner while on a pleasure trip along the banks of the Rhine, that the importance of the method began to be realized. The manner in which this was brought about has been well told by F. Fouqué, in the *Revue des deux Mondes*, a somewhat free translation of a portion of whose article is here given.

> Arriving at Bonn he [Sorby] made the acquaintance of a student in the Prussian School of Mines, named Zirkel, by whom he was accompanied and guided on several excursions. Together they visited the Eifel, the Siebengebirge, and the environs of Lake Laach. Each day, during the journey, an animated conversation was carried on on the nature of the volcanic rocks, their mineral composition, and the marvellous details of structure revealed in them by the microscope. Sorby set forth with clearness and warmth the magnificent results of his studies. In the evenings, after the day's excursions, the conversations were continued, and finally, on returning to Bonn, the extempore teacher placed before his young auditor some microscopic preparations which he had brought and let him see for himself the importance of the matter that had been the subject of their long discussions. When they separated, he left in Zirkel an enthusiastic disciple who thenceforth devoted himself entirely to the study of microscopical geology, who was soon to progress along this path, from discovery to discovery, to gather about him a multitude of followers, and become one of the most celebrated of German scientists.

So much in earnest was Zirkel that he at once proceeded to the laboratory of the Reichsanstalt at Vienna, and, as a result of his work during the winter, prepared a memoir only second in importance to Sorby's own paper. This, containing descriptions of the microscopical characters of thirty-nine very typical rocks, was read before the Vienna Academy on March 12, 1863. Zirkel continued energetically his labors in the new field of research, and published in succession a number of valuable papers in which minerals like leucite, nepheline, apatite, sphene, etc., were for the first time recognized as rock-constituents. He found a worthy coadjutor, as enthusiastic as himself, in his brother-in-law, Hermann Vogelsang, whose *Philosophie der Geologie und Mikroskopische Gesteinstudien*, published in 1867, and illustrated by descriptions and drawings of rock sections, did much toward making the new method of research widely known. Other German investigators like Heinrich Fischer, Tschermak, Doelter, and Von Lasaulx soon took up the work, and numerous papers on the subject were published. By 1873 microscopical petrography had

dated November 6, 1850, and was published in Vol. VII, 1851, of the Quarterly Journal of the Geological Society of London.

become established as a recognized department of geological research. Zirkel that year gave to the world his *Mikroskopische Beschaffenheit der Mineralien und Gesteine,* full of very detailed descriptions of the minute characters of minerals and rocks, while Heinrich Rosenbusch, in his *Mikroskopische Physiographie der petrographische wichtigen Mineralien,* developed the optical principles on which mineral determinations must be made by the aid of the microscope. The publication in 1877 of Rosenbusch's *Mikroskopische Physiographie der massiger Gesteine,* and in 1879 of Fouqué and Michel-Levy's *Mineralogie micrographique des roches eruptives Françaises,* with its magnificent atlas of plates, was followed by numerous memoirs by these two authors, and by Lacroix, also of Paris, and showed how abundant was the harvest which had sprung from the seed sown by Sorby in 1850.

The utility of the method, it may be stated, is based upon the fact that all crystallized minerals possess the property of so acting on polarized light as to make their determination, however minute, a matter of comparative certainty. It has further the advantage that it is possible to determine the amount of alteration a rock has undergone, and even to trace these alterations through all their varying stages. Not merely then has the mineral nature of the densest rocks been now made plain, but the presence has been detected of numerous minerals, before quite unsuspected, as common constituents. Furthermore, it has been found that these minerals in rock masses undergo such molecular and chemical changes, that, while the entire mineral nature may be altered, the rock still retains its geological identity. Still again, with or without a mineralogical change, the entire structure may undergo alteration. Thus massive trappean rocks may have become converted into schists, gabbros or norites into diorites, granites into gneisses, peridotites into serpentines, etc. Indeed it is safe to say that no one method of research of modern times has done more to advance our knowledge of rock formation and deformation than that of micropetrology.

With this much in the way of introduction, there remains but to be noted the adoption of the new method by American workers.

Apparently the first reports of a microscopic examination of rocks in thin sections to appear *in print* in the United States were those of A. A. Julien and Charles E. Wright, which were included in Volume II of the *Reports of the State Geological Survey of Michigan.* The volume bears the date of 1873. Naturally little was accomplished beyond the determination of the mineral nature of the finer grained rocks. The determination of the exact nature and relations of the remarkable alteration to which these minerals had been subjected, and of the interesting series of transition rocks which had been the result, and the whole subject of the origin both of the rocks and ores, it was announced, must be deferred to some other time and channel. Wright's work was done under the direction of Professors Von Cotta and Kreischer, in Freiberg, Saxony, but the results were not received in season to be incorporated in full in the report.

At the Hartford meeting of the American Association for the Advancement of Science, August, 1874, Professor E. S. Dana presented a paper on the Triassic trap rocks of the Connecticut Valley, giving the results thus far obtained by himself and G. W. Hawes "of a preliminary microscopic examination of the rocks of more than 100 localities by means of thin sections prepared for this purpose, a method of investigation of the greatest value in the case of such rocks as these by means of which the mineral constituents may be definitely determined, instead of being guessed at from the results of a chemical analysis." So far as the author has information, this was the first public announcement of the introduction of the system into America. This paper was quickly followed by one by Persifor Frazer of Philadelphia, on the traps of the Mesozoic sandstone in York and Adams counties, Pennsylvania. Here also the microscope and thin sections were used and three plates showing microstructures from photographs reproduced by "the photozincograph process."

The first recognition of the method by the Government was that given in Volume VI of the monographs of the Fortieth Parallel Survey, under Clarence King. The work, however, was done in Germany by the Ferdinand Zirkel already mentioned. The quarto volume of 297 pages and 12 plates, in part colored to show the appearance of the rock as seen under the microscope, was issued in 1876 and did much to arouse interest in this line of work. In 1877, M. E. Wadsworth, then instructor in mathematics and mineralogy in Harvard University, published in the *Proceedings of the Boston Society of Natural History* the results of studies on the igneous rocks in the vicinity of Boston, following it up in 1879 with a paper on the classification of rocks, and later (1880) with some energetic reviews and criticisms of Zirkel's work. In 1878 there appeared Volume III of Hitchcock's *Geology of New Hampshire*, in which was included a chapter of 110 pages on lithology by the G. W. Hawes above mentioned, who then gave promise of being our first and ablest exponent of the new departure.[3] This work, which was accompanied

[3] Dr. Hawes died in 1882.

by 12 plates, in part colored, gave the first systematic account of the methods of study and the preparation of thin sections to appear in America.

In Chamberlin's reports on the geology of Wisconsin (1877-1883) were papers by R. D. Irving, A. A. Julien, Arthur Wichman (of Leipzig), and C. E. Wright, on the microstructure of rocks of the Lake Superior region. Although the volumes bear the date as given above, the work of Wichman seems to have been performed in 1876 and presumably that of the others mentioned considerably antedated publication, as the survey appropriations ceased in 1879.

Newton's *Geology of the Black Hills,* published in 1880, contained an important chapter on the microscopic petrography by J. H. Caswell. The report by G. F. Becker on the *Geology of the Comstock Lode,* 1882, likewise contained a chapter on the microscopic structure of the rocks, accompanied by five beautiful lithograph plates in part colored. *The Report of the Tenth Census on Building Stones,* 1883, contained a chapter on microstructures, illustrated by photomicrographs reproduced by the heliotype process.

From this time the development was rapid and new workers came forward in considerable numbers. For some years it was necessary that students seek their training in German universities, particularly those of Heidelberg and Leipzig, under Professors Rosenbusch and Zirkel. All this is now changed, and ample facilities are furnished by nearly all of the leading American universities. Among those workers and teachers not already mentioned who have been instrumental in bringing the science to its present high degree of perfection, the names of Frank Adams, Whitman Cross, J. S. Diller, J. P. Iddings, J. F. Kemp, L. V. Pirsson, H. S. Washington, George H. Williams, J. F. Williams, and J. E. Wolff appear most prominently.[4]

[4] For details of foreign workers and their accomplishment, see W. Cross on the Development of Systematic Petrography in the Nineteenth Century, Jour. of Geology, Vol. X, 1902, pp. 331-376 and 451-499.

CHAPTER XV.

How Old Is It?[1]

> "Some drill and bore
> The solid earth and from the strata there
> Extract a register by which we learn
> That he who made it and revealed its date
> To Moses, was mistaken in its age."
>
> *Cowper.*

THE problem of the age of the earth has been attacked from several standpoints, and the results are widely variable; so variable, indeed, as to appear at first almost absurd, and it is only when one considers the magnitude of the problem, and how insignificant are the available data, that they can be given the respect they deserve.

In the early days, particularly during the latter part of the eighteenth century, before geology as a science had become established, the cosmogonists worked out theories of their own regardless of facts, hampered only by the limits of their imaginations.

These cosmogonists were divided into three schools: (1) those who believed that the earth has existed in its present form for all time, (2) those who believed that matter, but not the form, has thus existed, and (3) those who believed that both the matter and the form are due to a spiritual cause, to direct creation.

While the cosmogonist has by no means as yet disappeared, his numbers have been so far reduced as to give us little cause for concern. It is safe to say that no man of learning holds today the opinion that the world has existed in its present condition from eternity.

While, on the other hand, there are still many men of learning who regard the world, both in matter and in form as a product of divine creation, there are none whose opinions are worthy of consideration who hold that this creation dates back only some 5,000 or 6,000 years, these being the figures derived from a consideration

[1] This subject can scarcely be limited to America or American workers. It is, however, one of great popular interest and it has been thought advisable to give the accompanying résumé compiled from readily available sources.

of the Mosaic account as given in Genesis. I think it probable that were those among geologists, physicists, and astronomers who have considered the matter, called upon today to subscribe to any one of the three doctrines already noted, they would accept, though with many mental reservations, the second, to the effect that the matter but not the form of the earth has existed from eternity.

And this view does not necessarily carry with it any disregard for the scriptural statement that "In the beginning God created the heavens and the earth, the sea and all that in them is." If one stands today and looks down the long straight street, he notes that the lines of houses on either hand approach one another nearer and nearer as the distance increases, until the street itself is quite pinched out in the acute angle known as the vanishing point. But if from the same point he looks with an instrument like a telescope, so designed that rays before unavailable are focused on the retina, the vanishing point is pushed back to a distance dependent upon the perfection of the instrument.

Research has made available to our use facts before as unavailable as were the light rays without the lens, and the vanishing point called eternity is being pushed constantly back, but never done away with. The term *in the beginning* now means simply the limit of our reason, or understanding.

As already stated, the problem of the antiquity of the earth has been attacked from several different standpoints. Each of these involves certain assumptions, and the acceptance of the results must necessarily be governed by the acceptance of these primary assumptions.

Those whose opinions, or assumptions, are worthy of serious consideration, may be divided likewise into three schools: (1) the Cataclysmal, (2) the Uniformitarian, and (3) the Evolutionist.

With the cataclysmal school naturally no limit of time could be reached. The earth as viewed by them was formed and transformed through a series of more or less violent cataclysms, the violence of each of which could be attested by earthquake shocks and the elevation of mountain ranges, but the time limits were wholly those of the imagination. The uniformitarians, on the other hand, assumed that the geological processes going on around us today are essentially the same, both in kind and rate of action, as those of the past. They were divided into two schools, according as their assumptions were based upon observations of the processes of sedimentation and erosion now going on, or upon data relative to the origin of the earth from the cooling of a molten magma. The evolutionist attacked the problem from the standpoint of the gradual change in the character of animal and vegetable life as shown by fossil remains. Inasmuch as this school assumes a fairly uniform rate of change, they too might well be grouped with the uniformitarians.

We will now consider some of the methods used and the results which have been reached through the consideration of the subject by the various workers and from these various standpoints.

"To understand aright the origin and progress of the dispute regarding the value of time in geological speculation," wrote Dr. Geikie,[2] "one must take note of the attitude maintained towards this subject by some of the early fathers of the science. Among these pioneers none has left his mark more deeply graven on the foundations of modern geology than James Hutton (1726-1797). To him, more than to any other writer of his day, do we owe the doctrine of the high antiquity of our globe. No one before him had ever seen so clearly the abundant and impressive proofs of this remote antiquity recorded in the rocks of the earth's crust. In these rocks he traced the operation of the same slow and quiet processes which he observed to be at work at present in gradually transforming the face of the existing continents. When he stood face to face with the proofs of decay among the mountains, there seems to have arisen uppermost in his mind the thought of the immense succession of ages which these proofs revealed to him.[3] His observant eye enabled him to see 'the operations of the surface wasting the solid body of the globe, and to read the unmeasurable course of time that must have flowed during those amazing operations, which the vulgar do not see, and which the learned seem to see without wonder.' In contemplating the stupendous results achieved by such apparently feeble forces, Hutton felt that one great objection he had to contend with in the reception of his theory, even by the scientific men of his day, lay in the inability or unwillingness of the human mind to admit such large demands as he made on the past. 'What more can we require?' he asks in summing up his conclusions; and he answers the question

[2] Trans. Brit. Assoc. for the Adv. of Science, 1899.

[3] One of the greatest of America's early geologists, when confronted by similar conditions, seems to have received impressions quite the opposite of those of Hutton. Referring to the comparatively rapid disintegration of the Connecticut greenstones he wrote: "The conclusion forces itself upon us that the period when this process began could not have been vastly remote, in other words, that the earth has not existed in its present form from eternity. Its precise age cannot indeed be determined by this chronometer, but I have often thought, that, judging from this alone, we should be led to conclude that Moses placed the date of the creation too far back, rather than not far enough." (Am. Jour. Sci., Vol. VI, 1823, p. 26.)

in these memorable words: 'What more can we require? Nothing but time. It is not any part of the process that will be disputed; but after allowing all the parts, the whole will be denied; and for what?—only because we are not disposed to allow that quantity of time which the ablution of so much wasted mountain might require.'

"Far as Hutton could follow the succession of events registered in the rocky crust of the globe, he found himself baffled by the closing in around him of that dark abysm of time into which neither eye nor imagination seemed able to penetrate. He well knew that, behind and beyond the ages recorded in the oldest of the Primitive rocks, there must have stretched a vast earlier time, of which no record met his view. He did not attempt to speculate beyond the limits of his evidence. 'I do not pretend,' he said, 'to describe the beginning of things; I take such as I find them at present, and from these I reason with regard to that which must have been.' In vain could he look, even among the oldest formations, for any sign of the infancy of the planet. He could only detect a repeated series of similar revolutions, the oldest of which was assuredly not the first in the terrestrial history, and he concluded, as the result of this physical inquiry, that we find no vestige of a beginning, no prospect of an end.

"This conclusion from strictly geological evidence has been impugned from the side of physics, and, as further developed by Playfair, has been declared to be contradicted by the principles of natural philosophy. But if it be considered on the basis of the evidence on which it was originally propounded, it was absolutely true in Hutton's time and remains true today."

But no attempt was made by Hutton to measure that antiquity by any of the chronological standards of human contrivance. He was content to realize for himself and to impress upon others that the history of the earth could not be understood, save by the admission that it occupied prolonged though indeterminate ages in its accomplishment. . . .

Playfair, from whose admirable "Illustrations of the Huttonian Theory" most geologists have derived all that they know directly of that theory, went a little further than his friend and master in dealing with the age of the earth. Not restricting himself, as Hutton did, to the testimony of the rocks, which showed neither vestige of a beginning nor prospect of an end, he called in the evidence of the cosmos outside the limits of our planet, and declared that in the firmament also no mark could be discovered of the commencement or termination of the present order, no symptom of infancy or old age, nor any sign by which the future or past duration of the universe might be estimated. He thus advanced beyond the strictly geological basis of reasoning, and committed himself to statements which, like some made also by Hutton, seem to have been suggested by certain deductions of the French mathematicians of his day regarding the stability of the planetary motions. His statements have been disproved by modern physics; distinct evidence, both from the earth and the cosmos, has been brought forward of progress from a beginning which can be conceived, through successive stages to an end which can be foreseen. But the disproof leaves Hutton's doctrine about the vastness of geological time exactly where it was.

Following Hutton and Playfair the next writer of importance to attack the problem, whose efforts need receive attention here, was Sir Charles Lyell, who began publishing the results of his observations and studies in 1830. Lyell was a follower of Hutton and, as Dr. Geikie has expressed it, "became the great high priest of uniformitarianism," he himself not adding much to geological history, but possessing a rare faculty for setting forth the results of others' labors in an attractive and instructive manner. He conceived the idea of basing a calculation on the rate of change, or modification of molluscan species, and divided geological time, beginning with the Cambrian, into twelve periods. Basing then his estimates on the supposed length of time necessary to bring about the change in species shown to have taken place during Tertiary times, he assigned a duration of 20,000,000 years to each of his twelve periods or 240,000,000 years for the lapse of time between the beginning of the Cambrian and the close of the Tertiary. Concerning these estimates Dr. Geikie further remarks:[4]

It is easy to understand how the Uniformitarian school, which sprang from the teaching of Hutton and Playfair, came to believe that the whole of eternity was at the disposal of geologists. In popular estimation, as the ancient science of astronomy was that of infinite distance, so the modern study of geology was the science of infinite time. It must be frankly conceded that geologists, believing themselves unfettered by any limits to their chronology, made ample use of their imagined liberty. Many of them, following the lead of Lyell, to whose writings in other respects modern geology owes so deep a debt of gratitude, became utterly reckless in their demands for time, demands which even the requirements of their own science, if they had adequately realized them, did not warrant. The older geologists had not attempted to express their vast periods in terms of years. The indefiniteness of their language fitly denoted the absence of any ascertainable limits to the successive ages with which they had to deal. And until some evidence should be discovered whereby these limits might be fixed and measured by human standards, no reproach could justly be brought against the geological terminology.

[4] *Op. cit.*, p. 4.

It was far more philosophical to be content, in the meanwhile, with indeterminate expressions, than from data of the weakest or most speculative kind to attempt to measure geological periods by a chronology of years or centuries.

It is not to be expected that the geologists would have to themselves so fruitful a field for speculation. In 1862 Sir William Thomson took up the subject from a physical standpoint, basing his arguments on the secular cooling of the earth. In this memoir he adopted the hypothesis of a solid sphere the thermal properties of which remain invariable, while it cools by conduction from an initial state of uniform temperature. His conclusions have been summed up as follows:[5]

Conceive a sphere having a uniform temperature initially, to cool in a medium which instantly dissipates all heat brought by conduction to its surface, thus keeping the surface at a constant temperature. Suppose we have given the initial excess of the sphere's temperature over that of the medium. Suppose also that the capacity of the mass of the sphere for diffusion of heat is known, and known to remain invariable during the process of cooling. This capacity is called diffusivity, and is a constant which can be observed. Then from these data the distribution of temperature at any future time can be assigned, and hence also the rate of temperature increase, or the temperature gradient, from the surface towards the center of the sphere can be computed. It is tolerably certain that the heat conducted from the interior to the surface of the earth does not set up any reaction which in any sensible degree retards the process of cooling. It escapes so freely that, for practical purposes, we may say it is instantly dissipated. Hence if we can assume that the earth had a specified uniform temperature at the initial epoch, and can assume its diffusivity to remain constant, the whole history of cooling is known as soon as we determine the diffusivity and the temperature gradient at any point. Now Sir William Thomson determined a value for the diffusivity from measurements of the seasonal variations of underground temperatures, and numerous observations of the increase of temperature with depth below the earth's surface gave an average value for the temperature gradient. From these elements and from an assumed initial temperature of 7,000°, he infers that geologic time is limited to something between twenty million and four hundred million years. He says: "We must allow very wide limits in such an estimate as I have attempted to make; but I think we may with much probability say that the consolidation cannot have taken place less than twenty million years ago, or we should have more underground heat than we actually have, nor more than four hundred million years ago, or we should not have so much as the least observed underground increment of temperature."

[5] R. S. Woodward, Mathematical Theories of the Earth.

The next physicist of note to enter the field was Dr. James Croll. He, in 1867, placed a limit of twenty-four million years upon the time necessary for the accumulation of the sedimentary rocks, but on the theory that the present known sedimentaries had passed at least twice through the cycle of destruction and deposition he multiplied these figures by three, giving 72,000,000 as the limit of time duration since the first deposition began.

Another English physicist and mathematician, Professor Tait, in consideration of the probable retarding action of the tides on the earth's rotation, and its rate of cooling, felt that the time which should be allowed the geologists is something less than 10,000,000 years, and further, that the time the sun has been illuminating the earth is probably not more than 15,000,000 or 20,000,000.

In 1879 T. Mellard Reade, an English engineer, took up the subject from the standpoint of the rate of chemical denudation. Estimating the average amount of rain-water draining from the areas of igneous rocks during all geological time at 28 inches a year, and the amount of carbonate and sulphate of lime it takes up from these rocks at 4 parts per 100,000, the annual yield from these areas would be 71.68 tons per square mile. Calling it 70 tons for simplicity, and assuming that one-tenth of the area of the land, or 5,100,000 square miles, is occupied by these igneous rocks, they would then lose 357,000,000 tons per annum by solution, or 1.813 tons per square mile: at 13½ feet to the ton, it would thus take 1,139,032 years to build up on the sea bottoms a deposit a foot in thickness from these accumulations; or, on the assumption that the limestone in the sedimentary rocks of the earth is 528 feet in thickness it would require in round numbers 600,000,000 years for the elimination of this material from the original igneous rocks of which it was a constituent. This he distributed as follows:

	Millions of years
Laurentian, Cambrian, Silurian	200
Old Red, Carboniferous, Permian, New Red	200
Jurassic, Wealden, Cretaceous, Eocene, Miocene, Pliocene, and Post Pliocene	200
	600

A. R. Wallace in 1881 in his work on *Island Life* devoted a chapter to the subject in hand. He discussed the estimates of both physicists and geologists, devoting particular attention to the work of Croll relative to the cause of the glacial epoch, to Lyell's estimates based on Tertiary deposits, and to Dana's expressed opinion that the Tertiary time was only one-fifteenth as long as the Mesozoic and Paleozoic together.[6] He thought that the entire period of continental denudation and marine sedimentation might be reduced to 28,000,-000 years. He, however, concluded with the remark that "the only value of such estimates is to define our notions of geological time, and to show that the enormous periods of hundreds of millions of years which have sometimes been indicated by geologists are neither necessary nor warranted by the facts at our command; while the result places us more in harmony with the calculations of physicists by leaving a very wide margin between geological time as defined by the fossiliferous rocks and that far more extensive period which includes all possibility of life upon the earth."

In 1886 Professor G. H. Darwin took up the subject in his address before the British Association for the Advancement of Science. Referring to the arguments of Sir William Thomson, based upon the hypothesis that the earth is a cooling globe, he went on to state that on the supposition that this is correct the solidification of the earth probably began at the center and spread to the surface. Under these conditions he believed it possible if not probable that after a firm crust had been formed the upper portion still retained some degree of viscosity. This being the case some tidal oscillations must take place in it, which, being subject to friction, would generate heat in the viscous portion and further, the diurnal rotation of the earth would be retarded. If this retardation resulted in a lengthening of the day from twenty-three to twenty-four hours the amount of heat generated would be equal to the amount lost at the present rate of radiation in 23,000,000 years. He then went on to state that our data for the average radiant of temperature may be somewhat fallacious, since it had recently been shown that the lower stratum of the ocean is at a temperature near that of freezing while the mean annual temperature of the earth's surface where borings have been made is at least 30° higher. It did not to him, therefore, seem impossible that the mean temperature gradient for the whole earth

[6] Dana made no attempt to determine the age of the earth in years. Beginning with the second edition of his Manual, however, he gave estimates on the *relative* lengths of the geological periods. The ratios first given for the Paleozoic, Mesozoic, and the two periods of the Cenozoic were 14:4:2:1. These he subsequently changed to 12:3:1 for the Paleozoic, Mesozoic and the whole of the Tertiary, or 15:1, as above.

differed sensibly from that shown by the borings already made. To his mind the argument based on the amount of radiated energy given out by the sun seemed by all means the strongest, and he quoted from Sir William Thomson to the effect that "it seems therefore on the whole most probable that the sun has not illuminated the earth for 100,000,000 years, and almost certain that it has not done so for 500,000,000 years. As for the future we may say with equal certainty that inhabitants of the earth cannot continue to enjoy the light and heat essential to their life for many millions years longer unless sources now unknown to us are preparing in the great storehouse of creation." He was inclined to protest against the precision with which Professor Tait sought to deduce from the various arguments that had been used and felt fully justified in following Sir William in his estimate of 100,000,000 years as a limit of all geological history showing a continuity of life.

The first prominent American geologist to enter the lists was Professor Alexander Winchell of Michigan. Like Lyell, Winchell advanced little that was new, but summed up the evidence presented by others and presented the subjects needing consideration under the following heads:

(1) The time required for the sun to contract from a nebulous condition.

(2) The time which the sun will require to cool from its present condition to a planetary body.

(3) The time required for the earth to cool from incipient incrustation to its present state.

(4) Relative time for the deposition of all the rocky sediments.

(5) Calculations based on the obliteration of the rotational effects of the upheaval of a continental mass.

(6) The time since the middle of the last glacial period, based on the theory that epochs of glaciation on the northern hemisphere have been caused by extreme eccentricity of the earth's orbit.

(7) Estimates based on rates of erosion and deposition.

(8) The rate of bluff recession and terrace formation.

(9) The decrease of temperature of ground covered by ice during the glacial period, as compared with temperature of the ground not chilled by the ice sheet.

He thought the evidence furnished by geological forces now in operation of equal value with that selected by the physicists, and succeeded in reconciling his conscience with a reduction of the figures offered by his predecessors to a limit of but 3,000,000 years. He added: "If our attempts to ascertain the age of the world, or the duration of any single period of its evolution, yield only uncertain results they suffice at least to demonstrate that geological history has limits far within the wild conceptions of a certain class of geologists. They show, if we may credit the indications here recorded most trustworthy, a restriction of the modern epoch within limits not exceeding one tenth or one twentieth the duration sometimes assigned to it."

At the meeting of the American Association for the Advancement of Science in 1890 Mr. C. D. Walcott, later secretary of the Smithsonian Institution, delivered an address on geological time as indicated by the sedimentary rocks of North America. In this he devoted his attention mainly to the 16,000 feet of deposits accumulated through both mechanical and chemical agencies during Paleozoic time only, and on the western side of the American continent, within an area called by him the Cordilleran Sea, comprising what are now portions of Arizona, Nevada, Utah, Montana, and Alberta. These deposits bore evidence to his mind of having been accumulated in a connected and continuous sea roughly computed to cover an area of some 400,000 square miles. From a consideration of the present relative rate of land denudation the world over he felt justified in assuming over the land areas contributory to this sea a minimum rate of mechanical denudation of considerably less than 1 foot in 1,000 years, and of chemical denudation at the rate of 113 tons per annum for each square mile. He found nothing to indicate any marked change in rate of deposition during pre-Cambrian, Paleozoic, or Mesozoic times, and, comparing the results of his own observations with those of previous workers in similar fields, felt justified in expressing the opinion that for the assumed 10,000 feet of mechanical sediments 1,200,000 years would then have been requisite and for the 6,000 feet of limestone 16,300,000 years, or a total of 17,500,000 years for the duration of Paleozoic time. Taking these figures in connection with those given by others, for other periods, he gave the following estimate:

Period.	*Time duration.*
Cenozoic, including Pleistocene	2,900,000 years
Mesozoic	7,240,000 years
Paleozoic	17,500,000 years
Algonkian	17,500,000 years
Archean	10,000,000 years
Total	55,140,000 years

"It is easy," he remarked in conclusion, "to vary these results by assuming different values for area and rate of denudation, the rate of deposition of carbonate of lime, etc.; but there remains after each attempt I have made that was based on reliable facts of thickness, extent and character of strata, a result that does not pass below 25 to 30 million years as a minimum and 60 to 70 million years as a maximum for post-Archæan geologic time. . . . In conclusion, geologic time is of great but not of indefinite duration. I believe that it can be measured by tens of millions, but not by single millions nor hundreds of millions of years."

In 1893 T. Mellard Reade, to whose work reference has been already made, returned to the problem.

Assuming the mean area of denudation throughout post-Archean times to have been one-third the entire land area, and the bulk of the post-Archean rocks equal to a segment of the entire land area of the globe two miles in thickness, and, further, that subaerial erosion had progressed at an average rate of 1 foot in 3,000 years, he reached conclusions as follows: 5280 x 2 x 3000 x 3 = 95,040,000 years, the time that has elapsed since the commencement of Cambrian era. These figures agree fairly well with those of the continental geologist, Lapparent, who, on the basis of the present rate of mechanical denudation and sedimentation, placed the limits of time since the first consolidation of the earth's crust as between 67,000,000 and 90,000,000 years.

This same year the subject was taken up by Clarence King, who used as the basis of his calculations data acquired by Carl Barus in researches on the latent heat of fusion, specific heats, and volume expansion between the solid and melted state of the rock diabase. His conclusions were to the effect that "we have no warrant for extending the earth's age beyond 24,000,000 years." Referring incidentally to the previous estimates of Kelvin and Tait he remarked:

Kelvin's comparison of the earth's present figure with that of a thousand millions of years ago when the terrestrial day would have been only half its present length is one of the most interesting. The earth, if then plastic, would have yielded to four times the present centrifugal force at the equator and shown a correspondingly greater flattening at the poles, and bulging at the equator and "therefore" (as Tait expresses it) "as its rate of rotation is undoubtedly becoming slower and slower it cannot have been many millions of years back when it became solid, else it would have solidified very much flatter than we find it." This implies that because a computed earlier and greater value of ellipticity does not exist it could never have existed; in other words, that terrestrial rigidity has been and is of such value that a form taken in the remote past by the solid earth would not be modified by the tidal retardation of rotation and its attendant change of centrifugal force.

There is in modern geology a growing body of evidence which is believed to prove the very general plasticity of the lithosphere, by which it may experience important deformations from very *slowly* applied stresses. So strongly has this belief taken root that many American geologists accept "isostasy" and consider it to be an expression of a fluid equilibrium for the earth.

And further:

If, as I hold, Kelvin's suggestions as to ellipticity and tidal retardation do not apply to an earth readily deformable by slow stress as this one evidently is, there remain but three earth-ages to be weighed—Kelvin's value from terrestrial refrigeration which this paper seeks to advance to a new precision, Helmholtz and Kelvin's age of the sun which must sharply limit the date of the redistributed earth crust, and the old stratigraphical method. From this point of view the conclusions of the earlier part of this paper become of interest. The earth's age, about twenty-four millions of years, accords with the fifteen or twenty millions found for the sun.

In 1895 G. K. Gilbert of the United States Geological Survey made an estimate of the years involved in North American Cretaceous time, founded upon the thickness and character of Cretaceous sediments in Colorado, and their calculated rate of deposition. He found here sandstones, shales, and limestones belonging to the Benton, Niobrara, and Pierre groups, occurring in characteristic rhythmical alternations. It appeared to him that this rhythm could not be due to other than conditions controlled or at least influenced by the precession of the equinoxes. This precessional period is about 26,000 years, but as the position of the perihelion also moves, the average period has been cut down to 21,000 years. This he adopted as the time unit corresponding to each sedimentary alteration of the calcareous portions of the deposits. From data of this nature, he calculated that the entire 3,900 feet of sediments belonging only to the Benton, Niobrara, and Pierre epoch alone would have been laid down during a period of about 20,000,000 years.[7]

In 1897 Lord Kelvin (Sir William Thomson) delivered an address before the Victoria Institute, entitled *The Age of the Earth as an Abode Fitted for Life*. In this address, taking into account the more recent work, including that of Barus, and founding his reckon-

[7] Jour. of Geol., Vol. III, 1895, pp. 121-127.

ing "on the very sure assumption that the material of our present solid earth all around its surface was at one time a white-hot liquid," taking into account also the constitution and history of our atmosphere, and the necessity of sunlight for the production and support of human life, he came to the conclusion that the consolidation of the earth was finished 20 to 25 million years ago, and the sun then gave heat enough to support some kind of vegetable or animal life upon its surface.

A later paper upon the subject is that of Professor Joly of the University of Dublin, bearing the date of 1899. Professor Joly is a uniformitarian and selected for the basis of his calculations the rate of removal from the land areas and its deposition in water, the alkali sodium. The quantity of sodium now in the sea (mainly as sodium chloride or common salt) and the amount annually removed from the land have been calculated. Conceiving then that the annual average in past ages is essentially the same as at present, he calculated the time that has elapsed since the earth assumed its present solid form and water condensed upon its surface, to be between 80 and 90 millions of years.

In 1910 Dr. George F. Becker of the U. S. Geological Survey read a paper bearing on this subject before the National Academy of Sciences, in which he reviewed and compared the results thus far obtained. His conclusions were in favor of figures between 60 and 65 millions of years.

Résumé.

In the following tables are given the various estimates by the authorities quoted, together with a few not before mentioned. They are divided into three groups, according to the limit set for "the beginning."

	Since life began.
Charles Lyell	240,000,000 years
Samuel Haughton	133,000,000 years
James Croll	60,000,000 years
Archibald Geikie	100,000,000 years
T. Mellard Reade	95,000,000 years
J. D. Dana	48,000,000 years
Joseph Le Conte	30,000,000 years
C. D. Walcott	28,000,000 years

	Since the ocean came into existence.
Samuel Haughton	200,000,000 years
James Croll	72,000,000 years
Charles Darwin	200,000,000 years
Alfred Wallace	28,000,000 years
C. D. Walcott	50,000,000 years
J. Joly	90,000,000 years

	Since the earth was in a molten condition.
A. de Lapparent	80,000,000 years
Alexander Winchell	3,000,000 years
William Thomson	100,000,000 years
George H. Darwin	57,000,000 years
Simon Newcomb	14,000,000 years
Clarence King	24,000,000 years

Concerning the validity of the various assumptions made and the value of the estimates one cannot do better than to quote once more from Dr. Geikie and from an article in which he is defending the estimates based upon stratigraphy and paleontology against those of the physicist.

Until, therefore, it can be shown that geologists and paleontologists have misinterpreted their records, they are surely well within their logical rights in claiming as much time for the history of this earth as the vast body of evidence accumulated by them demands. So far as I have been able to form an opinion, one hundred millions of years would suffice for that portion of the history which is registered in the stratified rocks of the crust. But if the paleontologists find such a period too narrow for their requirements, I can see no reason on the geological side why they should not be at liberty to enlarge it as far as they may find to be needful for the evolution of organized existence on the globe. As I have already remarked, it is not the length of time which interests us so much as the determination of the relative chronology of the events which were transacted within that time. As to the general succession of these events, there can be no dispute. We have traced its stages from the bottom of the oldest rocks up to the surface of the present continents and the floor of the present seas. We know that these stages have followed each other in orderly advance, and that geological time, whatever limits may be assigned to it, has sufficed for the passage of the long stately procession.

We may, therefore, well leave the dispute about the age of the earth to the decision of the future.[8]

[8] A recent (1917) masterly résumé of the subject from all standpoints and particularly from that of radioactivity has led the late Professor Barrell to the conclusion that 700 million years have elapsed since the beginning of the Cambrian period. See Bull. Geol. Soc. of Am., Vol. XXVIII, No. 4, December, 1917, pp. 745-904.

FINALE.

Should it be asked along what lines American geologists particularly distinguished themselves during the period under review, one might reply: In studies tending toward the solution of, first, the fundamental problems of continental uplift and depression as made by Dana; second, in those relating to the physics and structure of mountain ranges, made by the Rogers brothers, Le Conte, and Dana; third, in those relating to glaciers and glaciation by Agassiz and the elder Hitchcock, and later by Chamberlin; fourth, in those relating to isostasy and physiography, made by Dutton, Gilbert, and Powell in the arid regions; and fifth, in those relating to vertebrate evolution, made by Leidy, Cope, and Marsh.

In the making of these summations it is not intended to ignore the workers of other nationalities, nor to claim that the work of any one of the individuals mentioned was wholly original and the credit belongs to him alone. It is rather an exemplification of the old saying that "an idea belongs to him who puts it to best use."

The branch of geology which can most properly be claimed as of American origin and development is that relating to the study of land forms originating through subaerial erosion and uplift. This is sometimes spoken of as the new geology, or, more technically, physiography. It is an outgrowth almost wholly of work in the West by Dutton, Gilbert, and Powell, ably supplemented in the East by the eminent physiographer, W. M. Davis, and hence does not come wholly within the limits of this history. It is a line of investigation for which the American continent has unquestionably afforded opportunities unexampled elsewhere. To appreciate its scope and bearing one has to remember that earth history, as studied by the earlier geologists, was limited largely to a consideration of the relative ages of the stratified rocks as shown by their included fossils. Gaps in the history due to a lack of fossiliferous strata were beyond remedy—were comparable with lost pages in a book. It is now realized, however, that the effects of contemporaneous erosion on the existing land areas of any period form also important and intelligible data, and may be made to supply, in part, at least, the missing pages.

APPENDIX

LETTERS CONCERNING MATTERS REFERRED TO IN THE TEXT

APPENDIX

28th Dec. 1843.
Philadelphia.

James Hall, Esq.
My dear Sir:

. . . My notes were in a great measure, compiled between 1830 and 1834. After the State took in hand, very commendably, an actual detailed survey, under every possible advantage, my chief inducement in pursuing a laborious examination, at my own private charge, no longer existed, and I left it in better hands. I am very glad it so occurred, for in the outset of the investigation, without any guidance deserving of mention, some of my notions were rather crude. I had been led to these, in a great measure, by the prevailing opinions of the day, among the best geologists on the other side the Atlantic, as to the supposed different origin and age of anthracite and bituminous coals;—the one being invariably placed with the transition (now Silurian), the other with the secondary formations.

However, we all got right at last, and the old red sandstone helped me wonderfully. . . .

My own time, when I devote any to geology, has been devoted to an object of some magnitude, on which I make progress by slow degrees. In fact, I have two undertakings on my hands, of some magnitude. One of them is the ascertainment and a concise notice of every deposit of *coal* and lignite formation, upon our globe. I was almost accidentally led into this subject by a little controversy as to the amazing amount of this valuable mineral in Europe as well as here. I pursue the inquiry as to coal *everywhere;* and the map in illustration will elucidate probably some useful facts and principles which as yet are not manifest, for the want of information. All this is scattered through such a mass of published matter, in various languages, that it is laborious though very interesting.

A friend in this city has lately published a pamphlet suggesting that coal formed certain *zone* round the globe. As he theorized before getting together the facts and details, it may be very probably, his scheme will not *fit*. This, of course, can only be acquired, as I am now doing, by an accumulation of adequate data.

My other undertaking is somewhat colossal; for it comprehends the entire illustration of the present terrestrial globe. To do this requires a vast deal of labor and hard reading. If I were in Europe I could do with much less difficulty, as I should have access to every known geological map; while here, you know, we are very deficient in those illustrations. Whether I shall be able to bring it to maturity is uncertain. If therefore, you have any aid at command, you know my wants, and I shall be glad of any assistance.

It strikes me that no geological draught of this nature can be a correct representation, otherwise than upon a *globe*. It is impossible to represent zones, or true forms, and areas, on any flat projection yet contrived, or ever will be. . . .

(*Signed*) Richard C. Taylor.

Hall to Dr. John Torrey.

Albany, April 2, 1842.

. . . I learned from Dr. Emmons that you had informed him that Mr. Lyell had engaged with Wyley and Putnam of New York to publish an edition of his Elements, with notes and additions on American geology. Having had occasion to speak of this to several friends I am anxious to know definitely the ground for my assertion. I have condemned most unqualifiedly such a course in Mr. Lyell,—a course which I should not have anticipated, and which from my intercourse I thought him incapable of as in all cases he said that he intended to publish nothing till the Reports of our Geology were published. . . .

Boston, Nov. 1, 1845.

. . . .

I have seen no Reports of Lyell's Lectures in any newspaper. He is lecturing on miscellaneous topics relating to evidences and causes of Geological Change, in which he figures somewhat largely in Paleontology. On the whole he succeeds very well. He is in much better spirits and of a much more tolerable demeanor than formerly. He made a hasty tour through Maine and to White Mts., and made some interesting discoveries in the way of fossil bones and shells on the banks of the Kennebec. He intimates that Dr. J's spectacles must have been a little obscure when he examined this region, and that a great deal is to be learned in the region of Augusta.

I am no great judge of his boom geologically but in other matters I think he has exhibited himself as a very sensible man, if he thinks as he wrote. He says he did not intend this as a geological affair, though he could not avoid saying something about it, but that he was taunted with the assertion that he dared not print the opinions he had uttered respecting America and the American people, until he was obliged to do it. He has expressed some little anxiety to know whether or not you were satisfied with the use he had made of your name and labors.

. . . .

(*Signed*) A. A. Gould.

Prof. James Hall.

Boston, Mayday, 1849.

Dear Hall—

Alas, the squabbles! shall we ever emerge from them. Four weeks ago I made a business of seeing everybody I supposed were either aggrievers or aggrieved; and then determined to mind my own business, and let difficulties alone. On Saturday evening Foster and Whitney made their appearance, stating that in consequence of information received from Washington giving them warning that the western men were about taking measures to have Dr. Jackson removed for inefficiency, idleness, disagreeable behaviour and other obnoxious reasons,[1] and that they would be involved as aiders and abettors of his misconduct,—they had resolved that a separation must take place—that they would resign—show him the reasons they should be obliged to assign if called upon—and that the only terms on which they could continue on the Survey would be that Jackson himself should resign in their favor, and thereby save himself from falling into the hands of western men, who would do all they could to throw discredit on his labors and injure him in every way. They read the reasons to me—some of them appeared frivolous and irrelevant, others formidable. I expressed my regret—urged them to deal openly and fairly with Jackson—saw no more till Tuesday A.M. when I received a note from Mrs. J. requesting to see me. I siezed the opportunity as one in which I might possibly redeem my character with her and went to see her at Roxbury—found her sick from the excitement of the day before, for it seems she had been present during a most furious scene the day before, many hours in length. Jackson of course denied everything and called all sorts of names. I had been meditating during my ride, and thought I might make a proposition which should render a separation, if take place it must, harmless to both parties. I then hurried home and arrived just as Jackson was about mailing his resignation—got him to withhold it until I could see Foster and converse with him. Two other gents were there and he proposed to refer the whole matter to us. We declined acting as arbitrators but that we would tell him what seemed to us best under the circumstances. They met at my room with Foster, Whitney, Jay and

[1] The attempt was even made to remove him last year—but Mr. Walker would not hear complaints of his officers from anybody—and the appropriation was left out (for the Survey) in Bill and was only restored by J. being accidentally in Washington—and this is one of the proofs that F. & W. could not have been conspiring to take the Survey.

Thayer, his assistants. I made my proposition—it was at once rejected and a separation declared to be a sine qua non—after some little show and feeling too of resentment on my part it was proposed that each in order should give his opinion on the state of the case. It became perfectly evident to myself and to Dr. J.'s friend, Mr. Abbott. that the difficulty did not originate with them, that McNair and Wilson and a whole series of men on the Lakes had set the ball in motion and that F. and W. indeed had it not in their power to stay proceedings. But even if they had any improper motives we felt that the only way to save Dr. J. was to put himself in their hands and obtain pledges from them they would guard his reputation and give him the full credit of all the previous labors on the Survey. The testimony of every man of the assistants as to J's conduct and neglect of the Survey was such that we felt that no man could pass under such imputations publicly and not come off greatly injured by them even though false. Three of these men had become disconnected with the Survey and had never before opened their mouths to their nearest friends—they are men of character and veracity. F. & W. on the other hand preferred to have their resignation accepted, and pledge themselves never to be again connected with the Survey in any capacity. No one seemed to have any object in view, in their statements, but to set Jackson's danger before us, should an investigation take place. The contract was therefore drawn up securing Jackson from harm in the most stringent and satisfactory manner—and he chose to sign it. This occurred on the day before I saw you. Every man was pledged to secrecy, and it was hoped that all knowledge of the movement would be confined to ourselves. On the next day news of the resignation had reached some of Dr. J's friends, among others J. L. Hayes of Portsmouth, who was to have accompanied J. as assistant. They at once raised a breeze, and as neither he nor we were at liberty to divulge what had been told us, they pursuaded him to withdraw his resignation and to accept those of F. & W. At the same time calling these assistants of whom he had previously always spoken in high terms, all manner of approbrious names. Foster, meantime, had gone post haste to Wash. and arrived the morning Mr. Ewing left for Ohio—found that J's removal had been decided on and that Mr. E. had given his chief clerk special order not to allow it to pass. The case now is, as I understand it, that Dr. J. in the belief that there is no cause for complaint, will demand an investigation—that there is a strong demonstration against him from the Lakes and the Land Office—that almost every man who has known him on the Lakes will testify against him as having wasted time and money—that there is no prospect of his continuing on the Survey—and that he will come out of this little less than ruined. His manner of conducting his affairs is such that he would spoil even a good cause. Mr. Ewing is thoroughly acquainted with all these delinquencies and peculiarities. Jackson cannot feel that he has done wrong—and you know he can in any case—but I do not believe that his assertions can stand against the testimony of all his assistants. As to anything he can show as the result of his *own* labors, he really has nothing of any value. He has received $500. per annum for the use of his laboratory and the analysis of the minerals, and not a single specimen has yet been analysed. He has not yet made any Report, though repeatedly urged to do so by the department. As to any conspiracy to dethrone him, I do not believe, from what I know of the circumstances, that it is possible—though I told F. that the chain of events was such that he could not escape that imputation. I may have been deceived. F. has taken some steps I fear which he should not have done—but it is to be remembered that he has been greatly exasperated by what he considers the bad faith of J. and now feels himself called upon to make his own character good. Should the Survey fall into other hands than those of F. and W. Dr. J. would fare badly. I do not believe you will do Dr. J. or the rights of superior, any good by interposing in this case. You know it is a law in all Gov't. appointments that the subs should inform of any delinquencies of their superiors on penalty of being considered as accessories. J. and W. are now in Washington and Mr. Ewing has returned, and the matter will be speedily settled I doubt not. It is a most unfortunate affair—and though others with less grounds for judging than I have, differed from me, and Dr. J. has preferred their advice, yet the more I know and think, the more I am satisfied that his most judicious course would have been to abide by our contract. Should my fears prove groundless I shall certainly feel highly gratified. I shall make inquiries respecting your sister on the earliest occasion. How is it about that set of N. Y. reports. I wish to know *immediately*.

A. A. G.

My dear Agassiz:

I have received your favor relative to the parallelism of European and American geological formations.

I still hold essentially the same views I held some years since. In England very little is known of the equivalents of our Potsdam sandstone and formerly no equivalent was shown. More recently it would appear that the "Lingula flags" striped stones and quartz rock are in some degree equivalent to these ancient rocks. The Potsdam and Calciferous sandstones also, so far as we have an equivalent, correspond to the Zone Primordiale of Barrande and to the lower, Lingula and the Oleolus sandstones of Russia and Sweden.

The Llandeilo Flags and Bala limestone of the Silurian system never appeared to me to reach lower than a partial equivalent of our Trenton limestone and Utica slate with perhaps some portions of Hudson River Group. But the later discoveries in Scotland of what have been termed the Durness and Assynt limestone as well as some other beds of fossils which have an equivalent in fauna with our Black River, Birdseye and

Chazy limestones, so that in Great Britain the series can be pretty well made out.

Our series of Chazy, Birdseye, Black River and Trenton limestones with the Hudson River Group, seem to be the representative of Barrande's second period or second fauna.

The Caradoc sandstone, as originally propounded by Murchison, embraced the Hudson River group and some portions of the Clinton group of New York, but the later investigations have I think shown some of these English localities to lie between true Caradoc and the Wenlock proper.

Wenlock is clearly an equivalent of our Niagara group, while the place of our Clinton group in the European series when recognizable is between true Caradoc and Wenlock.

Our Onondaga Salt group is not recognized, nor do I know any European equivalent. The tendency in England is to identify the rocks of this age and our Lower Helderburg limestones with the Ludlow rocks of England.

The Pterygotus and Eurypterus occur in the upper beds of Onondaga Salt group, and these fossils in England occur in the uppermost beds of the Silurian system according to Murchison. If this be true they have in England no representative of our Lower Helderberg group and a very feeble representation of the Onondaga Salt group. Still I think the Upper Silurian limestones of the continent of Europe show evidences of the fauna of both the Niagara and the Lower Helderberg periods. The Oriskany sandstone is not known in Great Britain and but meagrely on the continent.

The Devonian system in Europe in all its subdivisions is the equivalent of our Upper Helderberg limestones, Hamilton, Portage and Chemung groups and we may include the sandstone of the Catskill Mts.

The fossils of the Ludlow rocks are in so many instances generically identical with our Hamilton forms while some species are undistinguishable in the two formations, that I can only reconcile the apparent contradiction by supposing that there is yet no clearly determined sequence between the Wenlock limestone and Old Red sandstone in Great Britain.

If you will take the trouble to read a chapter which I furnished for Foster and Whitney's Report, Vol. II, pp. 285 to 318, you will see a review of the subject up to that time. The table of equivalents on page 318 will be instructive as showing my views at that time. If you will compare that table with the one I now send you you will see that the principal modifications occur from later discoveries in Europe. My mark on page 318, that of the seventy species of fossils in beds below the Trenton limestone, may be modified since the discovery of Maclurea and associated fossils in Scotland which clearly mark the period intervening between the Calciferous sandstone and Trenton limestone.

J. H.

Cambridge, Nov. 17, 1849.

My dear Hall:

I have just had a sight of a monstrous map published under the title of *Forster's Complete Geological Chart.* I do not know its author, but I am so painfully struck with the crudeness of this production that I hasten to write to you to ask if you will not join me in exposing publicly a work so full of false and antiquated views most childishly misrepresented, that its mere circulation would be considered abroad as a disgrace to American geologists if they were not to protest against it before it is puffed out in the newspapers. This is the more necessary as it is a very showy sheet with tolerably well drawn figures, which might easily mislead ignorant men or directors of schools not conversant with the subject. Passing over the fact that in its general outlines it is framed upon statements which 25 years ago were already antiquated, there is one feature in it which should make it worthless for any American student, that no reference whatsoever is made to the geology of this continent, and that not one of the numerous beautiful results derived from the surveys of the different states is introduced, so that it could not even be used as an introduction to the study of those more recent investigations. But even if in our days of rapid progress it was allowed to publish elementary works representing our science as it was taught in the days of our childhood, it should at least give the facts then known correctly and in their natural relations. I have, however, never seen more grotesque combinations than the illustrations there introduced among the tertiary beds where the Ichthyosauri and Plesiosauri of the Lias parade among the Pachyderms of Montmartre in as ridiculous a manner as a narrative of the battles between Napoleon and Alexander the Great of Macedonia would appear in a textbook of universal history, were the scene transferred to China to render the farce more complete. It is a circumstance very much to be lamented that such nonsense has been compiled and probably published at great expense in such showy form, when it were so easy to construct correct diagrams or to republish some of the better ones already in existence. As I do not know what can best be done to stop the circulation of such scientific forgeries I send you this note requesting you to circulate my opinion respecting the merits of that production in whatever form you please, were it even publishing this letter in its whole extent. I have alluded only to the general features of the map and the absurdities respecting the introduction of fossil remains in periods during which they never appeared, and need not allude to similar gross misrepresentations respecting the succession of the rocks which must have struck you, as a practical geologist, at first sight, such as a regular bed of porphyry overlying the entire formation of the coal. Do you know who can be the author of such a masterpiece of absurdity? which to summarily characterize it in a few words *is below all criticism* and about which I would never have lost a word were it not for the fear that it might have a wide circulation artificially called forth.

Sincerely your friend,

(*Signed*) L. Agassiz.

Friend Dana:

To say of the book under consideration [*i.e.* Marcou's] that the conceit and impudence of the thing is only equalled by the ignorance displayed, would prove nothing, and therefore it is as well to go at once to the work. It is hardly necessary to say anything of the Introduction, though its omissions and blunderings are obvious to the most casual glance.

After Maclure the author speaks of Conrad and Lea (p. 14), celebrated conchologists of Philadelphia, and leaves out Morton altogether. "Finally Jackson and Alger published in 1828 a geological description of Nova Scotia, etc." and "Such nearly was the condition of geology in America when Murchison published his celebrated book entitled 'The Silurian System' "—leaving it here to be inferred that from 1828 to 1839 when Murchison was read in this country nothing had been done by the geologists in America—leaving out of view the publication of the Final Report on the Geology of New Jersey, the first final report on Massachusetts, the Annual Reports of Pennsylvania, New York, Ohio, Michigan, Tennessee and Maine, which gave altogether a pretty general view of the geology of a great part of the United States.

On page 15—"Troost, Vanuxem and Eaton were also among the first to compare the American formations with those of Europe; and *laid the true foundations on which all the geological maps and memoirs published on this side of the Atlantic for sixteen years, have been constructed.*" Now as one example in point. Troost identified the Silurian of Tennessee with the Carboniferous of Europe, and Eaton the Silurian of New York with New Red Sandstone. Giving every credit due to these named geologists for their labors which in many respects had important results; nothing can be more utterly false than the above quoted assertion, since what they did could not by any possibility have served as a foundation for the maps and memoirs subsequently published. Mr. Vanuxem did indeed in the division of the Cretaceous, identify that formation of New Jersey with the same in Europe, and had from the beginning a more clear idea of the age of our geological formations as compared with Europe than any other geologist. The map of Byram Lawrence, which is next mentioned in terms of his praise is a map essentially copied from one made by Dr. Owen and which has been published in the Transactions of the Geological Society of London.

On page 16, alluding to the surety of the Vernuil comparisons, the author says,—"The different groups of the Palaeozoic rocks are *now* positively distinguished, and connected with the great geological epochs of Europe." The parallelism was established long before for nearly all the formations, and that part where there was difficulty, viz., Devonian and Lower Carboniferous, is now worse confounded than before, as will be seen in Marcou's own book. The author's travels for three years, page 17, would if truly represented, give a sorry figure before his arrogant assumptions.

Page 20—"Lower Silurian." "The first strata are thick beds of very hard sandstone, rose colored, or whitish gray."

The sandstone is usually thin bedded and over a large part of the west is extremely friable, and often in such a state as to be shoveled up like sand. It is gray or grayish brown or reddish brown, but never "rose colored."

Then comes a series of strata of compact limestone, blue, often blackish." This may be true of the localities seen by Mr. Marcou, but the prevailing color often over hundreds of miles in the west, is gray, or yellowish gray, with scarcely a trace of blue or black.

On page 21, he again calls it "a very hard sandstone of sub-crystalline texture and very diffuse stratification."

J. H.

University of Arizona.
Arizona School of Mines.

Tucson, Arizona, Oct. 12, 1899.

Prof. Geo. P. Merrill,
Head Curator of Geology,
U. S. National Museum,
Washington.

Dear Prof. Merrill:

I take pleasure in complying as far as possible with your wish expressed Oct. 2nd.

When Marcou returned from the exploration field-work with Whipple he took the fossils abroad to Paris with him, and on the demand of the Secretary of War was supposed to have sent them all back. But whether he did or not, all that he did send were turned over to me for description. I took them to James Hall at Albany under an arrangement for their comparison and description.

When the fossils were unpacked and laid out on the table Hall said at once—this, this and that, etc. are not American but European and Jurassic types, from the Jura, not from the West. When the specimens were compared and these European specimens were selected out and placed aside there were not many specimens left but all that there were were returned to the Smithsonian. Prof. Hall was indignant at the apparent intended deception of sending us over true Jurassic types as coming from the route of the Expedition. I wrote to Marcou for an explanation and had his reply to the effect that in the "tirage" of packing up the Jurassic types got mingled with his collections.

In those days there was little or no room to unpack and display and arrange collections. The rocks and silicified wood specimens of Whipple's expedition were unpacked in a sort of gallery or loft above the platform of the lecture room (or one of the rooms).

As regards the types and collections made by myself in California, a list of which is given in Vol. V (my report) Appendix, in order to keep them together as a unit in which I took some pride, for my rock samples were generally of uniform size and well trimmed, I had a black walnut case of drawers made in New Haven and sent on to the Institution in which I arranged the collection including the Eocene fossil shells and other fossils. This was intended by me as a temporary placing of the collection until cases belonging to the Institution could be provided.

I was distressed upon a visit to Washington sometime after to find that the case had been accidentally overturned in moving it. It used to be in the Document room.—The top was split and all the contents jumbled together. I cannot positively remember now whether I tried to replace the labels and specimens or how it was done, if ever, but I do know that sometime after the case (which I wanted to recover) was in use for birds eggs and Prof. Baird preferred to pay for it or to get another for me. This, however, has never been done.

I was also astonished in looking through the Museum collections at Columbia Coll. N. Y. to see the series of very remarkable concretions which I collected on the Colorado Desert, some of which I had figured in the Report (Vol. V) in the cases. Wondering how they got there I enquired of Prof. Baird and he told me that an arrangement had been made with Newberry to look over the mass of the material which had accumulated with permission to him to take such duplicate material as he liked.

I have never since seen anything of my collection and do not know where the Eocene types are. Those are important for they are from a different horizon than those collected by Gabb, which he insisted were Cretaceous while mine were certainly Eocene. He no doubt thought that my series was from the same bed in which he obtained his. You have heard of the shield with two sides!

But in those days we did not have a National Museum. Certainly not of minerals or fossils. Prof. Baird and I were in close sympathy in desiring such a Museum and also in getting a fair display of our mineral resources.

In 1873 I was commissioned to Vienna to represent our Centennial. One of my duties was to secure as far as possible desirable exhibits for Philadelphia. The Swedish exhibits of iron ores were superb. I secured them for our exhibition as a gift, as also some other exhibits. By the way, one great mass of copper ore weighing two tons, when being loaded for transportation, rolled off its pedestal, crashed through the floor of the building and lies today, I suppose, in the soft alluvium of the Danube. Sweden at Phila. made yet another fine display and Prof. Baird and I secured it for the prospective U. S. National Museum. The other such from Vienna belonged to the U. S. Centennial Commission but was finally transferred to the Smithsonian so that we were well loaded with Swedish iron ores. I think that I had all the specimens from Vienna labelled with white paint numbers. They were packed and shipped after I had left Vienna.

When H. D. Pratt conceived the idea of getting a Governmental direct participation at the Exhibition we were in close correspondence and consultation about it.

We did more than I can now explain to promote the official recognition of the Centennial Commission and of the work by the U. S. Pratt was head of the Inst. Diplomatic Bureau in the State Dept. We arranged to get good Gail Grant—Resident—to go and drive the initial stake at Phila. and the act was by prearrangement wired to me and made much of at the time the work most needed it. Excuse this digression, it may never again be pertinent and possible. The fact was cabled to me in Vienna and was announced all over Europe. Secretary Fish you know was not a warm advocate of the recognition by the U. S. of the Phila. Expos. hence the more need of government exhibits.

Well, the U. S. needed a mineral display. Prof. Baird wanted one. Prof. Henry realized that we should have one and that it fell to the Smithsonian to secure one. In an interview with Prof. Henry on the subject he asked me if I thought it possible to secure a creditable display. I assured him that I could get one but must have the control and the responsibility. He gave it to me and I well remember his saying that if you can succeed in the allotted time you will accomplish a great and creditable work.

The result more than satisfied us both. I hope you saw it. I regret that the pictures taken of it are buried somewhere in the War Department. I had one but lost it. I never have had time to write up a Report on it. Dewey did something in that direction.

However, the point is as to the distribution of that collection. We toiled for months in the winter packing and shipping. It was as cold as the Arctic Zone and no fires allowed. Ink was solid for a month or more. Car load after car load was sent out to Wash. and contents dumped into the old Army Med. building or Hospital. We got mineral exhibits for all countries, labelled or otherwise. We did the best we could to save all. The hoped for Museum seemed secure, all except a place or a building.

After the stress of packing and shipping I went on to Burlington and was there some months and had the Doulton pulpit etc. set up. The Terra Cotta replica from the Albert Memorial pottery etc. Could not do much with the minerals as there was no suitable place. I could not remain as I had hoped, for want of any suitable building or appropriation.

Horan can tell you how I lived at the building and I reckon that some of my furniture is there yet. You will find some few things of mine possibly—some englazed white tiles, a snake jug, and one of the broken Doulton faience vases.

I fear that many of the fine large masses of ores from Nevada are lost for want of identification. We labelled all we could with the red serial number. I wanted large masses. I had a plan of a vast hall on ground floor for minerals, marbles, etc.,—something imposing. Witness the big blocks of iron ore!

Excuse this scrawl. I have written without stopping and fear that I weary you.

Yours very truly,

(*Signed*) W. P. Blake.

Marcou's version of the affair.

Notes on Explorations for Railroad routes from the Mississippi River to the Pacific Ocean, made under the orders of the War Department in 1853-54; by Jules Marcou.

Cambridge, Mass., 30 April, 1885.

In the middle of May, 1853, I received from the Smithsonian Institution a letter written under the direction of Prof. Joseph Henry, of which the following is an extract:

Smithsonian Institution,
Washington, D. C.
14 May, 1853.

. . . The Congress in its last sitting has ordered the exploration of the country from the Mississippi to the Pacific Ocean; the Commission for that purpose will be obliged to be divided into several parties. A geologist is asked for; he will receive besides his travelling expenses and board, $100. a month. The expedition will start in one month or six weeks, and the exploration will last one year or 18 months. The geological observations and all the collections will belong to the geologist employed, with the only condition not to publish anything anywhere else until the publication of the Report by the Congress.

What do you say of it? It will be a splendid occasion to complete your geological map of the United States [then under the press], and also to draw a section of the country from the Atlantic to the Pacific.

You can choose your road; it is to say, you can go with the party by the Upper Missouri and Oregon, or by New Mexico, or if you prefer by the intermediate road of Great Salt Lake.

Please write directly if you accept the offer; and if you decline will you be kind enough to indicate any other geologist who may be able to fill up the place. . . .

To Jules Marcou, Esq.,
Harrison Square, Dorchester, Mass.

Having accepted with the condition that I have chosen to go with the party near the thirty-fifth parallel of north latitude, I received the following order:

Washington, D. C.
May 28th, 1853.

Circular:
Sir:

You will please repair without delay to Napoleon, Arkansas, where further instructions will await you.

You will commence from that place an official journal of your operations and each day note whatever you may deem of interest with reference to the expedition with which you are connected.

An account memorandum of your strictly necessary travelling expenses will facilitate the settlement of your accounts.

I am, Sir,

Very respectfully,
Your obd't servant,

(*Signed*) A. W. Whipple,
1st Lieut.,
U. S. Corp Engrs.

To Jules Marcou,
Geologist and Mining Engineer,
Pacific Railway Survey,
Harrison Square, Dorchester, Mass.

Two days after receiving the above letter I left Boston on the 2nd of June en route to the Pacific shores. At Napoleon, Arkansas, I found the following order, sent forward from Dorchester after my departure.

Washington, May 30th, 1853.

Circular:
Sir:

Upon your arrival at Napoleon, Ark., you will please proceed as expeditiously as possible to Fort Smith, Ark. Upon this portion of the route each officer of the scientific corps will please make such reconnoisances and notes as the rapidity of the journey will admit.

Very respectfully,
Your ob't servant,

(*Signed*) A. W. WHIPPLE,
1st Lieut., U. S. C. Engrs.

To Jules Marcou,
Geologist, Pacific Railroad Survey.

From the 16th of June at Little Rock, Ark., I kept a record of the geological observations I was able to make in a field book, until we reached the Pacific Ocean at San Pedro near Los Angeles, Cal., the 24th of March, 1854, the party having disbanded the day before at Los Angeles.

During the five weeks stay of the party at Albuquerque (from the 5th of October to the 10th of November) New Mexico, I explored the part of New Mexico between Albuquerque, Cigeras, San Antonio, ascended to the top of the Sierra da Sandia, then to Galisteo, Pecos village, Santa Fe, San Felipe and Bernallilo.

Brev. Capt. John Pope of the Topographical Engineers in garrison at Albuquerque, having received, during our stay there, order for an exploration along the 32nd parallel of latitude from the Rio Grande to Preston, Tex., asked me some directions, how to collect and observe the geology of his contemplated expedition, which I gladly gave him.

Finally, having obtained a two weeks leave from Lieut. Whipple, I hastened to visit from the 1st to the 11th of April the gold mines on the Juba and Feather rivers in California.

As soon as I was in possession of my specimens I wrote a geological resume for the octavo edition of the Pacific Railroad explorations, which appeared in Whipple's "Report of Explorations for a Railway route, near the thirty-fifth parallel of Latitude;" 8vo Washington, 1855, forming Chapter VI under the title "Resume of a geological reconnoisance extending from Napoleon at the junction of the Arkansas with the Mississippi, to the Pueblo de Los Angeles in California." House Document 129, 1a, page 40 to 48. All geological discoveries along our road, until then entirely unknown, were put into this Resume.

I also wrote a geological resume of the road explored by Capt. Pope, using his notes and specimens which he placed in my possession early in September 1854. This report was published in the octavo edition of Brevet Captain Pope's "Report of Explorations of a route for the Pacific Railroad near the thirty-second parallel of Latitude, from the Red River to the Rio Grande." Washington, 1855; forming Chapter XIII under the title: "Geological notes of a survey of the country comprised between Preston, Red River, and El Paso, Rio Grande del Norte." House Document 129, 1c. pages 125 to 128.

Sickness contracted on crossing the Isthmus of Panama obliged me to seek a winter residence under the mild climate of France, and having obtained leave from my superior, Lieutenant Whipple, I was on the point of embarking when the very morning of going on board I learned that Jefferson Davis, then Secretary of War, objected to my going, and overruling the lease given me both by Lieutenants Whipple and Pope, ordered me to stay or give up my notes. Alternative to which it was impossible for me to submit, as all my luggage was already on board the Cunard steamer. So I sent my resignation to Lieutenant Whipple and started for Europe where I delivered my notes and specimens to the American legation in France.

I did not complain, nor ask any compensation, notwithstanding I was solicited to do so, and submitted to the arbitrary act of Jefferson Davis who tried on two lieutenants and me his overbearing manners and tyrannical power, which a few years later made him so conspicuous as the leader of the Rebellion against the United States.

Two years after the publication of the Pacific Railroad Exploration edition 8vo and edition 4to, I replaced as well as I could from memory and the printed field-notes, in my "Geology of North America, with two Reports on the prairies of Arkansas and Texas, the Rocky Mountains of New Mexico, and the Sierra Nevada of California, originally made for the United States Government," Zurich, Switzerland, in 4to, 1858, all the geological results and principal observations of my explorations by the thirty-fifth parallel of Latitude.

Madison, March 4, 1857.

Prof. Hall,
Dear Sir:

A bill has just passed the Legislature appointing yourself, myself and Daniels a geological commission to make a geological and agricultural survey of state, and appropriating $6000. per annum, for *six years,* for that purpose. The commission are to agree among themselves as to the division of labor, etc., and the Governor contracts with each to perform such labor at a salary not exceeding $2000. per annum or in that proportion for a part of the year. The appropriation of $6000. a year is to cover all expenses except printing the reports.

The connection of Daniels with the survey and his removal by Barston you are, I suppose, familiar with. He has, since his removal, been very active as a politician, so much so that his own party fear to appoint him again. I have been assured by those friendly to him that he could not have been appointed in any other way, and it is expected by all that you will employ a portion of the time in the field. Under no other consideration would the bill have passed. Daniels would, I suppose, like to have it understood that he is to do this work, and be in fact *the Geologist,* leaving to yourself the determination of the fossils, etc. This would not be satisfactory to the people of the state, who look to you for the determination of questions in relation to the geology of the state, which they believe no other man is competent to determine. I mention these things, for I suppose that Daniels will be in Albany in a few days for the purpose of conferring with you in the matter. I think it would be well if you were to contract with the Governor for all of the time you can possibly spare through the entire *six years,* not confining yourself to the Paleontology. If it were understood that Daniels were to do the work and be the responsible head I think the next Legislature would repeal the law if it can be done. (I am told, however, that the Governor can contract, under the present law, for the full term of six years, which contract cannot be *destroyed* by Legislative action.)

Could you not come out to Madison in the course of a week when we could have the whole matter arranged?

I should desire myself to have simply the agricultural and assaying department. To carry this out as I want to do it will require an expenditure of $2000. per annum including assistance and everything. This was the original designation in the bill.

I have written this much hastily and *confidentially,* in order that you may act more understandingly in the matter than you might be able to do from an interview with Daniels alone. Daniels and myself, however, so far as I am aware, are on perfectly friendly terms. I should not, however, implicitly rely on everything he says in relation to *himself,* and I think he is wanting, to some extent, in the confidence of the people of the state, in part, perhaps, from political considerations. The present arrangement, is, I think, a very satisfactory one, you being regarded as *the geologist* and occupying a portion of the year in the field.

I should like to hear from you very soon for should anything occur to prevent your connection with the survey it would have much influence on the course I should pursue.

Yours truly,

(*Signed*) E. S. Carr.

Albany, Dec. 20, 1865.

Dear Sir:

When I saw you last summer you advised me to write to the Governor elect after the November elections, in reference to the Geological Survey of Wisconsin, and my relations therewith. My absence upon some geological explorations in Georgia and Alabama have prevented me from doing so at an earlier date.

The Geological Survey of Wisconsin was suspended by an Act or Resolution of the Legislature in 1862, just after the publication of the first volume of the Report, and while I was going on with the preparation of the second volume.

I held a contract under the law authorizing the survey, a place I had not sought, and was authorized and directed by a subsequent law to assume other responsibilities and perform other duties without additional compensation. These duties I performed to the best of my abilities, and as far as the means at my disposal permitted. The law was repealed in violation of its requirements and of the stipulations of my contract.

I had already up to that time incurred expenses in drawing and engraving for the second volume, and I afterwards continued to work under my contract to complete the work I had commenced.

I hold in my hands collections of specimens necessarily brought here for study and description. I have manuscript report, and maps of Mr. Whittlesey's work in the iron region of Lake Superior—my own manuscripts prepared for my department of the work—proofs of engraved plates, etc., which I wish to deliver to the State and receive the money due me on my contract and the expenses I have incurred.

Two years since, and also at the last session, a bill was brought in to accomplish this object. Two years since it was defeated by some means, and last year after having passed the House, was defeated I believe by inattention or neglect to call it up in the Senate.

Mr. Julius T. Clark has very kindly aided in this matter, and Ex-Governor Randall has written to Senator Hood in reference to the same. I will ask Mr. Clark to call and explain to you his views on the subject.

To me it is a matter of importance; not only in the want of money I have expended and that which is due me; but the prevention of publication and illustration of my work as I had been promised, and which every scientific man feels to be important to his reputation, is an injury to me.

I do not wish any longer to remain the custodian of collections and manuscripts belonging to the State, and on every account I am desirous of closing this matter if possible.

The Introduction to the Geological Report, volume I, contains the Laws, etc., regarding the Survey, and a duplicate of my contract is on file in the Executive Department. I will, however, inclose a copy of the contract.

Should the State at any time conclude to resume the Survey, these materials would be useful and important to the parties engaged, and save considerable labor and expense.

I shall feel greatly obliged if you can do anything in aid of my objects, which I conceive to be right and just.

I am, very respectfully,

Your obedient servant.

J— H—

Hon. Governor Elect,
Gov. Fairchild.

John S. Newberry to Hayden, February 10, 1858.

Steamer Explorer, Mohåve
Villages, Colorado River.

My dear Hayden:

From Stimpson's letter just received I infer that you are now in Washington and I hasten with the first news of your return to acknowledge my indebtedness for several kind letters which you sent to Cleveland for me before and since my departure. I was not ungrateful for your letters, nor did I undervalue them—but I was unable to write letters—from the state of my health until I returned to C. from Lake Superior and then you were—who knows where?—I therefore concluded to wait and write you on your return. This I now do with great pleasure, as I am assured of your safe return from a region just now full of perils and as Baird writes me, having been as usual successful in collecting a large amount of interesting material.

I should be very happy to be one of your pleasant circle at the Smithsonian this winter, to compare notes, discuss the material, and above all, help you so far as in me lies, in the study of the fossil plants.—"But so the fates have not decreed."—I am doomed to pass the entire winter and spring doing the hardest kind of field duty with few of its pleasures or rewards. Day after day as we slowly crawl along up the muddy Colorado —confined to a little tucked up over-loaded, over-crowded steamer with no retreat from the cold, heat, wind or drifting sand, and nothing but the monotony of an absolute desert to feast our eyes upon, with nothing but bacon and beans and rice and bread *and sand*—or rather *sand and bacon,* etc., to eat, sleeping on shore on a sand drift, eyes, nose, mouth, ears, clothes and bed filled with sand—with almost everyone discontented and cross. I sometimes *almost* envy you who are reposing in your "otium cum dig.," studying abundant material, eating comforting food, sleeping on good beds, washing clean and dressing neatly every day, and having a good time generally. I only hope you appreciate your advantages, with sometimes pity on poor wretches who are not so fortunate.

However, there is another side to the picture—and I am not sorry to be here—nor wish to return till the work is done. The plan of the expedition has been all broken up by the Morman troubles—and we shall only be ready to commence work with all our resources and comfortably a month from now. My health is excellent and I shall at least be able to make a *complete* geological map of the river banks. Everything which we get is interesting and much of it I know is new.

The survey and mapping of the river is slow, monotonous work but it is going on very successfully and will soon be done. Then for land travel and more variety. I, thanks to the kindness of Lieut. —— suffer few of the evils or discomforts I have enumerated, my greatest source of regret being that my days which I now value so highly should so many of them profit me or the world so little.

We expect to leave San Francisco on the steamer of July 5, at farthest. Do not fail to write me there, care of Dr. Ayre, when you receive this.

Give my love to all who love me. Ah who does in all W. Be a good boy—Dont get tight. And when you read this, as always think of me as one of your best friends.

Most truly yours,

(*Signed*) John S. Newberry.

P.S. I am sorry this sheet is disfigured by this caricature of a Mohave belle but paper is very scarce and I must use up every scrap.

Albany, March 2, 1858.

My dear Sir:

I write you in great haste to inform you that our Permian matter is perhaps lost. Hawn and Swallow have acted very meanly about it. Hawn sent some of these fossils to me, and some to Swallow long since without ever dreaming what formation they belonged to. Since that time Swallow read a paper before the Montreal meeting of the Assoc. on the Kansas formation without ever mentioning the word *Permian.* I wrote Hawn several times frankly telling him I believed the rocks from which he obtained his fossils must be of Permian age, and requested him to get Swallow to let me have the fossils sent him from the same formation, if he (Hawn) wished us to describe them, otherwise, if he wanted Swallow to investigate them I would send him those in my possession,—stating that whoever took them should have the whole collection.

Since that time Hawn wrote me that he had written Swallow, requesting him to send me all the fossils in the collection he (Hawn) had sent him *that were from formations newer than Carboniferous.* Subsequently I received a letter from Hawn in which he stated that he had received a letter from Prof. Swallow, stating that the fossils *were all Carboniferous,* which I firmly believe was Swallow's opinion until he learned from Hawn what I had said about them.

Fearing someone might get hold of the fact and bring it out I wrote Hawn to be very cautious about the matter, and not to say much about it. He replied there was no danger, as no one had ever seen the fossils but Swallow, and he regarded them as being all Carboniferous. I believe I can show in at least six or eight,—certainly three or four of Hawn's letters, where he gave me free privilege to work up and publish these fossils—indeed he rather urged me to do it.

Since I returned from W. I wrote Hawn we had taken up his fossils in connection with Lieut. W.'s collection, and that *I was now more than ever satisfied they must be Permian.* He wrote back *from Columbia, Mo.,* that Shumard and Swallow had looked over his whole collection and *pronounced them to be all Carboniferous*—adding that Swallow *appeared to take very little interest in them.*

It was this letter from Hawn, stating that he had taken his collections to Columbia, which caused me to fear the thing might leak out, and to send on to Dr. Leidy a letter announcing the probable existence of Permian rocks in Kansas. Whether Leidy has yet read this letter before the Academy or not I do not know. If he has, and it can be out soon, it will perhaps be ahead of them yet.

This evening, to my utter astonishment, I received from Hawn a letter stating that supposing I had not time to do anything with these fossils, he had submitted the whole collection to Prof. Swallow who had *without any hints from him* determined them to be Permian, and written out descriptions of the fossils and read them before the St. Louis Academy of Science, and that the descriptions will be ready for distribution to the scientific world soon.

What can honorable men think of such a course? Is it not as clear as daylight that they were availing themselves of our study of them?

(*Signed*) F. B. MEEK.

B. F. Shumard to F. B. Meek.

[a]AUSTIN, April 5, 1860.

MY DEAR M.: Your kind letter reached me to-day, and I can not sufficiently thank you for the friendly feeling that prompted you to write it and for the course you have taken in refusing testimonials to the aspirant who desires to supplant me in the place I now occupy. I shall speak to you unreservedly, for I have had too many assurances of your friendship to doubt it. Of the qualifications of the person alluded to, to take charge of a work so important as the survey of this State, I need not inform you. What he knows of geology has not been gathered from study, but from conversations with geologists. Thus he at first made the Coal Measures of Fort Belknap Tertiary and wrote a long article which was published in the Texas papers! He then took some of the same fossils that he relied upon to prove their Tertiary age to the North, submitted them to "my friend Mr. Meek," returned to Texas, and shortly after published a learned (?) article in which he referred these beds to the Coal Measures, their true age. I am aware that he professes to be a friend of mine, but I can cry, "save me from such friends." He says truly that I am or have been in trouble, but all of the trouble has been caused either by himself or his friends. He spent some five or six weeks here last winter during the session of the legislature, and it is believed here that he used every endeavor to get either my position or that of one of my assistants, either of whom are much better geologists than he. Governor Runnels refused him the appointment solely on the ground of incompetency, and insisted, for the same reason, that I should not give him the place of assistant. He has been a politician all his life and for many years edited a leading political paper in this State. No one in this State believes him to be a geologist, although for political reasons some profess to think so. I believe conscientiously that if the geological survey of this State is abolished, it will be done through the maneuvering of Dr. M.'s friends or himself. It would have been abolished last winter had the legislature entertained the opinion that Dr. M. would be placed in charge of the survey. The people throughout the State feel a great interest in the survey. But the important work will assuredly cease with the next legislature if Houston makes the change. I do not know what encouragement Gov. Houston has given Dr. M. He (Houston) has removed every one of Runnel's appointees, except myself, and but for the interference of some of Houston's warmest admirers I should have shared the same fate ere this. I have had indirect assurances from influential Houston men that I am not to be disturbed. It may be, however, that he would like to remove me simply from his hatred to Runnels, and that he would like to shield himself behind testimonials in Dr. M's favor from such men as Professors Henry, Bache, and yourself. Or it may be that Dr. M. wishes the testimonials merely to induce Houston to give him the place. Of one thing I am quite certain, and that is, any testimonials he may succeed in procuring will be employed to the injury of the survey. . . .

I am sorry that I have had to say unkind things of anyone, but in the matter of the Texas survey the case requires it. I shall feel much obliged to you if you will communicate the contents of this letter to Professor Henry, to whom I am under *many* obligations for *many* favors. I shall strive to merit the good opinion that he entertains of me.

. . . .

(*Signed*) B. F. SHUMARD.

Salem, Mass.
Oct. 24, 1864.

PROF. JAMES HALL,
Dear Sir:

I have been much interested in reading the introduction to Vol. 3 of the Palaeontology of New York, and particularly with your observations upon mountains: but one thing I do not understand. You say in a note to p. 71—"This mode of depression . . . offers a satisfactory explanation

. . . of the difference of slope on the two sides of the anticlinals" etc. Now while I can see that compression of the upper surface might take place, I do not see why it should produce waves *steeper on one side than on the other*, or at any rate how it should do so to the extent and with the regularity described by Prof. Rogers in his Geol. of Penn. By the way *is* that regularity in flexures to be found on the ground, that Prof. Rogers describes?

My desire to understand correctly your excellent paper, has induced me to address a person with whom I have no acquaintance.

Respectfully yours,

(*Signed*) GEORGE L. VOSE,
Salem, Mass.

Albany, Nov. 5, 1864.

DEAR SIR:

. . . .

In reference to the point stated in your letter I will give you my explanation as it occurred to me at the time. The depression arising from accumulation will be the greatest in the center of the map or along the line of greatest thickness, thus, probably together with a general contraction of the crust, causing the plication. It appeared to me that the longest slopes would be toward the center of the great synclinal, while the outer slopes would be shorter and steeper from the general movement toward the center, this movement being in fact equivalent to a pressure from the northwest against the base of the synclinal.

Notwithstanding this agreement it may be found that I have invented an explanation for a condition which is far from universal, for I do not think that this difference in the slopes occurs with that regularity which we have been taught to believe, nor do I think it possible theoretically that such regularity could exist when we consider the inequalities of the thickness and nature of the sediment, and sometimes the sudden changes noticed in the material of these sediments.

The point to which you refer is one that I have intended to investigate and to discuss the principle at some time, but an accumulation of labor in the line of palaeontology will keep me occupied for sometime. My 4th volume is in press and some where about one hundred pages in type.

If I can sustain the great principle which I advocate viz., that mountains are not produced by upheaval but by accumulation and continental elevation I shall feel that I have done something to advance the Science of Geology in true principles. I feel quite sure of its ultimate adoption because I feel quite sure that it is the only true explanation—the only mode of making mountain ranges, for they cannot be made without material and no imaginary upheavals will ever explain their existence.

It is a gratification for me to know that any one is reading on this subject with sufficient courage to ask a question or make a criticism and it gives me pleasure to reply to your inquiry,

And I remain

Very sincerely,
Your obedient servant,
(*Signed*) JAMES HALL.

George S. Vose, Esq.,
Salem, Mass.

Tucson, Arizona,
May 18, 1906.

PROF. GEO. P. MERRILL,
Head Curator of Geology,
U. S. National Museum.
Washington, D. C.

My dear Prof. Merrill:

Please accept my sincere thanks for the copy of your "Contributions to the History of American Geology" and for the measure of recognition of my work which you have accorded me.

The ammonite referred to, on p. 516, was not, as represented, a fossil of uncertain locality but was truly found *in place* in the slates of the American River near Colfax, as I correctly stated. Brewer seems to have confused the idea of this specimen with one which was of uncertain locality, but which had its due weight in evidence. This error is pardonable but the claim that the fossils were not found by me *so early as the date* I claimed, and do claim for them, is unpardonable, and I am astonished that Brewer should think it. It is a mendacious misrepresentation, unworthy of its author, and a blot upon the page of your book.

Whitney's claim must rest upon such evidence as he secured in the northern counties (largely anticipated by Trask) and not upon the fossils I found and brought to notice from the central gold-belt. I appeal to the printed records.

But I will not now thrust this controversy and injustice upon you further. It needs the vituperative genius of some of our contemporary senators—Bailey, for example—to stigmatize Whitney's disposition and actions as they deserve. It was infamous.

Yours very truly,
(*Signed*) WM. P. BLAKE.

St. Louis, Sept. 4, 1871.

PROF. JAMES HALL,

Dear Sir:

I little thought that you would be one to do me an injury, or become the agent of another to do it. I am not fully satisfied now that you would do it, unless you were imposed upon and influenced by misstatements. Your resolution in the Scientific Association is construed to carry the idea that the work of the Geol. Survey of Mo. was not being properly conducted. The same one which I think influenced you to do the sad job in the A. A. A. S. has worked hard to get me out of the Survey—I mean Prof. Swallow, and the work is at last accomplished, for I am discharged, and the survey work is about suspended. Dr. Norwood, my assistant is fixing up some work in Madison Co., but must return in a few days to his university duties, as he will not longer be connected with the uncertain work of a survey of Mo. But it will be of no use to Prof. Swallow for the board dare not, in my opinion, appoint him to the position of State Geologist. He is too unpopular in the State. He has not worked to the satisfaction of the people. He has bored the Legislature with the assertion that he had a large amount of very valuable unpublished material on hand, when in fact he has not a line of unpublished MS. in his possession unless he told Dr. Norwood and myself a lie when we went to his house and took away all the MS. and other articles belonging to the State, for he said he had given up all he had except his note books—things which I did not want.

He represented to me when he wanted me to appoint him my assistant, that he had a great deal of valuable material and valuable notes on hand and I then believed he had, but since then am led reluctantly to doubt his word. When I nominated him, the members of the Mining Bureau would not confirm him for the reason, as they said, that they had been "Swallowed enough." They did not believe he had this "valuable material" on hand and directed me to make careful examination and report to them at an adjourned meeting what this "valuable material" consisted in. If I found a large amount of valuable MS. some said they would second my nomination.

I told Prof. S. that the board would not confirm him unless he was able to show that he had done something during his eight years' work which had resulted in value, and desired him to show all that had been done. I think he did so; showed everything he had. All the MS. which he could muster consisted in his Report on 147 species of fossils and about ten pages on five counties which had been examined and reported on by his assistants.

I subsequently was informed that the descriptions of the fossils had been published in the Proc. of St. Louis Acad. of Science, but he never told me so, but carried the idea that it was unpublished MS. and that he once had a great deal more, but was not then able to lay his hand on it. He has never been able to do so since. I would suggest that the memorializing committee, if they would not wish to be placed in a ridiculous attitude before the people of Mo., examine this *valuable material* before they press the Legislature to publish it. The reports of Dr. Shumard, Locke, Meek and Broadhead were in my hands and I had arranged to publish them, not however till the authors living had revisited the territory and amended their reports. By reference to my Report of progress which I sent you, you will see that I gave Prof. S. full credit for all he handed over, and even for the ten pages which he did not, and also suggested in that report that the amended reports of those formerly engaged in the Survey should be published. The maps which he exhibited at the Meeting—if a large "Folio Map of the State," has been colored since I took away his assistant's reports &c. He begged me to leave that map and I did so. It had no geological coloring on it then. It may be correctly colored. He must have done it recently and for the occasion. Were you deceived by him, provided my statements are true?

Respectfully,

(*Signed*) A. D. Hager.

Albany, Sept. 10, 1871.

Albert D. Hager, Esq.,
State Geologist &c., St. Louis, Mo.
Dr. Sir:

I have your favor of the 4th inst. which sets out with the charge that I have done you an injury, but you qualify it by saying that I have been deceived, etc.

I may have been, for I have trusted many persons and have been often deceived. However, there was no intention of doing you an injury and when the matter was discussed by the committee, it was suggested that I should write you a letter saying all this, but I was compelled to be absent on official business after my return from Indianapolis, and returned only on Friday to find your letter of the 4th inst.

The appointment of this committee could have had nothing to do with your being discharged, or even by inference simply that the work of the present survey was not properly conducted. Certainly no sane board would have discharged a man from office simply because a committee had been appointed by a scientific association to memorialize the Legislature to permit one who had formerly occupied the position of State Geologist to publish the results of his labor to the time of the suspension of his work by the events of the war; and that is the form of the resolution—asking simple justice and nothing more. It was an act which I would do for any in a similar position.

But you give me too much credit: the resolution did not originate with me. A person who now holds the position of State Geologist in another state came to me and asked if I would introduce such a resolution. If Prof. Swallow has such material in his hands, it is but an act of justice that he be allowed to publish it. I feel this from personal experience. If he has nothing to publish, as you say he has not, there is no harm done except to himself, for if he asks this privilege and it is granted, he is thereafter responsible to the Legislature and the public for what may be done or left undone.

No harm can come to you from allowing Prof. Swallow to publish. You then have a clear starting point, and what you do thereafter is your own work, and you can never be charged with revenging him. Were I in your position, I would join in asking this from the Legislature.

With regard to the present or past of Prof. Swallow, his popularity or unpopularity, with his giving up his manuscript, or whether he told you the truth or otherwise, I have nothing to do and will have nothing to do. It is a simple matter of justice, and there lies at the bottom of this a principle which has been so often violated, that numerous feuds have grown out of it and are likely to continue.

The states of the West have been unfortunate in their surveys or in their management of them. But I hold that in such a case as a suspension of a work of this kind, the smallest concession a Legislature can make is to give the author a chance to publish the results of his investigations.

So far as I am concerned, I disclaim any intention of injuring you personally, or the survey in your charge, nor do I think that this action properly understood, can be so construed. Had I not received your letter I should have written you today, disclaiming for myself and for the committee any thought or intention of doing you an injury.

I shall feel greatly obliged if you will show this letter to the Governor, or to some member of the Board controlling the survey.

I am very sincerely and respectfully yours &c

(*Signed*) JAMES HALL.

Washington, D. C., March 12th, 1877.

DR. J. S. NEWBERRY,
Columbia College,
Cor. 49th St. and 4th Av., New York.

Dear Sir:

Mr. Newton has failed to get an order for the publication of his Report, through influences that I need not mention, as you will fully understand the matter.

I have very carefully followed Mr. Newton in the preparation of his Report, and I know that it will be a very valuable contribution to Science as a monograph of the geology of that region. Its publication is due to Science, and also to the industry and ability of Mr. Newton. The Secretary of the Interior has authority to order its publication, but it will be necessary to pay out of the "contingent fund" of the Interior Department for the reproduction of some of the Plates and the Map, to the amount of two or three thousand dollars.

In behalf of Mr. Newton, I beg of you Doctor, to help us with your influence in presenting the matter to the new Secretary of the Interior,

Mr. Schurz. A good, strong letter from yourself, and from any person in New York whom you think would have any influence with Mr. Schurz, would be of great assistance.

I simply make this suggestion leaving you to act as you may deem wise.

I am with respect
Your obedient servant,
(*Signed*) J. W. POWELL.

[A "good strong letter" was written by Newberry, and apparently it was effective. G. P. M.]

Agreement between James Hall and I. A. Lapham.

This agreement made this first day of March A. D. 1853 by and between James Hall of Albany, N. Y., and I. A. Lapham of Milwaukee, Wis., witnesseth: that said Hall agrees to prepare a work to be called American Paleontology—et cetera, based upon manuscripts now placed in his hands by said Lapham (which MSS. embraces descriptions of about two thousand species) and to procure the publication thereof upon the most favorable terms, as the joint work of said Hall and Lapham; that in case no publisher is found to assume the expense of the publication, then the work is to be published at the joint and equal expense of the said parties hereunto; that said work and publication is to be completed within one year from the date hereof; and that all proceeds and profits resulting from said publication are to be divided equally, between the parties hereunto.

Witness, our hands, on the day and year first above written.

(*Signed*) I. A. LAPHAM
JAMES HALL

In presence of
F. B. MEEK.

[Lesley's standing with at least one of his contemporaries is shown in the following letter of James Hall to William A. Ingham, Esq. (June 8, 1874). G. P. M.]

I have just learned from my friend Prof. Barker of the University of Penna. that you are a member of the board, which by a recent act of the Pennsylvania Legislature is directed to appoint a State Geologist. Until today I had not supposed that anyone among the geologists whom I know would become a competitor with Prof. Lesley for the position.

And without disparagement to anyone else, whether an applicant for the place or otherwise, I would beg leave to say that Prof. Lesley's abilities and qualifications for the position are so far superior to those of anyone whom I know that I could not for a moment entertain a doubt of his selection for the place were the matter referred to a number of scientific men. The structure and relations of the formations composing the Appalachian Chain involve questions of the highest importance to the Science of Geology as well as others equally important in an economical aspect. I do not hesitate to say that no other man within the range of my acquaintance is so well prepared, or at all equally prepared for the solution of these problems. I have myself had considerable experience in Geological Surveys and have been connected with Geological Surveys of New York with its commencement, but for the solution of these questions in Pennsylvania I would never for one moment entertain the proposition of placing myself in competition with Prof. Lesley. I have many years since had the pleasure of discussing the Scientific ability of Prof. Lesley with Prof. Bache, Prof. Agassiz, Prof. Henry and Prof. Frazer, and his name was placed foremost amongst all as a Geologist. At a later period when the question of the appointment of a professor of Geology at the Cornell University was under consideration by those having the matter in charge I decided that if Prof. Lesley would accept the position there would be no competition. I wish to express myself in the most unqualified manner in favor of the appointment of Prof. Lesley as State Geologist of Pennsylvania, his native State. I feel quite sure that no other appointment would be so acceptable to the best Geologist of our Country, or receive so fully and cordially the assent and support of the best men in other departments of Science. It is right for me to say that Prof. Lesley has not addressed me in any manner upon this subject, and I write from my own interest in having the Pennsylvania survey properly and efficiently organized and administered throughout.

My Dear Dr. Merrill:

. . . .

Lesley's assistants did practically as they pleased but without exception worked hard. Every man had full credit for all he did, good or bad. But Lesley never exposed one who did poor work. If the report was very bad Lesley wrote a new one and published it under the other man's name. I know of two cases. He asked me to write a volume once "to relieve a man," but I slid out. He was a curious compound. One day I was correcting proof in Harrisburg, and a postal came demanding why I had invaded Platt's district and had described the choicest bit of his area. My letter of instructions was in N. Y. but Lesley's report to the Commissioners was at hand. I copied his statement to the Commissioners and sent it to him. By return mail came a postal, "Wrong as usual." Who could harbor ill-will against such a man? The Frenchman once said, "You can apologize but you can't unkick me." Lesley could come as near to unkicking a man as anybody I ever knew. Of malice he knew nothing: his resentment was as transient as the morning dew. Love once kindled in his heart was unquenchable. One day I was endeavouring to dissuade him from putting into a place of some prominence one who had proved himself, to say the least, harmful. Lesley looked at me a moment. "Stevenson, when he was baby I used to dance him on my foot. No matter what he does, I love him."

Yes, Lesley was nobility incarnate. If J. H. had lived in Philadelphia among men, and not in Albany among politicians, he would have been a very different man. But he never met anybody: he was isolated. Whenever any busybody in science found no devil's work to do elsewhere, he would put a suspicion into H's mind respecting some friend and make the poor man miserable. If H. had had half a chance he would have been a lovable man. You think I was wrong in calling him child-like. A man possessing his confidence could do anything with him. He was the most tractable man I ever met. . . .

Some other day I'll write you again about something else.

Sincerely yours,

J. J. S.

Feb. 24, 1869.

My dear Hayden:

At the last meeting your No. 2 was referred to the Secretaries with power to act.

I have sent on the two pictures to Osborne.

Le Conte read your Mss. through carefully and returned it with a note saying that it contained such important matter that it was a pity it had not been put into a more agreeable dress. He objected to pp. 88 and 89, and somewhat to the recapitulation of services, with which the Society has nothing to do and which is no part of a scientific memoir, although it will do very well in a book or a coal report.

I not only agree with him but feel a still greater repugnance to both these than he does. In fact I have no hesitation in drawing my pencil across them. And I send back the Mss. to you, that is, from p. 85 to 96, that you may see what is stricken out.

I have been charmed by your narration. But I must say your style of narrating is slovenly in the extreme and when it gets into print exactly in its present condition I shall blush for the belles lettres reputation of the society. You have not only written it straight ahead and at railroad speed from your notes, but somebody has copied it and has punctuated it into an unintelligible mass in some places. A hundred words (vincula) have been left out. There is absolutely no system of subdivision *"at all, at all*:" I wish to heavens you would give me a very little liberty to correct the proof. If you will, you shall have the revises sent you regularly, so that you can still have command of the text so far as any corrections of mine are concerned.

Frye says your paper is a mass of words, all stuff. But you know what *his* style of criticism is,—he would only read a page or two. So Le Conte and I act without him. Kendall is a man of delicate taste and is offended by slovenly writing—but he yields to our judgment of the great essential worth of your work. For my part I admire the steady tramp of your narrative, and I think a *very* little attention to the proof-reading will satisfy everybody.

Ever yours,

(*Signed*) LESLEY.

N. B. One or two of your observations are as surprising as they are important for the higher geological questions of the day.

Albany, May 9, 1853.

MR. HAYDEN.
Dear Sir:

I have just received your telegraphic despatch. Mr. Meek left this morning and I hope will be on the ground before the boat leaves.

I will mention now one point that may be of interest to Mr. Meek and yourself. It may sometimes be necessary that the direction of your expedition and the affairs connected with it be under the direction of one of the party. I presume you can have no objection to Mr. Meek exercising this prerogative from his greater age and his experience in the field. I presume you will have no difficulty, however, in arranging this matter between yourselves. Still I think were I in your place I should tender the direction of all matters to Mr. Meek as soon as he joins you. I think you both together can work to far more advantage than one alone and with strayers only.

I hope you may have a pleasant and successful journey.

Sincerely your friend,

(*Signed*) JAMES HALL.

St. Louis, May 16th, 1853.

PROF. HALL,
Dear Sir:

I received your letter dated May 9th, this morning. It was the first word I have had from you since leaving Cleavland. Mr. Meek arrived here 1 o'clock Sunday morning. We have today been making our purchases and shall be ready for the boat which will not start until the eighteenth. I had already adopted your suggestion of giving the whole charge of things to Mr. Meek. He has had experience in such matters and I have had none. He will be more cool and calm, and will make his past experience available. I had made up my mind from all the information I could receive, that the Land Route by Fort Laramie would be the best, and I know that it would not cost half as much, still there are reasons which Mr. Meek has for going up on the boat which he will tell you. The expense will be very great, 1000 or 1200 dollars this fall. I fear the collection will not equal the expense by far. Mr. Meek has refused here the most flattering offers to leave you and join Evans & Major Stevens. He has certainly acted nobly, truly so, showed himself to be a man of the sternest integrity. He deserves much credit, and should have the credit of this expedition. I shall labor with all diligence and accomplish all I can, but it will take most of this season at least to make up my account of expedition. Therefore I shall refuse to receive any credit for what may be done this season, however successful the expedition may be. With the expense necessarily incurred this summer, I wish to ask you whether I can spend the coming year there? It is an entirely new field.

. . . .

Drs. Norwood, Evans and Shumard are here. Shumard & Evans go to Bad Lands. Dr. Owen will be here soon. Major Stevens and his party are here. All are most violently opposed to our going to Bad Lands. Mr. Meek will tell you more fully. Prof. Swallow is here also.

. . . .

(*Signed*) F. V. Hayden.

St. Louis, May 19, 1853.

My dear Sir:

I now sit down after a world of perplexity to write you a hasty line. I did not write sooner because I wished to wait until I had settled, in my own mind, the propriety of going on with the expedition, under the circumstances in which I found myself placed. On arriving here, I found Drs. Evans and Shumard were going to stop on their way to Oregon to make a collection of fossils at the Mauvaises Terres. I called upon Major Stevens and found him very much offended at you, for sending out the expedition, but upon being assured by Mr. Hayden and myself that you know nothing about Evans' intention to visit that region this season, he appeared to view the matter in a different light. I have since read to him a portion of your letter of the 9th inst. and he said he was satisfied and took back all he had said about you.

Being at a loss how to proceed I telegraphed you, and waited some time for an answer, but have received none. After thinking the matter over I concluded the best thing I could do would be to submit the whole matter to Prof. Agassiz and Dr. Engelman, and be governed by their advice, for I am well satisfied they are your sincere friends. They consulted about it for some time and had a conference with Dr. Evans, after which they advised me to abandon the expedition, on the conditions that you should be compensated for the amount you have expended and that Mr. Hayden and myself should be taken into their corps as assistants. (They were anxious to engage me.) Prof. Agassiz said their reasons for this advice were that they, as your friends, considered it of more importance to you that the whole matter should be amicably settled, and your money paid back to you, than the value of my services for the coming year would be to you in New York. I showed Prof. Agassiz the obligation between you and myself, and told him I felt a great deal of delicacy about such a step. He answered that the relations between you and himself were such that he would not hesitate to take it upon himself to act for you in your absence, in a matter of this kind, and that he would see you on his way to the east and explain the whole matter to you so that you would be satisfied. Accordingly this proposition was submitted to Maj. Stevens and Dr. Evans, and after some consultation they answered that they declined it as far as Mr. Hayden was concerned, but that they would agree to all the other terms. Upon some little reflection I declined their proposition. Early on the following morning a different proposition was made Mr. Hayden and myself which we promptly declined and it was immediately withdrawn. This left the whole matter exactly where we started. After waiting another day to see Prof. Agassiz again we were advised by him and Dr. Engelman to go on with our expedition as first projected by you; and it is now our determination to start up on the company's Steam Boat on Saturday next, unless we should in the meantime receive instructions from you not to do so.

The company talk very fair and appear willing to give us all the necessary facilities. I have had several conversations with Mr. Culberson who appears to be quite a gentleman. He says they will furnish us men at $25. per month. Horses at from $60. to $75. each. A suitable boat for $50. He thinks we had better take two men and two horses with carts and proceed to the locality, and after making our collection we can send one of the men back to the fort after an ox team which will be able to bring in at least two thousand pounds weight.

Evans has not got a troop of U. S. soldiers but only takes out some five or six men and will only be on the ground a few weeks, after which he starts on to Oregon and Dr. Shumard returns with their collection. Every person says the field is large enough for any quantity of collectors. Dr. Evans says he would like to see the locality well searched by as many American collectors as may wish to go there.

I must confess that the last four or five days have been passed by me under extremely unpleasant circumstances; but it has been a great relief to me to be able to avail myself of the advice of two such high-minded and distinguished gentlemen as your friends, Dr. Engelman and Prof. Agassiz.

Prof. A. will see you on his way to the east, and will give you a more detailed account of my course in this matter.

I delivered your message to Prof. Swallow and Dr. Litton. They have not yet received your letter and do not know your terms. Dr. Norwood is here. I have much more to say but have not time to write. I will drop you a line occasionally as we pass up the river.

In great haste I remain,

Sincerely yours, etc.

(*Signed*) F. B. MEEK.

Prof. James Hall,
Albany, New York.

Missouri River S. B. Robt. Campbell,
22nd May, 1853.

MY DEAR SIR:

I intended to write you again before leaving St. Louis, but I had so many things to attend to that I had no time to do so. We left St. Louis at 12 oclock M. yesterday and are now about sixty miles up the Missouri. Inasmuch as the company had received no letters of credit for us from the house in New York before we left we had to get Dr. Engleman to go down with us and vouch for the payment by you of all our expenses at Fort Pierre as well as for our passage up and the freight on our provisions, etc. Our outfit, including our own clothing and traveling expenses, together with our boarding here, exhausted all the funds you paid us excepting about twenty-five dollars. You paid me at Albany, before I left, including the ten dollars you handed me that morning, and the ten dollars Mrs. Hall let me have a few days before, one hundred and twenty dollars; and Mr. Hayden paid me at St. Louis, of your money, one hundred and forty-five dollars and fifty cents, making in all two hundred and sixty-five dollars and fifty cents ($265.50) I keep an account of all I pay out, and Mr. Hayden has kept an account of all his own expenditures. We found upon inquiry that it would be necessary for us to have guns, and consequently purchased two at ($15.) fifteen dollars each. The gunsmith obligated himself in writing to take them back on our return at ($12.) twelve dollars each. Our expenses at the hotel were more ($18.70) (eighteen dollars and seventy cents) than we should have spent in that way. I would have taken boarding at cheaper house had it not been for the fact that we had no idea of staying so long. I arrived at St. Louis on Saturday the 14th inst. and the boat was to have left on the 15th. But they postponed their departure from day to day until the 21st. We also had to purchase our provisions, etc., under very unfavourable circumstances, as it was only two days before the boat started that it was definitely determined that we should go. The company charge six cents per pound on all our things excepting our baggage. They charge Evans and

Stevens ten cents per pound to Fort Union; and an Indian agent on board tells me he pays the company $7000. freight on the goods he now has on board for Fort Union. Our whole outfit weighs about 1900 pounds the freight upon which will amount to one hundred and fourteen dollars. I found it would be cheaper to pay even this enormous charge than to depend upon purchasing anything at the Fort, for I am informed that they sell flour there at from sixteen to eighteen cents per pound and everything else at corresponding rates.

I have had several conversations with Mr. Culbertson and others on the boat, who have been all over the Bad Lands; and they all agree that almost any quantity of bones may be collected there. They say these remains may be found over a considerable extent of country and that there is no danger of the locality being exhausted.

The Germans say that they are not going to the Bad Lands, but that they are going on to Fort Union, and thence on to the Rocky Mountains and perhaps to the Pacific. One of them is the Prince of Nassau and another his traveling companion. Two others appear to be servants. The Prince is quite a youth, not more than twenty years of age; and none of them appear to have any taste for natural history.

Evans' party consists of himself, Dr. Shumard, Mr. Hendrick and an artist from St. Louis. They only expect to stay about three weeks at the Bad Lands. Shumard, Hendricks and the artist are to return, and Evans is to go on to Oregon. There are altogether some eight or ten of Stevens' party on board. Five or six of them are soldiers and the others are topographical surveyors, etc. They go immediately on to Fort Union where they are to be joined by Gov. Stevens who will cross over from St. Paul with the larger part of his corps.

There is only one white lady on board. She is the wife of one of the Indian agents. The wives of several traders are on board, and all of them are squaws or half breeds. Culbertson's wife is a full blood squaw!!!

The boat has only landed three times (to take on wood) since we left St. Louis, and as there are some two hundred hired men on board they only stop a few minutes at a time, consequently Mr. Hayden and myself have had but little chance to be on shore. We made, however, good use of these few opportunities to collect botanical specimens, and have already collected some forty or fifty species. We are the only collectors on board, and the passengers turn out and help us. We have no opportunity to study the geology of the country as the boat always stops at wood yards which are generally on low alluvial bottoms some distance from the bluffs. The river sometimes runs close along under these bluffs which appear to be, and I am told are, heavy bedded carboniferous limestone, and sometimes rise to the height of 450 to 500 feet above the river. The slope or Talus is generally covered with a dense growth of oak, elm, maple, poplar, etc. whilst the rugged cliffs which tower high above are studded with stunted cedars and a few deciduous trees. The most conspicuous plants now in flower along the river banks are the anemone cylindrica, evigerou bellidifolia, senecio lobatus, etc. When we get higher up the Capt. tells us they will have to stop and cut their wood, which will give us a better chance to collect plants and fossils.

From the best estimate I can form the balance of the expense of the expedition will not cost less than a thousand dollars, and perhaps a little more. I will write perhaps once more, as we go up, and once by the boat as she comes back. After that they tell me it will be uncertain whether we will be able to communicate with our friends in the states or not. If you write direct to us at Fort Pierre, care of Pierre Choteau & Co., St. Louis.

Respectfully,

(*Signed*) F. B. MEEK.

James Hall.

P. S. Dr. Evans wishes you to send him a copy of your chart. The enclosed ten dollars is to pay for it. Direct it as follows: Per Adams & Co.'s Express. Dr. John Evans, U. S. Geologist, care of Allan, McKinlay & Co., Oregon City, Oregon Territory.

F. B. M.

Missouri River, May 25, 1853.

MY DEAR SIR:

I have only time to scribble you a few lines, merely to inform you that we have progressed this far on our journey, in good health and spirits. A fire, however, occurred on the boat yesterday evening which, had it not been immediately extinguished, might have blasted all our hopes of ever seeing the Mauvaises Terres. It originated amongst some barrels piled upon the bow of the boat. Some of the men had been sitting there, and had doubtless dropped fire from their tobacco pipes. Great alarm was created, because it was generally known that there were 10,000 pounds of powder in the hold of the boat immediately beneath the spot where the fire was burning. The boat was at the time under way, and nearly in the middle of the river. The men at first fled to the stern of the boat, but the coolness and courage of Capt. Sarpy soon restored order, and perhaps saved the lives of all on board.

It would be well for you to make such arrangements with the fur company as will enable us to draw enough money immediately on our arrival at St. Louis, when we come back, to pay our passage etc. from the place we take the steamboat to that city, and thence to Albany, as we will be out of funds.

Mr. Hayden would like to know if you still desire him to remain over the winter in this country.

In great haste, I remain sincerely yours,

(*Signed*) F. B. MEEK.

Prof. James Hall.

P. S. Dr. Evans offers to lend me money enough to pay our passage down, but it may be unnecessary to borrow it if we can draw funds as soon as we arrive at St. Louis.

F. B. M.

Missouri River, May 28, 1853.

MY DEAR SIR:

I avail myself of this opportunity to write you another hasty note. We are, as you may see, traveling at a very slow rate. The river is high and very rapid, and owing to the numerous snags and ever-shifting sandbars we have to stop during the night. Our boat even when in the best part of the river and running under the most favorable circumstances only makes about 3½ to 4 miles an hour.

The scenery, for some distance back, is exceedingly monotonous, as nothing is to be seen but low alluvial bottoms, and islands clothed with a dense growth of gigantic sycamores, cottonwoods, oaks, maples, etc. Once or twice yesterday evening and this morning we had a distant view of low bluffs (150 in height) which are destitute of timber excepting a few scattering clumps of scrubby oaks. This morning we passed near some low bluffs on our left, which present beautiful grassy slopes, and along the base we saw some low exposures of a buff colored rock, regularly stratified and having much the appearance of some magnesian beds we had seen in the Carboniferous limestone at several points lower down the river.

Yesterday evening we had a distant view of a Kickapoo village on a high rolling prairie; and along the river bank near the same place we saw several men and boys of this tribe nearly in a state of nudity, lazily lounging about on the green grassy sward, in the shade of some tall sycamores and spreading elms. These were the first Indians we have seen.

Last night when the boat stopped Mr. Hayden and myself lit our lantern and went ashore to collect botanical specimens, but as we had previously collected nearly every species now in flower in these low bottoms we found but one or two species new to us.

In my last I estimated the cost of the remaining portion of the expedition at about one thousand dollars. Since writing that letter I have had several conversations with Mr. Culbertson and Capt. Sarpy on the subject, and from the best estimate I can make I am inclined to think it will not be so much. They agree to furnish six horses and three carts (able to bring away 2400 lbs.) with all necessary harness etc. for three hundred and twenty-five dollars; three men for four months at twenty-five dollars per month each, and a suitable boat for fifty dollars—so that the expenses may be approximately set down as follows:

6 horses and 3 carts, harness, etc. for three months.	$325.
3 men, four months, at $25. per month each.	300.
1 boat.	50.
	$675.

This arrangement, I find, will be cheaper, and much more expeditious than to take an ox team, as I at first intended. We could probably get along while at the Bad Lands with two men, but as we would have to pay for the horses ($70.00 each) in case the Indians should steal them, I think it would be better to take a third man whose business it will be to watch them constantly. Mr. Culbertson tells us we could not come down in a boat with less than three men, one good pilot and two stout men to row. He agrees to furnish us a guide who can speak the Indian language and who is well acquainted with the country; and a pilot well acquainted with the river. We could purchase good horses at seventy dollars each, but they will not agree to take them back at the end of the season, so I find it will be cheaper to hire them.

Respectfully yours,

(*Signed*) F. B. MEEK.

Missouri River, June 2, 1853.

MY DEAR SIR:

When I wrote you from St. Joseph I was under the impression I would not have another opportunity to address you again before the return of the boat,—I have just learned, however, that we will stop at Kanesville soon, where I can mail another letter to you.

Owing to the fact that our boat generally stops where there is a wide alluvial bottom between the river and the bluffs, we have had but few opportunities to examine the geology of the country. The first drift deposits of any extent noticed by us are at St. Joseph, where the hills are about 200 to 250 feet in height, and appear to be almost entirely composed of clay and sand. We have not yet seen any boulders or gravel. We stopped for the night at the base of a hill some 300 feet in height, about 15 miles from St. Joseph. On ascending this hill by candlelight we found at an elevation of some 40 feet above the river, an exposure of heavy bedded compact fragmentary limestone, eight to ten feet in thickness, in which we saw no fossils excepting a few crinoid columns. Along the steep slope, between this and the top of the hill, amongst the loose earth and broken fragments of limestone we found the following fossils, viz. Two species of Productus—one like P. costatus and another like P. cora; an Orthis like O. untraculum of Owens' report; an Atryp like one of the smooth species in your collection from Crittenden Co., Ky.; and a species of smooth spinifer; also some fine specimens of the Chonetes figured and described by Owen under the name of C. granulifera. Drs. Evans and

Shumard ascended the hill at another place and found amongst the fragments of limestone on the summit, some curious Aviculoid and Pectenoid fossils, together with several Mytilus-like and Modiola-like shells, similar to some of the Chemung species.

We have seen at several points higher up the river, low exposures along the right shores of stratified rocks, which appear to be sandstone, with beds of bluish shaly rock beneath. These rocks look like those of the coal measures, but as we had no opportunity to examine them we do not know what they are.

Mr. Hayden and myself have already collected about 250 species of plants—many of them by candlelight. When the boat stops for the night the men usually make a large bright fire out in the woods, the light of which attracts thousands of insects, by which means we have collected two or three bottles full of specimens.

It is beyond the power of either pen or pencil to convey to one who has never seen similar scenery, any correct idea of the beauty of this country. The Missouri River winds about in a great valley, from 8 to 20 miles in width, which is bounded on either side by hills from 200 to 250 feet in height. The summit of these hills is the common elevation of the country after leaving the valley. The bottom land is almost perfectly level, or only *very* gradually sloping from the hills to the river, and is rarely cut up with sloughs. Often these bottoms and the low islands are covered with a heavy growth of trees, far over which we occasionally catch from the hurricane deck a distant view of the dim outlines of the naked bluffs. At other places these immense bottoms as well as the bluffs or hills, are destitute of trees, but thickly covered with luxuriant growth of prairie grass. But more frequently the eye wanders over a vast expanse of level prairie with here and there long, narrow belts of scattering trees and island-like groves beyond which the view is bounded by dimly seen ranges of hills, or less elevated rolling prairie. Although the prairie grass is not less than two feet high, when we look at the prairies and distant hills the whole surface appears as smooth and soft as velvet. All these objects when viewed in a proper light, present a scene which is beautiful beyond description. I am taking sketches as we go along, but these views are so vast and the objects generally so distant that the drawing has to be made on a very small scale so that it is very difficult to produce that peculiar mingling and contrast of light and shade, so pleasing to the eye, when we look upon the landscape itself.

I thought when I saw the land along the St. Peters in Minnesota I could boast of having seen the finest soil in the world. I am now convinced, however, it will not compare with that of the prairie land along the valley of the Missouri, in Nebraska and Iowa. The river banks along these bottom prairies are constantly caving in, and show a soil rich and black with organic matter as much as 6 to 8 feet below the surface. The manure the N. Y. farmer often hauls from the city to a distance of 8 or 10 miles into the country is scarcely more rich in organic matter than we here see the subsoil four or five feet below the surface. I am not speaking here of a few isolated spots, but of vast prairies, thousands of acres in extent.

We see Indians along the shores occasionally and when the boat runs close enough the passengers amuse themselves by throwing to them crackers, and pieces of bread. Though an amusing scramble takes place, I notice they always hand the bread and crackers to an old man when there is one in the party, who I suppose divides it equally amongst them. Some of these Indians are mounted on good horses and cut quite a figure as they sweep across the prairies with their long hair and trappings flying in the wind.

They say it is 1600 miles from St. Louis to Ft. Pierre so if we get safely back to Albany we will have traveled between six and seven thousand miles.

Very respectfully yours,

(*Signed*) F. B. MEEK.

Prof. James Hall,
Albany, N. Y.

P. S. I have just learned that Dr. Evans is going to get off at the Council Bluffs and go up by land and leave Shumard to come on on the boat. I am at a loss to know what is his object in so doing.

F. B. M.

P. S. No. 2. Dr. E. has just told me he is going to stop at the Bluffs to see about getting men and horses. He has just loaned me fifty dollars.

Missouri River, June 7, 1853.

MY DEAR SIR:

As we are just meeting the Mackinac boats of the fur company on their way down I avail myself of the opportunity to write you once more. We are now about 30 or 40 miles above Blackbird hills, which are near a creek of the same name laid down on Owen's section of this river. On our left we have a bluff about 150 feet in height, the base of which is a soft, heavy bedded rock, having all the external appearances, as seen from the boat, of the Potsdam sandstone (No. 1 of Owen). It is striped with alternate bands of yellowish brown and white, and is so friable that the bank swallows perforate it. This rock only rises 30 to 40 feet above the water and is surmounted with a heavy deposit of siliceous marl which is often seen, at various places along the river, as much as 200 to 250 feet above the water. Owen considers it the equivalent of the Loess of the Rhine. This marl is often seen washed into curious ridges and conical hills which are frequently destitute of vegetation. Banvard's Panorama gives a very incorrect idea of the scenery along this part of the Missouri.

On our right we have a level bottom prairie which stretches far away to the east, and is bounded by a dimly seen range of hills or bluffs of the same elevation as those on our left. The valley here, and indeed for a long distance below, is about 20 to 30 miles in width and as level as a floor. The bottoms are generally only elevated about 10 to 15 feet above the present stage of water, and are sometimes inundated. The river is constantly cutting away these bottoms and changing its course.

Dr. Evans left the boat at the Council Bluffs, and went up by land. He will meet us at Sargents Bluffs this evening, from which point he and Mr. Culbertson will go on to Fort Pierre by land with a drove of horses. They will reach that place some 4 or 5 days in advance of us. Evans says he will not leave the Fort until the boat comes, but I would not be astonished if he would push immediately on to the Bad Lands and leave Shumard to come on with his provisions. If so, he will be there a few days ahead of us. I think, however, as they only have two or three weeks to spend in that region and we can be there over three months, that we can make the most extensive collection. I did think of going up (or sending Mr. Hayden) by land from Sargents Bluff to Fort Pierre, and thence immediately to the Mauvaises Terres. But as this would only stimulate Evans to greater exertions, we concluded we had better not do so, especially as he could employ a strong force at the Fort, if he chooses.

The scenery along in this region is very fine. I am taking sketches. All on board in good health. We do not run at night, and have to stop to cut wood two or three times a day, consequently we make slow progress. The Capt. says we will reach Fort Pierre in 10 days.

Do not fail to write us to care of the Fur Company thru F. B. Meek or F. V. Hayden, Fort Pierre, Nebraska, care of P. Choteau & Co., St. Louis, Missouri.

Respectfully,

(*Signed*) F. B. MEEK.

June 11th, 1853. 30 miles above
mouth Vermilion River.

When I commenced this letter I was under the impression that some piles of flood wood seen at a long distance ahead of us were the Mackinac boats on their way down. We have not yet met them, but are expecting to do so very soon. Dr. Evans came on board again at Sargents Bluffs, but will start from some point (I forget the name) which we will probably reach some time today. The small streams are all very much swollen so that they will not be able to travel fast enough to beat the boat up more than three days. Mr. Culbertson will also go up by land, and has agreed to have our horses, carts, men and all in readiness by the time we reach F. P.

Maj. Vaughan, an Indian agent who has on board eight thousand dollars worth of presents for the Sioux Indians, near Fort Pierre, agrees to warn them not to molest us at the Bad Lands. He says he will tell them the Great Father has given us leave to travel in their country, and if they steal our horses or give us any trouble he will send them no more presents.

We stopped yesterday at the mouth of Vermilion River for Maj. Vaughan to distribute presents to a band of Yankton Sioux. As the boat approached they all assembled on the bank, men, women and children, together with a troop of gaunt, wolfish looking dogs. The men were painted and decked in their best, and were really fine looking fellows. As soon as the boat came near the shore they ranged themselves along the bank and fired a salute of several rounds. Maj. V. went ashore and was introduced to them by the interpreter and at the same time presented them, in the name of the Great Father, with some sacks of beans, flour etc., together with a keg of powder and a bag of ½ oz. balls. There was no chief present, but an old fellow came forward and delivered a short harangue, the substance of which was that they were glad to see their Father (the Agt.). They had always been poor, and needy, and were never more so than at present. They wished their father to tell the Great Father they were thankful for the present he had sent them. They were glad to see he had not forgotten them, etc. At the end of this harangue the Braves all responded *How! How!* Maj. V. told them the Great Father was their friend and would never forget them,—that he was anxious to keep liquor dealers and bad men out of their country, and that if they would drive all such bad white men away the Great Father would be glad, etc. To this they responded *How! How!* This ceremony occupied but a few minutes and we soon started, leaving our red friends preparing for a great feast.

This morning we killed a fine elk (Cerious canadensis, Say) in the river. It was not far from shore, and could have easily made its escape, but the noble animal lingered for its young fawn which was so frightened that it could scarcely swim and as the boat approached a dozen or more rifles were leveled upon her. All fired about the same time, so that no one could tell who was the lucky marksman. The poor animal struggled for a moment and floated off. The yawl was lowered and she was soon taken up. The young one could have escaped but it continued to swim round the dead body of its Dame until it was caught.

We are all in good health. We have collected about 350 species of plants, some of which are new to us.

Yours respectfully,

(*Signed*) F. B. Meek.

Prof. James Hall.

The Prince of Nassau started by land from Dorrian's Hills, leaving his men aboard.

June 12. Dr. Evans did not leave at Dorrian's Hills, and is going up all the way by boat.

Fort Pierre, June 19, 1853.

My dear Sir:

The Mackinac boats have not yet started, but will do so in a few days. We arrived here this morning at 7 oclock and our things are now on shore, but owing to the fact that all the teams and men are engaged in hauling up the goods of the Company, we will probably not be able to open our boxes until tomorrow. We were within two miles of this place last night before dark, but were compelled to tie up by a tornado which came very near sinking the boat. The Indians, or at least some bands of them, are not very well disposed towards the whites at this time. Some of them do not come in to meet the Agent, and refuse to accept their annuities. They have sent word to the Company that they will not allow the boat to go higher up than Ft. Clark, and that they will not allow Gov. Stevens' party to pass through their country. Our men, horses, carts, etc. will all be ready by tomorrow or the day after. Our things will be taken to the fort this evening when we will go immediately to work to separating what we are going to take to the Bad Lands from what we expect to use on our way down. Two of our men are good guides and interpreters. One of them is a half-breed who was raised amongst the Indians and is said to be better acquainted with their habits and customs than any other person in the country. He has hunted all over the Bad Lands. He will take two of his own horses and his squaw along, and Culbertson says if we have any fighting to do he will be the last man to leave us.

Drs. Evans and Shumard will start about the same time we do and have expressed a desire to have us camp near them during their stay at the Bad Lands. They say deer, buffalo, antelope elk and mountain sheep are very abundant there and that our guide can kill more meat than our party can use, though I do not think it prudent to rely upon this means of supply.

I hope you will excuse this incoherent letter for I am in a great hurry and have to write on a table in the cabin with about two hundred Indian chiefs and braves seated in rows on each side of me. They have come on board by invitation from the captain to a feast. They are elegantly dressed and their bearing is noble and dignified. One old fellow has just presented a fine buffalo robe to Capt. Sarpy. He first spread it down on the floor and made the Capt. sit down upon it, when he commenced a long speech which he wound up by presenting the robe, and telling the Capt. that he looked upon him as only a little inferior to the Great Father. They have given us all an invitation to a dog feast tonight. I would like to go but will not have time. The Capt., Dr. Evans & Maj. Vaughan, the Agt. will go, and they say they are going to eat some of the dog. I do not envy them their supper.

I have given the Company a draft on you for the amt. of the freight on our goods and for our passage, as follows:

Freight on 1871 lb. at 6¢ per lb.	$112.26
Passage for Mr. Hayden & myself ($50. each)	100.00
	$212.26

In great haste, yours,

(*Signed*) F. B. Meek.

P. S. We walked across the great bend and collected some fine cretaceous fossils—Baculites, Inoceranus, Circullea, Ammonites, Fusus, etc.

June 20th. We have all our things ready to start. Dr. Evans is not quite ready but will be by tomorrow. We have agreed to wait for him. I am much pleased with our men.

Ceres, April 24, 1855.

Mr. Hayden
My dear Sir—

I wrote a line to Dr. Newberry last night after arriving in Olean to ask you to wait for a letter from me. I had written and enclosed letters, from Albany, for you, to Cleveland. I have heard from Dr. Leidy who feels very much annoyed at the idea of the two German collectors going there and is as anxious as I am that a good collection shall be secured for our own country. Under all the circumstances I have concluded that it is better for Mr. Meek to accompany you. He has had experience in the kind of life and that will be of the first importance. His acquaintance with Indian character and customs is also important. Mr. Meek however will wish to remain at Albany as long as possible and you will perhaps do well to prepare yourself and leave for St. Louis to be there by the 10th. The boat leaving on the 20th according to Mr. Choteau, this will give you time to get prepared for the expedition. Mr. Meek will leave Albany about the 7th or 8th and arrive by the 15th or sooner and have five days. Even this time will be enough for all preparations. Still it may be well for you to have a few days to see some persons and when Mr. Meek arrives, you can together make your purchases of provisions, etc. for the voyage.

As to the matter of remaining there during the winter it can depend upon the success of the expedition, and the facility of getting down. This must in a great measure depend on what you may learn at St. Louis and at the fort. The result must be the finest collection ever obtained and now that I am fairly engaged in it I shall take all necessary steps to have it succeed. I regret to part with Mr. Meek for six months for I need his assistance here. Still I think it important that you have some one with you.

I shall enclose letters to you care of Dr. Engelman, St. Louis.

If there should be anything in the way of preparation which you have not done when you leave Cleveland I will have it ready and send by Mr. Meek. He will have little baggage and can take a few extra things if necessary. Should you decide, before leaving St. Louis to return this autumn, you had better go as light as possible, though the weight of your books will not add much, I suppose.

Dr. Leidy says, that crania and teeth are the most important. Bring no fragments of large bones, and none whatever unless they possess the articular extremities. Bring no large turtles or fragments of turtles at all. Only the perfect smaller ones. He says he has already 500 pounds of worthless fragments brought from there by different collectors, and they all came away loaded and could not take other things which they saw.

If you have not all the tools you want, I can get them in Albany when I return and send by Mr. Meek.

Should you open your box at Cleveland you will find in one pair of your boots a pair of James' slippers which will be useless to you and you can leave them with Dr. Newberry.

You can go to St. Louis by two routes, one by Chicago to Peru on the Illinois River and thence by boat to St. Louis; the other by Cincinnati and thence by boat. You will probably find one less expensive than the other. Of this you can learn in Cleveland.

You will prepare in Cleveland all that you can in the apparatus for your scientific objects and after Mr. Meek arrives in St. Louis there will be time to get provisions, etc.

Mr. Dana says he would like to get any crawfish or crustaceans of any kind you may meet with in your travels. You will give your first attention to the crania, teeth, etc., and leave other objects as of secondary importance. Should you conclude to remain through the winter, you will have abundance of time the next season as I doubt not you could then come down with Dr. Owen or some other parties who will go up to come away in October. This plan is worth considering as you could then get to Albany in time to attend the course of lectures there for the winter and there is also now a second course in the spring. A plan of this kind would give you an opportunity of taking a greater range of country and of embracing more subjects. Were I as free as you are I should be inclined to do so unless the expense of remaining over winter is too great, and that it would be economy to go out again in the spring.

You will find it much pleasanter to have Mr. Meek for a companion and I shall now feel less solicitous than if you had started entirely alone.

Give my kind regard to Dr. Newberry, and believe me

Very sincerely yours,

(*Signed*) James Hall.

Should you receive this in time, direct a letter to me at Olean. I shall leave there next Saturday or Monday and Buffalo a day or two later. I will inquire also in Buffalo.

G. K. Warren to Hayden.

Washington, D. C., Feb. 15, 1856.

Dear Sir:—

I have the following propositions to make to you.

1st. As you have valuable information in relation to the Sioux country, its topography, facilities for campaigning, etc., etc.—I will give you $200.00 and pay your traveling expenses from St. Louis here and back on condition of your furnishing me with it, and such preliminary sketch of the country as you can prepare, to be done as soon as possible.

2nd. If, instead, you like better, I will accept your services for the present year beginning January first at $1000. per annum. I shall still want the above information immediately. You will then return with me to the Sioux country next summer, and whatever collections you may make during this time are to be for the government of the U. S. You will have to furnish your own subsistence except when in the field and then you must furnish your own horse.

Neither of these propositions, of course, contemplate any control over collections heretofore made by you.

3rd. Did my public funds justify I would gladly date your time from the 1st of July last at $100. per m. and continue it so with the understanding that the U. S. was to have the collections and results made by you since that date, and if I should succeed in getting more aid from Congress this proposition will be made to you, to take the place of either of the others in case you should accept either, and then any money paid you will be considered as going to defray my obligations of this 3rd proposition.

But in any case if you wish to present a report illustrative of your labors on the geology of the portions of Nebraska Ter. explored by you, I will gladly transmit to the Department with my report and allow it to take the same chance of publication by Congress. I will also aid you to a limited extent in transporting your collections to such points for examination and comparison as may be proper. I have not money now to authorize any illustrations, but if you should gain the assistance of any person through his professional interest to make such, I will endeavor, should my future means justify to have them recompensed, the result of such labor then becoming the property of the United States.

I regret that my present limited means prevents my doing more for our country, for science and for yourself.

Yours respectfully,

(*Signed*) G. K. Warren,
Lt. Top. Eng. C.

Washington, D. C. Nov. 4, 1856.

DEAR SIR:

The article which appeared in the National Intelligencer of Dec. 2nd at the head of the column of local items, has been called to my attention by officers in the War Department as deserving notice at my hands. It is considered as unjust to me and to that Department that the *conduct* of the exploration should be attributed to you and to *enrich* the Smithsonian Institution, and the mention made of me in the last part of the notice might by the public as well be considered as applying to Capt. Pope's collection as to mine.

If this was the only instance in which these newspaper publications seem to ignore me as the head of this particular branch of the exploration I should have nothing more to say. But the notice that came out in the Missouri Republican the day after our arrival in St. Louis over the signature of H. who was represented to be one of my party. You are mentioned as if you were entirely distinct from my exploration. This article appeared subsequently in several papers and from it at the time a telegraphic dispatch was correctly made out which was published in all the papers and which deprived me of all credit in the collection we had made.

To go still further back, in an article published in a Rochester paper it is stated you had received the appointment of Naturalist and Geologist (I think) to my party *from the Topographical Bureau,* from which it might be inferred that you were independent of me for your situation.

These things separately might be regarded as accidental and as such I was inclined to consider them, but taken collectively they tend to deprive me of any credit in your labors as my assistant. It would be so easy for me in the prospect of such a consummation to employ another assistant in your place that I cannot suppose you or your friends intend to do it; and I hope you will not in the future publish anything or allow anyone else to do so in any way effecting the results of our exploration without my consent.

You know that I labored nearly as hard in making the actual collection of natural history objects as you did; most of the large animals were killed or skinned and brought in by me; many of the birds I prepared entirely; my men did a great deal; many of the articles necessary to their preservation I purchased at my own expense, my accounts having been stopped. If I had employed you especially as naturalist that amount would be stopped too. I only employed you on my right to engage *three* assistants one of whom was to have been the astronomer which position I filled myself on purpose to leave the place open to you. All the transportation etc. of your branch I provided for. I gave up half our boat and all our comfort to the collection, and all of us rowed like laborers to bring it a portion of the way down the Missouri River. You may be sure it was no agreeable thing to me to see in a paper like the National Intelligencer that you had "conducted" the exploration, when all this labor had been actually performed by me and at my own risk and responsibility; to see that Capt. Pope and yourself were held up as deserving the thanks of the scientific world for your efforts, and I merely mentioned, and, in such a way as to be essentially "damned with faint praise." Ask Prof. Baird if I do not at least deserve as honorable mention for services as Capt. Pope?

I wish you would not send anything of our collection from the Smithsonian Institution except through me, and I believe Prof. Baird understands that it is there on deposit and if it should incommode him I will endeavor to find storage for it elsewhere where it will perhaps be better for you. It is my wish to give you full justice for what you have done on my exploration, and I am sure you deserve all the credit in it that the exertions of any one man could win.

Yours truly,

(*Signed*) G. K. Warren,
Lt. Top. Engs.

Washington, D. C., May 6, 1857.

Sir:

On the expedition to the Black Hills of Nebraska the command of which has been intrusted to me by instructions from the War Department of May 6th, you are hereby appointed Physician and Geologist from this date at a salary of $125. per month.

In the event of your acceptance of this position it is expressly understood that no communication of the results of the exploration shall be communicated to any one except through me, to whom all reports will be made. Please inform me of your acceptance or otherwise in writing.

Very respectfully,
Your Obt. Serv't,

(*Signed*) G. K. Warren.
Lt. Top. Engrs.

Washington, May 18, 1857.

Dear Sir:

The Secretary of War has appointed a physician to accompany our expedition as "assistant physician and geologist" but as he is a good physician and no geologist, and you are more particularly devoted to geology I have concluded by consent, and I think as will be more agreeably to you, to give him the title of Physician to the expedition, and you that of Geologist, thus making you distinct and independent of each other and you will of course take precedence of him in rank on the exploration.

You need not therefore prepare any medicines or instruments of surgery, any further than you wish to practice on yourself. You will, however, prepare the natural history outfit, and leave full directions to me of what you have done.

That list of birds was not among the papers you sent back to Prof. Baird. Please return to him as soon as possible.

I expect a proof of that map tomorrow.

Yours truly,

(*Signed*) G. K. Warren
Lt. Top. Engrs.

Washington, D. C.
April 22nd 1859.

Sir:

You are hereby appointed as an assistant upon the expedition under my command for the exploration of the headwaters of the Missouri and Yellowstone Rivers.

You will be informed when it will be necessary for you to be in St. Louis, in order to leave with the expedition.

Precise instructions as to your duties cannot be given until the organization of the party is completed. Your services will probably be required as Naturalist.

The interests of the survey may, however, render your assistance necessary in other departments of the work.

You will not be desired to procure any article of outfit, for the use of the Government; the equipment for this object will be furnished you in the field. Your compensation will be at the time of your departure from Washington, until you are relieved from duty with the expedition. In addition to this you will be paid for the actual necessary transportation of yourself and baggage from Washington to the field and back, and while on duty in the field, will be entitled to subsistence (one army ration per day) and transportation; the weight of your baggage during this time being limited of necessity to fifty pounds.

An established rule of the Department requires that each assistant shall distinctly understand, in accepting his appointment, that the specimens, notes, sketches, memoranda and all the materials collected or prepared by him, shall be considered the property of the Government and shall be turned over to the chief of the expedition at the conclusion of the work; and that he shall not publish, or furnish to other parties for publication, either during the progress of the exploration, or after its conclusion, any of the information or results that he may have procured while engaged upon the duty—excepting through the chief of the party, or such other persons as the War Department may indicate.

Immediately upon the receipt of this letter you will transmit to me, at this place, your acceptance or non-acceptance of the appointment tendered to you. Instructions as to the time of leaving this city, will be given you by Lieut. Henry E. Maynardier, Asst. Topog. Eng. of the Exped'n.

Very respectfully Yr. Obdt. Sv't.

(*Signed*) W. F. Raynolds,
Topo. Eng. C.

Dr. F. V. Hayden,
Washington, D. C.

Prof. S. F. Baird to F. V. Hayden.

Washington, March 9, 1867.

Dear Dr.

A late appropriation bill provides that certain unexpended balances of appropriations made for Nebraska Territory and inapplicable now that it has become a state, may be expended for a Geological Survey under direction of Commissioner Lane's (?) office. I read the bill at breakfast this morning and immediately went to Mr. Wilson and presented your name and claim as a candidate. He asked me to write him out what I said and I will do so at once and send him.

If you want the place you had better come on at once and see about it. If you had not carried off my copy of your geology of the Upper Missouri I might perhaps have clenched the matter by giving it to Mr. Wilson on the spot. You should send or bring *two* copies at once; one to replace mine that you cabbaged, the other for Mr. Wilson.

Yours truly,

(*Signed*) S. F. Baird.

Gather up copies of the two (?) memoirs; as many of your other articles, proofs of Leidy's new plates, etc., and send me at once. I will add copy of Pal. Upper M. and send all to Wilson. Get copy of Warren's report if you can. A copy of your geological colored map would be well to show.

Black Buttes, Wyoming, Aug. 17, 1872.

Philadelphia Academy of Sciences
Philadelphia Penna.

I have discovered in southern Wyoming the following species:

Loxolophodon Cope. Incis 1, one canine tusk, pm4, with one crescent and inner tuberculi, molars 2, size gigantic. L. *cornutus,* horns tripedral, cylindric; nasals with short convex lobes. L. *furcatus,* nasals with long spatulate lobes. L. *pressicornis,* horns compressed subacuminate.

(*Signed*) EDWARD D. COPE,
U. S. Geological Survey.

New Haven, Nov. 12, 1873.

DEAR DR. HAYDEN:

. . . .

You are right in employing Prof. Cope. Marsh could be of no service to you. I wish that they would stop their race, and work quietly. I have told Marsh more than once that it would be vastly more for his reputation among zoologists to describe one species thoroughly, than to be the first to name a hundred. These short, insufficient descriptions must make an earnest zoologist of Europe curse American science. I think that more quiet work may now be looked for.

. . . .

Yours very truly,
(*Signed*) J. D. DANA.

E. D. Cope to F. V. Hayden.

Haddonfield, N. J. 12/1, 1873.

DEAR DR.

Yours of yesterday came to hand and I proceed to notice one or two points and ask for more definite information.

As to the matter of prompt publication, I am not "alarmed," but I have had experience enough to show its value. I have been anticipated in generalizations by Flower and Huxley; in species in paleontology by Leidy and Marsh, and in reptiles and fishes by Peters and Günther; in evolution by Spencer, Mandsley and Huxley. Also the reverse of all these with the same men; i.e. I have anticipated them often in important points. The differences of date are often very small, so I follow Sir David Brewster's advice to publish a good thing at once *when it is finished.* This is what you are pleased to call a *"race,"* but is no race in study. I study as slowly as though no other man were working in the same field, for I presume no man of sense will injure himself by publishing trash. Such a "race" as this will go on as long as there are two men (or more) in the same field and the defeated party will always (if he be foolish) cry out like the school-boy or the politician "You didn't play fair." Such is the origin of the cry from "The Dismal Swamp."

Now for the future; your monthly publication will supply all the necessities of prompt issue. Until that is ready to be issued (should it not come out in a month or two), should I have anything important to publish I can write a note to you submitting it as a survey report and receive one authorizing publication from you, if proper, and preface my article with them. This would credit the contents to you and the article would bear your official stamp. Thus the work would be secured by early publication, and be secured *as the survey work*. This you have often done before and would perhaps be willing to do again until your monthly periodical begins its issue.

I am willing to work with and for any institution that is endeavoring to *build up* a good thing in this country. This you are doing and I will do what I can towards its success. Educational influences of a high grade are too few in this country, and those that exist should prosper and grow. I am glad to hear of your monthly periodical; I hope it will have considerable scope and embrace such a variety as to be largely taken by the people.

I am desirous of adapting myself to your plans, for you have the disposition, time and ability to carry them to a successful issue. For myself, I am a student, and never will have time or ability for positions involving much abandonment of study. I will therefore gladly work under others, provided I can *live,* scientifically and physically; i.e. by securing my work and securing some pay.

If you choose I will preface the paper in Journal of the A. N. S. by a note from you, as you have done sometimes with Meek's work.

I hope the boxes will come on eer long, so that I can get to work on some of the Cretaceous, etc.

Yours truly,

(*Signed*) Ed. D. Cope.

Yale College, New Haven,
Dec. 3rd, 1873.

My dear Hayden:

I have been intending to come on to Washington to see you about a matter of importance, but I am at present so busy with plans for our new museum that I must try to make a letter answer, at least to open the subject.

During our conversation last spring in Washington and here you fully agreed with me as to Cope's character, and his work. You assured me especially that you would not allow him to put anything of a personal or controversial nature affecting myself in his Report which you were to publish. On this assurance I fully relied.

On my return here a short time since I found your 8vo Report for

1872, which you had been kind enough to send me. I was greatly surprised and pained to find that Cope had not only published in it many of my discoveries as his own (although he had not the slightest claim to them) but had also inserted all the dates which I had proved to be false, and even charged me with making mistakes on the points where I had exposed his own blunders.

Feeling confident that these portions of his Report—which are calculated to do me much injury and retard science—were inserted against your wish, and without your knowledge, I write to you frankly about the subject, after consulting with a number of prominent scientific men here and elsewhere as to what should be done.

I understand, of course, that your 8v. Report is now published, and cannot be changed. I ask, however, in simple justice,

1st. That nothing of the same objectionable nature be inserted in your 4to Report.

2nd. That some fair correction of the more important wrong statements in the 8vo. vols. be there published.

Omitting many minor points, the following especially should be set right in your final Report.

1st. The two *Pterodactyles* which I described as *P. occidentales* and *Pingens,* and which Cope redescribed in your Report for 1871, p. 337, with no reference to my work, should be credited to me. You will remember the case doubtless, and that I showed you a certificate from Lesley that Cope's paper on these fossils was not *printed* until several days after he and many others had received my paper. Cope has fully acknowledged that I have priority.

2nd. Cope's claims to the discovery of *Dinocerata* (or as he calls them *Proboscidia*) made in your Vol. for 1872, p. 545, 644, etc. should be withdrawn, as I described the first species (*Tinoceras anceps*) in June 1871 (Am. Journ. Sci., II, p. 35), a year before Cope ever saw a specimen.

3rd. The discovery of *Quadrumana* likewise claimed by Cope in the same vol. (p. 545, 546, ec.) should be disavowed likewise, as Limnotherium, one of the typical forms, was described by me in June 1871 (Am. J. S., II, p. 43) more than a year before Cope wrote a word about Bridger fossils. I also announced the determination of their genus as *Quadrumanous* Oct. 8th, 1872, or before the date even claimed by Cope for any similar determination.

Here are three important discoveries which I made and published, each of which Cope endeavors to deprive me of. The wide circulation of your 8vo Reports will aid him greatly in this, and to have the attempt continued in your 4to Vol. would be a most serious wrong to me, as well as a lasting blemish on your Report.

You can easily settle the matter now, and as I wish well to you and your survey, I write first to you frankly on the subject. If you will try to put yourself in my place you can easily imagine how strongly I feel on this point.

If you have any doubt about the injustice I wish to avoid I am willing to have Leidy, Baird or Gill examine the three cases I have referred to above, and decide the question of priority. Cope has changed the dates, and suppressed some of the most important facts in his statement of the case, as given in your last 8vo. Vol. I ask only simple justice.

Please give this matter your serious attention, and inform me of your decision as soon as you can conveniently, as my future action will depend largely on your reply.

If necessary I will come on and see you about the matter immediately, but I am very busy just now, and should prefer not to come to Washington until later in the winter when I hope to spend some time there.

With kind regards to Mrs. Hayden, I remain,

Yours very truly,

(*Signed*) O. C. MARSH.

St. Louis, Sept. 21, 1875.

MY DEAR DOCTOR:

As I have travelled eastward as fast as the mails to this place I have delayed writing to you, using what little time I had to relieve the anxiety of my family and in seeing my friends en route. We have forwarded such full accounts to the Press that by this time you must have a pretty good idea of our brush with the Indians. We were certainly in a tight place, and I for one saw but little chance of escape. What with want of food and water, losing strength every minute, and no apparent egress, things looked desperate. To follow the trail into the narrow canon would have been certain death and we had tried every other direction but were cut off by canons. Mr. Gardner's topographical knowledge came to our rescue and found the place of our escape. To have saved the property was a physical impossibility even if the question of saving life be left out. The mules were so exhausted that they could never have carried their burdens to water. This was the united opinion of the packers whose opinion was taken on this point. There was no rout about our retreat although under fire all the time. The topographical and geological notes were collected, all monies, accounts, vouchers, &c were brought off and are now in a place of safety.

The boys as a body did splendidly, and what was in my power to do, I did willingly for my own credit and that of the Survey to which I have been attached so long, and I wish to thank you most sincerely for allowing me to come out this year for severe as was the suffering, I have gained an unexpected experience this trip that I would not now give up for anything.

You have probably heard before this that the parties are all reported safe. We were quite anxious about Jackson as we knew he was going into the region we had just left. Mr. Gardner sent two men to head him off but it appears they trailed him many days and then lost his tracks, owing to rain or some other cause.

We are somewhat chagrined at the opinion of the Indian department to the effect that the Indians only wished to frighten us. Had the gentlemen who expressed this taken part in the fight I think they would have formed a different conclusion, and it is rather rough that men who fought hard for their lives should receive such a verdict. But as the Utes are well known to be in a very restless state over the bad faith of the country in the San Juan treaty it is wisest for that Bureau to adopt such a tone.

. . . .

Very sincerely yours,

(*Signed*) ROBT. ADAMS, JR.

L. Lesquereux to F. V. Hayden.

Columbus, Ohio, Apr. 15, 1873.

MY DEAR SIR:

I had yesterday a very kind note from Prof. Meek on account of my remarks on the distribution of the lignitic strata of Coalville. As well you know, Mr. Meek has made a very careful exploration of that locality and of course his views and conclusions shall exclude suppositions or theoretical considerations which are based upon uncertain evidence. If, therefore, you have nothing to say against it, I shall omit my remarks on Coalville or modify them in such a way that they do not be in contradiction with Prof. Meek's geological exposition. This, however, and the superposition of some thousands of feet of Cretaceous measures, or rather of measures marked with marine cretaceous fossils, do not change my views on the age of the lignitic formations. In the lignitic formation, as in the Carboniferous formation, as also in the coal formation of Richmond, etc., botanical paleontology will be always in many points in discordance with animal paleontology; as the one represents land or atmosphere influence which cannot be recognized by the other, and vice versa. I have therefore objected always to that separation of the so-called *Sub-Carboniferous* which though truly Carboniferous has thousands of feet too of strata with Devonian animal fossils overlaying beds of coal and shale with true Carboniferous plants. I have not, however, put this supposition: (that all our lignite beds are true Tertiary, or that there is not a Cretaceous coal bed) in my report. And this is of no moment in the present case. Of course, you will desire to dispense of opinions which you may recognize as unright or as based upon unreliable data and therefore that chapter on Coalville may be omitted if you think it proper to do it. I suppose that I may correct all this upon proof sheets. I am now sorting specimens to separate those which may be used for illustration.

Very truly yours,

(*Signed*) L. Lesquereux.

Columbus, O., April 16, 1873.

Dr. F. V. Hayden:

I wrote you yesterday a note about Prof. F. B. Meek's letter. I send you now two pages of Mss. which, if you will think it proper, may be substituted to what is said of Coalville in my report. As my opinion expressed on that locality is essentially based upon your researches and remarks you will better know than myself if what I have said is wrong.

The more I consider the question, the less I am disposed to admit of a lignitic Cretaceous formation. Had we not Black butte with its so *positively cretaceous* fossils 150 feet higher than the coal and with its roof full of remains of tertiary plants, we could think of a division of the lignitic. But now it is not possible to separate Black butte from Carbon, Golden, Marshall, Raton, etc., etc., as it has a comparatively large number of species of fossil plants identical with those of these localities; and still less possible to consider the whole of the lignitic as cretaceous. And now, if you admit Black butte as tertiary through its cretaceous remains, how can you come to separate into the cretaceous Coalville and Bear river which have the same kind of coal strata and where, for certain, fossil plants of the tertiary types will be found hereafter? and where already the fucoids of the lower Eocene have been found in profusion.

Yours very truly,

(*Signed*) L. Lesquereux.

---

42 Garden St., Cambridge, Mass.
Jan. 27, 1874.

F. V. Hayden, *U. S. Geologist*,
Washington.

Dear Sir:

I thank you for "The Extinct Vertebrate fauna" and "No. 1 Bulletin of the U. S. Geological and Geograph. Survey." Will you allow me a few remarks.

First.—It is to be regretted that you expend so much on map making. Topography is a far more expensive business than geology, and has little to do with it; at least for a reconnoisance. Unless you want to publish a geological survey like the English, on a very large scale but such a survey is neither wanted or even possible in the U. S. Territories for one century or more to come; besides the Geological Survey of Great Britain is strictly confined to geology, and has nothing to do with the Ordnance Survey.

Second.—Among your collaborators, the number of geologists is very limited, and the few employed are entirely unknown in science. For a geological survey, I think, the main force ought to be geologists; and some of them at least, if not all, having a good record and standing in the scientific world. The number of entomologists and botanists is so great, that it looks more like a natural history concern than a Geological Survey. When the first expeditions explored the country, 15 or 20 years ago, it was all right to collect and publish every bug, bird and plant; but now it is otherwise; botany and zoology ought to be let alone, they do not require the aid of State appropriations. It is, I think, always regretable to mix and drown geology and paleontology with other branches of natural history. More than two thirds of the money granted in the U. S. by states or the U. S. governments for geological surveys, have been swallowed by topographers, botanists and zoologists.

Third.—You have created as part of your geological survey a branch of honorary members. Do you not think highly proper to put in it, all the geologists employed in the field by the U. S. Government before the beginning of your survey in 1867? It will show your desire to do justice to your predecessors west of the Mississippi River. We are few, very few indeed, and we had a pretty hard time, at least if I should judge of the others by myself. Several are already dead, such as: Edwin James, D. D. Owen, Dr. Benjamin Shumard, Dr. George G. Shumard, and Dr. Evans. Such a branch of your survey will be a sort of geological association of the old compaigners in the Rocky Mountains.

For making those suggestions, I have no particular plan of my own to further, for I desire nothing, but the progress of science and a good feeling among co-workers in the same field of science.

Very truly yours,

(*Signed*) Jules Marcou.

P. S.—In reading over my letter, I beg you will excuse the abruptness of my style owing to my imperfect knowledge of English.

Contracts between Hall and Meek.

This Memorandum of an agreement made and entered into this ninth day of May, 1853, between Prof. James Hall, New York State Geologist and Palaeontologist, party of the first part, and F. B. Meek of the second part, both of the county of Albany and State of New York,

Witnesseth, That the said F. B. Meek obligates himself to serve the said James Hall as a draughtsman assistant, &c., for the term of four years from the tenth of next June. For which services the said James Hall hereby obligates himself to pay the said F. B. Meek the sum of four hundred dollars ($400) per annum, and his boarding and lodging, and a suit of good fossils, consisting of as many species as can be furnished from his duplicates.

It is likewise agreed and understood that the said F. B. Meek is to be allowed to spend at least four months of each year in the field collecting fossils and in making geological investigations; and that while so occupied the said James Hall is to pay said Meek's traveling expenses, boarding and lodging.

(*Signed*) JAMES HALL.
(*Signed*) F. B. MEEK.

This memorandum of an agreement made and entered into this 13th day of June, 1854, between Prof. James Hall, State Geologist and Palaeontologist of the State of New York, and F. B. Meek, both of the city and county of Albany and State of New York,

Witnesseth—That the said James Hall hereby agrees to release the sd. F. B. Meek from an agreement entered into between them (Sd. Hall and Meek) on the 9th day of June, 1853, by which sd. Meek bound himself to serve the sd. James Hall as an assistant and draughtsman for the term of four years from that date.

The terms upon which the sd. James Hall releases sd. Meek are as follows, viz. That the sd. Meek is to be allowed to accept an appointment now offered him in the Geological Survey of the State of Missouri, and enter immediately upon his duties in that capacity. It is understood however that the sd. Meek is to return to Albany, N. Y., about the first of January of each year for three successive years, to serve the sd. James Hall as an assistant and draughtsman until the last day of May of each year, unless in consequence of the suspension or completion of sd. geological survey of Missouri, in the meantime, sd. Meek's time and attention should be required in Missouri in assisting with the final reports of the survey; or unless sd. Meek should have an appointment or situation offered him which would render it inconvenient for him to leave the west, in which cases sd. Meek is to be considered released from this and all previous engagements with Prof. Hall, Provided either of these contingencies should occur after the first two of the three years specified in this contract, and that the above named appointment or situation should be a separate and distinct one from that upon which sd. Meek is now about to enter.

It is agreed and understood that for the above named four months' services of each of the three (3) years the sd. James Hall is to pay the said F. B. Meek whatever sums of money may be deducted from his salary as an assistant in the Missouri Survey in consequence of the above named four months annual absence from his duties in the sd. Survey.

It is likewise agreed and understood that in addition to the aforementioned compensation the sd. James Hall is to give the sd. Meek his boarding and lodging during the four months of each year he shall be occupied with him.

It is also further understood that in addition to the other compensation, in case the terms are fulfilled during the entire term of three (3) years, the said James Hall is also to give the sd. Meek a good suit of the fossils characterizing the rocks of the New York system; and in case the sd. Meek should be released as hereinbefore mentioned from the last 4 months services he is to receive a suit of fossils proportioned in value to the length of time he has been occupied with Prof. Hall.

(*Signed*) F. B. MEEK.
(*Signed*) JAMES HALL.

[On returning from the Bad Lands trip it would appear that Meek wished to sever relations with Hall and accept a position in Missouri, for under date of May 19, 1854, he writes Hall giving details of a proposed division of his time between New York and Missouri, and finally asking Hall "to give me an honorable release from my contract, on some terms, at once, so that I may telegraph Prof. Swallow's family. . . . I make this request as a great favor and with full consciousness that you have an unquestionable legal right to demand of me services for the entire term of our contract." G. P. M.]

James Hall to A. Litton.

Albany, May 25, 1854.

DEAR SIR:

The state of the correspondence between Mr. Meek, Prof. Swallow and yourself seems to render it necessary to make some decision as soon as possible, and I shall avail myself of the occasion to give you a short statement of the relations existing between Mr. Meek and myself. After some negotiation in which I stated to Mr. Meek my objects in employing him, he proposed certain terms to which I agreed, giving him, indeed, more than he offered to come for, and he came here in June, 1852. For nearly the entire first year his services were of little value to me, as he made no drawings till the spring of 1853. Last year he proposed an agreement and afterwards reduced the same to writing in which he stipulated to engage for four years on certain terms. He declined at that time my offer to engage year by year, which would have left him free to make a new engagement every year. Afterwards, at his earnest solicitation, I consented to give him the opportunity of going to the Mauvaises Terres, which journey he thought would renovate his health and he would then feel like going forward with our engagement.

Some overtures of Prof. Swallow, however, rendered him a little discontented as I judged, and finally the offer to engage in the Missouri Survey induced him to urge a release from our engagement. This, however, was finally declined. At the present time as I understand it, Prof. Swallow proposes to pay him $1000 per annum and his travelling expenses and give him an opportunity of doing more for his own advancement than he can do here.

To this proposition Mr. Meek declined to accede unless I will agree to release him from his engagement with me. In order to gratify Mr. Meek and if possible accommodate Prof. Swallow, I suggested an arrangement by which a part of his time could be given to Missouri and a part to me. To this Mr. Meek at first very cordially assented and wrote to Prof. Swallow to that effect. Since then he finds some objections to the place. I learn since I returned from Ohio that he has received a letter from you in the absence of Prof. Swallow in reply to his proposal and stating that you have no doubt Prof. Swallow will accede to the terms, &c.

You can well understand why, after Mr. Meek has become acquainted with my work and knows my collection to which he has had unrestricted access since he has been with me, I should feel disinclined to cancel the arrangement which was designed to extend to the completion of the Palaeontology of N. Y. Whatever Mr. Meek may think I feel quite sure that the two years here have given him that knowledge which he will find available to his own advancement, and notwithstanding our agreement I am disposed to yield, as far as consistent with the progress of my own work. I cannot now give the time to disciplining and instructing another assistant, nor shall I undertake the task. The sum offered by Prof. Swallow is much more than I am able to afford, but if I am compelled to do so I shall nevertheless pay it, and I have that confidence in Mr. Meek that I do not think he will break the engagement existing between us without my consent.

In the present state of the matter it appears to me that Prof. Swallow should propose definitely to Mr. Meek a certain sum for a portion of the year, say six or seven months and this will give time for field work and reports.

I presume this time is as much as he would be actually required to be occupied in the Survey if he were altogether in Missouri, and therefore I think Prof. Swallow should be willing to allow a sum considerably more than that proportion according to time.

In the commencement of the N. Y. Survey our time was six months in the field and making the Annual Reports, after which the time of Principals and Assistants was at their own disposal.

I presume Prof. Swallow will have no objections to this definite proposition so that Mr. Meek and myself can make arrangements for the future as far as relates to my work here.

Mr. Meek is now engaged in completing some drawings which are necessary for the engraver and some time must necessarily elapse before he can leave under any circumstances.

Hoping that you may soon have an opportunity of communicating with Prof. Swallow, I shall wait any changes until I hear from you.

I remain very sincerely

Yours &c,

(*Signed*) JAMES HALL.

[June 21, 1854, Meek again writes from St. Louis and says, "I am now satisfied that the arrangement for me to return every year and spend a portion of my time with you at Albany will be the most agreeable to me." G. P. M.]

War Department, Washington City, June 18th, 1874.

4926.

To the Honorable

The Secretary of the Interior,

SIR:

In view of the duplication of Geographical and Geological Surveys of Western territories, under the Interior and this Department, during the last summer, I have the honor to submit for your consideration, that in order to prevent such repetition during this season, an understanding should be had between the two Departments, as to the field within which each should confine the Surveying parties under its order. You will perceive from the plat, which is sent you herewith, the areas which have been surveyed by parties under this Department, and that following the evident design of their surveys, it is desirable that the same parties should connect these areas, by a completion of the survey of sections 51, 52, 60, 61, 67, 68 and 69, a part of each of which has already, as you will observe, been surveyed by the War Department parties, under the direction of Mr. Clarence King, and Lieutenant G. M. Wheeler.

It will be further noticed that the operations of this Department have been in the main South of the Union Pacific and Central Pacific Railways, and in consideration of the above and of the desirableness of preventing in the future duplication of surveys by either Department, I have to suggest that it may be acceptable to you that the parties under this Department be required to confine their operations during the present season south of the railway, first mentioned, and that the parties under your direction be required to conduct their operations northward of that line, In further connection with this matter, I have to say, that there appears to be a harmonious co-operation between the survey of the Public lands, under your direction, and the Geodetic surveys under this Department, and that in furtherance of the interests of both, a certain interchange of information may be made, The details furnished by the Land Office to this Department of its valley survey prevent the necessity of a detailed survey of these valleys by the Officers of this Department, when the requirement of our triangulation obliges the latter to pass over them, and on the other hand the Geographical positions determined by the Officers of this Department, will enable the Public Land Surveyors, to correct and adjust the standard meridians or bases township and sectional lines:—Should further checks be required, the establishment of meridians and bases, can be made at points remote from those which have been carried from the valley of the Mississippi westward, and from the Pacific eastward.

An early reply is requested.

Very respectfully, Your obedient Servant,

(*Signed*) WM. W. BELKNAP.
Secretary of War.

James Hall to Sterry Hunt.

Albany, May 2, 1872.

. . . .

My principal points of opposition to the Taconic were the bringing it westward to the Hudson River and carrying it beneath the Potsdam, and I think you will so find by looking at what I have written.

The Geol. map of N. Y. in its eastern part, colored by Emmons and Mather, shows their views in 1843. Later, in connection with his agricultural work Dr. Emmons compiled a new geol. map and colored 3000 copies but I do not think they were distributed very extensively. Many of them are still in the office of Secretary of Regents and I will endeavour to procure a copy for you in a day or two.

This gives Emmons latest views of the Taconic, a view followed by Billings[2] in his review of the Graptolitic Zones in Europe and America.

I became satisfied that the solution of the difficulty had not been reached long before Sir William or Mr. Billings suspected anything of the kind. Also for that reason as I said before, withheld publication of my sections which had been completed in 1848. And as I told you before, I never saw the Canadian rocks till 1854 and never at any time gave an opinion about the Philipsburgh fossils.

If the controversy is to be reopened and Sir William chooses to stand

2 Canadian Naturalist and Geologist, Vol. VI, pp. 344-348.

behind Billings, which I can scarcely believe then I may be induced to reply, but I have never yet replied to any of Billings' vituperations.

I have not yet read Billings' article in the last Naturalist on Taconic. I am not strong enough to take it up just yet, but I have glanced at it just enough to see its drift.

I see that we are likely to have Taconic over again from another source. What has become of the result of Sir. William's investigations in eastern N. Y. and western Penna.

James Hall to J. D. Dana.

Albany, March 25, 1888.

My dear Dana:

I received yours of the 22nd on Saturday, P. M.

The Taconic question had not been discussed among the New York geologists at their spring and autumn meetings so far as I recollect the matter until 1840 and 1841. (It may possibly have begun in the autumn of 1839.) The question rested almost entirely with Dr. Emmons, Mr. Vanuxem and Mr. Mather, but it was not till 1841 that the subject assumed any definite form, and became the special subject of discussion. Mr. Vanuxem never assented to Dr. Emmons' views while Mr. Mather accepted them and gave his own interpretation. This you already know from the reports of these gentlemen.

In 1841 the Taconic question was discussed at the Philadelphia meeting of the Geologists, and then, for the first time so far as I know, Prof. Rogers took a prominent part in opposition to Prof. Emmons' views.

The figures you refer to on page 148 are so much like the blackboard illustrations of Prof. Rogers that they may very well have been reproduced from those used in his illustrations at Philadelphia and again in Boston the next year.

Up to this time I had taken no prominent part in these discussions.

On the second page of your letter you ask the time when Emmons *first made known his views to the New York geologists.* I have no positive recollection of the subject of a Taconic system coming forward to the notice of the New York geologists before the spring of 1840, and from that time onward. It may have been talked of earlier. I do not think it entered into the discussions of the geologists at their first meeting in Philadelphia, further than what might have taken place in a conversational way. Had the question been seriously discussed I think Prof. Hitchcock would have alluded to it in his address in 1841.

In these earlier discussions it always seemed to me that Emmons's views were vague and uncertain, and that he had no really well defined ideas of the limitation to his system or any clear conception of the nature of the rocks constituting it. This seems to me pretty well shown by his own list of the specimens which he arranged in the State Cabinet of Natural History at Albany.

James Hall to J. P. Lesley.

Albany, Dec. 5, 1861.

My dear Lesley:

I have just received your favor of the 3d. You may have been surprised that I should have asked you the questions that I did, but I wished to have your impressions independently of anything I might say.

So fully have I been impressed that the views of Sir W. E. Logan were his own and independent of all former questions or discussions upon the subject that I think two years since he would have resented any intimation to the contrary. Not only have the independence of these observations and conclusions been proclaimed but they have come to us fortified by the chemical results of Mr. Hunt. Not only has Sir William traced these fossiliferous unaltered rocks till they become the folded and metamorphic masses of the Green Mountain range, but Mr. Hunt has shown by chemical analyses that these unaltered fossiliferous rocks have the same chemical composition even to the presence of rare and remarkable minerals.

In a conversation with Sir William which I had in Sept. 1860 he then assured me that there was no possibility of there being any break, overturn or other condition that could in the least interfere with his conclusions that these were one and the same groups. In November I had the same assurances from him after he had revisited and reexamined Quebec—and it was not until late in December that he wrote me that there must be a break, upon paleontological grounds, but that it did not affect Hunt's chemical results or generally alter it. Now I cannot see but both Logan and Hunt must give up that their mode of working and their results amount to nothing. I am sorry for this for you know how eager I have been to seize upon every evidence that mineral, lithological or other characters were reliable in the tracing of strata, yielding the precedence of paleontology, or at least claiming no more for it than I was willing to yield to chemistry or mineralogy.

But this is not all. In a letter received from Sir W. E. Logan a few days before I wrote you there is the following paragraph. He complains that Sir Roderick Murchison in his address before the Geological Section of the British Association has given me undue credit in reference to the determination of the Quebec horizon, and charges that his own errors are due to the influence of my paleontological opinions and that he must make the correction publicly.

I wrote him remonstrating against this course. In return I have received from him a reply in which he says it is already in the Am. Journ. for January. I think Sir William cannot see that by these admissions he takes out his props to his physical and topographical geology everywhere in Canada and renders it as worthless as he confesses this great work on the Green Mountain range to be—and not only this, confesses the worthlessness of his chemical geology too, and he leaves his enemies, if he has any, to take advantage of the circumstances.

There is no doubt of the fact that these shales are lower than those above the Trenton limestone. They are not in fact what have been termed Hudson River Group or No. 3, but a lower set of shales. Still again they *are* Hudson River group, for the shales along the Hudson Valley in the typical localities are not *No. 3* or above limestone No. 2. The truth is that the typical localities of *Hudson River Valley* afford none of the fossils of No. 3 which are found in that rock in the unaltered regions.

The old question of the Hudson River and Taconic comes up again. Beginning as I did in the Pal. N. Y. in 1843, I found the Hudson River group already designated and defined by my colleagues in the Survey, as it did not extend into my district. The question at once arose whether the Taconic slates were distinct from the Hudson River group, for Emmons as you will remember did not claim the extension of his Taconic system towards the Hudson River. In order to settle this question for myself I made sections from the Hudson River across the country to Mass., and from Lake Champlain through Vermont. I could make nothing more than Hudson River shales of the entire series of shales but I recognized the Potsdam sandstone is immediately beneath them in several places in Vermont and have cited Scolithes lineatus from the same rock on the borders of Mass.

At the meetings of the Association of American Geologists in 1845 and 1846 I showed my sections and they were fully discussed by Rogers and by Emmons. These sections were engraved and copies were in the hands of Adams, Hitchcock, Thompson of Vermont, and Dr. —— of Washington. I am not aware that I ever advocated any other views.

That I omitted to have these rocks westward was the great error, but this I believed had been done by my colleagues, and the only contested ground was between the Hudson River in the Hoosac River Valley and Taconic System and this ground I examined.

Now it seems to me we shall be compelled to come back to the description of Hudson River group for the older shales and that too includes all the Taconic slates as well and some other term must be adopted for the shales No. 3.

You are at liberty to make use of any of the facts I have stated, but not to Sir William's views till you see the Journal. In the meantime I will send you a copy of one of my sections that you may see what I had done in 1844 and 1845. Of course I do not now believe in the steep dips and vertical strata. I am sorry not to have sent you the plates of Vol. III before now but the lithographer has not printed any extra plates as he had agreed to and I fear I may not get them at all. If so I am at the loss of printing 300 copies of the text.

James Hall to Sterry Hunt.

Albany, Oct. 18, 1861.

In regard to Emmons' sections, do you think them possible as he represents them, taking into account the physics, or physical laws alone? I contend that they are not possible sections.

Have you ever compared Eaton's sections with Emmons'? You should recollect that Eaton taught Emmons his geology and he had none other until he went into the N. Y. survey.

Eaton had a theory that in every group or system of strata there was a carboniferous, a quartzose and a calcareous formation—the carboniferous being often black slates only. Here is his section.

Quartzose formation.		The localities cited are Canada to Georgia and their particular localities.
	1. Calciferous sand rock.	
	2. Sparry lime rock.	
	3. First grawacke.	
	4. Argillite.	
	5. Granular primary limerock.	
	6. Granular quartz.	
	7. Talcose rock, Hornblende rock, etc.	

You can compare these with Emmons—and then recollect that Eaton taught this for more than ten—yes, even fifteen years, and I learned from him to believe the slates etc. on the east of the Hudson were older than the calciferous sand rock. One sees how soon the opinions of an author are lost sight of, and though Emmons has simply followed Eaton, proposing only the name of a system for these strata, yet he has the credit of original ideas in regard to the matter of their position. Were I to attempt to show that Emmons has only borrowed Eaton's ideas I should be charged now with invidious motives since it turns out that these slates are primordial in spite of what has been said. And this brings me to say that in 1844 I printed a section across Vermont recognizing the Potsdam at the disputed Snake Mt.—submitted it to Rogers and afterwards to Adams and yielded my opinions, and afterwards when Sir W. E. Logan took the same view of these rocks I abandoned my work altogether. That this section was engraved and printed I will some time show you, and since it has the imprint of Endicott who has been dead fifteen years it will not be regarded as a recent publication.

It is very true I followed Hisinger's section and we had not evidence to the contrary until recently.

James Hall to Prof. Dana.

Mar. 25, 1888.

It has never been possible for me to find time to review and revise anything I may have written on the subject or to reply to any of the charges and insinuations coming from various sources. The most malicious insinuation, coming too from a letter of Dr. Emmons himself, is in relation to his geol. map which he says was suppressed. Such a thing would have been impossible. Everything published went into the charge of the Secretary of State. This map was colored by his own son and was so delivered, and it was not until 1855 when the storage rooms were cleared out preparatory to using them for a census department, that the boxes containing these maps came to light. Copies were brought to me and at my suggestion they were sent to the State Library, and they have been freely distributed from there for more than thirty years past. At the time when Dr. Emmons wrote his letter there were hundreds of copies in the schools and academies of New York, and as many more in the hands of amateurs or collectors of books and maps. Probably very few persons even fully understood the purpose of the map.

While upon this subject, which I may never touch again, I may allude to what Dr. Emmons wrote in his letter to Marcou about the treatment he received on account of his advocacy of the Taconic system. This treatment of which he complained had nothing whatever to do with the Taconic question, but solely with his course in helping to involve Prof. Agassiz and myself in an annoying and expensive litigation. I know that Prof. Henry once refused to speak to Dr. Emmons on that account.

. . . .

The taconic question, I fear, is one never to be settled. This unhappy business of identifying formations many miles away, and bringing into it all sorts of heterogeneous materials must always leave room for doubt and dissension.

Very sincerely yours,

(*Signed*) James Hall.

Albany, March 17th, 1889.

Prof. R. Ellsworth Call,
Dear Sir:

I have received your favor of the 10th in reference to the Emmons Geol. Map. It is by no means a recent find. In 1855 or beginning of 1856 in cleaning out some storage rooms in the upper part of the State Hall some boxes of maps were found. Some of the copies were sent to Dr. Woolworth, Secy. of the Board of Regents who in turn sent them to me for explanation. I gave him the proper explanation and advised him to have the boxes sent to the State Library, and since there were frequent inquiries for a geol. map of the state, to send out these maps in response. This has been done since 1856 and some hundred have been distributed in that way. I have had plenty of them in the past but I may not be able to send you one just now. Will look for a copy. If you do not receive a copy after a reasonable time please write me again.

Very respectfully yours,

(*Signed*) JAMES HALL.

P. S. I have omitted to state that there were a great number of these maps in the State Library, but in the removal to the New Capitol they were mislaid or covered up so as to be inaccessible at present.

H. per J. N.

James Hall to Edward Hitchcock.

Albany, February 12, 1841.

MY DEAR SIR:

Through the kindness of my friend Mr. G. B. Emerson of Boston I am in possession of a copy of your final report on the Geology of Massachusetts. I have received much pleasure from the perusal. I need not say that which will be universally conceded, that it is a monument to your ability and perseverance which remain while the rocks of Massachusetts are unchanged.

I regret that in your allusion to my report of 1839 (1840, p. 432) you have entirely misunderstood me. It would appear that I advocate a theory which would attempt to explain all that you denominate "diluvial phenomena" by supposing an immense inland sea, etc. Now the supposition of an immense inland sea, as I have explained, or as I intended to explain it, could only have existed after some excavating power had operated to produce nearly the present configuration of the country with the exception of a few water courses which I suppose to have been excavated in the passage of this water to the sea. It is very strange that when I state that the limit of this inland sea could never have been much more than 1000 feet above tide water and at the same time knowing that the tops of mountains in New York more than 3000 feet above the water exhibited marks of abrasion, and that immense numbers of boulders from these mountains and from similar ones in Canada exist in our older drift deposits—it is strange that possessing such knowledge I could say anything which implied a belief that diluvial phenomena were all produced by this agency. It would carry the presumption that I was quite ignorant of diluvial phenomena in Massachusetts, Maine and in Canada in many places far higher than the limits I had supposed this continent to have, and extending far beyond any point where I had supposed it to have produced any influence. Perhaps it may be impossible to prove all that I have said, and in the haste in which it was written may have said what has conveyed an impression that I did not intend. Many object to the possibility of such a phenomena, but when we have now such immense inland lakes whose bottoms are far lower than the level of the ocean and no known outlet or communication at a lower level than the present one is it not a fair inference that a still larger one may have existed—but I do not cite this as a proof—there are too many stronger evidences than this analogy.

How shall we explain the very well defined terrace along the shores of Lake Erie about 1500 feet above the level of the present lake—if we attribute the ridge bordering Ontario to its former greater elevation shall we not also that of Erie?—and if we admit this where shall we find the barrier for its waters till we come east to the primary range or westward to the Mississippi—for if we raise the water to this elevation it will surely flow by way of the Mississippi.

I regard truth far more than theory and shall be very happy to know the truth regarding this matter in which I have theorized a little, but if any other more satisfactory explanation has been given I am not aware of it. I am fully inclined to admit the agency of glacial action whenever it can explain phenomena, but when our drift is sorted and regularly stratified I imagine this action alone to be insufficient. And I am not aware that the theory of glacial action will explain the formation of those deep and narrow gorges which intersect the rocks of western New York, and in many of which streams or lakes exist. These gorges are from 100 to 600 or 800 feet deep and from a few hundred feet to 3 miles wide.

At the time I wrote the article in question I had formed no satisfactory conclusion regarding the great cause of diluvial action—or the cause producing drift, and in the first paragraph of page 432, report of 1840, in alluding to the physical geography, valleys, etc., I say, "these phenomena may have been caused in part by a sudden submergence, and the rapid passage of a wave over the surface." Absurd as this may appear perhaps, when we now have a more rational explanation, it still may not be more absurd than other theories—and certainly this had no reference to an inland sea,—but to that I attribute only some of the "subsequent changes."

. . . .

With sentiments of the highest regard,

I remain, yours very truly,

(*Signed*) James Hall.

INDEX.

PAGE